MAGNETISM IN METALS AND METALLIC COMPOUNDS

MAGNETISM IN METALS AND METALLIC COMPOUNDS

Edited by

Jan T. Łopuszański, Andrzej Pękalski, and Jerzy Przystawa

Institute of Theoretical Physics
University of Wrocław
Poland

PLENUM PRESS · NEW YORK AND LONDON

Library of Congress Cataloging in Publication Data

Winter School of Theoretical Physics, 11th, Karpacz, 1974.
Magnetism in metals and metallic compounds.

Includes bibliographical references and index.
1. Magnetism—Congresses. 2. Metals—Congresses. 3. Nuclear magnetism—Congresses. I. Łopuszański, Jan. II. Pękalski, Andrzej. III. Przystawa, Jerzy. IV. Title.
QC750.W56 1974   538   75-20369
ISBN 978-1-4757-0018-3   ISBN 978-1-4757-0016-9 (eBook)
DOI 10.1007/978-1-4757-0016-9

Proceedings of the 11th Winter School of Theoretical Physics organized by the University of Wrocław and held at Karpacz, Poland, February 19–March 4, 1974

© 1976 Plenum Press, New York
Softcover reprint of the hardcover 1st edition 1976

A Division of Plenum Publishing Corporation
227 West 17th Street, New York, N.Y. 10011

United Kingdom edition published by Plenum Press, London
A Division of Plenum Publishing Company, Ltd.
Davis House (4th Floor), 8 Scrubs Lane, Harlesden, London, NW10 6SE, England

Since 1964 Institute of Theoretical Physics of the University of Wrocław has organized the following Winter Schools:

1964 - Various Problems of Theoretical Physics

1965 - Applications of Group Theory in Physics

1966 - Statistical Physics and Many-Body Problems

1967 - Functional Methods in Quantum Field Theory and Statistical Mechanics

1968 - Axiomatic Approach to Quantum Field Theory and to Many-Body Problems

1969 - Magnetic Phenomena in Solids

1970 - Liquid Helium and Many-Body Problems

1971 - New Developments in Relativistic Quantum Field Theory

1972 - Theory of Metals and Many-Body Problems

1973 - Recent Developments in Relativistic Quantum Field Theory and Their Applications

1974 - Magnetism in Metals and Metallic Compounds

FOREWORD

The Proceedings presented here contain the notes of lectures
delivered during the Eleventh Winter School of Theoretical Physics,
held at Karpacz, Poland, February 19 - March 4, 1974.

The School was primarily devoted to new concepts in the theory
of magnetism in metals, alloys, and metallic compounds, but, as can
be seen from the table of contents of the book, other topics of the
theory of magnetism were also discussed in the course of the lectures. The organizers agreed to such a broadening of the scope in
order to satisfy particular requests from the Polish participants
for whose benefit the School was organized. These "local" interests
are clearly reflected in the Proceedings and are responsible for a
certain inhomogeneity of the topics selected for presentation.
Nevertheless, we have a strong hope that these materials will be
interesting to many physicists, not only in Poland, for the subjects
discussed here are important not only on the local level, as the
lectures contain quite fresh, unpublished results or excellent up-
to-date reviews.

The first part of the volume contains lectures directly corresponding to the title of the School, i.e., selected topics of the
theory of metallic magnetism, with slight bias toward rare earth
and actinide metals and their compounds. In the second half we
have collected the topics more loosely connected with the main-
stream, such as statistical and thermodynamic aspects of various
models, spin-phonon interaction, and others.

The School was very well organized by members of the Institute
of Theoretical Physics of the University of Wrocław in the persons
of J.M. Kowalski (Technical University of Wrocław), J. Lorenc, A.
Ogielski, T. Paszkiewicz, A. Pekalski, A. Podolska, J. Przystawa,
Z. Strycharski, W. Suski (Institute of Low Temperature, PAN),
and K. Walasek. But our thanks are first of all due to all lec-
turers for the good physics they taught us, for their readiness to

discuss any subject at any time, for their contribution to the warm
and friendly atmosphere of the School, and, last but not least, for
excellent preparation of the lecture notes, which are being published
virtually in the form in which they were submitted. Finally, we must
acknowledge the financial support of the University of Wrocław and
the Polish Ministry of Science, Higher Education, and Technology.

Professor Z. M. Galasiewicz
Head, Department of Solid
   State Theory
University of Wrocław

PREFACE

The present volume contains lectures and seminars given at the Eleventh Winter School of Theoretical Physics organized in Karpacz in 1974 by the Institute of Theoretical Physics of the University of Wrocław. The winter schools in Karpacz have been held each year since 1964. They are devoted alternatingly to a review of recent developments in quantum field theory and in solid state physics. The purpose of these schools is to give a uniform and consistent account of these topics to Polish post-graduate students. The reviews are presented by outstanding specialists from abroad and from Poland, either in the form of a series of lectures or in the form of more or less informal seminars. The schools also serve the purpose of establishing a closer contact between physicists from the West and from the East. The beautiful scenery of the Sudeten Mountains, the friendly and quiet atmosphere of the University Resort Houses, and the comfortable though simple living conditions provide a good setting for encouraging personal contacts, informal discussions, and scientific collaboration among experts and novices, the older and the young. It is this aspect that we, the organizers of the schools, enjoy and appreciate most.

Professor J. T. Łopuszański
Director, Institute of
   Theoretical Physics
University of Wrocław

CONTENTS

A Many Electron Theory of Localised Moments
        in Metals   1
  Kenneth W. H. Stevens

Magnetic Properties of a Dense Interacting
        Electron Gas   15
  M. L. Glasser

Conceptual Phase Diagrams and Their
        Applications to Itinerant
        Electron Magnetism   35
  John B. Goodenough

Narrow Band Magnetism of Alloys   91
  John W. Schweitzer

The Static and Dynamic Impurity Spin
        Susceptibility of Kondo Alloys 131
  Wolfgang Götze

Magnon Interaction in the Itinerant Electron
        Model of Ferromagnetism 159
  Janusz Morkowski

Studies of Ferro- and Paramagnetic Nickel by
        Positron Annihilation 177
  Henryk Stachowiak

Theory of Magnetic Properties of Rare Earths
        and Their Compounds 193
  Yu. P. Irkhin

Exchange Interaction in the Heavy Rare Earth
        Metals Calculated from Energy Bands 203
  Per-Anker Lindgård

Lattice Effects and the Magnetic Behaviour
            of Rare Earth Compounds   225
    Bernard R. Cooper

Elementary Excitations of High Degree Pair
            Interactions in Rare Earth Compounds 265
    Peter M. Levy

Magnetic Properties of Actinides and Their
            Compounds   287
    Daniel J. Lam

Exploratory Band Structure Calculations for
            Actinide Compounds   295
    Harold L. Davis

Magnetism in Actinides   319
    Bernard Coqblin and R. Jullien

Thermodynamic Properties of the One-Dimensional
            Hubbard Model   351
    Uwe Brandt

Weak Self-Consistent Approximation Scheme   363
    Hajo Leschke

On the Connection between Equation of Motion
            Technique and Perturbation Theory
            Approach to the Evaluation of Double
            Time Green's Function   375
    Calogero R. Natoli and J. Ranninger

First Order Green Function Theory of a
            Heisenberg Ferromagnet   385
    Adolf Kühnel

Spin-One Lattice-Gas Model   393
    Jean Sivardière

Dipolar and Quadrupolar Ordering in
            Magnetic Crystals   411
    Jean Sivardière

One and Two Magnon Light Scattering in
            Insulating Antiferromagnets and
            Its Possible Relevance for the
            Investigation of the Short Range
            Interatomic Correlations   435
    Calogero R. Natoli and J. Ranninger

Two-Fluid Hydrodynamic Description of
        Magnetic Crystals   463
    Charles P. Enz

Impurity Ferromagnetism in Non-magnetic
        Semiconductors   501
    A. A. Abrikosov and Yu. V. Kopayev

Phonon Hydrodynamics in Solids   527
    Armin Thellung

The Spin-Phonon Interaction in Anharmonic
        Ferromagnetic Crystals   543
    N. M. Plakida and H. Konwent

The Symmetry Properties of Magnetoelastic
        Waves and the Determination of
        Selection Rules for Magnon-Phonon
        Interactions   573
    Arthur P. Cracknell

Sound Absorption in Magnetics   601
    V. G. Kamensky

Contributors   611

Index   615

A MANY ELECTRON THEORY
OF LOCALISED MOMENTS IN METALS

K.W.H. STEVENS

*Department of Physics, University of Nottingham,
Nottingham, England*

1. INTRODUCTION

The Hamiltonian of a typical solid is so complicated that one
does not expect to get very far with exact methods. On the other
hand, if cne tries to split it into an unperturbed part and a
perturbation (so that approximation methods can be used) it is
very desirable to have the unperturbed part giving a fairly accurate description, so that the effect of the perturbation is
small. It is also implied that the unperturbed problem is itself tractable, for there is no point in having an unperturbed
Hamiltonian unless one can discuss its properties fairly readily.

A very common choice of unperturbed Hamiltonian is one in
which the electrons move independently in a periodic potential
which extends throughout the lattice. (There is also the need
to see that the electrons obey the exclusion principle). The
resulting theories, which can be described as band theories,
have been extremely successful, particularly in describing metals, where the conduction properties readily emerge. They are
not, however, universally successful, and problems seem to arise whenever there appear to be localised magnetic moments. For
example, the rare earth metals seem to contain electrons in partially filled $4f$ shells which behave very much like $4f$ electrons
of the free ions. Their orbits and spins are coupled according
to Hund's rules for atoms, so that $\sum_i \ell_i = L$ and $\sum_i s_i = S$, and

then L and S are coupled to form J. Further, the J's at adjacent sites are only very weakly coupled, so that for many purposes the associated moments can be regarded as independent of

1

one another. It is difficult to see how these properties can
be understood with independent electrons which move throughout
the crystal in Bloch-like wavefunctions. Of course the rare
earth metals do contain electrons in Bloch-like wavefunctions,
derived from the outer shell electrons of the rare earth atoms,
and these are mainly responsible for the conduction. But the
point I wish to make is that the standard band theory type of
approach is unlikely to provide a good starting approximation
for the magnetic electrons.

For some time I have been studying the potential of an approach to magnetic insulators which is not based on an independent electron type of unperturbed Hamiltonian, but which uses
an essentially many body unperturbed Hamiltonian [1]. The subsequent perturbation treatment has presented a number of problems,
but enough progress has been made with these, and enough new results have emerged to make me feel that the new approach has
some value. In particular it now seems that the problem of describing localised moments in periodic structures (insulators) is
really only a problem if one starts with independent electron
unperturbed Hamiltonians. With suitable many-electron Hamiltonians the problem disappears, and although new problems do appear
they seem to be tractable.

The extension of this approach to conductors has presented an
interesting challenge, and during the summer of 1973, Dr.E.
Pytte and I gave it a good deal of attention. We believe we
have found ways of making extensions, and it is these which I
would like to tell you about. The theory is still in an initial form, as you will see, and quite a lot remains to be done.
Nevertheless, it does seem to have promise, apart from the fact
that we do not know of any better approach to localised moments
in conductors!

2. THE MANY-ELECTRON UNPERTURBED HAMILTONIAN

There is little problem about writing down the Hamiltonian
of a metal, provided it is assumed that the lattice is regular
and fixed. It typically has the form

$$\mathcal{H} \equiv \sum_i \left(\frac{p^2}{2m}\right)_i - \sum_{i,N} \left(\frac{Ze^2}{r_{iN}}\right) + \sum_{i<j} \left(\frac{e^2}{r_{ij}}\right) + \dots \,,$$

where the first term describes the kinetic energies of the electrons, the second term their electrostatic energies in the
fields of the nuclei, the third their mutual Coulomb repulsive
energies, and ... represents additional terms, such as spin-orbit interactions, nuclear hyperfine interactions, etc., which
can be given in detail later, as needed. It is perhaps worth
noting in passsing that, superficially, there is very little
difference between the Hamiltonian of one metal and the Hamiltonian of another. Yet their physical properties may differ
very considerably.

The usual way to obtain an unperturbed Hamiltonian is to drop $\sum_r e^2/r_{ij}$ and the ... terms and to add an effective potential, $\sum_i V_i$, chosen to simulate $\sum e^2/r_{ij}$. The unperturbed Hamiltonian becomes

$$\mathcal{H}_0 \equiv \sum_i \left[\frac{p^2}{2m} - \sum_N \frac{Ze^2}{r_N} + V(r) \right]_i \, ,$$

which is a sum of commuting one-electron Hamiltonians, and the perturbation is the difference between $\mathcal{H}$ and $\mathcal{H}_0$. From the point of view of second quantisation, $\sum e^2/r_{ij}$, being a two-electron operator, is quite different from $\sum V_i$, which is a one-electron operator. Nevertheless, this way of obtaining an unperturbed Hamiltonian is very plausible, and it has been extremely successful in many cases.

In order to prepare the background for a different approach I shall argue that its success primarily stems from an attempt to match the energies of the unperturbed Hamiltonian with those of the actual Hamiltonian, and that it is far less clear, particularly with magnetic systems, that there is a good match between states. I believe that traditionally much more emphasis has been placed on getting energies right than on getting states right. If this assessment is correct, I would further argue that the position is rapidly changing, in that much of experimental physics is now devoted to discovering the properties of states, so that, increasingly, theoreticians will have to be more concerned with states than energies. In support of this contention I would direct your attention to the enormous growth of interest in group theory, which to me is indicative of a concern with the symmetry of states, with little to do with energies. In fact, of course, if one knows the states of a system one can readily find the energies, from the expectation values of the Hamiltonian, recalling that there is usually no great difficulty in writing down the Hamiltonian. (The difficulty is usually in writing down a suitable unperturbed Hamiltonian, which is not necessary if the states are known!).

The method I shall describe is one in which states are chosen, on the basis of prior knowledge about the system under consideration. In general these states will not be eigenstates of the Hamiltonian, but one hopes they are close to eigenstates, in which case a correction procedure is required. The way to obtain this is to define an unperturbed Hamiltonian having the chosen states as eigenstates. The difference between the actual Hamiltonian and the defined unperturbed Hamiltonian can then be treated by standard perturbation theory, with convergence on the properties of $\mathcal{H}$

Let us ask what the low-lying states of a typical rare earth metal will be like? Each rare earth atom will have a number of electrons paired off in inner closed shells, it will have a well defined number of electrons in $4f$ states, arranged in accordance

with Hund's rules (for the moment we ignore crystal field effects), and all the atoms will be immersed in a sea of conduction electrons coming from the outer shells of the atoms. As
an example we might take erbium. Then the experimental evidence
is that each atom has eleven $4f$ electrons, arranged to have $L =$
6, $S = 3/2$ and $J = 15/2$. If there are N rare earth atoms in the
lattice, the ground manifold of states of the whole crystal has
degeneracy $(16)^N$, owing to the $4f$ electrons. There is additional
near-degeneracy owing to the conduction electrons, which are $3N$ in
number (three for each rare earth atom).

One of the problems with many-electron systems is that of allowing for the Exclusion Principle. The simplest way of dealing
with this is to use the methods of second quantisation, which
automatically take care of the antisymmetry of the many-electron
wavefunctions. However, in order to have simple commutation
rules it is essential that all the one-electron orbitals be mutually orthogonal. So our description of the rare earth metal
must be such that all the orbitals into which electrons are
placed are orthogonal to one another. The simplest way to ensure this is as follows. Suppose that a one-electron approximation has been introduced, and that the corresponding band theory has been worked out. Then the states will fall into bands,
based on $1s$, $2s$, $2p$, ..., $4f$, ... and so on. The states in each
band are mutually orthogonal, and so are the states in different
bands. In other words the band states form a complete set of
orthogonal functions (as do the eigenfunctions of any one-electron Hamiltonian). However, the band wavefunctions are far from
localised, for they extend right through the crystal. If,
though, one follows the prescription for forming Wannier functions [2], which is a little tricky in the case of degenerate
bands [3], and defines, for a given band,

$$\psi(\mathbf{R}_n) = \frac{1}{\sqrt{N}} \sum_{\mathbf{k}} e^{i\mathbf{k}\cdot\mathbf{R}_n}\psi_{\mathbf{k}},$$

where $\mathbf{R}_n$ ranges over the lattice sites, the result is a set of
localised orthogonal functions. Clearly the Wannier functions
formed from a given band are orthogonal to all the functions in
the other bands, and thus to any Wannier functions formed from
them. In describing the electrons in the localised $4f$ states
we shall therefore regard them as in $4f$ Wannier functions formed
from the $4f$ band states. (They have slightly different symmetry
properties from atomic $4f$ orbitals). The $3N$ conduction electrons
will be regarded as in band states, bands other than $4f$, so their
states will be orthogonal to the $4f$ states. The electrons in
filled inner shells can either be regarded as filling the low-
lying bands or as filling $1s$, $2s$, ... Wannier states, for the
two descriptions are mathematically identical.

For each orthogonal one-electron function annihilation and
creation operators are defined, satisfying the Fermi commutation
rules. We shall use c to denote a conduction state, so that $c_{\mathbf{k}}^{*}$

creates an electron in the k-th conduction state, where k describes the momentum of the state and its band index and m_s. The
lattice points will be labelled A,B,C, etc., so $a_{m,\sigma}^*$ creates an
electron in a $4f$ Wannier state at site A with $m_\ell = m$ and $m_s =$
σ. Similarly $b_{m',\sigma'}$ annihilates a $4f$ electron at B with $m_\ell =$
m', $m_s = \sigma$. It is convenient to distinguish between bands which
are filled and bands which are empty in the low-lying states of
the crystal, and we use f for filled and e for empty. (f must
not be confused with the $4f$ states).

Suppose then we operate on the vacuum state with a string of
eleven a^* type operators. The result is a state in which there
are eleven electrons in the $4f$ states at A. To describe any of
the states in $J = 15/2$ it will, in general, be necessary to take
a linear combination of products of eleven a^* operators. We
shall not pause here to see in detail how this is done, but it
is clearly possible. To create eleven more electrons, at B, we
next operate with a string of eleven b^* type operators (or a linear combination of a string of eleven b^*). Proceeding in this way
we can build states in which every site has eleven $4f$ electrons,
with the states at each site being drawn from $J = 15/2$. They
will not, however, describe states of the crystal, because the
electrons in closed inner shells and the conduction electrons
are so far omitted. The conduction electrons are easily incorporated by multiplying by a string of $3N$ c_k^* and the filled
shells by a string of f^*, where every f_+^* is accompanied by f_-^*.
The value of the Fermi operators is now apparent, for although
the final states are really quite complicated many-electron
states, with their use the antisymmetry requirement is automatically satisfied, and the need to have $J = 15/2$ at each site is
readily incorporated. The totality of states formed in the
above way will be referred to as the ground manifold. It is the
product space of the spaces of the many $J = 15/2$ with the conduction band space.

There are also many excited manifolds, which can similarly
be described. For example, the $4f$ electrons at site A can be
rearranged to belong to some J other than $15/2$ by operating with
a suitable string of a^*a operators. (The number of a^* operators
must be the same as the number of a so that the number of $4f$ electrons is unaltered). Again, one might wish to take an electron out of the $4f$ shell at B and place it in the conduction band.
This is done by operating with c_k^*b. Every conceivable excited
state can be reached by applying a suitable string of operators
to any chosen ground state.

What I hope is now clear is that the above technique enables
one to write down any specific many-electron state. If, probably from experiment, we know broadly which states are occupied in
the ground state, we can write down these states quite compact-
ly. Similarly for excited states.

The states constructed according to the above scheme are mutually orthogonal and form a complete set. They have not been
obtained as eigenstates of an operator, though their properties
suggest that they could have been. We shall pursue this,

defining the unperturbed Hamiltonian to have these states as
eigenstates. Further, if these states were true eigenstates of
the full Hamiltonian, their eigenvalues would be the expectation
values of the Hamiltonian taken over these states.

Given a state $|n\rangle$, its projection operator P_n is defined to
be $|n\rangle\langle n|$. The operator $\sum_n \lambda_n P_n$, where n ranges over a complete
set, has the state $|m\rangle$ as an eigenstate with eigenvalue λ_m. I
therefore define an unperturbed Hamiltonian by

$$\mathcal{H}_0 \equiv \sum_n |n\rangle\langle n|\mathcal{H}|n\rangle\langle n|,$$

where the summation is over all the many-electron states, and a
specific choice $\lambda_n = \langle n|\mathcal{H}|n\rangle$, has been made. If the chosen
states were indeed eigenstates of $\mathcal{H}$, then $\mathcal{H}_0$ would coincide with
$\mathcal{H}$. Since they are unlikely to be exact eigenstates of $\mathcal{H}$, $\mathcal{H}_0$
differs from $\mathcal{H}$, by an amount which depends on how far wrong the
states are. Since we know $\mathcal{H}$ and have defined $\mathcal{H}_0$, the difference, $\mathcal{H} - \mathcal{H}_0$, can be treated as a perturbation on $\mathcal{H}_0$. At first
sight $\mathcal{H}_0$ looks formidable, because of the presence of the projection operators. In fact this is not at all the case, because
projection operators are particularly simple to work with. So
we have achieved one important goal, that the unperturbed Hamiltonian be tractable. It has a number of other attractive features. It is invariant under interchange of electrons and its
eigenstates are all antisymmetric with respect to interchanges
of electrons, as follows from the way in which it is constructed. It is invariant under all translational symmetry operations which leave the lattice invariant. Under a typical trans-
lation $|n\rangle$ may go into $|m\rangle$. Since $\mathcal{H}$ is invariant, $\langle n|\mathcal{H}|n\rangle \rightarrow \langle m|$
$|\mathcal{H}|m\rangle$ and $|n\rangle\langle n| \rightarrow |m\rangle\langle m|$. It follows that $\mathcal{H}_0$ is invariant.
Nevertheless, $\mathcal{H}_0$ can be expected to be quite a complicated oper-
ator if written in terms of electron position and momentum variables, and far as I can see it is not a sum of commuting one-
electron operators. My uncertainty on this point arises because
it appears to be a formidable task to express $\mathcal{H}_0$ in these variables, and as it is quite unnecessary to do so, except to answer
this sort of question, I have not tried very hard!

Since $\mathcal{H}_0$ is defined, the properties of $\mathcal{H}$, which are the main
interest, are only found during the course of the perturbation
theory. Nevertheless, it is of interest to examine some of the
detailed properties of $\mathcal{H}_0$, and particularly the question of
whether the states in the ground manifold, which are certainly
eigenstates of $\mathcal{H}_0$, are degenerate. If they are then $\langle n|\mathcal{H}|n\rangle$
takes the same value for every state in the ground manifold.

To examine such a question, and indeed for the rest of the discussion, it is advantageous to take $\mathcal{H}$ in its second quantised
form. This is a standard piece of theory [4,5] so I shall not
spend time on a detailed account of it. What we primarily need
to know is that the one-electron operators (denoted by h) give
expressions of the form $\langle\alpha|h|\beta\rangle\alpha^*\beta$ and the two-electron opera-

tors (denoted by g) give expressions like $\langle \alpha\beta | g | \delta,\gamma \rangle \alpha^* \beta^* \gamma \delta$. If now $\alpha^* \beta$ is applied to a state $|n\rangle$ an electron is removed from β and placed in α. The new state is orthogonal to $|n\rangle$, unless $\alpha = \beta$. Thus the only one-electron terms which contribute to $\langle n | \mathcal{H} | n \rangle$ are of type $\langle \alpha | h | \alpha \rangle \alpha^* \alpha$. That is, they involve diagonal matrix elements and number operators. Similarly, the only two-electron operators are of types $\langle \alpha\beta | g | \alpha\beta \rangle \alpha^* \beta^* \beta \alpha$ and $\langle \alpha\beta | g | \beta\alpha \rangle \alpha^* \beta^* \alpha\beta$. One has diagonal and other exchange-like matrix elements, and they both involve number operators. It is now readily seen that the ground manifold is not degenerate. For example, states which differ in having a conduction electron moved from one conduction state to another will differ both in their kinetic and Coulomb energy terms. On the other hand, states which differ because $4f$ electrons have been rearranged, while still preserving J, have identical kinetic energies because $\langle a_{m\sigma} | p^2/2m | a_{m\sigma} \rangle$ is independent of m. (Strictly true only in spherical symmetry). In general, though, these states have different electrostatic energies, and the matrix elements which enter are of considerable interest because many of them can be interpreted as part of what is usually called the crystal field. This concept was not, of course, introduced into the initial formalism, because the crystal field is not part of the many-electron Hamiltonian. Nevertheless, it is widely used in discussing rare earth metals, and one may anticipate that something which corresponds to it will appear as part of the discussion. So it is important to be able to recognise it when it does emerge.

A matrix element like $\langle a_m | Ze^2/r_N | a_m \rangle$ is the Coulomb energy of a $4f$ electron in orbit $m_\ell = m$ at nucleus A in the field of the nucleus at R_N. Provided that R_N is different from A the appearance of such a matrix element can be interpreted as a crystal field-like energy. Similarly, $\langle a_m, f_B | e^2/r_{ij} | a_m, f_B \rangle$ is the coulomb energy of a $4f$ electron at A in the field of an electron in a filled orbit at B. It is also crystal field-like. On the other hand $\langle a_m b_{m'} | e^2/r_{ij} | a_m, b_{m'} \rangle$ is the electrostatic energy of two electrons, one in a $4f$ orbit at A and the other in a $4f$ orbit at B. Such an element would not be regarded as crystal field-like, but of multipole interaction type.

The electrostatic terms contribute a variety of matrix elements to $\langle n | \mathcal{H} | n \rangle$, and their detailed examination does indeed show that a crystal field is, effectively, present. So also are multipole interactions between $4f$ shells and some exchange interactions associated with matrix elements of form $\langle a_m b_{m'} | e^2/r_{ij} | b_{m'} a_{\dot{m}} \rangle$. The answer to the question of whether or not the ground manifold is degenerate is that it is not. Similarly, the excited manifolds are also not degenerate.

3. PERTURBATION THEORY

From the point of view of perturbation theory it is important to know how the splittings within manifolds compare with the spacing between adjacent manifolds. I shall assume that the splittings within manifolds are small compared with the splittings between manifolds, also that the off-diagonal matrix elements of $\mathcal{H} - \mathcal{H}_0$ between manifolds are small compared with the separations between manifolds. It is not necessary to assume, though, that they are small compared with the splittings within manifolds. It will therefore be essential to use a form of perturbation theory suitable for degenerate or nearly degenerate manifolds. For this reason it is convenient to distinguish between two kinds of off-diagonal matrix elements. Those which connect distinct manifolds will be described as off-diagonal, while those which connect states in the same manifold will be described as semi-diagonal.

The perturbation treatment will be taken in two stages. The first will concentrate on reducing the effects of the off-diagonal elements and will proceed broadly as follows. Suppose one has a matrix

$$A \equiv \left[\begin{array}{c|c} X & Y \\ \hline Y^* & Z \end{array} \right] ,$$

where X and Z correspond to distinct, well separated manifolds, and Y and Y^* represent the off-diagonal matrices which couple them. A unitary transformation is sought which much reduces the elements in the off-diagonal positions, so that they can then be safely neglected. The study of the eigenvalues of A is then reduced to the study of the matrices which are then found in the X and Z positions. We may say that before the transformation there were matrix elements coupling the X and Z manifolds. After the transformation the new X and Z manifolds are decoupled.

The following analysis represents the above decoupling procedure in the context of the present problem. Let P_0 be the projection operator of the ground manifold and P_N the projection operator of an excited manifold at Δ_N. Then $P_0\mathcal{H}P_0$ corresponds to X, $P_N\mathcal{H}P_N \equiv Z$ and $P_0\mathcal{H}P_N \equiv \epsilon Y$ where ultimately ϵ will be set to unity. Thus

$$A \equiv P_0\mathcal{H}P_0 + P_N\mathcal{H}P_N + P_0\mathcal{H}P_N + P_N\mathcal{H}P_0,$$

and

$$\mathcal{H}(\epsilon) \equiv (P_0\mathcal{H}P_0 + P_N\mathcal{H}P_N + \dots) + \epsilon(P_0\mathcal{H}P_N + P_N\mathcal{H}P_0) + \dots .$$

We seek $T \equiv \exp(i\epsilon S) \simeq 1 + i\epsilon S - \frac{1}{2}\epsilon^2 S^2 \dots$ such that $T^*\mathcal{H}(\epsilon)T$ has all its off-diagonal elements of order less than ϵ. The condition is

$$P_0\mathcal{H}P_N - i(P_0 S P_N \mathcal{H}P_0 - P_0\mathcal{H}P_N S P_0) = 0,$$

which, to a good enough approximation, is satisfied with

$$P_0 S P_0 = P_0, \qquad P_N S P_N = P_N,$$

and

$$P_0 S P_N = - i \, \frac{P_0 \mathcal{H} P_N}{\Delta_N} \, .$$

After some manipulations it then follows that the transformed Hamiltonian, $T^* \mathcal{H} T$, within P_0, is

$$\tilde{\mathcal{H}} \equiv P_0 \mathcal{H} P_0 - \sum_N \frac{P_0 \mathcal{H} P_N \mathcal{H} P_0}{\Delta_N} \, .$$

What we have therefore achieved is the reduction of the off-diagonal matrix elements to negligible size at the cost of introducing new diagonal and semi-diagonal elements, the terms

$$- \sum_N \frac{P_0 \mathcal{H} P_N \mathcal{H} P_0}{\Delta_N} \, ,$$

into P_0.

There will be many such new terms, and it is of interest to examine some of them in more detail. Suppose we pick from $\mathcal{H}$ a one-electron term which transfers an electron from a $4f$ orbital at A into an unoccupied state in the conduction band. It will be of the form $\langle c_k | h | a_{m\sigma} \rangle c_k^* a_{m\sigma}$ and it will couple P_0 to a particular excited P_N. To return to P_0, again using a one-electron operator, a conduction electron, which may or may not be identical with the one just created, will have to be removed and placed in a $4f$ orbit at A, which again may or may not be $a_{m\sigma}$. It will, though, have to give a final state which belongs to $J = 15/2$. The contribution of this process will be

$$- \frac{\langle a_{m'\sigma'} | h | c_{k'} \rangle \langle c_k | h | a_{m\sigma} \rangle}{\Delta_N} \, a_{m'\sigma'}^* c_{k'}^* c_k^* a_{m\sigma},$$

where it is understood that the evaluation of $a_{m'\sigma'}^* a_{m\sigma}$ is to be done within $J = 15/2$. Then $a_{m'\sigma'}^* c_{k'} c_k^* a_{m\sigma}$ is clearly an operator within P_0. It can be given a physical meaning, as a scattering process in which a conduction electron is scattered from k to k' with a possible change in the $4f^{11}$ state at A from one in $J = 15/2$ to another, also in $J = 15/2$. Similar terms will already be present in $P_0 \mathcal{H} P_0$, from the two-electron terms, so these new contributions can be regarded as modifying scattering terms which are already present. (They are known as Ruderman-Kittel interactions).

A rather similar process is one in which a $4f$ electron is transferred to an excited conduction band state which is unoccupied in the ground manifold. Then on returning to the ground manifold this particular state must be emptied, so the effective operator within P_0 is $a_{m'\sigma'}^* e_k e_k^* a_{m\sigma}$. Further, $e_k e_k^*$ is unity, so the net contribution is

$$- \frac{\langle a_{m'\sigma'} | h | e_k \rangle \langle e_k | h | a_{m\sigma} \rangle}{\Delta} \, a_{m'\sigma'}^* a_{m\sigma}$$

which can be interpreted as an effective crystal field acting on the $J = 15/2$ states at A, provided $\sigma = \sigma'$, as will be the case for most terms in h.

The excited manifolds can be classified according to the number of electrons which must be excited to reach them. Those with one excited electron can be reached by both the one- and two-electron terms in $\mathcal{H}$, those with two only by the two-electron terms, and those with more than two cannot be reached, and therefore need not be considered (except in higher order). All the processes will result in operators within P_0 of which there are many different sorts. However, there will be no combinations with more than four starred and four unstarred operators.

It is instructive to consider one of these, the form $a^*a^*a^*a^*$ $aaaa$, which can only have arisen by using, twice, from $\mathcal{H}$ the two-electron terms of type a^*a^*aa (second-order Coulomb interactions within the $4f$ shell). Now it is one of the features of second quantisation that the formalism does not depend on the number of electrons present. We can therefore readily compare the present treatment with a similar treatment of an isolated atom at A. This is obtained by dropping, from $\mathcal{H}$, all the nuclear attracting terms except that from the nucleus at A, by modifying the conduction band states to be outer atomic orbitals and by not using the $4f$ orbitals at sites other than A. Terms of the form $a^*a^*a^*a^*aaaa$ will again be expected, arising from second-order Coulomb interactions within the $4f$ shell. They will be similar to those in the metal, but with different Δ_N. However, in the case of the atom the Coulomb interactions can only mix states of the same J, and due to them the ground multiplet may be a linear combination of terms of different L and S. With the correct starting combinations there should be no terms of type $a^*a^*a^*a^*aaaa$. I have assumed that knowing the J values in a metal uniquely determines the $(2J + 1)$ states which span it. This is ensured when the $4f$ shell is more than half full by the rule of maximum L,S and J, but when it is less than half full the rule is maximum L and S and minimum J, and the states are not uniquely determined. One can anticipate that the $a^*a^*a^*a^*$ $aaaa$ term will be absent in the second half of the $4f$ shell, and in the first half if there are such terms it should be possible to eliminate them by a better choice of initial states. There will be no comparable terms of type $c^*c^*c^*c^*cccc$ because the whole of the occupied conduction bands are regarded as part of P_0, and c^*c^*cc is regarded as semi-diagonal.

4. DECOUPLING WITHIN P_0

I now come to the second stage in the perturbation theory. The expression for $\tilde{\mathcal{H}}$ is an operator within P_0, which is the projection operator for the Hilbert space made up of the product of the $4f$ space and the conduction band space. $\tilde{\mathcal{H}}$ contains operators connecting the two sub-spaces. The question arises as to whether or not it is possible to find a transformation which decouples the two sub-spaces. At first sight this appears to be a problem similar to that already considered, decoupling the ground and excited manifolds. There, is however, a very important difference, in that previously the states in the ground and excited manifolds differed by energies of order Δ, which could be used as an expansion parameter, for $T \equiv \exp(i\varepsilon S) = 1 + i\varepsilon S + \dots$ was really an expansion in powers of $1/\Delta$. Now there is nothing comparable with Δ.

$\tilde{\mathcal{H}}$ can be split into two parts, $\tilde{\mathcal{H}}_0$ composed entirely of number operators ($a_{m\sigma}{}^*a_{m\sigma}$, $c_k{}^*c_k$, etc.) and $\varepsilon\tilde{V}$ composed of scattering terms, of which a typical one is

$$\{U_{mm'kk'}a_{m\sigma}{}^*a_{m'\sigma'}c_k{}^*c_{k'} + \bar{U}_{mm'kk'}c_{k'}{}^*c_k a_{m'\sigma'}{}^*a_{m\sigma}\}$$

The usual way of reducing the order of the perturbation in $\tilde{\mathcal{H}}_0 + \varepsilon\tilde{V}$ is to apply

$$e^{i\varepsilon S} \simeq 1 + i\varepsilon S + \dots \;,$$

and choose S so that the terms linear in ε disappear [6]. From the expansion

$$(1 - i\varepsilon S + \dots)(\mathcal{H}_0 + \varepsilon V)(1 + i\varepsilon S - \dots),$$

the condition is

$$V - i(S\mathcal{H}_0 - \mathcal{H}_0 S) = 0.$$

In general this does not determine S as an operator, but it gives its matrix elements in a representation in which $\mathcal{H}_0$ is diagonal. With near degeneracies in $\mathcal{H}_0 + \varepsilon V$, some of its eigenstates may differ significantly from those of $\mathcal{H}_0$, so one wants to avoid, if possible, using eigenstates of $\mathcal{H}_0$. In the present example we do not aim to eliminate the whole of εV, but only those parts in which conduction electron are scattered. A term like $a_{m\sigma}{}^*a_{m'\sigma'}c_k{}^*c_k$ is now left, while $a_{m\sigma}{}^*a_{m\sigma}c_k{}^*c_{k'}$ is to be removed. Now

$$[a_{m\sigma}{}^*a_{m'\sigma'}c_k{}^*c_{k'}, \mathcal{H}_0] = a_{m\sigma}{}^*a_{m'\sigma'}c_k{}^*c_{k'} \times \left\{ \begin{array}{c} \text{function of} \\ \text{number operators} \end{array} \right\},$$

so iS should contain

$$X_{mm'kk'} a_{m\sigma}^* a_{m'\sigma'} c_k^* c_{k'} - \text{conjugate complex,}$$

where

$$U_{mm'kk'} - X_{mm'kk'} \langle \text{function of number operator} \rangle = 0,$$

and $\langle \ldots \rangle$ denotes an average over some distribution of occupation numbers yet to be determined. Setting $i(S\mathcal{H}_0 - \tilde{\mathcal{H}}_0 S) = \tilde{V}_1$ ($\neq \tilde{V}$) the transformed Hamiltonian, $\mathcal{H}_{II}$, to second order in ε is:

$$\mathcal{H}_{II} \equiv \tilde{\mathcal{H}}_0 + \varepsilon(\tilde{V} - \tilde{V}_1) + \frac{i\varepsilon^2}{2} \{S(\tilde{V}_1 - \tilde{V}) - (\tilde{V}_1 - \tilde{V})S\}$$

$$- \frac{i\varepsilon^2}{2} \{S\tilde{V}_1 - \tilde{V}_1 S\}.$$

You may have been wondering why I have thought it necessary to go through this piece of analysis. There are several reasons. First, it is by no means obvious that the $4f$ and the conduction spaces can be satisfactorily decoupled, and the only way in which to find out is to try and do it. The simplest approximate decoupling is the following. To obtain an effective Hamiltonian in the $4f$ space from $\mathcal{H}$ replace all pairs of $c_k^* c_{k'}$ by their average values, which are to be determined self-consistently from an effective Hamiltonian in the conduction space. This is similarly obtained by replacing $a_{m\sigma}^* a_{m'\sigma'}$ by a mean value, determined from the effective Hamiltonian in the $4f$ space. I have wanted to avoid too early a use of this type of decoupling lest it be too crude. It seems highly desirable for the effective Hamiltonian in the $4f$ space to contain what are usually called exchange interactions via the conduction electrons, and these would not be included in the simple decoupling. It has therefore seemed essential to go the next order, $\mathcal{H}_{II}$, before decoupling (which appears to be inevitable at some stage).

Suppose then that $\mathcal{H}_{II}$ has been decoupled in this way, and let us examine $\mathcal{H}_{J-J}$, the effective Hamiltonian coupling the $4f$ shells. Apart from mean values of conduction operators (e.g. $\langle c_k^* c_{k'} \rangle$) it contains $a_{m\sigma}^* a_{m'\sigma'}$ and $b_{m\sigma}^* b_{m'\sigma'}$ pairs, and also averages of pairs, coming from the transformation. If one had an isolated atom, $a_{m\sigma}^* a_{m'\sigma'}$ could be replaced by an equivalent operator within $J = 15/2$, in which case it would be an expression in J_x, J_y and J_z. Something similar is possible in the present case. We replace the $4f$ sub-space, in which $\mathcal{H}_{J-J}$ operates and which is composed of antisymmetric many-electron wavefunctions, by another space of the same dimensions, $(2J + 1)^N$, which is the direct product space of N angular momentum spaces, each of which corresponds to $J = 15/2$ and is associated with a particular lattice point. Then the $a_{m\sigma}^* a_{m'\sigma'}$, $b^* b$, etc., can be

written as equivalent operators in this space, following exactly
the same rules as for the isolated atoms. The result is an effective spin-Hamiltonian containing crystal field-like terms
(single site operators) and exchange and multipole interactions
(linking two or more sites). It will therefore be a generalised
version of the effective Hamiltonians which are frequently postulated for rare earth metals. However, we now have a rather
better idea of where the various expressions could have come
from and, of course, this formulation contains the mean values,
which are yet to be determined and which are commonly omitted
in phenomenological Hamiltonians. Also it is now apparent that
it will not be a particularly simple task to determine them, for
it is by no means easy to diagonalise a generalised exchange
plus crystal field Hamiltonian, even with fixed parameters. A
discussion of this would take me outside the scope of this set
of lectures and I hope you will not mind if I do no more than
point out the link with spin-wave theory. What one has to do
is a spin-wave theory from which the required averages can then
be found.

5. CONCLUSION

What I have tried to do is to pass from the many-electron
Hamiltonian of rare earth metals to an effective Hamiltonian in
a $(2J + 1)^N$ spin-space, tracing out the course which a full
theory might have to follow. Various approximations have been
made, of which the decoupling of the $4f$ and conduction space
presents the most problems. It seems, though, that the effective spin-Hamiltonian will be much like that written down on
phenomenological grounds, except that the present development
suggests that the parameters which are usually introduced should
not be regarded as constants but will have to be determined
self-consistently.

A more subtle point is one that arises from the various
transformations. At the begining a^* created an electron in a
$4f$ Wannier function. After a transformation of $\mathcal{H}$ by e^{iS}, a^*
continues to appear in the transformed Hamiltonian. Its commutation rules with the other operators, b^*, etc., are unaltered, but its meaning has changed. It no longer creates an electron in a Wannier $4f$, but in a much more complicated orbit. As
decoupling by successive transformations proceeds, so the meanings of a^*, etc., get increasingly obscure, and at the final
stage, when the equivalent angular momentum operators are introduced, the meanings of J_x, J_y and J_z at site A are almost completely lost. Because crystal field operators are typically
single site and involve J_x, J_y and J_z, but based on true $4f$ orbitals, it is convenient to call expressions like $a_{m\sigma}{}^* a_{m'\sigma'}$ in
the decoupled Hamiltonian a 'crystal field term'. But this can
be most misleading, because by no means all the $a_{m\sigma}{}^* a_{m'\sigma'}$ pairs
are accompanied by matrix elements which can be given simple electrostatic interpretations. (For example, $p^2/2m$ appears in
many terms). Of course, all these rather difficult points are

missed if one just writes down an effective Hamiltonian on symmetry or other arguments. This does not matter very much provided one does not attach too much meaning to the magnitudes which the parameters are found to require (though we have seen that they may vary with the occupation numbers, that is with temperature). But as a result of the above considerations I would hesitate to estimate a crystal field strength in a phenomenological spin-Hamiltonian of a rare earth metal using simple electrostatic arguments!

As I said early on, and as I think you will now appreciate, there are various directions in which this theory needs further development. Nevertheless, I think it is on the right lines and I hope it has an interesting future. Even at this stage I have found it helpful in orienting my thinking about rare earth metals!

REFERENCES

1. Stevens, K.W.H. (1972). *J. Phys.*, **C5**, 1360, 2217.
2. Wannier, G. (1937). *Phys. Rev.*, **52**, 191.
3. Kohn, W. (1973). *Phys. Rev.*, **B7**, 4388.
4. Ziman, J.M. (1970). *Elements of Advanced Quantum Theory, Vol. 1*, (Cambridge University Press).
5. White, R.M. (1970). *Quantum Theory of Magnetism*, (McGraw-Hill, New York).
6. Schrieffer, J.R. and Wolf, P.A. (1966). *Phys. Rev.*, **149**, 491.

MAGNETIC PROPERTIES OF A
DENSE INTERACTING ELECTRON GAS

M.L. GLASSER†

Battelle Memorial Institute,
505 *King Avenue, Columbus, Ohio* 43201

The electron gas is an extremely reasonable and versatile model to use when one is trying to understand the nature of 'shiny' things such as plasmas and metals. It is for this reason so much attention and effort has been devoted to the calculation of its physical properties, particularly the ground state energy. Unfortunately, exact results have been obtained only for densities higher or lower than the electron densities of real metals, but plausible approximation and interpolation schemes are presently available for the metallic range. Hence, the ground state properties of the electron fluid (in a uniformly charged positive background) are reasonable accurately known as a function of density [1,2].

Since magnetic fields (both internally and externally generated) play a key rôle in experimental investigation of metals, it is imperative to gain some understanding of the behaviour of the ground state energy as one moves from the $H = 0$ axis in the magnetic field-density plane, and to have an idea of the phase boundaries existing in this plane. The purpose of these lectures is to present the results of a preliminary investigation of this problem for the case of a high density electron fluid.

The organization of these talks is as follows: we first review the quantum mechanics of free electrons in a magnetic field and calculate the auxiliary quantities needed; next, we describe a procedure for obtaining the principal terms in the ground state energy on the basis of the many body formalism given by

---

† Present address: Department of Applied Mathematics, University of Waterloo, Waterloo, Ontario N2L 3Gl, Canada.

Martin and Schwinger [3]. Thirdly, we summarize the results of
the calculation obtained in this way for the cases of weak, intermediate and 'super-strong' magnetic fields; finally, some
interesting features of the results will be discussed along with
suggestions for future investigation. For the most part, the
mathematical details of the calculations will be omitted as well
as references and comparison with previous work; these will be
presented at length in a review article which is in preparation.

We begin by setting some notation. Our system is a collection of N electrons of charge e and effective mass $m*$ contained
in an indefinitely large box of volume V neutralized by a uniform positively charged background, and in the presence of an
externally applied uniform magnetic field $\vec{H} = \vec{\nabla} \times \vec{A}$. Thus, the
system is characterized by the two parameters ($\hbar = 2m = 1$)

$$r_{\mathrm{S}} = \left(\frac{3V}{4\pi N}\right)^{1/3}, \qquad H,$$

and we are concerned with the ground state energy per unit volume $E_0(r_{\mathrm{S}}, H)$. In addition, we shall make much use of the thermodynamic Gibbs' potential

$$W = \ln \mathrm{Tr} \, \{\exp(-\beta(\mathcal{H} - \zeta N))$$

where ζ is the chemical potential, $\beta = 1/kT$ is the inverse temperature and $\mathcal{H}$ is the Hamiltonian of the system.

The Hamiltonian for a free electron $H = 0$ is

$$h = \frac{p^2}{2m*} \, ,$$

and describes rectilinear motion corresponding to plane waves.
When $H \neq 0$, this becomes

$$h = \frac{\pi^2}{2m*} + a\mu_0 * \vec{H} \cdot \vec{\sigma},$$

$$\pi = \vec{p} + \frac{e}{c}\vec{A}, \qquad \mu_0* = \frac{e}{2m*c} \, ,$$

$$a = \frac{gm*}{2m} \, ,$$

(we shall use units such that $\hbar = 1$) and corresponds to helical
paths whose wave functions $\psi_\sigma(\vec{r}, t)$ are plane waves along the
field and oscillator functions normal to it. Fortunately, it
is not necessary to deal with these complicated functions individually, but only a single function of space and time: the (retarded) single particle Green's function. In our case, this is
defined by

$$iG_<(1,2) = iG_<(\vec{r}_1 t_1, \vec{r}_2 t_2) = \qquad\qquad \text{(Contd)}$$

(Contd)

$$
= \begin{cases}
0, & t_2 < t_1, \\[2ex]
\dfrac{e^{i\zeta(t_1-t_2)}}{\mathrm{Tr}\ e^{-\beta(\mathcal{H}-\zeta N)}}\ \mathrm{Tr}\ \{e^{-\beta(\mathcal{H}-\zeta N)}\psi_{\sigma_1}(\vec{r}_1 t_1)\psi_{\sigma_2}{}^+(\vec{r}_2 t_2)\} & t_2 > t_1,
\end{cases}
$$

and similarly for the advanced Green's function $G_>(1,2)$. For non-interacting electrons these functions are the outgoing and incoming solutions to the equation

$$
\left[i\,\frac{\partial}{\partial t_1} + \zeta - h_1\right]G^0(1,2) = \delta(\vec{r}_1 - \vec{r}_2)\delta(t_2 - t_1).
$$

From the invariance of the trace under cyclic permutations, we find

$$
G_<(1,2) = -\ G_>(\vec{r}_1 t_1 + i\beta, 2),
$$

which, when h is time-independent, leads to the spectral representation

$$
G_<(1,2) = -\int\frac{d\omega}{2\pi i}\ e^{-i\omega(t_1-t_2)}\frac{A(\vec{r}_1,\vec{r}_2,\omega)}{1 + e^{\beta\omega}}.
$$

The spectral density $A(\vec{r}_1,\vec{r}_2,\omega)$ is temperature independent and obeys the simple equation (for free electrons)

$$
\left[i\,\frac{\partial}{\partial t} + \zeta - h_1\right]A^0(\vec{r}_1,\vec{r}_2,t) = 0.
$$

In our case, this can be solved by Fourier analysis, whence we find

$$
A^{\sigma_1\sigma_2}(\vec{r}_1,\vec{r}_2,t)
$$

$$
= \delta_{\sigma_1\sigma_2}C(\vec{r}_1,\vec{r}_2)\int\frac{d\vec{p}}{(2\pi)^3}\ e^{i\vec{p}(\vec{r}_1-\vec{r}_2)}\exp\left[i\left\{\zeta - a\mu_0{}^*H\sigma_1 - \frac{p_z^2}{2m^*}\right\}t\right]
$$

$$
\times\ \sec(\mu_0{}^*Ht)\exp\left[-i\,\frac{p_x^2 + p_y^2}{2m^*\mu_0{}^*H}\tan\mu_0{}^*Ht\right],
$$

where C is a unitary factor which depends on the choice of gauge and is unity for $\vec{r}_1 = \vec{r}_2$. Therefore, after some mathematical manipulation, we have

$$G_<^{\sigma\sigma'}(\vec{r},t,\vec{r}'t')$$

$$= iC(\vec{r},\vec{r}')\delta_{\sigma\sigma'}\int_{c-i\infty}^{c+i\infty}\frac{ds}{2\pi i}\frac{\pi e^{s\beta\zeta}}{\sin[\pi(s+i\tau/\beta)]}\,\text{sech}(\mu_0{}^*H\beta s)$$

$$\times\int\frac{d\vec{p}}{(2\pi)^3}\,e^{i\vec{p}\cdot(\vec{r}-\vec{r}')}\exp\left[-\beta\left(a\mu_0{}^*H\sigma+\frac{p_z^2}{2m^*}\right)s\right.$$

$$\left.-\frac{\bar{p}^2}{2m^*\mu_0{}^*H}\tanh(\mu_0{}^*H\beta s)\right]\,,\tag{1}$$

$$\bar{p}=(p_x,p_y,0),\qquad\tau=t-t',$$

The s-integral is an inverse Laplace transform and we have used the identity

$$\int_{-\infty}^{\infty}d\omega\,\frac{e^{\beta\omega}}{e^{\beta(\omega-\zeta)}+1}=\frac{\pi e^{p\zeta}}{\beta\sin(\pi p/\beta)}\,.$$

By inspection, it can be seen that the internal energy of the non-interacting system of electrons is given by

$$E=-\frac{i}{V}\,\text{Tr}_\sigma\int d\vec{r}\,\lim_{\vec{r}\to\vec{r}'}[hG_<^{\sigma\sigma'}(\vec{r}t,\vec{r}'t^+)]_{t=0}\,,\tag{2}$$

and the density by

$$\rho=\frac{N}{V}=-\frac{i}{V}\,\text{Tr}_{\vec{\sigma}}\int dr G_<^{\sigma\sigma'}(\vec{r}t,\vec{r}t^+)|_{t=0}\,,\tag{3}$$

from which we can determine $\zeta(\rho)$, and hence all the thermodynamic properties. The ground state energy is the $\beta\to\infty$ limit of E. As an exercise, the reader is invited to check that the proper results are obtained in the $H=0$ limit.

To investigate the effects of electron interactions, we go back to the Gibbs free energy W from which the internal energy is given by

$$E=-\frac{1}{V}\frac{\partial W}{\partial\beta}\bigg|_{\beta\zeta,V}=\frac{\text{Tr}\{\mathcal{H}e^{-\beta(\mathcal{H}-N\zeta)}\}}{V\text{Tr}\{e^{-\beta(\mathcal{H}-N\zeta)}\}}\equiv\frac{1}{V}\langle\mathcal{H}\rangle.$$

We now have

$$\mathcal{H} = \mathcal{H}_0 + \mathcal{H}_c(e^2),$$

where

$$\mathcal{H}_0 = \sum_{i=1}^{N} h_i,$$

and

$$\mathcal{H}_c(x) = \tfrac{1}{2}\iint d\vec{r}_1 d\vec{r}_2 \psi^+(\vec{r}_1 t)\psi^+(\vec{r}_2 t)v_x(r_{12})\psi(\vec{r}_2 t)\psi(\vec{r}_1 t),$$

$$v_x(r) = \frac{x}{r} .$$

Formally, we find

$$\frac{\partial W}{\partial x} = -\frac{\beta}{x}\langle \mathcal{H}_c(x)\rangle,$$

so $(\mathcal{H}_c(0) = 0)$

$$W = W_0 - \beta\int_0^{e^2}\frac{dx}{x}\langle \mathcal{H}_c(x)\rangle.$$

We now consider the two-particle Green's function:

$$G^x(1,2;1',2')$$

$$= e^{i\zeta(t_1-t_1'+t_2-t_2')}\langle T\psi(\vec{r}_1 t_1)\psi(\vec{r}_2 t_2)\psi^+(\vec{r}_1' t_1')\psi^+(\vec{r}_2' t_2')\rangle,$$

where the superscript x denotes that the Hamiltonian involves $\mathcal{H}_c(x)$ and T is the (Fermion) time ordering symbol. In particular,

$$\langle \psi^+(\vec{r}_1 t)\psi^+(\vec{r}_2 t)\psi(\vec{r}_2 t)\psi(\vec{r}_1 t)\rangle = -G^x(1,2;1^+,2^+)\big|_{t_1=t_2} .$$

Therefore,

$$W = W_0 + \tfrac{1}{2}\beta\int d\vec{r}_1 d\vec{r}_2 \int_0^{e^2}\frac{dx}{x}v_x(r12)G^x(1,2;1^+,2^+)\big|_{t_1=t_2} \qquad (4)$$

so that once we have the two-particle Green's function, all the thermodynamic properties of the system can be calculated. Unfortunately, G^x is beyond our grasp as yet and we must resort to some approximations.

The simplest of these is the Hartree-Fock approximation (HF):

$$G^X(1,2;1',2') = G^0(1,1')G^0(2,2') - G^0(1,2')G^0(2,1')$$

(in our case $G^0 = G_<$). After inserting this into (4), we find the term

$$W = \tfrac{1}{2}\beta \iint d\vec{r}_1 d\vec{r}_2 \int_0^{e^2} \frac{dx}{x}\, v_x(r_{12}) G^\sigma(1,1^+) G^0(2,2^+),$$

which corresponds to the energy

$$E' = \frac{1}{2}\frac{e^2}{V}\iint d\vec{r}_1 d\vec{r}_2 \frac{\rho(\vec{r}_1)\rho(\vec{r}_2)}{r_{12}},$$

where $\rho(r) = |\psi(\vec{r})|^2$, and this exactly cancels the interaction of the positively charged background with itself and with the electrons. Therefore, in the HF approximation, we have

$$W_{HF} = W_0 + W_{ex},$$

$$W_{ex} = -\tfrac{1}{2}e^2\beta \mathrm{Tr}_\sigma \iint \frac{d\vec{r}_1 d\vec{r}_2}{r_{12}}\, G_<^{\sigma_1\sigma_2}(\vec{r}_1\sigma;\vec{r}_2,\sigma^+) G_<^{\sigma_2\sigma_1}(\vec{r}_2,\sigma;\vec{r}_1\sigma^+),$$

(5)

where the latter term is the *exchange* energy.

Following Wigner, it has become customary to call the difference between the exact and the Hartree-Fock energies the *correlation* energy. Hence, the correlation term in the Gibbs free energy is

$$W_c = W - (W_0 + W_{ex}) = W_c^{(1)} + W_c^{(2)} - W_{ex},$$

where

$$W_c^{(1)} = \tfrac{1}{2}\beta \iint d\vec{r}_1 d\vec{r}_2 \int_0^{e^2}\frac{dx}{x}\, v_x(r_{12})[G^X(1,2;1^+,2^+)$$

$$- G^X(1,1^+)G^X(2,2^+)]\big|_{t_1=t_2},$$

$$W_c^{(2)} = \tfrac{1}{2}\beta \iint d\vec{r}_1 d\vec{r}_2 \int_0^{e^2}\frac{dx}{x}\, v_x(r_{12})[G^X(1,1^+)G^X(2,2^+)$$

$$- G^0(1,1^+)G^0(2,2^+)]\big|_{t_1=t_2}.$$

If the system is subjected to a weak external potential $U(1)$, the electrons will redistribute themselves in such a way as to screen it partially and a test charge at the point $\vec{r}_2 t_2$ will experience a different potential $V(2)$. We define an *inverse dielectric function*

$$K_\chi(1,2) = \frac{\delta V(2)}{\delta U(1)} \; .$$

As shown by Martin and Schwinger

$$K_\chi(1,3) = \delta(1,3) + i\int d\vec{r}_2 v_\chi(r_{12})[G^\chi(3,2;3^+,2^+)$$

$$- G^\chi(3,3^+)G^\chi(2,2^+)].$$

For a homogeneous system, K_χ depends only on $t_1 - t_2$ and on the vector $\vec{r}_1 - \vec{r}_2$ and it is convenient to consider its Fourier transform with respect to these variables: $K_\chi(\vec{p},\omega)$. In terms of this quantity $W_c^{(1)}$ can be written

$$W_c^{(1)} = \frac{1}{2i} \beta V \int \frac{d\vec{p}}{(2\pi)^3} \int \frac{d\omega}{2\pi} \int_0^{e^2} \frac{dx}{x} [K_\chi(\vec{p},\omega) - 1].$$

However, $K_\chi(\vec{p},\omega)$ has the spectral representation

$$K_\chi(\vec{p},\omega) - 1 = \int \frac{d\omega'}{2\pi} \left\{ \frac{\mathcal{P}}{\omega - \omega'} - \pi i\coth(\tfrac{1}{2}\beta\omega)\delta(\omega - \omega') \right\} A_\chi(\vec{p},\omega'),$$

where the spectral density obeys the sum rule

$$\int \frac{d\omega}{2\pi} A_\chi(\vec{p},\omega) = 0.$$

For a uniform system, the spectral density in this case is simply the imaginary part of the reciprocal of the longitudinal dielectric function. Consequently, we have

$$W_c^{(1)} = \tfrac{1}{2}\beta V \int_0^{e^2} \frac{dx}{x} \int \frac{d\vec{p}}{(2\pi)^3} \int \frac{d\omega}{2\pi} \coth(\tfrac{1}{2}\beta\omega) A_\chi(\vec{p},\omega).$$

In the presence of the perturbation $U(1)$, the exact single particle Green's function obeys the integro-differential equation

$$\left[i\frac{\partial}{\partial t} + \varsigma - h - V(1) - i\int v_\chi(r_{12})\frac{\delta V(3)}{\delta U(3)}\frac{\delta}{\delta V(3)}\right]G^\chi(1,1') = \delta(1,1'),$$

which is intractible. By neglecting the complicated integral term, we find the much simpler Hartree equation

$$\left[i\,\frac{\partial}{\partial t}\,-\,h\,+\,\zeta\,-\,v(1)\right]G(1,1') \;=\; \delta(1,1').$$

In this approximation, $K_\chi(1,2)$ can be easily calculated and the result is called the *random phase approximation*. Physically, it corresponds to assuming that in the screening process, all that can occur is the creation and subsequent annihilation of electron-hole pairs. In terms of the ground state energy, it corresponds to the classical ring diagram approximation. Since it is clear that in this approximation $G^\chi(1,2) \;=\; G^0(1,2)$, the term $W_c{}^{(2)}$ vanishes. Therefore, in terms of the respective spectral densities for the exchange and random phase energies we have

$$W_c \;=\; -\,\tfrac{1}{4}\beta V\int_0^{e^2}\frac{dx}{x}\int\frac{d\vec{p}}{(2\pi)^3}\int\frac{d\omega}{2\pi}\,\coth\,(\tfrac{1}{2}\omega\beta)\,[A_\chi^{RPA}(\vec{p},\omega)\,-\,A_\chi^{HF}(\vec{p},\omega)].$$

The expression in square brackets is precisely the discontinuity of the quantity

$$A(\vec{p},\omega) \;=\; \left\{\left[1\,+\,\frac{x}{e^2}\,\alpha(\vec{p},\,-\,i\omega)\right]^{-1}\,-\,\left[1\,-\,\frac{x}{e^2}\,\alpha(\vec{p},\,-\,i\omega)\right]\right\}$$

across the real ω axis where $\alpha(\vec{p},\omega)$, the dynamic longitudinal polarizabiltiy for a free electron gas in a magnetic field, has a branch cut. The latter quantity has been thoroughly studied by Horing, Glasser and Kaplan, and others [4,5,6].

The frequency integral in W_c can be expressed as a contour integral:

$$-\,\frac{1}{2\pi i}\int_C d\xi\coth(\tfrac{1}{2}\beta\xi)A(\vec{p},\xi)$$

ξ-plane

The quantity A is analytic off the real axis and vanishes as ξ^{-2} at infinity, so the contour can be deformed into a path circling the imaginary axis, thereby enclosing the simple poles of coth $(\tfrac{1}{2}\beta\xi)$ at $\xi_n = 2\pi ni/\beta$, $n = \pm1,\pm2,\dots$. Hence, by residues we obtain

$$-\frac{2}{\beta} \sum_{-\infty}^{\infty}{}' \; A(\vec{p},\xi n).$$

After the trivial coupling constant integration, we obtain

$$W_c = -\frac{1}{2}V\int\frac{d\vec{p}}{(2\pi)3} \sum_{-\infty}^{\infty}{}' \{\ln[1 + \alpha(\vec{p}, -i\xi_n)] - \alpha(\vec{p}, -i\xi_n)\}.$$

Finally, at low temperatures,

$$\frac{1}{\beta} \sum_{u=1}^{\infty} F\left(\frac{2\pi u}{\beta}\right) \to \frac{1}{2\pi}\int_0^{\infty} d\omega F(\omega),$$

and after an integration by parts, the low temperature correlation Gibbs' free energy is

$$W_c = -\frac{1}{2}\beta V\int\frac{d\vec{p}}{(2\pi)3}\int_{-\infty}^{\infty} \frac{d\omega}{2\pi} \; \frac{\omega\alpha(\vec{p},\omega)\alpha'(\vec{p},\omega)}{1 + \alpha(\vec{p},\omega)} \; , \tag{6}$$

where the prime denotes integration with respect to the second argument. From the work of Horing [4] we have

$$\alpha(\vec{p},\omega) = -\frac{4\pi e^2}{p^2}\int\frac{ds}{2\pi i} \; \frac{e^{s\varepsilon\beta}}{\sin(s\pi)} \; \eta(s)B(\vec{p},\omega,s), \tag{7}$$

where

$$\eta(s) = 2\left(\frac{m^*}{2\pi}\right)^{3/2}(\mu_0^*H)(\beta s)^{-\frac{1}{2}} \frac{\cosh(\mu_0^*Ha\beta s)}{\sinh(\mu_0^*H\beta s)} \; , \tag{7a}$$

is the Laplace transform of the Landau density of states and

$$B(\vec{p},\omega,s) = \exp\left(-\frac{\beta p_z^2 s}{8m^*}\right)\exp-\left(\frac{\bar{p}^2}{4m^*\mu_0^*H}\right)\coth(\mu_0^*H\beta s)$$

$$\times\int_0^{\infty} dt e^{-|\omega|t}\left\{\exp\left(-\left(\frac{p_z^2}{8m^*\beta s}\right)(2t - i\beta s)^2\right]\right. \tag{7b}$$

$$\times\exp\left[\left(\frac{\bar{p}^2}{4m^*\mu_0^*H}\right)\frac{\cos(\mu_0^*H(2t - i\beta s))}{\sinh(\mu_0^*H\beta s)}\right] - (i \to -i)\right\}.$$

To gain an idea of how the calculations are organized, we

consider in detail the evaluation of the chemical potential from the identity (2) for the case on non-interacting electrons. By inserting (1) into (2) we obtain for $\beta \to \infty$

$$\rho = \frac{1}{4}(\mu_0^* H)\left(\frac{2m^*}{\pi}\right)^{3/2} \int_{c-i\infty}^{c-i\infty} \frac{ds}{2\pi i} \frac{e^{s\zeta}}{s^{3/2}} \frac{\cosh(a\mu_0^* Hs)}{\sinh(\mu_0^* Hs)} \quad . \tag{8}$$

The integrand of the complicated integral on the right hand side of this expression has a structure which is characteristic of all the quantities we shall consider: there is a branch point at $s = 0$ and isolated singularities along the imaginary s-axis at the points $s_n = n\pi i/\mu_0^* H$, $n = \pm 1, \pm 2, \ldots$. When H is not too large, the branch cut, as we shall see, determines the 'monotonic' field dependence of the property in question (in this case ξ) whilst the isolated singularities give rise to the so called de Haas-van Alphen oscillations. On the other hand, when H grows very large, the isolated singularities merge with the branch point and the oscillations disappear.

In the former case, by closing the contour to the left by a semicircle indented so as to avoid the negative s-axis, we find

$$\rho = \frac{1}{4} \mu_0^* H \left(\frac{2m^*}{\pi}\right)^{3/2} \left[\rho_B + \sum_{-\infty}^{\infty}{}' \rho_n\right],$$

where

$$\rho_B = \frac{1}{2\pi i}\int_{\Gamma} ds \ s^{-3/2} e^{s\xi} \frac{\cosh(a\mu_0^* Hs)}{\sinh(\mu_0^* Hs)} \quad ,$$

and

$$\rho_n = \frac{1}{2\pi i}\oint_{C_n} ds \ s^{-3/2} \frac{\cosh(a\mu_0^* Hs)}{\sinh(\mu_0^* Hs)} \ e^{s\zeta}.$$

In these expressions Γ is a loop enclosing the negative real s-axis in the positive sense and C_n is a small circle about s_n. For simplicity, let $\eta = \zeta/\mu_0^* H \gg 1$ and make the substitution $\mu_0^* Hs = t$. Then by using Hankel's formula

$$\frac{1}{2\pi i}\int_{\Gamma} dt \ t^{-v} e^{\eta t} = \frac{\eta^{v-1}}{\Gamma(v)}$$

we find

$$\rho_B = \frac{(\mu_0^* H)^{1/2} \eta^{3/2}}{\Gamma(5/2)} \left[1 + \frac{3}{8}\left(a^2 - \frac{1}{3}\right)\eta^{-2} + \ldots \right] \quad .$$

Note that in this case the singularities s_n are simple poles, so by using the residue theorem we find

$$\rho_n = \frac{\mu_0^* H}{(n\pi i)^{3/2}} (-1)^n \cos(a\pi n) e^{n\pi i \eta},$$

and therefore

$$\rho = \frac{(2m^* \zeta)^{3/2}}{3\pi^2} \left[1 + \frac{3}{8}\left(a^2 - \frac{1}{3} \right) \left(\frac{\mu_0^* H}{\zeta} \right)^2 + \cdots \right.$$

$$\left. + \frac{3}{2\pi} \left(\frac{\mu_0^* H}{\zeta} \right)^{3/2} \sum_1^\infty \frac{(-1)^n \cos(a\pi n)}{n^{3/2}} \sin\left[\frac{n\pi\zeta}{\mu_0^* H} - \frac{\pi}{4} \right] \right].$$

This must be solved to obtain ζ in terms of H (and ρ); the solution is plotted schematically for the case $m^* = m$, $a = 1$ below.

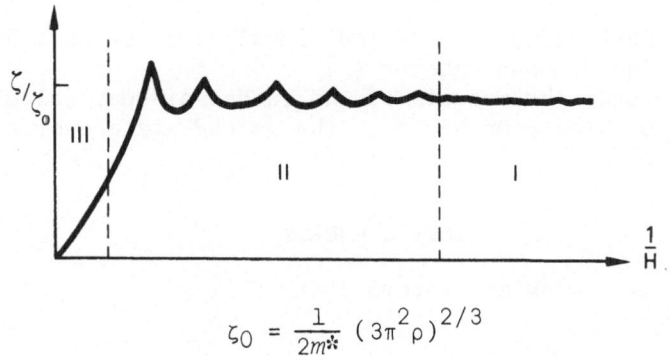

$$\zeta_0 = \frac{1}{2m^*} (3\pi^2 \rho)^{2/3}$$

As suggested by our examination of the integrand above, we can distinguish three regions of field strength:

I. *Weak Fields*: ζ differs little from the zero field value,

$$\frac{\zeta}{\zeta_0} \cong 1 + \frac{3}{8}\left[\left(\frac{gm^*}{2m} \right)^2 - \frac{1}{3} \right] \left(\frac{\mu_0^* H}{\zeta_0} \right)^2 + \cdots ,$$

where $(\mu_0^* H/\zeta_0) \ll 1$;

II. *Intermediate Fields*: ζ displays perceptible de Hass-van Alphen oscillations although its mean value is close to ζ_0;

III. *High Field Quantum Limit*: (HFQL): ζ is a monotonically decreasing function of H. The electrons are frozen in-to the lowest Landau level whose degeneracy increases with

H so the electrons can give up their kinetic energy along the field and ζ becomes small algebraically.

In the latter case, the above treatment is no longer applicable, but we may simple replace the hyberbolic functions in (8) by exponentials (thus emphasising the discrete Landau level structure) to obtain

$$\rho \cong \frac{1}{4}(\mu_0^*H)\left(\frac{2m^*}{\pi}\right)^{3/2}\zeta^{1/2}\int_{c-i\infty}^{c+i\infty}\frac{ds}{2\pi i}\,s^{-3/2}e^s\exp\left[\left(\frac{a\mu_0^*Hs}{\zeta}\right)-\left(\frac{\mu_0^*Hs}{\zeta}\right)\right]$$

$$=\begin{cases}\mu_0^*H(m^*)^{3/2}\pi^{-2}[2\zeta+(a-1)\mu_0^*H]^{1/2}, & 2\zeta+(a-1)\mu_0^*H>1,\\[2mm]0, & 2\zeta+(a-1)\mu_0^*H<1.\end{cases}$$

Thus, for $a=1$, for example

$$\zeta \cong \frac{\rho^2\pi^4}{2(m^*)^3(\mu_0^*H)^2}$$

Other cases must be considered individually; a few have been discussed by Anselm and Asherov [7].

We now present the results of the rather arduous calculations above for the three components of the ground state energy in turn.

KINETIC ENERGY

Weak and Intermediate Fields

$$E_k = \frac{(2m^*)^{3/2}\zeta^{5/2}}{5\pi^2}\left\{1-\frac{5}{8}\left(a^2-\frac{1}{3}\right)\left(\frac{\mu_0^*H}{\zeta}\right)^2+\dots\right\}$$

$$+\frac{5}{2\pi^2}\left(\frac{\mu_0^*H}{\zeta}\right)^{5/2}\sum_{n=1}^{\infty}\frac{(-1)^n\cos(n\pi a)}{n^{5/2}}$$

$$\times\left\{\left(\frac{n\pi\zeta}{\mu_0^*H}\right)\sin\left(\frac{n\pi\zeta}{\mu_0^*H}-\frac{\pi}{4}\right)+\cos\left(\frac{n\pi\zeta}{\mu_0^*H}-\frac{\pi}{4}\right)\right\}.$$

High Field Quantum Limit

$$E_K = \left(\frac{\mu_0^*H}{6\pi^2}\right)(2m^*\zeta)^{3/2}.$$

EXCHANGE ENERGY

Weak and Intermediate Fields

$$E_{ex} = \frac{e^2 (m^*)^2 \zeta^2}{\pi 3}$$

$$\times \left\{ 1 - \left[\left(a^2 - \frac{1}{3} \right) + \frac{1}{18} + c + \ln 2 - \frac{1}{18} \ln(\infty) \right] \left(\frac{\mu_0^* H}{\zeta} \right)^2 + \ldots \right.$$

$$- \frac{2}{\pi} \left(\frac{\mu_0^* H}{\zeta} \right)^{1/2} \sum_{n=1}^{\infty} \frac{(-1)^n \cos(n\pi\zeta)}{n^{3/2}}$$

$$\left. \times \left[n\pi \cos \left(\frac{n\pi\zeta}{\mu_0^* H} - \frac{\pi}{4} \right) + \ln \left(\frac{4n\pi\zeta}{\mu_0^* H} \right) \sin \left(\frac{n\pi\zeta}{\mu_0^* H} - \frac{\pi}{4} \right) \right] \right\} .$$

High Field Quantum Limit

$$E_{ex} = - \frac{e^2 (m^* \mu_0^* H)^2}{8\pi 4} F\left(\frac{2\zeta}{\mu_0^* H} \right) , \qquad F(x) \cong x \left(1 - \frac{7}{6} x \ln x \right) .$$

CORRELATION ENERGY

Weak and Intermediate Fields

$$E_c = - \frac{2(m^*)^2 e^4 (2m^* \zeta^3)^{1/2}}{6\pi 4} (1 - \ln 2) \ln r_s - \frac{e^2 (m^*)^2 \zeta^2}{18\pi 3} - \ln(\infty) \left(\frac{\mu_0^* H}{\zeta} \right)^2$$

$$- \frac{(m^*)^2 e^4 (2m^* \zeta^3)^{1/2}}{8\pi 4} \left[(1 - \ln 4) \left(a^2 - \frac{1}{3} \right) - \frac{1}{9} \right] \ln(r_s) \left(\frac{\mu_0^* H}{\zeta} \right)^2$$

$$+ \left\{ \frac{(m^*)^2 e^2 \zeta^2}{36\pi 3} \ln r_s \right.$$

$$+ \frac{(m^*)^2 e^2 \zeta^2}{36\pi 3} \left[\ln \left(\frac{4}{\pi} \left(\frac{4}{9\pi} \right)^{1/3} \right) + 2 - \frac{16}{3\pi} \right] \right\} \left(\frac{\mu_0^* H}{\zeta} \right)^2$$

$$+ \frac{m^* e^4 (2m^* \zeta)^{3/2}}{3\pi 5} (1 + 3\ln 2) \ln(r_s) \left(\frac{\mu_0^* H}{\zeta} \right)^{3/2} \times \qquad \text{(Contd)}$$

(Contd) $\qquad \times \sum_{n=1}^{\infty} \frac{(-1)^n \cos(n\pi a)}{n^{3/2}} \sin\left(\frac{n\pi\zeta}{\mu_0 *H} - \frac{\pi}{4}\right)$.

High Field Quantum Limit

$$E_c = \frac{e^2 (m*)^2}{8\pi 3} (\mu_0 *H)^2 r \ln r, \qquad r = \frac{e^4 m*}{2\zeta} .$$

The expressions given above are exact in the sense that they represent all the terms in the ground state energy (per particle) which are singular in r_s, i.e. those through $\ln r_s$. Note that in the exchange and correlation energies there appear terms proportional to $\ln(\infty)$ and that these exactly cancel. The infinity appearing here represents the range of the Coulomb interaction. This is a very interesting result and has the following significance. Shortly after Overhauser proposed that the ground state of the electrons in the alkali metals might actually be a spin density wave, Mermin and Celli investigated the Hartree-Fock equations for an electron gas in a magnetic field and proved that they were unstable; that is the Hartree-Fock ground state was not the lowest state for the system. However, they left open the question of the precise origin of this instability and whether it might be removed by the inclusion of correlation effects. Our results provide a direct answer to these questions: The instability is due to the long range of the inter-particle interaction and is exactly cancelled by the ring diagram correlation.

The equations given above are not yet in usable form because they involve the exact chemical potential which must be determined anew from the relation

$$N = \beta^{-1} \left(\frac{\partial W}{\partial \zeta}\right)_{\beta, v}$$

each time a new term is added to the energy. For example, if we consider only the kinetic energy, then we have already found that

$$\frac{\zeta}{\zeta_0} = 1 + \frac{3}{8}\left[\left(\frac{gm*}{2m}\right)^2 - \frac{1}{3}\right]\left(\frac{\mu_0 *H}{\zeta_0}\right)^2 + \cdots ,$$

and therefore the kinetic energy is

$$E_k = \frac{(2m*)^{3/2} \zeta_0^{5/2}}{5\pi^2} \left\{1 - \frac{5}{4}\left(a^2 - \frac{1}{3}\right)\left(\frac{\mu_0 *H}{\zeta_0}\right)^2 + \cdots \right.$$

$$\left. + \frac{5}{2\pi 2}\left(\frac{\mu_0 *H}{\zeta_0}\right)^{5/2} \sum_{n=1}^{\infty} \frac{(-1)^n \cos(n\pi a)}{n^{5/2}} \times \right.$$

(Contd)

(Contd) $\times \left[\left(\dfrac{n\pi \zeta_0}{\mu_0 {*} H} \right) \sin \left(\dfrac{n\pi \zeta_0}{\mu_0 {*} H} - \dfrac{\pi}{4} \right) + \cos \left(\dfrac{n\pi \zeta_0}{\mu_0 {*} H} - \dfrac{\pi}{4} \right) \right] \Big\}$.

Similarly, in the high field quantum limit,

$$\zeta \cong \frac{\rho^4 \pi^4}{2(m{*})^3 (\mu_0 {*} H)^2} ,$$

so

$$E_k \cong \frac{\rho^3}{6\pi^2 (m{*})^3 (\mu_0 {*} H)^2} .$$

In the case of weak and intermediate magnetic fields, denoting by ζ' the corrected zero field chemical potential, we shall have a relation of the form

$$\zeta \doteq \zeta' \left[1 + \gamma \left(\frac{\mu_0 {*} H}{\zeta'} \right)^2 + \dots \right] .$$

Then, in the above expression for the terms in the energy, ζ must be replaced by the full expression above in the field independent term, but merely by ζ' in the field dependent terms, to give a result correct to leading order in the magnetic field. We illustrate this for the Hartree-Fock approximation in zero magnetic field.

We have $W = W_K + W_{ex}$ and correspondingly

$$N = N_K + N_{ex}, \qquad E = E_K + E_{ex}.$$

We have seen that

$$N_k = (2m{*}\zeta)^{3/2} \frac{V}{3\pi^2} , \qquad E_k = (2m{*})^{3/2} \frac{\zeta^{5/2}}{5\pi^2} .$$

But, in addition, it is straightforward to show that

$$N_{ex} = 2(em{*})^2 \frac{\zeta V}{\pi^3} , \qquad E_{ex} = (em{*})^2 \frac{\zeta^2}{\pi^3} .$$

Hence the chemical potential is to be determined from the relation

$$\rho = \frac{(2m{*})^{3/2}}{3\pi^2} \zeta^{3/2} + \frac{2(em{*})^2}{\pi^3} \zeta$$

If, in the spirit of the Hartree-Fock approximation we regard e^2 (or equivalently r_s) as small, we may set

$$\zeta \cong \zeta_0[1 + e^2\delta]$$

and linearize the resulting equation to obtain δ. This gives

$$\zeta \cong \zeta_0\left[1 - \frac{e^2}{\pi}\left(\frac{2m^*}{\zeta_0}\right)^{\frac{1}{2}}\right]$$

Consequently

$$E_k \cong \frac{(2m^*)^{3/2}}{5\pi^2}\zeta_0^{5/2} - \frac{(2m^*)^2\zeta_0^2 e^2}{2\pi^3}$$

whilst

$$E_{ex} \cong \frac{(2m^*)^2 e^2 \zeta_0^2}{4\pi^3}$$

Dividing by the density to obtain the energy per particle, taking $m^* = m$, and using atomic units, we obtain

$$E_{HF} = E_k + E_{ex} = \frac{2.21}{r_s^2} - \frac{0.916}{r_s} \quad,$$

which recovers the 'text book' result. Note that in the finite temperature formalism we are using, the kinetic energy actually includes a contribution to exchange owing to the correction to the chemical potential. This feature, which persists with respect to all the higher order terms in the energy, makes a direct comparision with the usual zero temperature formalism very difficult.

As a second example, we consider the case of the High Field Quantum Limit. Using atomic units and introducing the quantity $h = \frac{1}{2}\mu_0 H/R$, where R is the Rydberg, and the quantity $p = 2\zeta/\mu_0 H$, in the HFQL we have $hr_s^2 > 1.84$. In these terms the kinetic contributions to the energy and density are

$$E_k = \frac{h^{5/2}}{6\pi^2}p^{3/2},$$

$$\rho_k = \frac{h^{3/2}p^{1/2}}{\pi^2} \quad,$$

whilst the corresponding exchange contributions are

$$E_{ex} = - \frac{h^2 p}{4\pi^4} \, ,$$

$$\rho_{ex} = \frac{h}{2\pi^4} (2 - C - \ln p),$$

where $C = 0.577$ is Euler's constant. The chemical potential or p is determined by the condition

$$\rho_k + \rho_{ex} = \frac{3}{4\pi r_s^3} \, ,$$

or

$$F(p) \equiv (hp)^{\frac{1}{2}} \pi^2 + 2 - C - \ln p = \frac{3\pi^3}{4hr_s^3} \, .$$

The expression on the right hand side has a minimum value $F_0 = 6.616 + \ln h$ at the argument $p_0 = 0.041/h$. Hence, if the density is greater than the critical value $\rho_c = hF_0/\pi^4$, we have the approximate solution

$$p = 4.15 \exp\left(- \frac{3\pi^3}{4hr_s^3}\right) \, .$$

If the density falls below the critical value, our approximations become inapplicable and other terms must be included in the energy. When our solution is inserted into the expressions for the energy we find

$$\left| \frac{E_{ex}}{e_0} \right| \cong \left(\frac{3h}{2\pi^2} \right) p^{-\frac{1}{2}} \, .$$

Hence, if $p < 9h^2/4\pi^4$, the interaction energy exceeds the kinetic energy which signals the onset of an ordered state in which the exchange energy is maximised. One possible configuration corresponds to an ordered arrangement of charged rods parallel to the field [8]. In such high magnetic fields, the electron spins are all aligned parallel so that the exchange energy, which is essentially the energy of interaction amongst the electrons having parallel spins, is the dominant term in this part of the energy. Indeed, it can be shown explicitly that in this case the correlation energy is negligible.

Finally we present the leading terms for the ground state energy per electron on the weak field case for a free electron gas ($m^* = m$) in atomic units. Because of the mixing brought about by the chemical potential the division into kinetic, exchange and correlation terms is merely suggestive

$$\zeta \cong \frac{3.69}{r_s^2} \{1 - 0.330 \; r_s + 0.006 \; r_s^2 \ln r\}$$

$$- \frac{0.625}{r_s^2} \left(\frac{\mu_0 H}{\zeta_0}\right)^2 \{1 - 0.013 \; r_s + 0.231 \; r_s^2 \ln r_s\},$$

$$\varepsilon \cong \varepsilon_k + \varepsilon_{ex} + \varepsilon_c,$$

$$\varepsilon_k = \frac{2.21}{r_s^2} - \frac{0.435}{r_s^5} \left(\frac{\mu_0 H}{\zeta_0}\right)^2,$$

$$\varepsilon_{ex} = - \frac{0.916}{r_s} - \frac{0.081}{r_s^4} \left(\frac{\mu_0 H}{\zeta_0}\right)^2,$$

$$\varepsilon_c = 0.062 \; \ln r_s + \frac{0.005}{r_s^4} \left(\frac{\mu_0 H}{\zeta_0}\right)^2 \ln r_s.$$

Similarly, we have in the intermediate field range the de Haas-van Alphen terms (again for the energy per electron)

$$\varepsilon_k^{osc} \cong \frac{0.561}{r_s^2 \eta^{5/2}} \sum_{n=1}^{\infty} n^{-5/2} [n\pi\eta\sin(n\pi\eta - \tfrac{1}{4}\pi) + \cos(n\pi\eta - \tfrac{1}{4}\pi)],$$

$$\varepsilon_{ex}^{osc} \cong - \frac{0.586}{r_s \eta^{1/2}} \sum_{n=1}^{\infty} n^{-3/2} [n\pi\cos(n\pi\eta - \tfrac{1}{4}\pi)$$

$$+ \ln(4n\pi\eta)\sin(n\pi\eta - \tfrac{1}{4}\pi)],$$

$$\varepsilon_c^{osc} \cong \frac{0.197}{\eta^{3/2}} (\ln r_s) \sum_{n=1}^{\infty} n^{-3/2}\sin(n\pi\eta - \tfrac{1}{4}\pi),$$

where $\eta = \zeta_0/\mu_0 H \gg 1$.

There are two outstanding problems which remain to be investigated: to extend these results to higher values of r_s, and to include relativistic effects. The calculations presented here can, in a straightforward manner, be continued to provide an expansion of the ground state energy through terms, say, of order r_s^2, but from the zero field calculations we know that such a series converges very poorly. In the extreme low density case, where at $H = 0$ the electrons form a Wigner lattice, the effect of a magnetic field is negligible, except that it will enhance the localization of the electrons and thus push the critical value of r_s down somewhat. At intermediate densities a modified procedure, such as Hubbard's, will have to be introduced, and at high magnetic fields this will present a very complicated problem.

The inclusion of relativistic effects presents an attractive field for investigation. For example, owing to the magnetic field, the spin-orbit interaction should have a number of interesting consequences, and it has been shown that owing to current-current interactions amongst the electrons, the latter respond to B rather than H, which, as is known from the work of Shoenberg, shows up in an interesting way in the de Haas-van Alphen effect.

To conclude, our preliminary considerations indicate that the H-r_s plane for the electron gas consists of at least three phases as sketched below.

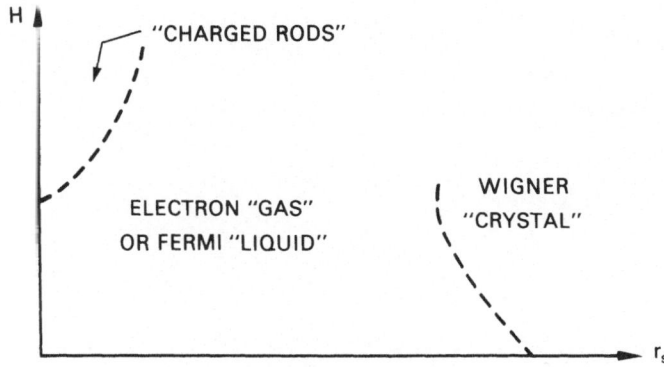

When relativistic effects are included this simple picture should become much more complicated.

REFERENCES

1. Pines, D. (1963). *Elementary Excitations in Solids*, (W.A. Benjamin, Inc., New York), chapter 3.
2. Vashishta, P. and Singwi, K.S. (1972). *Phys. Rev.*, **B6**, 875.
3. Martin, P.C. and Schwinger, J. (1959). *Phys. Rev.* **115**, 1342.
4. Horing, N.J. (1965). *Ann. Phys. (N.Y.)*, **31**, 1.
5. Glasser, M.L. and Kaplan, J.I. (1972), *Ann. Phys. (N.Y.)*, **73**, 1.
6. Hebborn, J.E. and March, N.H. (1970). *Adv. Phys.*, **19**, 175.
7. Anselm, A.I. and Asherov, B.M. (1960). *Fiz. Tverd. Tela*, **2**, 2821.
8. Kaplan, J.I. and Glasser, M.L. (1972). *Phys. Rev. Lett.*, **28**, 1077.

CONCEPTUAL PHASE DIAGRAMS AND THEIR APPLICATIONS TO ITINERANT ELECTRON MAGNETISM†

JOHN B. GOODENOUGH

Lincoln Laboratory, Massachusetts Institute of Technology, Lexington, Massachusetts 02173, USA

1. INTRODUCTION

An adequate description of the magnetic properties of a solid requires not only a knowledge of the relative energies of the atomic outer-electron states, but also the location of a given electron state within a conceptual phase diagram of electronic properties. In the last few years the physical ideas essential to this task have been identified. My intent in this paper is to review briefly these ideas and to illustrate their usefulness. First, I develop the phase diagrams. Then I review briefly some complex magnetic and crystallographic properties exhibited by several metallic compounds and show how their magnetic order and the magnitudes of the observed atomic moments can be derived in terms of the phase diagrams, known variations across the Periodic Table, and qualitative arguments about the energy bands. These examples illustrate how, from a small set of physical ideas and the phase diagrams, it is possible to develop experimental strategies for controlled modification of known magnetic materials and for a search for new magnetic materials. Although development of fruitful experimental strategies is adequate justification for inclusion of a qualitative discussion in a theoretical conference, my hope is that a heightened confidence in the sufficiency of the physical ideas discussed will also aid the theorists' search for simplified techniques with which to calculate the observed properties from first principles.

† This work was sponsored by the Department of the Air Force.

2. CONCEPTUAL PHASE DIAGRAMS

2.1 PRELIMINARY CONSIDERATIONS

Of particular interest to solid state theory are the physical properties imparted by the electrons outside of closed atomic shells. There are two limiting theories of the outer electrons in solids: crystal-field theory and band theory. Crystal field theory rests on the assumption that the outer electrons remain *localized* to discrete atomic positions even after the atoms have been brought together to form a crystalline array. It is therefore applicable only to those electrons that are strongly bound to the atomic nucleus and experience weak interatomic interactions. Band theory, on the other hand, rests on the assumption that each outer electron is *itinerant*, being shared equally by all the like atoms of a periodic array.

The limiting example of localized electrons is the free atom. From the successive ionization energies of a free atom, it is apparent that the addition of an electron to a partially occupied manifold costs an energy U or U', the electrostatic energies required to add an electron to a half filled or an empty orbital, respectively. The energy difference $J^{intra} \sim (U - U')$ is responsible for Hund's highest multiplicity rule for the free atom.

If the atoms form a periodic array, as in a crystal, interatomic interactions broaden the free-atom energy level into an energy band. To lowest order in tight-binding theory, the width of a band is

$$w_b \approx 2zb, \tag{1}$$

where z is the number of nearest like neighbours, b is the nearest-neighbour, spin-independent transfer integral

$$b_{ij} = (\psi_i, \mathcal{H}'\psi) \sim \varepsilon_{ij}(\psi_i, \psi_j), \tag{2}$$

$\mathcal{H}'$ is the perturbation of the electron potential at the atom at position R_j by the presence of a like atom at position R_i, and the ψ are localized wave functions for an electron at R_i or R_j.

Electrons in partially filled bands experience electron-phonon interactions. For broad bands, where the time

$$\tau \approx \frac{\hbar}{w_b} \tag{3}$$

is short compared to the period ω_R^{-1} for an optical-mode vibration, electron mobilities are described by conventional collison theory. However, below a critical band width, $\tau > \omega_R^{-1}$ and electrons become trapped by crystalline deformations. If the number of charge carriers is small, the deformations are local

and randomized. The charged particle, though 'dressed' by the
local deformation, is mobile. However, the mobility is reduced
by a factor $\exp(-\varepsilon_A/kT)$, where ε_A is the thermal energy required
to make equal the energy at an occupied site and a near neigh-
bour empty site. Such a 'dressed' particle is called a small
polaron, and the charge-carrier mobility is described by diffusion theory. On the other hand, if the number of charge carriers is large, as in a half-filled band, a cooperative distortion
of the crystal takes place. Such cooperative distortion may induce electron ordering into distinguishable lattice positions,
as in low temperature Fe_3O_4, or disproportionation: $2M^{m+} \rightarrow$
$M^{(m+1)+} + M^{(m-1)+}$, as in $Tl^+Tl^{3+}F_4$ and $Pd^{2+}Pd^{4+}F_6$. In the thal-
lium compound, a half filled $6s$ band disproportionates into fil-
led $6s^2$ and empty $6s^0$ cores, thereby quenching any spontaneous
magnetism. In PdF_6, on the other hand, disproportion does
not cost the energy U, but the smaller energy U'. The $4d$ orbitals are split by crystalline fields, and two-fold degenerate
orbitals of E_g symmetry (e_g orbitals) contain one electron per
atom, making them one quarter filled. In place of itinerant
electron ferromagnetism, disproportionation stabilizes triplet
atomic states on half the atoms: $2e_g^1 \rightarrow e_g^2 + e_g^0$.

Conventional band theory includes the assumption that $U << w_b$,
and a $U = 0$ is taken for the zero-order problem. The Hartree-
Fock hypothesis is that the wave function Ψ for a system of N
itinerant electrons may be represented by a *single* determinantal
wave function D whose elements are one-electron wave functions
$\phi_i(\tau_j)$, where τ_j stands for the three spatial variables r_j plus
the spin variable σ_j. The function D antisymmetrizes the Hartree one-electron product function $\prod_i \phi_i(\tau_j)$, thereby introduc-

ing an accidental correlation between electrons of parallel spin,
but it fails to require that electrons of either spin should keep away
from one another. This is the principal shortcoming of the Hartree-Fock hypothesis. The difference $E(D) - E(\Psi_g)$ between the
energy of a D and of an exact ground state wave function Ψ_g is
a measure of the accuracy of D as an approximation of Ψ_g. If
$E(D)$ is stationary, this energy difference is called the *correlation energy*. This difference in energy for the ground determinantal wave function D_g is primarily due to neglect of the
Coulomb energy U arising from interatomic interactions between
electrons of anti-parallel spin. From the e^2/r_{mn} character of
this energy, the energy U is larger the more tightly bound are
the outer electrons.

The outer s and p electrons are loosely bound to their atomic
nuclei ($U \approx$ 2-5 eV), and they interact strongly with the neighbouring atoms of a crystalline array (w_b > 10 eV) to provide the
primary binding energy of the crystal. Therefore, the condition
$U << w_b$ is generally met, and conventional band theory is applicable. Since U is small, the condition U/w_b > 1 for spontaneous
magnetism requires band widths that are smaller than the critical value for strong, cooperative electron-phonon interactions.

Therefore, outer s and p electrons do not exhibit spontaneous
magnetism, unless localized to impurity atoms in one-electron
donor states. Narrow s bands disproportionate, and p bands order into bonding and non-bonding orbitals to give the (8-N) rule
for crystal symmetry.

Outer $4f$ electrons, on the other hand, are tightly bound to
their atomic nuclei ($U > 20$ eV) and are screened from the neighbouring atoms by $5s^2 5p^6$ core electrons, which keep $w_b < 1$ eV.
Therefore, the strong-correlation condition $U \ll w_b$ is always
fulfilled, and partially filled $4f$ cores exhibit atomic-like
properties that are but weakly perturbed by the crystalline
fields. Even in insulators, where the energy gap E_g between
valence and conduction bands is large, the $4f$ electrons have a
$U > U' > E_g$. This is why the $4f$ core states correspond to one,
or at most two, possible valence states. If the energy of one
$4f^n$ manifold falls in the gap E_g, the $4f^{n-1}$ manifold is also
stable; if no $4f^n$ manifold falls in the gap, only one core state
is possible, as can be seen from figure 1. However, electrons
may be added to a $5d$ conduction band, as in the case of metallic
$Eu_{1-x}Gd_x S$, which contains $4f^7$ cores at both cations.

Outer d electrons are intermediate in character ($U \approx w_b$).
Except in fluorides and a few oxides, the $4d$ and $5d$ electrons

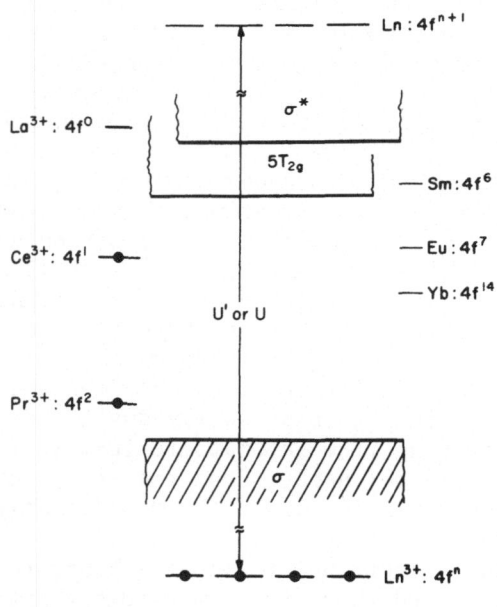

Ln = Nd, Pm; Gd, Tb, Dy, Ho, Er; Lu

Figure 1 - Locations of the $4f^n$-manifold energies relative to the conduction band and valence band edges in
rare earth oxides.

are itinerant or form closed-core states ($U < w_b$). The $3d$ electrons, on the other hand, may be strongly correlated ($U \gg w_b$), as in the antiferromagnetic insulator MnO, itinerant ($U \ll w_b$) as in superconducting TiO, or intermediate ($U \approx w_b$) as in the ferromagnetic metal CrO_2. It is just this ambivalent character of the $3d$ electrons that gives the rich variety of physical properties manifest by first-row transition elements and their alloys.

The intermediate character of the $3d$ electrons forces us to ask, and at the same time permits us to investigate experimentally, the following fundamental question: Is there one thermodynamic state over the entire range of interatomic interactions and a gradual transition in physical properties on passing from the weak interaction limit, where crystal-field theory is applicable, to the strong interaction limit where band theory is applicable; or are there at least two distinguishable thermodynamic states, one in which a given set of electrons are strongly correlated (localized) and another in which they are weakly correlated, with or without spontaneous magnetism? The first order electronic transition in NiS suggests that, where the atoms are not clamped, a first order lattice dilatation occurs from weak to strong correlations [1]. The transition metal nitrides and carbides discussed in this paper provide additional support for this conclusion.

Data analysis is aided by a conceptual phase diagram in terms of three variables: the temperature T, the occupancy numbers n_ℓ of the partially filled d orbitals ($0 \leqslant n_\ell \leqslant 2$ because of the spin degeneracy), and the theoretical parameter b of equations (1,2), which defines the strength of the interactions between localized, crystal-field electrons at nearest-neighbour-like atoms in equivalent lattice sites. Although b is not measurable directly, it is generally obvious how to order the b in any set of isostructural compounds, since the relative magnitudes of the overlap integrals (ψ_i, ψ_j) can be estimated. Therefore, from a conceptual phase diagram, it is possible to deduce how the electronic properties vary on passing between members of a group of isostructural compounds, and even to some extent on passing from structure to structure.

The energy U would appear explicitly in the phase diagrams if a normalized energy b/U, or $w_b/U \approx 2zb/U$, were used in place of b. However, use of such a normalized energy obscures the sharpness of the transition region $U \approx w_b$ as a function of observable parameters. The Coulomb energy U is a function of the transfer energy b through a screening parameter $\xi(b)$, which increases with b:

$$U = (|\phi_m(1)|^2, V|\phi_m(2)|^2), \tag{4}$$

$$V = \left(\frac{e^2}{r_{12}}\right)\exp(-\xi r_{12}). \tag{5}$$

Therefore, U decreases with increasing b, and the ratio b/U increases unambiguously with increasing b. However, at the transition from strong to weak correlations, the screening parameter ξ changes particularly rapidly with b, which makes b/U increase abruptly in a narrow interval of b. It is just this feature that makes possible a first-order transition, the degree of electron correlation and electron localization increasing significantly with small increases in lattice parameter. It is for this reason that I choose the energy b rather than a normalized parameter b/U or w_b/U for the third variable of a conceptual phase diagram.

1.2   CONCEPTUAL PHASE DIAGRAM FOR $n_\ell = 1$

An orbital occupancy number $n_\ell = 1$ corresponds to each relevant orbital being half-filled, and therefore to an integral number of electrons per like atom on equivalent lattice sites. Consider half-filled $3d$ orbitals and a Bravais lattice containing N atoms. In the broad-band limit $U \ll w_b$, the $3d$ orbitals form a half filled band of itinerant electron states, and the crystal is metallic. In the narrow-band limit $U \gg w_b$, the single atom ground configuration d^n is separated from the d^{n+1} ground level by an energy U, as shown in figures 2(a,b). Electron excitations to excited states of the d^n manifold are important for spectroscopy, but not for the present discussion, so d^n

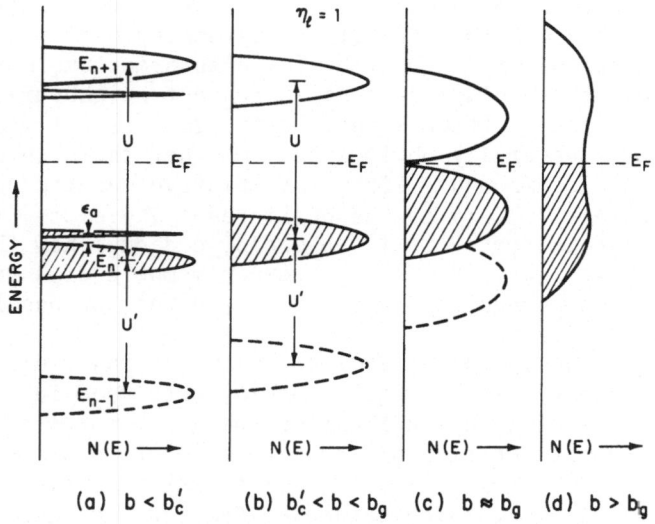

Figure 2 - Energy density of states $N(E)$ vs energy for a half-filled d band for four different values of the nearest neighbour transfer energy integral b. Small-polaron activation energy is ε_a.

excited-state energies are not shown in figure 2. The single-atom energy levels are broadened into narrow bands of width w_b as a result of virtual electron transfer excitations, which cost an energy $(U - w_b)$. As the intermediate range $(U \approx w_b)$ is approached with increasing b, the parameter U decreases sharply, see equations (4,5).

If U is small in the narrow-band limit, as is the case for outer s and p electrons, figure 2 is not applicable. Band splitting with decreasing b is not accomplished by the stabilization of free-atom configurations and spontaneous magnetism, but by a cooperative crystallographic distortion and disproportionation. However, figure 2 does apply to $3d$ electrons, which are our concern in this paper. For these electrons, the critical bandwidth for strong electron-phonon interactions in partially filled bands occurs at a $b_c' < b_g$, where b_g marks the critical bandwidth for a semiconductor-to-metal transition, see figure 2(c). Therefore, the electron-correlation splitting quenches the possibility of a cooperative crystal distortion at b_c'. However, if a small concentration of charge carriers is injected, holes or electrons, small-polaron formation occurs for $b < b_c'$. The extremely narrow small-polaron bands are separated from the band edge by an energy ε_a, as indicated in figure 2(a).

Since splitting of the half-filled band in figure 2 is due to correlations between electrons of opposite spin, spontaneous antiferromagnetism must be associated with the domain $b < b_g$. Consider, for example, a two-sublattice array of magnetic atoms in which each magnetic atom has all its nearest-neighbour magnetic atoms belonging to the other sublattice. In the strong-correlation limit, the dominant interactions between magnetic electrons on neighbouring atoms are generally via electron-transfer excited states. From figure 2(a) and second-order perturbation theory, the stabilization energy from such a superexchange interaction is

$$\Delta\varepsilon_{ij} = - \frac{t_{ij}^2}{(U - w_b)} , \qquad (6)$$

where t_{ij} is the spin-dependent transfer integral. It is related to the integral b_{ij} of equation (2) through the matrix that rotates through an angle θ a local spin at R_i having the direction components α, β to the direction of the spin at R_j having the components α', β'

$$\alpha = \cos(\tfrac{1}{2}\theta)\alpha' + \sin(\tfrac{1}{2}\theta)\beta',$$

$$\beta = - \sin(\tfrac{1}{2}\theta)\alpha' + \cos(\tfrac{1}{2}\theta)\beta'. \qquad (7)$$

For half-filled orbitals, the Pauli exclusion principle only allows electron transfer if the spins are antiparallel, so

$$t_{ij}^2 = b_{ij}^2 \sin^2 \tfrac{1}{2}\theta = \tfrac{1}{2}b_{ij}^2 - \tfrac{1}{2}b_{ij}^2 \cos\theta. \tag{8}$$

Comparison with the Heisenberg isotropic exchange operator

$$\mathcal{H} = - \sum_{ij} J_{ij}s_i \cdot s_j \tag{9}$$

gives the superexchange contribution

$$J_{ij}^S = - \frac{2b_{ij}^2}{4S^2 U} \tag{10}$$

where S is the net atomic spin of a magnetic atom. The antiferromagnetic ordering temperature T_N is proportional to the exchange-energy parameter:

$$kT_N \sim zJ_{ij}S(S+1) \approx zJ_{ij}^S S2 \sim \frac{zb^2}{U} , \tag{11}$$

where b/U increases unambiguously with increasing b. Therefore,

$$\frac{dT_N}{db} > 0 \quad \text{for} \quad b < b_c, \tag{12}$$

where $b < b_c$ defines the domain within which the superexchange perturbation expansion converges. From transport measurements on NiO, it appears that $b_c' < b_c$ for the $3d$ electrons [2]. For $b > b_c$, the interatomic interactions are comparable to, or greater than, the intra-atomic interactions. In this limit, the band criterion for spontaneous magnetism applies, and T_N increases with the correlation energy U. Therefore, the T_N vs b curve should exhibit a sharp maximum at $b \approx b_c$. Moreover, this maximum is larger the greater the value of U in the strong-correlation limit. Since the radial extension of the d wave functions decreases with increasing effective nuclear charge, the maximum possible magnetic ordering temperature should increase on passing to heavier atoms of a transition series for the same valence state.

The most straightforward application of these ideas is to the case of a half-filled two-fold degenerate band in a simple cubic array of like transition metal atoms. Orbitals of E_g symmetry in an ABX_3 cubic perovskite, for example, correspond to this case.

In the domain $U > w_b$, the relative stability of antiferromagnetic vs ferromagnetic order can be appreciated from the energy density of states shown in figure 3. As in figure 2(a), the energy E_2 of the two-electron state is separated from that of the

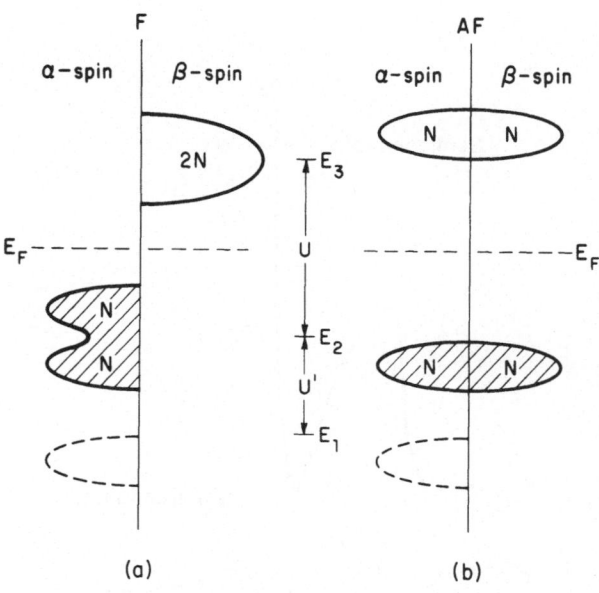

Figure 3 - Energy density of states vs energy for states of α and β spin and a half-filled ($n_\ell = 1$), two-fold degenerate band on a simple cubic array with $b_c < b < b_g$: (a) ferromagnetic and (b) antiferromagnetic ordering.

three-electron state E_3 by an energy U, since addition of a third electron per atom requires occupancy of an orbital that is already occupied. However, E_1 and E_2 are separated by a $U' < U$, since the addition of a second electron per atom can be accomplished by occupying empty orbitals. Moreover, within a band of allowed energies, the bonding orbitals have energies below the reference energy E_i and the anti-bonding orbitals have energies above it. The counterpart of an occupied bonding state of α spin is an anti-bonding state of β spin, of an occupied anti-bonding state of α spin is a bonding state of β spin, and vice versa. Each atom carries an atomic moment as a result of the relation $U > w_b$. Figure 3(a) represents ferromagnetic coupling and figure 3(b) antiferromagnetic coupling. Ferromagnetic coupling requires that only states of one spin can be occupied. For a half-filled band, all the states of this spin are occupied, both bonding and anti-bonding states. Therefore, no inter-atomic binding energy can be achieved by ferromagnetic coupling. Antiferromagnetic coupling, on the other hand, allows states of both α and β spin to be occupied, and all the bonding states are just filled if $n_\ell = 1$. Therefore, binding energy can be gained by antiferromagnetic coupling, and the antiferromagnetic state is stabilized at lowest temperature for all $b < b_m$, where $b_m > b_c$. Antiferromagnetic coupling of all nearest neighbours of a simple cubic array is known as Type G antiferromagnetic order.

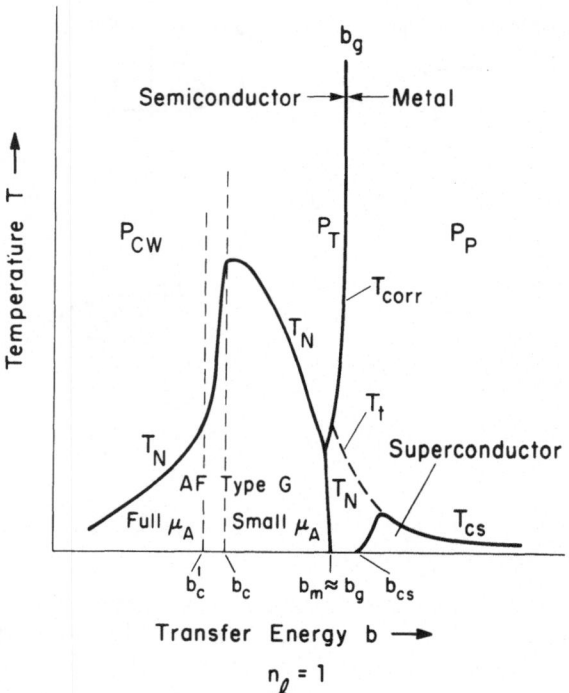

Figure 4 – Conceptual phase diagram of temperature vs transfer energy for a band occupation number $n_\ell = 1$, a Bravais lattice, and a free atom $U \approx 10$ eV.

Figure 4 shows a schematic plot of the Néel temperature T_N vs b. It is intersected by the line $b = b_g$, which increases somewhat with rising temperature T because of the spin entropy associated with the magnetic states. Unless a spin-density wave is stabilized by 'nesting' of a Fermi surface in a Brillouin zone, spontaneous magnetism may be expected to disappear abruptly at a b_m close to the value of b at which T_N and b_g intersect.

Electron-transfer excited states introduce spin pairing, which reduces the atomic moment μ_A observed by neutrons below T_N. In the strong-correlation domain $b \ll b_g$, the atomic moments are reduced from their crystal-field values by only a small fraction of a Bohr magneton. However, as $b \to b_g$ important reductions in the atomic moment are observed.

In the broad-band limit $b \gg b_g$, a half-filled band gives a weak, temperature-independent susceptibility (Pauli paramagnetism $\chi = 2\mu_B^2 N(E_F)$), and a transition from the normal to a superconducting state may occur below a temperature T_{cs}. As b decreases toward b_g, the energy U suppresses T_{cs} for $b < b_{cs}$ and introduces an enhancement of the susceptibility χ via an enhancement of $N(E_F)$, the density of states at the Fermi energy.

This is an effective-mass enhancement that is also reflected in the electronic specific heat [3]. The weak, temperature-independent susceptibility becomes temperature dependent, but the apparent Curie-Weiss law gives a μ_{eff} and a Weiss constant $|\theta|$ that are too large to be meaningfully interpreted by crystal-field and superexchange theories [4].

Also shown in figure 4 is a dashed line T_t marking possible crystallographic distortions that suppress the appearance of spontaneous magnetism at temperatures $T < T_t$. This phenomenon appears to be well illustrated by NiS, which undergoes a transformation from the high-temperature, hexagonal $B8_1$ structure of NiAs to the peculiar millerite structure if slow cooled below 485°C. There is no spontaneous magnetism associated with the millerite phase. If quenched to room temperature from above T_t, NiS retains the $B8_1$ structure and, on further cooling, undergoes a first-order dilatation (no change in crystal symmetry) at a Néel temperature $T_N \approx 260°$K [5]. Spontaneous magnetism is suppressed by a hydrostatic pressure $P > 20$ kbar [6], which demonstrates that $b \approx b_m$ and that on cooling through $b = b_m$ the larger potential energy provided by the electron correlations results in a reduced kinetic energy, as required by the virial theorem. A dilatation reflects reduced kinetic energy, since binding is caused by electron transfer. In NiS, the localized-electron configuration $t_2^6 e^2$ for octahedral site $Ni^{2+}:d^8$ ions is transformed, via the large covalent mixing with $S^{2-}:3p$ orbitals, to a $t_2^6 \sigma^{*2}$ configuration, where σ^{*2} denotes a half-filled, two-fold degenerate band. Intra-atomic exchange stabilization associated with localized atomic moments increases the effective nuclear charge seen by the electrons; hence the increase in potential energy. It also reduces the radial extension of the electrons, thus decreasing the overlap integrals $b_{ij} \approx \epsilon_{ij}(\psi_i, \psi_j)$ responsible for binding and the magnitude of the kinetic energy. Because the atoms are not clamped, a first-order transition is possible. This observation is consistent with two distinguishable thermodynamic states, one containing strongly correlated electrons, the other weakly correlated electrons.

In VO_2, which has the tetragonal rutile structure above a first-order semiconductor-to-metal transition at $T_t = 67°$C and a monoclinic structure below it, the vanadium d^1 configurations become spin paired in V-V homopolar bonds at $T < T_t$ [7]. Above T_t, the V-V interactions are strong enough to induce band formation, and below T_t band uncrossing leaves a half filled, nondegenerate d band that becomes split in two by V-V pairing [8]. Because $b_m < b < b_{cs}$, a band splitting that stabilizes occupied states relative to unoccupied states is energetically favored over either spontaneous antiferromagnetism or a superconducting state at lowest temperatures.

1.3 CONCEPTUAL PHASE DIAGRAM FOR $n_\ell = \frac{1}{2}$

If two-fold degenerate bands of our simple cubic array contain only one electron per atom, the energy density of states

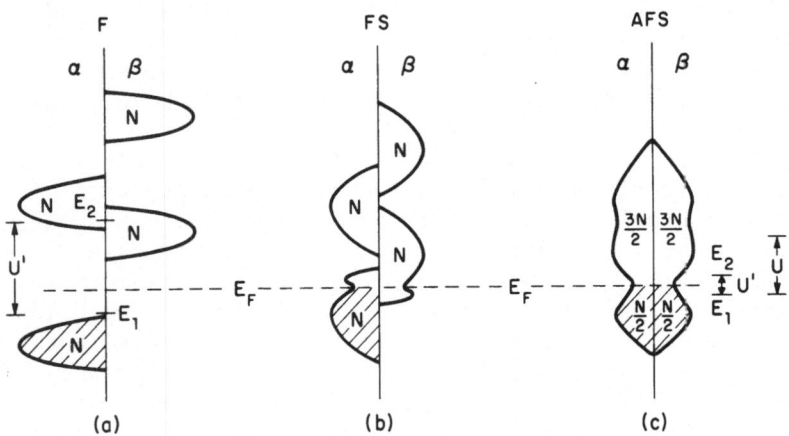

Figure 5 - Energy density of states vs energy for states of α and β spin and a quarter filled ($n_\ell = \frac{1}{2}$), two-fold degenerate band on a simple cubic array with (a) $b_c < b \gtrless b_m$; (b) $b_g < b < b_m$; and (c) $b \approx b_m$.

is as shown in figure 5. It is immediately apparent that, even in the ferromagnetic state, only bonding states need be occupied. Therefore, either ferromagnetic or antiferromagnetic coupling allows binding. In such a case, the ferromagnetic correlations implicit to the Hartree-Fock hypothesis will stabilize ferromagnetic order by the Weiss molecular field energy kT_w.

This same conclusion may be derived from the superexchange perturbation theory. The Pauli exclusion principle does not inhibit transfer of electrons of either spin, since one orbital state is empty. Therefore, it is necessary to go to third-order theory. The intra-atomic exchange energy $J^{\text{intra}} \sim (U - U')$ stabilizes ferromagnetic order, and the Curie temperature varies as

$$kT_c \sim zb^2 \frac{(U - U')}{U'^2} . \tag{13}$$

Thus T_c increases monotonically with b as long as the perturbation expansion is valid. However, in the simple cubic array electron-phonon interactions induce a Jahn-Teller distortion below a $T_t > T_c$. The distortion to tetragonal ($c/a < 1$) symmetry distinguishes interactions between half-filled orbitals ($n_\ell = 1$) along the c-axis from interactions between an empty and a half filled orbital ($n_\ell = \frac{1}{2}$) within basal planes. As a result, magnetic ordering below T_t is antiferromagnetic Type A at $T_N < T_t$. Type A order has ferromagnetic coupling in the basal planes and antiferromagnetic coupling along the c-axis. This situation is reflected in the phase diagram of figure 6.

In the interval $b_c' \approx b_c < b < b_m$, on the other hand, there

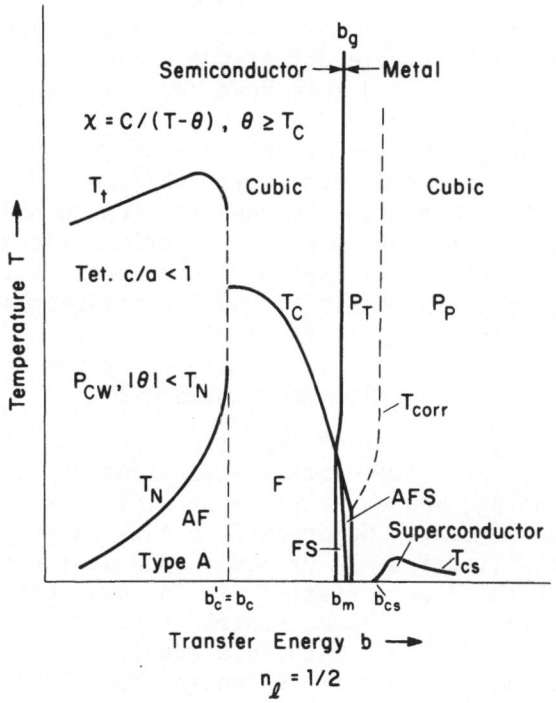

Figure 6 – Conceptual phase diagram of temperature vs transfer energy for a quarter filled ($n_\ell = \frac{1}{2}$), two-fold degenerate band at a simple cubic array and a free atom $U \approx 1C$ eV.

is no Jahn-Teller distortion, and the coupling is ferromagnetic. According to the Stoner theory of itinerant electron ferromagnetism, the Curie temperature varies as

$$kT_c \sim \varepsilon_F \left\{ \frac{kT_W}{\varepsilon_F} - \frac{2}{3} \right\}^{\frac{1}{2}}, \tag{14}$$

where $\varepsilon_F \approx w_b$ is the Fermi energy measured from the bottom (or the top for holes) of the band. The Weiss molecular field energy kT_W is an effective Coulomb energy U_{eff} that changes more rapidly with b than ε_F, and $dT_c/db < 0$. As in the case $n_\ell = 1$, a maximum T_c is anticipated at $b \approx b_c$, and this maximum should be larger the greater U' in the strong-correlation limit.

As $b > b_c$ increases, U' decreases and w_b increases as indicated in figure 5(b,c). The spontaneous magnetization is reduced for values of b larger than that where the spin up and spin down bands begin to overlap ($b > b_g$). With no overlap, spin waves are generated by the thermal excitation of down spin states. Where spin down electrons are present in the ground state, a

standing ferromagnetic spin density wave (FS) may be stabilized
by interactions with the Fermi surface. The net magnetization
of the FS would be the simple difference between the population
densities of the spin up and spin down bands multiplied by the
Bohr magneton μ_B.

For a large enough overlap of the spin-up and spin-down bands,
an antiferromagnetic spin-density wave (AFS) may be stabilized
relative to the FS. Metamagnetic behavior (FS to AFS transitions as a function of T and applied magnetic field H) would be
observed in the transitional region. This complex transition
from spontaneous ferromagnetism to Pauli paramagnetism at $b \approx b_m$
is also shown schematically in figure 6.

2.4 ATOMIC MOMENTS

Figure 5(a) contains another important concept. At $E_2 = E_1
+ U'$, the α spin states are destabilized relative to E_2 because
they are antibonding, whereas the β spin states are stabilized
because they are bonding. Superposed on this is the Weiss molecular field stabilization kT_w of the α spin states and destabi-
lization $-kT_w$ of the β spin states. It follows that an added
electron enters the α spin band, contributing to the ferromagnetic moment, only if $2kT_w > w_b$. This condition is met only in
the strong-correlation limit $b < b_c$ where $kT_w \to J$intra $\sim (U - U')$
and $w_b << U$. In the range $b_c < b < b_m$ we may anticipate a $kT_w <
(U - U') \approx w_b$, and any ferromagnetic moment is reduced. This is
a particularly important result for the theory of itinerant-
electron ferromagnetism. It may be generalized to the following
statement about the *magnitudes of spin only atomic moments in
ferromagnets:*

For all $b < b_g$ and $n_\ell \leqslant \frac{1}{2}$ or for $b < b_c$ and $\frac{1}{2} < n_\ell < 1$,

$$\mu_A^F = \nu n_\ell \mu_B, \tag{15}$$

whereas for $b_c < b < b_g$ and $\frac{1}{2} < n_\ell < 1$

$$\mu_A^F = \nu(1 - n_\ell)\mu_B, \tag{16}$$

and for $b_c < b < b_g$ and $1 < n_\ell < 3/2$

$$\mu_A^F = \nu(n_\ell - 1)\mu_B, \tag{17}$$

whereas for all $b < b_g$ and $n_\ell > 3/2$ or for $b < b_c$ and
$1 < n_\ell < 3/2$

$$\mu_A^F = \nu(2 - n_\ell)\mu_B, \tag{18}$$

where ν is the degeneracy of the partially filled atomic
orbitals.

In the domain $b_c < b < b_m$, it follows from equations (16,17) that spontaneous ferromagnetism would have an atomic moment μ_A^F that vanishes as $n_\ell \rightarrow 1$. However, zero moment represents a paramagnetic state, and from figure 4 an antiferromagnetic state is more stable at $n_\ell = 1$ if $b < b_m$. Therefore, we may anticipate a transition from ferromagnetism at that value of n_ℓ where the atomic moment of the antiferromagnetic state μ_A^{AF} is

$$\mu_A^{AF} = \mu_A^F. \tag{19}$$

For a two-sublattice structure in which all ν orbitals are active in nearest-neighbour interactions, as occurs in the simple cubic array of atoms with partially filled e_g orbitals ($\nu = 2$), the atomic moment of an antiferromagnetic state is

$$\mu_A^{AF} \quad \begin{aligned} &= \nu(n_\ell - \delta)\mu_B \qquad \text{if } n_\ell < 1, \\[2ex] &= \nu(2 - n_\ell - \delta)\mu_B \quad \text{if } n_\ell > 1, \end{aligned} \tag{20}$$

where δ is a fraction that measures the reduction in atomic moment due to admixture of spin paired electron-transfer excited states. Substitution of equations (16,17,20) into equations (19) yields the critical values n_ℓ^t at which the ferromagnetic-to-antiferromagnetic transition would occur:

$$n_\ell^t \quad \begin{aligned} &= \tfrac{1}{2}(1 + \delta) \text{ if } n_\ell < \tfrac{1}{2}, \\[2ex] &= \tfrac{1}{2}(3 - \delta) \text{ if } n_\ell > \tfrac{1}{2}. \end{aligned} \tag{21}$$

Experimentally, a $\delta \approx 0.1$ is found where $b < b_c$ occurs; a $\delta \approx 0.5 \pm 0.2$ is found in the domain $b_c < b < b_m$. Localized moments in other orbitals induce a smaller δ; and even where $b \gtrsim b_m$, they may induce an atomic moment from spin paired bonding electrons ($\delta = 0$) to produce a $\delta \approx 0.9$. For $\delta \approx 0.5$, equation (21) gives antiferromagnetic order in the domain

$$\frac{3}{4} < n_\ell < \frac{5}{4} . \tag{22}$$

These conclusions are summarized in figure 7, where the left-hand and bottom scales give μ_A^F/μ_B vs z_d, the number of d electrons per atom in a five-fold degenerate d band, whereas the right-hand and top scales give μ_A^F/μ_B in units of ν vs n_ℓ. If the order is ferromagnetic, a quantitative spin-only contribution to the moment can be deduced from a knowledge of the magni-

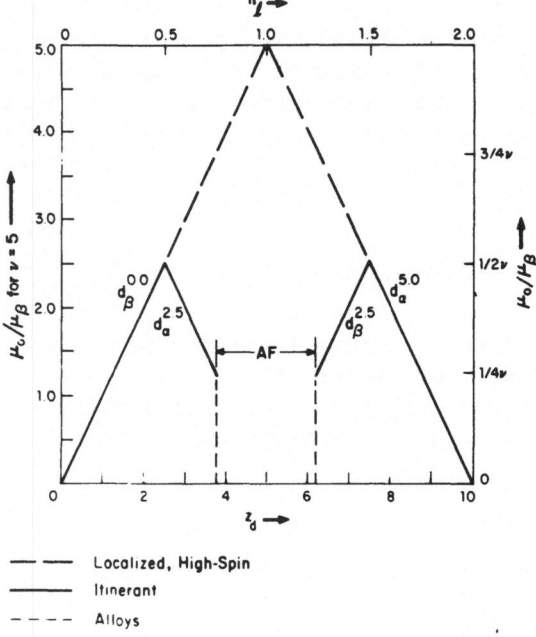

Figure 7 - Spin only spontaneous atomic moment at $T = 0°K$
vs $3d$ electron/atom ratio z_d for itinerant $3d$ electrons
in ferromagnets.

tude of b relative to b_c and b_m. Conversely, from a knowledge
of the moment, it is possible to deduce the relative magnitude
of the transfer energy integral b. If the order is antiferromagnetic, on the other hand, there is an uncertainty in the
magnitude of the atomic moment because of the fraction δ.

2.5 MAGNETIC ORDER

Empty orbitals ($n_\ell = 0$) and full orbitals ($n_\ell = 2$) can have
no spontaneous magnetism. At lowest temperatures, small concentrations of mobile charge carriers are trapped at the isolated
impurities or lattice defects that create them. For $b > b_c'$
they occupy molecular-orbital trap states; for $b < b_c'$ they occupy small-polaron states that may move among the atom's nearest
neighbour to the defect site. For n_ℓ near 0 or 2, there is no
long-range magnetic order; but as the concentration increases,
these isolated moments begin to interact with one another. Beyond a critical concentration, either long-range magnetic order,
if $b < b_m$, or impurity-band formation, if $b > b_m$, is established.
In the range $b_c < b < b_g$, the long-range order would be ferro-
magnetic outside the interval $\frac{1}{2}(1 + \delta) < n_\ell < \frac{1}{2}(3 - \delta)$, antiferromagnetic within it, as discussed above and represented schem-
atically in figure 8. However, a complex ferromagnetic → FS →
AFS → paramagnetic sequence occurs as b increases through b_m.
If $b < b_c$, on the other hand, the magnetic order may be more
complex. If $b_c' < b < b_c$, the site-transfer time $\tau \approx \hbar/b$ of

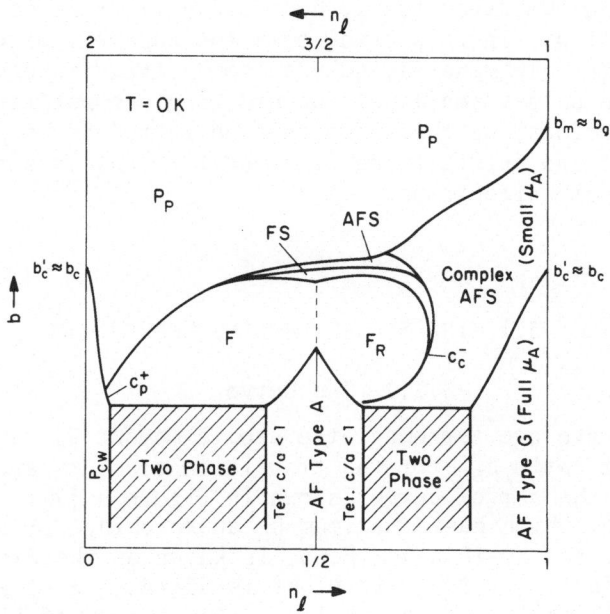

Figure 8 - Conceptual phase diagram of transfer energy b vs band occupation number n_ℓ at $T = 0°K$ for a two-fold degenerate band at a simple cubic array and a free-atom $U \approx 10$ eV.

equation (3) is short relative to the relaxation time of the atomic moment orientation. In this case, the mobile electron spins, which conserve their spins during transfer, couple ferromagnetically any stationary spins at the atoms over which they move. This is called double exchange. From equation (7) and first-order perturbation theory, applicable for mobile carriers (see figure 2), the double exchange stabilization energy is

$$\mathcal{H}_{ex}^{D} = - zcb\cos\tfrac{1}{2}\theta, \tag{23}$$

where c is the concentration of mobile carriers. Near $n_\ell = 1$, an antiferromagnetic superexchange interaction varying as $\cos\theta$ and a double exchange interaction from the mobile electrons combine to stabilize an angle between neighbouring spins that falls between 0 and π. Such a cant angle can be accommodated either by a spiral spin configuration (AFS) or by a canted spin configuration (FS). Metamagnetic AFS $\rightleftarrows$ FS transitions may be encountered at critical temperatures and applied external fields. Mobile spins trapped at lattice impurities or defects create ferromagnetic clusters. Only those that move through the bulk can stabilize long-range ferromagnetic, FS, or AFS order.

If $b < b_c'$, on the other hand, the activation energy in the mobility also reduces τ so as to make it too long for a mobile

spin to retain the direction it had before it transferred to its
present position. In this case, only the superexchange interactions remain. In general, several competitive interactions
are operative unless small-polaron ordering, cooperative Jahn-
Teller distortions, or their combination introduce cooperative
ferromagnetic and antiferromagnetic interactions in different
crystallographic directions.

3. APPLICATIONS

3.1 THE SLATER-PAULING CURVE

3.1.1 The Curve

A plot of average ferromagnetic atomic moment $\bar{\mu}_A$ vs outer
electron/atom ratio z_{sd} for the transition elements and their
alloys gives the set of curves shown in figure 9 [9]. The up-
permost curve, which has a maximum $\bar{\mu}_A$ approaching 2.5 μ_B at
$z_{sd} \approx 8.5$ for the 50-50 alloy FeCo, is known as the Slater-Pauling curve. For $z_{sd} > 8.5$, its slope is $d\bar{\mu}_A/dz_{sd} \approx -1$ μ_B; and
for $z_{sd} < 8.5$, it is $d\bar{\mu}_A/dz_{sd} \approx 1$ μ_B. Any theory of itinerant
electron ferromagnetism must be able to predict this curve.

3.1.2 The Model

Three unknown parameters immediately confront us: (1) the
number n_s of electrons per atom in broad, primarily $4s$ bands
overlap the narrower $3d$ bands; (2) the spectroscopic splitting
factor g for the atomic magnetic moment $\mu_A = gS\mu_B$, where S is
the effective net spin of the atom; and (3) the small broad-band
contribution $\delta_s\mu_B$ to the ferromagnetic moment per atom that is

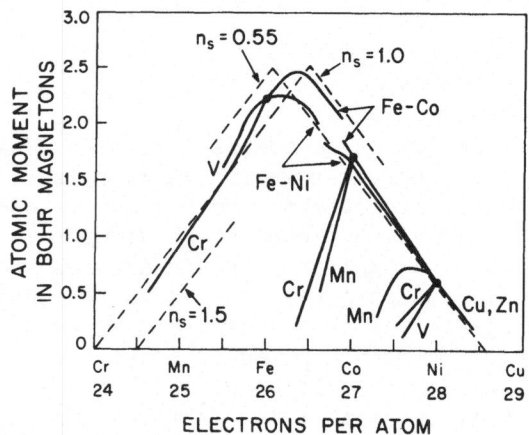

Figure 9 - Average atomic moment vs outer electron/atom
ratio for first-row transition elements and their alloys.

induced by the molecular field. For a Weiss molecular field
$H = WM$, the induced broad-band contribution is $\delta_S \mu_B = \chi_S \bar{W}\bar{\mu}_A$,
where the broad-band susceptibility χ_S and the Weiss constant W
have a product $\chi_S W \sim 0.1$. Therefore, we estimate a

$$\delta_S = \chi_S W \left(\frac{\bar{\mu}_A}{\mu_B}\right) \sim 0.1 \left(\frac{\bar{\mu}_A}{\mu_B}\right) . \qquad (24)$$

Itinerant electrons have a g-factor close to the spin only value
$g = 2.0$, and small deviations from this value are obscured by
uncertainties in the parameter n_S. Therefore, we assume a $g =$
2.0. Because the $4s$ band is broad relative to the $3d$ bands, we do not
anticipate large changes in n_S on passing through a phase change.
Moreover, the relative stabilities of the $3d$ electrons increase
with atomic number, so that systematic changes in n_S on traversing a long period of the periodic table should not be large.
Therefore, severe constraints are placed on the adjustable parameter n_S. Detailed band calculations together with the Stoner-
Wohlfarth interpretation of the ferromagnetism in Co and Ni give
$(n_S + \delta_S) = 0.6$ and 0.75 in these elements. At the face centered cubic (fcc) to body centered cubic (bcc) transitions in the
Fe-Co and Fe-Ni alloys, a $\Delta n_S \approx 0.15$ accounts for the discontinuity $\Delta\mu_A$ in figure 9.

Were the $3d$ bands five-fold degenerate, the Slater-Pauling
curve would follow directly from figure 7, since $z_d = z_{sd} - n_S$.
However, the $3d$ bands are not degenerate, and it is not possible
to use a rigid band approximation for the alloys, even for alloys containing nearest-neighbour elements in the periodic table.
However, the success of the degenerate band approximation indicates that figure 7 provides an essential concept for the interpretation of itinerant-electron ferromagnetism.

In order to modify the degenerate-band approximation, it is
logical to consider the effect of the crystalline fields. Except for low temperature cobalt, which is hexagonal close packed
(hcp), the elements and the alloys of figure 9 are all cubic.
At the centre Γ of the Brillouin zone, cubic crystalline fields
split the three atomic t_{2g} orbitals directed toward nearest
neighbours from the two atomic e_g orbitals directed toward next
nearest neighbours. Conventional band theory ($U = 0$) gives $3d$
band widths that are considerably larger than the cubic-field
splitting at Γ. However, where spontaneous magnetism occurs,
near neighbour interactions dominate the $3d$ band structure.
Therefore, neglecting spin-orbit coupling, we make as a zero
order approximation the assumption that two types of $3d$ band
states, though overlapping in energy, are distinguishable: t_2
bands containing six states per atom and e bands containing four
states per atom. We will demonstrate how it is possible to obtain a fair quantitative estimate of the magnetic order and the
magnitudes of the atomic moments for even complex alloy systems
from this zero-order approximation and figures 7,8. In the case

of alloys, the predominance of near-neighbour interactions is
particularly striking, and it is necessary to consider explicitly the formation of molecular-cluster orbitals. Moreover,
atomic charge neutrality is preserved in the alloys by appropri-
ately distributing the $3d$ electron charge density.

3.1.3 The Elements

In order to apply figure 7 to the transition elements, it is
necessary not only to place b in the phase diagram of figure 8,
but also to know n_ℓ for each of the bands t_2 and e. For this
task, we first note that the effective nuclear charge seen by
the $3d$ electrons increases with atomic number on traversing the
first long period, and it increases faster than the effective
nuclear charge of the $4s$ electrons. It follows that U for the
free atom increases with atomic number, as prescribed by equation (4), while the b_{ij} for the $3d$ electrons decrease for any
isostructural series. As b/U increases on going from ferromagnetic Ni to paramagnetic Ti, both b_e and $b_t > b_e$ move unambiguously to larger b in figure 8. This conclusion is consistent
with a lack of any spontaneous magnetism in the elements lighter
than Cr, spontaneous ferromagnetism in Fe, Co, and Ni, and unusual magnetic properties in Cr and Mn. Therefore, we assume
that $b_c < b \lesssim b_m$ for all the magnetic transition elements.

A distinction must also be made between the face centred cubic (fcc) and body centred cubic (bcc) structures. In the latter, the nearest neighbour distances are longer and the next
nearest neighbour distances are shorter than in the close packed
phases. It follows that the e and t_2 bands are more similar in
the bcc structure. In the fcc structure, strongly correlated e
electrons are more probable.

In this connection, it is interesting to compare the magnetic
properties for the two forms of iron. An atomic moment $\mu_{Fe} =
2.2$ μ_B for bcc iron is consistent with $n_s \approx 1$, $z_d \approx 7$, $\delta_s \approx 0.2$,
an $n_\ell \lesssim 3/2$ for both the e and the t_2 bands, and at least b_t located in region F_R of figure 8, which corresponds to itinerant
electron ferromagnetism. For example, a configuration Fe:$e^3 t_2^4 s^1$
having strongly correlated e electrons and $b_c < b_t < b_m$ would
give an e-electron contribution and a t_2-electron contribution
of $1\mu_B$ each to the atomic moment. A small contribution from the
$4s$ band susceptibility in the molecular field would account for
the additional 0.2 μ_B per iron atom. Although we do not need
to assume a splitting of the e^3 and e^4 manifolds in bcc iron,
in fcc iron we do. In the fcc phase, which may be stabilized
to low temperatures as precipitate particles in a fcc matrix,
the atomic moment appears to decrease from about $2\mu_B$ at high
temperatures to about 0.7 μ_B in an antiferromagnetic, low-temperature phase [10]. This observation can be understood if the e
electrons are strongly correlated and the configuration Fe:e^4
$t_2^3 s^1$ is only slightly more stable than the configuration Fe:e^3
$t_2^4 s^1$. At low temperatures, a half filled t_2 band gives antiferromagnetic coupling with $\mu_{Fe} = 3(1 - \delta)\mu_B$, and a large

$\delta \approx 0.75$ indicates a $b_t \to b_m$. As the temperature is raised, the configuration Fe:$e^3t_2^4s^1$ becomes increasingly populated. As a solute in a magnetic alloy, internal molecular fields would stabilize the $e^3t_2^4$ configuration relative to the $e^4t_2^3$ configuration.

In fcc cobalt, spontaneous ferromagnetism in the t_2 bands creates an internal molecular field that stabilizes the e^3 configuration relative to e^4, and therefore $n_\ell \gtrsim 3/2$ for both the e and t_2 bands. In hexagonal cobalt, even a filled a_1 band leaves the two e bands with $n_\ell \gtrsim 3/2$, since $n_s \approx 0.7$. Therefore, spontaneous ferromagnetism gives a $\mu_{Co} = (10 - z_d^{Co} + \delta_s)\mu_B = (1 + n_s + \delta_s)\mu_B$ for both close packed phases. Similarly, a configuration Ni:$e^4t_2^{5+\lambda}s^{1-\lambda}$ gives a $\mu_{Ni} = (10 - z_d^{Ni} + \delta_s)\mu_B = (n_s + \delta_s)\mu_B$.

In manganese, on the other hand, an $n_s \gtrsim 1$ gives, according to figure 7, an average $n_\ell \lesssim 1.2$. Thus, in fcc γ-Mn the t_2 bands should stabilize antiferromagnetic coupling between nearest neighbours. Since antiferromagnetic coupling between all nearest neighbours cannot propagate in a fcc array, ferromagnetic (001) layers coupled antiparallel provides the best collinear spin compromise. Such a tetragonal magnetic configuration would be reflected in a tetragonal ($c/a < 1$) crystal structure, t_2-electron bonding occuring between nearest neighbours coupled antiparallel. The magnetic moment would be $\mu_{Mn} \approx (4 - 2\delta)\mu_B$. Extrapolation of fcc manganese alloys to pure manganese indicates [11] that high temperature γ-Mn would undergo a fc tet $\rightleftarrows$ fcc transition at a $T_N = 600°K$, that the magnetic order would be tetragonal ($c/a < 1$) as predicted, and that $\mu_{Mn} \approx 2.4 \mu_B$, corresponding to a $\delta \approx 0.8$ and therefore a $b \gtrsim b_m$ for the two bonding t_2 orbitals $d_{yz}d_{zx}$. This, in turn, is consistent with a $b_t \to b_m$ in γ-Fe. Moreover, a $b_t \approx b_m$ in fcc γ-Mn indicates that more complex crystallographic distortions may be anticipated, since $n_\ell = 1$ for the t_2 electrons. This is, in fact, what is found, α-Mn and β-Mn representing extremely complex magnetic structures [12].

Chromium, with approximately half-filled $3d$ bands and $b \gtrsim b_m$ for next-nearest neighbours of a bcc array, exhibits Pauli paramagnetism, an AFS with a $\mu_{Cr} \approx 0.4 \mu_B$, and a T_N that is extremely sensitive to impurities [13]. Spontaneous magnetism and the magnetic order depend on 'nesting' of the Fermi surface in the Brilloiun zone as well as on a $b \approx b_m$ [14].

3.1.4 Substitutional Binary Alloys

3.1.4.1 *Non-Transition Metal Alloys*

An isolated solute atom interacts with the nearest-neighbour solvent (host) atoms to create molecular-cluster orbitals. These interactions create more stable bonding states that are split from less stable antibonding states. If a non-transition metal atom having v' outer valence electrons is substituted into a ferromagnetic solvent, n_s' and $(v' - n_s')$ antibonding states are, respectively, lifted in energy from the antibonding 4s and 3d bands. A corresponding number of bonding states are stabilized. Removal of antibonding molecular-cluster states does not change the total number of antibonding electrons, which are responsible for the atomic moment; therefore, it does not change the average moment per solvent atom provided an $n_\ell < 3/2$ is maintained in all 3d bands. However, if an $n_\ell > 3/2$ occurs, removal of antibonding states from this band is equivalent to pairing spins at the neighbouring solvent atoms.

In order to apply this reasoning, let us first present the expressions for pure-solvent, itinerant-electron atomic moments:

$$\mu_A^0 = (z_{sd}^0 - n_s^0 - 5 + \delta_s^0)\mu_B$$

$$\text{(25)}$$

if $5 < (z_{sd}^0 - n_s^0) < 7.5$ and all $n_\ell \leqslant 3/2$,

$$\mu_A^0 = (10 - z_{sd}^0 + n_s^0 + \delta_s^0)\mu_B$$

$$\text{(26)}$$

if $(z_{sd}^0 - n_s^0) > 7.5$ and all $n_\ell \geqslant 3/2$,

As we have seen, bcc iron has $z_{sd}^0 = 8$, $n_s^0 \approx 1$ and $\delta_s^0 \approx 0.2$, giving $\mu_{Fe} = 2.2\ \mu_B$, whereas fcc nickel has $z_{sd}^0 = 10$ and $n_s^0 + \delta_s \approx 0.6$, giving $\mu_{Ni} \approx 0.6\ \mu_B$. Similarly, fcc cobalt has $z_{sd}^0 = 9$ and $n_s^0 + \delta_s^0 \approx 0.75$, giving $\mu_{Co} \approx 1.75\ \mu_B$. Extension to the non-transition metal alloys gives the mean moments

$$\bar{\mu}_A = (1 - c)(z_{sd}^0 - n_s - 5 + \delta_s)\mu_B$$

$$\approx (1 - c)\mu_A^0 - c\Lambda \qquad \text{(27)}$$

provided $5 < (z_{sd}^0 - n_s^0 + c(v' - n_s')) < 7.5$ and $n_\ell \leqslant 3/2$ for all solvent 3d bands. Here

$$\Lambda = \frac{d(n_s - \delta_s)}{dc} \qquad \text{(28)}$$

is a small fraction that measures changes with concentration in 4s-band population and magnetization. On the other hand,

$$\bar{\mu}_A = \{(1 - c)(10 - z_{sd}^0 + n_s + \delta_s) - c(v' - n_s')\}\mu_B$$

$$\approx \mu_A^0 - c(v' + m + \Lambda')\mu_B, \qquad \text{(29)}$$

provided $(z_{sd}^0 - n_s) > 7.5$ and $n_\ell \geqslant 3/2$ for all solvent $3d$ bands. Here $m = 10 - z_{sd}^0$ and

$$\Lambda' = (n_s - n_s') + \frac{d(n_s + \delta_s)}{dc} \; . \tag{30}$$

Experimentally, equation (27) is found to apply where iron is a solvent: $\mu_A^0 = \mu_{Fe} = 2.2 \; \mu_B$, whereas equation (29) applies where Co ($m = 1$) or Ni ($m = 0$) are the solvent, [15]. This situation is very satisfactory.

3.1.4.2 *Transition Metal Alloys*

In the case of transition metal solute atoms, both their $3d$ and their valence electrons participate in the formation of molecular-cluster orbitals. The stability of the $3d$ bands relative to the $4s$ band increases with atomic number Z, which is why, for the elements, n_s^0 decreases with increasing z_{sd}^0. Since $n_s \approx n_s'$ and the $4s$ band is much broader than the $3d$ bands, electrons in the molecular-cluster $3d$ orbitals distribute themselves so as to preserve charge neutrality. Therefore, it is meaningful and useful to speak of effective atomic configurations for the constituents of an alloy.

Although antibonding molecular-cluster orbitals are relatively unstable, this instability decreases with the overlap integral and the energy separation of the interacting orbitals. Therefore, antibonding molecular-cluster $3d$ orbitals may have energies that are overlapped by the solvent t_2 band. Moreover, as the concentration of solute atoms increases, solute-solute interactions cause the cluster orbital energies to be broadened into energy bands that overlap the solvent t_2 bands and need not be distinguished from them.

Transition metal bonding electrons give no contribution to a ferromagnetic moment, while the antibonding $3d$ electrons are spontaneously magnetic, whether they occupy molecular-cluster or solvent-band states. Since e-orbital interactions are weaker than t_2-orbital interactions, the antibonding molecular-cluster orbitals first occupied are those formed from e orbitals.

From these considerations, it follows that equations (27,29) become modified to

$$\bar{\mu}_A \approx (z_{sd} - n_s - 5 + \delta_s)\mu_B \tag{31}$$

provided $5 < (z_{sd} - n_s) < 7.5$ and $n_\ell \leqslant 3/2$ for all $3d$ orbitals; but if $n_\ell > 3/2$ for all solvent $3d$ orbitals and $(z_{sd}^0 - n_s) > 7.5$

$$\bar{\mu}_A \approx \mu_A^0 + c\{2n_d' - (v' + m + \Lambda')\}\mu_B$$

$$\text{if } n_d' < (v' - n_s') \leqslant 5, \tag{32a}$$

$$\bar{\mu}_A \approx (10 - z_{sd} + n_s + \delta_s)\mu_B \quad \text{if } n_d' \geq 7.5, \tag{32b}$$

where v' is $z_{sd}{}^0$ for the solute atom. Equation (32a) applies to lighter solute atoms in a heavier-atom solvent, so the n_d' molecular-cluster antibonding electrons give a solute atom moment $\mu' = n_d'\mu_B$. In fcc alloys, strong correlations separate the e^2 configuration from the e^3 configuration by a relatively large energy U, and only the e^2 molecular cluster configuration is occupied before the t_2 orbitals. Therefore, the n_d' electrons at the solute atom behave like localized electrons in a tetrahedral crystalline field: only e orbitals are occupied if $n_d' \leq 2$, but a high spin configuration having two e and $(n_d' - 2)t_2$ electrons of parallel spin is found where $2 < n_d' < 5$. In the latter case, where half-filled t_2 orbitals are at the solute, antiferromagnetic solute-solute interactions can stabilize homopolar-bond formation between the two like atoms. Such a bond formation between like atoms is stronger than the ferromagnetic interactions between unlike nearest-neighbour atoms, so equation (32a) only applies, in this case, for dilute solutions where solute-solute interactions are not present. The moments at solute atoms coupled antiparallel are reduced from $n_d'\mu_B$ by $(1 - \delta)\mu_B$ for each solute orbital participating in a homopolar bond.

The magnitudes of the individual atomic moments in a given alloy can be obtained from the additional requirement of charge neutrality, which is accomplished by the $3d$ electron distribution.

Because the relative stability of the atomic $3d$ orbitals and their occupancy increases with atomic number Z, the magnitude of any n_d' must increase with the solute atomic number Z' for a given solvent Z, and with Z for a given Z'.

Where equation (31) applies,

$$\frac{d\bar{\mu}_A}{dz_{sd}} = (1 + \tilde{\Lambda}')\mu_B, \tag{33}$$

and where equation (32b) applies,

$$\frac{d\bar{\mu}_A}{dz_{sd}} = -(1 + \tilde{\Lambda}')\mu_B, \tag{34}$$

where

$$\tilde{\Lambda} = \frac{d(\delta_s - n_s)}{dz_{sd}} \quad \text{and} \quad \tilde{\Lambda}' = -\frac{d(n_s + \delta_s)}{dz_{sd}} \tag{35}$$

are positive fractions. Equations (33,34) provide the basis for the Slater-Pauling curve. Where equation (32a) applies, or where the conditions for ferromagnetic order are not fulfilled, deviations from the Slater-Pauling curve are found.

The two equations (32a,b) should apply to alloys having, as a solvent, Co, Ni, or Cu. If we assume, in zero-order approximation, a $\Lambda' \approx 0$, the curves of figure 9 indicate that $n_d' = 0$ at V solutes, $n_d' \approx 0.2$ and 0.8 for Cr and Ni, respectively, and $n_d' \approx 2.8$ and 4.7 for Mn in Co and Ni, respectively. Furthermore, antiferromagnetic Mn-Mn interactions have been observed directly in the Co-Mn and Ni-Mn systems [16]. These are responsible for the curvature in figure 9 of the $\bar{\mu}_A$ vs z_{sd} curve for the Ni-Mn system. Because $5 < n_d' < 7.5$ for Fe in nickel, the condition $n_\ell \geqslant 3/2$ is not fulfilled for Fe-Fe interactions, and equation (32b) only applies for dilute concentrations of Fe in nickel, see equation (38). At this point in the discussion, it is sufficient to point out that n_d' varies qualitatively with Z and Z' as required from the systematics of the periodic table.

With the exception of the Fe-Ni system, alloys having iron as a solvent should be described by equation (31). So long as correlations among the e electrons maintain an e^3 configuration for both Fe and Co atoms in the Fe-Co system, the e bands have $n_\ell = 3/2$ and equation (31) should apply for the iron-rich alloys, equation (32b) for the cobalt-rich alloys. Experimentally, $\bar{\mu}_A = (2.2 + c)\mu_B$ for a cobalt concentration $c < 0.33$, in excellent agreement with equation (31) for a constant $n_s \approx 1$ and $\tilde{\Lambda} = 0$. For $c > 0.5$ equation (32) applies if $n_s \approx 0.85$ and $\tilde{\Lambda}' = 0$ in the bcc phase. In the interval $0.33 < c < 0.5$, a small decrease in n_s appears to accompany increasing atomic order. Moreover, a combination of neutron-diffraction and magnetometer data [17] indicate that in the ordered 50-50 alloy FeCo, one sublattice contains $2.9 \, \mu_B$ and the other $2.0 \, \mu_B$. We can identify the larger moment with the iron sublattice, the smaller with the cobalt sublattice, since charge neutrality requires placing one more antibonding t_2 electron on a cobalt than on an iron atom and $n_\ell > 3/2$.

For larger concentrations c of Ti, V, or Cr in bcc iron, the condition $n_\ell < 3/2$ for all the $3d$ bands is fulfilled. Moreover, charge neutrality requires that the antibonding electrons responsible for spontaneous ferromagnetism be associated with the heavier Fe atoms. Therefore,

$$\mu_{Ti} = \mu_V = \mu_{Cr} = 0\mu_B, \tag{36}$$

and from equation (31)

$$\mu_{Fe} = \frac{\bar{\mu}_A}{(1 - c)} = \frac{(z_{sd} - n_s - 5 + \delta_s)\mu_B}{(1 - c)} . \tag{37}$$

A 50-50 FeTi alloy would have $z_{sd} = 6$, so that for $n_s > 1$ only bonding orbitals are occupied and $\mu_{Fe} = 0\mu_B$. Experimentally,

the 50-50, ordered FeTi alloy has no spontaneous magnetism [18].
A 50-50 FeV alloy, on the other hand, would have z_{sd} = 6.5 and
c = 0.5. Therefore, for $n_s \approx 1$, a $\mu_{Fe} \approx 1.1$ μ_B is predicted.
Neutron-diffraction data [18] for an ordered alloy gave 0.9 μ_B
on one sublattice and 0.1 μ_B on the other hand, which is in excellent agreement if allowance is made for a small amount of
atomic disorder in the sample. Figure 9 shows that the Fe-Cr
alloys follow equation (33) with a reasonable value for $\tilde{\Lambda}$ at
larger Cr concentrations, but require a $\tilde{\Lambda} \approx 0.45$ for small c.
A plausible explanation for this can be found. Because the
antibonding molecular cluster states are unstable in dilute alloys, five $3d$ electrons per Cr atom enter the antibonding $3d$
orbitals at near-neighbour iron atoms, and an n_ℓ > 3/2 may occur
locally.

Difficulty with the value of n_ℓ for different $3d$ orbitals introduces problems in the Fe-Ni alloys also. Although concentrations c < 0.33 of iron in nickel give a $\bar{\mu}_A$ that is well described
by equation (32b) if $\Lambda' \approx 0.1$ in equation (34), a $\Lambda' \approx -0.13$
must be chosen for larger values of c. Moreover, deviations
from the Slater-Pauling curve are found for all values of c in
the bcc alloys. These difficulties require an analysis of n_ℓ
for all the $3d$ orbitals. Introduction of an isolated Fe atom
into fcc nickel creates molecular-cluster orbitals, and an
$n_\ell \geqslant 3/2$ in every Fe-Ni bond allows holes only in the spin down
states. Charge neutrality places three holes at an Fe atom to
one hole at a Ni atom, and therefore a $\mu_{Fe} \approx (3 + \delta_s)\mu_B$ and a
$\mu_{Ni} \approx (n_s + \delta_s)\mu_B$, in accord with equation (32b). However, as
c increases, Fe-Fe interactions are introduced. Since n_ℓ < 3/2
for the orbitals participating in Fe-Fe interactions, the holes
add as in elemental iron, where $\mu_{Fe} \approx (2 + \delta_s)\mu_B$. Therefore,
for larger c and random atomic distribution,

$$\mu_{Fe} \approx (3 - c + \delta_s)\mu_B \quad \text{and} \quad \mu_{Ni} = (n_s + \delta_s)\mu_B. \quad (38)$$

These values should hold in the bcc phase as well, provided allowance is made for a discontinuous increase in n_s on passing
from the fcc to the bcc phase. Therefore,

$$\bar{\mu}_A \approx \{n_s + (3 - n_s)c - c^2 + \delta_s\}\mu_B \quad (39a)$$

for larger c changes, as $c \to 0$, to the values given by equation
(32b)

$$\bar{\mu}_A \approx (10 - z_{sd} + n_s + \delta_s)\mu_B = (n_s + 2c + \delta_s)\mu_B. \quad (39b)$$

For larger c in the bcc phase, where $n_s \approx 1$, equation (39a) reduces to

$$\bar{\mu}_A = (2 - c'^2 + \delta_s)\mu_B, \quad \text{(Contd)}$$

(Contd) $$c' = (1 - c) = \tfrac{1}{2}(z_{sd} - 8),\tag{40}$$

which is in striking agreement with figure 9. Moreover, from a combination of neutron-diffraction and magnetometer data [19], a $\mu_{Fe} = (2.60 \pm 0.1)\mu_B$ and a $\mu_{Ni} = (0.67 \pm 0.1)\mu_B$ were deduced for $c = 0.50$, which is in excellent agreement with equation (38). The change in slope in $\bar{\mu}_A$ vs c at $c = 0.33$ may be attributed to the onset of short range order that minimizes the number of Fe-Fe interactions. In this case, we should expect to find for $c < 0.33$

$$(3 - c + \delta_s)\mu_B < \mu_{Fe} < (3 + \delta_s)\mu_B,\tag{41}$$

and the neutron data [19] give $\mu_{Fe} = (2.91 \pm 0.2)\mu_B$ and $\mu_{Ni} = (0.62 \pm 0.02)\mu_B$ for $c = 0.257$. Since $\delta_s \approx 0.1$, the data are in good accord with equation (41), although still within experimental error for equation (38).

3.2 ORDERED ALLOYS

3.2.1 Manganese-Copper and Manganese-Gold Alloys

Several ordered alloys are known in the Mn-Au system: MnAu, $MnAu_2$, $MnAu_3$ and $MnAu_4$, in all of which the manganese atoms carry a spontaneous atomic moment. Similarly, even dilute concentrations of manganese in copper are characterized by localized atomic moments on the manganese atoms. Also, the ferromagnetic Heusler alloys Cu_2MnM, where M is a group III, IV or V element, have been known for a long time.

The characteristic feature of both Cu and Au is the outer electron configuration $d^{10}s^1$. We must anticipate that manganese substitutes with the configuration d^6s^1, each Mn atom stabilizing occupied, bonding molecular-cluster s orbitals and destabilizing empty, antibonding molecular-cluster s orbitals. Each isolated Mn atom also creates molecular-cluster $3d$ orbitals. Because the Mn atom is lighter than copper or gold atoms and because of charge neutrality considerations, the $3d$ holes are associated with the Mn atoms. In a binary Cu or Au alloy, the $3d$ states of an isolated manganese atom may all be considered antibonding, so four $3d$ holes per Mn:d^6 atom leaves $n_\ell > 3/2$. Introduction of near neighbour Mn-Mn interactions, on the other hand, introduces a condensation of one $3d$ hole per manganese $3d$ orbital participating in Mn-Mn σ-bonding. This condensation creates an $n_\ell = 1$ for any molecular Mn-Mn σ-bonding orbital, and antiferromagnetic Mn-Mn coupling follows. Formation of homopolar σ bonds between manganese nearest neighbours tends to spin pair the bonding electrons, but intra-atomic exchange with localized spins in other orbitals of each Mn atom induces some contribution to the atomic moments from the bonding electrons. It follows that isolated manganese atoms would experience only

ferromagnetic Mn-Cu-Mn or Mn-Au-Mn interactions and have an
atomic moment

$$\mu_{Mn}{}^F = (4 + \delta_S)\mu_B, \tag{42}$$

whereas nearest-neighbour Mn atoms would exhibit antiferromagnetic Mn-Mn interactions and have a

$$\mu_{Mn}{}^{AF} = (4 - \nu\delta)\mu_B, \tag{43}$$

where ν is the number of $3d$ orbitals per Mn atom participating
in antiferromagnetic coupling.

This prediction is confirmed experimentally [20]. Ferromagnetic $MnAu_4$, for example, contains isolated manganese atoms
having atomic moments $\mu_{Mn} = 4.15 \mu_B$. The Heusler alloys provide
an even more telling illustration.

The crystal structure of the ternary Heusler alloys Cu_2MnM is
shown in figure 10. It consists of two inter-penetrating, simple
cubic (sc) arrays, as in the bcc structure. One of the sc arrays is occupied by Cu atoms, the other by Mn and M atoms ordered as in the NaCl structure. This ordered structure is ferromagnetic. However, the spontaneous magnetization is extremely
sensitive to the thermal history of the specimen because any
atomic disorder results in antiferromagnetic Mn-Mn interactions.
The maximum spontaneous magnetization in these alloys is about
$4.1 \mu_B$ per molecule.

Interpretation of the Heusler alloys must include a consider-

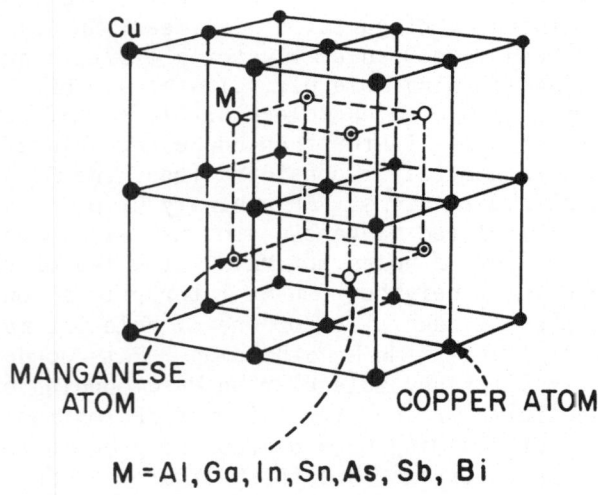

Figure 10 - Crystal structure of the ternary Heusler alloys Cu_2MnM.

ation of the outer s and p electrons at the M atoms. The symmetry of the eight Cu nearest neighbours to an M atom is conduc-
ive to (sp^3) hybridization at the lighter M atoms. The filled
$3d^{10}$ cores at the Cu atoms participate little in M-Cu bonding.
The Cu outer s and p electrons interact so as to create eight
s-p bonding states per M atom within the Cu_2M subarray, which
means that the bonding bands are not completely filled for M
atoms from groups III-V of the periodic table. Moreover, interactions with the next-near-neighbour Mn atoms is of secondary
importance, and the Mn atoms appear to retain an $e^2t_2^4 4s^1$ configuration with strongly correlated e_g electrons and antibonding
t_{2g} electrons in t_2 bands having $n_\ell > 3.2$. The Cu:d^{10} core sup-
plies the bonding electrons. Because the ratio of broad-band to
narrow-band electrons at the Mn atoms is not changed significantly by the presence of the M atoms, equations (42,43) apply.

3.2.2 Iron-Aluminium and Iron-Silicon Alloys

Substitution of Al and Si into bcc iron gives an initial
variation of $\bar{\mu}_A$ that is well described by equation (27). However, as the solute concentration exceeds $c \approx 0.15$, the effective parameter Λ begins to increase, reaching a $\Lambda \gtrsim 0.3$ in the
ordered alloys Fe_3Al and Fe_3Si. Moreover, the magnetic properties of alloys in the interval $0.25 < c < 0.5$ are quite complex,
and the ordered 50-50 alloy FeAl exhibits no spontaneous magnetism [21]. These observations indicate that at least one of our
assumptions breaks down at higher solute concentrations.

In the ordered alloy FeAl, which has the CsCl structure, symmetry permits covalent mixing of the σ-bonding t_2^3s orbitals at
Fe atoms and the hybrid (sp^3) orbitals at Al atoms. Moreover,
the energies of the orbitals on the respective atoms are comparable, so that the bonding states include a strong contribution from both atoms. Therefore, the t_2 orbitals at the iron
atoms participate in the formation of broad, bonding σ bands
of itinerant electron states $(b > b_m)$, and these σ bands are
only partially filled. The antibonding σ^* bands are therefore
empty. Since the more electropositive Al atoms each contribute
only three outer electrons to these bonding bands, the outer
electron configurations become $Fe:e^4(\sigma^4) - Al:(\sigma^3)$, where, the
parentheses enclose itinerant electrons. Filled e bands on the
iron subarray cannot support spontaneous magnetism; nor can partially filled σ bands having a $b > b_m$. Therefore, we may conclude that the observation of no spontaneous magnetism in ordered FeAl signals a breakdown of the assumption that the Fermi
energy falls between the correlation-split e^3 and e^4 bands. The
e^3 and e^4 bands are split by a relatively small energy $U' < U$
that must decrease with reduced intra-atomic exchange stabilization as spontaneous t_2-electron magnetism disappears, so breakdown of this assumption is a consequence of the disappearance
of anti-bonding t_2 electrons, which occurs where molecular orbital
formation between solvent $3d$ orbitals and solute p orbitals raises
the energies of the antibonding states above the Fermi energy.

In the ordered alloys Fe_3Al and Fe_3Si, two types of iron atoms can be distinguished. The structure is similar to that shown in figure 10, Fe_{II} atoms replacing the sc Cu array, Fe_I atoms the Mn array, and M = Al or Si. Polarized neutron data [22] from an ordered Fe_3Al specimen gave average site moments $\mu_{FeI} \approx 2.14\,\mu_B$, $\mu_{FeII} \approx 1.46\,\mu_B$ and $\mu_{Al} \approx 0.1\,\mu_B$, the moment on the aluminium sites presumably signalling some atomic disorder in the sample. What is striking is that the Fe_I atoms, which have only Fe_{II}-atom near neighbours, retain a $\mu_{Fe} = (2 + \delta_s)\mu_B$. Only the Fe_{II} atoms, which bond with four nearest neighbour Al atoms, have a reduced moment.

In Fe_3Al, the nearest neighbour Al-Fe_{II} interactions can create occupied, bonding s states and unoccupied, antibonding s states that leave one p electron per Al atom for spin pairing with the Fe_{II}:t_2 electrons. Such interactions would also quench the conduction band contribution $\delta_s\mu_B$ to μ_{FeII}. Since there are two Fe_{II} atoms for every Al atom, the result would be a $\mu_{FeII} \approx 1.5\,\mu_B$. The excellent agreement with experiment obtained by such a naïve argument suggests that the number of occupied Fe_{II}:t_2-Al:p bonding states removed from the Fe_I-Fe_{II} t_2 bands is just sufficient to maintain charge neutrality. In a random alloy, the 'extra' electron per Al atom may participate in Al-Al bonding; but in a domain of short range order, spin pairing between Al:p and Fe:t_2 electrons appears to occur.

3.2.3 Ordered MPd$_3$ and MPt$_3$ Alloys

The ordered MPd_3 and MPt_3 alloys, like the ordered MNi_3 alloy, consists of a sc array of M atoms with Ni, Pd, or Pt at the face centers. Since charge neutrality maintains n_s holes in the $4d$ and $5d$ bands of Pd or Pt, equation (32) should provide the atomic moments for M = Cr, Mn, Fe, or Cr even though the host $4d$ and $5d$ bands have no spontaneous magnetism. Moreover, an $n_\ell > 3/2$ in the $4d$ and $5d$ bands introduces ferromagnetic coupling through the host atoms, even though their induced atomic moments may not be saturated:

$$\delta_{sd}\mu_B < \mu_{Pt} < \mu_{Pd} < (n_s + \delta_s)\mu_B. \tag{44}$$

As in Ni alloys, n_d' must increase with M-atom atomic number. In the nickel alloys, n_d' does not reach saturation, i.e. equivalent to n for an atomic configuration $d^{n_s}1$ to give $\mu_M = (10 - n + \delta_s)\mu_B$, until M = Fe, as was discussed above. In the Pd and Pt alloys, n_d' is larger, reaching saturation at M = Mn. The observed moments are [23]

$$\mu_{Cr} \approx (n_d' + \delta_s)\mu_B \approx (2.4 \pm 0.2)\mu_B, \tag{45}$$

$$\mu_{Mn} \approx 4\mu_B, \qquad \mu_{Fe} \approx 3\mu_B, \qquad \mu_{Co} \approx 2\mu_B, \tag{46}$$

$$\mu_{Pt} \approx 0.15\,\mu_B \quad \text{and} \quad \mu_{Pd} \approx (0.40 \pm 0.05)\mu_B, \tag{47}$$

3.3 NITRIDES AND CARBIDES $M^cXM_3'^f$

3.3.1 Essential Features of the Energy Bands

The nitrides and carbides $M^cXM_3'^f$, where X = N or C, are ordered interstitial compounds. The M^c atoms form a sc array with
M^f atoms at the face centers, as in the ordered MPd_3 and MPt_3
alloys, and the X atom occupies the octahedral interstice at the
body center position. This structure is illustrated in figure 11.
Introduction of the interstial X atom tends to stabilize many
$M:M_3'$ ratios that do not exist as ordered MM_3' alloys. Compounds have been formed with M' = Cr, Mn, Fe, Co and Ni. Most
extensively studied are the compounds $M^cXMn_3^f$ and $M^cXFe_3^f$. Carbides of these two classes are known for M^c from groups II, III,
and IV of the periodic table. Nitrides are known that have M^c
from groups I-V and from the heavier elements of the three transition series. Fruchart and co-workers at Bellevue, France have
made an extensive and systematic study of the magnetic and crystallographic properties of these compounds and their pseudo-bin-
ary allyos. The extremely complex interplay between magnetic order and structural changes as a function of composition provide
an important testing ground for the ideas that have been developed thus far in this paper.

The essential features of the energy bands can be illustrated
by qualitative curves of the energy density of states. Consider, for example, the energy density of states for fcc iron illustrated in figure 12. From magnetic data, we concluded that,
at lowest temperatures, the low spin $Fe:e^4t_2^3s^1$ atomic configuration is a little more stable than the intermediate spin configuration $Fe:e^3t_2^4s^1$. We also concluded that the e^3 and e^4
configurations could be represented by narrow bands of width w_b
that are split from one another by a correlation energy $U' > w_b$.

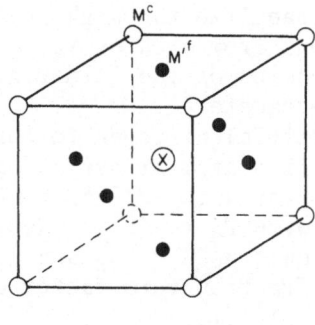

X = C or N,   M' = Mn or Fe

Figure 11 - Crystal structure of the $M^cXM_3'^f$ nitrides and
carbides.·

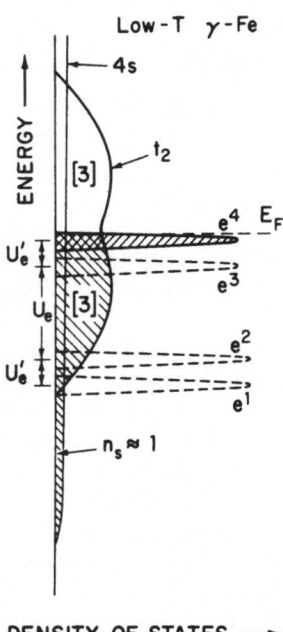

Figure 12 - Schematic energy density of states for low temperature γ-iron Fe:$e^4 t_2{}^3 4s^1$.

At higher temperatures, the greater entropy associated with the $e^3 t_2{}^4$ configuration ($\mu_{Fe} \approx 2\mu_B$ vs $\mu_{Fe} \approx 0.7\ \mu_B$) tends to raise the e^4-band energies relative to the t_2-band energies. In addition, and perhaps more importantly, a larger lattice parameter at higher temperatures increases the energies U and U', thereby shifting the e^3 and e^4 configurations to higher energies relative to the Fermi energy E_F. The energy $U > U'$ separates the e^2 and e^3 configurations, and sufficiently large increase in lattice parameter would stabilize the high-spin configuration e^2. The shift of electrons from e bands to t_2 bands is cooperatively induced by a narrowing of the itinerant electron t_2 bands with increasing lattice parameter.

Introduction of interstitial nitrogen to form $Fe^c N Fe_3{}^f$ creates two distinguishable iron atom arrays and increases the lattice parameter. Magnetometer data [24] give a ferromagnetic molecular moment, $\mu_m \approx 9\ \mu_B$ with $T_c = 765°K$. Neutron diffraction data [25] give atomic moments $\mu_{Fe}{}^c \approx 3\mu_B$ and $\mu_{Fe}{}^f \approx 2\mu_B$ for Fe^c and Fe^f, and electron diffraction form factor data [26] indicate a valence formula $Fe^0 N^{-1}(Fe^{+1/3})_3$.

The point symmetry about an Fe^c atom remains cubic, so the corner atom atomic configurations continue to consist of e_g and t_{2g} electrons. Observation of a $\mu_{Fe}{}^c \approx 3\mu_B$ can only be reconciled with a high-spin configuration Fe^c:$e^2 t_2{}^{5+\varepsilon} s^{1-\varepsilon}$, so we conclude that expansion of the lattice parameter has increased U^c

sufficiently to do this.

The Fe^f point symmetry is tetragonal, and the two e_g orbitals (d_{z^2} and $d_{x^2-y^2}$) become a_{1g}, which is directed parallel to the unique Fe^f-N axis, and b_{1g}, which lies in the face perpendicular to this axis and is directed toward next near neighbour Fe^f atoms. The three t_{2g} orbitals (d_{yz}, d_{zx}, d_{xy}) become two e_g orbitals directed toward nearest neighbour Fe^f atoms and a b_{2g} orbital directed toward nearest neighbour Fe^c atoms. The p orbitals of the interstitial N atom σ-bond with the Fe^f:p orbital directed toward the N atom to form filled, bonding σ-band states that are split, via the translational symmetry, from empty, antibonding σ^*-band states by a discrete energy gap. Such molecular orbital formation may leave the X atom essentially neutral. The Fe^f:a_1 orbitals also interact with the N:p orbitals, but because only bonding states are occupied, there is no a_1-electron contribution to the magnetic moment. However, the b_1^1 configuration is split from the b_1^2 configuration by a large energy $U_f <$ U_c, comparable to the energy U_c splitting the high spin e^2 configuration from the intermediate spin configuration e^3 at an Fe^c atom. Therefore, the Fe^f-atom configuration becomes Fe^f:(a_1^0) $b_1^1 b_2^{1+\lambda+\epsilon} e^{4-\lambda} s^{1-\epsilon}(p^1)$, and the b_{1g} electrons contribute $1\mu_B$ to μ_{Fe}^f. The degenerate t_2 bands of fcc iron are split into Fe^f:e bands and Fe_3^f:b_2-Fe^c:t_2 bands that contain a total of $4(1 - \epsilon)$ holes per molecule, corresponding to an average $\bar{n}_\ell = 20/24 = 5/6$ $> 3/2$ for $\epsilon \approx 0$. Therefore, ferromagnetic coupling between nearest neighbour iron atoms (Fe^f-Fe^f and Fe^f-Fe^c) is predicted, each hole contributing $1\mu_B$ to the magnetization, and charge parity requires that the $4(1 - \epsilon)\mu_B$ holes per molecule contribute $(1 - \epsilon)\mu_B$ per atom to both μ^c and μ^f. Since the high-spin e^2 configuration at an Fe^c atom gives $2\mu_B$, the predicted moments are

$$\mu_{Fe}^c = (3 - \epsilon + \delta_s)\mu_B \quad \text{and} \quad \mu_{Fe}^f = (2 - \epsilon + \delta_s)\mu_B, \quad (48)$$

which agree with the measured moments provided

$$\epsilon \approx \delta_s. \quad (49)$$

The corresponding band structure is illustrated schematically in figure 13.

Four conclusions emerge from this exercise:

Conclusion (1): Any covalent mixing between nitrogen s and p and metal-atom $4s$ orbitals has little influence on the total population of the half-filled $4s$ bands;

Conclusion (2): The existence of a low-spin $\rightleftarrows$ high-spin transformation on increasing the cubic lattice parameter a_0 of fcc iron from ca 3.64 Å to the 3.795 Å of $Fe^cNFe_3^f$ shows that there is a critical value of a_0 at which such a transition takes place. The additional intra-atomic ex-

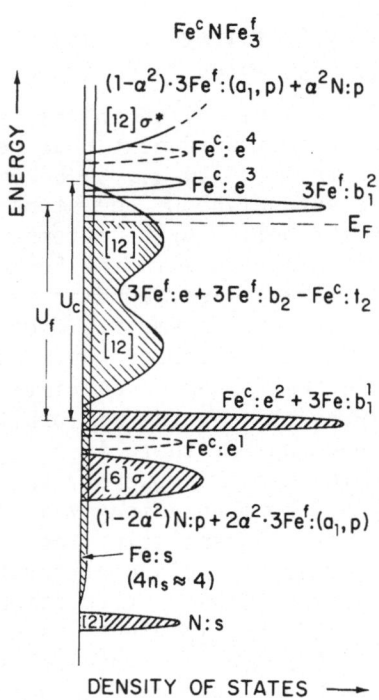

Figure 13 - Schematic energy density of states for Fe^c NFe_3^f corresponding to outer-electron configurations $Fe^c: e^2 t_2{}^5 4s^1$ and $Fe^f: (a_1{}^0) b_1{}^1 b_2{}^{1+\lambda} e^{4-\lambda} s^1 (p^1)$.

change stabilization of the two-fold degenerate e^2 configuration at an Fe^c atom insures a $U_c > U_f$ and therefore a transition $b_1{}^2 \to b_1{}^1$ at a smaller a_0 than the transition e^3 or $e^4 \to e^2$;

Conclusion (3): Stabilization of the $M^c X M_3{}'^f$ structure is achieved by the formation of filled σ-bands and empty σ^*-bands. Therefore, this feature should be common to all the alloys $M^c X Fe_3{}^f$ and $M^c X Mn_3{}^f$,

Conclusion (4): Atomic configurations provide a convenient short-hand notation for keeping track of the charge neutrality conditions and band occupancies.

3.3.2 The System $Fe_{1-x}{}^c Ni_x{}^c N Fe_3{}^f$

Substitution of Ni for the Fe^c atom in the cubic system $Fe_{1-x}{}^c Ni_x{}^c N Fe_3{}^f$ reduces the ferromagnetic Curie temperature and the molecular moment [27]:

$$T_c = (765 - 125x)\,^\circ K \quad \text{and} \quad \mu_m \approx (9 - 2x)\mu_B. \qquad (50)$$

Relative to the Fermi energy, the $3d$-electron energies at a Ni atom are more stable than those at an Fe atom. Therefore, the $Ni^c:e^4$ band would lie below E_F in figure 13 and the population of the $3Fe^f:b_2$-$Ni^c:t_2$ bands would remain unchanged ($n_\ell > 3/2$). Moreover, the condition of charge neutrality would distribute $n_s = (1 - \varepsilon)$ d-band holes at the Ni^c atoms while leaving the number of holes at the Fe atoms unchanged. Consequently, from equations (48,49)

$$\mu_{Ni}^c = (1 - \varepsilon + \delta_s)\mu_B \approx 1\mu_B, \tag{51}$$

$$\mu_m = (1 - x)\mu_{Fe}^c + x\mu_{Ni}^c + 3\mu_{Fe}^f \approx (9 - 2x)\mu_B, \tag{52}$$

and $d\mu_m/d(2x) \approx -1\mu_B$ is similar to the Slater-Pauling curve.

3.3.3 The System $Mn^cN_{1-x}Mn_3^f$, $0 \leqslant x \leqslant 0.2$

The nitrogen X-ray structure factor [28] indicates that nitrogen is essentially neutral in metallic $Mn^cNMn_3^f$, and neutron diffraction data together with magnetometer data reveal antiferromagnetic coupling between Mn^c and Mn^f atoms with [29]

$$\mu_{Mn}^c \approx 3.9\ \mu_B, \qquad \mu_{Mn}^f \approx 0.9\ \mu_B,$$

$$\mu_m = (\mu_{Mn}^c - 3\mu_{Mn}^f) \approx 1.2\ \mu_B. \tag{53}$$

Removal of x nitrogen atoms increases the Curie temperature and decreases the molecular moment [30]:

$$T_c = (740 + 175x)°K \quad \text{and} \quad \mu_m = (1.2 - 3.5x)\mu_B. \tag{54}$$

As in $Fe^cNFe_3^f$, the $N:p^3$ orbitals form a filled, bonding band with $Mn^f:p$ orbitals, and covalent mixing with the $Mn^f:a_1$ orbitals effectively lifts them above the Fermi energy as empty, antibonding states. Although the energy bands for $Mn^cNMn_3^f$ are similar to those shown in figure 13 for $Fe^cNFe_3^f$, their relative positions are shifted. Since the $3d$ bands are less stable relative to the $4s$ band in the lighter element manganese, the broad $4s$ band overlaps the $3d$ bands to a larger extent, and an $n_s \approx 1$ should be maintained even though the total number of outer electrons is decreased. Moreover, the $3d$ electrons at the lighter Mn atoms experience a smaller effective nuclear charge, which is why their relative stability is less. Therefore, they have a larger radial extension and the energies $U_f < U_c$ are lower for the Mn atoms. This means that the $Mn^f:b_1^2$ and $Mn^c:e^3,e^4$ levels are also more stable relative to the broader $3d$ bands. In fact, the measured atomic moments for $Mn^cNMn_3^f$ indicate that these

JOHN B. GOODENOUGH

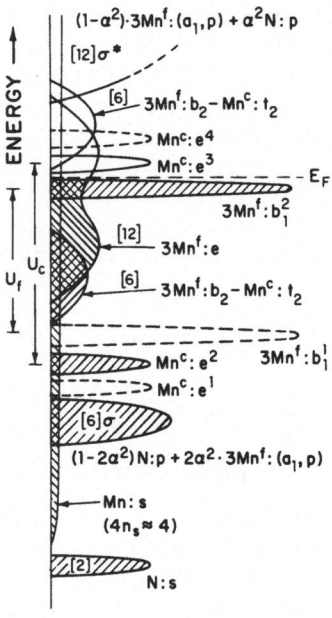

Figure 14 - Schematic energy density of states for Mn^c NMn_3^f corresponding to outer-electron configurations $Mn^c : e^2 t_2^4 4s^1$ and $Mn^f : (a_1^0) b_1^2 b_2^{2/3} e^{2+(1/3)} 4s^1 (p^1)$.

levels are shifted sufficiently to place E_F between the $Mn^f : b_1^2$ and $Mn^c : e^3, e^4$ levels, as shown in figure 14. If $n_s \approx 1$, this leaves four electrons/Mn^c atom and three electrons/Mn^f atom in the broader $3d$ bands.

Antiferromagnetic coupling between Mn^c and Mn^f atoms indicates an $n_\ell = 1$, corresponding to six electrons per molecule, for the $Mn^c : t_2$-$3Mn^f : b_2$ bands. Charge neutrality proportions these electrons as $Mn^c : t_2^4$ and $Mn^f : b_2^{2/3}$, so that the contribution to the atomic moments from these electrons is $(2 - 3\delta')\mu_B$ for Mn^c and $((2/3) - \delta')\mu_B$ for Mn^f atoms. Because charge neutrality has placed one spin pair on each Mn^c atom, δ' is not to be confused with the δ of equations (20,43). Each Mn^c atom bonds to nearest neighbours via three t_2 orbitals, and $\delta \approx \delta' + 0.33$.

Of the remaining $3d$ electrons, one electron per Mn^f atom participates in the σ-bands, which leaves $7/3$ electron per Mn^f atom in the two-fold degenerate $Mn^f : e$ bands, where two electrons per Mn^f atom corresponds to $n_\ell = 1$. Therefore, ferromagnetic coupling amongst the Mn^f atoms means a contribution to the atomic moment from only the antibonding e electrons, and hence $(1/3)\mu_B$.

In shorthand notation, for $n_s = 1$ the electron configurations at the atoms are:

$$N:s^2(p^3), \quad Mn^c:e^2t_2{}^4s^1, \quad Mn^f:(a_1{}^0)b_1{}^2b_2{}^{2/3}e^{2+(1/3)}s^1(p^1), \quad (55)$$

and the corresponding atomic moments are

$$\mu_{Mn}{}^c \approx (4 - 3\delta' + \delta_s)\mu_B, \qquad \mu_{Mn}{}^f \approx (1 - \delta' - \delta_s)\mu_B, \qquad (56)$$

where $2\mu_B$ are contributed to the $\mu_{Mn}{}^c$ from the high-spin $Mn^c:e^2$ configuration. A molecular moment $\mu_m = (\mu_{Mn}{}^c - 3\mu_{Mn}{}^f) = 1.2 \ \mu_B$ requires a $\delta_s{}^0 \approx 0.05$, where the superscript 0 refers to $x = 0$. The atomic moments then follow for a

$$\delta' \approx \delta_s{}^0 \approx 0.05. \qquad (57)$$

This is equivalent to a $\delta \approx 0.4$ in equations (20,43).

If x nitrogen atoms are removed, equation (55) is altered only by $Mn^f:(a_1{}^0)(p^1) \rightarrow Mn^f:(1 - x)(a_1{}^0)xa_1{}^1(p^{(1-x)})$, and each $a_1{}^1$ configuration adds $1\mu_B$ to the average Mn^f moment. Moreover, from equation (24) $\delta_s \sim \mu_m$, and a $\mu_m \rightarrow 0$ at $x = 1/3$ makes

$$\delta_s = \delta_s{}^0(1 - 3x). \qquad (58)$$

It follows that

$$\mu_m = (\mu_{Mn}{}^c - 3\mu_{Mn}{}^f) \approx [(1 + 4\delta_s{}^0) - (3 + 12\delta_s{}^0)x]\mu_B$$

$$\approx (1.2 - 3.6 \ x)\mu_B, \qquad (59)$$

which is in excellent agreement with equation (54). From these considerations, we may add to Conclusion (2) the following:

> *Conclusion* (2'): In the nitrides of manganese, $M^cNMn_3{}^f$, a low-spin $\rightleftarrows$ high-spin transformation $b_1{}^2 \rightarrow b_1{}^1$ can be expected at an $a_0 > 3.857$ Å, the room temperature lattice parameter of $Mn^cNMn_3{}^f$.

3.3.4 The Compounds $Ni^cNMn_3{}^f$

$Ni^cNMn_3{}^f$ exhibits a first-order cubic $\rightleftarrows$ cubic transition at a magnetic ordering temperature $T_C = 266°K$, the cubic lattice parameter increasing 0.004₅ Å on cooling through T_C [31]. From neutron diffraction data [32],

$$\mu_{Ni}{}^c = 0\mu_B \quad \text{and} \quad \mu_{Mn}{}^f \approx 0.98 \ \mu_B. \tag{60}$$

The manganese moments lie in (111) planes, and all nearest neighbour Mn^f atoms have moments at 120° to one another. Below a second-order transition at $T_t = 180°K$, the Mn^f moments are parallel to $\langle 110 \rangle$ directions. This triangular configuration, illustrated in figure 15(a), is referred to by its symmetry notation Γ_{5g}. In the interval $T_t < T < T_c$, the moments rotate continuously with temperature through an angle of 90°, and a small canting out of the (111) plane gives a weak ferromagnetism. This triangular spin configuration, illustrated in figure 15(b). is identified as Γ_{4g}.

The outer electron configurations,

$$Ni^c\!:e^4 t_2{}^{6-\varepsilon} 4s^\varepsilon \quad \text{and} \quad Mn^f\!:(a_1{}^0)b_1{}^2 b_2{}^{(\varepsilon/3)+x}e^2+y 4s^1(p^1), \tag{61}$$

give $(x + y) = 1 - (\varepsilon/3)$ antibonding electrons per molecule in the $Ni^c\!:t_2\text{-}3Mn^f\!:b_2$ and $Mn^f\!:e$ bands. The greater stability of the nickel $3d$ bands places all the antibonding electrons on the Mn^f array, which is consistent with a $\mu_{Ni}{}^c = 0\mu_B$. Because there is an important antiferromagnetic component to each $Mn^f\text{-}Mn^f$ interaction, the $\mu_{Mn}{}^f$ must contain a contribution $2(1 - \delta)\mu_B$ from the bonding e electrons as well as the $(x + y)\mu_B$ from the antibonding electrons:

$$\mu_{Mn}{}^f = (x + y + 2(1 - \delta))\mu_B = (3 - \frac{\varepsilon}{3} - 2\delta)\mu_B, \tag{62}$$

where both ε and δ are large fractions. For $0.6 < \varepsilon < 1$, a $\mu_{Mn}{}^f \approx 1\mu_B$ requires $0.9 > \delta \gtrsim 0.83$, a reasonable range of

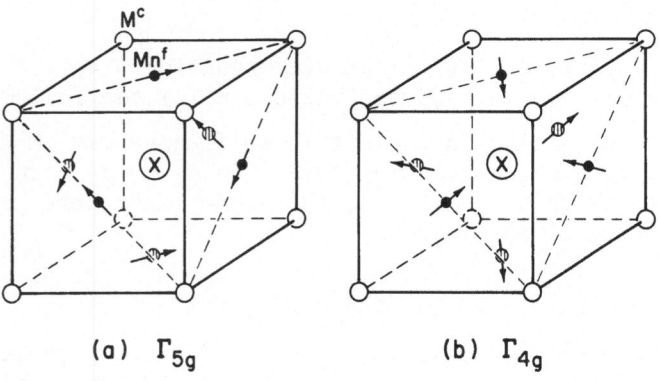

(a) Γ_{5g} (b) Γ_{4g}

Figure 15 - The cubic Γ_{5g} and Γ_{4g} magnetic structures found for several $M^cNMn_3{}^f$ compounds.

values characteristic of an e band having $b > b_m$ for the bonding electrons.

The dilatation on cooling through T_N is characteristic of a first order phase change on passing from itinerant electron $(b > b_m)$ to strongly correlated $(b < b_m)$ behaviour. Since the Γ_{5g} antiferromagnetic order induces a contribution to the magnetic moment from the bonding e electrons, this condition is fulfilled.

Observation of ferromagnetic Mn^f-Ni^c-Mn^f interactions indicates an $x > y$ in the magnetically ordered state. However, retention of cubic symmetry below T_N and the Γ_{5g} magnetic order seem to imply a $y > 0$. Were $y = 0$, the e bands would have $n_\ell = 1$ and competitive, antiferromagnetic Mn^f-Mn^f interactions would be expected to give collinear spin antiferromagnetism with a deformation to tetragonal $(c/a < 1)$ symmetry via strong exchange striction, as in γ-Mn.

3.3.5 The Compounds $Cu^cNMn_3^f$ and $Ag^cNMn_3^f$

Although $Cu^cNMn_3^f$ and $Ag^cNMn_3^f$ have the same number of outer electrons, their low temperature properties are quite different. $Ag^cNMn_3^f$ undergoes a first-order $\Gamma_{5g} \rightleftarrows$ paramagnetic transition at a $T_N = 290°K$ and a second-order $\Gamma_{4g} \rightleftarrows \Gamma_{5g}$ transition at $55°K$ [33], analgous to $Ni^cNMn_3^f$. However, in the silver compound a large

$$\mu_{Mn}^f = 3.1 \ \mu_B \tag{63}$$

signals a high-spin b_1^1 configuration, so that the electron configurations become

$$Ag^c : d^{10}5s^1 \quad \text{and} \quad Mn^f : (a_1^0)b_1^1b_2^{(2-y)}e^{2+y}4s^1(p^1). \tag{64}$$

It follows that

$$\mu_{Mn}^f = [1 + (2y) + 2(1 - \delta)]\mu_B. \tag{65}$$

A $y = 1$ corresponds to $n_\ell = 3/2$ in the e bands and should give ferromagnetic Mn^f-Mn^f interactions. From equation (21), a Γ_{5g} antiferromagnetic order requires $1 + \frac{1}{2}y < \frac{1}{2}(3 - \delta)$, or a $y < (1 - \delta)$; and a $\mu_{Mn}^f = 3.1 \ \mu_B$ then requires a $\delta < 0.475$. Therefore, reconciliation of the large μ_{Mn} and the Γ_{5g} magnetic order with the model of this paper seems to require: (1) a high-spin b_1^1 configuration; and (2) a small δ, characteristic of $b \approx b_c$. These requirements are compatible with the large lattice parameter $(a_0 = 4.025 \ \text{Å})$. From Conclusion (2') above, a transition from the low-spin b_1^2 to the high-spin b_1^1 configuration must occur at larger lattice parameters. Moreover, the larger lattice parameter plus the enhanced intra-atomic exchange with the

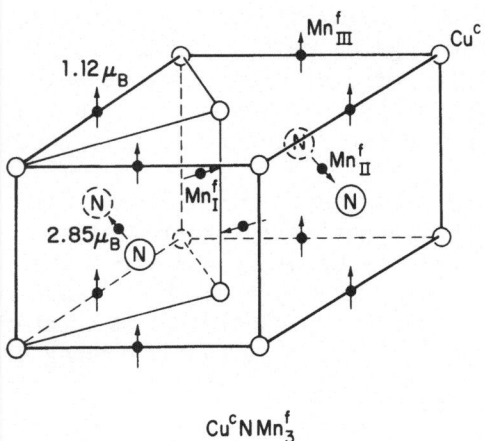

$Cu^cN Mn_3^f$

Figure 16 - Magnetic order found in tetragonal ($c/a < 1$) $Cu^cNMn_3^f$.

localized b_1^1 electrons would narrow the broader $3d$ bands toward $b \approx b_c$. Finally, a $d^{10-\epsilon}$ configuration may occur at the silver, as is indicated by the tendency of an Ag^+-ion $4d$ core to undergo a spontaneous deformation, via hybridization with $5s$ orbitals, to allow for shorter Ag-O bonding [34].

The compound $Cu^cNMn_3^f$ has a smaller lattice parameter ($a_0 = 3.906$ Å), but this parameter is nevertheless larger than those found in $Mn^cNMn_3^f$, $Ni^cNMn_3^f$, and $Zn^cNMn_3^f$. The copper compound undergoes a first order tetragonal ($c/a < 1$) $\rightleftarrows$ cubic transition at a Curie temperature $T_c = 143°K$. The tetragonal unit cell and complex ferrimagnetic order below T_c are shown in figure 16. The measured atomic moments are [35]

$$\mu_{Cu}{}^c = 0\mu_B, \qquad \mu_{Mn}{}^f(I,II) \approx 2.85 \ \mu_B,$$

$$\tag{66}$$

$$\mu_{Mn}{}^f(III) \approx 1.12 \ \mu_B.$$

From the magnitudes of the μ_{Mn}^f, it is clear that the high-spin configuration b_1^1 is stabilized at the $Mn_{I,II}^f$ atoms, a low-spin configuration b_1^2 at the Mn_{III}^f atoms. The tetragonal ($c/a < 1$) distortion can therefore be interpreted to be the result of ordering of high-spin and low-spin configurations. This observation pin points the critical room temperature lattice parameter in the nitrides for the low spin $\rightleftarrows$ high spin ($b_1^2 \rightleftarrows b_1^1$) transition anticipated in Conclusion (2') above as

$$a_{oc} \approx 3.905 \ \text{Å}. \tag{67}$$

The tetragonal distortion creates orthorhombic point symmetry
at the $Mn_{I,II}^f$ atoms, which lifts the e_g orbital degeneracy. If
the c-axis is taken as the z-axis, the two distinguishable orbitals are the d_{xy} directed toward near neighbour $Mn_{I,II}^f$ atoms of
a basal plane and d_{iz}, where $i = x$ or y, directed toward nearest
Mn_{III}^f atoms. The outer electron configurations become

$$Cu^c:d^{10}4s^1p^0 \qquad \text{and} \qquad Mn_{III}^f:(a_1^0)b_1^2b_2^1e^24s^1(p^1),$$

$$(68)$$

$$Mn_{I,II}^f:(a_1^0)b_1^{2-\zeta}b_2^1d_{iz}^{1+\zeta-\eta}d_{xy}^1\eta 4s^1(p^1).$$

The 90° magnetic ordering introduces contributions $(1 - \delta)\mu_B$ to
μ_{Mn}^f form bonding $3d$ electrons. The tetragonal $c/a < 1$ distortion makes the contribution from $Mn_{I,II}^f:d_{xy}$ orbitals larger
than that from the $2Mn_{I,II}^f:d_{iz}-Mn_{III}^f:e$ orbitals. The measured
$\mu_{Mn}^f(III)$ is compatible with neglect of the latter, in which
case

$$\mu_{Cu}^c = 0\mu_B, \qquad \mu_{Mn}^f(I,II) \approx (2 + 2\zeta - \delta)\mu_B,$$

$$(69)$$

$$\mu_{Mn}^f(III) \approx (1 + \delta_s)\mu_B.$$

In view of equation (21) and the existence of 90° coupling for
all Mn^f-Mn^f interactions, it is reasonable to set $(1 + \eta) \approx$
$\frac{1}{2}(2 + \zeta - \eta) < \frac{1}{2}(3 - \delta)$. From equations (66,69), $(2\zeta - \delta) \approx$
0.85. These relationships call for $\eta \lesssim 0.233$, $\zeta \lesssim 0.7$, $\delta \lesssim 0.55$.
Larger values of these parameters would presumably lead to fer-
romagnetic coupling. Although possible, such a large value for
the $(1 - \delta)\mu_B \approx 0.45 \mu_B$ contribution seems somewhat surprising
even though the lattice parameter is large enough to insure a
$b_{xy} < b_m$.
    In general, Mn-Cu-Mn interactions are ferromagnetic, as in
the Heusler alloys. From the model, this follows if the $Cu:d^{10}$
core electrons are considered part of the $3d$ bands and n_ℓ is calculated on that basis. However, if covalent mixing with $Cu:4p$
orbitals were more important than that with $Cu:d^{10}$ core orbitals,
an antiferromagnetic $Mn:d^1-Cu:p^0-Mn:d^1$ would occur. Since the
larger lattice parameter introduced by interstitial nitrogen
weakens coupling via d^{10} relative to that via p^0, the antiferromagnetic and ferromagnetic components to the $Mn:d^1-Cu-Mn:d^1$ interaction may be comparable in the nitrides. The magnetic order
of figure 16 implies antiferromagnetic $Mn_{I,II}^f:b_2^1-Cu-Mn_{I,II}^f:b_2^1$
interactions are present, and these could be responsible for
stabilizing the complex ferrimagnetic order instead of ferromagnetic order

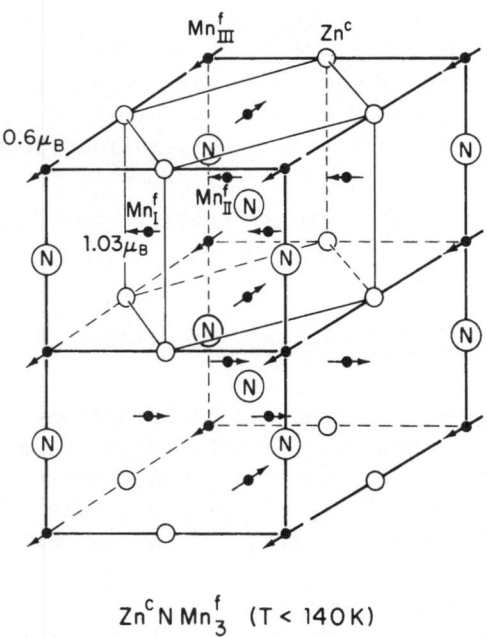

$Zn^cN Mn_3^f$ (T < 140K)

Figure 17 - Magnetic order found in tetragonal ($c/a < 1$) $Zn^cNMn_3^f$.

3.3.6   The Compounds $M^cNMn_3^f$, M = Zn, Ga and In

Like $Ni^cNMn_3^f$ and $Ag^cNMn_3^f$, the three compounds $Zn^cNMn_3^f$, $Ga^cNMn_3^f$ and $In^cNMn_3^f$ undergo $\Gamma \rightleftarrows$ paramagnetic transitions: and in at least the first two the transition is first order, the lattice parameter increasing on cooling through T_N. At room temperature, the cubic lattice parameters are, respectively, 3.900, 3.885, and 4.021 Å. Therefore, from equation (67) low-spin configurations are anticipated for Zn^c and Ga^c, a high-spin configuration for In^c.

In all three compounds, the d^{10} cores at the M^c atoms are too tightly bound to play a role. Occupancy of $M^c{:}p$ orbitals permits the formation of directed bonds between M^c and Mn^f atoms. Formation of such bonds would be reflected in a crystalline deformation to lower symmetry and spin pairing of $M^c{:}p$ and $Mn^f{:}$ (b_2,p) electrons.

In the case of $Zn^cNMn_3^f$, two magnetic phases are observed [36]: (1) the Γ_{5g} magnetic order of figure 15 within the interval $T_t < T < T_N = 183°K$; and (2) the tetragonal magnetic order of figure 17 below $T_t = 140°K$. A lattice dilatation is found only in the interval $T_t < T < T_N$ where the magnetic order is Γ_{5g}. The measured [35,37] atomic moments are

$$\mu_{Mn}{}^f \approx 1.21 \ \mu_B \tag{70}$$

for the cubic phase, and

$$\mu_{Mn}^{f}(I,II) \approx 1.03 \ \mu_{B} \qquad \text{and} \qquad \mu_{Mn}^{f}(III) \approx 0.6 \ \mu_{B} \qquad (71)$$

for the tetragonal phase.

The cubic phase contains no directed Zn^{c}-Mn^{f} bonds, and the configurations

$$Zn^{c}:4s^{2-\lambda}p^{\lambda} \qquad \text{and} \qquad Mn^{f}:(a_1^{0})b_1^{2}b_2^{1-\zeta}e^{2+\zeta}4s^{1}(p^{1}) \qquad (72)$$

give the predicted moment

$$\mu_{MN}^{f} = [1 + 2(1 - \delta)]\mu_{B}, \qquad (73)$$

where a $\delta \approx 0.9$ is characteristic of a $b > b_m$ for the bonding e orbitals. The tetragonal phase, on the other hand, would be represented by

$$Zn^{c}:4s^{2-\lambda}(p_x p_y)^{\lambda} \text{ and } Mn_{I,II}^{f}:(a_1^{0})b_1^{2}b_2^{1-\eta}d_{iz}^{1}d_{xy}^{1+\eta}4s^{1}(p^{1}),$$
$$(74)$$
$$Mn_{III}^{f}:(a_1^{0})b_1^{2}b_2^{1-\lambda-\xi}e^{2+\xi}4s^{1}(p^{1})(p_x p_y) ,$$

where the electrons in brackets are spin paired by molecular-orbital formation with $N:2p$ or $M^{c}:p$ orbitals. The corresponding atomic moments are

$$\mu_{Mn}^{f}(I,II) \approx 1\mu_{B} \qquad \text{and} \qquad \mu_{Mn}^{f}(III) \approx (1 - \lambda)\mu_{B}, \qquad (75)$$

if the small contributions $(1 - \delta)\mu_{B} < 0.05 \ \mu_{B}$ from bonding electrons are neglected for the case of 90° coupling

The model is also consistent with the complex magnetic order found in the tetragonal phase. Whereas $Mn_{III}^{f}:b_2^{1-\lambda}$-$M^{c}:p^{\lambda}$-$Mn_{III}^{f}:b_2^{1-\lambda}$ interactions would be ferromagnetic, $Mn_{I,II}^{f}:b_2^{1}$-$M^{c}:p^{0}$-$Mn_{I,II}^{f}:b_2^{1}$ interactions would be antiferromagnetic. Moreover, the character of the Mn_{I}^{f}-Mn_{II}^{f} coupling depends on the magnitude of η, whereas the $Mn_{I,II}^{f}$ coupling depends on the magnitude of $\frac{1}{2}\xi$. The larger this number of antibonding electrons per orbital, the greater the probability of ferromagnetic Mn^{f}-Mn^{f} interactions.

The tetragonal phase in $Zn^{c}NMn_3^{f}$ is apparently stabilized in order to allow directed-bond formation. In $Ga^{c}NMn_3^{f}$ and $In^{c}NMn_3^{f}$, where $M^{c}:p$ electrons are necessarily present, the tetragonal phase is not competitive with the cubic phase having Γ_{5g} magnetic order. Without the formation of directed bonds, the $Mn^{f}:b_2$ electrons presumably do not spin-pair with the $Ga:p$

electrons, and the electron configurations for $Ga^cNMn_3^f$ are

$$Ga^c:4s^2p^1 \quad \text{and} \quad Mn^f:(a_1^0)b_1^2b_2^{1-\zeta}e^{2+\zeta}4s^1(p^1), \quad (76)$$

which gives

$$\mu_{Mn}^f = [1 + 2(1 - \delta)]\mu_B \approx 1.2 \ \mu_B \qquad (77)$$

for a $\delta \approx 0.9$, as was used for M = Zn. This prediction is to be compared with an experimental [38] $\mu_{Mn}^f \approx 1.17 \ \mu_B$ at $T = 4.2°K$.

The value of μ_{Mn}^f for the indium compound has not been measured, but from the model a larger μ_{Mn}^f is anticipated because the larger lattice parameter should induce both a high-spin b_1^1 configuration and a smaller δ.

3.3.7 The Compounds $M^cNMn_3^f$ where M = Ge, Sn, As, Sb

Covalent mixing between $M^c:p$ and $Mn^f:b_2$ and p orbitals becomes more important for M atoms from groups IV,V of the periodic table. A large occupancy of the bonding p states may be reflected in distortions characteristic of directed bonds: (sp^3) bonds for the lighter elements Ge and As where the hybridization energy is not too large, and p bonds for the heavier elements Sn and Sb where hybridization is energetically more costly. However, formation of directed bonds means the stabilization of filled bonding bands that are primarily $M^c:s^2p^6$ in character. In this case, the σ-bonding $Mn^f:b_2$ and π-bonding $Mn^f:b_1$ orbitals are not so much spin paired by band formation, as emptied, since they contribute primarily to the empty antibonding bands. Moreover, the bond formation utilizes covalent mixing with $Mn^f:p$ orbitals, so the broad Mn:4s conduction band remains about half filled: $4s^1$. The particular p orbitals involved can be obtained from the magnetic order and/or the structure.

Both $Ge^cNMn_3^f$ and $As^cNMn_3^f$ have T_4-tetragonal structures [39], although the germanium compound exhibits a $T_4 \rightleftarrows$ cubic transition at $T_t \approx 510°K$. The T_4 structure may be derived from the cubic structure by opposite rotations, about c-axis Mn_{II}^f-N-Mn_{II}^f chains, of successive NMn_3^f octahedra along the c-axis. These rotations create a tetragonal unit cell having $c \approx 2a_0$ and $a \approx \sqrt{2}a_0$. The rotations, through about 17° in $Ge^cNMn_I^f$ also create four shortest M^c-Mn_I^f distances tetrahedrally coordinated about M^c. Each Mn_I^f atom has two nearest neighbour M^c atoms. This T_4 structure thus satisfies the crystallographic criteria $M^c:(sp^3)$ bonds coupling covalently with $Mn_I^f:b_2p_1p_2$ orbitals, where p_1 and p_2 are directed toward nearest neighbour M^c atoms within the pseudo-cubic (110) and (101) planes. The corresponding outer electron configurations for the T_4 phase of $Ge^cNMn_3^f$ are

$$\text{Ge}^c:(s^1p^3) \qquad \text{and} \qquad \text{Mn}_{II}{}^f:(a_1{}^0)b_1{}^2b_2{}^1e^24s^1(p^1),$$

$$(78)$$

$$\text{Mn}_I{}^f:(a_1{}^0)b_1{}^1(b_2{}^0)d_{iz}{}^1d_{xy}{}^14s^1(p^1)(p_1{}^1p_2{}^1),$$

where a high-spin $b_1{}^1$ configuration is assumed for the $\text{Mn}_I{}^f$ atoms, but not for the $\text{Mn}_{II}{}^f$ atoms, because the room temperature lattice parameter $a_0 = 3.911$ Å is similar to that for $\text{Cu}^c\text{NMn}_3{}^f$ ($a_0 = 3.906$ Å). The $\text{Mn}_I{}^f:(b_2{}^0p_1{}^1p_2{}^1)$ notation describes filled, bonding s-p states associated with the Ge^c atom and empty, antibonding $\text{Mn}_I{}^f:b_2$ bands. However, the bonding states obviously contain considerable $\text{Mn}_3{}^f:b_2$ character.

Equation (78) is misleading in one important respect. Because the directed bond states have more Ge character, since this is the more electronegative atom, the $\text{Mn}_I{}^f:(b_2{}^0)(p_1{}^1p_2{}^1)$ electrons are shifted toward the Ge atoms. Since no similar shift occurs from the $\text{Mn}_{II}{}^f$ atoms towards Ge atoms, charge parity among the Mn^f atoms requires some shift of narrow band $3d$ electrons from $\text{Mn}_{II}{}^f$ to $\text{Mn}_I{}^f$ atoms. This transfer, or shift, occurs within the $2\text{Mn}_I{}^f:d_{iz}\text{-Mn}_{II}{}^f:e$ bands. Since these are half filled, the shift should have little influence on the atomic moments.

Half-filled $2\text{Mn}_I{}^f:d_{iz}\text{-Mn}_{II}{}^f:e$ and $\text{Mn}_I{}^f:d_{xy}$ orbitals call for competitive, antiferromagnetic $\text{Mn}^f\text{-Mn}^f$ interactions and therefore a complex magnetic order. Individual atomic moments have not been measured, but a ferrimagnetic molecular moment that is unsaturated at 26.6 kOe has been extrapolated [39] to 0°K and infinite applied field to $\mu_{m\infty} \approx 0.79~\mu_B$.

$\text{Ge}^c\text{NMn}_3{}^f$ undergoes a complex series of transitions $T_4 \gtrless T_1 \gtrless \Gamma_{5g} \gtrless$ paramagnetic within the narrow temperature interval $510 < T < 525$°K. Presumably the formation of directed bonds is no longer maintained at higher temperatures. $\text{As}^c\text{NMn}_3{}^f$, on the other hand, retains the T_4 structure to highest temperatures, which is consistent with the more electronegative character of As stablizing the bonding bands corresponding to formal valence state $\text{As}^{3-}: s^2p^6$.

The tetragonal ($c/a > 1$) T_1 structure occurs below 237°K in $\text{Sn}^c\text{NMn}_3{}^f$ and below 360°K in $\text{Sb}^c\text{NMn}_3{}^f$. Since ($sp^3$) hybridization is energetically more costly in these heavier M^c atoms, it is reasonable to suppose that the distortion is due to the formation of directed $M^c\text{-Mn}^f$ bonds involving only $M^c:p$ orbitals. This type of deformation is to be distinguished from the tetragonal ($c/a < 1$) distortion observed in $\text{Cu}^c\text{NMn}_3{}^f$.

$\text{Sn}^c\text{NMn}_3{}^f$ exhibits a series of crystallographic and magnetic order phase changes with increasing temperature [39,40]. Within the tetragonal ($c/a > 1$) T_1 phase below 237°K, there is a first order change at 184°K. In the interval $237 < T < 357$°K, a cubic phase having the Γ_{5g} structure is found; and in the interval $357 < T < T_N = 465$°K a tetragonal ($c/a < 1$) structure occurs. The paramagnetic phase is cubic. Preliminary neutron diffraction data [41] for the tetragonal ($c/a > 1$) T_1 phase gives

$$\mu_{Mn}{}^f(III) \approx - 3.5 \ \mu_B \qquad \text{and}$$

(79)

$$\mu_{Mn}{}^f(I,II) \approx 1.75 \left[1 \pm \sin\left(\frac{2\pi c}{k}\right) \right] \mu_B$$

where k apparently decreases with increasing temperature above 184°K. At 77°K, $k = 0.205 \ c$ and at 200°K, $k = 0.175 \ c$. The minus and plus signs refer to the orientations of the moments along the c-axis.

A high atomic moment indicates little M^c:p-Mn^f:b_2p directed-bond formation, a low atomic moment implies the opposite. Therefore, in the T_1 phase of $Sn^cNMn_3{}^f$ the directed bonds must be to $Mn_{I,II}{}^f$ atoms only, which places them in the pseudo-cubic (100) and (010) planes. Moreover, a cubic room temperature lattice parameter $a_0 = 4.06$ Å insures a high-spin $b_1{}^1$ configuration at all the Mn^f atoms. This reasoning leads to the outer electron configurations

$$Sn^c{:}s^2(p_1{}^1p_2{}^1p_3{}^0) \quad \text{and} \quad Mn_{III}{}^f{:}(a_1{}^0)b_1{}^1b_2{}^1e^24s^1(p^1)(p_x{}^{\frac{1}{2}}p_y{}^{\frac{1}{2}}),$$

(80)

$$Mn_{I,II}{}^f {:}(a_1{}^0)b_1{}^1(b_2{}^0)d_{iz}{}^1d_{xy}{}^24s^1(p^1)(p_1{}^{\frac{1}{2}}p_2{}^{\frac{1}{2}}),$$

where the empty $p_3{}^0$ orbital, which is alternately directed along (100) and (010) axes, neither stabilizes nor spin pairs the $Mn_{I,II}{}^f$:$b_1{}^1$ electron via π-bond formation. However, interactions between the M^c:p_3 and $Mn_{III}{}^f$:p_xp_y states are assumed to form partial occupancy being just that required to preserve charge parity among the Mn^f atoms. Without charge transfer between $Mn_{III}{}^f$ and $Mn_{I,II}{}^f$ atoms, the atomic moments can be obtained directly from equation (80) as

$$\mu_{Mn}{}^f(III) = (4 - 2\delta)\mu_B \qquad \text{and} \qquad \mu_{Mn}{}^f(I,II) = (2 - \delta)\mu_B, \text{(81)}$$

where δ refers to the half-filled $Mn_{III}{}^f$:e-$Mn_{I,II}{}^f$:d_{iz} band responsible for antiferromagnetic $Mn_{III}{}^f$-$Mn_{I,II}{}^f$ coupling. The filled $Mn_{I,II}{}^f$:d_{xy} band introduces no competitive coupling. It is apparent from equation (79) that an antiferromagnetic spin density wave is superposed on this basic magnetic structure. This feature is outside the scope of our model. It originates, presumably, in the overlapping conduction bands, which include the partially filled, bonding M^c:p_3-$Mn_{III}{}^f$:p_xp_y bands as well as the partially filled $4s$ band. However, comparision of equations (79,81) gives a $\delta \approx 0.25$, which is reasonable for an $a_0 = 4.06$ Å.

Neutron-diffraction data [41] for the T_1 phase of $Sb^cNMn_3{}^f$ reveal a strikingly different magnetic order, which is shown in figure 18. The experimental moments are

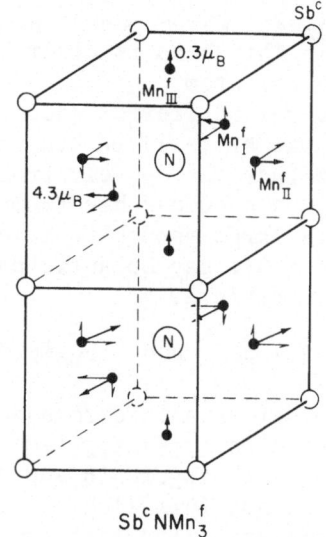

$$Sb^cNMn_3^f$$

Figure 18 - Magnetic order found in T_1-tetragonal
$(c/a > 1)$ $Sb^cNMn_3^f$.

$$\mu_{Mn}^f(III) \approx 0.3 \ \mu_B \qquad \text{and} \qquad \mu_{Mn}^f(I,II) \approx 4.3 \ \mu_B, \qquad (82)$$

with the $\mu_{Mn}^f(I,II)$ having a 1.3 μ_B component parallel to [001]
and a 4.1 μ_B component in the basal plane. Here the directed
bond orbitals responsible for the T_1 symmetry involve M^c-Mn_{III}^f
bonding rather than M^c-$Mn_{I,II}^f$ bonding as in $Sn^cNMn_3^f$. Thus the
existence of a tetragonal $(c/a > 1)$ distortion can be accomplished by more than one directed bond configuration.

Since the Sb^c:p bonds are formed primarily with the Mn_{III}^f
atoms, the outer electron configuration would appear to be

$$Sb^c:s^2(p^3) \qquad \text{and} \qquad Mn_{III}^f:(a_1^0)(b_1^0)(b_2^0)e^34s^1(p^1)(p^2),$$
$$(83)$$

$$Mn_{I,II}^f:(a_1^0)b_1^1b_2^1d_{iz}^{1.5}d_{xy}^14s^1(p^1)(p^{0.5}),$$

where the bonding p_xp_y bands are primarily Sb^c:$p_x^2p_y^2$ in character, which is why the Mn_{III}^f:b_2 orbitals are represented as
empty, antibonding orbitals. Similarly, the Sb^c:p orbitals, which
interact with the $Mn_{I,II}^f$:p orbitals and π-bond with the Mn_{III}^f:b_1
orbitals, from bonding band states that are primarily Sb^c:p_z in character. Therfore, the Mn_{III}^f:b_1 orbitals are also represented as
empty, antibonding orbitals. As in the case of equation (78),
equation (33) is misleading as it stands, because there is a
greater transfer of charge form Mn_{III}^f than from $Mn_{I,II}^f$ atoms
to the more electronegative Sb^c atoms via the p bands. Therefore, charge must transfer from the $Mn_{I,II}^f$ atoms to the Mn_{III}^f
atoms so as to preserve charge parity among the Mn^f atoms. This

transfer takes place within the $Mn_{III}{}^f:e$-$2Mn_{I,II}{}^f:d_{iz}$ bands to produce the distribution $d_{iz}{}^1e^4$, which means that the $2\mu_B$ per molecule spin density within this band is distributed as $1\mu_B$ per $Mn_{I,II}{}^f$ atom and $0\mu_B$ per $Mn_{III}{}^f$ atom.

Configuration (83) calls for antiferromagnetic $Mn_I{}^f$-$Mn_{II}{}^f$ superexchange interactions via half-filled d_{xy} orbitals, in agreement with figure 18, so that the d_{xy}-electron contribution to $\mu_{Mn}{}^f(I,II)$ is $(1-\delta)\mu_B$. A pseudo-cubic room temperature lattice parameter $a_0 \approx 4.2$ Å is large enough to make $b \lesssim b_c$ for the d_{xy} electrons, so a small $\delta \approx 0.1$ may be anticipated. The predicted atomic moments are, therefore,

$$\mu_{Mn}{}^f(I,II) = (4-\delta)\mu_B \approx 3.9\ \mu_B \quad \text{and} \quad \mu_{Mn}{}^f(III) \approx 0\mu_B, \quad (84)$$

and the expected magnetic order is antiferromagnetic $Mn_I{}^f$-$Mn_{II}{}^f$ interactions and ferromagnetic $Mn_{I,II}{}^f$-$Mn_{III}{}^f$-$Mn_{I,II}{}^f$ interactions. However, comparision with figure 18 and equation (82) shows that this prediction is not fulfilled. A complex spin density wave, presumably reflecting indirect coupling via conduction electrons, is superposed to give a more complex magnetic order and a contribution $\delta_s\mu_B \approx 0.3\ \mu_B$ to each atom from the conduction electrons. This feature does not follow from the qualitative considerations of this paper. It is, however, perhaps significant that superposed spin density waves are found where there is considerable d-orbital mixing into the conduction bands, as is implied by the $(b_2{}^0)$ and $(b_1{}^0)$ antibonding states in configurations (80,83).

3.3.8   The System $Mn^cN_{1-x}C_xMn_3{}^f$, $0 \leqslant x < 0.7$

Substitution of nitrogen by carbon in ferrimagnetic, cubic $Mn^cN_{1-x}C_xMn_3{}^f$ reduces the molecular moment and increases the Curie temperature [30,42]:

$$T_c = (740 + 310x)^\circ K \quad \text{and} \quad \mu_m = (\mu_{Mn}{}^c - 3\mu_{Mn}{}^f) \approx 1.2\ (1-x)\mu_B,$$

$$(85)$$

$$\mu_{Mn}{}^c \approx 3.9\ \mu_B \quad \text{and} \quad \mu_{Mn}{}^f \approx (0.9 + 0.4)\mu_B.$$

Interpretation of these results follows immediately from the model. Whereas the formation of directed N-$Mn^f p$ bands has been designated $N:2s^2(p^3)$, $Mn^f:a_1{}^04s^1(p^1)$, the C-$Mn^f p$ bands must be characterized by $C:s^2(p^2)$, $Mn^f:a_1{}^04s^1(p^{2/3})$ in this notation. Covalent mixing between carbon p and $Mn^f:a_1$ orbitals is still strong enough to lift the antibonding a_1 bands above the Fermi energy. As in the nitrides, the primacy of these C-Mn^f bands should be maintained throughout all $M^cCMn_3{}^f$ compounds. It follows that, for the system $Mn^cN_{1-x}C_xMn_3{}^f$, the outer electron configurations are:

$$N:s^2(p^3), \qquad C:s^2(p^2), \qquad Mn^c:e^2t_2{}^44s^1, \qquad \text{(Contd)}$$

(Contd) $Mn^f:(a_1^0)b_1^2b_2^{2/3}e^{2+(1+x)/3}4s^1(p^{1-(x/3)})$,   (86)

which gives

$$\mu_{Mn}^c = (4 - 3\delta + \delta_s)\mu_B \quad \text{and} \quad \mu_{Mn}^f = [1 + \frac{x}{3} - \delta' - \delta_s]\mu_B \quad (87)$$

where $b_2^{2/3}$ is taken so as to correspond to $n_\ell = 1$ in the $Mn^c:t_2$-$3Mn^f:b_2$ bands responsible for antiferromagnetic coupling between Mn^c and Mn^f atoms. The $Mn^f:e$ bands have an $n_\ell = 1.17 + 0.17\ x$, which means that the ferromagnetic component of the Mn^f-Mn^f interactions increases linearly with x, as does T_c. Since $\mu_m \to 0$ at $x = 1$, we must assume

$$\delta_s = \delta_s^0(1 - x) \approx 0.05\ (1 - x). \quad (88)$$

If $\delta \approx 0.05$ remains independent of x,

$$\mu_{Mn}^c \approx (3.9 - 0.05\ x)\mu_B \quad \text{and} \quad \mu_{Mn}^f \approx (0.9 + 0.383\ x)\mu_B, \quad (89)$$

$$\mu_m = (\mu_{Mn}^c - 3\mu_{Mn}^f) \approx 1.2\ (1 - x)\mu_B. \quad (90)$$

3.3.9 The Carbides $M^cCMn_3^f$, M = Zn, Al, Ga

The compound $Zn^cCMn_3^f$ exhibits a tetragonal ($c/a < 1$) $\rightleftharpoons$ cubic transition at $T_t = 233°K$. Below T_t the magnetic order assumes the complex ferromangetic spiral illustrated in figure 19. The atomic moments are [43]

$$\mu_{Mn}^f(III) \approx 1.6\ \mu_B \quad \text{and} \quad \mu_{Mn}^f(I,II) \approx 2.7\ \mu_B. \quad (91)$$

In the temperature interval $T_t < T < T_c = 380°K$, the compound is ferromagnetic with

$$\mu_{Mn}^f \approx 1.6\ \mu_B. \quad (92)$$

At room temperature, the cubic lattice parameter is $a_0 = 3.924$ Å, which is larger than the $a_{oc} \approx 3.905$ Å found in the nitrides, and the existence of at least some high-spin $Mn^f:b_1^1$ configurations is anticipated. In fact, the tetragonal ($c/a < 1$) distortion at low temperatures is similar to that found in $Cu^cNMn_3^f$ where there was an ordering of high-spin and low-spin $Mn^f:b_1^1$ and b_1^2 configurations. With this analogy in mind, we write

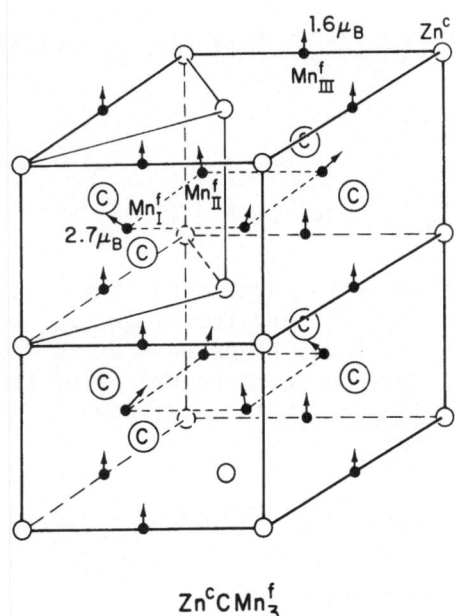

$$Zn^cCMn_3^f$$

Figure 19 - Magnetic order found in low temperature, tetragonal (c/a < 1) $Zn^cCMn_3^f$.

$$Zn^c:4s^2 \qquad \text{and} \qquad Mn^f:(a_1^0)b_1^{2-\xi}b_2^1e^{2+\xi}+(1/3)4s^1(p^{2/3}) \qquad (93)$$

for the ferromagnetic phase. This gives a

$$\mu_{Mn}^f = (1.33 + 2\xi + \delta_s)\mu_B, \qquad (94)$$

and from equation (21), a requirement for ferromagnetic Mn^f-Mn^f coupling that $(3 - \delta)/2 \lesssim 1 + (1/6) + (\xi/2)$, or that $\delta \gtrsim (2/3) - \xi$. A $\delta \approx 0.5$ seems reasonable for an a_0 = 3.924 Å, so ferromagnetic coupling is compatible with $\xi \gtrsim 1/6$ and hence

$$\mu_{Mn}^f > (1.67 + \delta_s)\mu_B. \qquad (95)$$

Though a little higher than the experimental value above 233°K, this moment is close to what might be anticipated for T = 0°K. Moreover, the existence of a $\xi \approx 0.2$ provides justification for interpreting the low temperature distortion to be owing to an electron ordering similar to that found in $Cu^cNMn_3^f$. Therefore, for $T < T_t$ we assume

$$Mn_{III}^f:(a_1^0)b_1^2b_2^1e^{2+(1/3)}4s^1(p^{2/3}), \qquad (96)$$

(Contd)

(Contd) $\mathrm{Mn_{I,II}}^f : (a_1{}^0) b_1{}^{2-\eta} b_2{}^1 d_{iz}{}^{1+(1/3)+\eta} d_{xy}{}^1 4s^1 (p^{2/3})$,

which provides antiferromagnetic $\mathrm{Mn_I}^f$-$\mathrm{Mn_{II}}^f$ interactions via half-filled d_{xy} orbitals and ferromagnetic $\mathrm{Mn_{I,II}}^f$-$\mathrm{Mn_{III}}^f$ interactions, as demanded by figure 19. This gives the atomic moments

$$\mu_{\mathrm{Mn}}{}^f(\mathrm{III}) = [1.33 + \delta_s + 2(1 - \delta)]\mu_B, \qquad (97)$$

where $(1 - \delta) \approx 0.1$ is a small contribution from the bonding electrons owing to canting of the $\mathrm{Mn_{I,II}}^f$ moments from the c-axis, and

$$\mu_{\mathrm{Mn}}{}^f(\mathrm{I,II}) = [1.33 + 2\eta + \delta_s + 2(1 - \delta)]\mu_B \qquad (98)$$

From equations (91,97,98), $\mu_{\mathrm{Mn}}{}^f(\mathrm{I,II}) - \mu_{\mathrm{Mn}}{}^f(\mathrm{III}) = 2\eta\mu_B \approx 1.1\mu_B$ gives an $\eta \approx 0.55$, which is quite consistent with the small value of ξ in the cubic phase. In the cubic phase, the total number of high-spin atoms per molecule is $3\xi > 0.5$, or at least half the number in the tetragonal phase.

This interpretation of $\mathrm{Zn^cCMn_3}^f$ seems to provide a critical lattice parameter for the low-spin $\not\gtrless$ high-spin $b_1{}^2 \not\gtrless b_1{}^1$ transition at Mn^f atoms that is larger for carbides than for nitrides. Although a smaller U_f in the carbides would not be unreasonable, the fact is that the high-spin and low-spin states have comparable energies over quite a range of lattice parameters. A study of the system $\mathrm{Ga_{1-x}}^c\mathrm{Zn_x}^c\mathrm{CMn_3}^f$ is instructive on this point.

As shown in figure 20, a ferromagnetic phase is stable over

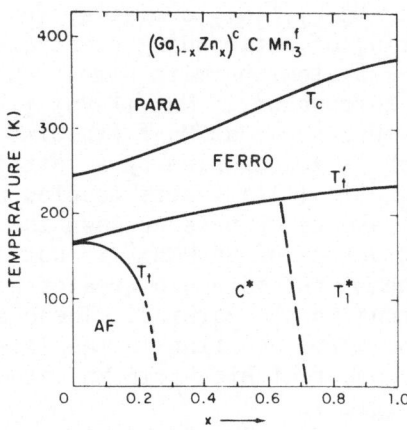

Figure 20 - Magnetic phase diagram for the system $\mathrm{Ga_{1-x}}^c\mathrm{Zn_x}^c\mathrm{CMn_3}^f$. (*Non-collinear ferrimagnetic structure of figure 19). (After [44]).

the entire compositional range [44], the Curie temperature decreasing from 380°K for $Zn^cCMn_3^f$ to 245°K for $Ga^cCMn_3^f$, and the
transition temperature T_t changing from 233°K to 163°K. As in
the nitrides, the $Ga:s^2p^1$ configuration is not expected to induce an effective spin pairing between the $Mn^f:b_2$ and $Ga:p$ electrons in the absence of directed-band formation. Therefore, in
the cubic phase configuration (93) is maintained at the Mn^f
atoms, and μ_{Mn}^f is given by equation (94). However, the magnitude of ξ is expected to increase with the lattice parameter,
and hence with x, whereas the magnitude of δ decreases with x.
Therefore, the condition $\delta \gtrsim (2/3) - \xi$ for ferromagnetic ordering may be maintained in $Ga^cCMn_3^f$ even though ξ is reduced to
nearly zero. An apparent reduction in μ_{Mn}^f of 0.3 μ_B in $Ga^cCMn_3^f$
presumably reflects a smaller ξ as well as a lower T_c. Moreover,
a $\xi \to 0$ in $Ga^cCMn_3^f$ is compatible with an $a_{oc} \approx 3.905$ Å.

The influence of lattice parameter on the number of high-spin
Mn^f atoms is particularly striking at lower temperatures ($T < T_t$).
As the Ga^c-atom concentration is increased, the tetragonal distortion associated with an ordering of high-spin and low-spin
ions decreases in magnitude, and apparently disappears even
though the peculiar magnetic order of figure 19 is maintained.
But from configuration (96), that magnetic order depends on an
ordering of antibonding e electrons among d_{iz} and d_{xy} orbitals
of the $Mn_{I,II}^f$ atoms and not on the distribution of high-spin
atoms. As η of (96) decreases, the tendency for ordering of high
spin atoms must also. Therefore, the fact that the magnetic order is independent of the distortion is quite consistent with the
model. Actually, exchange striction associated with the tetragonal magnetic order must produce a small, if unresolved, tetragonal distortion of the crystal.

Finally, a decrease in η of (96) with decreasing lattice parameter also means a decrease in the number of antibonding elec-
trons in the $Mn_{III}^f:e-Mn_{I,II}^f:d_{iz}$ orbitals, for which $n_\ell = 1.25 + \frac{1}{2}\eta$.
As $\eta \to 0$, ordering of antibonding electrons between d_{iz} and
d_{xy} orbitals on $Mn_{I,II}^f$ atoms permits almost enough antibonding
electrons to sustain ferromagnetic $Mn_{III}^f-Mn_{I,II}^f$ interactions.
Nevertheless, low temperature $Ga^cCMn_3^f$ exhibits a new kind of
magnetic order, which is accompanied by a lattice dilatation. It
consists of ferromagnetic (111) sheets coupled antiparallel to
near neighbour (111) sheets on passing down the [111] axis. The
fact that the Γ_{5g} structure of $Ga^cNMn_3^f$ is not stabilized is due
to the influence of the 1/3 per molecule greater concentration of
antibonding e electrons in the carbide. These apparently order
so as to give ferromagnetic coupling in the (111) planes. In
both compounds, the number of high-spin Mn^f atoms at low temperatures should be negligible.

In the aluminum compound $Al^cCMn_3^f$, which has a smaller lattice
parameter, an absence of high-spin Mn^f atoms would make $\xi \to 0$ in
equation (94), and

$$\mu_{Mn}^f = (1.33 + \delta_s)\mu_B \qquad (100)$$

for a ferromagnetic phase. In fact, $AlCMn_3^f$ has been reported [45] to be ferromagnetic to lowest temperatures with a μ_{Mn} = 1.26 μ_B and a T_c = 288°K. The condition $\delta \gtrsim (2/3) - \xi = 2/3$ for ferromagnetic order should be easily met because of the smaller lattice parameter of $Al^cCMn_3^f$.

3.3.10 Conclusions

The use of atomic configurations and near neighbour molecular orbital identification has proved to provide a convenient short hand notation for keeping track of charge neutrality conditions and band occupancies in the transition metal alloys. Although it has been successfully extended to more complex compounds, difficulties can arise where atoms more electronegative form directed bonds that introduce an anisotropic electron transfer from neighbouring atoms. However, it is possible to see from the notation where this is happening, and a qualitative redistribution of charge can generally be made to restore charge parity among like atoms.

Interpretation of the complex magnetic and structural data available for the $M^cXMn_3^f$ nitrides and carbides, which can be successfully extended to the systems $M_{1-x}^cM_c'^cXMn_3^f$, depends on seven major concepts:

(a) point symmetry for identifying molecular orbital formation among near neighbour atoms;

(b) relative atomic orbital stabilities to obtain band occupancies;

(c) outer electron configurations to keep track of band occupancies and charge neutrality or charge parity considerations;

(d) phase diagrams to provide the sign of the magnetic coupling as a function of band-occupancy number n_ℓ as well as the magnitude of the atomic moments as a function of the transfer energy b and correlation energy U;

(e) a critical lattice parameter for high spin $\gtrless$ low spin transitions, identified as $a_{oc} \approx 3.905$ Å in the $M^cXMn_3^f$ compounds, with the possibility of ordering high-spin and low-spin atoms via a cooperative lattice distortion if a_0 is in the neighbourhood of a_{oc};

(f) maintenance of a nearly constant occupancy of the broad $Mn:4s$ or $Fe:4s$ bands for all the nitrides and carbides $M^cXMn_3^f$ and $M^cXFe_3^f$,

(g) the formation of directed (sp^3) or p^n bonds where the s and p states of the M^c atom are stable enough to form filled bonding states, corresponding to the formal anionic valence state.

From the magnetic and crystallographic data for the nitrides M^cN Mn_3^f, it has been possible to show that the crystallographic distortion in $Cu^cNMn_3^f$ reflects ordering of high-spin and low-spin Mn^f atoms owing to an $a_0 \approx a_{oc}$, that the dilatation encountered with Γ_{5g} magnetic order is owing to electron correlation transition or exchange striction among the $3d$ electrons, that directed bond formation occurs with elements from groups IV,V of the periodic table, $(sp3)$ bonding occurring for intermediate weight Ge and As and p^n bonding occurring for the heavier Sn and Sb atoms. The octahedral coordination of the N and C atoms in the $M^cXM'_3^f$ compounds permits directed bonding with only the $2p$ orbitals, and the difference between the physical properties of the nitrides and carbides is essentially owing to a smaller p band occupancy in the carbides, which places an additional 1/3 electron per molecule into the $3d$ bands of the M'^f atoms. Because the number of electrons in the $3d$ bands responsible for M'^f-M'^f magnetic coupling corresponds, in many instances, to an $n_\ell \approx 1.25$, complex magnetic order transitions are common.

REFERENCES

1. Raccah, P.M. and Goodenough, J.B. (1967). *Phys. Rev.*, **155**, 932.
2. Bosman, A.J. and Van Daal, H.J. (1970). *Adv. Phys.*, **19**, 1.
3. Brinkman, W.F. and Rice, T.M. (1970). *Phys. Rev.*, **B2**, 4302.
4. Goodenough, J.B. (1973). Valence Bond Approach to Magnetic Semiconductors, in *New Developments in Semiconductors*, (eds. Wallace, P.R., Harris, R. and Zuckermann, M.J.), (Nordhoff International Publishing Co., Leyden), p. 107; Goodenough, J.B., Mott, N.F., Pouchard, M., Demazeau, G. and Hagenmuller, P. (1973). *Mater. Res. Bull.*, **8**, 647.
5. Sparks, J.T. and Komoto, T. (1968). *J. Appl. Phys.*, **38**, 715.
6. McWhan, D.B., Marezio, M., Remeika, J.P. and Dernier, P.D. (1972). *Phys. Rev.*, **B5**, 2552.
7. Magnéli, A. and Andersson, G. (1955). *Acta Chem. Scand.*, **9**, 1378; Goodenough, J.B. (1960). *Phys. Rev.*, **117**, 1442.
8. Goodenough, J.B. (1971). *J. Solid State Chem.*, **3**, 490; (1971). *Annu. Rev. Mat. Sci.*, **1**, 101.
9. Slater, J.C. (1937). *J. Appl. Phys.*, **8**, 385; Pauling, L. (1938). *Phys. Rev.*, **54**, 899.
10. Abrahams, S.C., Guttman, L. and Kasper, J.S. (1962). *Phys. Rev.*, **127**, 2052; Wilkinson, M.K. and Shull, C.G. (1956). *Phys. Rev.*, **103**, 516.
11. Meneghetti, D. and Sidhu, S.S. (1957). *Phys. Rev.*, **105**, 130; Bacon, G.E., Dunmur, I.W., Smith, J.H. and Street, R. (1957). *Proc. Roy. Soc.*, **A241**, 223.
12. Kasper, J.S. and Roberts, B.W. (1956). *Phys. Rev.*, **101**, 537; Kunutomi, N., Yamada, T., Nakai, Y. and Fujii, Y. (1969). *J. Appl. Phys.*, **40**, 1265.
13. DeVries, G. (1959). *J. Phys. Radium*, **20**, 438; Wilkinson, M.K., Wollan, E.O., Koehler, W.C. and Cable, J.W. (1962). *Phys. Rev.*, **127**, 2080; Werner, S.A. and Arrott, A. (1969). *J. Appl. Phys.*, **40**, 1447.

14. Lomer, W.M. (1962). *Proc. Phys. Soc.*, **80**, 489.
15. Bozorth, R.M. (1951). *Ferromagnetism*, (D. Van Nostrand, New York).
16. Kasper, J.S. and Kouvel, J.S. (1959). *J. Phys. Chem. Solids*, **11**, 231; Kouvel, J.S. and Graham, C.D., Jr., (1961). *J. Phys. Chem. Solids*, **21**, 57, and references therein.
17. Shull, C.G. (Private communication).
18. Nevitt, M.V. (1960). *J. Appl. Phys.*, **31**, 155; Chandross, R.J. and Shoemaker, D.P. (1962). *J. Phys. Soc. Jap. Suppl. B-III*, **17**, 16.
19. Shull, C.G. and Wilkinson, M.K. (1955). *Phys. Rev.*, **97**, 304.
20. Meyer, A.J.P. (1956). *C.R. Acad. Sci. Paris*, **242**, 2315; Bacon, G.E. and Street, R. (1958). *Proc. Phys. Soc.*, **72**, 470; Bozorth, R.M. (1951). *Ferromagnetism*, (D. Van Nostrand, New York), p. 328.
21. Oleś, A. (1965). *Acta Phys. Pol.*, **27**, 343.
22. Pickart, S.J. and Nathans, R. (1960). *J. Appl. Phys. Suppl.*, **31**, 372S.
23. Cable, J.W., Wollan, E.O., Koehler, W.C. and Wilkinson, M.K. (1962). *J. Appl. Phys. Suppl.*, **33**, 1340; Pickart, S.J. and Nathans, R. (1962). *J. Appl. Phys. Suppl.*, **33**, 1336.
24. Wiener, G.W. and Berger, J.A. (1955). *J. Met.*, **7**, 360.
25. Frazer, B.C. (1958). *Phys. Rev.*, **112**, 751.
26. Nagakura, S. (1968). *J. Phys. Soc. Jap.*, **25**, 488.
27. Goodenough, J.B., Wold, A. and Arnott, R.J. (1960). *J. Appl. Phys. Suppl.*, **31**, 342S.
28. Kuriyama, M., Husoya, S. and Suzuki, T. (1963). *Phys. Rev.*, **130**, 898; Barbéron, J., Madar, R., Fruchart, E., Lorthioir, G. and Fruchart, R. (1970). *Mater. Res. Bull.*, **5**, 903.
29. Takei, W.J., Shirane, G. and Frazer, B.C. (1960). *Phys. Rev.*, **119**, 122.
30. Bouchaud, J.-P. (1968). *Ann. Chim.*, **3**, 81.
31. Fruchart, D., Bertaut, E.F., Mader, R., Lorthioir, G. and Fruchart, R. (1971). *Solid State Commun.*, **9**, 1793.
32. Bertaut, E.F. and Fruchart, D. (1972). *Int. J. Magn.*, **2**, 259.
33. Fruchart, D., Bertaut, E.F., Fruchart, E., Barbéron, M., Lorthioir, G. and Fruchart, F. (In press).
34. Hong, H.Y.-P., Kafalas, J.A. and Goodenough, J.B. *J. Solid State Chem.*, (In press).
35. Bertaut, E.F., Fruchart, D. and Fruchart, R. (1971). *AIP Conference Proceedings: Magnetism and Magnetic Materials*, Section **31**, 1355.
36. Madar, R., Gilles, L., Roualt, A., Bouchaud, J.P., Fruchart, E., Lothioir, G. and Fruchart, R. (1967). *C.R. Acad. Sci.*, **264**, 308.
37. Fruchart, D., Bertaut, E.F., Madar, R. and Fruchart, R. (1971). *J. Phys. Colloque C1, Suppl. 2-3*, **32**, 876.
38. Bertaut, E.F., Fruchart, D., Bouchaud, J.-P. and Fruchart, R. (1968). *Solid State Commun.*, **6**, 251.

39. Barbéron, M. (1973). (Thèse), (Faculté des Sciences d'Orsay, Université de Paris).

40. Fruchart, R., Madar, R., Barbéron, M., Fruchart, E. and Lorthioir, M.-G. (1971). *J. Phys. Colloque C1, Suppl. 2-3*, **32**, 981.

41. Fruchart, D. and Bertaut, E.F. (Quoted in reference [*39*]).

42. Bouchaud, J.-P. and Fruchart, R. (1964). *Bull. Soc. Chim.*, 1579.

43. Fruchart, D., Bertaut, E.F., Lorthioir, G. and Fruchart, E. (In press).

44. Fruchart, E., Lorthioir, G. and Fruchart, R. (1973). *Mater. Res. Bull.*, **8**, 21.

45. Lowe, L. and Hyers, H.P. (1957). *Phil. Mag.* [8], **2**, 554.

NARROW BAND MAGNETISM OF ALLOYS

JOHN W. SCHWEITZER†

*Department of Physics and Astronomy,
University of Iowa, Iowa City, Iowa 52242, USA*

1. INTRODUCTION

The purpose of these lectures is to discuss the use of the coherent potential approximation (CPA) for the study of disordered magnetic alloys from the itinerant point of view. The CPA has contributed significantly to the theory of concentrated non-magnetic alloys described by one-electron Hamiltonians. For these alloys the CPA scheme gives a practical treatment of the disorder which yields reasonable results for arbitrary concentrations and scattering strengths. The extension to magnetic alloys necessarily involves many body effects which have yet to be systematically studied. At the present stage one relies on highly simplified models and straightforward extensions of the CPA as formulated in the context of one-electron Hamiltonians. While this does not permit a detailed confrontation of theory and experiment, the studies can be made tractable and the satisfactory features and defects can be recognized.

2. MODEL

In these discussions we assume a disordered substitutional binary alloy A_xB_{1-x} to be described by the non-degenerate narrow energy band model Hamiltonian

† Supported in part by the National Science Foundation under Grant No. GH34359.

$$H = \sum_{\substack{i,j,\sigma \\ (i \neq j)}} t_{ij} c_{i\sigma}^\dagger c_{j\sigma} + \sum_{i,\sigma} \varepsilon_i n_{i\sigma} + \frac{1}{2} \sum_{i,\sigma} U_i n_{i\sigma} n_{i-\sigma}. \qquad (1)$$

This is the so-called Hubbard Hamiltonian [1] in a Wannier representation. For the generalization to the case of an $A_x B_{1-x}$ alloy [2], the atomic level and the intra-atomic Coulomb repulsion ε_i and U_i take on values ε_A or ε_B and U_A or U_B depending on whether the i-th site is occupied by a type A or B atom. The transfer integral t_{ij} is assumed to be independent of the species of atoms occupying the i and j sites.

It is hoped that this model might contain a qualitative account of the electronic structure of alloys where both constituents are transition metals. Clearly it ignores complications present in real transition metal alloys such as d-band degeneracy, interband interactions, and inter-atomic exchange. Also we have neglected the disorder in the transfer term which is no doubt present in real alloys.

3. THE COHERENT POTENTIAL APPROXIMATION (CPA) [3-5]

Before proceeding further, let us give a brief review of the usual CPA for a one-electron Hamiltonian with cell-localized disorder with respect to the Wannier representation:

$$H = \sum_{\substack{i,j,\sigma \\ (i \neq j)}} t_{ij} c_{i\sigma}^\dagger c_{j\sigma} + \sum_{i,\sigma} \varepsilon_i n_{i\sigma}, \qquad \varepsilon_i = \varepsilon_A \text{ or } \varepsilon_B. \qquad (2)$$

For the $A_x B_{1-x}$ alloy we assume the sites are occupied in a random way with concentrations x and $1 - x$ for A and B type atoms respectively. This defines a whole ensemble of possible arrays of atoms. One is interested in the average of observables over this ensemble. Since the configuration averaged equilibrium properties are determined by the averaged Green function, the problem is one of obtaining the configuration averaged Green function.

For a given configuration and complex ω, one introduces the Green function (operator in this case)

$$G(\omega) = (\omega - H)^{-1}, \qquad (3)$$

where H is now restricted to be the Hamiltonian for one electron. Next one introduces the so-called coherent potential $\Sigma(\omega)$, a complex function of the ω variable to be determined by the prescription that will follow. This coherent potential is used to rewrite the Hamiltonian as

$$H = \bar{H} + V, \qquad (4)$$

where

$$\bar{H} = \sum_{\substack{i,j,\sigma \\ (i \neq j)}} t_{ij} c_{i\sigma}^\dagger c_{j\sigma} + \sum_{i,\sigma} \Sigma(\omega) n_{i\sigma}, \tag{5}$$

$$V = \sum_{i,\sigma} [\varepsilon_i - \Sigma(\omega)] n_{i\sigma}. \tag{6}$$

$\bar{H}$ is an effective medium Hamiltonian which has the full period-icity of the lattice, since we have assumed that t_{ij} is independ-ent of atomic configuration. The scattering potential V rela-tive to this effective medium defines a T-matrix (again an opera-tor in this case),

$$G = \bar{G} + \bar{G}T\bar{G}, \tag{7}$$

$$T = V + V\bar{G}T, \tag{8}$$

where

$$\bar{G} = (\omega - \bar{H})^{-1}.$$

The cell-localized character of the potential V,

$$V = \sum_n V_n, \tag{9}$$

is exploited to decompose the T-matrix into a sum of contribu-tions from each site,

$$T = \sum_n Q_n, \tag{10}$$

$$Q_n = V_n + V_n \bar{G}T \tag{11}$$

$$= V_n + V_n \bar{G}Q_n + V_n \bar{G} \sum_{m \neq n} Q_m. \tag{12}$$

It is seen that Q_n is not a single-site scattering matrix since it involves all the other sites as well as the n-th site. Re-writing once more one finds

$$Q_n = (1 - V_n \bar{G})^{-1} V_n \left(1 + \bar{G} \sum_{m \neq n} Q_m \right). \tag{13}$$

In this form we easily identify a scattering matrix T_n associated with the n-th site,

$$T_n = (1 - V_n \bar{G})^{-1} V_n, \tag{14}$$

and a modifying factor

$$(1 + \bar{G} \sum_{m \neq n} Q_m)$$

due to all the other sites.

The configuration averaged Green function $\mathfrak{G}(\omega)$ is obtained by averaging equation (7):

$$\mathfrak{G}(\omega) \equiv \langle G \rangle_{avg} = \bar{G} + \bar{G} \langle T \rangle_{avg} \bar{G}. \tag{15}$$

From the decomposition of the T-matrix one has

$$\langle T \rangle_{avg} = \sum_n \langle Q_n \rangle_{avg} \tag{16}$$

$$\langle Q_n \rangle_{avg} = \langle T_n (1 + \bar{G} \sum_{m \neq n} Q_m) \rangle_{avg}. \tag{17}$$

This last equation can be rewritten as

$$\langle Q_n \rangle_{avg} = \langle T_n \rangle_{avg} (1 + \bar{G} \sum_{m \neq n} \langle Q_m \rangle_{avg})$$

$$+ \langle (T_n - \langle T_n \rangle_{avg}) \bar{G} \sum_{m \neq n} (Q_m - \langle Q_m \rangle_{avg}) \rangle. \tag{18}$$

The equations up to this point have all been exact. The crucial approximation which makes things tractable is to neglect the second term in equation (18):

$$\langle Q_n \rangle_{avg} \simeq \langle T_n \rangle_{avg} (1 + \bar{G} \sum_{m \neq n} \langle Q_m \rangle_{avg}). \tag{19}$$

This approximation, known as the single-site approximation, consists in neglecting all statistical correlations between the n-th site and all the other sites. $\langle Q_n \rangle_{avg}$ is approximated by the average n-th site scattering matrix multiplied by a factor which gives an averaged modification due to the multiple scattering. Within this approximation the configuration averaged Green function is given by

$$\mathbf{G}(\omega) \simeq \bar{G} + \bar{G} \sum_j [\langle T_j \rangle_{\text{avg}} (1 + \bar{G} \sum_{m \neq j} \langle Q_m \rangle_{\text{avg}})] \bar{G}. \qquad (20)$$

The choice for the coherent potential $\Sigma(\omega)$ now seems obvious. One chooses $\Sigma(\omega)$ such that

$$\langle T_j \rangle_{\text{avg}} = 0, \qquad (21)$$

for in that case

$$\mathbf{G}(\omega) \simeq \bar{G}(\omega). \qquad (22)$$

We see that the effective medium is chosen such that on the average a single site produces no further scattering. One refers to equations (19,21) as the usual coherent potential approximation.

Let us now review the results. The configuration averaged Green function is given by $\bar{G}(\omega)$ which has the full symmetry of the lattice, and is therefore diagonal in the Bloch representation:

$$\mathbf{G}_{\mathbf{k}\mathbf{k}'}^{\sigma}(\omega) \simeq \langle \mathbf{k}\sigma | \bar{G}(\omega) | \mathbf{k}'\sigma \rangle = \frac{\delta_{\mathbf{k}\mathbf{k}'}}{\omega - \varepsilon_{\mathbf{k}} - \Sigma(\omega)}, \qquad (23)$$

where

$$\varepsilon_{\mathbf{k}} = N^{-1} \sum_{\substack{i,j \\ (i \neq j)}} t_{ij} e^{-i\mathbf{k}\cdot(R_i - R_j)}. \qquad (24)$$

Here we use the notation $|\mathbf{k}\sigma\rangle$ for the Bloch states which are related to the Wannier states $|j\sigma\rangle$ by

$$|\mathbf{k}\sigma\rangle = N^{-\frac{1}{2}} \sum_j e^{i\mathbf{k}\cdot R_j} |j\sigma\rangle. \qquad (25)$$

The coherent potential is determined by the CPA condition, equation (21). The explicit equation for $\Sigma(\omega)$ is easily obtained by noting from the definition of T_j, equation (14), that it is a diagonal matrix in the Wannier representation, with diagonal elements

$$\langle i\sigma | T_j | i\sigma \rangle = \frac{[\varepsilon_j - \Sigma(\omega)]\delta_{ij}}{1 - [\varepsilon_j - \Sigma(\omega)]\langle j\sigma | \bar{G}(\omega) | j\sigma \rangle} \qquad (26)$$

Since for a random $A_x B_{1-x}$ alloy

$$\langle T_j \rangle_{avg} = x T_A + (1 - x) T_B, \tag{27}$$

the CPA condition is just

$$\frac{x[\varepsilon_A - \Sigma(\omega)]}{1 - [\varepsilon_A - \Sigma(\omega)]\bar{F}(\omega)} + \frac{(1 - x)[\varepsilon_B - \Sigma(\omega)]}{1 - [\varepsilon_B - \Sigma(\omega)]\bar{F}(\omega)} = 0, \tag{28}$$

where

$$\bar{F}(\omega) \equiv \langle j\sigma | \bar{G}(\omega) | j\sigma \rangle = N^{-1} \sum_{\mathbf{k}} \langle \mathbf{k}\sigma | \bar{G}(\omega) | \mathbf{k}\sigma \rangle$$

$$= N^{-1} \sum_{\mathbf{k}} \frac{1}{\omega - \varepsilon_{\mathbf{k}} - \Sigma(\omega)} = F^0(\omega - \Sigma(\omega)), \tag{29}$$

where

$$F^0(\omega) = N^{-1} \sum_{\mathbf{k}} \frac{1}{\omega - \varepsilon_{\mathbf{k}}} . \tag{30}$$

We see from equation (23) that the coherent potential $\Sigma(\omega)$ is the self energy of the configuration averaged Green function. It can be shown [6] that there exists a solution to equation (28) for $\Sigma(\omega)$ which satisfies the analyticity requirements for a self energy. (The k-independence of $\Sigma(\omega)$ is a shortcoming of the single site approximation).

A given alloy $A_x B_{1-x}$ is specified in this simplest of models for disordered alloys by the concentration x, the energies ε_A and ε_B, and the band structure $\varepsilon_{\mathbf{k}}$. As a measure of the disorder one defines

$$\delta = \frac{\varepsilon_B - \varepsilon_A}{w} , \tag{31}$$

where w is half the band width associated with $\varepsilon_{\mathbf{k}}$. From equation (28) it is clear that $\Sigma(\omega)$ depends on $\varepsilon_{\mathbf{k}}$ only through the density of states

$$\rho^0(\varepsilon) = N^{-1} \sum_{\mathbf{k}} \delta(\varepsilon - \varepsilon_{\mathbf{k}}), \tag{32}$$

since by equation (30)

$$F^0(\omega) = \int_{-\infty}^{+\infty} \frac{\rho^0(\varepsilon)}{\omega - \varepsilon} \, d\varepsilon. \tag{33}$$

The CPA yields a quite satisfactory approach to this model for disordered alloys. We shall cite some of the results of this approach:

(1) The CPA gives the exact results for the dilute alloy limit defined by $x \ll 1$ or $(1 - x) \ll 1$;

(2) The CPA reduces to the virtual crystal approximation (same as the rigid band approximation for this model) in the limit of weak scattering defined by $\delta \ll 1$;

(3) The CPA gives reasonable results for arbitrary concentrations and scattering strengths.

This last vague statement will be illustrated by considering some results for a particular choice for $\rho^0(\varepsilon)$, namely

$$\rho^0(\varepsilon) = \frac{2}{\pi} (1 - \varepsilon^2)^{\frac{1}{2}}, \qquad |\varepsilon| \leqslant 1, \tag{34}$$

where the energy is in units of half the band width ($w = 1$). Figures 1,2, taken from Velicky, Kirkpatrick, and Ehrenreich [4], show the CPA result for the density of states for the alloy,

$$\rho(\varepsilon) = - \frac{1}{\pi} N^{-1} \sum_{\mathbf{k}} \mathrm{Im} G_{\mathbf{kk}}(\varepsilon + i0^+), \tag{35}$$

for a variety of concentrations and disorder parameters δ. (Here $\varepsilon_A = \frac{1}{2}\delta$ and $\varepsilon_B = -\frac{1}{2}\delta$). In figure 1 the virtual crystal results are shown for comparison. In the CPA, if the disorder parameter δ is sufficiently large the band will split. This is seen to occur at a reasonable value for δ. Although there are no band edge tailing effects because of the single site approximation character of the CPA, it can be shown [7] that the CPA preserves the first eight moments of the density of states.

Figure 3, again from Velicky, Kirkpatrick and Ehrenreich, [4], shows the partial densities of states associated with A and B atoms. One defines a $\rho_A(\varepsilon)$ which gives the contribution of the A sites per A site to the total density of states, and likewise a $\rho_B(\varepsilon)$.

$$\rho(\varepsilon) = x\rho_A(\varepsilon) + (1 - x)\rho_B(\varepsilon). \tag{36}$$

To calculate $\rho_A(\varepsilon)$ one considers the case where an A atom is located at the i-th site of the alloy. Within the CPA this situation is described by the effective Hamiltonian

$$H_i^A = \sum_{\sigma} \varepsilon_A n_{i\sigma} + \sum_{\substack{\ell,j,\sigma \\ (\ell \neq j)}} t_{\ell j} c_{\ell\sigma}^+ c_{j\sigma} + \sum_{j \neq i,\sigma} \Sigma(\omega) n_{j\sigma}. \tag{37}$$

The Green function in the Wannier representation for that i-th

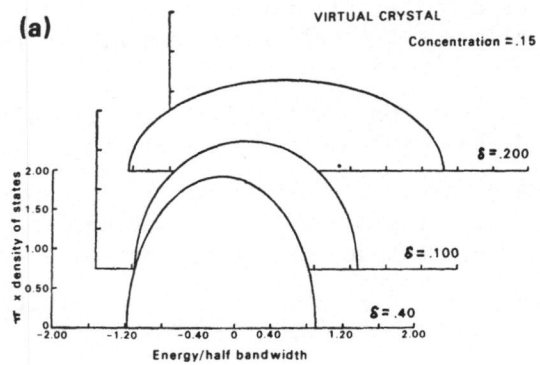

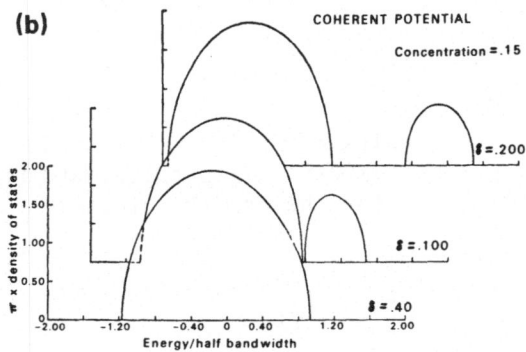

Figure 1 - Density of states for A_xB_{1-x} alloy as calculated with model $\rho^0(\varepsilon)$ in (a) the virtual crystal approximation, and (b) the coherent potential approximation with $x = 0.15$ and $\delta = (\varepsilon_A - \varepsilon_B)/\omega$ having values 0.4, 1.0, and 2.0. [After Velicky et $al.$, (1968). $Phys.$ $Rev.$, **175**, 747].

site is easily determined to be

$$\langle i\sigma | G^A(\omega) | i\sigma \rangle = \frac{F^0(\omega - \Sigma(\omega))}{1 - [\varepsilon_A - \Sigma(\omega)]F^0(\omega - \Sigma(\omega))} . \qquad (38)$$

Then the density of states contributed by that A site is just

$$\rho_A(\varepsilon) = -\frac{1}{\pi} \text{Im} \left\{ \frac{F^0(\varepsilon + i0^+ - \Sigma(\varepsilon + i0^+))}{1 - [\varepsilon_A - \Sigma(\varepsilon + i0^+)]F^0(\varepsilon + i0^+ - \Sigma(\varepsilon + i0^+))} \right\}, \qquad (39)$$

and similarly for that contributed by a B site. To check that equation (36) is satisfied one can show that the CPA condition, equation (28), is equivalent to

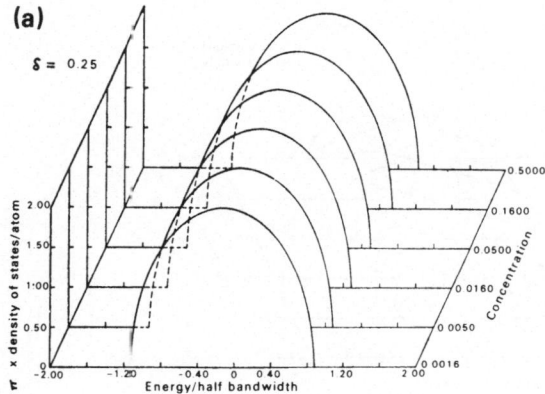

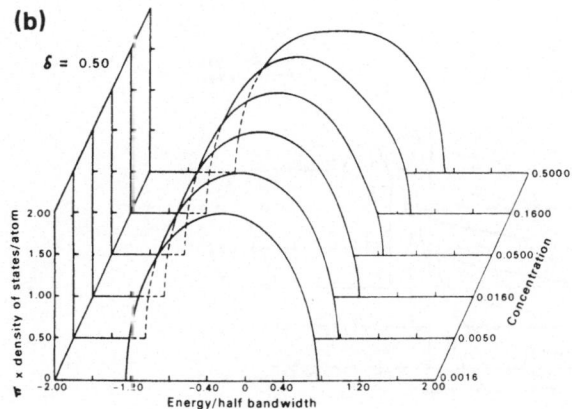

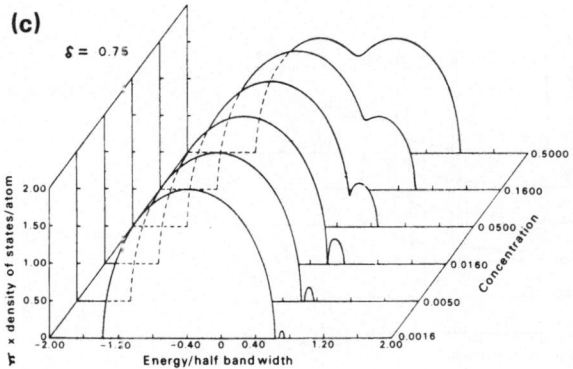

Figure 2 - Density of states for A_xB_{1-x} alloy as calculated with model $\rho^0(\varepsilon)$ in the coherent potential approximation for various concentrations x and disorder parameters $\delta = (\varepsilon_A - \varepsilon_B)/\omega$. [After Velicky *et al.* (1968). *Phys. Rev.*, **175**, 747].

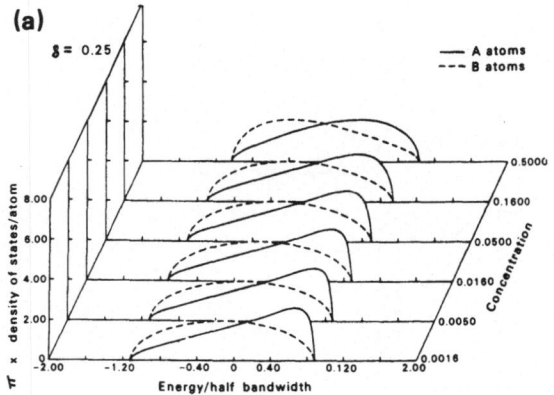

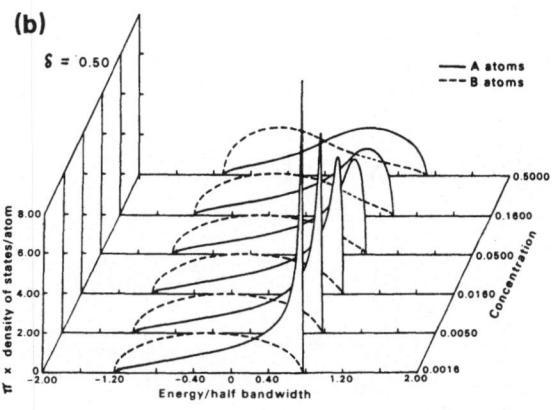

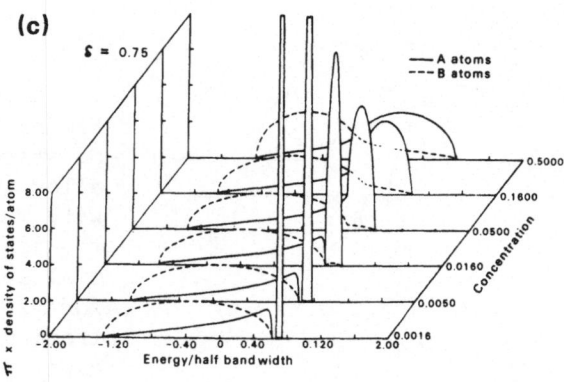

$$F^0(\omega - \Sigma(\omega)) = \frac{xF^0(\omega - \Sigma(\omega))}{1 - [\epsilon_A - \Sigma(\omega)]F^0(\omega - \Sigma(\omega))}$$

$$+ \frac{(1 - x)F^0(\omega - \Sigma(\omega))}{1 - [\epsilon_B - \Sigma(\omega)]F^0(\omega - \Sigma(\omega))} . \qquad (40)$$

Therefore equation (36) follows directly. Note that figure 3 shows quite dramatically the inequivalence of the A and B sites.

4. MAGNETIC PHASES OF THE HUBBARD MODEL FOR A PERFECT CRYSTAL

Let us now review some results for the ground state magnetic properties of the Hubbard model in the case of a perfect crystal (equation (1) without disorder). Unfortunately very little is rigorously known about the ground state magnetic properties of this model (except for the one-dimensional form which we shall not consider here), even though it is the simplest model that incorporates both the itineracy of the electrons and the correlation between electrons on the same site. One has at present only the rigorous low density results of Kanamori [8], the exact results of Nagaoka [9] for nearest neighbour tight binding cubic structures having one electron more or less than the number of sites and infinite Coulomb repulsion, some numerical solutions for high temperatures [10,11], and many approximate treatments the validity of which are difficult to judge.

One does not even know with certainty whether the ground state of this single band Hubbard model is ferromagnetic for any physical range of parameters. For example, consider the case where the transfer integral t_{ij} is non-vanishing only between nearest neighbour sites,

$$t_{ij} = \begin{cases} t < 0, & i \text{ and } j \text{ nearest neighbours,} \\ 0, & \text{otherwise,} \end{cases} \qquad (41)$$

and the crystal structure is simple cubic (sc). The model is characterized by the ratio of the transfer energy to the Coulomb repulsion energy and the electron concentration. The Hartree-Fock treatment of the Coulomb repulsion, which consists of the replacement

Figure 3 (Opposite) - Component densities of states in the coherent potential approximation for the same values of x and δ as in figure 2. [After Velicky *et al.* (1968). *Phys. Rev.*, **175**, 747].

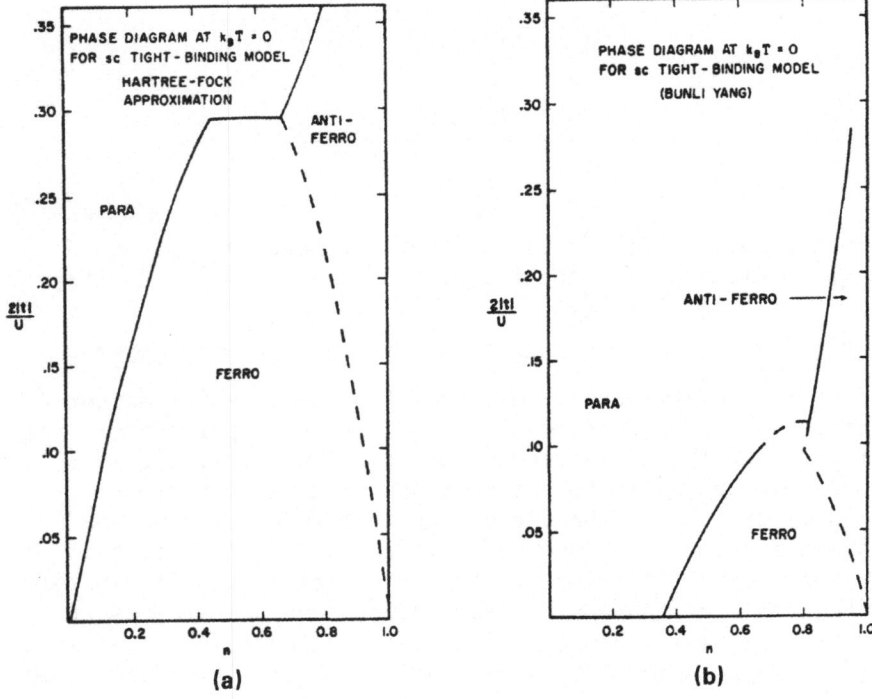

Figure 4

$$\tfrac{1}{2}U \sum_{i,\sigma} n_{i\sigma}n_{i-\sigma} \xrightarrow{\text{H-F}} U \sum_{i,\sigma} \langle n_{i-\sigma}\rangle n_{i\sigma}, \qquad (42)$$

yields [12] the phase diagram shown in figure 4(a) for the plane
of $2|t|/U$ versus the number of electrons per site denoted by n.
Here only paramagnetic, ferromagnetic, and antiferromagnetic
solutions are considered and the stable phase is taken to be the
phase of lowest energy. The solid curves mark second order
transition boundaries between the phases, and the dashed curve
marks a first order transition which is a discontinuity in the
magnetization. The diagram is symmetric about $n = 1$. One sees
that except for an electron concentration near 0,1, or 2, the
ferromagnetic state is stable if the Coulomb repulsion is of the
order of the band width (here equal to $6|t|$) or greater.

We know that this is not the correct phase diagram. The Hartree-Fock approximation does not include correlation effects
that tend to keep anti-parallel electrons away from each other.
Hence it seriously over-estimates the energy of the paramagnetic
phase and thereby does not provide a stringent enough condition
on ferromagnetism. What then is a reasonable phase diagram?
Rigorous low density t-matrix calculations [8,13] have shown
that ferromagnetism is not stable for $n \gtrsim 0.3$ and $n \gtrsim 1.7$ for
this example. Nagaoka [9] has given us the exact result that

for infinite Coulomb repulsion the ferromagnetic state with maximum total spin is the ground state for the sc example if there are $N \pm 1$ electrons, where N is the number of sites. While this result must be used with care since as $N \to \infty$ one electron more or less is indistinguishable from $n = 1$, it seems reasonable that this result can be extended to values of n which differ by finite amount from $n = 1$, as argued by Nagaoka [9]. Therefore one expects ferromagnetic solutions along the $2|t|/U = 0$ axis for some finite range of n about $n = 1$. For $n = 1$, when U is infinite all phases have the same energy, but when U is finite it is very likely that the ground state is antiferromagnetic. This is because there can be virtual hopping onto a site occupied by an opposite spin electron giving a reduction of energy of order t^2/U. This can not occur in the ferromagnetic phase because of the exclusion principle.

All these considerations plus some features of high temperature numerical studies suggest a phase diagram something like figure 4(b). This diagram is one that was calculated by Bunli Yang [14] using a rather elaborate approximation scheme. It is presented here merely as a likely diagram on the basis of available information†. The condition on the parameters for a stable ferromagnetic phase is seen to be much more restrictive than the Hartree-Fock condition. The Hartree-Fock approximation is only reasonable when the Coulomb repulsion is weak. Clearly correlations tend to destroy ferromagnetism, yet it is our conviction that a ferromagnetic phase does exist for this model for some band shapes.

It is unfortunate that correlation effects play a crucial rôle in determining the magnetic properties, for if we use the Hartree-Fock approximation for the Coulomb repulsion the CPA can be directly applied to the alloy generalization. This is owing to the fact that the Hartree-Fock replacement, equation (42), yields a one-electron Hamiltonian with site-diagonal disorder. Our model for the alloy, equation (1), is replaced by

$$H^{\text{H-F}} = \sum_{\substack{i,j,\sigma \\ (i \neq j)}} t_{ij} c_{i\sigma}{}^{\dagger} c_{j\sigma} + \sum_{i,\sigma} (\varepsilon_i + U_i \langle n_{i-\sigma} \rangle) n_{i\sigma}, \qquad (43)$$

which is nearly identical to the Hamiltonian considered in section 3 where the usual CPA was discussed. The only differences are the spin dependence of the disorder and the implied self-consistency, where the average $\langle n_{i-\sigma} \rangle$ must be determined so that it depends only on the type of atom occupying the i-th site.

---

† Recently Visscher [11] has argued that the system will become an inhomogeneous mixture of ferromagnetic and antiferromagnetic phases for a range of electron concentrations near $n = 1$ for U finite. This is owing to the neglect of inter-atomic Coulomb repulsion which is an unphysical feature of the model. We do not consider such phases in our diagram.

Indeed most studies of non-dilute magnetic alloys have used the Hartree-Fock approximation. We shall note some of the results of these studies [15-17] in section 5.

It is clear that our ability to treat correlation effects in the Hubbard model is uncertain. Nevertheless, we shall consider some approximate schemes to include correlation which can be used with the CPA for the alloy problem. The feature that these approximations have in common with the Hartree-Fock approximation is their single-site character. By this we mean that for a given configuration of A and B atoms the one-electron Green function [18] in the Wannier representation $G_{ij}^{\sigma}(\omega)$ is determined by an equation of the form

$$G_{ij}^{\sigma}(\omega) = g_i^{\sigma}(\omega)\delta_{ij} + g_i^{\sigma}(\omega) \sum_{\ell \neq i} t_{i\ell} G_{\ell j}^{\sigma}(\omega), \qquad (44)$$

where $g_i^{\sigma}(\omega)$, an 'atomic limit' Green function, is assumed to depend only on the type of atom occupying the i-th site. An equation of this form is not exact for the Hubbard model but does occur in these approximation schemes. Let us consider some examples:

(1) Hartree-Fock Approximation:

$$g_i^{\sigma}(\omega) = \frac{1}{\omega - \varepsilon_i - U_i\langle n_{i-\sigma}\rangle} ; \qquad (45)$$

(2) First Hubbard Approximation [1]:

$$g_i^{\sigma}(\omega) = \frac{1 - \langle n_{i-\sigma}\rangle}{\omega - \varepsilon_i} + \frac{\langle n_{i-\sigma}\rangle}{\omega - \varepsilon_i - U_i} \qquad (46)$$

(3) Hubbard Alloy Analogy Approximation (without resonance broadening) [19,20]:

$$g_i^{\sigma}(\omega) = \frac{1}{\omega - \varepsilon_i - \Delta\varepsilon_i^{\sigma}(\omega)} , \qquad (47)$$

where

$$\Delta\varepsilon_i^{\sigma}(\omega) = \frac{U_i\langle n_{i-\sigma}\rangle}{1 - (U_i - \Delta\varepsilon_i^{\sigma}(\omega))G_{ii}^{\sigma}(\omega)} ; \qquad (48)$$

(4) Improved First Hubbard Approximation (Harris and Lange [21], Roth [22], Tahir-Kheli and Jarrett [23] for $U \gg t_{ij}$:

$$g_i{}^\sigma(\omega) = \frac{1 - \langle n_{i-\sigma} \rangle}{\omega - \varepsilon_i - \langle n_{i-\sigma} \rangle W_{i-\sigma}}$$

$$+ \frac{\langle n_{i-\sigma} \rangle}{\omega - \varepsilon_i - U_i - (1 - \langle n_{i-\sigma} \rangle) W_{i-\sigma}} , \qquad (49)$$

where

$$\langle n_{i\sigma} \rangle (1 - \langle n_{i\sigma} \rangle) W_{i\sigma} = - \sum_{j \neq i} t_{ij} \langle c_{i\sigma}{}^\dagger c_{j\sigma} (1 - 2n_{i-\sigma}) \rangle . \qquad (50)$$

If our convictions about the magnetic properties of the Hubbard model are correct, approximations (2,3) are inadequate for a discussion of magnetic alloys. They are too restrictive towards the ferromagnetic phase. For the perfect crystal, approximation (2) can have ferromagnetic solutions, but not for simple band shapes. For example, it has no ferromagnetic solutions for the sc case previously considered. Approximation (3) was developed to correct the defect in the first Hubbard approximation that the single-particle spectrum has a gap for any U, no matter how small. It does this, but it is even more unfavourable to ferromagnetism, and probably does not have ferromagnetic solutions for any band shape [20]. Approximation (4) seems to give reasonable results in the limited cases where it has been applied [22,24,25]. Since it shares with the first Hubbard approximation the defect of a gap for any U, it should only be used for $U \gg t_{ij}$, the strongly correlated limit where one expects a gap.

Approximation (4), as represented by equations (44,49,50), is very similar to the 'two-pole approximation' of Roth [22], except for the neglect of some small terms that are inconsistent with the single-site character [26]. It is seen that this approximation differs from the first Hubbard approximation by the spin-dependent band shifts caused by the $W_{i-\sigma}$ energy. These band shifts were first indicated by the work of Harris and Lange [21] who studied the frequency moments of the spectral weight function for the one-electron Green function. The Roth [22] approximation was obtained by a decoupling scheme for the Green function equations of motion which is related to a variational principle. Similar results were obtained by Tahir-Kheli and Jarrett [23] using a procedure for conserving the first four moments of the spectral weight function. It is clear that these spin-dependent shifts must be included in any discussion of the question of magnetic stability.

Let us here briefly note some results from our study [25] of the Hubbard model (perfect crystal) where we used approximation (4). These studies were limited to simple cubic (sc), body centred cubic (bcc), and face centred (fcc) structures with a tight binding nearest neighbour band. The Coulomb repulsion was taken to be infinite, and solutions of ferromagnetic and antiferromagnetic symmetry were sought for the sc and bcc cases, and solutions of ferromagnetic symmetry for the fcc case. All results are for zero temperature. Figure 5 shows the density of states associated with ε_k for these structures. Figure 6-10

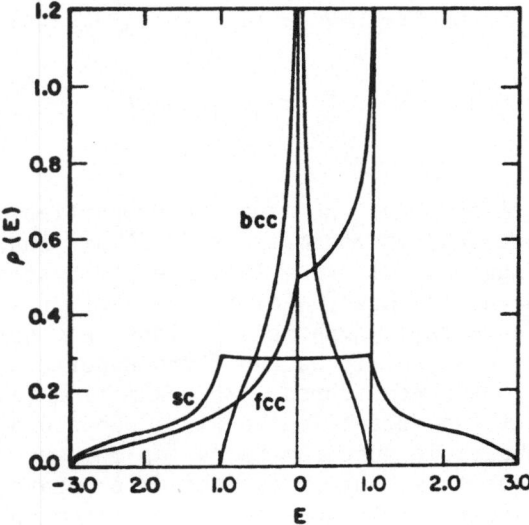

Figure 5 - Tight binding nearest neighbour density of states for simple cubic, body centred cubic, and face centred cubic structures as a function of energy. All densities have been normalized to unity.

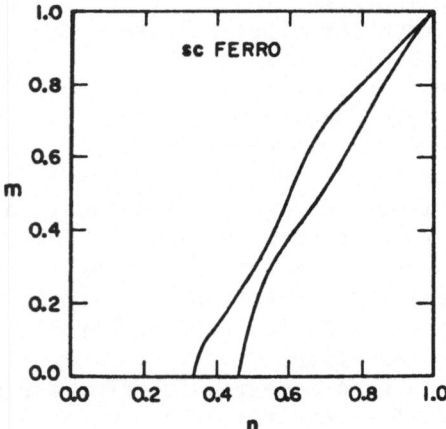

Figure 6 - Magnetization per site at absolute zero as a function of electron concentration for the simple cubic lattice.

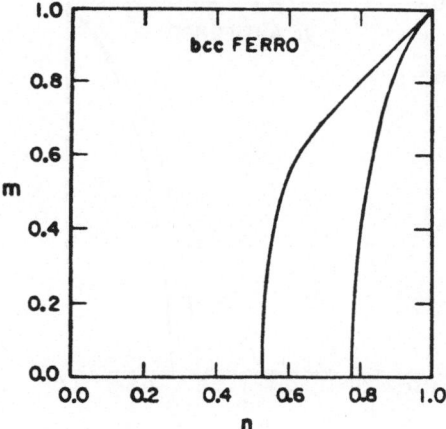

Figure 7 - Magnetization per site at absolute zero as a function of electron concentration for the body centred cubic lattice.

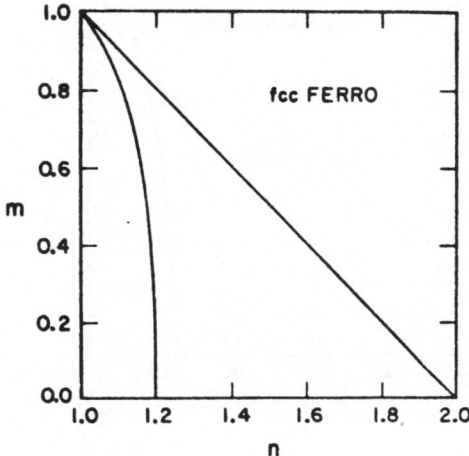

Figure 8 - Magnetization per site at absolute zero as a function of electron concentration for the face centred cubic lattice. No magnetic solutions occur until the band is half filled.

show the magnetic solutions that one obtains for infinite Coulomb repulsion ($U = \infty$). These are plots of the magnetization ($m = |\langle n_\sigma \rangle - \langle n_{-\sigma} \rangle|$) as a function of electron concentration ($n = \langle n_\sigma \rangle + \langle n_{-\sigma} \rangle$) for the ferromagnetic solutions and plots of sub-lattice magnetization for the antiferromagnetic solutions. Figures 11-13 are calculated energies as a function of electron concentration.

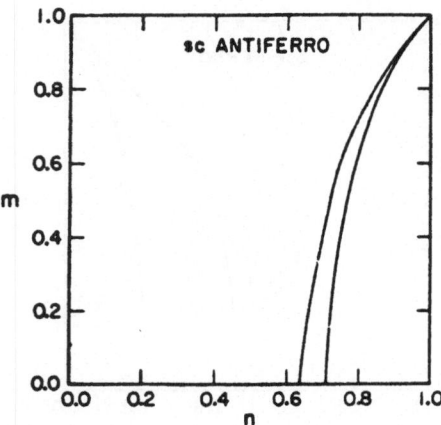

Figure 9 - Sublattice magnetization per site at absolute zero as a function of electron concentration for the simple cubic lattice.

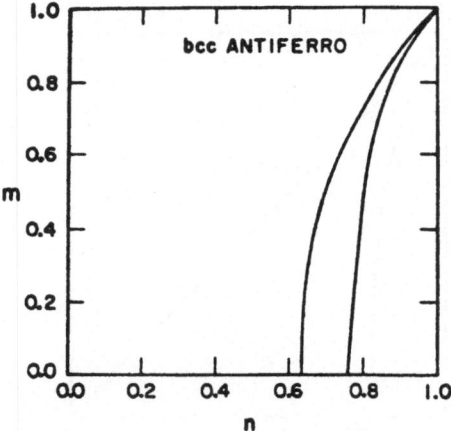

Figure 10 - Sublattice magnetization per site at absolute zero as a function of electron concentration for the body centred lattice.

Since the sc and bcc magnetization curves are symmetric about $n = 1$, only results for $n \leqslant 1$ are shown. There is a range of electron concentrations where ferromagnetic solutions exist, and over part of this range there are two solutions with ferromagnetic symmetry. The solution of larger magnetization becomes saturated ($m = n$) near $n = 1$. For example, in the sc case this solution exists for $0.35 \leqslant n \leqslant 1.65$ and is saturated for $0.75 \leqslant n \leqslant 1.25$. The antiferromagnetic solutions are similar, except that they do not saturate for $n < 1$. For the fcc case the solutions are not symmetric about $n = 1$. When

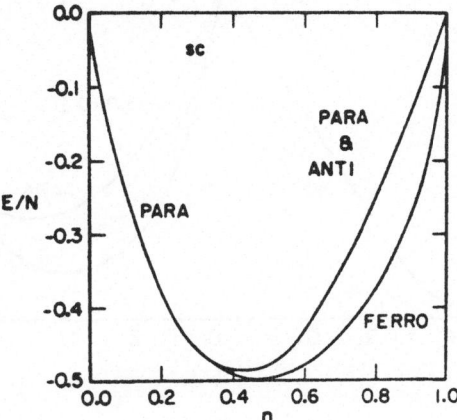

Figure 11 - Energy per site at absolute zero as a function of electron concentration for simple cubic lattice. The curve marked PARA is the paramagnetic energy. The lowest energy antiferromagnetic configuration is labelled ANTI, and the lowest energy ferromagnetic configuration is labelled FERRO. The scale of energy is such that the transfer energy between states at neighbouring sites is $-\frac{1}{2}$.

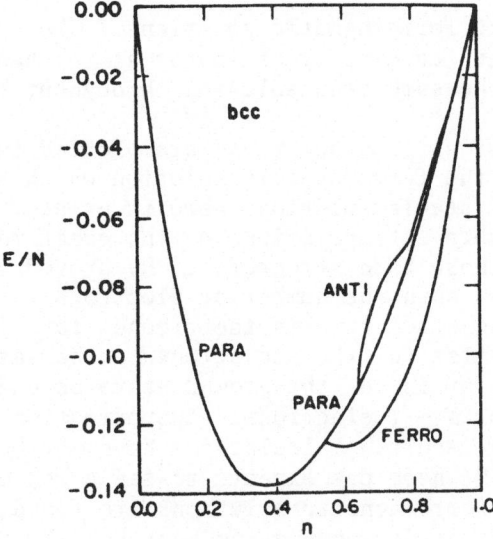

Figure 12 - Energy per site at absolute zero as a function of electron concentration for the body centred cubic lattice. The curve marked PARA is the paramagnetic energy. The lowest energy antiferromagnetic configuration is labelled ANTI, and the lowest energy ferromagnetic energy is labelled FERRO. The scale of energy is such that the transfer energy between states at neighbouring sites is $-\frac{1}{8}$.

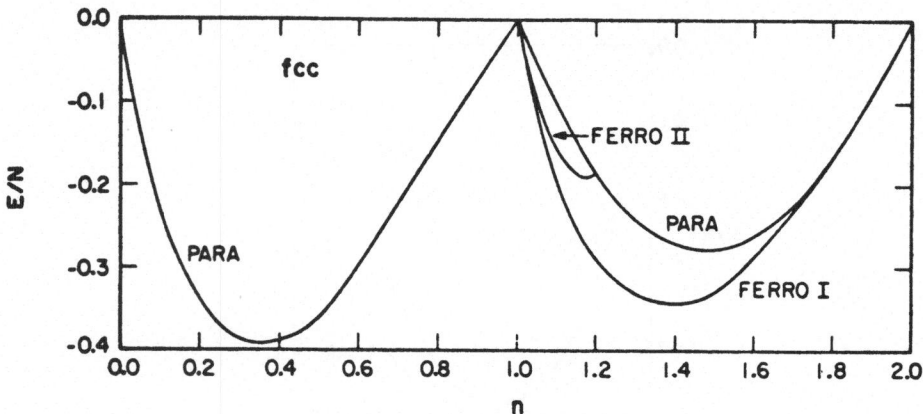

Figure 13 - Energy per site at absolute zero as a function of electron concentration for the face centred cubic
lattice. The curve marked PARA is the paramagnetic energy. The curve labelled FERRO I is the energy of the
saturated configuration, and the curve labelled FERRO II
is the energy of the configuration of smallest non-zero
magnetization. For $n > 1$ the energies plotted are $E/N -
\frac{1}{2}I(n - 1)$. The scale of energy is such that the trans-
fer energy between states at neighbouring sites is $-\frac{1}{4}$.

$n < 1$ there are no ferromagnetic solutions. The logarithmic
singularity in the fcc density of states at the band edge is
responsible for the saturated solution throughout the range of
$n \geqslant 1$.

The energies of various solutions are plotted in figures 11-
13. We see that the ferromagnetic solution which saturates is
the energetically stable solution where it exists. These results are reasonable extrapolations of the exact results of
Nagaoka [9] for these band structures. He proved in the case
of infinite U that when the number of electrons in $N \pm 1$ the
saturated ferromagnetic state is the ground state for the sc
and bcc cases, whilst for the fcc structure the saturated ferromagnetic state is or is not the ground state depending on whether
there are $N + 1$ or $N - 1$ electrons. Approximation (4) is seen
to reproduce this result. Calculations have not been done for
large but finite U where one expects to see a reduction in the
range of n where ferromagnetic solutions are found, and an instability towards antiferromagnetism near $n = 1$. Such calculations are much more difficult than they are in the infinite U
case. For infinite U doubly occupied sites are forbidden when
$n < 1$, whilst for finite U one must calculate the probability
for such doubly occupied sites.

In summary, our studies of the physical content of approximation (4) suggests that it provides an acceptable treatment, when
U is very large, for the correlation effects which have a major

influence on the stability of the ferromagnetic phase. Furthermore it can be used with the CPA for the alloy problem.

5. CPA APPLIED TO MAGNETIC ALLOYS

In this section we discuss the magnetic alloy generalization
within the CPA formalism. For approximations of the Coulomb
correlations that have the single-site character described by
equation (44), one can introduce an effective one-electron Hamiltonian to be used to calculate one-electron properties:

$$\tilde{H} = \sum_{\substack{i,j,\sigma \\ (i \neq j)}} t_{ij} c_{i\sigma}^{\dagger} c_{j\sigma} + \sum_{i,\sigma} \tilde{\varepsilon}_{i\sigma}(\omega) n_{i\sigma}, \tag{51}$$

where the energy-dependent effective potential $\tilde{\varepsilon}_{i\sigma}(\omega)$ is given
by

$$\tilde{\varepsilon}_{i\sigma}(\omega) = \omega - [g_i^{\sigma}(\omega)]^{-1}. \tag{52}$$

We assume that $\tilde{\varepsilon}_{i\sigma}(\omega)$ will be self-consistently determined such
that it depends only on the type of atom occupying the i-th site.
Note that this restricts our discussions to paramagnetism and
ferromagnetism. We will not consider the necessary generalization for antiferromagnetism.

The CPA formalism discussed in section 3 can be straightfor-
wardly applied to this Hamiltonian. The only difference is that
the disorder energies are spin-dependent and in general energy-
dependent. The CPA configuration averaged one-electron Green
function is

$$G_{kk'}{}^{\sigma}(\omega) = \frac{\delta_{kk'}}{\omega - \varepsilon_k - \Sigma^{\sigma}(\omega)} \tag{53}$$

where the coherent potential $\Sigma^{\sigma}(\omega)$ satisfies, for a concentration x of A atoms, the equation

$$x\tilde{\varepsilon}_{A\sigma}(\omega) + (1 - x)\tilde{\varepsilon}_{B\sigma}(\omega) - \Sigma^{\sigma}(\omega)$$

$$= [\tilde{\varepsilon}_{A\sigma}(\omega) - \Sigma^{\sigma}(\omega)][\tilde{\varepsilon}_{B\sigma}(\omega) - \Sigma^{\sigma}(\omega)]F^0(\omega - \Sigma^{\sigma}(\omega)), \tag{54}$$

which is the generalization of equation (28) written in a more
convenient form.

In the Hartree-Fock approximation $\tilde{\varepsilon}_{i\sigma}$ is energy independent.

$$\tilde{\varepsilon}_{i\sigma} = \varepsilon_i + U_i \langle n_{i-\sigma} \rangle \qquad \text{(Hartree-Fock)}, \tag{55}$$

$$\tilde{\varepsilon}_{i\sigma} = \begin{cases} \varepsilon_A + U_A n_{A-\sigma} & \text{for A sites,} \\ \varepsilon_B + U_B n_{B-\sigma} & \text{for B sites.} \end{cases} \tag{56}$$

The expectation values for the number of electrons of spin σ on an A site and on a B site, $n_{A\sigma}$ and $n_{B\sigma}$, are evaluated by standard Green function techniques [18] using the Green functions that describe atoms of the required type occupying the site. These Green functions, which we denote by $G_A{}^\sigma(\omega)$ and $G_B{}^\sigma(\omega)$, are given by a generalization of equation (38). For R = A or B,

$$G_R{}^\sigma(\omega) = \frac{F^0(\omega - \Sigma^\sigma(\omega))}{1 - [\tilde{\varepsilon}_{R\sigma}(\omega) - \Sigma^\sigma(\omega)]F^0(\omega - \Sigma^\sigma(\omega))}, \tag{57}$$

which in the Hartree-Fock approximation is just

$$G_R{}^\sigma(\omega) = \frac{F^0(\omega - \Sigma^\sigma(\omega))}{1 - [\varepsilon_R + U_R n_{R-\sigma} - \Sigma^\sigma(\omega)]F^0(\omega - \Sigma^\sigma(\omega))}, \tag{58}$$

Then

$$n_{R\sigma} = -\frac{1}{\pi} \int_{-\infty}^{\varepsilon_F} d\varepsilon \, \mathrm{Im} G_R{}^\sigma(\varepsilon + i0^+), \tag{59}$$

where ε_F is the Fermi energy. This completes the formalism in the Hartree-Fock approximation. The numbers $n_{A\sigma}$ and $n_{B\sigma}$ for both spin orientations are determined by solving the self-consistency condition equation (59) with the coherent potential $\Sigma^\sigma(\omega)$ simultaneously determined by equation (54). The band structure associated with ε_k enters only through $F^0(\omega)$, and the parameters are the concentration x of A atoms, the Fermi energy ε_F, and the energies ε_A, ε_B, U_A, and U_B. For finite temperatures equation (59) is replaced by

$$n_{R\sigma} = -\frac{1}{\pi} \int_{-\infty}^{+\infty} d\varepsilon f(\varepsilon) \, \mathrm{Im} G_R{}^\sigma(\varepsilon + i0^+), \tag{60}$$

where $f(\varepsilon)$ is the Fermi distribution function.

This Hartree-Fock plus CPA yields a generalization of the Stoner theory to $A_x B_{1-x}$ alloys where the average magnetic moments of the A and B type atoms are determined. These moments can be studied as a function of temperature and alloy concentration, and applications can be made to real alloys. The first attempt to apply this theory semi-quantitatively to ferromagnetic alloys was by Hasegawa and Kanamori [16]. In a series of papers they applied the theory to fcc $Ni_{1-x}Fe_x$, $Ni_{1-x}Co_x$, $Ni_{1-x}Mn_x$, $Ni_{1-x}Cr_x$, and bcc $Fe_{1-x}Ni_x$, $Fe_{1-x}Co_x$, $Fe_{1-x}Mn_x$, and $Fe_{1-x}Cr_x$. It is be-

yond the scope of these lectures to analyze these applications.
There are many interesting considerations associated with applying this over-simplified model to real transition metal alloys.
How does one choose a consistent set of parameters? How does
one choose a model density of states $\rho^0(\epsilon)$ which is sufficiently
realistic, yet one for which the calculations are tractable?
How does one take into account the five-fold degeneracy of the
actual d band? One is referred to Hasegawa and Kanamori [16]
for the discussion of these topics and the results of their calculations. However, to give some indication of this important
work we show in figures 14,15 the parameters used for the calculations and some comparisons with experiments for the magnitude
of the magnetic moments as a function of alloy concentration.
The agreement with experiment is quite encouraging.

The Hartree-Fock plus CPA has also been applied to enhanced
paramagnetic alloys. Levin, Bass, and Bennemann [17] found good
agreement with experiment for the susceptibility as a function
of alloy concentration for Pt_xPd_{1-x}, Rh_xPd_{1-x}, Ni_xRh_{1-x}, and
Ni_xPd_{1-x} in the paramagnetic state. We will not illustrate these
results, but mention them as an example of the applicability of
the theory.

The CPA represents an important advance in the theory of
itinerant magnetic alloys. The older discussions based on the
rigid band approximation could not confront the experimental observations for ferromagnetic A_xB_{1-x} alloys that show, in general,
moments of different magnitude and sometimes sign associated
with the A and B atoms. The CPA plus Hartree-Fock approximation
treatment by Hasegawa and Kanamori [16] gives a good description
of these alloys. However, as pointed out by Fukuyama [27],
there is a fundamental inconsistency in the use of the Hartree-
Fock approximation with the CPA. The success of the CPA is owed to its
ability to take into account strong disorder scattering. In this application of the CPA the strong disorder is primarily the result of the
intra-atomic Coulomb repulsion as it enters the expression for $\tilde{\epsilon}_{i\sigma}$.
But the Hartree-Fock approximation should be valid only if the Coulomb
repulsion is weak, and then the CPA is not required to treat the disorder. It is apparent that correlation effects should be examined.

Fukuyama and Ehrenreich [20] have applied the Hubbard alloy
analogy approximation for the Coulomb correlation together with
the CPA to this model. They consider the special case where the
disorder is so strong that only the sites occupied by one of the
two types of atoms are accessible to the electrons. Their result for the paramagnetic susceptibility showed no ferromagnetic
instability for any band structure. As we previously noted,
this treatment of correlations is so unfavourable to ferromagnetism that it has no ferromagnetic solutions when there is no
disorder, and it is very likely that this result remains when
there is arbitrary disorder. There has also been some work using the first Hubbard approximation by Esterling and Tahir-Kheli
[28]. Since they consider disorder in the transfer integrals as

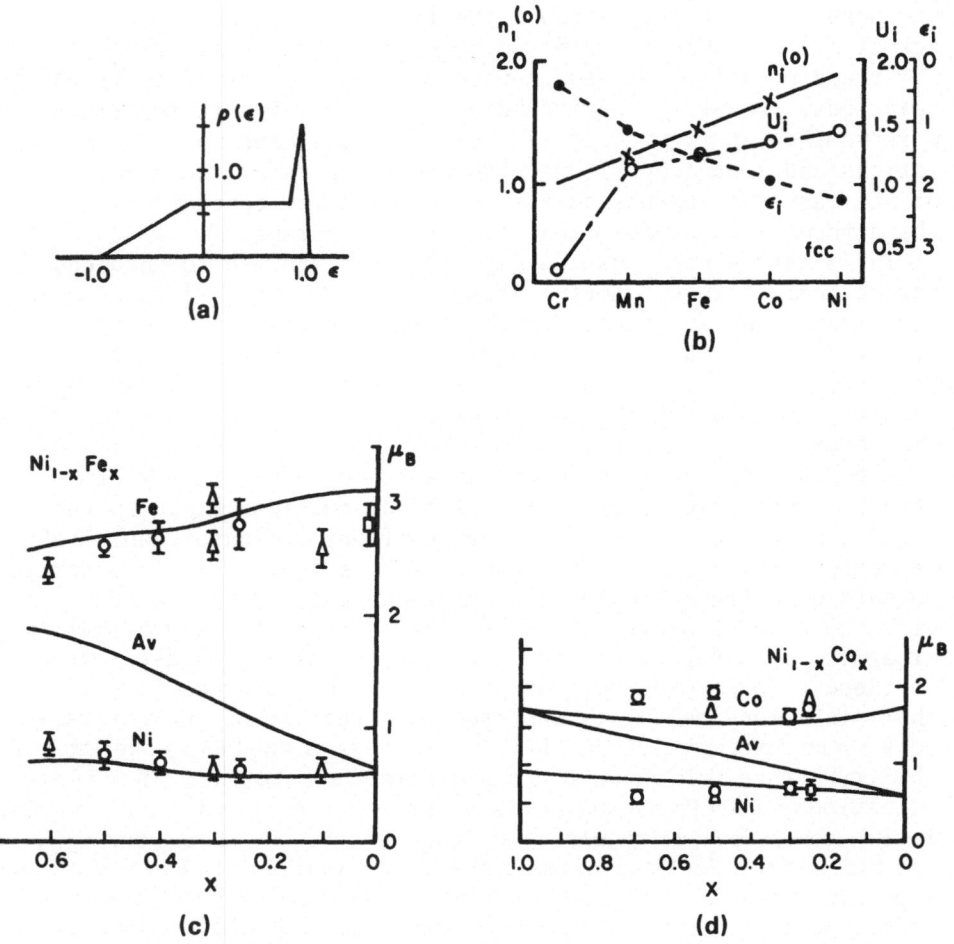

Figure 14 - Application of the coherent potential approximation plus Hartree-Fock approximation to fcc ferromagnetic alloys by Hasegawa and Kanamori. Shown are: (a) the
model density of states $\rho^0(\varepsilon)$; (b) the parameters assumed;
and comparisons of calculated results (lines) with neutron
diffraction data for (c) $Ni_{1-x}Fe_x$ and (d) $Ni_{1-x}Co_x$. [After
Hasegawa, H. and Kanamori, J. (1972). *J. Phys. Soc. Jap.*,
33, 1599].

well, this work goes behond the usual CPA. Let us merely note
that ferromagnetism is most unlikely within the framework of the
first Hubbard approximation. In contrast Brouers, Lederer, and
Heritier [29] have shown that the Nagaoka exact results for $U = \infty$
and $N \pm 1$ electrons remain unchanged when one introduces disorder. This suggests the desirability of applying approximation
(4) discussed in section 4. Hopefully this approximation can
give some insights into the effects of strong correlations and
disorders in the Hubbard model.

We shall now discuss our limited application [30] of approximation (4) with the CPA. We restrict our considerations to
cases where U_i is either zero or infinite. It is expected that
the results for infinite U_i will be representative of the situation where the ratio of the band width to the Coulomb repulsion
is small, and therefore not devoid of physical content. Hence
with these limited choices we will be able to comment on the
situation where the Coulomb repulsion between electrons is negligible for sites occupied by one type of atom, but very strong
for the other sites. This is just the Wolff model [31] for
local magnetic moments generalized to arbitrary concentrations
of magnetic impurities. Also we will be able to comment on the
case where the Coulomb repulsion between electrons is large on
all sites. Now clearly for $U_i = 0$,

$$\tilde{\varepsilon}_{i\sigma}(\omega) = \varepsilon_i \qquad (U_i = 0). \qquad (61)$$

For U_i infinite, approximation (4), as described by equations
(49,50), yield for $\tilde{\varepsilon}_{i\sigma}(\omega)$, according to equation (52), the following:

$$\tilde{\varepsilon}_{i\sigma}(\omega) = \frac{\varepsilon_i + \langle n_{i-\sigma}\rangle (W_{i-\sigma} - \omega)}{1 - \langle n_{i-\sigma}\rangle} \qquad (U_i = \infty), \qquad (62)$$

$$\langle n_{i\sigma}\rangle (1 - \langle n_{i\sigma}\rangle) W_{i\sigma} = - \sum_{\substack{j \neq i}} t_{ij}\langle c_{i\sigma}{}^+ c_{j\sigma}\rangle \qquad (U_i = \infty). \qquad (63)$$

The simplification seen in these expressions is the result of
two electrons on the i-th site being rigorously forbidden when
U_i is infinite. We are thus restricted to $\sum_\sigma \langle n_{i\sigma}\rangle \leqslant 1$, but this
is without loss of generality, because we can view the model as
describing holes if we change the signs of the transfer integrals
and ε [9,25].

The energy shift $W_{i\sigma}$ and $\langle n_{i\sigma}\rangle$ must be self-consistently determined to depend only on the type of atom occupying the i-th
site. Recall for an atom of type R = A or B,

$$n_{R\sigma} = - \frac{1}{\pi}\int_{-\infty}^{\varepsilon_F} d\varepsilon \, \mathrm{Im} G_R{}^\sigma(\varepsilon + i0^+), \qquad (64)$$

where

$$G_R{}^\sigma(\omega) = \frac{F^0(\omega - \Sigma^\sigma(\omega))}{1 - [\tilde{\varepsilon}_{R\sigma}(\omega) - \Sigma^\sigma(\omega)]F^0(\omega - \Sigma^\sigma(\omega))} \ . \tag{65}$$

Now, from equation (44) one sees that

$$\sum_{j \neq i} t_{ij} G_{ji}{}^\sigma(\omega) = [g_i{}^\sigma(\omega)]^{-1} G_{ii}{}^\sigma(\omega) - 1 \tag{66}$$

$$= [\omega - \tilde{\varepsilon}_{i\sigma}(\omega)] G_{ii}{}^\sigma(\omega) - 1. \tag{67}$$

Hence the usual Green function techniques yield

$$n_{R\sigma}(1 - n_{R\sigma}) W_{R\sigma} \tag{68}$$

$$= \frac{1}{\pi} \int_{-\infty}^{\varepsilon_F} d\varepsilon \, \mathrm{Im}\{[\varepsilon - \tilde{\varepsilon}_{R\sigma}(\varepsilon + i0^+)] G_R{}^\sigma(\varepsilon + i0^+)\}.$$

Figure 15 - Application of the coherent potential approximation plus Hartree-Fock approximation to bcc ferromagnetic alloys by Hasegawa and Kanamori. Shown are: (a) the model density of states $\rho^0(\varepsilon)$; (b) the parameters assumed; and comparisons of calculated results (lines) with neutron diffraction data for (c) $Fe_{1-x}Co_x$, (d) $Fe_{1-x}Ni_x$, and (e) $Fe_{1-x}Cr_x$. [After Hasegawa, H. and Kanamori, J. (1972). *J. Phys. Soc. Jap.*, **33**, 1607].

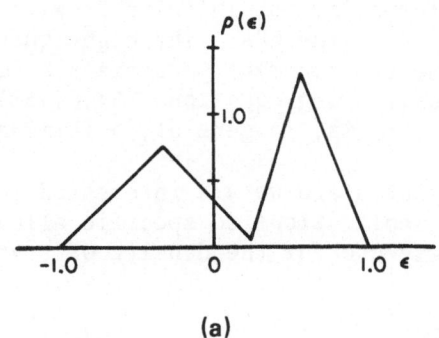

(a)

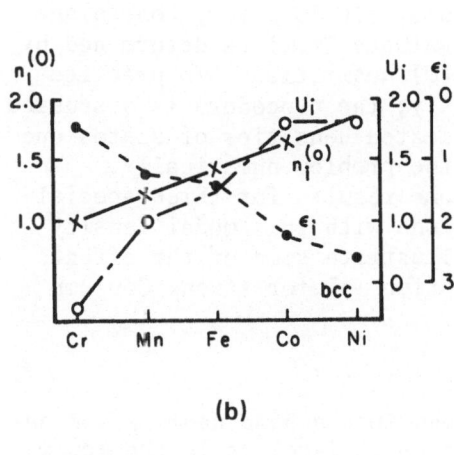

(b)

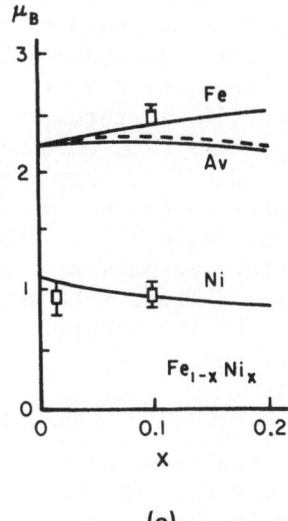

(c)

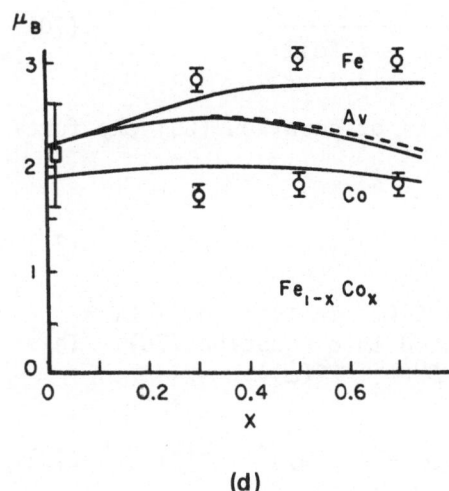

(d)

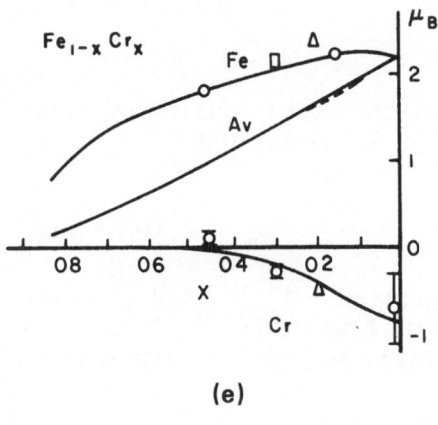

(e)

This completes the formalism for infinite U_i within the scheme of approximation (4) plus the CPA. The eight quantities $n_{R\sigma}$ and $W_{R\sigma}$ together with the two coherent potentials $\Sigma^\sigma(\omega)$ must be determined simultaneously using equations (54,64,68) and the definitions (61,62,65). This is, in general, a formidable numerical task.

Since at the present stage we are interested in qualitative aspects rather than applications to specific alloys, we will for the sake of simplicity use for the density of states associated with ε_k the form

$$\rho^0(\varepsilon) = \frac{2}{\pi} (1 - \varepsilon^2)^{\frac{1}{2}}, \qquad |\varepsilon| \leqslant 1. \tag{69}$$

This is the same form that was used in section 3 to illustrate the CPA for one-electron Hamitonians. It is a very convenient choice since one can explicitly evaluate $\Sigma^\sigma(\omega)$ as determined by equation (54) in terms of the $\tilde{\varepsilon}_{R\sigma}(\omega)$ quantities. In practice one evaluates $F^0(\omega - \Sigma^\sigma(\omega))$ directly; the procedure is discussed in reference [4]. For more complicated densities of states one would have to do this portion of the problem numerically.

In the following we describe some results for three special cases where we have made calculations with this model density of states. These will serve to illustrate some of the effects produced by the CAP plus approximation (4) for strong Coulomb correlations.

CASE A: $\delta \equiv \varepsilon_B - \varepsilon_A \gg 1$

Here the disorder splits the band into a band near ε_A and another near ε_B. We assume that the Fermi level is in the lower ε_A band in which case the B sites are unoccupied. For energies near ε_A the CPA condition equation (54) reduces to

$$\Sigma^\sigma(\omega) = \tilde{\varepsilon}_{A\sigma}(\omega) - \frac{1 - x}{F^0(\omega - \Sigma^\sigma(\omega))}. \tag{70}$$

For the model density of states $\rho^0(\varepsilon)$ of equation (69) the function $F^0(\omega)$ is just

$$F^0(\omega) = 2[\omega - (\omega^2 - 1)^{\frac{1}{2}}]. \tag{71}$$

Equation (71) is used to solve for $\Sigma^\sigma(\omega)$ in terms of $F^0(\omega - \Sigma^\sigma(\omega))$ and this result is substituted into equation (70). This yields an explicit expression for $F^0(\omega - \Sigma^\sigma(\omega))$:

$$F^0(\omega - \Sigma^\sigma(\sigma)) = 2\{(\omega - \tilde{\varepsilon}_{A\sigma}(\omega)) - [(\omega - \tilde{\varepsilon}_{A\sigma}(\omega))^2 - x]^{\frac{1}{2}}\}. \tag{72}$$

The component density of states for the A sites,

$$\rho_{A\sigma}(\varepsilon) \equiv -\frac{1}{\pi} \, \mathrm{Im} G_A{}^\sigma(\varepsilon + i0^+) = \frac{2}{\pi x} \, [x - (\varepsilon - \tilde{\varepsilon}_{A\sigma}(\varepsilon))^2]^{\frac{1}{2}}, \qquad (73)$$

where equation (65) for $G_A{}^\sigma(\omega)$ has been used. This expression
for $\rho_{A\sigma}(\varepsilon)$ is seen to differ with the density of states for the
perfect A type crystal by having its width narrowed by a factor
$x^{\frac{1}{2}}$ whilst keeping the same normalization.

The ferromagnetic solutions obtained as one varies the Fermi
energy across the band are shown in figure 16 where the moment
of an A atom, $m_A = n_{A\uparrow} - n_{A\downarrow}$, is plotted as a function of $n_A =
n_{A\uparrow} + n_{A\downarrow}$, the average occupation number for an A site. The
dashed curves are Hartree-Fock results for various values of
$U_A x^{-\frac{1}{2}}$ which is the single parameter that characterizes the Hartree-Fock solutions. For $U x^{-\frac{1}{2}} < \frac{1}{2}\pi$ there are no ferromagnetic
solutions. The solutions obtained when correlation effects are
included within approximation (4), which consists of using equa-
tion (62) for $\tilde{\varepsilon}_{A\sigma}(\varepsilon)$, are shown by the solid curve in figure 16.
For $n_A \lesssim 0.72$ there are no ferromagnetic solutions. The onset
of magnetic solutions at $n_A \approx 0.72$ is discontinuous (i.e., the
moment as a function of increasing n_A first appears with a finite value near the saturation value) and for $0.72 < n_A < 1$ there
are in fact two ferromagnetic solutions with slightly different
moments for each value of n_A. The solution of larger magnetization becomes saturated at $n_A \approx 0.77$. The scale in figure 16
is too small for the solutions to appear as separate curves.

These solutions for $U_A = \infty$ are independent of the concentra-

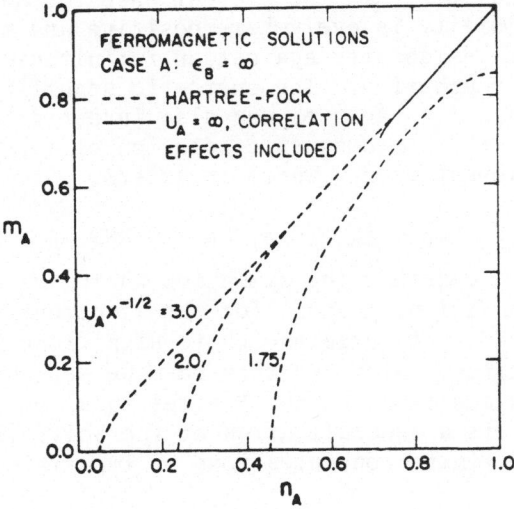

Figure 16 - Magnetization as a function of electron concentration for the case where the B sites are not accessible.

tion x (i.e., the m_A for a given n_A does not depend on x). This result is not general to the approximation scheme but is owing to our particular choice for $\rho^0(\varepsilon)$ which leads to equation (73) for $\rho_{A\sigma}(\varepsilon)$. Upon substitution for $\tilde{\varepsilon}_{A\sigma}(\varepsilon)$ from equation (62) one obtains

$$\rho_{A\sigma}(\varepsilon) = \frac{2}{\pi(1 - n_{A-\sigma})x}$$

$$\times [(1 - n_{A-\sigma})^2 x - (\varepsilon - \varepsilon_A - n_{A-\sigma}W_{A-\sigma})^2]^{\frac{1}{2}}, \qquad (74)$$

when $|\varepsilon - \varepsilon_A - n_{A-\sigma}W_{A-\sigma}| \leqslant (1 - n_{A-\sigma})x^{\frac{1}{2}}$ and zero otherwise. For the perfect crystal our approximate treatment of the Coulomb correlations for infinite U_A is seen to produce a band shift $n_{A-\sigma}W_{A-\sigma}$, a band narrowing by $(1 - n_{A-\sigma})$, and a change in normalization from 1 to $(1 - n_{A-\sigma})$ as the other $n_{A-\sigma}$ states are displaced to infinite energy. The dependence on the concentration x for this particular $\rho^0(\varepsilon)$ is only through a further narrowing by an $x^{\frac{1}{2}}$ factor. Hence x can be absorbed into a change in the energy scale. For general band shapes there can be non-trivial x-dependencies; however, to the extent that the major effect of the disorder is to narrow the band with dilution rather than to distort its shape, the solutions will be insensitive to x in our approximation with U_A infinite.

Also the discontinuous onset of ferromagnetic solutions is not a general feature of the approximation scheme. Recall the solutions shown in figures 6,7 for the sc and bcc tight binding nearest neighbour model densities of states did not show this discontinuous behaviour. In the present case the uniform paramagnetic susecptibility is everywhere positive and finite, and hence there is no instability against an infinitesimal uniform magnetic field perturbation. The energetic stability of one solution with respect to another was also investigated. For a given value of n_A the ferromagnetic solution of larger magnetization has the lowest energy where it exists.

CASE B: $\delta \equiv \varepsilon_B - \varepsilon_A = 0.8$, $U_B = 0$ AND $U_A = \infty$

In this case we consider the situation where the disorder parameter δ is sufficiently small for the electrons to have access to the B sites, yet large enough to significantly modify the density of states. Also for this case we set the Coulomb repulsion between electrons on the B sites equal to zero. When $U_B = 0$, the model is a generalization of the Wolff [31] magnetic impurity model to finite concentrations of impurities. Since $U_B = 0$,

$$\tilde{\varepsilon}_{B\sigma} = \varepsilon_B, \qquad (75)$$

and since U_A is taken to be infinite, $\tilde{\varepsilon}_{A\sigma}(\varepsilon)$ is given by equa-

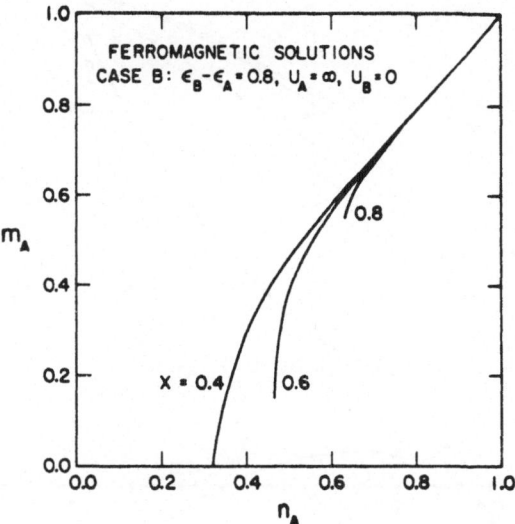

Figure 17 - Magnetization as a function of electron density for A sites for several alloy compositions.

tion (62). The self-consistency problem of determining $n_{A\sigma}$, $W_{A\sigma}$, and $n_{B\sigma}$ was solved by an iterative procedure.

Figure 17 shows ferromagnetic solutions for several concentrations x as curves of A site magnetization as a function of the A site electron concentration (m_A versus n_A). For $x = 1.0$ one of course obtains the solid curve of figure 16, since it also describes a perfect A type crystal. One sees that the range of n_A where ferromagnetic solutions exist is extended to smaller values of n_A with decreasing concentration x of A type atoms. In contrast for $\delta = -0.8$ (i.e., $\varepsilon_A - \varepsilon_B = 0.8$), the range of ferromagnetic solutions is reduced as x decreases. Since the occurrence of magnetic solutions depends primarily on the shape of the partial density of states $\rho_{A\sigma}(\varepsilon)$, one can understand this behaviour in terms of the distortion of the A band caused by the disorder and its dependence on the concentration x of A atoms. In figures 18-20 one finds a more complete picture of these ferromagnetic solutions. These figures show as a function of ε_F the partial electron concentrations and magnetizations n_A, m_A, n_B, and m_B, and the average electron concentration and magnetization, $\bar{n}$ and $\bar{m}$, defined by

$$\bar{n} = xn_A + (1 - x)n_B,$$

$$\bar{m} = xm_A + (1 - x)m_B.$$

(76)

The curves show the onset of ferromagnetic solutions with increasing ε_F, but were not continued to larger values of ε_F where

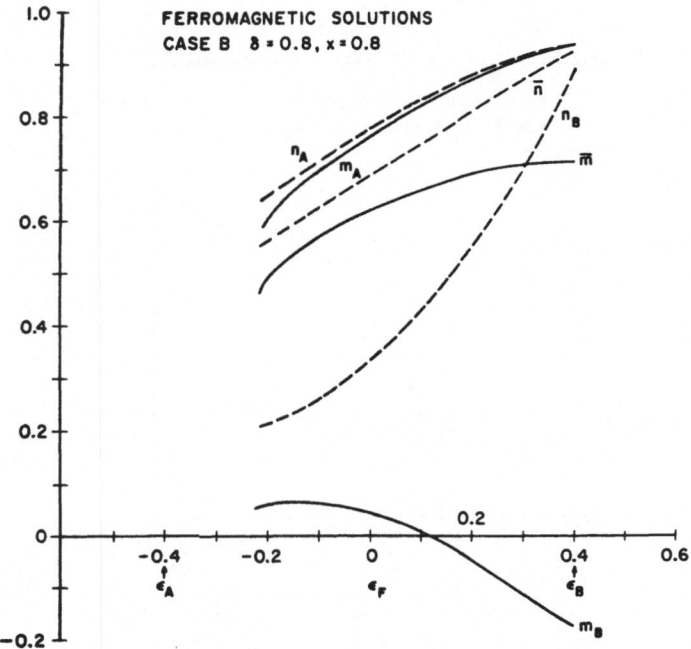

Figure 18 – Component and average magnetization and electron densities for ferromagnetic solutions (solutions
continue to larger values of the Fermi energy).

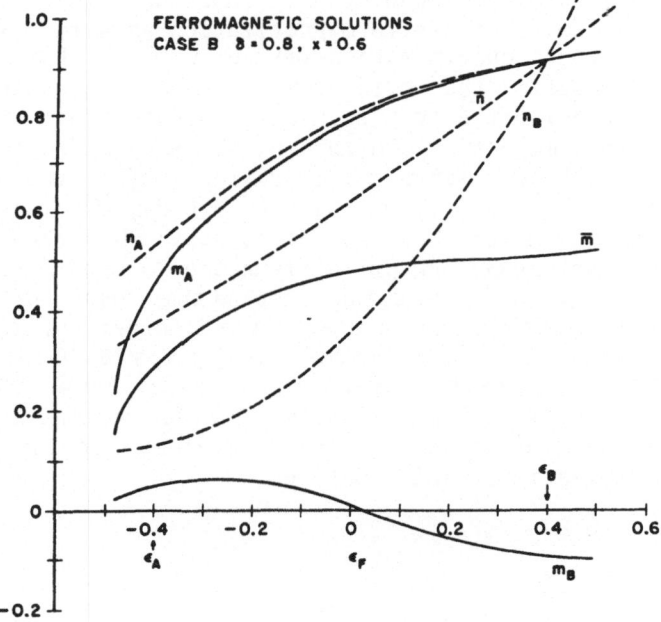

Figure 19 – Component and average magnetizations and electron densities for ferromagnetic solutions (solutions continue to larger values of ε_F).

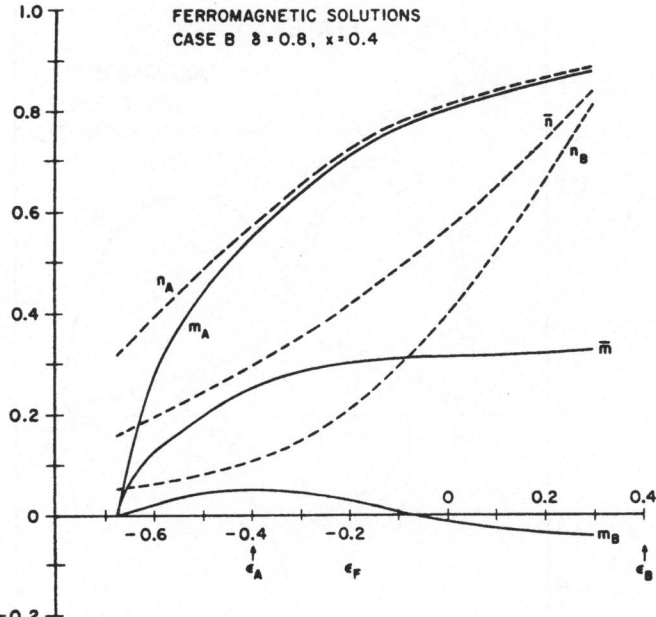

Figure 20 - Component and average magnetizations and electron densities for ferromagnetic solutions (solutions continue to larger values of ε_F).

n_A will eventually reach 1 and n_B reach 2. We plan to extend
our search for ferromagnetic solutions to higher values of ε_F
in the near future.

Figures 21,22 show representative curves for the partial
densities of states. Figure 21 corresponds to a solution where
the A site magnetization is small compared with the A site electron concentration, whilst figure 22 corresponds to a solution
where the A site magnetization is near saturation. As expected
for our treatment of correlations, one sees a narrowing of the
A minority spin band which becomes more pronounced as saturation
is approached.

CASE C: $\delta \equiv \varepsilon_B - \varepsilon_A = 0.8$, $U_A = U_B = \infty$

Because the intra-atomic Coulomb repulsion is infinite on
both the A and B sites, there can be no doubly occupied sites.
This is in contrast to case B where this restriction only held
for the A sites. Now both $\tilde{\varepsilon}_{A\sigma}(\varepsilon)$ and $\tilde{\varepsilon}_{B\sigma}(\varepsilon)$ are given by equation (62) in our treatment, and one must use an iterative pro-
cedure to determine $n_{A\sigma}$, $n_{B\sigma}$, $W_{A\sigma}$, and $W_{B\sigma}$ simultaneously. We
have just begun to investigate the ferromagnetic solutions.
Figure 23 shows magnetic solutions for a concentration $x = 0.4$
of A atoms. Again this figure shows the onset of magnetic solutions as a function of increasing ε_F, but does not follow these
solutions to arbitrarily large values of ε_F. In this case n_A

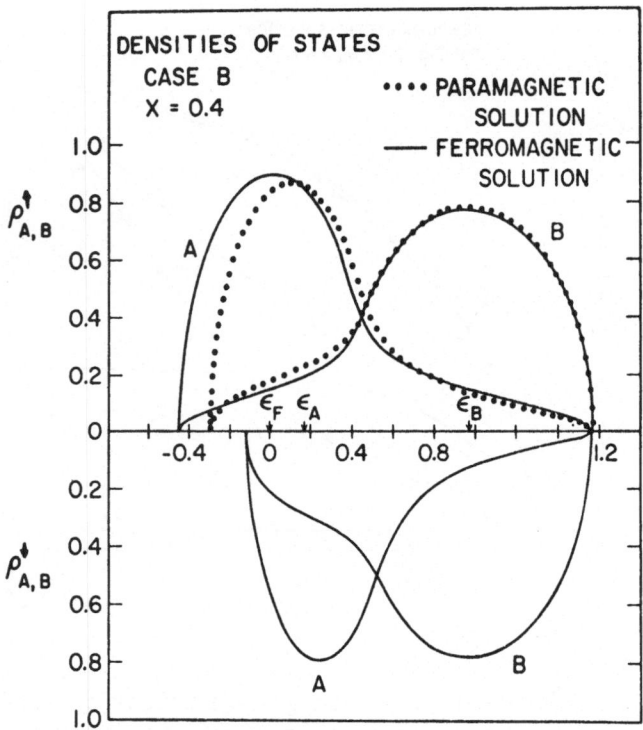

Figure 21 - Component densities of states for paramag-
netic and ferromagnetic solutions where $\varepsilon_B - \varepsilon_A = 0.8$
and $\varepsilon_F - \varepsilon_A = 0.18$.

and n_B approach 1. It will be interesting to see if m_B reverses
its sign again at larger ε_F and approaches 1 with m_A, as would
be expected from the Nagaoka exact results. Also there may be
other solutions with ferromagnetic symmetry.

In figures 24-26 partial densities of states are shown for
three values of the ε_F in the case where $x = 0.4$. Figure 24
shows the situation for $\varepsilon_F = -0.4 = \varepsilon_A$. Here one sees that both
the m_A and m_B are just saturated, since the minority bands are
displaced such that the lower edge is at the Fermi level. Figure 25 for $\varepsilon_F = 0.38$ shows the majority spin A band at the point
of splitting into two bands. In figure 26, for $\varepsilon_F = 0.5$ the
splitting has led to a finite gap. These graphs of the partial
density of states illustrate the variety of complicated behaviour that can result.

Finally there are several observations that should be made
concerning these results obtained using the CPA together with
our approximate treatment of strong correlations. Many of the
results can be reproduced by a Hartree-Fock treatment where the
U_i is understood as an effective intra-atomic Coulomb repulsion
which is reduced from the actual value to account for correlations neglected in the Hartree-Fock approximation. However,

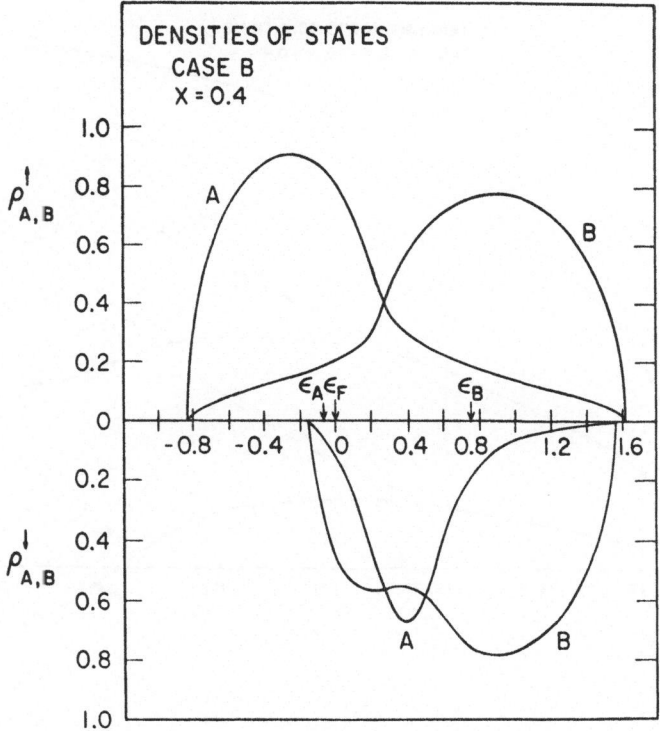

Figure 22 - Component densities of states for ferromag-
netic solution where $\varepsilon_B - \varepsilon_A = 0.8$ and $\varepsilon_F - \varepsilon_A = 0.07$.

this effective U_i would need to depend on various quantities
such as $n_{A\sigma}$, $n_{B\sigma}$, $m_{A\sigma}$, and $m_{B\sigma}$; and even then the Hartree-Fock
approximation could not produce the band narrowing seen in our
treatment of correlations. It is interesting to note that in
Hartree-Fock treatments the occurrence of ferromagnetism is
strongly dependent on the values of the partial densities of
states at the Fermi level, whilst in our treatment of correlations integrated properties of these partial densities of
states are the important parameters. This might help explain
the remarkable successes in describing experiments by using
the CPA and Hartree-Fock approximation with very simple densities of states. We hope to apply our treatment to a system
like $Co_xFe_{1-x}S_2$ for which the simple model discussed might be
reasonable.

6. CONCLUDING REMARKS

We have tried to indicate how the CPA can be used to study
itinerant ferromagnetic alloys with random disorder. The discussions were based on the single narrow band Hubbard model with
no disorder in the electron transfer term. Unfortunately, the
nature of the ground state of this model is not known. It may

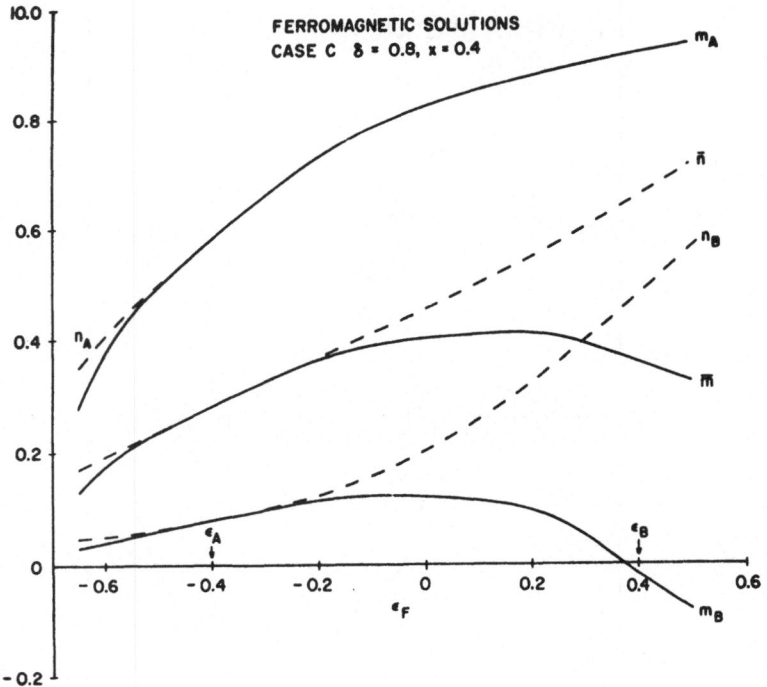

Figure 23 – Component and average magnetizations and electron densities for ferromagnetic solutions (solutions continue to larger values of the Fermi energy).

never have a ferromagnetic ground state, and hence be inappropriate for discussions of ferromagnetic alloys [33]. Nevertheless, it seems worthwhile to continue studies with this model until these questions are completely clarified.

We noted that applications of the Hartree-Fock approximation plus the CPA were able to obtain good agreement with many experimental observations on transition metal alloys. This is very encouraging, since one might have expected that magnetic properties would depend strongly on the neighbouring environment of component atoms. Such behaviour could not be described by the usual CPA, since atoms of a given type are all treated equally.

We have described our programme for investigating strong Coulomb correlation effects within the CPA framework for disordered alloys. This work is continuing and applications should be attempted in the near future. The need for an investigation of correlation effects was suggested by the inconsistency of treating the Coulomb repulsion as weak by using the Hartree-Fock approximation and then requiring the CPA to treat the strong disorder scattering primarily associated with the Coulomb repulsion. The treatment of correlation effects is a difficult problem. We have used a scheme which might contain much of the

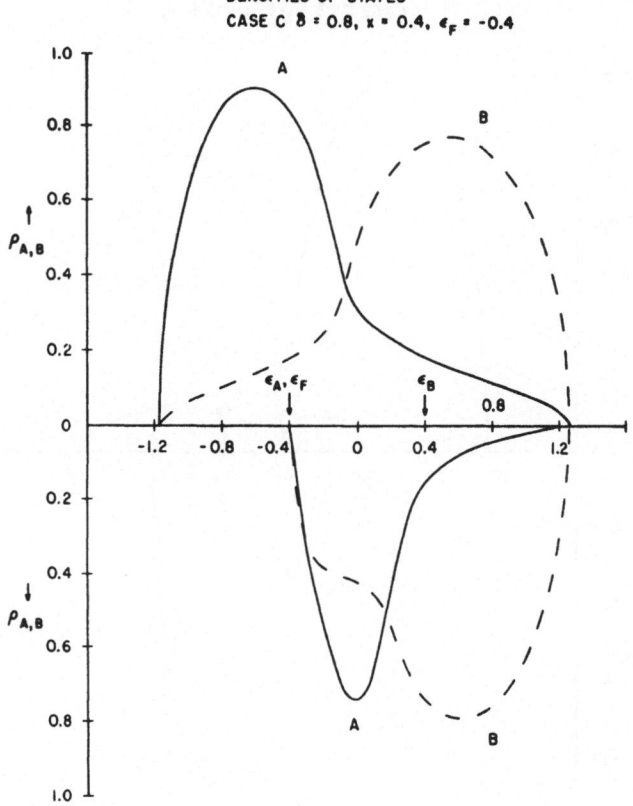

Figure 24 - Component densities of states for ferromagnetic solution.

essential physics in the limit of strong Coulomb repulsion.
It is necessarily a highly simplified scheme in order to make
the calculations tractable. It is hoped that at least it provides some insights into inadequacies of the Hartree-Fock scheme.

We would be remiss if we did not note that whilst the CPA for
one-electron Hamiltonians gives the correct dilute alloy limit,
none of the treatments for the Hubbard model described in these
notes produce the correct behaviour for a dilute magnetic alloy.
Where local moments are likely to exist these treatments will
tend to produce ferromagnetic solutions. Fluctuation effects
responsible for local moment behaviour are not taken into account.

Finally, we would like to note that no attempt was made to
give a complete review of all the contributions in this active
field of research. The omissions are too numerous to list. For
example, there has been extensive work on non-uniform and dynamical effects, and on work that attempts to go beyond the single-
site CPA.

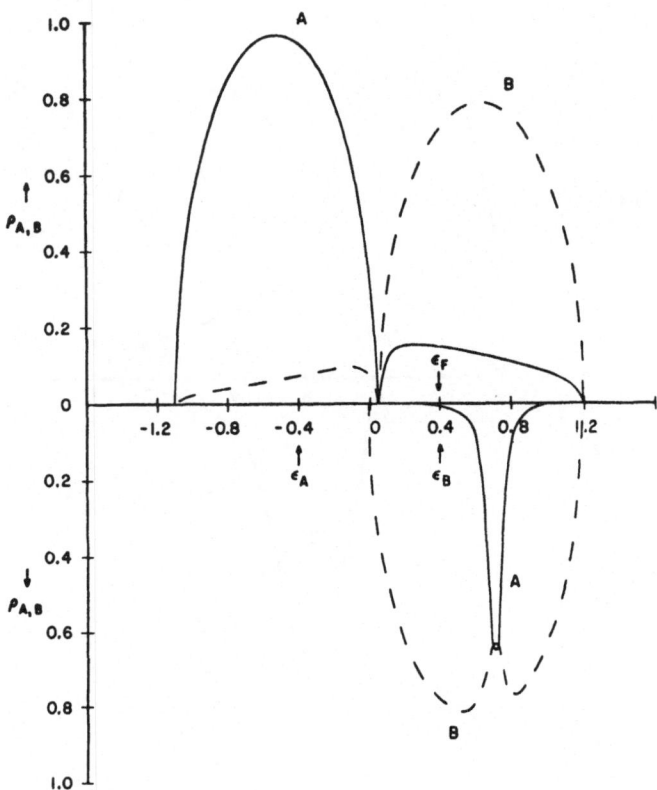

Figure 25 - Component densities of states for ferromagnetic solution.

ACKNOWLEDGMENTS

The author wishes to acknowledge the significant contributions to this work by G.F. Abito and J.S. Meyer.

REFERENCES

1. Hubbard, J. (1963). *Proc. Roy. Soc.*, **A276**, 238.
2. Hasegawa, H. and Kanamori, J. (1971). *J. Phys. Soc. Jap.*,
   31, 382.
3. Soven, P. (1967). *Phys. Rev.*, **156**, 809; (1969). *Phys. Rev.*,
   178, 1136.
4. Velický, B., Kirkpatrick, S. and Ehrenreich, H. (1968).
   Phys. Rev., **175**, 747.
5. Ehrenreich, H. (1972). *Lectures at the Winter College on
   Electrons in Crystalline Solids, International Centre for
   Theoretical Physics, Trieste.*

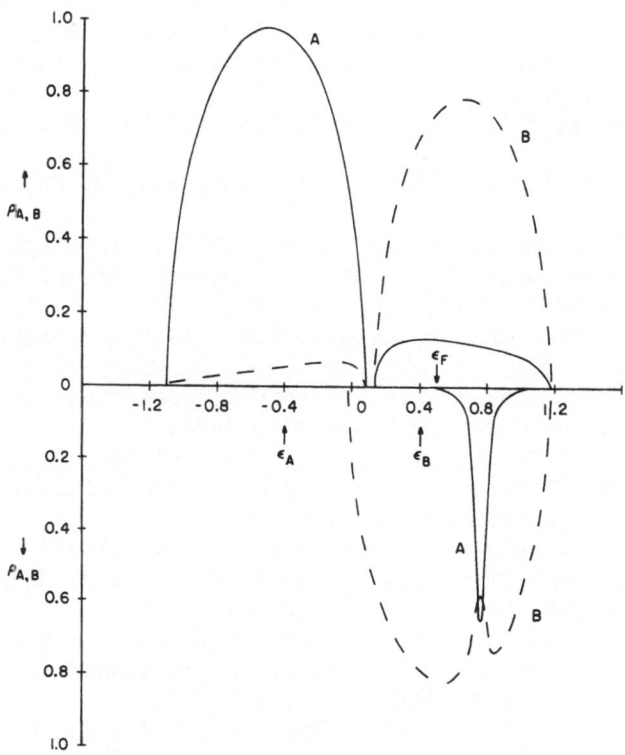

DENSITIES OF STATES
CASE C $\delta = 0.8$, $x = 0.4$, $\epsilon_F = 0.5$

Figure 26 - Component densities of states for ferromagnetic solution.

6. Müller-Hartmann, E. (Preprint).
7. Blackman, J.A., Esterling, D.M. and Berk, N.F. (1971).
   Phys. Rev., **B4**, 2412.
8. Kanamori, J. (1963). *Prog. Theor. Phys.*, **30**, 275.
9. Nagaoka, Y. (1966). *Phys. Rev.*, **147**, 392.
10. Beni, G., Pincus, P. and Hone, Daniel (1973). *Phys. Rev.*,
    B8, 3389.
11. Visscher, P.B. (Preprint).
12. Penn, D.R. (1966). *Phys. Rev.*, **142**, 350.
13. Caron, L.G. and Kemeny, G. (1971). *Phys. Rev.*, **B4**, 150.
14. Yang, Bunli (1973). (Thesis), (University of Illinois),
    (unpublished).
15. Roth, L.M. (1970). *Phys. Lett.*, **A31**, 440.
16. Hasegawa, H. and Kanamori, J. (1971): *J. Phys. Soc. Jap.*
    31, 382; (1972): *J. Phys. Soc. Jap.*, **32**, 1665; **33**, 1599,
    1607.
17. Levin, K., Bass, R. and Benneman, K.H. (1971). *Phys. Rev.*
    Lett., **27**, 589; (1972). *Phys. Rev.*, **B6**, 1865.
18. Zubarev, D.N. (1960). *Sov. Phys. Usp.*, **3**, 320.

19. Hubbard, J. (1964). *Proc. Roy. Soc.*, **A281**, 401.
20. Fukuyama, H. and Ehrenreich, H. (1973). *Phys. Rev.*, **B7**, 3266.
21. Harris, A.B. and Lange, (1967). *Phys. Rev.*, **157**, 295.
22. Roth, L.M. (1968). *Phys. Rev. Lett.*, **20**, 431; (1969). *Phys. Rev.*, **184**, 451.
23. Tahir-Kheli, R.A. and Jarrett, H.S. (1969). *Phys. Rev.*, **180**, 544.
24. Nolting, W. (1972). *Phys. Lett.*, **A38**, 417; Z. *Phys.*, **255**, 25.
25. Meyer, J.S. and Schweitzer, J.W. (1972). *AIP Conference Proceedings No.10: Magnetism and Magnetic Materials, 1972*, p. 526; (1973). *Phys. Rev.*, **B7**, 4253.
26. Faulkner, D.H. and Schweitzer, J.W. (1972). *J. Phys. Chem. Sol.*, **33**, 1685.
27. Fukuyama, H. (1972). *AIP Conference Proceedings No.10: Magnetism and Magnetic Materials, 1972*, p. 1127.
28. Esterling, D.M. and Tahir-Kheli, R.A. (1973). In *Amorphous Magnetism*, (eds. Hooper, H.O. and de Graaf, A.M.), (Plenum Press, London and New York), p. 365.
29. Brouers, F., Lederer, P. and Héritier, M. (1973). In *Amorphous Magnetism*, (eds. Hooper, H.O. and de Graaf, A.M.), (Plenum Press, London and New York), p. 151.
30. Abito, G.F. and Schweitzer, J.W. *Conference on Magnetism and Magnetic Materials, 1973*; (to be published).
31. Wolff, P.A. (1961). *Phys. Rev.*, **124**, 1030.
32. Jarrett, H.S., Cloud, W.H., Bouchard, R.J., Butler, S.R., Frederick, G.G. and Gillson, J.L. (1968). *Phys. Rev. Lett.*, **21**, 617.
33. Izuyama, T. (1972). *Prog. Theor. Phys.*, **48**, 1106; (1973). *Prog. Theor. Phys.*, **49**, 1066, 1068, 1772; (1973). *Prog. Theor. Phys.*, **50**, 1265.

THE STATIC AND DYNAMIC IMPURITY
SPIN SUSCEPTIBILITY OF KONDO ALLOYS

WOLFGANG GÖTZE

Physik-Department der Technischen Universität,
München, West Germany

and

Max-Planck-Institut für Physik, München,
West Germany

INTRODUCTION

In this lecture we want to discuss some properties of a single magnetic impurity imbedded in a metal matrix. We assume that the impurity is described completely by its spin operators S_z, $S_\pm = (S_x \pm S_y)/\sqrt{2}$. For the sake of simplicity we assume the spin to be $\frac{1}{2}$. The metal we will approximate as a free Fermi gas of conduction electrons, and as the interaction between impurity and matrix we will use a simplified exchange contact Hamiltonian. Such a model presents a reasonable description of simple metals like copper or gold containing a small fraction of magnetic ions like Fe, Mn or Cr. Direct measurements of magnetic properties can be performed by determining the spin polarization in an external static magnetic field, or by measuring hyperfine splittings for the impurity nucleus. Hyperfine techniques also provide information about the dynamical behaviour of the impurity spin. In the following we will focus our attention on the static zero field susceptibility and on the zero field excitation spectrum of the impurity.

The relevant low lying excitations of a degenerate Fermi system can be described approximately as excitations of a one-dimensional boson field. In the first section this spin boson model will be derived. In the second section the infra-red divergencies associated with the emission of soft bosons will be discussed. In the last section it will be shown how the soft boson emission and absorption yields a transformation of the spin $\frac{1}{2}$ impurity to a spin zero impurity complex.

1. THE SPIN BOSON MODEL

If the Zeeman energy of the impurity in an external magnetic field in the z-direction is denoted by B_i the impurity Hamiltonian reads

$$H_i = - B_i S_z. \tag{1}$$

Hence one gets for the motion of the spin operators the standard result

$$S_z(t) = S_z, \tag{2a}$$

$$S_\pm(t) = \exp(\mp iB_i t)S_\pm. \tag{2b}$$

S_z is a constant of motion and $S_\pm$ oscillate with the Larmor frequency. Remember that for spin $\frac{1}{2}$ one has the following equations

$$S_\pm S_\mp = \tfrac{1}{4} \pm \tfrac{1}{2}S_z, \qquad S_z S_\pm = - S_\pm S_z = \pm \tfrac{1}{2}S_\pm, \tag{3a}$$

$$S_+ S_+ = S_- S_- = 0, \qquad S_z S_z = \tfrac{1}{4}. \tag{3b}$$

The s-wave part of the conduction electron Hamiltonian reads

$$H_\ell = \sum_\sigma \int d\varepsilon \; \varepsilon_\sigma c_{\varepsilon\sigma}{}^* c_{\varepsilon\sigma}. \tag{4}$$

Here $c_{\varepsilon\sigma}{}^*$ and $c_{\varepsilon\sigma}$ denote respectively electron creation and annihilation operators for s-wave states with kinetic energy ε and spin $\sigma = \pm\frac{1}{2}$; $\varepsilon_\sigma = \varepsilon - \sigma B_\ell$, where B_ℓ is the conduction electron Zeeman energy. One has the usual anticommutation relations

$$[c_{\varepsilon\sigma}, c_{\varepsilon'\sigma'}]_+ = 0, \qquad [c_{\varepsilon\sigma}, c_{\varepsilon'\sigma'}{}^*]_+ = \delta_{\sigma\sigma'}\delta(\varepsilon - \varepsilon'). \tag{5}$$

The energy is normalized so that $\varepsilon = 0$ marks the Fermi level; ε runs from the bottom of the conduction band, ε_-, to the top, ε_+. The free motion of the electron is given by

$$c_{\varepsilon\sigma}(t) = \exp(- i\varepsilon_\sigma t)c_{\varepsilon\sigma}. \tag{6}$$

The level occupation is determined by the Fermi distribution

$$\langle c_{\varepsilon\sigma}^{*} c_{\varepsilon'\sigma'} \rangle = \delta_{\sigma\sigma'} \delta(\varepsilon - \varepsilon') \left[\exp\left(\frac{\varepsilon_{\sigma}}{T}\right) + 1 \right]^{-1}, \qquad (7)$$

where T denotes the temperature.

The most general interaction of the impurity spin with the conduction electrons, which is invariant under spin rotations, reads

$$H_{i\ell} = \sum_{\sigma\sigma'} \int d\varepsilon \int d\varepsilon' c_{\varepsilon\sigma}^{*} J(\varepsilon,\varepsilon') \vec{s}_{\sigma\sigma'} \cdot \vec{S} c_{\varepsilon'\sigma'}; \qquad (8)$$

here $J^{*}(\varepsilon,\varepsilon') = J(\varepsilon',\varepsilon)$ and s_i denote the 2×2 electron spin matrices. Since only low energy phenomena are of interest one can approximate $J(\varepsilon,\varepsilon')$ by a separable potential

$$J(\varepsilon,\varepsilon') = J v^{*}(\varepsilon) v(\varepsilon'); \qquad |v(0)|^2 = 1, \qquad (8b)$$

where J is dimensionless and $v(\varepsilon)$ is a cut off function of form factor. It is convenient to distinguish between longitudinal and transverse coupling, i.e. we write

$$H_{i\ell} = H_{i\ell}^{\parallel} + H_{i\ell}^{\perp} \qquad (8c)$$

$$H_{i\ell}^{\parallel} = J^{\parallel} S_z \sum_{\sigma\sigma'} c_{\sigma}^{*} s^{z}_{\sigma\sigma'} c_{\sigma'} \qquad (8d)$$

$$H_{i\ell}^{\parallel} = J^{\perp} S_+ \sum_{\sigma\sigma'} c_{\sigma}^{*} s^{-}_{\sigma\sigma'} C_{\sigma'} + \text{c.c.}, \qquad (8e)$$

where c_{σ} abbreviates the Wannier-operator

$$C_{\sigma} = \int d\varepsilon v(\varepsilon) c_{\varepsilon\sigma}. \qquad (8f)$$

($J^{\parallel} = J^{\perp} = J$ for the experimentally interesting case of isotropic coupling).

Let us write the total Hamiltonian in the form

$$H = H^{\parallel} + H^{\perp}, \qquad (9a)$$

$$H^{\parallel} = H_i + H_{\ell} + H_{i\ell}^{\parallel}, \qquad (9b)$$

$$H^{\perp} = H_{i\ell}^{\perp}. \qquad (9c)$$

S_z commutes with $H^{\parallel}$. Hence the eigenstates of $H^{\parallel}$ are classified according to the two impurity spin positions up and down. Then $H^{\parallel}$ describes the motion of up-/down-spin electrons in a separable potential of strength $\pm S_z J^{\parallel}$. The ground state will be a

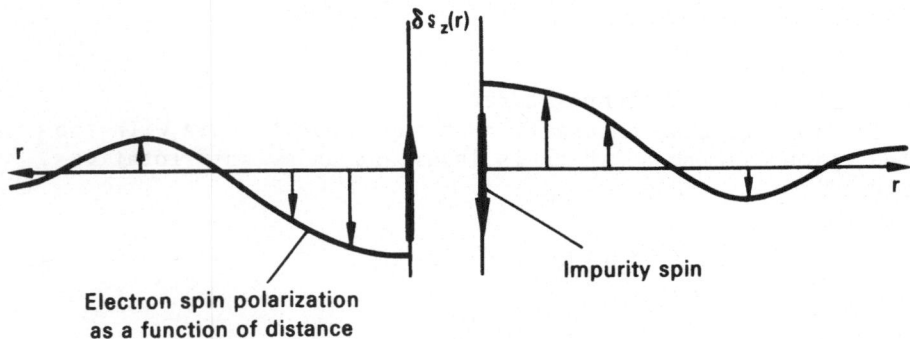

Figure 1

Friedel polarization cloud for each spin orientation. Provided
we have a symmetric band, the polarization of up and down spin
electrons will be of equal magnitude and opposite sign. Hence
there will be no density fluctuation but a spin polarization as
indicated in figure 1.

For $B = 0$ the ground state of $H^{\parallel}$ is degenerate: impurity
spin up and down surrounded by polarization clouds. Excited
states are given by the excitations of the polarization cloud.
Operator $H^{\perp}$ induces transitions between the degenerate states;
the impurity spin can flip under simultaneous emission of
particle-hole excitations of the electron gas.

To simplify the discussion of H let us introduce the operators

$$b_{\varepsilon\sigma}{}^* = \frac{i}{\sqrt{\varepsilon}}\int d\bar{\varepsilon}\, c_{\bar{\varepsilon}+\varepsilon\sigma}{}^* c_{\bar{\varepsilon}\sigma}, \tag{10a}$$

$$b_{\varepsilon\sigma} = \frac{-i}{\sqrt{\varepsilon}}\int d\bar{\varepsilon}\, c_{\bar{\varepsilon}-\varepsilon\sigma}{}^* c_{\bar{\varepsilon}\sigma}. \tag{10b}$$

Here $\varepsilon \geqslant 0$ and the integrals are extended over all values of $\bar{\varepsilon}$
for which the corresponding operators are defined. These opera-
tors create particle-hole excitations of the electron gas. With
(5) one evaluates the commutation relations

$$[b_{\varepsilon\sigma}{}^*, b_{\varepsilon'\sigma'}{}^*] = [b_{\varepsilon\sigma}, b_{\varepsilon'\sigma'}] = 0, \tag{11a}$$

$$[b_{\varepsilon\sigma}, b_{\varepsilon'\sigma'}{}^*] = \frac{\delta_{\sigma\sigma'}}{\sqrt{\varepsilon\varepsilon'}}\left[\int_{\varepsilon_-}^{\varepsilon_-+\varepsilon} d\bar{\varepsilon}\, c_{\bar{\varepsilon}+\varepsilon'-\varepsilon\sigma}{}^* c_{\bar{\varepsilon}\sigma} \right.$$

$$\left. - \int_{\varepsilon_+-\varepsilon'}^{\varepsilon_+} d\bar{\varepsilon}\, c_{\bar{\varepsilon}+\varepsilon'-\varepsilon\sigma}{}^* c_{\bar{\varepsilon}\sigma}\right], \tag{11b}$$

$$[b_{\varepsilon\sigma}, c_{\sigma'}{}^*] = \frac{-i\delta_{\sigma\sigma'}}{\sqrt{\varepsilon}} \int d\bar{\varepsilon} v^*(\bar{\varepsilon} + \varepsilon) c_{\bar{\varepsilon}\sigma}{}^*, \tag{11c}$$

$$[b_{\varepsilon\sigma}{}^*, c_{\sigma'}{}^*] = \frac{i\delta_{\sigma\sigma'}}{\sqrt{\varepsilon}} \int d\bar{\varepsilon} v^*(\bar{\varepsilon} - \varepsilon) c_{\bar{\varepsilon}\sigma}{}^*, \tag{11d}$$

$$[b_{\varepsilon\sigma}, H_\ell] = \varepsilon b_{\varepsilon\sigma}{}^*, \tag{11e}$$

$$[b_{\varepsilon\sigma}{}^*, H_\ell] = - \varepsilon b_{\varepsilon\sigma}{}^*. \tag{11f}$$

Now let us make two approximation assumptions. *First*, we assume there exists a characteristic energy D of the order of the band width such that

$$c_{\varepsilon\sigma}|\psi\rangle = 0 \quad \text{for} \quad \varepsilon > D \sim \tfrac{1}{2}\varepsilon_+ \ ,$$

$$c_{\varepsilon\sigma}{}^*|\psi\rangle = 0 \quad \text{for} \quad \varepsilon < -D \sim \tfrac{1}{2}\varepsilon_- \ ,$$

for all states $|\psi\rangle$ relevant to the discussion of our system. This means that all excitations of our system (owing to the external magnetic field, owing to the temperature, and owing to the interactions $J^{||}$, $J^{\perp}$) do not extend in energy further than D. Thus one can replace the operators in (11b) by their free free gas expectation values

$$c_{\bar{\varepsilon}+\varepsilon'-\varepsilon\sigma}{}^* c_{\bar{\varepsilon}\sigma} = \delta(\varepsilon - \varepsilon') \quad \text{for} \quad \varepsilon_- + \varepsilon < - D,$$

$$c_{\bar{\varepsilon}+\varepsilon'-\varepsilon\sigma}{}^* c_{\bar{\varepsilon}\sigma} = 0 \quad\quad \text{for} \quad \varepsilon_+ - \varepsilon' > D,$$

and this yields the Bose commutation relations

$$[b_{\varepsilon\sigma}, b_{\varepsilon'\sigma'}{}^*] = \delta_{\sigma\sigma'}\delta(\varepsilon - \varepsilon'). \tag{12}$$

The particle-hole excitations are described approximately by boson excitations. *Second*, we assume that all relevant states $|\psi\rangle$ can be obtained by applying the operators $b_{\varepsilon\sigma}$, $b_{\varepsilon'\sigma'}{}^*$, $S_{\pm}$, S_z on the ground states $|0\rangle$ of the non-interacting system. This means that the already mentioned algebra of operators is irreducible, and, according to Schur's lemma, that any operator X, obeying

$$[b_{\varepsilon\sigma}, X] = [b_{\varepsilon'\sigma'}{}^*, X] = [S_{\pm}, X] = [S_z, X] = 0 \tag{13a}$$

is a multiple of the unity

$$X = \lambda 1. \tag{13b}$$

The constant λ can be worked out by calculating some convenient matrix element $\langle\psi|X|\psi\rangle = \lambda$. As an example let us consider

$$X = \sum_\sigma \int d\varepsilon \ \varepsilon b_{\varepsilon\sigma}{}^* b_{\varepsilon\sigma} - H_\ell.$$

Because of (11e,f,12) X obeys (13a). Thus the electron kinetic energy is expressible in terms of Bose operators

$$H_\ell = \sum_\sigma \int d\varepsilon \ \varepsilon b_{\varepsilon\sigma}{}^* b_{\varepsilon\sigma} + \lambda_\ell. \tag{14}$$

Another example is

$$X_\sigma = c_\sigma{}^* \exp\left\{ i \int d\varepsilon \ \frac{(b_{\varepsilon\sigma}{}^* + b_{\varepsilon\sigma})}{\sqrt{\varepsilon}} \right\}.$$

If we choose $v(\varepsilon + \bar{\varepsilon}) \approx v(\varepsilon)$ in (8f) for the relevant energies ε, $\bar{\varepsilon}$ one uses (11c,d,12) to verify (13a). Consequently

$$c_\sigma = \lambda_\sigma \exp\left\{ i \int d\varepsilon \ \frac{b_{\varepsilon\sigma}{}^* + b_{\varepsilon\sigma}}{\sqrt{\varepsilon}} \right\}. \tag{15}$$

Notice that the localized Fermi operator c_σ has been expressed in terms of Bose operators. Equation (15) is an approximation valid for expectation values of low energy states. In this approximation the creation of localized fermions can be expressed as a superposition of density and spin density changes.

Before re-writing the Hamiltonian H in terms of bosons, let us take out the expectation value of the free electron polarization from (8d). Because of (7) this term would yield an $\varepsilon = 0$ singularity for $\langle b_\varepsilon \rangle$ which we want to avoid this way.

$$\sum_{\sigma\sigma'} \langle c_\sigma{}^* s_{\sigma\sigma'}{}^z c_\sigma \rangle_{H_\ell} = \sum_{\sigma\sigma'} \int d\varepsilon \int d\varepsilon' \ v^*(\varepsilon) v(\varepsilon') \langle c_{\varepsilon\sigma}{}^* c_{\varepsilon'\sigma'} \rangle s_{\sigma\sigma'}{}^z$$

$$= \frac{1}{2} B_\ell + o\left(\frac{B_\ell}{D}, \frac{I}{D}\right).$$

So we obtain

$$H^{||} = \int d\epsilon \omega_\epsilon b_\epsilon^{d*} b_\epsilon^{d} - B S_z +$$

$$+ \int d\epsilon \omega_\epsilon b_\epsilon^{s*} b_\epsilon^{s} + i J^{||} S_z \int d\epsilon (\rho_\epsilon \sqrt{\tfrac{1}{2}\omega_\epsilon})(b_\epsilon^{s} - b_\epsilon^{s*}), \qquad (16a)$$

$$H^{\perp} = \frac{J^{\perp}}{\sqrt{2}} (S_+ B_q^{-} + S_- B_q^{+}); \qquad q = 2,$$

$$B_q^{\mp} = \exp\left\{ \pm iq \int d\epsilon \left[\frac{\rho\epsilon}{\sqrt{2\omega_\epsilon}} \right] (b_\epsilon^{s} + b_\epsilon^{s*}). \right. \qquad (16b)$$

Here

$$b_\epsilon^{d} = \frac{1}{\sqrt{2}} (b_{\epsilon\frac{1}{2}} + b_{\epsilon-\frac{1}{2}}),$$

$$b_\epsilon^{s} = \frac{1}{\sqrt{2}} (b_{\epsilon\frac{1}{2}} - b_{\epsilon-\frac{1}{2}}), \qquad (17)$$

denote boson annihilation operators for density and spin density
excitations respectively. $B = B_1 - \frac{1}{2} J^{||} B_\ell$ denotes the Zeeman en-
ergy of the impurity containing the first order g-shift correc-
tion due to the electron polarization. $\omega_\epsilon = \epsilon$ for the metal;
but to avoid low energy divergencies in the following formulae
we write $\omega_\epsilon = (\epsilon^2 + \Delta^2)^{\frac{1}{2}}$ and consider the limit $\Delta \to 0$. The situation $\Delta \neq 0$ roughly corresponds to a semiconductor with gap Δ be-
tween valence and conduction band. The uninteresting constant
λ_ℓ has been dropped. The constants λ_σ will be shown in the following (see end of next section) to be equal to one and have not
been written therefore. Form factors ρ_ϵ have been introduced to
provide a cut off at energies of order D; $\rho_{\epsilon=0} = 1$.

The Hamiltonian (16) defines the spin boson model for a magnetic impurity in a metal. The metal is described by a one-dimensional Bose field of spin density waves of energy ω_ϵ. Actually there is the second Bose field of density waves given by
the first term in equation (16a). Since this field does not
couple to the impurity, we can forget about it in the following.
The impurity is a localized system of two levels: spin up with
energy $-B$ and spin down with energy $+B$. The Bose field experi-
ences an external source given by the last term in equation
(16a). The strength of the source is $J^{||} S_z$; it depends on the
state of the localized impurity. $H^{\perp}$ describes the fact that the
impurity can flip by emission and absorption of bosons. The
question is: how does the two level system change if the coupling to the bosons is taken into account? How does the Boson
field dress the bare spin $\frac{1}{2}$ impurity? What is the ground state
and the excitation spectrum of the interacting spin boson system?

The so called s-d Hamiltonian (8) can be derived [1] from the
more fundamental Anderson Hamiltonian for magnetic impurities.
This derivation brings out that the strong Coulomb repulsion for
localized electrons is the origin of the exchange coupling J.

The idea of describing one-dimensional Fermi systems by bosons
is owing to Tomonaga [2]. Therefore we will refer to the Bose
excitations discussed before as Tomonagons. Schotte and Schotte
[3] adapted Tomonaga's considerations to describe impurities in
metals; in particular they proved equation (15) and formulated
the spin boson or Tomonaga model (16). The Hamiltonian (16) is
equivalent to (8) provided the coupling constants $J^\parallel$, $J^\perp$ and the
other relevant energies are small enough; in practice J is of
the order 0.2 or smaller. The advantage of (16) is its simplicity compared with (8) and its analogy with models studied in
other branches of theoretical physics. Its disadvantage is its
not being manifestly spin rotational invariant.

2. THE SOFT BOSON SINGULARITIES

To discuss the physics described by $H^\parallel$ it is most convenient
to introduce new operators

$$a_\varepsilon = b_\varepsilon^s - iJ^\parallel S_z \frac{\rho_\varepsilon}{\sqrt{2\omega_\varepsilon}}, \tag{18a}$$

$$a_\varepsilon^* = b_\varepsilon^{s*} + iJ^\parallel S_z \frac{\rho_\varepsilon}{\sqrt{2\omega_\varepsilon}}, \tag{18b}$$

which again obey Bose commutation relations. Dropping the first
term in (16a) and dropping the uninteresting additive constant
$J^{\parallel 2}S_z^2 \int d\varepsilon(\rho_\varepsilon^2/2\omega_\varepsilon)$, one gets then

$$H^\parallel = - B\tau_0 + \int d\varepsilon \omega_\varepsilon a_\varepsilon^* a_\varepsilon. \tag{19}$$

For later use we have written $\tau_0 = S_z$. The operator (19) represents a system consisting of a free Bose field and a free spin.
Two eigenstates of $H^\parallel$ are the degenerate 'vacuum' $|x\rangle$, $x = \pm\frac{1}{2}$,
characterized by

$$\tau_0|x\rangle = x|x\rangle, \qquad a_\varepsilon|x\rangle = 0. \tag{20}$$

Excitations are given by applying a_ε^* operators on $|x\rangle$. The
states $|x\rangle$ have been discussed in connection with figure 1.
The polarization cloud is obtained from (18)

$$\langle x|b_\varepsilon^s|x\rangle = iJ^\parallel x \frac{\rho_\varepsilon}{\sqrt{2\omega_\varepsilon}}, \tag{21a}$$

$$\langle x|b_\varepsilon^d|x\rangle = 0. \tag{21b}$$

Equations (18) are a canonical transformation. To calculate this explicitly we check that from

$$u = \exp\left\{ i \int d\epsilon \alpha_\epsilon (b_\epsilon {}^{S*} + b_\epsilon {}^{S}) \right\}$$ (22a)

for $\alpha = \alpha^*$ it follows that $u^{-1} = u^*$ and

$$b_\epsilon {}^{S} u = u(b_\epsilon {}^{S} + i\alpha_\epsilon).$$ (22b)

Choosing

$$\alpha_\epsilon = - J\|S_z \frac{\rho_\epsilon}{\sqrt{2\omega_\epsilon}} ,$$ (22c)

we find indeed

$$u^* b_\epsilon {}^{S} u = \alpha_\epsilon.$$ (22d)

Hence one can write $|x\rangle = u^* |x\rangle_0$, where $|x\rangle_0$ are the states corresponding to (20) for the non-interacting system ($J\| = 0$):

$$|x\rangle = \exp\left\{ ix J\| \int d\epsilon \frac{\rho_\epsilon}{\sqrt{2\omega_\epsilon}} (b_\epsilon {}^{S*} + b_\epsilon {}^{S}) \right\} |x\rangle_0.$$ (23)

Consequently, $|x\rangle$ is a coherent superposition of many boson excitations. The impurity is surrounded by a polarization (21) or a coherent field of virtual bosons in the same way a charge is surrounded by a Coulomb field (coherent superposition of photons), a nucleon is surrounded by a Yukawa field (coherent superposition of pions), or a defect in an elastic medium is s surrounded by a deformation field (coherent superposition of phonons). Making use of the identity

$$e^{(A+B)} = e^A e^B e^{-\frac{1}{2}[A,B]},$$ (24)

which is valid whenever $[A,B]$ commutes with A as well as with B, we get from (23,12)

$$|x\rangle = e^{-\frac{1}{8}W} \sum_{n=0}^{\infty} \left[ix J\| \int d\epsilon \frac{\rho_\epsilon}{\sqrt{2\omega_\epsilon}} b_\epsilon {}^{S*} \right]^n \frac{1}{n!} |x\rangle_0.$$ (25a)

Here W is the Debye-Waller exponent

$$W = J\|^2 \int d\epsilon \frac{\rho_\epsilon^2}{2\omega_\epsilon} = \tfrac{1}{2} J\|^2 \log\left(\frac{2D}{\Delta}\right)$$ (25b)

(if we use a cut off D and $\Delta \ll D$). Hence the probability w_n to find n Tomonagons of energy ε in state $|x\rangle$ is given by the Poisson distribution

$$w_n = e^{-\frac{1}{8}W} \left(\frac{|J|^2 \rho_\varepsilon^2}{8\omega_\varepsilon} \right)^n \frac{1}{n!} , \qquad (25c)$$

$$e^{-\frac{1}{8}W} = 1 - \frac{W}{8} + \frac{W^2}{128} - \dots . \qquad (25d)$$

Let us consider now the limit $\Delta \to 0$, i.e. the boson field with mass zero. Then W in (25b) diverges logarithmically and the Debye-Waller factor approaches zero according to

$$e^{-\frac{1}{8}W} = \left(\frac{\Delta}{2D} \right)^{(\frac{1}{4}J|I)^2} . \qquad (26)$$

Owing to the very soft bosons the perturbation expansion (25d) shows logarithic divergences; the leading term is finite, the next one diverges like $\log\Delta$, the third one like $(\log\Delta)^2$, etc.. The perturbation expansion is impossible, since the probability w_n of finding n bosons in the coherent state $|x\rangle$ is zero. Owing to the $\varepsilon \to 0$ divergence of the W-integral for massless bosons, the operator u in (22a) is not well defined. Meaningfull results can be obtained only by taking $\Delta \neq 0$, working out the physical result without using perturbation theory for the construction of the coherent state, and then taking the limit $\Delta \to 0$ in the final formulae. These logarithmic divergences (25b) are a common feature of boson fields with zero mass, in particular of metals. Correct handling of the coherent states turns the logarithmic singularities into algebraic ones of type (26).

To see the physical relevance of the soft boson singularities let us calculate the probability $\sigma(\omega)$, that a γ-quantum with energy ω and definite helicity propagating in the z direction is absorbed by the impurity. Up to the electromagnetic coupling constant this quantity is given by the absorptive part of the transverse impurity susceptibility [4]

$$\sigma(\omega) = \int dt \exp\{i\omega t\} \langle S_+(t)S \rangle . \qquad (27)$$

For a free spin we get from (2b)

$$\sigma_0(\omega) = \pi p \delta(\omega - B_i), \qquad (28a)$$

where p is the population of the spin up state (see (3a))

$$p = \langle S_+ S_- \rangle = \frac{1}{4} + \frac{1}{2} \langle S_z \rangle . \qquad (28b)$$

The γ-quantum is absorbed if its energy ω is in resonance with the energy B_i which is necessary to flip the spin from the up into the down position.

The spin operators $S_\pm$ do not commute with the bosons, but we get from (18a,36)

$$[a_\epsilon, S_\pm] = \mp i J \| S_\pm \frac{\rho_\epsilon}{\sqrt{2\omega_\epsilon}} \ .$$

Introducing new operators

$$\tau_\pm = S_\pm \exp\left\{ \pm i J \| \int d\epsilon \ \frac{\rho_\epsilon}{\sqrt{2\omega_\epsilon}} (a_\epsilon + a_\epsilon^*) \right\}, \tag{29}$$

one finds, that the τ_0, τ_+, τ_- obey the same spin $\frac{1}{2}$ algebra (3) as the original spin operators, and they commute with the a_ϵ, a_ϵ^*. From (23) we find

$$\tau_\mp |\mp\tfrac{1}{2}\rangle = 0, \qquad \tau_\mp |\pm\tfrac{1}{2}\rangle = \frac{1}{\sqrt{2}} |\mp\tfrac{1}{2}\rangle, \tag{30}$$

and thus the $\tau_\pm$ are the operators flipping the impurity states dressed by the polarization cloud. The equation of motion $i\partial_t \tau_\pm = [\tau_\pm, H\|]$ yields

$$\tau_\pm(t) = \exp\{\mp i B t\} \tau_\pm. \tag{31}$$

These operators $\tau_\pm$ precess with the renormalized Larmor-frequency B. Since the τ motion is not coupled to the boson motion one gets in (27)

$$\langle S_+(t)S_-\rangle = p \exp\{-Bt\}\langle B_q^-(t)B_q\rangle; \qquad q = -J\|. \tag{32}$$

Here $p = \langle \tau_-\tau_+\rangle$ is the spin up population, and according to (29, 18) the B_q operators (16b) read

$$B_q^\mp = \exp\left\{ \pm iq \int d\epsilon \ \frac{\rho_\epsilon}{\sqrt{2\omega_\epsilon}} (a_\epsilon + a_\epsilon^*) \right\} \tag{33}$$

For the Hamiltonian (19) one gets, as usual,

$$a_\epsilon(t) = \exp\{-i\omega_\epsilon t\} a_\epsilon, \tag{34a}$$

$$\langle a_\varepsilon{}^* a_{\varepsilon'} \rangle = \delta(\varepsilon - \varepsilon')\left[\exp\left(\frac{\omega_\varepsilon}{T}\right) - 1\right]^{-1}. \qquad (34b)$$

For a single harmonic oscillator one verifies easily

$$\langle \exp\{-ik(a + a^*)(t)\}\exp\{ik(a + a^*)\}\rangle$$

$$= \exp\{k^2\langle((a + a^*)(t) - (a + a^*))(a + a^*)\rangle\},$$

and since the various oscillators entering

$$H_q(t) = \langle B_q{}^-(t)B_q \rangle \qquad (35a)$$

are independent, it follows from (33) that

$$H_q(t) = \exp\{-q^2 f(t)\}, \qquad (35b)$$

with

$$f(t) = -\frac{1}{2}\int d\varepsilon\int d\varepsilon'\, \frac{\rho_\varepsilon \rho_{\varepsilon'}}{\sqrt{\omega_\varepsilon \omega_{\varepsilon'}}}\, \langle [(e^{-i\omega_\varepsilon t} - 1)a_\varepsilon$$

$$+ (e^{i\omega_\varepsilon t} - 1)a_\varepsilon{}^*](a_{\varepsilon'} + a_{\varepsilon'}{}^*)\rangle \qquad (35c)$$

$$= \frac{1}{2}\int d\varepsilon\, \frac{\rho_\varepsilon{}^2}{\omega_\varepsilon}\, [(1 - e^{i\omega_\varepsilon t}) + e^{\omega_\varepsilon/T}(1 - e^{-i\omega_\varepsilon t})](e^{\omega_\varepsilon/T} - 1)^{-1}.$$

Let us concentrate on the zero temperature limit. Then one finds

$$f_0(t) = \int d\varepsilon\, \frac{\rho_\varepsilon{}^2}{2\omega_\varepsilon}\, (1 - \exp\{-i\omega_\varepsilon t\}) \qquad (35d)$$

$$= \frac{1}{2}\, \log(1 + itD). \qquad (35e)$$

and the last equation holds for $\omega_\varepsilon = \varepsilon$ and $\rho_\varepsilon = \exp\{-\varepsilon/2D\}$.

Formulae (35) have to be substituted into (32) and then into (27). The standard procedure consists of taking the time independent term in (35c) as the Debye-Waller factor (25b) in front of the integrals and then to perform a Taylor expansion of (35c). In this manner one finds

$$\sigma(\omega) = e^{-W}\left[\delta(\omega - B) + \frac{1}{2}|J|^2\int d\varepsilon\, \frac{\rho_\varepsilon{}^2}{\omega_\varepsilon}\, \delta(\omega - B - \omega_\varepsilon)\right] \qquad \text{(Contd)}$$

(Contd)

$$+ \tfrac{1}{8}J|^4 \int d\varepsilon_1 \int d\varepsilon_2 \, \frac{\rho_{\varepsilon_1}{}^2 \rho_{\varepsilon_2}{}^2}{\omega_{\varepsilon_1}\omega_{\varepsilon_2}} \, \delta(\omega - B - \omega_{\varepsilon_1} - \omega_{\varepsilon_2}) + \dots \Bigg]$$

$$= e^{-W}\Bigg[\delta(\omega - B) + \frac{\tfrac{1}{2}J|^2}{((\omega - B)^2 - \Delta^2)^{\frac{1}{2}}} \, \theta(\omega - B - \Delta) + \dots \Bigg] . \quad (36)$$

The first term in the preceeding formula describes the resonance absorption of the γ-quantum at the Zeeman energy B. This infinitely sharp resonance occurs as for a free spin, but its total absorption strength is reduced by the Debye-Waller factor e^{-W}. The second term represents the γ-absorption accompanied by the additional emission of one Tomonagon. This process contributes a smooth tail extending from $\omega = B$ to higher frequencies. The third term represents the γ-absorption accompanied by the additional emission of two Tomonagons etc. (see figure 2). If one increases the temperature the Debye-Waller factor goes down exponentially, the inelastic background increases and anti-Stokes absorption for $\omega < B$ becomes possible.

Let us consider now the limit of zero mass bosons, i.e. the case of an impurity in a metal. $\Delta = 0$ yields a $J|^2/2(\omega - B)$ tail for the one boson processes and this implies a total cross section which shows the logarithmic soft boson singularity mentioned in connection with formulae (25). The two Tomonagon contribution diverges even more strongly. On the other hand, the Debye-Waller factor tends to zero according to (26), and so no process with a finite number of bosons involved contributes to $\sigma(\omega)$. Perturbation expansion of finite order is not adequate to calculate $\sigma(\omega)$. Formula (35d) is well defined, though, yielding in (35b)

$$H_q(t) = (1 + itD)^{-\frac{1}{2}q^2}, \quad (36a)$$

which has the Fourier transform

$$H_q(\omega) = \left[\frac{2\pi\omega\theta(\omega)e^{-\omega/D}}{\Gamma(\tfrac{1}{2}q^2)}\right]\left(\frac{D}{\omega}\right)^{2-\frac{1}{2}q^2}. \quad (36b)$$

Thus one gets for small ω and $J|$

$$\sigma(\omega) = (\pi D J|^2)\theta(\omega - B)\left[\frac{D}{\omega - B}\right]^{1-\frac{1}{2}J|^2} \quad (37)$$

The absorption cross section increases to infinity algebraically. This singularity is owing to the simultaneous emission of infinitely many soft Tomonagons during the spin flip. The total

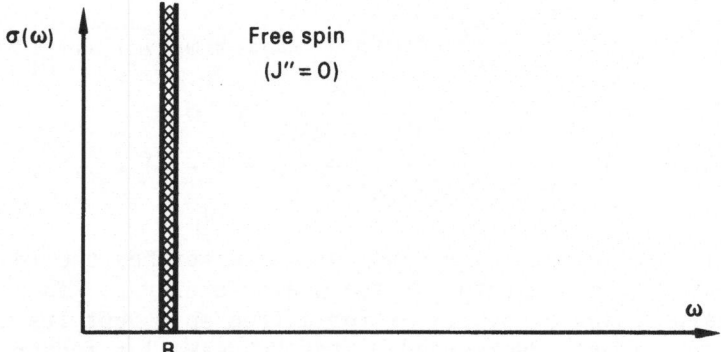

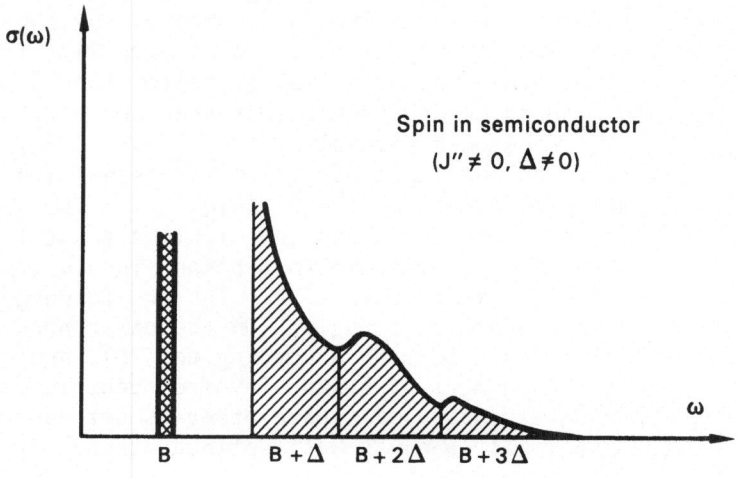

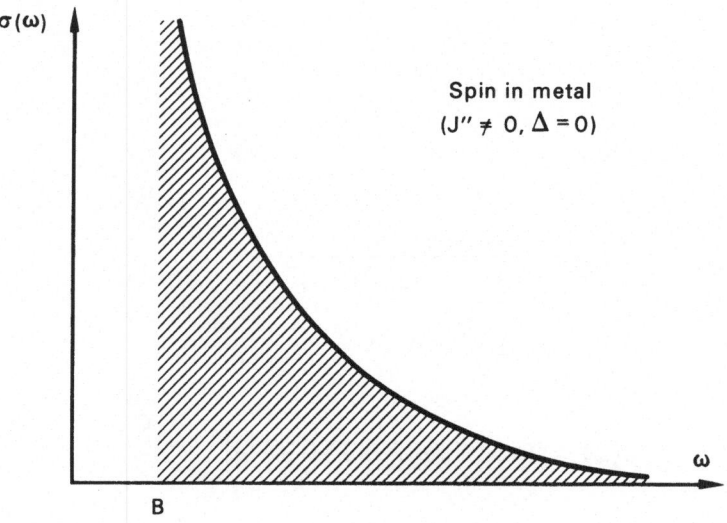

Figure 2

cross section is finite and agrees with the one for a free spin. The strong soft Tomonagon singularities similar to (37) are responsible for the anomalies observed experimentally for magnetic ions in metals.

Calculating $\langle c_\sigma^*(t) c_{-\sigma}(t) c_{-\sigma}^* c_\sigma \rangle$ with (8f,6,7) for a free Fermi gas, and using (15) to calculate the same expression with the aid of equations (35), yields $\lambda_\sigma = 1$.

The logarithmic soft boson singularities have first been found for Bremsstrahlung processes in electrodynamics. The correct treatment along the lines discussed above was originally due to Bloch and Nordsieck [5], and formulae similar to (37) can be found more or less explicitly in most textbooks on quantum electrodynamics. The relevance of these infrared singularities for the impurity problem in metals was first realized by de Dominicis and Nozières [6] in 1969. Originally the logarithmic divergences for the magnetic impurity problem were found by Kondo [7] in 1964, and therefore these divergences are usually referred to as Kondo singularities. Experimental observation of singularities like (37) has been reported for Bremsstrahlung tails and X-ray thresholds in metals. For magnetic impurities the preceeding formulae have no direct relevance because of the important rôle played by the transverse interaction $H^\perp$.

3. THE KONDO SUSCEPTIBILITY

Before starting an evaluation of the susceptibility of a magnetic impurity in a metal, we have to remember the standard formulae [8] of linear response theory. The longitudinal dynamical susceptibility $\chi(\omega)$ is the coefficient of proportionality between the change of the spin polarization $\delta\langle S_z \rangle$ and the change of the magnetic field δB:

$$\delta\langle S_z \rangle(\omega) = \chi(\omega)\delta B(\omega) + O(\delta B^2), \qquad (38a)$$

where the magnetic field varies monochromatically with frequency ω. The zero frequency limit is the static susceptibility

$$\chi^0 = \chi(\omega = 0) = \frac{\partial\langle S_z \rangle}{\partial B} . \qquad (38b)$$

The dynamical susceptibility is given by a correlation function. In the Zubarev notation one defines the Laplace transform

$$\langle\langle A;B \rangle\rangle_z = -i\int_0^\infty dt e^{izt}\langle A(t)B - BA(t)\rangle, \qquad \text{Im} z > 0, \qquad (39a)$$

and gets $\chi(\omega) = -\langle\langle S_z, S_z \rangle\rangle_{\omega+i0}$. Such functions can be written as spectral integrals

$$F(z) = \int d\left(\frac{\omega}{\pi}\right) \frac{F''(\omega)}{\omega - z} \tag{39b}$$

The spectral function of the Zubarev function is the Fourier transform of the commutator

$$\langle\langle A;B\rangle\rangle_\omega'' = -\frac{1}{2}\int dt e^{i\omega t}\langle A(t)B - BA(t)\rangle. \tag{39c}$$

For zero frequency one gets from (38b,39b) the sum rule

$$\chi^0 = \int d\left(\frac{\omega}{\pi}\right) \frac{\chi''(\omega)}{\omega}. \tag{40}$$

It is often more convenient to use Kubo's relaxation function

$$\phi(z) = \left[\frac{\chi(z)}{\chi^0} - 1\right]z^{-1}. \tag{41a}$$

Its spectral function reads

$$\phi''(\omega) = \frac{\chi''(\omega)}{\chi^0\omega}, \tag{41b}$$

it is symmetric in ω, non-negative, and normalized according to (40)

$$1 = \int d\left(\frac{\omega}{\pi}\right)\phi''(\omega). \tag{41c}$$

The fluctuation dissipation theorem reads

$$\langle(S_z - \langle S_z\rangle)^2\rangle = \int d\left(\frac{\omega}{2\pi}\right)\coth\left(\frac{\omega}{2T}\right)\chi''(\omega). \tag{(42)}$$

The energy the system absorbs from the external field is given by

$$\delta E = \int d\left(\frac{\omega}{2\pi}\right)\omega\chi''(\omega)|\delta B(\omega)|^2. \tag{43}$$

The last equation shows, that $\phi''(\omega)$ is the normalized longitudinal excitation spectrum of the impurity. If $\phi''(\omega) = 0$ no excitations of energy ω can be made by δB; if $\phi''(\omega)$ is big, states with energy ω can be excited easily. The fluctuation dissipation theorem can be re-written by using $S_z^2 = \frac{1}{4}$ and (41b, c)

$$(1 - 4\langle S_z\rangle^2)(\chi^0)^{-1} = 4T + 2\int d\left(\frac{\omega}{\pi}\right)\varphi(\omega,T)\phi''(\omega), \qquad (44a)$$

with the non-negative function

$$\varphi(\omega,T) = \omega\left[\coth\left(\frac{\omega}{2T}\right) - \frac{2T}{\omega}\right]$$

$$= \begin{cases} \dfrac{\omega^2}{6T} & \text{for} \quad |\omega| \ll T, \\[3mm] |\omega| & \text{for} \quad |\omega| \gg T. \end{cases} \qquad (44b)$$

Let us illustrate these equations (44) for $B = 0$ by four elementary examples. First, for a spin with $J^\perp = 0$, in particular for a free spin, S_z is a constant of motion and thus $\chi(z) = 0$ for $z \neq 0$. Hence $\phi''(\omega) = 0$ for $\omega \neq 0$ and (41c) requires

$$\phi''(\omega) = \pi\delta(\omega); \qquad (J^\perp = 0). \qquad (45a)$$

Equations (44) yield the Curie law

$$\chi^0_c = \frac{1}{4T} . \qquad (45b)$$

The ground state is degenerate. An arbitrarily small field will polarize the system completely at $T = 0$, hence the zero temperature susceptibility diverges. This divergence is the experimental characterization of a magnetic impurity. At non-zero temperature one has to polarize against the thermal fluctuations and so χ^0_c decreases. Generally we have

$$\chi^0 \leqslant \chi^0_c, \qquad (45c)$$

and the equality holds only if (45a) is valid.

Second, we consider the classical limit; i.e. we assume the thermal energy to be large compared with the excitation energies of the impurity

$$\phi''(\omega) \approx 0 \quad \text{for} \quad \omega \gtrsim T. \qquad (45d)$$

Again one obtains the Curie law (45b).

Third, let us assume our spin to be bounded in some molecule. Then there is a gap ω_0 in the excitation spectrum

$$\phi''(\omega) = 0 \quad \text{for} \quad \omega \leqslant \omega_0. \qquad (46a)$$

One gets the finite zero temperature susceptibility

$$\chi^0(T = 0) = \frac{1}{4\pi} \int_{\omega_0}^{\infty} d\omega\, \omega\phi''(\omega). \tag{46b}$$

If $\phi''(\omega)$ is independent of temperature $\chi^0(T) - \chi^0(T = 0) \propto \exp(-\omega_0/T)$.

Fourth, we consider a Lorentzian of width Γ,

$$\phi''(\omega) = \frac{\Gamma}{\omega^2 + \Gamma^2}. \tag{47a}$$

Such a form implies via (41a,38a) that spin deviations die out exponentially in time,

$$\delta\langle S_z\rangle(t) = \delta\langle S_z\rangle(t = 0)\exp\left(-\frac{t}{T_1}\right), \tag{47b}$$

where the decay time reads $T_1 = 1/\Gamma$. T_1 is the longitudinal relaxation time and equation (47b) is the first Bloch equation for the spin motion. With (47a) the RHS of (44a) diverges logarithmically, implying $\chi^0 = 0$. Actually, one has to introduce some high frequency cut off D, writing, e.g., in (47a)

$$\Gamma = \left(\frac{D^2}{T_1}\right)(\omega^2 + D^2)^{-1}. \tag{47c}$$

For $\Gamma \ll D$ we can write (47a) but replace $d\omega$ by $d\omega D^2/(\omega^2 + D^2)$. This simplifies the algebra slightly. Remembering

$$\coth z - \frac{1}{z} = \left[\psi\left(1 + \frac{iz}{\pi}\right) - \psi\left(1 - \frac{iz}{\pi}\right)\right]\frac{1}{\pi i}, \tag{47d}$$

where $\psi(z) = d\log\Gamma(z)/dz$ is the digamma function one finds from (44a)

$$\frac{1}{\chi^0} = 4T + \frac{\Gamma D^2}{D^2 - \Gamma^2}\left[\psi\left(1 + \frac{D}{2\pi T}\right) - \psi\left(1 + \frac{\Gamma}{2\pi T}\right)\right]. \tag{47e}$$

Using the asymptotic expansion

$$\psi(z) = \log z - \frac{1}{2z} + O(z^{-2}), \tag{47f}$$

one gets for low temperatures

$$\chi^0 = \frac{D}{4\Gamma}\left[T + \frac{1}{\pi}\log\left(\frac{D}{\Gamma}\right)\right]^{-1}; \qquad T \ll \Gamma \ll D. \tag{47g}$$

Again the zero temperature susceptibility is finite; it varies according to a Curie-Weiss law $\chi^0 = C/(T + \theta)$ with an enhanced Curie-constant. For intermediate temperatures the susceptibility looks similar, but the Curie constant is $\frac{1}{4}$ and θ varies slightly with temperatures,

$$\chi^0 = \frac{1}{4}\left[T + \frac{\Gamma}{\pi}\left(\log\left(\frac{D}{2\pi T}\right) - \psi(1)\right)\right]^{-1}; \quad \Gamma \ll T \ll D. \quad (47h)$$

Equations (44) show that only an impurity with a degenerate groundstate having the trivial excitation spectrum (45a) can have a divergent zero temperature susceptibility. If a molecule is formed, or a spin fluctuation (47a) is built up, χ^0 remains finite.

From (45a) we expect $\phi(z)$ to have a resonance, and so it is advisable to introduce a function characterizing position and width of this resonance. Since $\phi''(\omega) \geqslant 0$ the dispersion relation (39b) yields $\phi(z) \neq 0$ for $\mathrm{Im}\,z \neq 0$, and we introduce a function $N(z)$ by the definition

$$\phi(z) = - \frac{1}{z + \dfrac{N(z)}{\chi^0}}; \quad (48)$$

$N(z)$ is holomorphic off the real axis. Because of (41c), $N(z) = O(z-1)$ for large frequencies, and hence it can be represented by a spectral integral like (39b). $N''(\omega)$ is even and non-negative. $N''(\omega)$ is called the relaxation or noise spectrum. Let us assume $N''(\omega)$ to be given as a function of frequency ω, field B and temperature T. $N(z)$ then follows from (39b) and thus $\phi''(\omega)$ is given depending on $\xi = 1/\chi^0$. Then (44a) yields the transcendental equation ($\eta = \langle S_z \rangle$):

$$\xi(1 - 4\eta) = F(\xi; T, N''_{T,B}(\omega)), \quad (49)$$

yielding $\xi(B;\eta,T)$. For $B = 0$ we have $\eta = 0$, and (49) fixes the static zero field susceptibility. For $B \neq 0$ (38b) yields a differential equation to determine the spin polarization η. By inventing some approximation for $N''(\omega)$ one thus obtains the dynamical susceptibility, the static susceptibiltiy and the spin polarization.

To illustrate (49) for $B = 0$ let us assume the phenomenological Bloch equation (47a) or (47b) to be valid. This means

$$N''(\omega) = \gamma. \quad (50a)$$

The relaxation spectrum is a white one; up to the cut off D it does not depend on frequency. $N'(\omega) = O(D^{-1}) \approx 0$. The relaxation rate reads

$$\frac{1}{T_1} = \frac{\gamma}{\chi^0} \; . \tag{50b}$$

With $\Gamma = \gamma/\chi^0$ and $\Gamma \ll D$, equation (49) follows from (47e) to

$$\xi = F(\xi) = 4T + \frac{4\xi\gamma}{\pi}\left[\psi\left(1 + \frac{D}{2\pi T}\right) - \psi\left(1 + \frac{\xi\gamma}{2\pi}\right)\right] \; . \tag{50c}$$

Obviously $\xi = \alpha T$ is not a solution of this equation (if γ is independent of T) i.e. χ^0 does not diverge in Curie fashion. Equation (50c) has a non-zero zero temperature solution $\xi_0 \neq 0$

$$\xi_0 = \frac{4\xi_0\gamma}{\pi} \log\left(\frac{D}{\gamma\xi_0}\right) \; ,$$

i.e.,

$$\chi^0(T = 0) = \frac{\gamma}{D} \exp\left(\frac{\pi}{4\gamma}\right) \; . \tag{50d}$$

Later we will see in connection with figure 3, that (50d) is the limit of the only solution of (50c) for $T \to 0$. Hence a white relaxation spectrum with $\gamma(T \to 0) \neq 0$ implies a non-divergent susceptibility. According to (47a) the excitation spectrum does not contain a $\delta(\omega)$ peak, and hence the ground state is not degenerate. Therefore a non-zero energy is necessary to achieve polarization, and χ^0 does not diverge. For low temperatures one gets from (50c)

$$\chi^0(T) = \chi^0(T = 0) - bT^2, \tag{50e}$$

with some constant b.

Let us now calculate $N''(\omega)$ for the spin boson Hamiltonian $H = H^\| + H^\perp$. $H^\|$ in (19) does not yield relaxation since it commutes with $S_z = \tau_0$. The interaction (16b) can be written with (29,33) as

$$\boxed{H^\perp = \frac{J^\perp}{\sqrt{2}}\,(\tau_+ B_q^- + \tau_- B_q^+), \qquad q = 2 - J^\|,} \tag{51}$$

and it yields a change of τ_0 with the velocity $j = [\tau_0, H^\perp]$, where

$$j = \frac{J^\perp}{\sqrt{2}}\,(\tau_+ B_q^- - \tau_- B_q^+). \tag{52}$$

In the first Born approximation N will be proportional to $(J^\perp)^2$.

The Heisenberg equation for the Zubarev function (39a) reads

$$z\langle\langle A^*;B\rangle\rangle_z = \langle [A^*,B]\rangle - \langle\langle [A,H]^*;B\rangle\rangle_z$$

$$= \langle [A^*,B]\rangle - \langle\langle A^*;[B,H]^*\rangle\rangle_z, \qquad (53)$$

and hence

$$z\chi(z) = \langle\langle j^*;S_z\rangle\rangle_z$$

$$= (\langle [j^*,S_z]\rangle - \langle\langle j^*;j\rangle\rangle_z)z^{-1}.$$

For $z = 0$ we find $\langle [j^*,S_z]\rangle = \langle\langle j^*;j\rangle\rangle_{z=0}$ and so

$$z\chi(z) = (\langle\langle j^*;j\rangle\rangle_{z=0} - \langle\langle j^*;j\rangle\rangle_z)z^{-1}. \qquad (54a)$$

The RHS is proportional to $(J^\perp)^2$. On the other hand one gets from (48,41a) the expansion

$$z\chi(z) = \chi^0(z\phi(z) + 1) = N(z) + O(N^2). \qquad (54b)$$

Comparision of (54a) with (54b) and using (39c) yields in leading second order

$$N''(\omega) = \int dt\exp(i\omega t) \frac{\langle j^*(t)j - jj^*(t)\rangle}{2\omega}. \qquad (55a)$$

The correlation function can be calculated with $H^{||}$ and factorizes in the motion of the τ- and a-operators, and hence with (31) we find

$$N''(\omega) = \frac{J^{\perp 2}}{4\omega}\int dt\exp(i\omega t)$$

$$\times\{\exp(iBt)(\langle\tau_-\tau_+\rangle\langle B_q^+(t)B_q^-\rangle - \langle\tau_+\tau_-\rangle\langle B_q^-B_q^+(t)\rangle$$

$$+ \exp(-iBt)(\langle\tau_+\tau_-\rangle\langle B_q^-(t)B_q^+\rangle - \langle\tau_-\tau_+\rangle\langle B_q^+B_q^-(t)\rangle\}.$$

Specializing for zero field, equation (3a) yields $\langle\tau_-\tau_+\rangle = \langle\tau_+\tau_-\rangle = \frac{1}{2}$, and we get with notation (35a,b)

$$N''(\omega) = \frac{(J^\perp)^2}{8}\frac{H_q(\omega) - H_q(-\omega)}{\omega}; \qquad q = 2 - J^{||}. \qquad (56a)$$

With (36b) one finds

$$N''(\omega) = \left[\frac{\pi(\frac{1}{2}J^\perp)^2}{\Gamma(2 - 2\alpha)}\right]\left[\frac{D^2}{\omega^2}\right]^\alpha \; ; \quad \alpha = J^\parallel(1 - \tfrac{1}{4}J^\parallel). \qquad (56b)$$

If the temperature is non-zero, ω^2 has to be replaced by $\omega^2 +$ $(2\pi T)^2$. [10]. Formula (56b) is the relaxation spectrum taking into account the soft Tomonagon singularities as discussed in the preceding section.

Had we performed an expansion with respect to $J^\parallel$ we would have obtained in leading order

$$N^\parallel(\omega) = \pi(\tfrac{1}{2}J^\perp)^2. \qquad (57a)$$

(for all temperatures; put $J^\parallel = 0$ in (56b)). This is a white relaxation spectrum as discussed in equations (47). Hence Bloch's equation is valid, and using (45b,50b) we find the famous Korringa law for the spin relaxation in metals,

$$\frac{1}{T_1} = \pi J^{\perp 2} T. \qquad (57b)$$

Because of the Curie divergence of χ^0 the relaxation rate slows down linearly with decreasing temperature. Actually, one should calculate $\chi^0(T)$ from (49). Then one does not find a divergent χ^0 but the limiting value (50d). Since $J < 0.2$ this value is very large and (57b) is valid for all practical purposes. In next order (expand (56b)) one gets

$$N''(\omega) = \pi(\tfrac{1}{2}J^\perp)^2\left[1 + J^\parallel \log\left(\frac{D^2}{\omega^2}\right)\right] , \qquad (57c)$$

a meaningless result because of the logarithmic divergence. In perturbation theoretical language (57a) corresponds to a spin flip accompanied by the emission of one spin density wave, and (57c) is the Mott-Sommerfeld Bremsstrahlung correction due to the emission of two soft bosons. In section 2 we have seen that such an expansion is not possible, no process with only a finite number of Tomonagons involved is important. One has to calculate the coherent emission of Tomonagons to get the meaningful but singular result (56b).

Owing to the result of interference effects, $N''(\omega)$ — contrary to the example discussed in section 2 — depends critically on the sign of the exchange coupling $J^\parallel$. If we have ferromagnetic coupling ($J^\parallel < 0$) the interference is destructive and the relaxation rate is much smaller than the Korringa value (57a). We expect the

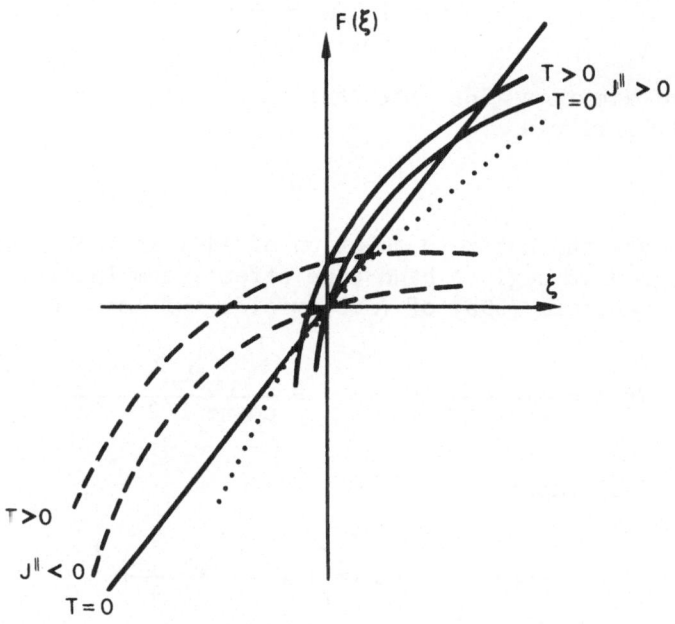

Figure 3

solution of (49) to bring out a χ^0 larger than (50d). If we have anti-ferromagnetic coupling ($J\|> 0$) the interference is constructive, and the relaxation rate $N''(\omega)$ exceeds the Korringa value by a very large margin. Hence we expect a χ^0 much smaller than (50d). To solve (49) one looks for the intersection of $F(\xi)$ vs ξ with the diagonal. The curves look as indicated in figure 3. $F(\omega)$ is convex and has two intersections; only the solutions $\xi \geqslant 0$ are allowed. For ferromagnetic coupling one finds indeed $\chi^0 = \infty$ as for a free spin. For $J\|> 0$ there are two solutions. To pick out the correct one we examine $T > 0$; then $F(\xi = 0) = 4T$ and the curves shift as indicated in figure 3. Hence $\xi(T = 0) = 0$ can not be obtained as zero temperature limit of a solution $\xi(T \to 0)$; only $\xi(T = 0) \neq 0$ is allowed. Keeping $J^{\perp}$ fixed but decreasing $J\|$, χ^0 increases. For a certain value one obtains the dotted curve in figure 3 having slope unity for $\xi = 0$. This $J\|$ marks the transition from the generalized antiferromagnetic behaviour ($\chi^0(T = 0) < \infty$) to the ferromagnetic behaviour ($\chi^0(T = 0) = \infty$). In the $J^{\perp}-J\|$ plane one gets a sort of phase diagram. From (44a, 48) we get for $T = 0$

$$F(\xi) = -\frac{2}{\pi} \, \mathrm{Im} \int d\omega \, \frac{\omega}{\omega + \xi N(\omega)} \, , \qquad (58)$$

and $\partial F/\partial \xi = 1$ for $\xi = 0$ yields

$$\frac{\pi}{4} = \int_0^\infty d\omega \, \frac{N''(\omega)}{\omega} \, . \tag{59a}$$

With approximation (56b) the phase separation curve (see figure 4) is the parabola

$$J^{||} = - 2J^{\perp^2} . \tag{59b}$$

To calculate the non-zero solution of (49) in the antiferromagnetic régime we neglect band edge effects completely. Then the Hilbert transform (39b) of (56b) yields for $\text{Im} z > 0$

$$N(z) = \frac{ia}{(-iz)^{2\alpha}} \; ; \qquad a = \frac{\pi(\tfrac{1}{2}J^{\perp})^2 D^{2\alpha}}{\cos\pi\alpha \, \Gamma(2 - 2\alpha)} \, . \tag{60a}$$

Equation (58) implies

$$F(\xi) = - \frac{2}{\pi} \int d\omega \, \text{Im}\left(1 + \frac{\xi a}{(-i\omega)^{2\alpha+1}}\right)^{-1}$$

$$= \frac{4}{\pi} \, \text{Im} \int_0^\infty d\omega \left(1 + \frac{(-i\omega)^{2\alpha+1}}{\xi a}\right)^{-1} = \qquad \text{(Contd)}$$

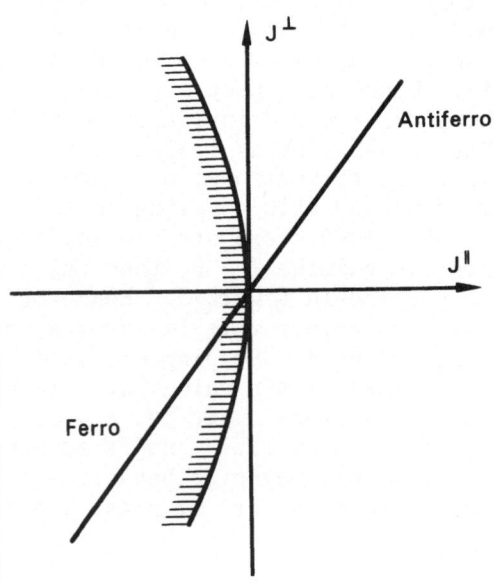

Figure 4

(Contd) $= \dfrac{4}{\pi} \, (\xi a)^{1/(1+2\alpha)} \displaystyle\int_0^\infty dx (1 + x^{2\alpha+1})^{-1}$

$$= \frac{4(\xi a)^{1/(1+2\alpha)}}{(1+2\alpha)\sin\!\left(\dfrac{\pi}{1+2\alpha}\right)}$$

$$\approx (\xi a)^{1/(1+2J^{\parallel})} \, \frac{2}{\pi J^{\parallel}} \quad \text{for } J^{\parallel} \ll 1. \tag{60b}$$

Equation (49) then gives the final approximate formula for the zero temperature susceptibility

$$\frac{1}{\chi^0(T=0)} = \frac{4D}{(1+2\alpha)\sin\!\left(\dfrac{\pi}{1+2\alpha}\right)}$$

$$\times \left[\frac{\pi J^{\perp 2}}{(1+2\alpha)\Gamma(2-2\alpha)\cos(\pi\alpha)\sin\!\left(\dfrac{\pi}{1+2\alpha}\right)}\right]^{1/2\alpha} \tag{60c}$$

$$\approx \frac{2D}{\pi J^{\parallel}}\left(\frac{J^{\perp 2}}{2J^{\parallel}}\right)^{1/2J^{\parallel}} .$$

This value depends very sensitively on $J^{\parallel}$. For $D \sim 2$ eV one gets $1/\chi^0 \approx 10$ mdeg if $J \approx 0.1$ or $1/\chi^0 \approx 10$ deg if $J \approx 0.2$.

For non-zero temperature (49) is solved graphically as indicated in figure 3. The result is sketched in the Curie plot of figure 5. The susceptibility for larger temperatures is given by a Curie-Weiss law with a renormalized Curie constant

$$\frac{1}{\chi^0} \approx 4.6 \ (T + \theta); \qquad T > \theta, \tag{61a}$$

at low temperature it bends to the finite value

$$\frac{1}{\chi^0} \approx 3.8 \ \theta, \qquad T < \frac{\theta}{5} . \tag{61b}$$

For ferromagnetic coupling χ^0 obeys a Curie law with a renormalized Curie constant. The relaxation rate — defined as half width of $\phi''(\omega)$ — does not slow down to zero as expected from the Korringa law (57b) since χ^0 does not diverge for antiferromagnetic coupling. It behaves as sketched in figure 6 and approaches the zero temperature limit

$$\frac{1}{T_1(T=0)} \approx 1.2 \ \theta. \tag{61c}$$

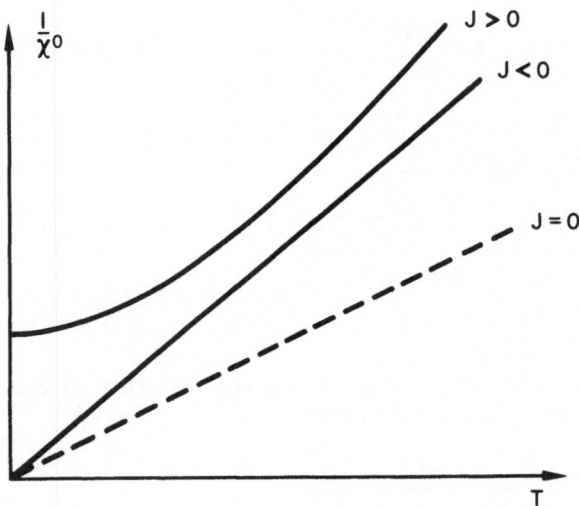

Figure 5

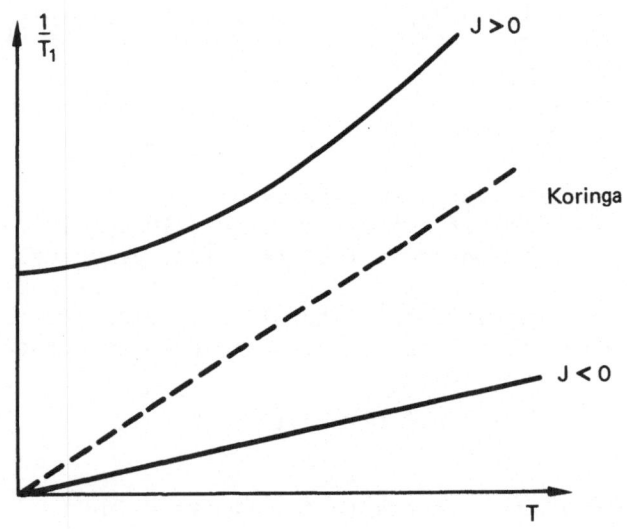

Figure 6

As a result we arrive at the following picture for a magnetic impurity coupled antiferromagnetically to the metal. Owing to the effective coupling of the spin to the soft excitations of the conduction electron gas, it flips around with a flipping spectrum extending continuously from 0 to θ. If one probes the impurity complex with energies (fields and temperatures) smaller than θ it looks like a non-magnetic system. Only if energies much larger than θ are applied can the bare impurity spin be pulled out of the local perturbation. For ferromagnetic coupling the impurity behaves like a slightly perturbed free spin. These conclusions agree with the experiments.

Anderson, Yuval and Hamann [9] in 1969 realized the importance of the soft boson singularity for the Kondo problem. For $J^{\parallel} = \frac{1}{2}$ they were able to calculate the static susceptibility and they gave qualitative arguments that for all $J^{\parallel} > 0$ χ^0 should behave as sketched in figure 5. The quantitative theory discussed above is due to Schlottmann and the author [10]. In the last reference one can find more details, in particular a discussion of the non-zero field problem and an analysis of the Cu:Fe data.

REFERENCES

1. Schrieffer, J.R. and Wolff, P.A. (1966). *Phys. Rev.*, **149**, 491.
2. Tomonaga, S. (1950). *Prog. Theor. Phys. (Kyoto)*, **5**, 544.
3. Schotte, H.D. and Schotte, U. (1969). *Phys. Rev.*, **182**, 479; (1970). *Z. Phys.*, **230**, 99.
4. Slichter, C.P. (1963). *Principles of Magnetic Resonance*, (Harper and Row, New York).
5. Bloch, F. and Nordsieck, A. (1937). *Phys. Rev.*, **52**, 54.
6. Nozières, P. and de Dominicis, C.I. (1969). *Phys. Rev.*, **178**, 1084.
7. Kondo, J. (1969). *Advances in Solid State Physics*, (Academic Press, New York), **23**, 183.
8. Zubarev, D.N. (1960). *Usp. Fiz. Nauk*, **71**, 71; Kadanoff, L.P. and Martin, P.C. (1963). *Ann. Phys. (N.Y.)*, **24**, 419.
9. Anderson, P.W. and Yuval, G. (1969). *Phys. Rev. Lett.*, **23**, 89; Anderson, P.W., Yuval, G. and Hamann, D.R. (1970). *Phys. Rev.*, **B1**, 4464.
10. Götze, W. and Schlottmann, P. (1974). *Solid State Commun.*, **13**, 17, 511, 861; *J. Low Temp. Phys.*, **16**, 87.

MAGNON INTERACTION IN THE ITINERANT ELECTRON MODEL OF FERROMAGNETISM

J. MORKOWSKI

*Ferromagnetics Laboratory, Institute of Physics
of the Polish Academy of Sciences, 60-179 Poznan,
Poland, ul. Smoluchowskiego 15*

1. INTRODUCTORY REMARKS

In the present lecture I shall present an elementary procedure which can be used to study various aspects of magnon-magnon interaction and also magnon-phonon interaction. No magnon-
electron interaction will be considered. The procedure can be
applied to calculate magnon relaxation times, to calculate the
correction to the spontaneous magnetization due to magnon interaction and, finally, we shall demonstrate how magnon interaction
leads to formation of a two-magnon bound state.

An interesting feature of the method described here is that
it reveals quite close analogies between properties of magnon
interaction in the itinerant as well as in the localized spin
models. It thus enables one to look into effects which are essentially model independent although a little bit different language is used in describing corresponding situations.

The method which I am going to present is based upon the Random Phase Approximation. It is well known that RPA is too crude
for calculating consistently even a relatively simple quantity
such as the magnon energy. However, a substantial amount of
theoretical work has been done recently on improving the calculations of the magnon energy in the long-wavelength limit beyond
RPA. There exist fairly reliable results, at least in the low
density limit, and we can use them to improve our treatment.
We shall comment upon this point more specifically later on in
the lecture.

2. MODEL ASSUMPTIONS

We consider the simplest possible model of itinerant electron
ferromagnetism, namely the Hubbard model [1]. It consists of a
single, narrow, partially occupied band. We consider the case
of strong ferromagnetism and assume that the spin down band is
occupied up to the Fermi energy ε_F and the spin up band is empty
in the ground state.

It is not obvious that the Hubbard model actually does have
a ferromagnetic ground state. The problem has been reviewed at
the present School by J.W. Schweitzer [2]. We may summarize the
discussion by stating that the present state of knowledge of the
subject justifies the assumption that for some region of the
$(W/I,n)$-plane (W/I is the ratio of the band width W to the intra-atomic Coulomb integral I, and n is the concentration of
itinerant electrons per atom) the paramagnetic state of the Hubbard model is unstable against formation of a ferromagnetic
ground state. It can also be argued that at a given W/I and n
the strong ferromagnetic state is more stable than the weak one
[3]†. Thus, there exists a region in the $(W/I,n)$-phase diagram,
for which the strong ferromagnetic state is the ground state of
the Hubbard Hamiltonian.

3. ELEMENTARY THEORY OF MAGNONS (RPA)

Let us write down the Hubbard Hamiltonian [1] in the usual
notation

$$\mathcal{H} = \sum_{k\sigma} \varepsilon_{k\sigma} a_{k\sigma}{}^{\dagger} a_{k\sigma} + \frac{I}{N} \sum_{kk'q} a_{k+q,+}{}^{\dagger} a_{k,q} a_{k'-q,-}{}^{\dagger} a_{k',-}, \quad (1)$$

and denote its ground state¶ by

$$|\phi_0\rangle = \prod_{\substack{k \\ \varepsilon_k < \varepsilon_F}} a_{k,-}{}^{\dagger}|0\rangle.$$

As is well known, the state $\beta_q{}^{\dagger}|\phi_0\rangle$, with $\beta_q{}^{\dagger}$ defined as [4,5]

$$\beta_q{}^{\dagger} = \sum_k b_{k+q,k} a_{k+q,+}{}^{\dagger} a_{k,-}, \quad (2)$$

is, within RPA, an eigenstate of the Hubbard Hamiltonian (1),
provided $b_{k+q,k}$ are given by

---

† See this reference, for example.
¶ Note, that in the strong ferromagnetic case the ground state
$|\phi_0\rangle$ is an *exact* ground state.

$$b_{k+q,k} = \frac{d_q}{\varepsilon_{k+q} - \varepsilon_k + \Delta - E_q} . \tag{3}$$

We take a constant magnetic field H along the positive direction of the z-axis ('+' means 'up' in our convention), thus $\varepsilon_{k\sigma} = \varepsilon_k + \sigma\mu_B H$, $\sigma = \pm 1$ and

$$\Delta = nI + 2\mu_B H. \tag{4}$$

d_q in equation (3) is a normalization constant and E_q is the excitation energy of the state $\beta_q^+|\sigma_0\rangle$, i.e. $\mathcal{H}\beta_q^+|\phi_0\rangle^{RPA} = \{(\text{ground state energy}) + E_q\}\beta_q^+|\phi_0\rangle$.

The state $\beta_q^+|\phi_0\rangle$ represents a propagating bound state of an electron and a hole of opposite spin, it corresponds thus, in a sense, to a propagation of a spin reversal or a change by unity of the total spin momentum of the system. Thus $\beta_q^+|\phi_0\rangle$ can be interpreted as a one magnon state, $\vec{q}$ is the magnon wave vector and E_q the magnon energy.

β_q^+, defined by (2), can be considered as the magnon creation operator. The operators β_q^+ for different q commute exactly, $[\beta_q^+, \beta_{q'}^+] = 0$, as is easy to see. When we calculate the commutators $[\beta_q, \beta_{q'}^+]$ we find that they are expressed in terms of the products $a_{k\sigma}^+ a_{k'\sigma}$. If we replace them by average values over the ground state, that means if we calculate the commutators within RPA, we find $[\beta_q, \beta_{q'}^+]_{RPA} = \delta_{qq'}$, provided the normalization constant d_q is suitably chosen. We determine d_q from the condition [5]

$$\sum_k |b_{k+q,k}|^2 n_k = 1, \tag{5}$$

where $n_k = 1$ for $\varepsilon_k < \varepsilon_F$, and $n_k = 0$ otherwise.

The magnon energy, in the RPA, is calculated from the equation

$$\frac{I}{N} \sum_k n_k(\varepsilon_{k+q} - \varepsilon_k + \Delta - E_q)^{-1} = 1. \tag{6}$$

In the long-wavelength (small q) limit we have for cubic crystals an approximate expression

$$E_q = 2\mu_B H + Dq^2 \tag{7a}$$

where

$$D = (6nN)^{-1} \sum_k n_k\left\{\nabla_k^2 \varepsilon_k - \frac{2}{nI} |\nabla_k \varepsilon_k|^2\right\} . \tag{7b}$$

The formula (7b) for the exchange stiffness constant D is actually a very crude one for applying to real itinerant electron ferromagnetics, we shall comment on it in the next section.

The operators $\beta_q{}^\dagger$ and their Hermitian adjoints β_q can be used to construct a theory of a system of interacting magnons which, from a formal point of view, has close analogies with the well developed theory of magnons for Heisenberg ferromagnets.

4. POSSIBLE IMPROVEMENTS OF THE THEORY BEYOND RPA

The improvements of the theory based on RPA can be introduced intuitively in at least two directions. First, in applying our theory to real systems we should replace the two crucial parameters, Δ and D, by appropriate values calculated by a more rigorous treatment. Second, we should introduce a more accurate definition of a magnon creation operator than by equation (2), perhaps by introducing terms of higher order with respect to the electron operators.

The exchange splitting parameter Δ given by equation (4) can be corrected by replacing the bare Coulomb integral I by an effective quantity, say I_{eff}, which somehow takes into account effects of many body interactions. Such an approach was proposed by Kanamori [6], who had shown that in the low-density limit we obtain roughly $I_{eff} = (1 + AIW^{-1})^{-1}I$, where A is a numerical constant of the order of magnitude of unity and W stands for the band width. Although the low density approximation is actually not accurate for electron densities allowing the existence of a ferromagnetic ground state on the $(W/I, n)$-phase diagram, Kanamori's arguments may still retain some value.

However, correcting Δ is not enough for calculating the magnon energy in better than RPA approximation. Much effort was spent upon calculating exactly the magnon energy in the asymptotic régime of small q vectors. The actual value of D in equation (7b) is a result of a competition between the first, positive term, which we may call the kinetic energy term and the second, which let us call the Coulomb term. The value of D depends on details of the band structure and is substantially modified by many-body effects. The leading many body corrections can be taken into account by replacing I in the formula (7b) by an effective, k-dependent interaction function, $I \rightarrow U(\vec{k})$. The interaction function was calculated in the low density approximation by Edwards [7] and, with band structure effects taken into account, by Callaway and co-workers [8-10]. Roughly speaking, the order of magnitude of the interaction function $U(\vec{k})$ is the same as the band width, as in Kanamori's theory. Roth [11] had shown by a variational procedure that $1/I$ in equation (7b) had to be replaced by $(I^{-1} + V^{-1})$, where

$$V \simeq \frac{2 - n}{n(1 - n)} \frac{1}{N} \sum_k \frac{(1 - n_k)}{(1 - n)} \varepsilon_k$$

is of the order of magnitude of the band width W. This result
also recovers the result of Kanamori, but is valid outside the
low density approximation.

Summing up what we have said, we can state that it may be not
too crude an approximation to start with formulae based on the
RPA and correct the final results by taking properly renormalized parameters Δ and D.

Let us return to the second above-mentioned possibility of
improving the theory by more accurate definition of the magnon
creation operators. It has been argued [12,11] that in order
to correct the one magnon state $\beta_q^\dagger |\phi_0\rangle$ for correlations neg-
lected in RPA the next approximation would be to add to $\beta_q^\dagger$ an
expression like

$$(\delta\beta)_q^\dagger = N^{-1} \sum_{kk'k''} A_{k,q} a_{k'+q,+}^\dagger {}^\dagger a_{k'+k''-k,-} a_{k'',-} {}^\dagger a_{k,-} . \qquad (8)$$

Roth [11] proposed a variational procedure for calculating $A_{k,q}$.
She had shown that in the first approximation $A_{k,q}$ can be taken
proportional to $(\vec{q}\cdot\nabla_k\varepsilon_k)$ in the long-wavelength limit. Thus, in
the limit of small q we would have a well defined expression for
$\beta_q^\dagger$ outside the RPA. However, taking $\beta_q^\dagger + (\delta\beta)_q^\dagger$ as the magnon
creation operator, although possible in principle, would lead
us to much more complicated calculations. Therefore, we decided
to ignore the correlation corrections and we will work within
the scheme of the RPA.

5. EFFECTIVE MAGNON HAMILTONIAN

We shall introduce an equivalent or effective Hamiltonian for
the system of magnons as an expansion in powers of the magnon
operators $\beta_q^\dagger$ and β_q.

We aim at formulating a simple formalism to treat excited
states of the Hamiltonian (1) which should be valid at small
number of electrons excited outside the ground state Fermi sea
$|\phi_0\rangle$. The simplest excited state, determined by the wave func-
tion $a_{k+q,\sigma}^\dagger a_{k,-}|\phi_0\rangle$, corresponds to the so-called Stoner elec-
tron-hole pair (the wave vectors $\vec{k}$ and $\vec{q}$ have to satisfy the
conditions $\varepsilon_k < \varepsilon_F$, $\varepsilon_{k+q} > \varepsilon_F$). The Stoner pairs without spin reversal ($\sigma = -1$) are of secondary importance to magnetic proper-
ties. They influence the temperature dependence of thermodynamic quantities and can be taken into account separately just
by introducing proper correction terms (e.g. they give a correction to the spontaneous magnetization, proportional to temperature squared [13], also a term proportional to T^2 appears in the
exchange stiffness constant D corrected for interaction of magnons with the Stoner pairs [14]).

The Stoner pairs with spin reversal, $a_{k+q,+}^\dagger a_{k,-}|\phi_0\rangle$ have the
excitation energy $\Delta + \varepsilon_{k+q} - \varepsilon_k$ which for small q is large compared with the corresponding magnon energy E_q, since $\Delta \gg E_q$ for
small q. Thus, at low temperature the population of free pairs

with spin reversal is much smaller then the population of magnons. These intuitive arguments make plausible the approach for
low-lying excited states which ignore free Stoner pairs. Therefore, it seems reasonable to interpret low energy excited states
in terms of magnons and to replace the Hubbard Hamiltonian by
an equivalent effective Hamiltonian of the system of magnons.
We may say that we divide the full Hamiltonian (1) into the Hamiltonian of free Stoner pairs, the effective Hamiltonian of magnons, and the residual magnon-electron and electron-electron interaction. This statement is formal, but it has a definite
meaning within the RPA. As we have explained above, we are interested here only in the effective magnon Hamiltonian.

We start with the RPA expressions for the magnon operators
$\beta_q^\dagger$ and their Hermitian adjoint β_q. We write down the formal
expansion of the effective magnon Hamiltonian $\mathcal{H}_e$ to determined,
in powers of operators $\beta_q^\dagger$, β_q. The form of the expansion is
restricted by the condition of hermiticity of $\mathcal{H}_e$ and by the requirement of conservation of the crystal momentum. The Hamiltonian (1) commutes with the total magnetic moment of itinerant
electrons $\sum\limits_{k\sigma} (-\mu_B)\sigma a_{k\sigma}^\dagger a_{k\sigma}$; thus the corresponding effective

Hamiltonian should conserve the number of magnons which is also
proportional to the total magnetic moment. Consequently, the
expression for the effective Hamiltonian should contain only
such products in which the numbers of creation and annihilation
operators are the same. The above remarks can be used to simplify the formal expression for the effective Hamiltonian. However, there is no ambiguity in the final effective Hamiltonian:
had we not made explicit use of the above general properties,
the coefficients of the spurious terms in the effective Hamiltonian would have ultimately vanished.

The most general effective magnon Hamiltonian corresponding
to (1), up to terms of the fourth order, and apart from a trivial constant, is

$$\mathcal{H}_e = \sum_q K_q \beta_q^\dagger \beta_q + \sum_{kk'q} \Gamma_{kk'q} \beta_{k+q}^\dagger \beta_{k'-q}^\dagger \beta_k \beta_{k'}. \qquad (9)$$

The expansion coefficients can be written as

$$K_q = [\beta_q, [\mathcal{H}_e, \beta_q^\dagger]], \qquad (10a)$$

$$\Gamma_{kk'q} = \tfrac{1}{4}[\beta_{k+q}, [\beta_{k'-q}, [[\mathcal{H}_e, \beta_{k'}^\dagger], \beta_k^\dagger]]]. \qquad (10b)$$

These relations are used to define the coefficients of the effective Hamiltonian $\mathcal{H}_e$, which is equivalent to the initial Hamiltonian $\mathcal{H}$ of the system of itinerant electrons, [5],

$$K_q = \langle \phi_0 | [\beta_q, [\mathcal{H}, \beta_q^\dagger]] | \phi_0 \rangle, \qquad (11a)$$

$$\Gamma_{kk'q} = \tfrac{1}{4}\langle\phi_0|[\beta_{k+q},[\beta_{k'-q},[[\mathcal{H},\beta_{k'}^\dagger],\beta_k^\dagger]]]|\phi_0\rangle. \qquad (11b)$$

In the last section some comments on the above procedure of calculating the coefficients of the effective Hamiltonian will be included.

Now we express the magnon operators β_q, $\beta_q^\dagger$ in terms of the electron operators $a_{k\sigma}$, $a_{k\sigma}^\dagger$ using (2) and, after calculating the ground state averages of products of electron operators, we obtain [5,16]

$$K_q = \sum_k (\varepsilon_{k+q} - \varepsilon_k + \Delta)|b_{k+q,k}|^2 n_k - \frac{N}{I}d_q^2 \qquad (12)$$

and

$$\Gamma_{kk'q} = \tfrac{1}{4}(\bar{\Gamma}_{kk'q} + \bar{\Gamma}_{kk'}{}^{k'-k-q} + \bar{\Gamma}_{k'k}{}^{-k'+k+q} + \bar{\Gamma}_{k'k}{}^{-q}), \qquad (13a)$$

where

$$\bar{\Gamma}_{kk'} = -\frac{I}{N}\sum_{pp'} b_{p+k',p}b_{p'+k+q,p'+q}$$

$$\times\, b_{p+k'-q,p}{}^\dagger b_{p'+k+q,p'}{}^\dagger n_p n_{p'}$$

$$+\frac{I}{N}\sum_{pp'} b_{p+k',p}b_{p+k+q,p+q}$$

$$\times\, b_{p'+k'-q,p'}{}^\dagger b_{p+k+q,p}{}^\dagger n_{p'} n_p. \qquad (13b)$$

Here $n_k = \langle\phi_0|a_{k-}^\dagger a_{k-}|\phi_0\rangle$ is equal to 1 for $\varepsilon_k < \varepsilon_k$, and 0 otherwise. If equations (3,5) are used, the right hand side of equation (12) reduces to E_q, which is the solution of equation (6). Thus K_q is equal to the magnon energy E_q, as it should be for consistency.

6. APPLICATIONS OF THE EFFECTIVE HAMILTONIAN

The method of the effective Hamiltonian can be used to study various aspects of magnon interaction, in particular we shall discuss magnon relaxation and bound states of two magnons.

It is also easy to generalize the effective Hamiltonian for magnetic systems of itinerant electrons determined by more complicated Hamiltonians than that of equation (1). In order to illustrate this possibility we take into account magnetic dipolar interaction between itinerant electrons in a single nar-

row band (this generalization is also relevant to the discussion
of magnon relaxation). Now the Hamiltonian of the system of
itinerant electrons $\mathcal{H}$ will be the sum of the right hand side of
equation (1) and the magnetic dipolar energy $\mathcal{H}_m$. The last contribution can be written as [15]

$$\mathcal{H}_m = N^{-1} \sum_{kk'q} \left[\sum_{\alpha,\beta} a_{k+q,\alpha}{}^\dagger \vec{\sigma}_{\alpha\beta} a_{k,\beta} \right] |F(\vec{q})|^2 \hat{D}_q$$

$$\times \left[\sum_{\alpha',\beta'} a_{k'-q,\alpha'}{}^\dagger \vec{\sigma}_{\alpha'\beta'} a_{k',\beta'} \right]. \qquad (14)$$

Here $F(\vec{q})$ is the magnetic form-factor

$$F(\vec{q}) = \int d\vec{r} |\phi(\vec{r})|^2 e^{i\vec{q}\cdot\vec{r}}, \qquad (15)$$

where $\phi(\vec{r})$ is the Wannier function. In deriving equation (14)
we assumed narrow band, i.e. well localized, Wannier functions
and we neglected 2-, 3- and 4-center integrals of the Wannier
functions. $\hat{D}_q$ is the usual dipolar dyadic

$$\hat{D}_q = \frac{N}{V} \int d\vec{r} e^{-i\vec{q}\cdot\vec{r}} \frac{1}{2} \frac{(2\mu_B)^2}{r^3} \left[1 - 3 \frac{\vec{r}\cdot\vec{r}}{r^2} \right], \qquad (16)$$

which can be easily calculated in the limit of small q; for results see, e.g., [5].
   The effective Hamiltonian will be now of the form

$$\mathcal{H}_e = \sum_q \{ K_q \beta_q{}^\dagger \beta_q + (L_q \beta_q \beta_{-q} + h.c.)$$

$$+ \sum_{qq'} (C_{qq'} \beta_{q+q'}{}^\dagger \beta_q \beta_{q'} + h.c.)$$

$$+ \sum_{kk'q} G_{kk'q} \beta_{k+q}{}^\dagger \beta_{k'-q}{}^\dagger \beta_k \beta_{k'}$$

$$+ \sum_{qq'q''} (F_{qq'q''} \beta_{q+q'+q''}{}^\dagger \beta_q \beta_{q'} \beta_{q''} + h.c.)\}, \qquad (17)$$

where the terms proportional to

$$L_q = \frac{1}{2} \langle \phi_0 | [[\mathcal{H}, \beta_q{}^\dagger], \beta_{-q}{}^\dagger] | \phi_0 \rangle, \qquad (18a)$$

$$C_{qq'} = \frac{1}{2} \langle \phi_0 | [\beta_{q+q'}, [[\mathcal{H}, \beta_q^\dagger], \beta_{q'}^\dagger]] | \phi_0 \rangle, \tag{18b}$$

$$F_{qq'q''} = \frac{1}{6} \langle \phi_0 | [\beta_{q+q'+q''}, [[[\mathcal{H}, \beta_{q''}^\dagger], \beta_{q'}^\dagger], \beta_q^\dagger]] | \phi_0 \rangle, \tag{18c}$$

are of dipolar origin. K_q is the sum of E_q and a dipolar correction, and similarly,

$$G_{kk'}{}^q = \Gamma_{kk'}{}^q + P_{kk'}{}^q, \tag{19}$$

where $\Gamma_{kk'}{}^q$ is given by equations (13), and $P_{kk'}{}^q$ represents a dipolar contribution. Explicit expressions for K_q, L_q, $C_{qq'}$, etc., in terms of sums of products of $b_{k+q,k}$ are given in [5, 16].

For any applications of the effective Hamiltonian (17) it is necessary to know explicitly its coefficients K_q, L_q, etc.. They depend on electronic band structure through relations (3). Expressions for K_q, L_q, etc., in terms of $b_{k+q,k}$, are rather complicated and analytical calculations can be practicably made only in the effective mass approximation for the electron energy, $\varepsilon_k = \hbar^2 k^2 / 2m^*$. For most purposes it is sufficient to know the coefficients of the effective Hamiltonian only in the limit of long wavelengths of magnon or small vectors (as compared with the lattice spacing). Thus in the present section we determine the effective Hamiltonian under the following assumptions: effective mass approximation for the electron energy, small magnon wave-vectors (the dipolar contribution will be given only up to terms of the order of magnitude of q^0, where q is the magnon wave vector).

The results are summarized as follows (for simplicity we assume that the ferromagnetic crystal is an ellipsoid of revolution along the z-axis, parallel to the magnetic field direction) [5,16,17]:

$$K_q = E_q - 2C\left[N_z - \frac{1}{3}\right] - 2C\left[\left(\frac{q_z}{q}\right)^2 - \frac{1}{3}\right] + O(Cq^2). \tag{20}$$

Here

$$C = 4\pi\mu_B{}^2 \frac{N}{V}, \tag{21}$$

N_z is the demagnetization factor for the z-direction, and E_q is given by the simplified expression (7a) with the exchange stiffness constant D calculated in accordance with the remarks made in section 4, or taken from experimental data.

$$L_q = -\frac{1}{2} C \frac{(q_+)^2}{q^2} + O(Cq^2), \qquad (q_\pm = q_x \pm iq_y); \tag{22}$$

$$C_{qq'} = - C(nN)^{\frac{1}{2}} \left[\left[\frac{q_z q_+}{q^2} \right] + \left[\frac{q_2' q_+'}{q'^2} \right] \right] + O(Cq^2, Cq'^2); \qquad (23)$$

(In the case of three-magnon relaxation processes small correction terms $O(Cq2, Cq'2)$ in $C_{qq'}$ can be of some importance, see [17] for a more accurate expression).

$$F_{qq'q''} = - \frac{1}{6N} C(f_q + f_{q'} + f_{q''}), \qquad (24a)$$

$$f_q = \left[\frac{q_-^2}{q^2} \right] + O(q^2); \qquad (24b)$$

$$P_{kk'q} = - \frac{1}{12N} C(4d_q + 4d_{k'-k-q} + d_{k+q} + d_{k'-q} + d_k + d_{k'}) \qquad (25a)$$

$$d_q = 1 - 3 \left[\frac{q_z}{q} \right]^2 + O(q^2). \qquad (25b)$$

The magnetic form factor $F(\vec{q})$ in equations (20-25) has been approximately replaced by unity.

$$\Gamma_{kk'q} = \frac{4}{5} IN^{-1} \eta^2 k_F^{-2}$$

$$\times \left\{ \vec{k} \cdot \vec{k}' + \chi \left[\frac{1}{2} \left(1 - \frac{6}{7} \eta \right) k^2 k'^2 - \frac{1}{2} \left(1 - \frac{66}{35} \eta \right) (\vec{k} \cdot \vec{k}')^2 \right. \right.$$

$$+ \left(1 - \frac{17}{35} \eta \right) (k'^2 (\vec{k} \cdot \vec{q}) - k^2 (\vec{k}' \cdot \vec{q}) - 2(\vec{k} \cdot \vec{q})(\vec{k}' \cdot \vec{q}))$$

$$+ \left(1 + \frac{1}{35} \eta \right) (\vec{k}' \cdot \vec{q} - \vec{k}\,\vec{q})(\vec{k} \cdot k') - \frac{18}{35} \eta q^2 (\vec{k} \cdot \vec{k}')$$

$$\left. \left. + \frac{1}{2} \left(\frac{\alpha k_F^2}{\varepsilon_F} - 1 + \frac{36}{35} \eta \right) (k^2 + k'^2)(\vec{k} \cdot \vec{k}') \right] \right\}, \qquad (26)$$

where

$$\chi = \frac{\hbar^2}{(m^* \Delta)} = 2 \left(\frac{\varepsilon_F}{\Delta} \right) k_F^{-2}, \qquad \eta = \frac{\varepsilon_F}{\Delta}.$$

Equations (20-26) determine the effective Hamiltonian completely. An important property of the effective Hamiltonian (17, 20-26) is its equivalence to the magnon Hamiltonian for the Heisenberg ferromagnets. It means that: (1) the form of the ef-

fective Hamiltonian (17) is the same as the magnon Hamiltonian
for the Heisenberg ferromagnets; (2) the coefficients of both
Hamiltonians are in first approximation the same functions of
magnon wave vectors.

The last property has to be formulated in more detail. The
terms of dipolar origin, if calculated in lowest order approximation with respect to magnon wave vectors, are the same in our
itinerant electron model as in the localized electron one. The
other terms, due to the intra-atomic Coulomb interaction I, are
in the lowest order the same functions of magnon wave vectors in
both models, only the proportionality factors being different.
In particular, the exchange stiffness constant D, equation (7b),
in our model depends on the Coulomb integral I, details of the
band shape and many body corrections (see section 4), whereas
in the localized electrons model D is simply proportional to the
Heisenberg exchange integral [18,19]. Similarly, the matrix
element for 2-body interaction of electrostatic origin, $\Gamma_{kk'q}$
is, in the first approximation, proportional to $\vec{k} \cdot \vec{k}'$, as is also
the case for the Heisenberg ferromagnets [18], although the proportionality factors are different in each case. It should be
mentioned here, that among higher order terms (i.e. terms of the
fourth order with respect to magnon wave vectors) in the right
hand side of equation (26) there appear spurious ones, which
lead to some unphysical consequences (see [16b] for details).
These spurious terms result from approximations used in our calculations, presumably the use of the RPA is responsible for
their presence.

As we said above, the effective magnon Hamiltonian in our
itinerant electron model is virtually the same as for the Heisenberg model. Thus properties of magnons are, in a sense, independent of the model of ferromagnetism, at least as far as
magnons of long wavelengths are concerned. Therefore, physical
consequences of the effective Hamiltonian (17) will correspond
to the well known results of the spin wave theory for Heisenberg ferromagnets (see, e.g. [19]). In the following we shall
briefly comment on some effects.

DIPOLAR CORRECTIONS TO THE MAGNON ENERGY

The bilinear terms in the effective Hamiltonian (17) can be
diagonalized using the well-known Holstein-Primakoff transformation [20]. Magnon energy, corrected for dipolar contributions,
will be given by the formula

$$\tilde{E}_q = (K_q{}^2 - 2|L_q|^2)^{\frac{1}{2}}, \tag{27}$$

which upon inserting (20,22) gives in the first approximation
the same expression as that known for the Heisenberg ferromagnets [20]. Differences between itinerant and localized electrons
model appear as small second order corrections which are negligible in most cases, see [21].

MAGNON RELAXATION

Relaxation processes to itinerant electron ferromagnets due to magnon-magnon interaction will be essentially the same as in the Heisenberg model, because of the equivalence of the magnon Hamiltonians. Thus, instead of a systematic discussion of the subject we shall quote here only some more important results (for details see [16,17]). Excellent reviews of the relaxation theory for localized electron ferromagnets can be found in [19, 22]. Many aspects of the theory described in [19,22] retain its validity for itinerant electron ferromagnets, provided the parameters of the theory are properly changed.

(a) Relaxation of Uniform Magnons

Relaxation of uniform magnons ($\vec{q} = 0$) is due to dipolar interaction. Dominant processes determining relaxation of $q = 0$ magnons are the four-magnon scattering processes ('2,2') and four four-magnon confluence processes ('3,1'), as depicted in figures 1 (c,d). Lower order processes, i.e. three-magnon ones are forbidden for $q = 0$ by the energy conservation condition. The inverse relaxation time of the uniform magnon due to four-magnon scattering ('2,2') and confluence ('3,1') processes in approximately proportional to temperature squared [16]. More accurately,

$$\frac{1}{\tau_0(2,2)} \sim T^2 + f_{2,2}\left(\frac{2\mu_B H}{k_B T}\right) \,,$$

$$\text{(28a)}$$

$$\frac{1}{\tau_0(3,1)} \sim T^2 + f_{3,1}\left(\frac{2\mu_B H}{k_B T}\right) \,,$$

where the corrections $f(2\mu_B H/k_B T)$ are negligible in the usual temperature range $2\mu_B H/k_B T \ll 1$. The total relaxation time τ_0 is given by

$$\frac{1}{\tau_0} = \frac{1}{\tau_0(2,2)} + \frac{1}{\tau_0(3,1)}. \qquad \text{(28b)}$$

(b) Relaxation of Non-Uniform Magnons

Non-uniform ($q \neq 0$) magnons relax by all three- and four-magnon processes of dipolar origin, as presented in figure 1(a-d), and by four-magnon '2,2' scattering processes (figure 1(c)) of electrostatic origin. For very small wave vectors q the total inverse relaxation time is dominated by processes coming from the dipolar interaction, whereas for large q the relaxation time is practically determined by the four-magnon '2,2' scattering

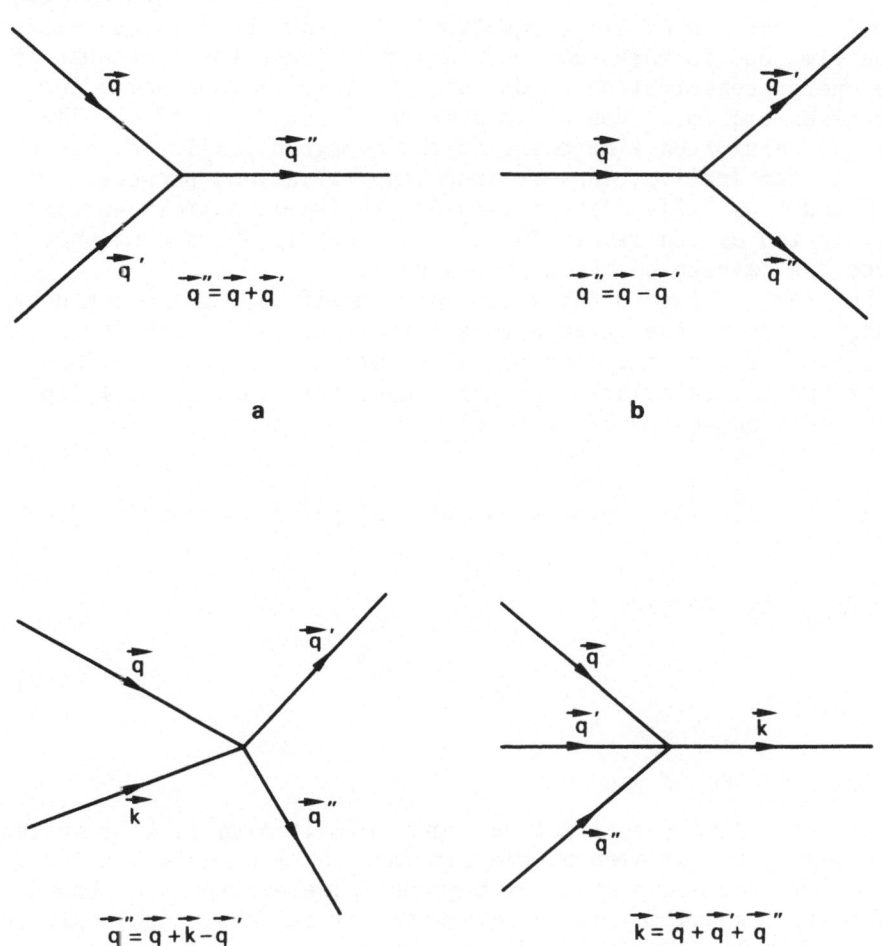

Figure 1 - Scattering processes contributing to magnon relaxation. (a) Three-magnon confluence processes '2,1';
(b) Three-magnon splitting processes '1,2'; (c) Four-magnon scattering processes '2,2'; (d) Four-magnon confluence
processes '3,1'.

processes due to electrostatic interaction, represented by the
term $\Gamma_{kk'}{}^{q}$ in the effective Hamiltonian (17).

For magnons with very small wave vectors q, leading contributions to the inverse relaxation times of the processes exhibited
in figures 1(a-d) are: $1/\tau_q{}^{(2,2)} \simeq 1/\tau_0{}^{(2,2)}$,

$$\frac{1}{\tau_q(3,1)} \simeq \frac{1}{\tau_0(3,1)} \; , \qquad \frac{1}{\tau_q(2,1)} \sim qT, \qquad \frac{1}{\tau_q(1,2)} \simeq 0.$$

Thus, for small enough q the inverse relaxation time $1/\tau_q$ is equal to the sum of $1/\tau_0$, equation (28), and the inverse relaxation time due to three-magnon confluence processes. Because of the energy conservation condition, the three-magnon splitting processes are forbidden up to some specified value of q. The inverse relaxation time owing to three-magnon confluence processes, is for small q, in very crude approximation, proportional to T and to q [17]. As q increases, the whole pattern becomes complicated as contributions to $1/\tau_q$ from all above mentioned processes have to be taken into account.

For large enough q the situation simplifies, as the dominant contribution to the inverse relaxation time $1/\tau_q$ is the term $1/\tau_q^{(2,2)e}$, due to the electrostatic interactions $\Gamma_{kk'}^q$. This contribution, calculated for the temperature range satisfying the condition $Dq^2 \ll k_B T$, is [16]

$$\frac{1}{\tau_q(2,2)e} \sim T^2 q^4 \left\{ \ln^2\left(\frac{Dq^2}{k_B T}\right) + \text{constant} \times \ln\left(\frac{Dq^2}{k_B T}\right) + \text{constant} \right\} . \quad (29a)$$

For $Dq^2 \gg k_B T$ we have [16]

$$\frac{1}{\tau_q(2,2)e} \sim T^{5/2} q^3 . \quad (29b)$$

7. BOUND STATES OF MAGNONS

The method of the effective magnon Hamiltonian is well suited to studying the problem of the interaction of magnons leading to formation of a bound state of magnons. The existence of bound states of magnons in the ferromagnetic linear chain of localized spins was proved by Bethe [23] many years ago. Later it was shown [24] that magnon bound states exist also in two- and three-dimensional Heisenberg ferromagnets, as well as in antiferromagnets. Recently some experimental evidence has become available in favour of the existence of magnon bound states [25].

We shall consider bound states of two magnons in our model of strong itinerant ferromagnet. As a magnon is equivalent to an electron-hole bound state, two-magnon states have to be constructed from two-electron and two-hole states. Thus we are faced with a four-body problem, which is quite involved even if RPA were to be adopted. However, within the framework of the method of the effective Hamiltonian, the two-magnon bound state problem is just a two-body problem, as in Heisenberg ferromagnets, and can be solved relatively easily by known methods (see, e.g. [24c]).

Let us consider two magnons having wave vectors $\frac{1}{2}K + k$ and $\frac{1}{2}K - k$. The total wave vector K is a constant of the motion and can be used as a quantum number for a general two-magnon state which we define as

$$|K\rangle = \sum_k g^0_k \beta_{\frac{1}{2}K+k}^\dagger \beta_{\frac{1}{2}K-k}^\dagger |\phi_0\rangle \quad (30)$$

We solve Schrödinger equation $\mathcal{H}_e|K\rangle = E|K\rangle$ with $\mathcal{H}_e$ given by equation (9) (we neglect the dipolar terms in the present problem). The Schrödinger equation is equivalent to the following integral equation for the amplitudes g_k

$$g_k(E_{\frac{1}{2}K+k} + E_{\frac{1}{2}K-k} - E) = \sum_q 2g_q \Gamma_{\frac{1}{2}K+q,\frac{1}{2}K-q}^{k-q}. \qquad (31)$$

If the interaction between magnons were to be ignored, i.e. $\Gamma_{kk'}^q$ replaced by zero, the trivial solutions of equation (31) would be δ-functions for the amplitudes g_k and $(E_{\frac{1}{2}K+k} + E_{\frac{1}{2}K-k})$ for the energy eigenvalues, corresponding to two non-interacting magnons. A solution of equations (31) for an energy eigenvalue $E = E(K)$ lying below the band of free two-magnon states ($E(K) < E_{\frac{1}{2}K+k} + E_{\frac{1}{2}K-k}$, for any k), if it exists, corresponds to a two-magnon bound state.

Equations (31) have been analysed numerically for a one-dimensional case, in the tight binding approximation for the electron energy [26]. The magnon energies were calculated from the equation (6), taking $\varepsilon_k = $ constant $- A \cos ak$. Then the coefficients $\Gamma_{kk'}^q$ of the effective Hamiltonian were computed from equations (13). Finally, the determinant of the system of linear equations (31) for fixed K is computed for a range of values of E, and a value $E = E(K)$ for which the determinant vanishes gives the energy of the bound state. The numerical analysis reveals that bound states do exist. The typical energy spectrum of the two-magnon bound states, taken from [26], is exhibited in figure 2. The following values of parameters were chosen for computations: $A = 0.04$ eV, $\Delta = 0.57$ eV, and the Fermi momentum $k_F = 1.9$ a^{-1}.

8. EFFECTIVE HAMILTONIAN FOR MAGNON-PHONON INTERACTION

The method of the effective Hamiltonian can be extended in a straight forward way to problems involving interaction of magnons with phonons. Let us consider an interaction between itinerant magnetic electrons and phonons of the usual form

$$\mathcal{H}' = \sum_{kq\mu\sigma} v_\mu(\vec{q})(\xi_{q\mu} + \xi_{-q,\mu}^\dagger)a_{k+q,\sigma}^\dagger a_{k\sigma} \qquad (32)$$

$\xi_{qu}^\dagger$ is the creation operator for a phonon of wave vector $\vec{q}$ and polarization μ, $v_\mu(\vec{q})$ is the coupling matrix element. Since in our model the magnon is a bound state of an electron and a hole, the interaction $\mathcal{H}'$ will result in some effective interaction between magnons and phonons.

It is easy to show that in our approximation scheme, based on RPA, the effective Hamiltonian for magnon-phonon interaction will be

$$\mathcal{H}_{m-mp} = \sum_{qq'\mu} C^\mu_{qq'}(\xi_{q'\mu} + \xi_{-q',\mu}^\dagger)\beta_{q+q'}^\dagger\beta \qquad (33)$$

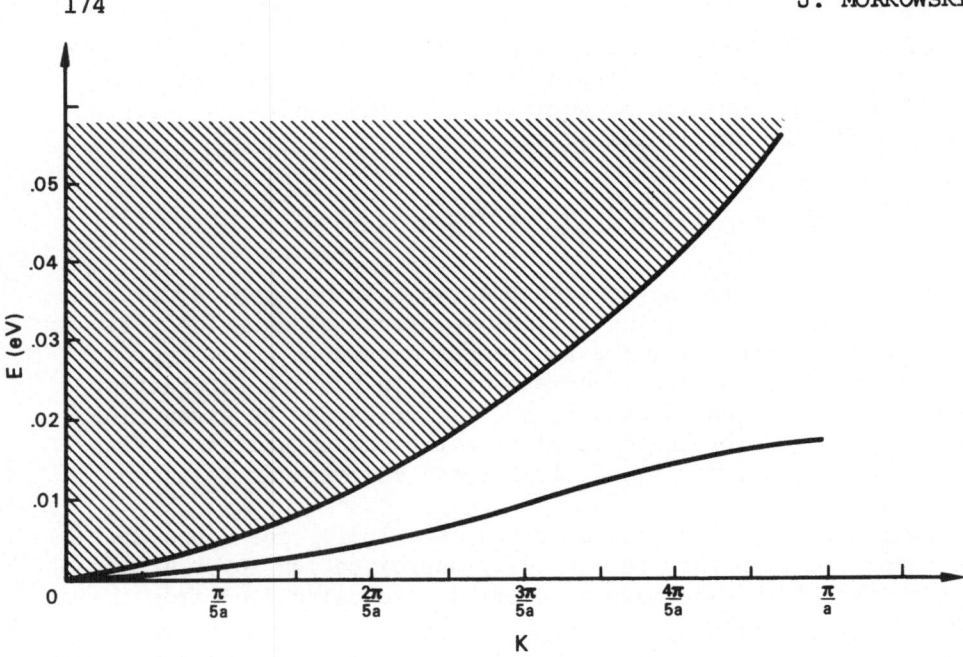

Figure 2 - The energy spectrum of the two-magnon bound state. The shaded area represents the quasi-continuum of free two-magnon states, its lower boundary is given by $\min (E_{\frac{1}{2}K+k} + E_{\frac{1}{2}K-k})$.

where

$$C^{\mu}{}_{qq'} = \langle \phi_0 | [[\beta_q{}^{\dagger}, [\xi_{q'\mu}{}^{\dagger} \mathcal{H}']], \beta_{q+q'}] | \phi_0 \rangle . \qquad (34)$$

Straightforward calculations give [27]

$$C^{\mu}{}_{qq'} = v_{\mu}(\vec{q}') \sum_k (b_{k+q,k} - b_{k+q+q',k+q'}) b_{k+q+q'}{}^{*}{}_{,k} n_k . \qquad (35)$$

The magnon-phonon effective Hamiltonian can be used, for instance, to calculate the magnon relaxation time owing to (magnon) → (magnon) + (phonon) scattering processes.

Assuming $v_{\mu}(\vec{q}) \sim q$, taking the Debye model and effective mass approximation for the Bloch energy of itinerant electrons, we obtain for $q \ll \hbar c_s/D$ [27]

$$\frac{1}{\tau_q{}^{(\text{m-ph})}} \sim q^4 \, \frac{\theta}{T} \, \frac{e^{\theta/T}}{(e^{\theta/T} - 1)^2} \qquad (36)$$

where $\theta = (\hbar c_s)^2/Dk_B$ and c_s is the velocity of sound. For $\theta \ll T$, $1/\tau_q{}^{(\text{m-mph})} \sim q^4 T$.

The result (36), taken from [27], is in disagreement with more recent calculations by a different method [28]. Because of different approximations adopted in [27,28] it is not easy to ascertain which of these results is accurate. An objection against the method used in [28] may be raised, in that it leads to finite relaxation time at $T = 0°K$; the physical origin of this result is difficult to understand.

9. CONCLUDING REMARKS

In the method of the effective magnon Hamiltonian, described in the present lecture, the system of itinerant electrons, which are fermions, has for some physical problems been replaced by a system of magnons, which are bosons. Our arguments have been based on RPA, and this approximation allowed us to calculate all quantities we needed in a systematic and simple way. We used an approximate correspondence between bosons and pairs of fermions. However, in establishing exact one-to-one correspondence between pairs of fermions and bosons it is necessary to take the Pauli principle properly into account. In our RPA treatment such kinematical effects have been neglected. The theory can be generalized using exact methods of transforming a Hamiltonian of fermions into a boson representation (see, for example, [29]).

REFERENCES

1. Hubbard, J. (1963). *Proc. Roy. Soc.*, **A277**, 237; (1964). P *Proc. Roy Soc.*, **A281**, 407.
2. Schweitzer, J.W. (1975). See this volume, p. 91.
3. See, e.g.: Nagaoka, Y. (1966). *Phys. Rev.*, **147**, 392.
4. Izuyama, T. (1960). *Prog. Theor. Phys.*, **23**, 969.
5. Morkowski, J. (1973). *Acta Phys. Polon.*, **A43**, 809.
6. Kanamori, J. (1963). *Prog. Theor. Phys.*, **30**, 275.
7. Edwards, D.M. (1967). *Proc. Roy. Soc.*, **A300**, 373; see also: Edwards, D.M. (1968). *J. Appl. Phys.*, **39**, 481; Edwards, D.M. and Fisher, B. (1971). *J. Physique, Coll. Suppl.*, **32**, C1-C697.
8. Callaway, J. (1968). *Phys. Rev.*, **170**, 576; (1969). *J. Appl. Phys.*, **40**, 110.
9. Young, W. and Callaway, J. (1970). *J. Phys. Chem. Solids*, **31**, 855; Young, W. (1970). *Phys. Rev.*, **B2**, 167.
10. Callaway, J. and Zhang, H.M. (1968). *Phys. Lett.*, **28A**, 2921; (1970). *Phys. Rev.*, **B1**, 305.
11. Roth, L.M. (1967). *J. Phys. Chem. Solids*, **28**, 1549; (1968). *J. Appl. Phys.*, **39**, 474.
12. Herring, C. (1966). Exchange Interaction Among Itinerant Electrons, in *Magnetism, Vol. IV*, (eds. Rado, G. and Suhl, H.).
13. Thompson, E.D., Wohlfarth, E.P. and Bryan, A.C. (1964).

Proc. Phys. Soc., **83**, 59.

14. Izuyama, T. and Kubo, R. (1964). *J. Appl. Phys. Suppl.*, **35**, 1074S.
15. Morkowski, J. (1968). *J. Appl. Phys.*, **39**, 476.
16. Morkowski, J., Król, Z. and Krompiewski, S. (*a*) (1972). *Phys. Lett.*, **39A**, 19; (*b*) (1973). *Acta Phys. Polon.*, **A43**, 817.
17. Morkowski, J. (1971). *J. Physique, Coll. Suppl.*, **32**, C1-C816.
18. Dyson, F.J. (1956). *Phys. Rev.*, **102**, 1217, 1230.
19. Keffer, F. (1966). Spin Waves, in *Encyclopedia of Physics, Vol. XVIII/2*, (ed. Flügge, S.), (Springer-Verlag).
20. Holstein, T. and Primakoff, H. (1940). *Phys. Rev.*, **58**, 1098.
21. Morkowski, J. (1967). *Acta Phys. Polon.*, **32**, 147.
22. Sparks, M. (1964). *Ferromagnetic Relaxation Theory*, (McGraw-Hill, London and New York).
23. Bethe, H.A. (1931). *Z. Phys.*, **71**, 205.
24. (*a*) Wortis, M. (1963). *Phys. Rev.*, **132**, 85; (*b*) Hanus, J. (1963). *Phys. Rev. Lett.*, **11**, 336; (*c*) Oguchi, T. (1971). *J. Phys. Soc. Jap.*, **31**, 394.
25. Torrance, J.B. Jr. and Tinkham, M. (1969). *Phys. Rev.*, **187**, 587, 595; see also: Elliott, R.J. and Smith, A.M. (1971). *J. Physique, Coll. Suppl.*, **32**, C1-C585.
26. Krompiewski, S., Morkowski, J. and Jezierski, A. (1974). *Phys. Lett.*, (in the press).
27. Morkowski, J. (1966). *Phys. Lett.*, **21**, 146.
28. George, P.K. (1970). *Physica*, **49**, 278.
29. Usui, T. (1960). *Prog. Theor. Phys.*, **23**, 787; Lam, J. (1972). *Kond. Mat.*, **15**, 46.

STUDIES OF FERRO- AND PARAMAGNETIC NICKEL
BY POSITRON ANNIHILATION[†]

HENRYK STACHOWIAK

Institute for Low Temperatures and Structure Research,
Polish Academy of Sciences, Wrocław, Poland

INTRODUCTION

This is an account of a work still unfinished concerning the investigation of the electronic structure of ferro- and paramagnetic nickel as well as the course of the Curie transition by means of positron annihilation [1]. The experimental curves for angular correlation of annihilation quanta in long slit geometry were obtained at the Technical University of Denmark. The interpretation goes on as a collaboration of the Institute of Experimental Physics of the University of Wrocław and the Institute for Low Temperatures and Structure Research of the Polish Academy of Sciences.

1. POSITRON ANNIHILATION IN METALS

This subject has been discussed in more detail in two previous lectures in Karpacz [2,3]. Here let us recall just a few basic facts:

A positron in a metal interacts strongly with the conduction electrons because of Coulomb attraction. So, in principle, we should describe the positron and the annihilating electron using the two-body wave function $\varphi_{\vec{k}}(\vec{r}_p, \vec{r}_e)$, where $\vec{k}$ is the electron momentum at large distance from the positron and $\vec{r}_p$, $\vec{r}_e$ are the positron and electron coordinates respectively. Then the amplitude $A(\vec{\ell})$ of electron-positron annihilation with emission of two photons with total momentum $\hbar\vec{\ell}$ is given by the formula

---

[†] This is not an original contribution, but an early account of a work still under way.

$$A(\vec{\ell}) \sim \int \varphi_{\vec{k}}^*(\vec{r},\vec{r}) e^{-i\vec{\ell}\cdot\vec{r}} d\vec{r}. \tag{1}$$

In the approximation where we neglect the k-dependence of $\varphi_{\vec{k}}^*(\vec{r},\vec{r})$ we get a good description of angular correlation results (but not life time measurements) assuming that

$$\varphi_{\vec{k}}^*(\vec{r},\vec{r}) = \varphi(\vec{r})\psi_{\vec{k}}^*(\vec{r}), \tag{2}$$

where $\varphi(\vec{r})$ is the positron ground state wave function, and $\psi_{\vec{k}}^*$ is the electron wave function. This approximation is not exact [3-6]. But it can be improved by introducing a k-dependent coefficient in the right hand side of (2). The positron is assumed to be in its ground state, since its lifetime is long enough, so it loses all its initial kinetic energy. One says that it is thermalized.

The electron density in k-space as seen by positron annihilation is given by the formula

$$\rho(\vec{k}) = \sum_{\substack{\ell \\ occupied}} \left| \int \varphi(\vec{r})\psi_{\vec{k}}^*(\vec{r}) e^{-i\vec{k}\cdot\vec{r}} d\vec{r} \right|^2. \tag{3}$$

But, unfortunately, $\rho(\vec{k})$ is not directly measured in the experiment. Using the so called long slit geometry we measure rather the quantity

$$N(k_z) = \int dk_x dk_y \rho(\vec{k}), \tag{4}$$

where the z-axis is connected with the apparatus. We used the word 'rather' because the finite resolution of the equipment introduces a smearing in the curves measured.

The samples investigated have the form of plates with the z-axis perpendicular to them. In order to study the anisotropy of the electronic structure we must investigate a set of single crystal plates cut in different crystallographical directions. Then $\rho(\vec{k})$ can be extracted out of the experimental curves $N(k)$ using Mijnarends equations [7].

$\rho(\vec{k})$ is obtained in the form of a series in lattice harmonics for the trivial representation:

$$\rho(\vec{k}) = \sum_{\ell,v} \rho_{\ell v}(\vec{k}) F_{\ell v}(\Omega), \tag{5}$$

where $F_{\ell v}(\Omega)$ are the suitable lattice harmonics, ℓ indicates the spherical harmonics out of which $F_{\ell v}$ is built, v is an additional index used in the case when several independent lattice harmonics correspond to the same ℓ, Ω gives the orientation of $\vec{k}$ with respect to the crystal axes.

The experimental curves $N_{\beta\alpha}(k_z)$ are measured for a set of polar and azimuthal angles β and α of the z-axis in a coordinate system connected with the crystal axes of the metal. Then we obtain the auxiliary functions $g_{\ell v}(k_z)$ from the set of equations

$$N_{\beta\alpha}(K_z) = \sum_{\ell,v} F_{\ell v}(\beta,\alpha) g_{\ell v}(k_z). \qquad (6)$$

Here in practice we must resign ourselves to a very crude approximation, allowing only for as many lattice harmonics as there are experimental curves at our disposal. Now the functions $\rho_{\ell v}$ in (5) are determined from Mijnarends' equation

$$\rho_\ell(p) = -\frac{1}{p}\left[\frac{dg_\ell(p)}{dp} - \frac{\ell(\ell+1)}{2p} g_\ell(p)\right.$$

$$\left. + \frac{1}{p^2}\int_0^p g_\ell(z)P_\ell''(zp^{-1})dz\right], \qquad (7)$$

where P'' denotes the second derivative of the Legendre function with respect to its argument.

2. EXPERIMENTAL RESULTS

Angular correlation curves have been obtained for samples of nickel oriented in the (100), (110) and (111) directions. Measurements have been performed at room temperature but also at 330°C and 380°C (below and above the Curie temperature equal to 361°C). The curves show a marked anisotropy (figure 1). Because of the finite number of counts registered, the experimental points show some dispersion (figure 2).

3. THE ELECTRONIC STRUCTURE OF FERROMAGNETIC NICKEL

The electronic structure of ferromagnetic nickel has attracted considerable interest. The total number of valence electrons is ten. In a free atom these electrons are distributed between $3d$ and $4s$ states. In the solid these states have only a conventional meaning, since the orbital quantum number is no longer conserved. Nevertheless one can expect that most electronic states are mainly of s character or of d character, hybridization being essential for only a limited part of the conduction electrons. A more sophisticated approach to how separate bands are seen by positrons can be found in [8].

Since there is no spin degeneracy in ferromagnetic nickel, the Fermi surface is different for spin up and spin down. The crystal lattice for nickel is fcc. The Brillouin zone is shown in figure 3 together with indices for symmetry points. For

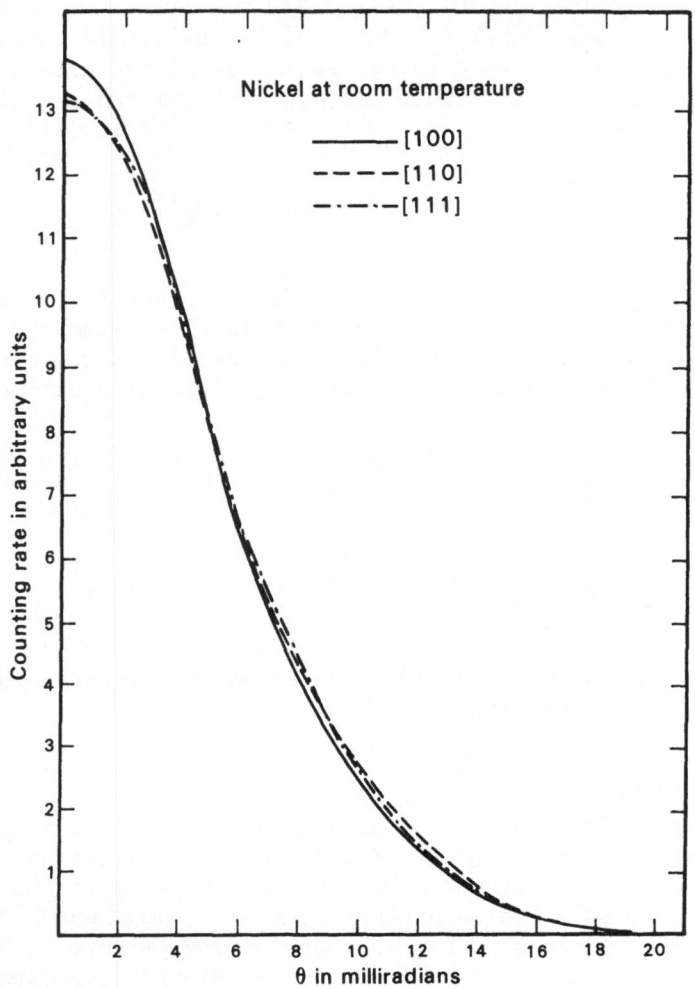

Figure 1 - Angular correlation curves for nickel at room
temperature. The momentum is shown in milliradians, the
number of counts in arbitrary units.

majority spin the Fermi surface exists only in the sixth zone
where it has an s character (figure 4) and is similar to that
of copper but considerably smaller. The lower zones are completely filled. For minority spin the sixth zone FS is closed,
but there exists also a closed FS in the fifth zone which has
a d character and a hole in the fourth zone around X. Alto-
gether, neglecting the holes in the fourth and eventually in
the third zone, we have approximately 1.4 electrons per atom in
the fifth zone and 0.6 in the sixth zone (compared to unity in
copper). So the minority spin fifth zone FS is a little bigger
than the sixth zone FS.

One should expect that when the temperature approaches the

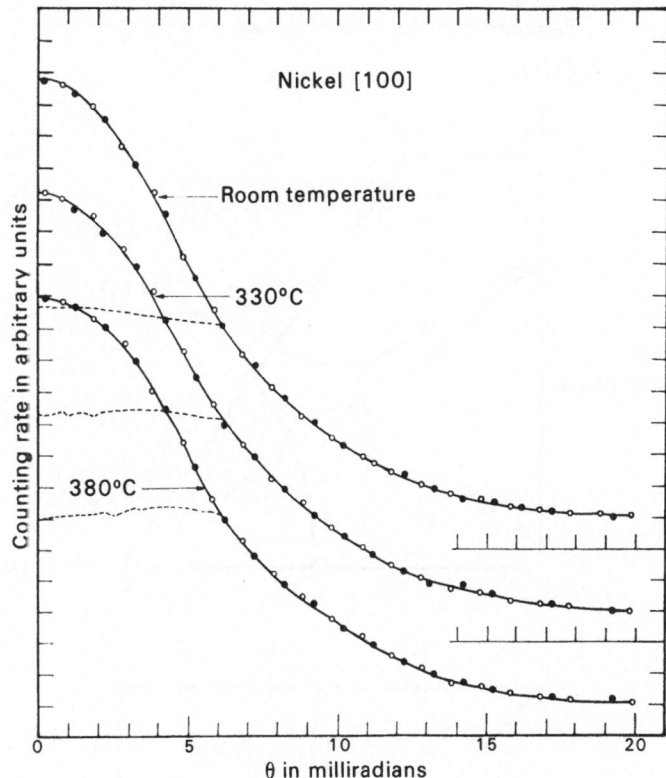

Figure 2 - Experimental results for the (100) direction.
The black and white circles correspond to positive and
negative k_z.

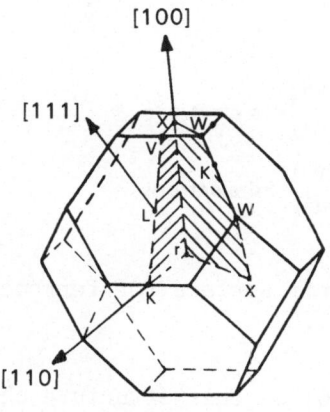

Brillouin zone for F.C.C. lattice

Figure 3 - The Brillouin zone for nickel.

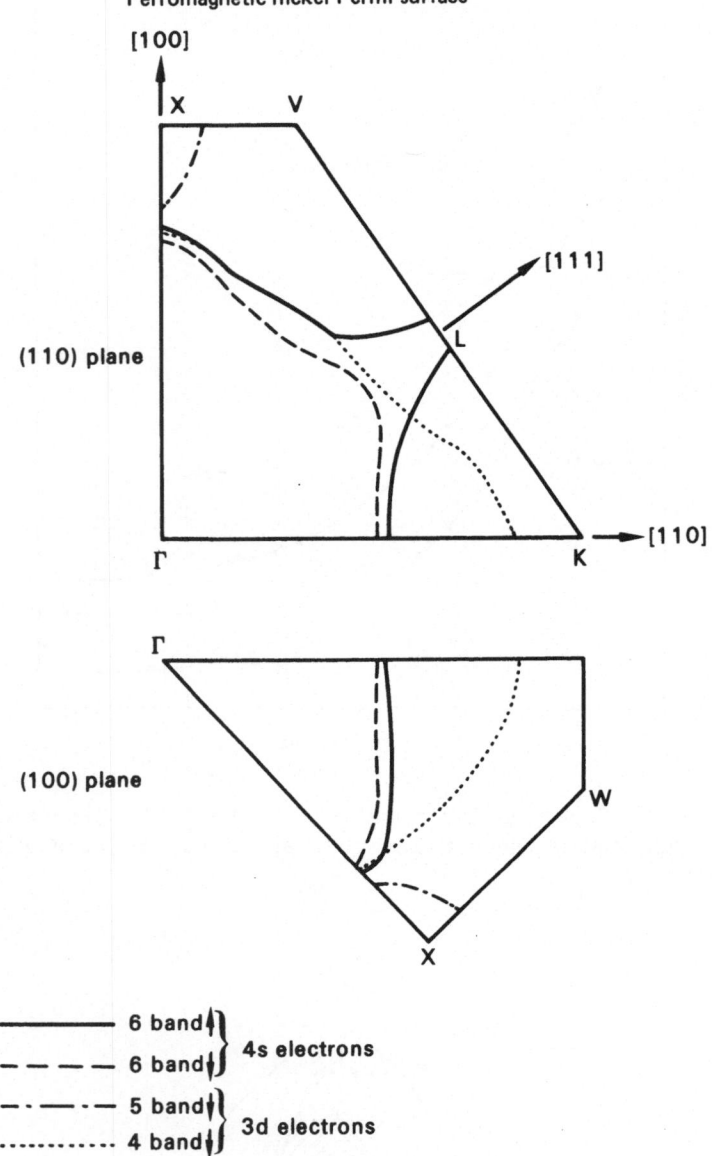

Figure 4 - The Fermi surface for ferromagnetic nickel
(according to [9]).

Curie temperature (361°C) this structure changes because of spin
disorder. As concerns paramagnetic nickel (above the Curie
point) we do not know about any earlier experimental studies nor
theoretical calculations of its Fermi surface.

4. INTERPRETATION OF THE ANNIHILATION CURVES FOR FERROMAGNETIC NICKEL

One way of interpreting the experimental results is to compute the anisotropic density of states $\rho(\vec{k})$ using Mijnarends' equations. Unfortunately, the number of directions investigated is not sufficient to obtain in this way strongly angular dependent features like necks in the Fermi surface. But other features like FS bumps or depressions can be obtained. The density of states along the main crystallographical directions is shown in figure 5. The results are more visible if one draws isodenses (curves of equal density) on planes in k-space (figure 6). The bump in the sixth band FS in the (100) direction and the depression in the (110) direction are well seen. An important bulging in the (111) direction is also visible, but of course one cannot describe a neck with only three lattice harmonics.

Now we shall use an alternative approach to Mijnarends' density of states. Assuming that the contribution from HMC is well described by three lattice harmonics one can split the total number of counts into two parts — one being the contribution from the first Brillouin zone, the other from HMC and core annihilation [10].

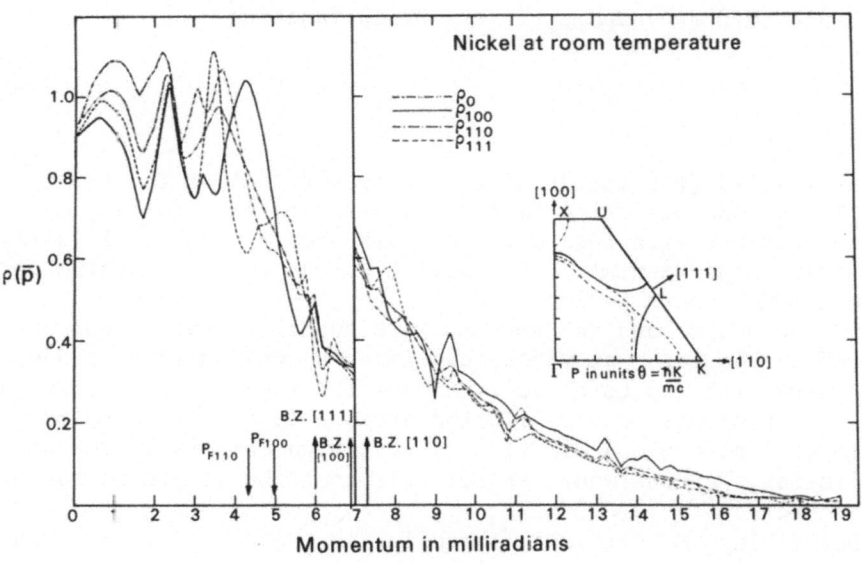

Figure 5 - The electronic density of states along the main crystallographical directions as obtained for ferromagnetic nickel (room temperature) using Mijnarends' equations.

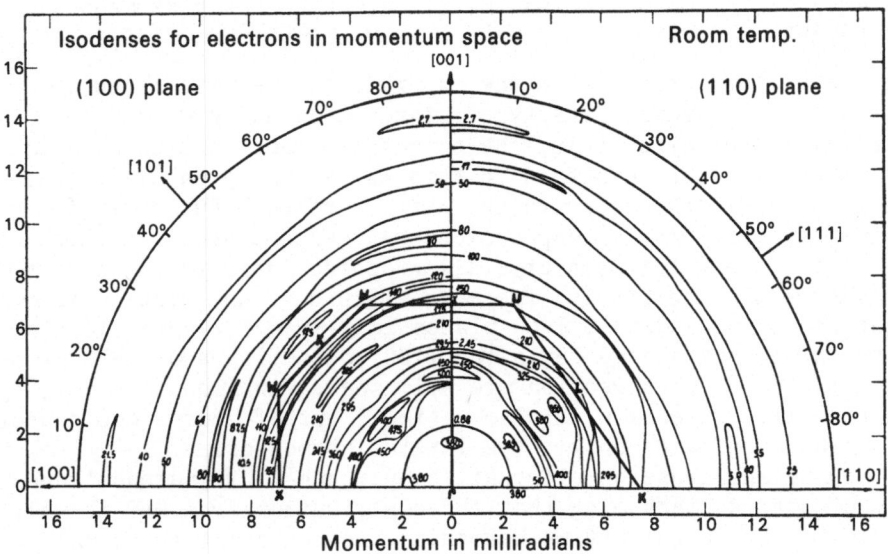

Figure 6 - Isodenses for ferromagnetic nickel (room temperature).

Since both $\varphi(\vec{r})$ and $\psi_{\vec{k}}(\vec{r})$ are Bloch functions

$$\varphi(\vec{r})\psi_{\vec{k}}(\vec{r}) = e^{ikr} \sum_{\vec{\ell}} a_{\vec{\ell}}(\vec{k})e^{i\ell Br}, \qquad (8)$$

a Fermi solid (the inside of the Fermi surface) in the first Brillouin zone will give rise to additional Fermi solids in k-space switched with regard to the main one by reciprocal lattice vectors. Experimentally this will give the higher momentum component (HMC) contribution.

On the other hand the positron wave function, though concentrated in the inter-atomic space, gives a non-vanishing product $\varphi(\vec{r})\psi_C(\vec{r})$ with the core electron wave functions $\psi_C(\vec{r})$. Core annihilation occurs in very limited areas around lattice sites, so it will give a contribution to very high momenta in the annihilation, but one would expect this contribution to be rather isotropic.

Describing HMC with only three lattice harmonics could seem to be a rather rough approximation, but let us remark here that usually this contribution is approximated by an isotropic Gaussian.

The result is shown in the next figures (figure 7). Here the theoretical curve corresponds to cross-sections of the sixth band FS. This way of computing theoretical curves was used for a long time [11] and is exact if $\varphi(\vec{r})$ is constant and $\psi_{\vec{k}}(\vec{r})$ are plane waves. The agreement between theory and experiment gets much better after subtracting HMC and core. The discrepancy

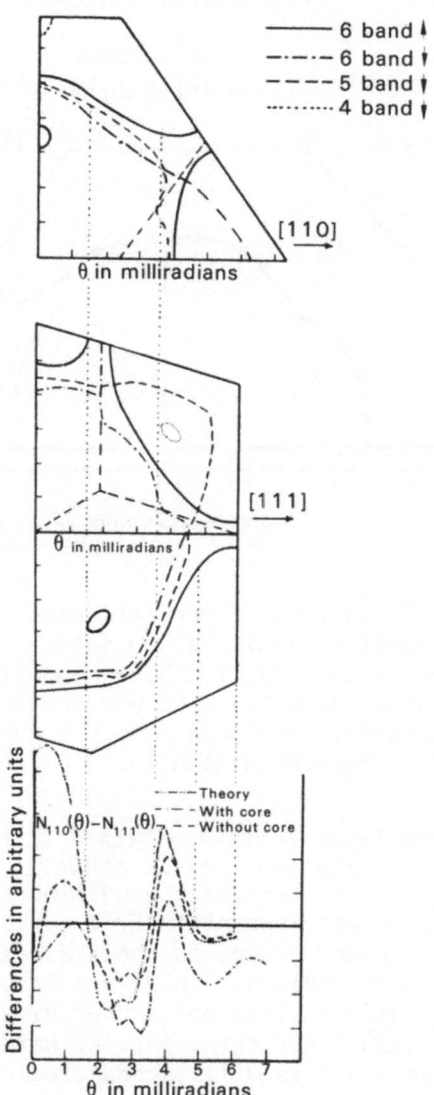

Figure 7 - Annihilation curve differences as follows from the sixth zone FS cross-sections and from experiment before and after subtracting the contribution from HMC and core annihilation.

still existing is probably partially owing to the contribution from other bands, particularly from the fifth band. But one should remember that the agreement is not very good even for copper in which FS exists only in the first zone [12].

Instead of using the cross-sectional area method one could determine the FS from the (smeared) discontinuity in the density

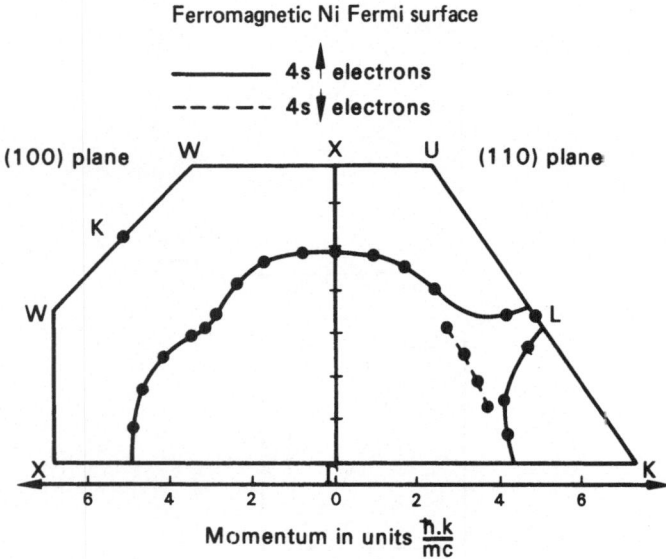

Figure 8 - Sixth zone FS for ferromagnetic nickel as follows from positron annihilation data. In the (111) direction the FS is shown for both majority and minority
spins, in other directions it was impossible to separate
the two surfaces. The points on the surface are shown as
obtained from experimental data.

of states. The Fermi surface obtained finally in this way from
our positron annihilation data is shown in figure 8. Because
the sixth band FS and the fifth band minority FS are very close
to each other, it was impossible from our data to determine the
FS in the fifth band, except in the K-direction where we found
the FS at 6.1 milliradians. These results, however, should be
treated with caution, since one cannot expect to describe a
neck properly using only three lattice harmonics.

Some hope was, and is still, connected with studies of positron annihilation in magnetized ferromagnetic nickel [13].
Positrons resulting from nuclear processes have an initial energy of 0.15-1.5 MeV. Their longitudinal polarization is equal
to v/c, where v is the positron velocity and c is the velocity
of light. This polarization is conserved, in principle, during
thermalization. By magnetizing ferromagnetic nickel one gets
the domains polarized in a definite direction, obtaining then
an excess of electrons with spin in a given direction. Since
positrons annihilate mainly with electrons of opposite spin, we
can in this way study separately the electronic structure for
both spins. Positron annihilations with electrons of parallel
spin result in three photons, and are less probable, besides
they are not registered in angular correlation experiments.

Repeated experiments on magnetized nickel did not lead to definite conclusions [14]. Probably the right experiments remain to
be done.

5. ELECTRONIC STRUCTURE OF NICKEL AT HIGHER TEMPERATURES

The annihilation curves obtained at 330°C and 380°C, i.e.
just below and just above the Curie temperature, have been interpreted in the way described above. The density of states
along the main crystallographical directions has been obtained
and is shown in figure 9 for 330°C and in figure 10 for 380°C.
It is visible, at first sight, that the features of the density
of states at the Fermi surface are smeared at 330°C and they become sharp again above the Curie temperature. The isodenses for
paramagnetic nickel are shown in figure 11.

In order to understand the results it would be good to get
an orientation on what should be the electronic structure of
paramagnetic nickel. Unfortunately, very little research has
been devoted to this problem. The main contribution in positron
annihilation should come from the sixth and fifth bands where
there are sheets of the Fermi surface. The lower bands are filled, except possibly for a small hole around X in the fourth and
third zones. These Fermi surfaces should have an analogy to
those of palladium, except that in Pd (figure 12) there are 1.64
electrons per atom in the fifth band, against 0.36 in the sixth
band, whilst in Ni there are approximately 1.4 electrons in the

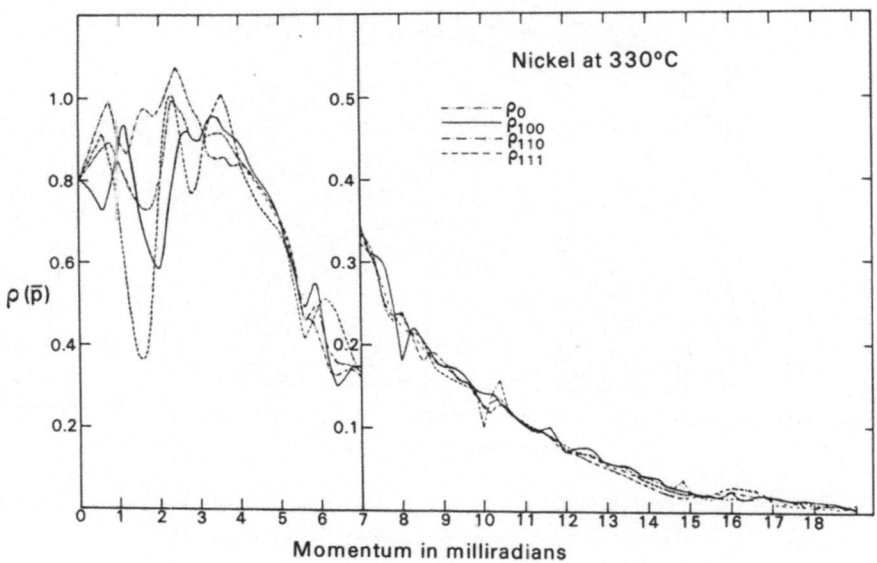

Figure 9 - The same as in figure 5 but for 330°C.

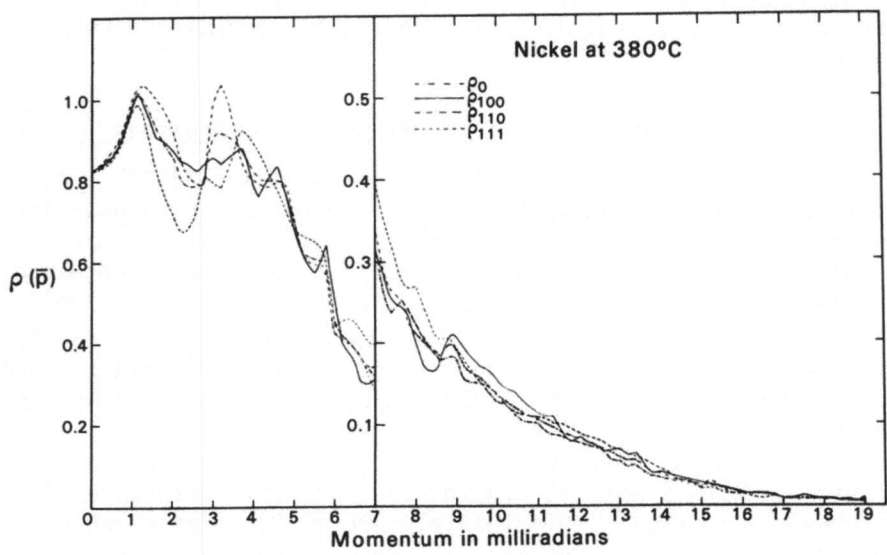

Figure 10 - The same as in figure 5 but for 380°C (para-magnetic nickel).

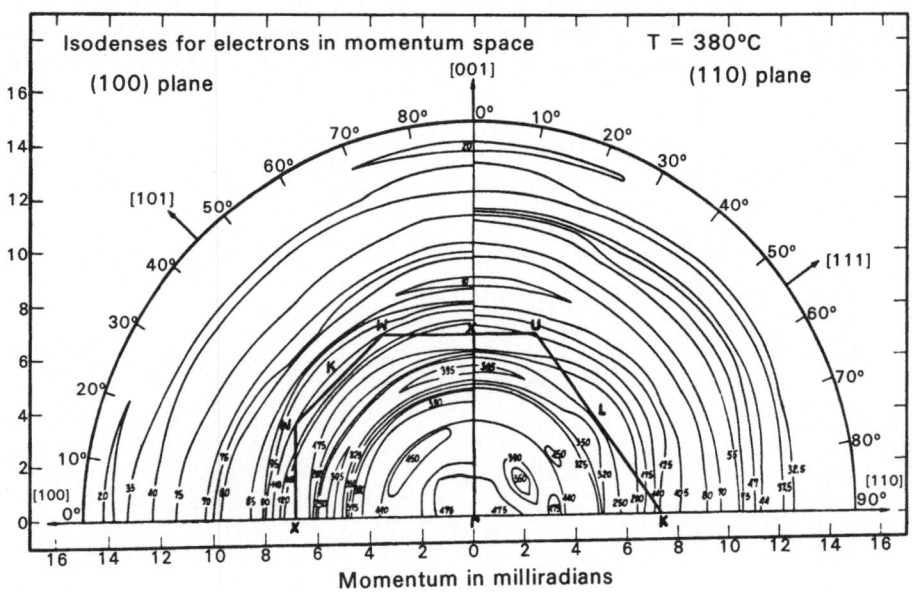

Figure 11 - Isodenses for paramagnetic nickel (380°C).

fifth band and 0.6 in the sixth band. So in comparison with palladium, the fifth band Fermi surface of nickel should be smaller and sixth band FS should be bigger. In comparison with

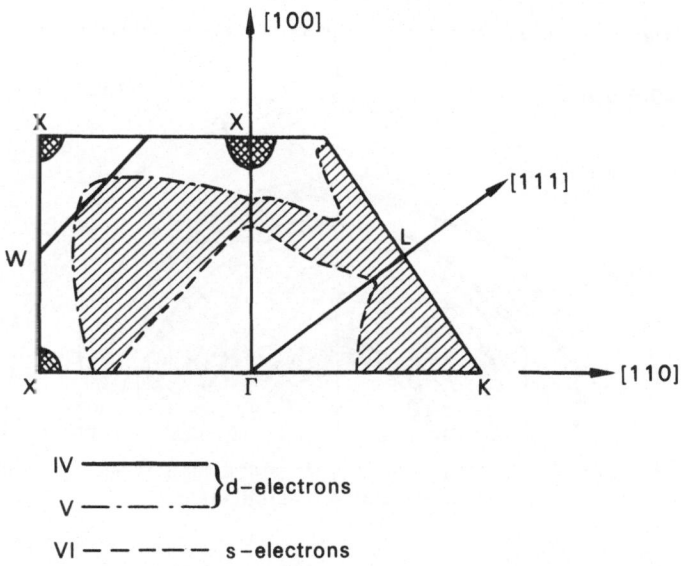

Figure 12 - The Fermi surface of palladium (according to
[15]).

the ferromagnetic state the occupied majority spin states in the
fifth zone should empty for momentum close to the Brillouin zone
boundary to the benefit of the fifth zone minority spin states,
whilst the fifth zone FS should become well separated both from
the sixth zone FS and the Brillouin zone boundary. What happens
in reality is shown in figure 13 and is in agreement with our
expectations. What is less expected is the decrease of the density of states inside the sixth band FS. This could result either
from an increase in the total number of electrons in the less
localized sixth zone or from a more pronounced delocalization of
the fifth band electrons in comparison with the ferromagnetic
state. Also the existence of a free Fermi surface for both spins
in the fifth zone should increase the enhancement owing to electron-positron interaction. In all these cases electrons inside
the sixth band FS would have more competition against annihilating with positrons than in the ferromagnetic state. But the
second case seems to us more probable: in the ferromagnetic
state, if the fifth zone electrons are more localized, their
kinetic energy increases somewhat, but then the separation of
minority and majority FS leads to smaller increase of the kinetic energy in comparison with what would happen without the additional localization — in short some balance should be established between these effects. In the paramagnetic state no additional localization is needed for minimizing the total energy.
    The Fermi surfaces obtained from the density of states dis-
continuities are shown in figure 14. For nickel in comparison

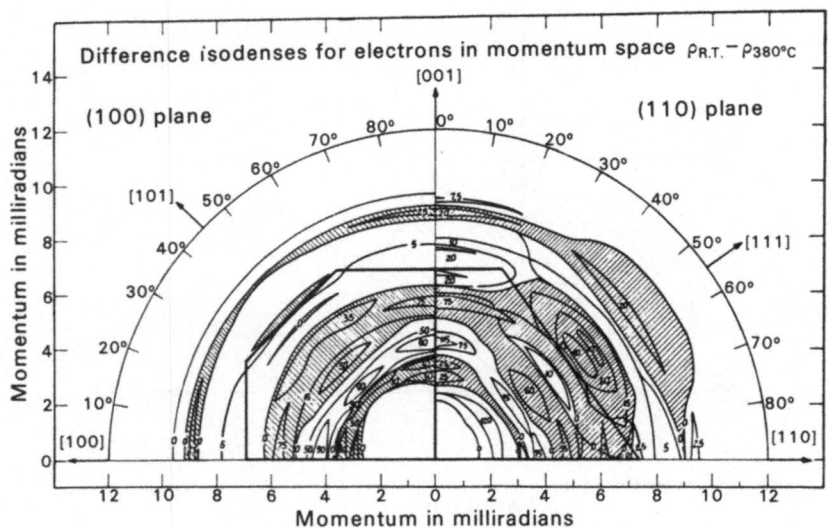

Figure 13 - Difference isodenses comparing the density of states at 380°C and at room temperature. The shaded areas show where the density of states has increased as the result of the transition from the ferromagnetic to the paramagnetic state.

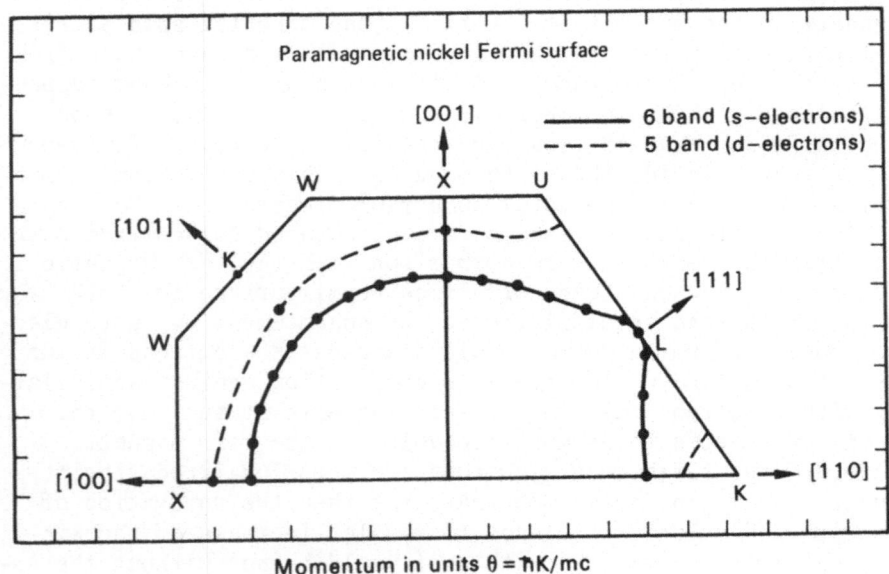

Figure 14 - Fermi surface of paramagnetic nickel as follows from positron annihilation. We are unable to decide whether the sixth zone FS has a neck or not.

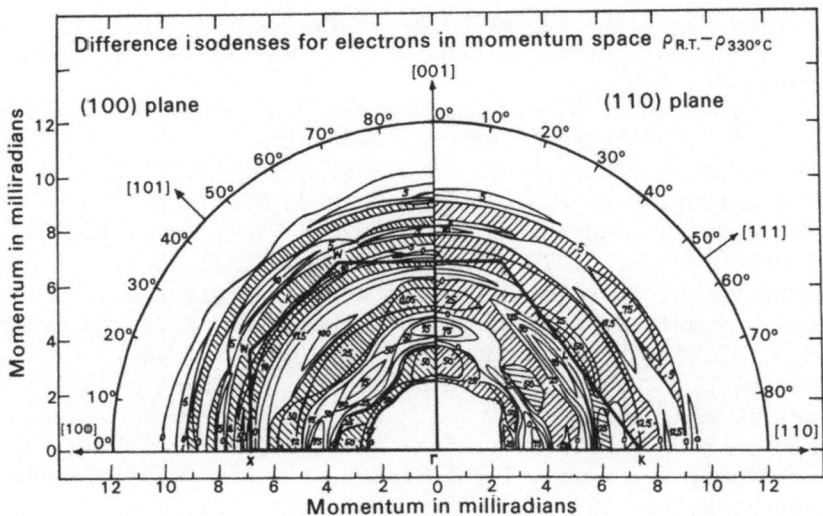

Figure 15 - Difference isodenses comparing the density of states at 330 C° and at room temperature.

with palladium let us point out that the points K and U are empty in the fifth zone and the Fermi surface is open and topologically similar to that of copper. As far as the sixth zone, is concerned, we cannot decide whether there is a neck in L or not. Probably there is, but then this FS, too, is similar to that of copper. As follows from our results, the sixth zone FS is more spherical than in the ferromagnetic case. Especially the bump towards X and the depression in the (110) direction become less pronounced.

Concerning the results for 330°C we are unable to reach any reasonable conclusion from them, except for the smearing of the density of states in the neighbourhood of the Fermi surfaces (figure 15) resulting from spin disorder.

ACKNOWLEDGMENT

The author wishes to thank Mr. W. Wierzchowski and Mrs. G. Kontrym-Sznajd for help in preparing this lecture.

REFERENCES

1. Kontrym-Sznajd, G., Petersen, K., Stachowiak, H., Thrane, N., Trumpy, G. and Wierzchowski, W. (To be published).
2. Stachowiak, H. (1969). *VI-th Winter School of Theoretical Physics in Karpacz*, (Institute of Theoretical Physics, University of Wrocław).
3. Stachowiak, H. (1972). IX-th Winter School of Theoretical

Physics in Karpacz, *Acta Univ. Wratislav.*, No.181, 81.

4. Kahana, S. (1963). *Phys. Rev.*, **129**, 1622.

5. Fujiwara, K., Hyodo, T. and Ohyama, J. (1972). *J. Phys. Soc. Jap.*, **33**, 1047.

6. Kontrym-Sznajd, G. and Stachowiak, H. (1973). *Third International Positron Conference, Helsinki*.

7. Mijnarends, P.E. (1967). *Phys. Rev.*, **160**, 512.

8. Mijnarends, P.E. (1973). *Physica*, **63**, 235.

9. Callaway, J. and Zhang, H.M. (1970). *Phys. Rev.*, **B1**, 305.

10. Stachowiak, H. (1970). *Phys. Stat. Sol.*, **41**, 599.

11. Schoenberg, D. (1965). *Low Temperature Physics LT9*, (eds. Daunt, J.G., Edwards, D.O., Milford, F.J. and Jaqub, M.), (Plenum Press, London and New York), p. 680.

12. Berko, S., Cushner, S. and Erskine, J.C. (1968). *Phys. Lett.*, **A27**, 668.

13. Sedov, V.L. (1968). *Usp. Fiz. Nauk*, **94**, 417.

14. Shiotani, N., Okada, T., Sekizawa, H., Mizoguchi, T. and Karasawa, T. (1973). *J. Phys. Soc. Jap.*, **35**, 456.

15. Mueller, F.M., Freeman, A.J., Dimmock, J.O. and Furdyna, A.M. (1970). *Phys. Rev.*, **B1**, 4617.

THEORY OF MAGNETIC PROPERTIES OF RARE EARTHS AND THEIR COMPOUNDS

Yu.P. IRKHIN

Institute of Metal Physics,
Sverdlovsk, USSR

1. INTRODUCTION: PECULIARITIES OF RARE EARTHS' MAGNETIC PROPERTIES AND ELECTRON STRUCTURE

The systematic research of Rare Earths (RE) metals and their compounds started about twenty years ago, i.e. considerably later than the study of iron group, and has given us a number of interesting results both in theory and experiment.

This rapid development has been caused by the remarkable RE properties that have proved valuable for practical applications. The more interesting amongst these properties are the existence of large atomic magnetic moments and large magnetic anisotropy.

From the theoretical point of view, RE are well suited for quantitative considerations due to the strong localisation of $4f$ magnetic electrons near RE ions in crystal structure. We can neglect the direct overlapping of $4f$ wave functions of neighbouring lattice points. Thus, the situation is opposite to the one existing in $3d$ magnetics, in which the development even of qualitative theory is hardly possible at present.

There are two main peculiarities in the theory of RE magnetics. The first is the necessity of a multi-electron treatment, particularly the unfilled shells' electrons' energy in $(4f^n \ldots \ell^{n'})$-configuration depends on the set Γ of multi-electron quantum numbers (including spin and orbital numbers)

$$E(4f^n \ldots \ell^{n'}) = E_\Gamma(4f^n \ldots \ell^{n'}). \tag{1}$$

The second essential peculiarity is the existence of large orbital contributions in RE magnetic moments. Here we have the

important distinction from the case of $3d$ magnetics having quenched orbital moments. These orbital contributions change drastically the nature of magnetic anisotropy and the effect of spin-
orbital interaction in RE compared with the $3d$ case.

In order to check the validity of the theory, it is interesting to compare the calculated dependence of some physical properties on the atomic number of RE with the experimental one. A
good example of such dependence is the well-known De Gennes formula for the paramagnetic Curie temperature

$$\theta_p = (g - 1)^2 I J (J + 1), \tag{2}$$

where g is the g-factor, J the $4f$ shell total momentum. The exchange integral I is constant for all RE, in good agreement
with experiment. This example demonstrates for us the possibility of constructing the theory with a small number of parameters.

2. METHOD

The method of RE theory must take into account two main features of $4f$ shell structure discussed earlier: multi-electron
aspects and large orbital moments. In this case, the most suitable representation is the second quantization. The Hamiltonian
we begin with is

$$\hat{\mathcal{H}} = \sum_{\lambda_1 \lambda_2 \lambda_1' \lambda_2'} Q(\lambda_1 \lambda_2 \lambda_1' \lambda_2') a_{\lambda_1}^\dagger a_{\lambda_2}^\dagger a_{\lambda_1'} a_{\lambda_2'}. \tag{3}$$

As is well-known, we can transform the exchange part of (3)
to the spin Hamiltonian in the simple case of one s electron
per atom [1].

The transformation formulae are ($\lambda = \sigma = \pm\frac{1}{2}$)

$$s_z = \tfrac{1}{2}(a_{\frac{1}{2}}^\dagger a_{\frac{1}{2}} - a_{-\frac{1}{2}}^\dagger a_{-\frac{1}{2}}), \qquad s^+ = a_{\frac{1}{2}}^\dagger a_{-\frac{1}{2}}, \tag{4}$$

$$s^- = a_{\frac{1}{2}}^\dagger a_{-\frac{1}{2}}, \qquad \sum_\sigma a_\sigma^\dagger a_\sigma = 1,$$

and we have

$$\mathcal{H}_{ex} - \tfrac{1}{2} \sum_{i,j} (\tfrac{1}{2} + 2 s_i s_j). \tag{5}$$

In our case of RE, it is necessary to generalize (4) for
$\ell \neq 0$ and more than one electron per ion.

The first step can be made in the following way. Let us
write down the expression for an arbitrary one-electron operat-
or in the second quantization representation ($\lambda = mc$)

$$\hat{B} = \sum_{\lambda\lambda'} B_{\lambda\lambda'} a_\lambda^\dagger a_{\lambda'}. \tag{6}$$

Let $\hat{B} = \hat{\ell}^\alpha s^\beta$ [2]. Then we have for $\ell = 1$ ($n_{m\sigma} = a_{m\sigma}^\dagger a_{m\sigma}$)

$$(\ell^z)^0(s^z)^0 = 1 = \sum_{m\sigma} n_{m\sigma}, \qquad s^z = \sum_{m\sigma} \sigma n_{m\sigma},$$

$$\ell^z = \sum_{m\sigma} m n_{m\sigma}, \qquad \ell^z s^z = \sum_{m\sigma} m\sigma n_{m\sigma}, \tag{7}$$

$$(\ell^z)^2 = \sum_{m\sigma} m^2 n_{m\sigma}, \qquad (\ell^z)^2 s^z = \sum_{m\sigma} m^2 \sigma n_{m\sigma}.$$

Equations (7) represent the system with six unknown quantities. The solution of (7) is

$$n_{\pm 1, \pm\frac{1}{2}} = \tfrac{1}{2}\ell^z(\ell^z \pm 1)(s \pm \tfrac{1}{2}),$$

$$n_{0, \pm\frac{1}{2}} = [1 - (\ell^z)^2](\tfrac{1}{2} \pm s^z). \tag{8}$$

The same method can be used for off-diagonal combinations determination

$$\ell^\pm = \sqrt{2} \sum_\sigma (a_{0\sigma}^\dagger a_{\mp 1\sigma} + a_{\pm 1\sigma}^\dagger a_{0\sigma}),$$

$$\ell^\pm s^z = \sum_\sigma \sigma(a_{0\sigma}^\dagger a_{\mp 1\sigma} + a_{\pm 1\sigma}^\dagger a_{0\sigma}),$$

$$\ell^\pm \ell^z = \mp \sqrt{2} \sum_\sigma a_{0\sigma}^\dagger a_{\mp 1\sigma}, \qquad \ell^\pm \ell^z s^z = \mp \sqrt{2} \sum_\sigma \sigma a_{0\sigma}^\dagger a_{\mp 1\sigma},$$

$$(\ell^\pm)^2 = \sqrt{2} \sum_\sigma a_{\pm 1\sigma}^\dagger a_{\mp 1\sigma}, \qquad (\ell^\pm)^2 s^z = \sqrt{2} \sum_\sigma \sigma a_{\pm 1\sigma}^\dagger a_{\mp 1\sigma}.$$

The next step is to generalize formulae (7-9) for the multi-electron configuration ℓ^n. For this purpose, the appropriate method is the Racah technique (see for example [3]) in second quantization form [4].

As is known, a multi-electron function can be constructed as a linear combination of one-electron function products

$$\psi_{\Gamma_n}(\ell^n) = \sum_{\Gamma_{n-1}\lambda} G_{S'L'} C_{L'M', \ell m} C_{S'\mu', \frac{1}{2}\sigma} \psi_{\Gamma_{n-1}'}(\ell^{n-1}) \psi_\lambda(r_i s_i), \tag{10}$$

where $\Gamma = SL\mu M$ (μ and M are z-projections of spin and orbital moments), $G_{S'L'}^{SL}$ the fractional parentage coefficient, and $C_{B\beta,D\delta}^{A\alpha}$ the Clebsch-Gordan coefficient. The Coulomb interaction between ℓ^n configuration electrons is diagonalized in Γ-representation.

In analogy with (10), we can introduce multi-electron operators

$$A_{\Gamma_n}{}^{\dagger}(\ell^n) = \frac{1}{\sqrt{n}} \sum_{\Gamma_{n-1}'\lambda} G_{S'L'}^{SL} C_{L'M'}^{LM},_{\ell m} C_{S'\mu'}^{S\mu},_{\frac{1}{2}\sigma} A_{\Gamma_{n-1}'}{}^{\dagger}(\ell^{n-1}) a_\lambda{}^{\dagger}. \quad (11)$$

Particularly in the case $s2$-configuration, we have from (11)

$$A_{0000}{}^{\dagger}(s^2) = \tfrac{1}{2}(a_{\frac{1}{2}}{}^{\dagger} a_{-\frac{1}{2}}{}^{\dagger} - a_{-\frac{1}{2}}{}^{\dagger} a_{\frac{1}{2}}{}^{\dagger}) = a_{\frac{1}{2}}{}^{\dagger} a_{-\frac{1}{2}}{}^{\dagger}. \quad (12)$$

The operators A_Γ play the same rôle for the momenta operators $\hat{S}, \hat{L}, \hat{J}$ as the usual $a_{m\sigma}$ operators for one-electron $\hat{s}, \hat{\ell}, \hat{j}$ moments.

The corresponding relation is

$$\hat{\vec{J}} = \sum_{M_J M_J'} (JM_J|\hat{\vec{J}}|JM_J') A_{JM_J}{}^{\dagger} A_{JM_J'}. \quad (13)$$

Then in analogy with the one-electron case, we can transform the Hamiltonian to the J-representation by using the relations between the components of the $(\vec{J})^p$ quantity ($p = 0,\ldots,2J$) and A_Γ. It can be shown that in the A_Γ-representation the Hamiltonian (3) has the form

$$\mathcal{H}(\ell_1{}^{n_1}, \ell_2{}^{n_2}) = \sum_{\Gamma_1\Gamma_2\Gamma_1'\Gamma_2'} \phi(\Gamma_1\Gamma_2\Gamma_1'\Gamma_2') A_{\Gamma_1}{}^{\dagger} A_{\Gamma_2}{}^{\dagger} A_{\Gamma_2'} A_{\Gamma_1'}. \quad (14)$$

As was demonstrated in [4], this method allows one to obtain, besides the usual Heisenberg product $(\vec{S}_1 \cdot \vec{S}_2)$, also the orbital dependent factor $(\vec{L}_1 \cdot \vec{L}_2)^p$ ($p = 0,\ldots,2\ell$) in the Russel-Saunders case.

The consideration of $(\vec{L}_1 \cdot \vec{L}_2)^p$ invariants is very fascinating from a theoretical point of view. But we must remember that in reality $3d$ electrons' orbital moments are quenched and average $\underline{L} = 0$. Moreover, there is no addition of one-electron momenta ℓ in the total moment $\vec{L}$ in $3d$ metals, where strong crystal field exists and $3d$ electrons are collectivized.

Furthermore, we proceed with our consideration for $4f$ electrons only

The exchange part of the s-f interaction Hamiltonian can be written in the form

$$\mathcal{H}(f^n, C) = - \sum_{\vec{\upsilon}\Gamma_1\Gamma_2\vec{k}\sigma\vec{k}'\sigma'} e^{i(\vec{k}-\vec{k}')\vec{\upsilon}} \times \qquad \text{(Contd)}$$

$$\times \langle \Gamma_1, \vec{k}\sigma | \sum_{i=1}^{n} \frac{e^2}{|\vec{r}_i - \vec{r}_c|} \, pic | \Gamma_2, \vec{k}'\sigma' \rangle A_{\vec{J}\Gamma_1} \dagger a_{\vec{k}\sigma} \dagger a_{\vec{k}'\sigma'} A_{\vec{J}\Gamma_2}. \qquad (15)$$

Here the index c designates conduction electrons with quantum numbers $\vec{k}$ (quasi-moment) and σ, ν-lattice point index, $\Gamma = SLJM_J$.

In the plane wave approximation for conduction electrons and for the lowest energy term $S_1 L_1 J_1 = S_2 L_2 J_2$ for f shells, we can calculate matrix elements in (15) and represent them through one-electron integrals. These last quantities depend on quasimoments $\vec{k}, \vec{k}'$ and spin and orbital projections. All angular dependencies are expressed in terms of $3j$-symbols, the summation of products of which can be carried out by use of the known relations [3]. Then we have for the two lowest terms $\ell = 0, 1$ in Legendre polynomial expansion of conduction electrons wave functions [4]

$$\hat{\mathcal{H}}_{ex}(f^n[SLJ], c) = \sum_{\vec{\nu}\vec{k}\sigma\vec{k}'\sigma'} e^{i(\vec{k}-\vec{k}')} \, a_{\vec{k}\sigma}\dagger a_{\vec{k}'\sigma'}$$

$$\times \left\{ A\left[\tfrac{1}{2}n + 2(g-1)\vec{s}\vec{J}_{\vec{\nu}}\right] + (B\langle\vec{k},\vec{k}'\vec{J}\rangle + iC[\vec{k},\vec{k}']\vec{J}_{\vec{\nu}}\right.$$

$$+ D(\vec{k}\vec{k}'))\delta_{\sigma\sigma'} + E\{(s\vec{J}_{\vec{\nu}}), \langle\vec{k},\vec{k}'\vec{J}_{\vec{\nu}}\rangle\} + iF\{(s\vec{J}_{\vec{\nu}}), [\vec{k}\vec{k}']\vec{J}_{\vec{\nu}}\}$$

$$+ G(\vec{s}\vec{J}_{\vec{\nu}})(\vec{k}\vec{k}') + iH[\vec{k}\vec{k}']\vec{s} + I[(\vec{s}\vec{k}')(\vec{k}\vec{J}_{\vec{\nu}})$$

$$\left. + (\vec{s}k)(k'\vec{J}_{\vec{\nu}})]\right\}. \qquad (16)$$

Here $\{X, Y\} = XY + YX$, $\langle\vec{k}, \vec{k}'\vec{J}\rangle = (\vec{k}\vec{J})(\vec{k}'\vec{J}) + (\vec{k}'\vec{J})(\vec{k}\vec{J})$ and $A, B, \dots$ are the one-electron integrals multiplied by some numerical coefficients.

The first term in (16) is the usual De Gennes exchange. The next terms are interesting for the anisotropy effects. We can see from (16) that there is a direct dependence between the electron wave vector $\vec{k}$ and the total moment $\vec{J}$ of $4f$ shells. Such dependence permits one to consider the effect of the Fermi surface shape on RE magnetic structure [5].

The most interesting consequences of (16) are: (1) anisotropic exchange; (2) anisotropic magnetic scattering. All coefficients in (16) can be determined either by direct calculations or by making use of the irreducible tensor operators technique [3,6]. It can be shown for example that

$$C = C_{SLJ}(kk') = \frac{4\pi N e^2}{V} G_{||}^{(2)}(k,k') \frac{g}{140} (g-2), \qquad (17)$$

where $G_{\parallel}^{(2)}(k,k')$ is the exchange integral with $\ell = \ell' = 1$, and N and V are the number of ions and the volume of our system. The term with C is especially interesting. It represents the orbital-orbital exchange $[\vec{k}\vec{k}'](g-2)\vec{J} \sim (\vec{\ell}_c\vec{L})$ ($\vec{\ell}_c$ is the conduction electron orbital moment).

3. THE PARAMAGNETIC SUSCEPTIBILITY ANISOTROPY AND THE ANISOTROPY CONSTANTS OF RE

The iron group magnetic anisotropy is usually small. The order of magnitude of $K_2 \approx 10^7$ erg/cm^3 for Co is the largest amongst all others. For RE, $K_2 \approx 10^8$ erg/cm^3. The reason for that lies in the different nature of magnetic anisotropy of $3d$ and $4f$ magnetics.

The orbital moments of $3d$ electrons are almost fully quenched and their spin moments can rotate nearly freely relative to the crystal axes. The small orbital moment appears to be only due to the spin-orbit interactions. Thus, we have in the $3d$ case $K_2(3d) \approx \lambda^2/\Delta E \approx 10^6$ erg/cm^3, where λ is the spin-orbital constant and ΔE is the energy difference of different d electron states in the crystal field.

On the contrary, in RE magnetics there exists a large orbital moment which is tied with spin rigidly. The total moment $\vec{J} = \vec{L} + \vec{S}$ is constant, and we can consider the spin-orbital interaction as an infinitely large quantity. The magnetic anisotropy energy is determined in this case by the magnitude of work needed to rotate the $\vec{J}$-vector from an easy to a hard direction.

There exist two main mechanisms of magnetic anisotropy in RE: (a) the crystal field; (b) anisotropic exchange.

(a) CRYSTAL FIELD THEORY

The energy of the non-spherical $4f^n$ electron state SLJ in a one-axial crystal field can be expressed in the form

$$V_2^0 = \alpha_{SLJ} \frac{e^2 Z'}{a} v_2^0 \frac{\overline{r_+^2}}{a^2} [3J_z^2 - J(J+1)],$$

$$v_2^0 = -1.035\left[1.633 - \frac{c}{a}\right],$$

(18)

Here eZ' is the ion charge, α_{SLJ} is the Stevens factor, $\overline{r_f^2}$ is the $4f$ orbit average square radius, c and a are the hexagonal parameters.

Making use of (18) it is easy to calculate the value of the paramagnetic Curie point anisotropy $\Delta\theta_p = \theta_p(a) - \theta_p(c)$ and the anisotropy constant K_2 (see Table I). The dependence on the periodic table number is determined mainly by the factor α_{SLJ}. The last quantity changes sign between f^{10} (Ho) and f^{11} (Er) configurations, in agreement with experiment. The single

TABLE I

		Gd	Tb	Dy	Ho	Er	Tm
$\Delta\theta\,^\circ K$	$\Delta\theta_p^{exp}$	≈ 0	44	48	15	-29	-58
	$\Delta\theta_p^{CF}$	0	44	38	15	-16	-44
	$\Delta\theta_p^{ex}$	0	48	40	21	-25	-24
	$\Delta\theta'_p^{ex}$	0	44	48	21	-23	-66
$K_2(0)$ $\times 10^{-8}$ erg/cm^3	K_2^{exp}	≈ 0	≈ -5.5	≈ -5	≈ -2.2	≈ 1.8	
	K_2^{CF}	0	-5.5	-5.1	-2	2	5.5
	K_2^{ex}	0	-5.5	-4.6	-1.4	1.1	1.9

unknown quantity in (18) is $\overline{r_f^2}$. Numerical coincidence of the theory with experimental data can be obtained with the value $\overline{r_f^2} = 1.5(\overline{r_f^2})_{F-W}$ where $(\overline{r_f^2})_{F-W}$ was calculated earlier in the well known Freeman-Watson paper. This means that there exists an essential expansion of $4f$ shells in metals, or another mechanism gives the important contribution.

(b)   ANISOTROPIC EXCHANGE

The Hamiltonian (16) consists of several terms, which give the anisotropic exchange in the second approximation. The most interesting is the following [6]:

$$\mathcal{H}_{ex}^{(2)} = \sum_{1,2} I(R_{12})(g - 1)D_1(\vec{R}_{12}\vec{J}_1)(\vec{R}_{12}\vec{J}_2), \qquad (19)$$

where $I(R_{12})$ is the effective exchange integral, and

$$D_1 = \left[\frac{2J + 1}{J(J + 1)}\right]^{\frac{1}{2}}\begin{Bmatrix} LJS \\ LJS \\ 211 \end{Bmatrix} \qquad (20)$$

changes the sign between f^{10}(Ho) and f^{11}(Er). The values $\Delta\theta_p^{ex}$
calculated with (19) are given in Table I.

As we can see from the table, the anisotropic exchange gives
satisfactory results from Tb up to Er and is in disagreement
with experiment for Tm. The crystal field results are more satisfactory in this case.

The anisotropic exchange theory can be improved, if we take
into account the symmetry of conduction electrons wave functions, the results are in this case [7]

$$\mathcal{H}_{ex}^{(2)'} = A_2 n \alpha_J J(2J - 1) P_2(\cos\vartheta),$$

$$A_2 \approx 10^{-16} \frac{V^2}{N^2} \omega k_F^4 F(2k_F R) I_{eff}^2. \tag{21}$$

Here n is the $4f$ electrons number, ϑ is the angle between J and
the c-axis, k_F is the Fermi wave vector, $F(y) = (y\cos y - \sin y)/y^4$,
ω is the nearest neighbors number, and I_{eff} is the effective
tive exchange integral.

The dependence of this last contribution on the periodic
table number is similar to that given by crystal field theory.

The final solution of the problem of the separation of different contributions in RE magnetic anisotropy can be obtained
from the concentration dependence study in RE alloys [8].

The last paper of Boutron [9] is interesting in this respect.
According to this paper, the exchange part of anisotropy can be
separated by use of the experimental data $\Delta\theta_p$ in Gd doped by
R = Tb, Dy, Ho Er, Tm. A good agreement with formula (19) was
obtained for R-dependence and the value of the exchange contribution was determined. The latter is about one half of the crystal field part and has the opposite sign. RE compounds magnetic properties were studied by many authors (as for the theory,
see for example Levy's papers [10]).

Finally, we shall consider shortly the popular RCo_5 compounds.

4. MAGNETIC ANISOTROPY IN RCo_5 COMPOUNDS

RCo_5 compounds are interesting as an example of materials in
which the crystal field plays the main rôle for R-ions. We can
neglect the R-R exchange interaction in these materials.

Standard crystal field theory calculations lead to the following result for the anisotropy constant

$$K_2 = - 3\left(\frac{4\pi}{5}\right)^{\frac{1}{2}}\left(16 \frac{2x^2 - 1}{(1 + x^2)^{5/2}} - 3\sqrt{3}\right)\frac{e^2 Z r_+^2}{a^3} \alpha_J J(J - \tfrac{1}{2}),$$

$$(x = c/a). \tag{22}$$

The geometrical factor in parentheses differs essentially from the case of a close-packed hexagonal lattice. This quantity is not equal to zero at any x-values. At $x = 0.8$ (the actual value for RCo_5 lattice) this factor is ≈ 1. For this reason we have the significantly larger theoretical value for K_2 than in pure RE metals. The value of K_2 calculated from (22) is two orders of magnitude larger than the experimental one.

The agreement with experiment can be obtained only with an account of the screening effect caused by the formation of localised states of conduction electrons near R^{3+}-ions [11].

REFERENCES

1. Dirac, P.A.M. (1958). *Principles of Quantum Mechanics*, (Oxford University Press).
2. Irkhin, Yu.P. and Abelsky, Sh.Sh. (1964). *Fiz. Tverd. Tela*, **6**, 1635.
3. Sobel'man, I.I. (1963). *Introduction to The Theory of Atomic Spectra*.
4. Irkhin, Yu. P. (1966). *Sov. Phys. JETP*, **50**, 379.
5. Dzialozhinsky, I.E. (1964). *Sov. Phys. JETP*, **47**, 336.
6. Irkhin, Yu. P., Druzhinin, V.V. and Kasakov, A.A. (1968). *Sov. Phys. JETP*, **54**, 1183.
7. Irkhin, Yu.P. and Karpenko, V.P. (1971). *Fiz. Tverd. Tela*, **13**, 3586.
8. Karpenko, V.P. and Irkhin, Yu.P. (1973). *Sov. Phys. JETP*, **64**, 756.
9. Boutron, P. (1973). *Intermag Report*, (Moscow).
10. Levy, P.M. (1964). *Phys. Rev.*, **135**, A155; (1970). *Phys. Rev.*, **B2**, 1429.
11. Irkhin, Yu.P., Zabolotsky, E.I., Rosenfeld, E.V. and Karpenko, V.P. (1973). *Fiz. Tverd. Tela*, **15**, 2963; (1974). *Fiz. Tverd. Tela*, **16**, (in press).

EXCHANGE INTERACTION
IN THE HEAVY RARE EARTH METALS
CALCULATED FROM ENERGY BANDS

PER-ANKER LINDGÅRD

Danish Atomic Energy Commission,
4000 Roskilde, Denmark

1. INTRODUCTION

The heavy rare earth metals were obtained in pure form and
as single crystals about ten years ago. This made a detailed
experimental investigation possible. Neutron scattering in
particular has been an important tool. As a result we have, by
now, obtained a very complete knowledge about the magnetic interactions. The experimental facts, which are reviewed in [1],
revealed that the magnetic properties are determined by an in-
tricate interplay of forces of similar magnitude. The dominant
is the indirect Ruderman-Kittel-Kasuya-Yosida (RKKY) exchange
interaction, which we shall attempt to calculate from first
principles, here. Of importance also is the crystal field anisotropy and magnetoelastic effects. The anisotropy of this
origin is of a single ion type. Recent neutron scattering
measurements [2] have shown also that two-ion anisotropy may be
of importance. There are numerous possibilities for anisotropy
of the interaction between the moments at different sites. As
we shall see, the RKKY interaction, which is mediated by the
conduction electrons, is anisotropic in the magnetically ordered
phase. The two-ion interaction, which is mediated by phonons,
is strongly anisotropic. The magnitude of the interaction between the spin system and the lattice is determined by the
coupling between the spin- and orbital-momentum of the electrons.
If the spin-orbit coupling and the orbital momentum is large we
must therefore expect large anisotropies both of single-ion and
two-ion nature. Also the RKKY interaction becomes anisotropic,
as discussed by Kaplan and Lyons [3].

In order to avoid the complications of anisotropy we shall
start by considering the RKKY interaction in a pure spin system
with no orbital effects. This is exemplified by gadolinium,
which has an ^8S ground state. The electronic configuration of
a Gd atom is a xenon core with seven $4f$ electrons and three $5d^1$
$6s^2$ outer electrons.

The basic interaction is between the localized $4f$ electrons
belonging to the inner shells of gadolinium and the conduction
electrons. Ruderman and Kittel assumed for simplicity that the
condition electrons were completely free (i.e. plane wave
states). We are now able to go a step further and treat the
conduction electrons in a more realistic fashion. A standard
technique is the augmented plane wave (APW) method [4,5].

2. THE AUGMENTED PLANE WAVE METHOD

In the APW method the electrons are supposed to move in a
simplified potential which is atomic-like inside a sphere (the
muffin tin (in two dimensions)) around each ion and constant
(= 0) between the spheres. The Schrödinger equation is then
solved numerically for this potential by the variational method.

The trial wave function is obtained by expanding the wave
functions inside the spheres in atomic-like functions and between the spheres in plane waves. The wave functions are matched
at the surface of the spheres and the coefficients in the expansion is determined by minimizing the energy.

The wave function for the electrons, the crystal wave function, is therefore

$$\psi_{\mathbf{k},E}(\mathbf{r}) = \sum_i A_i(\mathbf{k})\phi_{\mathbf{k}_i}(\mathbf{r}), \qquad (1)$$

where the coefficients $A_i(\mathbf{k})$ are to be determined variationally.
The sum is over a set of reciprocal lattice vectors τ_i where we
write $\mathbf{k}_i = \mathbf{k} + \tau_i$.

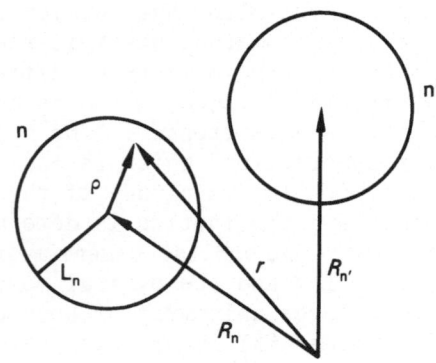

The augmented plane wave is outside the spheres

$$\phi_{k_i}(r) = \frac{1}{\sqrt{\Omega}} e^{ik_i \cdot r} \tag{2}$$

where Ω is the volume of the unit cell.

Inside the n-th sphere it is:

$$\phi_{k_i}(r) = e^{ik_i \cdot R_n} \sum_{l=0}^{\infty} \sum_{m=-l}^{l} A_{lm}(k) \Psi_{l,E'}(\rho) Y_{lm}(\hat{\rho}),$$

$$|\rho| = |r - R_n| < L_n, \tag{3}$$

where L_n is the radius of the sphere and R_n the vector to the centre. The number of l values to be included in the sum are not specified at this point. However, if we want to represent s, p, d or f character of the crystal wave function we must include $l = 0,1,2,3$. The wave function in the two regions (2,3) can be made to match at the sphere surface by choosing the expansion coefficients $A_{lm}(k)$ in (3). By expanding (2) in spherical harmonics and Bessel functions around the centre of the sphere,

$$\frac{1}{\sqrt{\Omega}} e^{ik_i \cdot r} = \frac{4\pi}{\sqrt{\Omega}} e^{ik_i \cdot R_n} \sum_{l=0}^{\infty} \sum_{m=-l}^{l} i^l j_l(k\rho) Y_{lm}^*(\hat{k}_i) Y_{lm}(\hat{\rho}), \tag{4}$$

and equating this with (3) at $\rho = L_n$ we find

$$A_{lm}(k) = \frac{4\pi}{\sqrt{\Omega}} \frac{i^l Y_{lm}^*(\hat{k}_i) j_l(kL_n)}{\Psi_{l,E'}(L_n)} . \tag{5}$$

The APW function $\phi_{k_i}(r)$ with this $A_{lm}(k)$ is called a basis function. It is continuous, but has a discontinuous slope at the sphere radius. The expansion coefficients $A_i(k)$ in (1) are found by minimizing the energy,

$$E_k = \frac{\langle \psi_k^* | H | \psi_k \rangle}{\langle \psi_k^* | \psi_k \rangle} . \tag{6}$$

This gives a secular equation for the determination of the $A_i(k)$. We shall not go further into this.

The $\Psi_{l,E'}(\rho)$ functions are the radial solution of the Schrödinger equation inside the sphere:

$$\left(-\frac{1}{r^2} \frac{d}{dr} \left[r^2 \frac{d}{dr} \right] + \frac{l(l+1)}{r^2} + V(r) - E' \right) \Psi_{l,E'}(r) = 0.$$

$\Psi_{l,E'}(\rho)$ must be regular at the centre ($\rho = 0$), but there is no boundary condition at $\rho = \infty$, and hence there exist solutions for all E'. This is a complication and E' must be chosen self-consistently according to (6). Several methods have been devised to do this practicably. Harmon [5] used a linearized APW method [6] to obtain the wavefunction for Gd, which we are going to use later. Also the crystal potential inside the sphere $V(r)$, must be chosen self-consistently. This is done by summing the contribution to the Coulomb potential from a large number of surrounding ions, including the conduction electron charge density. The exchange interaction may be included in the Slater $\rho^{1/3}$ approximation.

By carrying out this programme we are able to find a set of self-consistent energy bands E_k and the corresponding wave functions ψ_k for the conduction electrons. The variationally determined wavefunctions are presumably less reliable than the energies. Also they are more sensitive to the approximation made when constructing the muffin tin potential. However, we may expect them to be best near the atoms inside the spheres. Therefore they should be quite adequate in calculating the matrix element between the conduction electrons and the localized $4f$ electrons, which is relevant for the calculation for the RKKY interaction. The $4f$ electrons are well approximated by the atomic wavefunctions of Herman and Skillman [7].

3. THE RKKY INTERACTION WITH REALISTIC ENERGY BANDS

3.1 THE INTERACTION BETWEEN CONDUCTION ELECTRONS AND THE $4f$ ELECTRONS

By means of these realistic energy bands and wave functions we can proceed to calculate the RKKY interaction†

In the calculation of the energy bands we did not consider explicitly the interaction between two electrons, but rather the interaction between one electron and the average potential for all the other electrons. As a perturbation of this model we shall now consider the interaction between a conduction electron and a $4f$ electron. The direct interaction is the Coulomb interaction $v(\mathbf{r}_1 - \mathbf{r}_2) = e^2/|\mathbf{r}_1 - \mathbf{r}_2|$ between a conduction electron at $\mathbf{r}_1$ and a $4f$ electron at $\mathbf{r}_2$. In general, however, the potential is screened by the presence of the other electrons, in which case $v(\mathbf{r}_1 - \mathbf{r}_2)$ will be modified to, for example, the Yukawa potential $e^2 \exp\{-\kappa|\mathbf{r}_1 - \mathbf{r}_2|\}/|\mathbf{r}_1 - \mathbf{r}_2|$, where κ^{-1} is the screening length.

Since we are interested in the magnetic interaction, we shall only consider the exchange interaction, and furthermore, only the term which involves the scattering of a conduction electron on a $4f$ electron.

This is represented in terms of electron creation and anni-

---

† We follow closely the notation of A.J. Freeman in reference [1].

hilation operators, $c_k{}^+$ and c_k respectively, as follows:

$$v(r_1 - r_2) = \tfrac{1}{2} \sum_{\substack{k_i s_i \\ i=1,2,3,4}} \langle k_4 s_4, k_3 s_3 | v(r_1 - r_2) | k_2 s_2, k_1 s_1 \rangle$$

$$\times c_{k_4 s_4}{}^+ c_{k_3 s_3}{}^+ c_{k_2 s_2} c_{k_1 s_1}, \tag{7}$$

where $\langle | v | \rangle$ is the matrix element, k_i the crystal momentum and s_i the spin index. Since $v(r_1 - r_2)$ is independent of spin, the spin must be conserved in the scattering process. Let us assume that the $4f$ electrons are well approximated by localized atomic states $\phi_{4f}(r - R_n)$ at the site R_n and the conduction electron wave function is $\psi_{k,E}(r)$ in (1). Then the conduction electrons are scattered from one state of momentum k to another of k' whereas the localized electrons are scattered from one localized state to another, with or without spin flip. We can represent the change in the localized states by the change in the total local spin S_n instead of by means of the creation and annihilation operators in (7).

The perturbation of the single electron Hamiltonian which was used in the band calculation is therefore in this approximation for the s-f exchange interaction as follows

$$H_{sf}(R_n) = -\frac{1}{N} \sum_{k,k'} j_{sf}(k,k') e^{i(k-k') \cdot R_n}$$

$$\times \{ (c_{k\uparrow}{}^+ c_{k'\uparrow} - c_{k\downarrow}{}^+ c_{k'\downarrow}) S_n{}^z \tag{8}$$

$$+ c_{k\uparrow}{}^+ c_{k'\downarrow} S_n{}^- + c_{k\downarrow}{}^+ c_{k'\uparrow} S_n{}^+ \},$$

where the last line shows the spin flip scattering processes and the middle line the processes without spin flip.

The matrix element is

$$j_{sf}(k,k') = N \int dr_1 dr_2 \{ \phi_{4f}{}^*(r_1 - R_n) \psi_k{}^*(r_2) v(r_1 - r_2) \tag{9}$$

$$\times \phi_{4f}(r_2 - R_n) \psi_{k'}(r_1) \} e^{i(k'-k) \cdot R_n}.$$

$j_{sf}(k,k')$ is independent of the lattice site R_n since

$$\psi_k(r) = u_k(r) e^{ik \cdot r} = \psi_k(r - R_n) = u_k(r) e^{ik \cdot r} e^{-ik \cdot R_n},$$

according to Bloch's theorem. We shall assume that $\phi_{4f}(r - R_n)$

vanishes outside the muffin tin sphere around R_n and therefore
we only integrate (9) inside the sphere to obtain the generalized exchange integral $j_{sf}(k,k')$.

3.2 THE EFFECT OF ORBITAL MOMENT
OF THE $4f$ ELECTRONS

Let us generalize the interaction Hamiltonian (8) slightly.
In the presence of orbital momentum L for the $4f$ electrons the
total angular momentum $J = L + S$ ($J = |L \pm S|$ since L and S are
parallel, with + for the heavy and - for the light rare earth
metals). Then we can replace S in (9) by the spin projection
along J namely $(g - 1)J$, where g is the Landé factor. A proper
calculation of the orbital effects will give rise to a more complicated form for (8) as discussed by Kaplan and Lyons [3].
The effect is, however, small and will be neglected here.

3.3 THE EFFECT OF MAGNETIC ORDERING
OF THE LOCALIZED MOMENTS

If the localized moments are ordered throughout the crystal
they will give rise to a molecular magnetic field H_M which will
shift the energy of the otherwise degenerate spin-up and spin-
down electrons, i.e. $E_{k\uparrow} \neq E_{k\downarrow}$. This molecular field model is
equivalent to the rigid-band-shift model. The shift in energies
can be calculated exactly by diagonalizing the single electron
Hamiltonian and the molecular field term,

$$H = \sum_{k,s} E_{k,s} c_{k,s}^+ c_{k,s} + H_M. \qquad (10)$$

The molecular field is obtained by taking the thermal average
value of localized moments S_n in (8).

For the sake of generality we shall calculate the RKKY interaction for the conically ordered phase. The cone structure,
which contains as special cases both the ferromagnetic and the
spiral structure, is defined by the following parameterization
of the ionic moments:

$$\langle S_{R_n} \rangle = m(T)S\{\sin\theta\cos(Q \cdot R_n), \sin\theta\sin(Q \cdot R_n), \cos\theta\}, \qquad (11)$$

where $m(T)$ is the temperature dependent reduced magnetization,
θ is the cone angle and Q the spiral vector.

Using (8,11) we find the molecular field H_M to be used in
(10), which then can be diagonalized using standard techniques.
We find the new energies,

$$\varepsilon_{k,Q^\pm} = \varepsilon_p \pm ((\varepsilon_m - \Delta)^2 + \gamma^2)^{\frac{1}{2}}, \qquad (12)$$

where

$$\varepsilon_{\substack{p \\ m}} = \tfrac{1}{2}(E_{k-\frac{1}{2}Q} \pm E_{k+\frac{1}{2}Q}),$$

(13)

$$\Delta = Sm(T)\cos\theta j_{\mathrm{sf}}(k,k) \quad \text{and} \quad \gamma = Sm(T)\sin\theta j_{\mathrm{sf}}(k,k + Q),$$

the new wave functions are

$$\psi_{k,+} = \cos\phi|k - \tfrac{1}{2}Q,\uparrow\rangle + \sin\phi|k + \tfrac{1}{2}Q,\downarrow\rangle,$$ (14)

$$\psi_{k,-} = -\sin\phi|k - \tfrac{1}{2}Q,\uparrow\rangle + \cos\phi|k + \tfrac{1}{2}Q,\downarrow\rangle,$$

where

$$\tan\phi = \frac{\gamma}{\varepsilon_{k,Q}^{+} + \Delta - E_{k-\frac{1}{2}Q}}.$$ (15)

(12) shows the energies of the conduction electrons in the magnetically ordered phase.

For the ferromagnetic case $\theta = 0$ and $Q = 0$ and we find the rigid band model:

$$\varepsilon_k^{\uparrow\downarrow} = E_k \pm \Delta,$$ (16)

where Δ goes to zero when the magnetization vanishes at T_c.

For the spiral case we obtain the results discussed by Elliott and Wedgwood [8]. In this case, as in the general case of cone structure, the magnetic order produces gaps in the electron energy bands related to the spiral vector Q. This is of importance when calculating the temperature dependence of the spiral vector $Q(T)$, and in general the temperature dependence of the exchange interaction.

The magnetic order, the effect of which we have just included, is of course a consequence of the interaction between the local moments. In other words the interaction must be calculated self-consistently.

3.4 THE GENERALIZED RKKY INTERACTION IN THE ORDERED PHASE

We now proceed to calculate the RKKY interaction by taking into account the terms left in (8). $H_{\mathrm{sf}} - H_{\mathrm{M}}$ is not diagonal between the states (14), but the effect thereof can be found by second order perturbation theory.

The shift in energy is then, using (9,12):

$$\delta E^{(2)} = \sum_{n,n'} \sum_{i} \frac{\langle 0|H_{\mathrm{sf}}(R_n) - H_{\mathrm{M}}|i\rangle\langle i|H_{\mathrm{sf}}(R_{n'}) - H_{\mathrm{M}}|0\rangle}{(\varepsilon_0 - \varepsilon_i)},$$ (17)

where $|0\rangle$, $|i\rangle$ are the initial and intermediate states respectively and ε_0, ε_i the corresponding energies, from (12). We must remember that the electrons can be scattered only from an occupied state to an empty state according to the Pauli principle. This can be accounted for by the Fermi factors $f_k = (e^{(\varepsilon_k - E_F)/kT} + 1)^{-1}$. We shall assume that f_k is a step function, being 1 for energies smaller than the Fermi energy and 0 for larger energies†. We then find from (17) for the cone structure the following effective interaction between localized moments:

$$H_q = - J_q^{\parallel} S_q^z S_{-q}^z - \tfrac{1}{2} J_q^{\perp} (S_q^+ S_{-q}^- + S_q^- S_{-q}^+), \qquad (18)$$

where the wave vector dependent exchange interaction is

$$J_q^{\parallel} S_q^z S_{-q}^z =$$

$$\sum_{nn'} S_n^z S_{n'}^z e^{iq(R_n - R_{n'})} \frac{1}{2N} \sum_k f_k (1 - f_{k+q}) |j_{sf}(k, k+q)|^2$$

$$\times \left[(1 + \cos^2 2\phi) \left(\frac{1}{\varepsilon_k^+ - \varepsilon_{k+q}^+} + \frac{1}{\varepsilon_k^- - \varepsilon_{k+q}^-} \right) \right.$$

$$\left. + \sin^2 2\phi \left(\frac{1}{\varepsilon_k^+ - \varepsilon_{k+q}^-} + \frac{1}{\varepsilon_k^- - \varepsilon_{k+q}^+} \right) \right]$$

and   (19)

$$J_q^{\perp} (S_q^+ S_{-q}^- + S_q^- S_{-q}^+) =$$

$$\sum_{nn'} e^{iq(R_n - R_{n'})} \{ S_n^+ S_{n'}^- e^{iQ(R_n - R_{n'})} + S_n^- S_{n'}^+ e^{-iQ(R_n - R_{n'})} \}$$

$$\times \frac{1}{2N} \sum_k f_k (1 - f_{k+q}) |j_{sf}^*(k, k+Q+q) j_{sf}(k, k-Q+q)|$$

$$\times \left[\sin^2 2\phi \left(\frac{1}{\varepsilon_k^+ - \varepsilon_{k+q}^+} + \frac{1}{\varepsilon_k^- - \varepsilon_{k+q}^-} \right) \right.$$

$$\left. + (1 + \cos^2 2\phi) \left(\frac{1}{\varepsilon_k^+ - \varepsilon_{k+q}^-} + \frac{1}{\varepsilon_k^- - \varepsilon_{k+q}^+} \right) \right] .$$

---

† This is a very good approximation at all relevant temperatures.

For a ferromagnetic ordering $\phi = \theta = Q = 0$ from (13,15), and (19) reduces considerably. We find

$$J_q{}^{\|} = \frac{1}{N} \sum_k f_k(1 - f_{k+q})|j_{sf}(k,k + q)|^2$$

$$\times \left(\frac{1}{\varepsilon_k{}^{\uparrow} - \varepsilon_{k+q}{}^{\uparrow}} + \frac{1}{\varepsilon_k{}^{\downarrow} - \varepsilon_{k+q}{}^{\downarrow}} \right) , \qquad (20)$$

and

$$J_q{}^{\perp} = \frac{1}{N} \sum_k f_k(1 - f_{k+q})|j_{sf}(k,k + q)|^2$$

$$\times \left(\frac{1}{\varepsilon_k{}^{\uparrow} - \varepsilon_{k+q}{}^{\downarrow}} + \frac{1}{\varepsilon_k{}^{\downarrow} - \varepsilon_{k+q}{}^{\uparrow}} \right)$$

where $\varepsilon_k{}^{\uparrow\downarrow}$ are given by (16). This interaction is anisotropic, contrary to the paramagnetic RKKY interaction. For the ferromagnetic phase we obtain a $J_q{}^{\|}$ and a $J_q{}^{\perp}$ for the spin components parallel or perpendicular to the average moment direction. $J_q{}^{\|}$ involves only scattering of electrons with no spin-flip and $J_q{}^{\perp}$ only with spin-flip. $J_q{}^{\perp}$ can be measured directly by spin wave measurements, whereas $J_q{}^{\|}$ cannot be measured as a function of the wave vector. However, the magnetic contribution to the free energy is $-J_q{}^{\|}S_q{}^z S_q{}^z$ for $q = 0$. If $J_q{}^{\|}$ has a maximum for $q \neq 0$ it shows that, if for no other reasons, a non-ferromagnetic state would have lower free energy. However, it is necessary to calculate self-consistently the energy difference between the various phases.

3.5 THE MAGNITUDE OF THE s-f INTERACTION

Experimental information about the magnitude of the s-f interaction $j_{sf}(k,k')$ can be obtained directly by considering the polarization of the conduction electrons. This can be found either by measuring the total moment per atom or by means of an NMR technique for measuring the magnetic field, which the conduction electrons create at the nucleus.

For ferromagnetic ordering, the net polarization is given by the difference between the number of electrons with spin up and spin down. In the rigid band model ($T = 0$) this is to a good approximation:

$$n_{\uparrow} - n_{\downarrow} = \tfrac{1}{2}(\Delta_{\uparrow} - \Delta_{\downarrow})\rho(E_F)$$

$$= \Delta\rho(E_F) = Sm(T) \frac{1}{N} \sum_k j_{sf}(k,k)\rho(E_F), \qquad (21)$$

where $\Delta_{\uparrow,\downarrow}$ is the energy shift of the spin up and spin down electrons relative to the paramagnetic Fermi energy E_F, and $\rho(E_F)$ is the density of states at the Fermi energy. We obtained (21) by averaging over all momenta in (9a).

Since each unpaired electron contributes to the magnetic moment by $\frac{1}{2}g_S\mu_B = 1\mu_B$ we find for the average s-f interaction $(m(0) = 1)$

$$j_{sf}(0) = \frac{\Delta}{S} = \frac{2\delta M}{g_S\mu_B S\rho(E_F)} \, ,$$

where δM is the conduction electron polarization in μ_B and $g_S = 2$. From magnetization data [1] and a theory for the temperature dependence of the magnetization [9] we find the results given in table I.

The s-f interaction can be estimated from the ferromagnetic transition temperature as follows:

$$kT_c = 0.792 \, \frac{1}{3} \, J_0 \| J(J + 1), \tag{22}$$

where

$$J_0\| = [(g - 1)j_{sf}(0)]^2 \rho(E_F).$$

0.792 is a factor which corrects the molecular field value for T_c.

Having derived the expressions (19,20) for the indirect exchange interaction and estimated the interaction strength, we shall consider the actual calculation. The summation over the wave vectors k in (19,20) must be done numerically.

4. NUMERICAL METHODS

On the basis of the APW energy bands calculated by Louks [10] we can evaluate the sums in (19,20). The matrix element $j_{sf}(k,k + q)$ must be evaluated using the wave functions. The major contribution to the sum comes when the denominator is small. In other words, when the electrons are scattered from just below to just above the Fermi surface. This makes the numerical calculation difficult. A possible way is to sum over a very large number of k points and exclude the contribution when the denominator is smaller than a chosen number δ. This is called the root sampling method. This is a brute force principal value calculation (correct in the mathematical sense, if we let δ go to zero). However, it is very difficult to test the convergence of this procedure numerically. In fact the noise in the computer sets a limit for how small δ can be chosen and how fine a mesh of k points we can use —apart from the practical problem of increasing computer time. However, the method is

TABLE I

Data for the Heavy Rare Earth Metals

	J	L	S	g	$\rho(E_F)$ $\left(\dfrac{\text{States}}{\text{Ryd}}\right)$	E_F (Ryd)	δM (μ_B)	Δ (Ryd)	$j_{sf}(0)$ (δM)	$j_{sf}(0)$ (T_C)
Gd	7/2	0	7/2	2	25.6	0.440	0.55	0.021	0.006	0.006
Tb	6	3	3	3/2	28.0	0.509	0.41	0.015	0.005	0.006
Dy	15/2	5	5/2	4/3	27.7	0.513	0.41	0.015	0.006	0.007
Er	15/2	6	3/2	6/5	23.0	0.451	0.22	0.010	0.006	0.009

$\rho(E_F)$ is the calculated density of states, Δ is half the ferromagnetic splitting, and $j_{sf}(0)$ the deduced s-f interaction in Rydbergs. We notice it is almost independent of the elements. The values estimated from the ferromagnetic transition temperature are given in the last column.

simple and was used by Liu *et al.* [11] and in several of the results to be discussed [12]. The convergence seems to be good,
and the computing time reasonable with a mesh of 450,000 points
in the total Brillouin zone. These calculations were simplified
by the assumption that the matrix element $j_{sf}(k,k + q)$ was independent of k and only dependent on the difference q, i.e.
$j_{sf}(k,k + q) \sim j_{sf}(q)$.

In order to test the convergence and also to make it feasible
to include a **k** dependence of $j_{sf}(k,k + q)$ a different numerical
method was used. In this method the Brillouin zone is divided
into a relatively small number of micro-cells. Inside each cell
are the constant energy surfaces ε_k approximated by planes.
This makes it possible to integrate analytically inside each
micro-cell. The integrals are only divergent if the energy sur-
faces ε_k and ε_{k+q} are exactly parallel. This will occur very
rarely. This so called linearized method was developed for dens-
ity of state calculations by Gilat and Raubenheimer [11] and was
later simplified by Jepsen and Andersen, who used it for calculating Fermi surface areas. The sums in (19,20) are more complicated and have not previously been calculated using this
method. We shall therefore briefly describe it. The Brillouin
zone is divided into micro-cells of the shape of tetrahedra [14]
of a volume V as shown on figure 1. In each corner are the energies

$$\varepsilon_k{}^i = \varepsilon_1, \varepsilon_2, \varepsilon_3, \varepsilon_4 \quad \text{and} \quad \varepsilon_{k+q}{}^i = e_1, e_2, e_3, e_4.$$

Since the constant energy surfaces are approximated by planes
the constant energy difference $\omega = \varepsilon - e$ is also a plane. The
problem is therefore to integrate

$$I = \int_{\omega_{min}}^{\omega_{max}} \frac{1}{\omega} A(\omega)d\omega \tag{23}$$

over the part P of the tetrahedra for which $\varepsilon < E_F$ and $e > E_F$,
where the area of the constant energy difference plane inside
P is $A(\omega)$. P may be a complicated polyhedron because of the restrictions coming from the Fermi f_k factors in (19,20).
$f_k(1 - f_{k+q})$ can by symmetry considerations be replaced by
$\frac{1}{2}(f_k - f_{k+q})$. We do not use the latter form (although it simplifies the calculation considerably) because the result then
is given as the difference between two large numbers which may
be inaccurate numerically. For illustration we shall consider
the case where the condition "$\varepsilon < E_F$ and $e > E_F$" is fulfilled
for the whole tetrahedron.

The area $A(\omega)$ is then simply the area of a cut of the tetrahedron perpendicular to the ω planes. This area is clearly a
quadratic function of ω, being zero for ω outside the range
$\omega_{max} - \omega_{min}$. The area can easily be expressed by geometrical

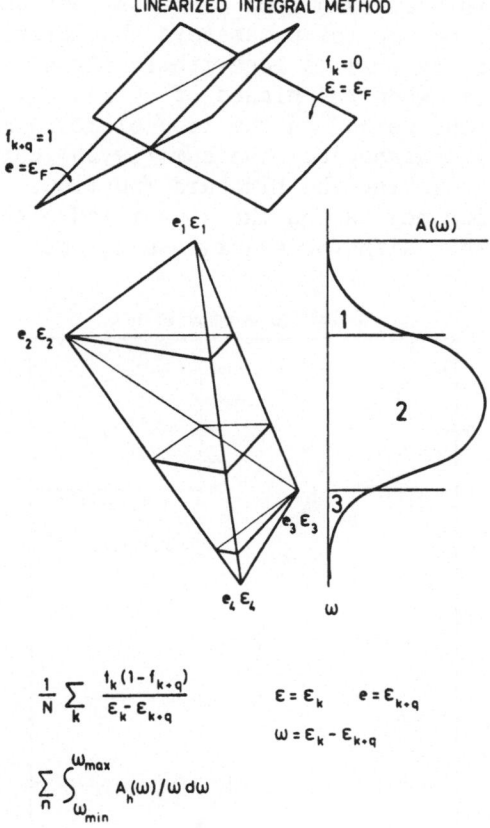

Figure 1 - At the top is shown the constant energy planes for $\varepsilon_k = \varepsilon_F$ and $\varepsilon_{k+q} = \varepsilon_F$. The Brillouin zone is divided into tetrahedra, as shown below, of constant volume V, here oriented so that the direction of increasing energy difference ω is vertical. The cut with the constant ω planes are shown. The area of these cuts are quadratic functions of ω in the regions 1, 2 and 3. The sum then reduces to the integral shown in the lowest line. In general the plane $\varepsilon_k = \varepsilon_F$ and $\varepsilon_{k+q} = \varepsilon_F$ may also cut the tetrahedra. In this case the integration must only be over the part P for which the $f_k(1 - f_{k+q})$ condition is fulfilled.

considerations in terms of the corner energies ε^i and e^i and V, it is not necessary to calculate the normal vector to the ω planes. Therefore, in the case where we must integrate over the whole tetrahedron (23) is simply

$$I = \sum_n \int_{\omega_{min}}^{\omega_{max}} [a_n(\varepsilon^i, e^i) + b_n(\varepsilon^i, e^i)\omega + c_n(\varepsilon^i, e^i)\omega^2] \frac{1}{\omega}\, d\omega, \quad (24)$$

where the sum is over each type of cross-section (triangle or square) and a_n, b_n, c_n the parameters characterizing this. We notice that the integral is logarithmically divergent when $\omega_{min} = \omega_{max}$, i.e. when the planes of ε_k and ε_{k+q} are parallel.

We can test the method on the free electron model where the energy bands are parabolic. The sum (19,20) can then be integrated exactly giving the Lindhard function. The result of the root sampling method and the linearized method is shown on figure 2, together with the exact result. We see that the

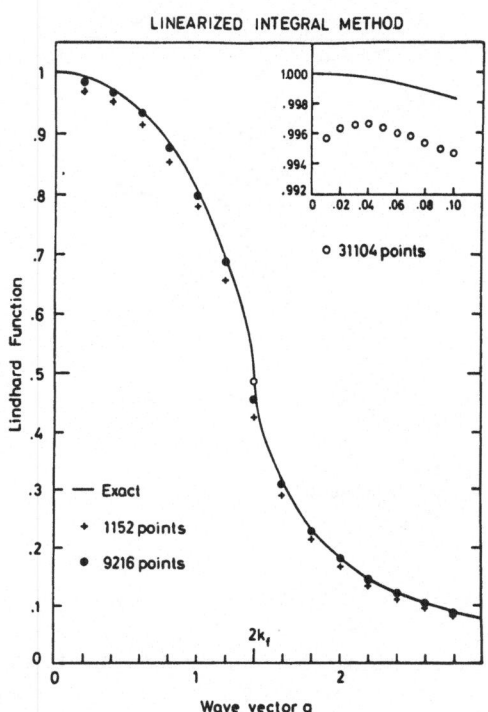

Figure 2(a) - The generalized susceptibility for free electrons. The points are the numerical results for the linearized integral method for meshes with 1,000, 9,000 and 30,000 points in the entire Brillouin zone (hcp) with $k_F = 0.7$ of the zone boundary wave vector ($\Gamma - K$). We notice a very good agreement with the theoretical Lindhard function already with the mesh with 9,000 points. The insert shows that the most difficult region for $q \to 0$ is reproduced well. The systematic deviation is due to the fact that the integration is performed in the inscribed polyhedra in the Fermi sphere. It has both convex and concave parts and the volume is better approximated by the polyhedra in a realistic system.

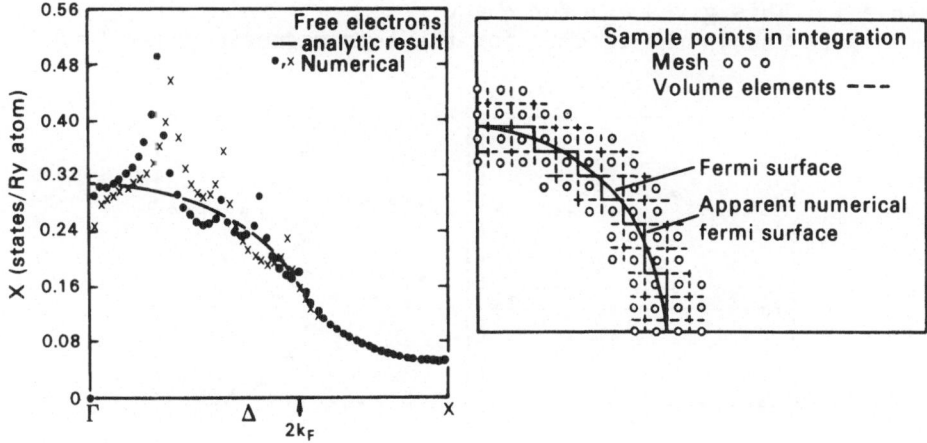

Figure 2(b) - The result of the root sampling method [16] in a coarse mesh of 27,000 points. We notice that spurious peaks occur because of the mesh for k values less than $2k_F$. The convergence is good in a mesh with 450,000 points, not shown [11,12].

linearized method gives an excellent result for only 9,000 k-points in the entire Brillouin zone.

5. RESULTS

Let us start by considering what effect the magnetic ordering has on the RKKY exchange interaction. That is the same as asking what is the intrinsic temperature dependence. The formulae were developed in (19,20). We shall only be interested in a qualitative answer, which will show the general magnitude and the direction of the effects. We therefore make the simplifying assumption that for this purpose we can consider the matrix element $j_{sf}(k,k + q)$ only to depend on the difference q. Our problem then reduces to calculating the electronic static susceptibility.

$$\chi_q^{\alpha\beta} \sim \frac{1}{N} \sum_k \frac{f_k(1 - f_{k+q})}{(\varepsilon_k^\alpha - \varepsilon_{k+q}^\beta)}$$

We determine the matrix element $|j_{sf}(q)|^2$ from experiment, by comparing $\chi_q^\perp$, the calculated sum without it, with the $J_q^\perp$ obtained from spin wave measurements. The matrix element is assumed to be insensitive to the magnetic structure and is used in obtaining the exchange interaction in other magnetic phases. The absolute scale of $J_q^\perp$ cannot be determined from the spin waves. The scale is found from the transition temperature T_N and coincidently from the conduction electron polarization

table I. This gives $J_q\parallel$ for $q = 0$.
   Figure 4 shows the results for the ferromagnetic phase $T = 0$

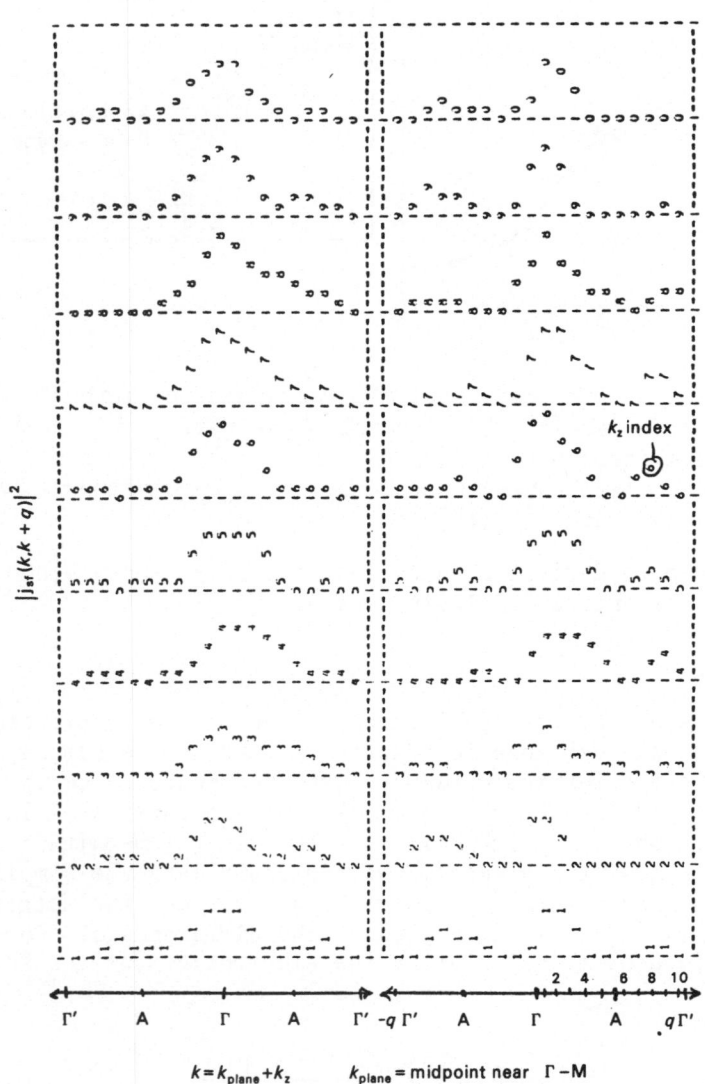

Figure 3 - Example of $\left|j_{sf}(k,k + q)\right|^2$ for Gd calculated
by the APW method by Harmon [5].

for Gd, Tb, Dy, and Er using the APW energy bands and the root
sampling method with 450,000 points in the entire Brillouin zone
(the linearized method was also used as a test, it gave essentially the identical result and is not shown). It is clear that
$J_q\parallel$ and $J_q^\perp$ differ significantly for all materials. The dots
show the points compared with the experimental $J_q^\perp$; the calculation was done for sixty equidistant q-values. For terbium the

experimental $J_q^\perp$ shows no maximum for $q \neq 0$, whereas the calculated $J_q^\parallel$ shows that Tb has a tendency to form a spiral structure even in the ferromagnetic phase. The enhancement of the
maximum for $q \neq 0$ is also evident for Dy and Er in which the
spiral region is large. The opposite effect occurs for Gd,
where $J_q^\parallel$ shows that Gd should not form a spiral phase, and nor
it does. Furthermore, it is clear that the maxima in $J_q^\parallel$ occur
at q-values very close to the experimental spiral vectors (indicated with an arrow) and that it is significantly displaced
from the peaks in $\chi(q)$, which is directly related to the presence
of flat parallel pieces of Fermi surface. The matrix element
thus plays an important rôle in determining the wave vector dependence of the exchange interaction. The semi-empirically found
wave vector dependence of the matrix element is very similar for
all materials, despite the rather different $\chi(q)$ functions.
This is encouraging for the present analysis. Overhauser [15]
has argued that the matrix element should follow the $4f$ form
factor. By extending his model to include the Bloch character
of the conduction electrons, we would expect a narrow central
peak originating from the conduction electrons. This is the
form found in figure 4.

The energy difference between the ferromagnetic and spiral
phases is, as judged from the $T = 0$ ferromagnetic data figure 4,
for Gd, Tb, Dy, and Er in per cent of the exchange energy: -14%,
+5%, +5%, +12%. This gives for Tb, Dy, and Er a stabilization
of the spiral phase by 10 K/ion times the reduced magnetization
squared. The magnetoelastic stabilization of the ferromagnetic
phase is for these materials at the ferromagnetic-spiral transition typically 1 K/ion.

The last column in figure 4 shows a calculation at half the
saturation moment of $\chi_0^\parallel(q)$ in the ferromagnetic phase and
$\chi_Q^\perp(q)$ in the spiral phase, with spiral vector Q. The contribu-
tion to the free energy is proportional to $-|j(Q)|^2\chi_Q^\perp(0)$.
$\chi_Q^\perp(0)$ as a function of the spiral vector Q follows closely that
of $\chi_0^\parallel(q)$ as a function of q, which shows that the most probable
spiral vector coincides with that found in the ferromagnetic
phase. The precise location is sensitive to the wave vector dependence of the matrix element.

The above simple calculation gave encouraging results and is
a natural extension of the calculation of the exchange interaction in the paramagnetic phase [13]. However, the next step
is to consider the matrix element more seriously. We shall do
this for the paramagnetic phase with no band splitting. Harmon
[5] has by means of the APW functions calculated $|j_{sf}(\mathbf{k},\mathbf{k} + \mathbf{q})|^2$
for the simplest material Gd. A few of the matrix elements are
shown on figure 3. They generally show the $\mathbf{q}$-dependence we anticipated, namely a sharp peak at $\mathbf{q} = 0$. On the other hand it
is clear that they are quite sensitive to the value of $\mathbf{k}$, and
irregularities occur as a function of $\mathbf{q}$, which comes from the
change in wave function character at band crossings.

It is therefore of importance to carry out the complete sum

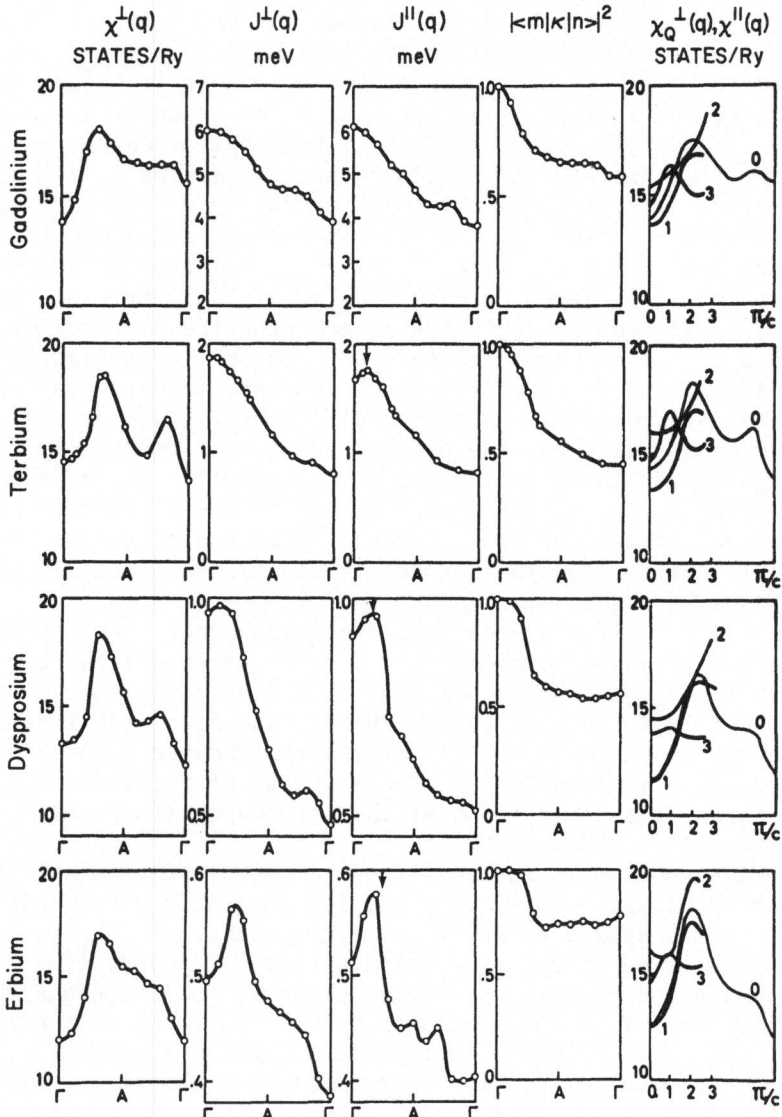

Figure 4 - The perpendicular susceptibility $\chi(0)^{\perp}$, the perpendicular experimental exchange interaction $J_q^{\perp}$, the calculated parallel exchange interaction $J_q^{\parallel}$ and the deduced matrix element $|j(q)|^2/|j(0)|^2 = |\langle m|\kappa|n\rangle|^2$ for the ferromagnetic phase (splitting: 0.008 Ryd). The last column shows $\chi_Q^{\perp}(q)$ in the spiral phase for $Q_0 = 0$, $Q_1 = \pi/6c$, $Q_2 = 2\pi/6c$ and $Q_3 = 3\pi/6c$ (splitting 0.004 Ryd); the corresponding ferromagnetic $\chi(q)^{\parallel}$ is also shown (marked 0).

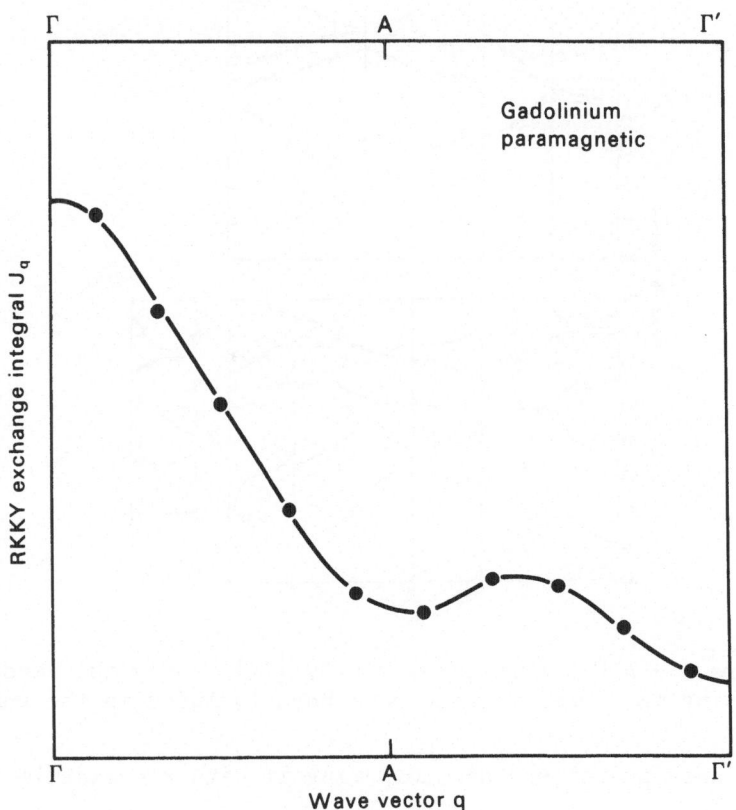

Figure 5 - Preliminary result for the calculated RKKY interaction, using Harmon's APW matrix element [5]. Only
scattering relevant for an extended zone has been included as a first approximation.

(20) including the k-dependence of $j_{sf}(k,k + q)$. Preliminary
results are shown on figure 5. The calculation is performed by
the linearized integral method (23) with 7,000 k points in the
entire Brillouin zone and with $j_{sf}(k,k + q)$ included rectangu-
larily at 1,250 k points. The result is the first direct calculation of the RKKY interaction for Gd with no adjustable parameters. The q-dependence of J_q is in satisfactory agreement
with that obtained experimentally from spin wave measurements,
shown as $J_q^\perp$ in figure 4. An important question to be investigated is if the major contribution to J_q comes from the part of
the sum for which $j_{sf}(k,k + q)$ is insensitive to k, or if both
the k- and q-dependence are equally important, the last case
would indicate that the matrix element is an important in determining the magnetic properties of the heavy rare earths as the
Fermi surface topology.

Work on these questions is in progress. A large number of
problems are waiting to be done in developing and refining the

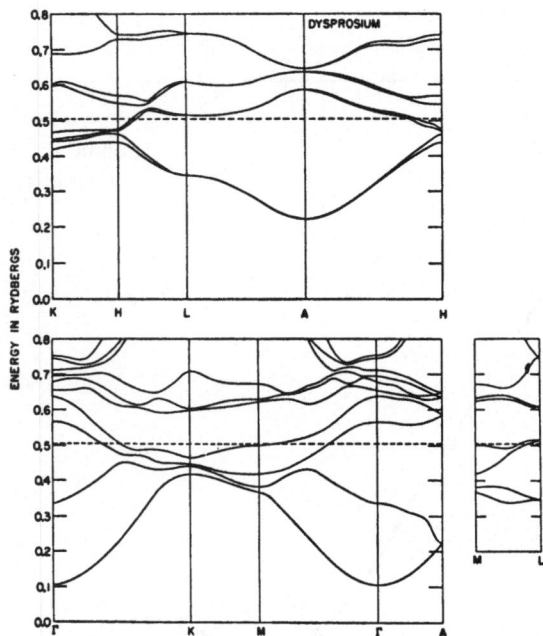

Figure 6 - APW energy bands for Dy [*11*]. Only the bands
crossing the Fermi surface have been included in the sum.

theory, here presented, and comparing it with the experimental
facts.

REFERENCES

1. Elliott, R.J. (ed.). (1972). *Magnetic Properties of Rare
   Earth Metals*, (Plenum Press, London).
2. Houman, J.G., Jensen, J. and Touborg, P. (1974). (To be
   published).
3. Kaplan, J.A. and Lyons, D.H. (1963). *Phys. Rev.*, **129**, 2072.
4. Louks, T.L. (1967). *Augmented Plane Wave Method*, (W.A.
   Benjamin, Inc., New York).
5. Harmon, B.N. (1973). *Conduction Electron Polarization,
   Spin Densities and The Neutron Magnetic Form Factor of
   Gadolinium*, (Thesis).
6. Koelling, D.D. (1972). *J. Phys. Chem. Sol.*, **33**, 1335.
7. Herman, F. and Skillman, S. (1963). *Atomic Structure
   Calculations*, (Prentice-Hall, Inc., Englewood Cliffs, New
   York).
8. Elliott, R.J. and Wedgwood, F.A. (1963). *Proc. Phys. Soc.*,
   81, 846.
9. Lindgård, P.A. and Danielsen, O. (1974). (To be published).
10. Keeton, S.C. and Louks, T.L. (1968). *Phys. Rev.*, **168**, 672.
11. Liu, S.H., Gupta, R.P. and Sinha, S.K. (1971). *Phys. Rev.*,

B4, 1100.
12. Lindgård, P.A. and Liu, S.H. (1973). *Proceedings of the International Conference on Magnetism, Moscow.*
13. Gilat, G. and Raubenheimer, L.J. (1966). *Phys. Rev.*, **144**, 390.
14. Jepsen, O. and Andersen, O.K. (1971). *Solid State Commun.*, **9**, 1763.
15. Overhauser, A.W. (1963). *J. Appl. Phys.*, **34**, 1019.
16. Evenson, W.E. and Liu, S.H. (1968). *Phys. Rev.*, **178**, 783.

LATTICE EFFECTS AND THE MAGNETIC BEHAVIOUR
OF RARE EARTH COMPOUNDS

BERNARD R. COOPER

*General Electric Research and Development Center,
Schenectady, New York* 12301, *USA*

1. INTRODUCTION

The fact that the rare earth ionic moment in a crystal typically has a substantial orbital contribution qualitatively affects the nature of the magnetic behaviour. Because there is a large orbital contribution to the total ionic moment, the effect of the crystal electric field is qualitatively more striking than for magnetic systems of the $3d$ transition elements. Also magnetoelastic effects and anisotropic exchange can be qualitatively important, or even dominant, for the magnetic behaviour.

With this in mind, I am going to talk about two types of rare earth systems, the behaviour of which have attracted much interest in recent years. First, I want to talk about singlet ground state magnetism. The induced magnetic ordering, and the associated behaviour of the collective magnetic excitations — the magnetic excitons, has been studied extensively in the rare earth monopnictides — especially the TbSb system, and in metallic Pr — both fcc and dhcp, as well as in Pr_3Tl which is essentially a type of fcc Pr which is experimentally better behaved. As I shall briefly review, the basic picture for the bulk magnetic properties and magnetic ordering of these materials was already well understood probably three or four years ago. On the other hand, important questions still remain open about the magnetic exciton behaviour and its relationship to the magnetic ordering process. Indeed the question of whether or not there is a soft magnetic mode involving excitation from the singlet crystal field ground state still remains open. I will therefore concentrate on discussing the magnetic exciton behaviour and its relationship to the magnetic ordering.

The second type of system I will talk about seems to involve behaviour that is intrinsically more involved and difficult to understand than the singlet ground state systems. This is the behaviour of the cerium monopnictides, especially CeSb and CeBi. Here, in addition to crystal field effects, a highly anisotropic interaction is present. Indeed, this interaction dominates the magnetic behaviour. It is especially appropriate to discuss the behaviour of these materials at the Winter School of the University of Wrocław, since the striking magnetic structures found would be unique, were they not also shown by some actinide compounds that have been studied at the Academy of Sciences Institute in Wrocław and at Argonne National Laboratory.

2. SINGLET GROUND STATE MAGNETISM

Figure 1 shows some typical crystal field level schemes for rare earth systems of interest. The symmetry type of each crystal field level is indicated on the left and the degeneracy on the right. For the systems of interest, the splitting from the singlet ground state to the first excited state is characteristically a few tens of degrees Kelvin.

A good deal of theoretical discussion has focussed on the two singlet level system, where the ground state $|0_c\rangle$ is a singlet, and the only excited state $|1_c\rangle$ is also a singlet at an energy Δ above $|0_c\rangle$. For such a system, all magnetic moments present are induced, that is they result from polarization or admixture effects of the states for zero magnetic field. Therefore it has been felt that the two singlet level system offered a simple model that could yield the essence of understanding induced magnetic

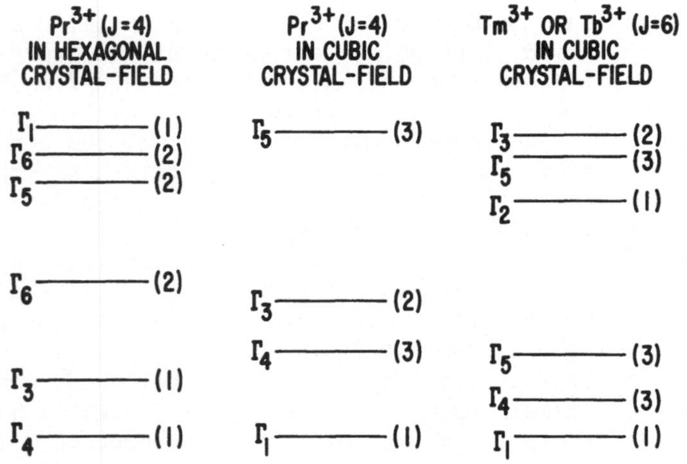

Figure 1 - Crystal field level schemes for rare earth ions in sites of hexagonal or cubic symmetry.

ordering. Whilst — as we shall see — the presence of additional excited states does have important effects, the motivation for
studying the two singlet state system is well justified, especially since, formally, this system is exactly equivalent to an
Ising system in a transverse magnetic field.

We consider a model Hamiltonian,

$$\mathcal{H} = \sum_i V_{ci} - \sum_{\langle i,j \rangle} J_{ij} \vec{J}_i \cdot \vec{J}_j - g\beta H \sum_i J_{iz}, \tag{1}$$

with crystal field, exchange and Zeeman terms.

In the molecular field picture, one can think of the molecular field as mixing some of the zero field singlet excited state
into the zero field singlet ground state to give a polarized
ground state possessing a magnetic moment. On the other hand,
the existence of such a polarized ground state is necessary in
order to have a molecular field. Thus there is a sort of self-
consistent, or 'boot strap' condition, that the ratio of exchange
interaction to crystal field splitting must exceed some critical
value to have magnetic ordering even at zero temperature.

Bleaney [1] pointed out that one way of seeing the approach
to this threshold is by looking at the susceptibility in the molecular field approximation. The crystal field only inverse susceptibility is,

$$\frac{1}{\chi} = \frac{\Delta}{2g^2\beta^2\alpha^2} \left[\tanh\left(\frac{\Delta}{2T}\right) \right]^{-1}, \tag{2}$$

where

$$\alpha \equiv \langle 0_c | J_z | 1_c \rangle. \tag{3}$$

In the presence of exchange in the molecular field approximation,
$1/\chi$ is given by (2) with the replacement,

$$\left[\tanh\left(\frac{\Delta}{2T}\right) \right]^{-1} \rightarrow \left[\tanh\left(\frac{\Delta}{2T}\right) \right]^{-1} - A, \tag{4}$$

with

$$A \equiv 4J(0) \frac{\alpha^2}{\Delta}, \tag{5a}$$

$$J(\vec{k}) \equiv \sum_j J_{ij} e^{i\vec{k} \cdot (\vec{r}_i - \vec{r}_j)}. \tag{5b}$$

So, as seen in figure 2, for ferromagnetic exchange, the $1/\chi$
versus T curve shifts rigidly downward until the critical value
of exchange, $A = 1$, is reached at which the susceptibility

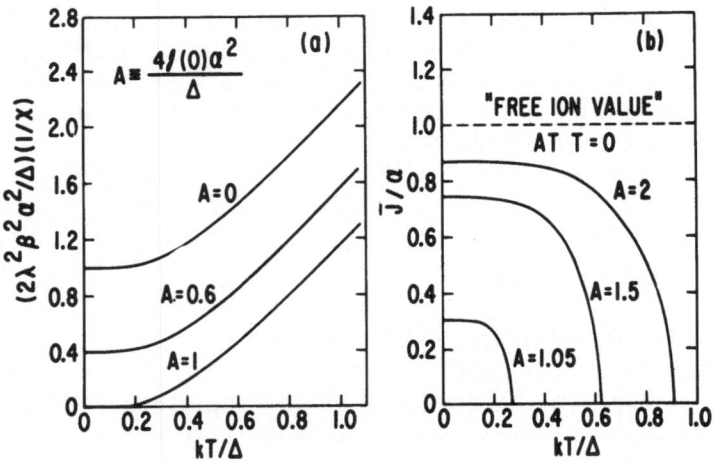

Figure 2 - (a) Inverse susceptibility, and (b) magnetization (in dimensionless form) versus temperature for the
two singlet level model for varying ferromagnetic exchange.
(After [2]).

diverges at zero temperature, signalling the onset of magnetic
ordering. Above the threshold value of exchange, the two key
qualitative features of the magnetic behaviour are [2] the abrupt
increase of the ordering temperature, changeing into a linear
increase for larger exchange; and the corresponding abrupt increase of the ordered moment at zero temperature, which approaches the free ion moment with increasing exchange. Trammell's [3] recognition, in the early 1960's, of systematic trends
[4] for the rare earth pnictides, related to this behaviour,
played a large rôle in arousing interest in singlet ground state
magnetism. Another characteristic feature of induced magnetic
ordering is the specific heat behaviour [2]. For an ordering
temperature lower than the crystal field splitting, the magnetic
ordering anomaly becomes quite inconspicuous compared to the
broad Schottky peak. This, of course, becomes increasingly true
as T_C falls, since the total entropy under the specific heat
curve remains constant.
    The molecular field theory can easily be carried over to cal-
culations involving the full crystal field level schemes. As
shown in figure 3, such a theory was quite successful in understanding the behaviour of the $Tb_zY_{1-z}Sb$ system [5]. Figure 3
shows the inverse susceptibility per Tb ion versus temperature
in $Tb_zY_{1-z}Sb$ as the Tb concentration is increased. For this system, to a very good approximation, the crystal field splitting
remains constant, and the molecular field exchange constant
varies linearly with Tb concentration. Since the exchange is
antiferromagnetic, the curves shift upward with increasing Tb
concentration. One can fit the behaviour, as shown, using a simple theory with only two adjustable parameters, one for the

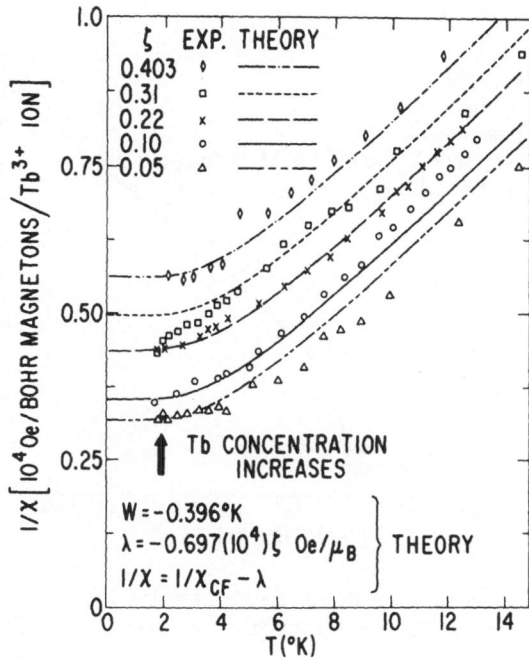

Figure 3 - Inverse susceptibility per Tb ion as a function of temperature for $Tb_\zeta Y_{1-\zeta} Sb$. Experiment is compared to molecular field theory for $\zeta \leqslant 0.403$. (After [5]).

crystal field splitting and one for the exchange. Furthermore, as shown in figure 4, this theory then yields the critical concentration for antiferromagnetic ordering at $T = 0$, and the variation of the Néel temperature with Tb concentration. The agreement between theory and experiment is extremely good; where I want to emphasize that at this point in the theory there are no adjustable parameters. There are also specific heat measurements [6] for this system that are consistent with the behaviour expected from a molecular field theory. There has also been a good deal of work on the fcc Pr metal system [7] and the Pr_3Tl system [8] which overall fits in with the bulk magnetic behaviour expected on the basis of a molecular field type theory. Thus as of a few years ago, one felt that the basic picture for the induced magnetization ordering process was well understood. On the hand, understanding of the collective excitation behaviour and the relationship of that behaviour to the magnetic ordering process has developed more slowly.

There was a rather substantial amount of theoretical work [9] on the magnetic exciton behaviour before any neutron scattering experiments were done. Much of this work concentrated on the two singlet level system.

The first work [10,11] studying the magnetic exciton behaviour used a Bogoliubov type of approximation to treat effective bosons.

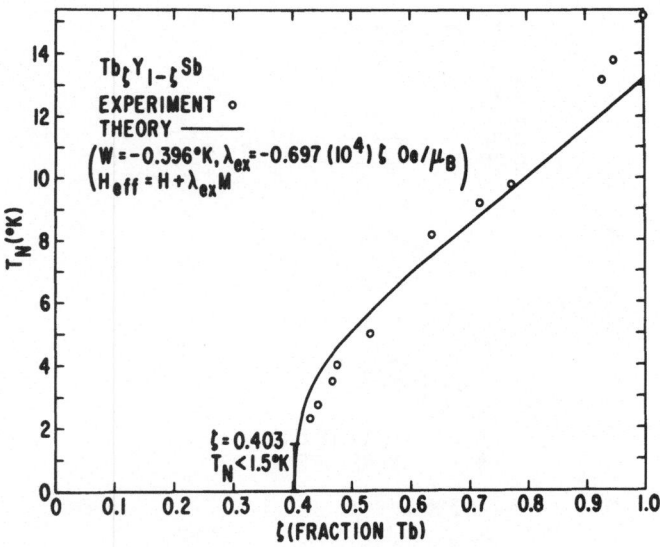

Figure 4 - Variation of Néel temperature with concentration of terbium in $Tb_\zeta Y_{1-\zeta} Sb$ for $\zeta \geqslant 0.403$. (After [5]).

The Bogoliubov type of approximation proceeds by assigning fermion operators to each molecular field energy level,

$$\mathcal{H}_0 = \sum_{\substack{i \\ sites}} \sum_{\substack{n \\ levels}} \varepsilon_n d_{in}{}^\dagger d_{in}; \tag{6}$$

and the correction giving the difference between the exact Hamiltonian and the molecular field Hamiltonian is

$$\mathcal{H}_1 = - \sum_{i,j} \sum_{n,m,n',m'} J_{ij} \langle n | \vec{\mathcal{J}}_i - \langle \vec{\mathcal{J}}_i \rangle | m \rangle$$

$$\cdot \langle n' | \vec{\mathcal{J}}_j - \langle \vec{\mathcal{J}}_j \rangle | m' \rangle d_{in}{}^\dagger d_{im} d_{jn'}{}^\dagger d_{jm'}, \tag{7}$$

It should be noted that the use of Fermi statistics is justified only at low temperatures. At high temperatures the fermion representation admits states with more than one of the single ion states occupied for a single ion. I will return to this point and a further discussion of the use of the fermion representation later. The one great advantage of the fermion representation and the boson representation derived from it, is that it is simple to treat all the crystal field levels for the level scheme of a rare earth ion such as Pr^{3+} or Tb^{3+}.

From the fermion Hamiltonian, one arrives at the boson picture by introducing operators

$$a_{in} \equiv d_{i0}^{\dagger} d_{in}, \qquad a_{in}^{\dagger} \equiv d_{in}^{\dagger} d_{i0}, \qquad n \neq 0, \qquad (8)$$

where 0 labels the molecular field ground state. The commutation relationships for the a and $a^{\dagger}$ are,

$$[a_{in}, a_{jn}] = [a_{in}^{\dagger}, a_{jn}^{\dagger}] = 0, \qquad (9a)$$

$$[a_{in}, a_{jn}^{\dagger}] = \delta_{ij} n_{i0} - \delta_{ij} n_{in} \approx \delta_{ij}, \qquad (9b)$$

so that for $T \approx 0$, the a and $a^{\dagger}$ are boson operators.

The Hamiltonian can then be written as a quadratic form in terms of these boson operators. Diagonalization of the quadratic boson Hamiltonian then proceeds by standard technique. Figure 5 shows the boson excitation behaviour for the two singlet model at $T = 0$ with increasing exchange. The characteristic feature is the presence of a soft mode at $\vec{q} = 0$, which goes un-

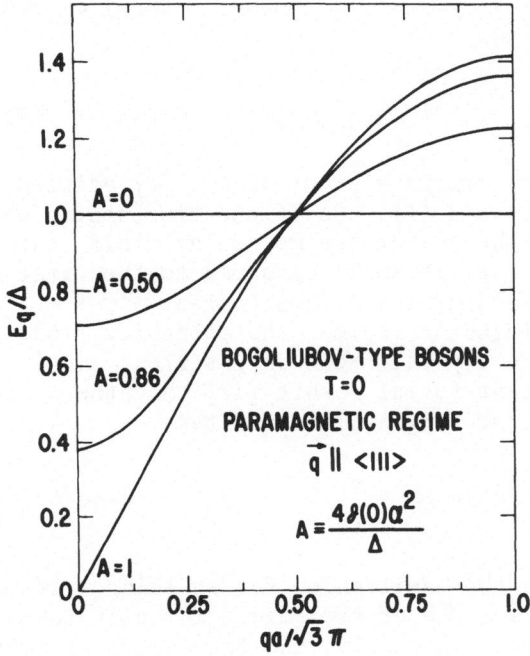

Figure 5 - Dispersion curve at $T = 0$ of elementary excitations in the two singlet level model for a simple cubic lattice with nearest neighbour ferromagnetic exchange. This is shown calculated in the Bogoliubov type of approximation for several values of A in the paramagnetic regime. (After [9]).

stable at a value of exchange equal to the critical value for
ferromagnetic ordering in the molecular field approximation.
This coincidence with the molecular field criterion for magnetic
ordering is a built in feature of the effective boson theory.

A more definitive study of the two singlet level system is
possible using a pseudo-spin $\frac{1}{2}$ representation [12,13]. This
makes use of the fact that any two level system can be represented exactly by spin $\frac{1}{2}$ operators. To treat the present problem,
one uses the molecular field states as the basis onto which one
projects the pseudo-spin Hamiltonian.

The Hamiltonian of interest has a crystal field and an exchange term:

$$\mathcal{H} = \sum_i V_{ci} - \sum_{\langle i,j \rangle} J_{ij} \vec{J}_i \cdot \vec{J}_j. \tag{10}$$

In the pseudo-spin representation this becomes,

$$\mathcal{H} = \sum_i \epsilon_0 S_{iz} - \sum_{\langle i,j \rangle} J_{ij}\{\alpha_{01}S_i + \alpha_{10}S_i^+ + \alpha_{00}S_i^- S_i^+ + \alpha_{11}S_i^+ S_i^-\}$$

$$\tag{11}$$

$$\times\{\alpha_{01}S_j^- + \alpha_{10}S_j^+ + \alpha_{00}S_j^- S_j^+ + \alpha_{11}S_j^+ S_j^-\},$$

where α_{mn} is the matrix element of $\vec{J} - \langle \vec{J} \rangle$ between the molecular
field states $|0\rangle$ and $|1\rangle$. So if the true states of the system
differ only slightly from the molecular field states, the lengthy
second term in (11) is small compared to the first term, which
is just the molecular field Hamiltonian.

In the paramagnetic régime, the molecular field states are
identical to the crystal field only states; and in that case we
have the important formal result [12] that the pseudo-spin Hamiltonian for the two singlet level system,

$$\text{paramagnetic } \mathcal{H} = \sum_i \Delta S_{iz} - 4 \sum_{\langle i,j \rangle} J_{ij} \alpha^2 S_{ix} S_{jx}, \tag{12}$$

is identical to the Hamiltonian for an Ising system in a transverse field, where the crystal field formally takes the rôle of
the applied field. Actually, one can always project the pseudo-
spin Hamiltonian onto the crystal field only states even in the
magnetically ordered régime, although in that case the true
states of the system differ substantially from the crystal field
only states. However, this means that the formal equivalence
with the Ising transverse field problem always holds true.

Once the Hamiltonian is in the pseudo-spin form, one finds
the excitation energies at $T = 0$ by treating the equations of motion of the Fourier transformed generating operators, $S^+(k)$ and
$S^-(k)$, in the usual way. The problem as always is to choose a
tractable, but physically meaningful way for decoupling the equa-

tions of motion. The simplest decoupling scheme is the random phase approximation, where $T = 0$, one replaces S_z in the equations of motion by its ground state expectation value:

$$\text{RPA:} \qquad S_{gz}S_f^{\pm} \xrightarrow[g \neq f]{} \langle S_z \rangle S_f^{\pm}. \qquad (13)$$

The RPA takes account of the fact that the true ground state of the system differs from the molecular field ground state. The expectation value of S_z is a measure of that difference. If the true ground state were the molecular field ground state, then the expectation value of S_z at $T = 0$ would be $-\frac{1}{2}$. Instead, $\langle S_z \rangle$ at $T = 0$ differs typically by about 4% from $-\frac{1}{2}$. This indicates the admixture of the molecular field excited state into the true ground state.

The RPA does not take account of correlations between excitations on neighbouring sites. Such correlation effects can be included [12] by using a two site correlation approximation (TSCA) for decoupling, where one replaces operators of the form

$$S^+(\vec{k} - \vec{k}_1 + \vec{k}_2)S^+(\vec{k}_1)S^-(\vec{k}_2)$$

by a linearization approximation taking all possible expectation values for two spin operators. So in the TSCA one includes correlation functions of the form

$$\varepsilon = - 2\langle S_z \rangle \frac{1}{N} \sum_k \gamma_k [\langle S_k^+ S_k^- \rangle + \langle S_k^+ S_{-k}^+ \rangle], \qquad (14a)$$

where

$$\gamma \equiv \frac{J(\vec{k})}{J(0)}. \qquad (14b)$$

Both the RPA and TSCA can be generalized to finite temperature by studying the equations of motion of the retarded time Green's functions corresponding to the generating operators. The same result is obtained from the zero temperature results by simply regarding $\langle S_z \rangle$ and $\langle S_k^+ S_k^- \rangle$, etc., as thermal expectation values rather than ground state expectation values.

I want to emphasize that a fully self-consistent RPA calculation involves taking summations over the excitation spectrum in order to determine $\langle S_z \rangle$ and the magnetization. I am emphasizing the two singlet level model, because that is the only model for which such fully self-consistent RPA calculations have been done; and this may be important for discussing the soft mode question in the critical regime. The set of self-consistent equations determining the excitation spectrum E_k and the magnetization $\langle J \rangle$ is,

$$a = - \frac{1}{2\langle S_z \rangle A}, \qquad (15)$$

$$\langle S_z \rangle = -\frac{1}{2} \frac{1}{(\phi_a + \psi_a)} , \tag{16}$$

$$\phi_a = \frac{1}{2N} \sum_k (1 - a^2 \gamma_k)^{-\frac{1}{2}} \coth(\tfrac{1}{2} E_k), \tag{17a}$$

$$\psi_a = \frac{1}{2N} \sum_k (1 - a^2 \gamma_k)^{-\frac{1}{2}} \coth(\tfrac{1}{2} E_k), \tag{17b}$$

$$\frac{E_k}{T} = \frac{(1 - a^2 \gamma_k)^{\frac{1}{2}}}{\frac{T_a}{\Delta}} , \tag{18}$$

where one defines,

$$J(\vec{k}) \equiv \sum_j J_{ij} e^{i\vec{k} \cdot \vec{r}_{ij}}, \tag{19a}$$

$$\gamma_k = \frac{J(\vec{k})}{J(0)} , \tag{19b}$$

$$\alpha = \langle 1_c | J_z | 0_c \rangle , \tag{19c}$$

$$A \equiv \frac{4 J(0) \alpha^2}{\Delta} . \tag{19d}$$

To solve these equations, one fixes T/Δ. Then for a series of values of a, one calculates ϕ_a and ψ_a, thereby determining $\langle S_z \rangle$ and hence A for each a. Thus we calculate a curve of A versus a for fixed T. Hence, for a specified A, this curve gives us the value of a for the fixed T, and we also have the value of $\langle S_z \rangle$ for that A. Then from (18) we have the self-consistent excitation spectrum with the corresponding self-consistent magnetization given by

$$\frac{\langle J \rangle}{\alpha} = -2 \langle S_z \rangle (1 - a^2)^{\frac{1}{2}}. \tag{20}$$

The collective excitation behaviour at $T = 0$ in the RPA is similar to that in figure 5 for the Bogoliubov type of approximation. The only real difference is that the critical value of A at which the $\vec{q} = 0$ soft mode goes unstable is about 4% larger than in the molecular field theory. This comes about because the true ground state of the system is not the molecular field ground state.

Figure 6 shows the behaviour at finite temperatures for the case where the exchange is only slightly larger than the critical value for ferromagnetism at zero temperature, giving a Curie

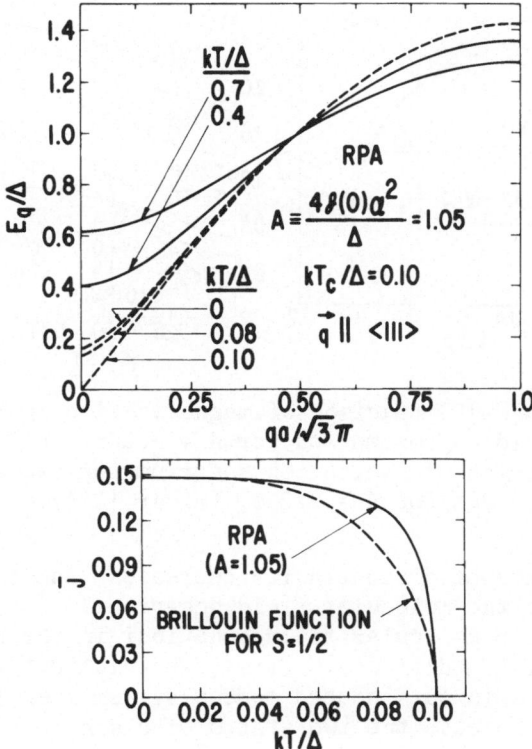

Figure 6 - (Upper) Change of dispersion curve as temperature is raised through T_C for A = 1.05, T_C/Δ = 0.1.
(Lower) Corresponding magnetization vs temperature. (After
[12]).

temperature equal to 0.1 of the crystal-field splitting. As T/Δ
increases from zero, the k = 0 mode energy drops toward zero.
The rate of decrease is, however, slow until T is close to the
Curie temperature, whereupon the $\vec{q}$ = 0 mode frequency drops to
zero precipitously. For T greater than T_C, the mode frequency
increases again. For high temperatures, the collective excitation energies approach Δ, and the dispersion tends to disappear.
From the self-consistent determination of the excitation behaviour, one also finds the temperature dependence of the magnetization. The most striking feature is the sharp drop in magnetization close to T_C.

As shown in figure 7, as T_C/Δ increases, the RPA magnetization shows double valued behaviour at high T in the ordered régime. This indicates that there is a first order transition at
T_C. This first order transition occurs [12] for values of T_C/Δ >
0.1.

The discontinuity in magnetization is most significant for
values of T_C/Δ near unity. Also, the excitation spectrum is considerably narrower than for lower T_C. As T_C/Δ increases still
further in the RPA, while the discontinuity in magnetization

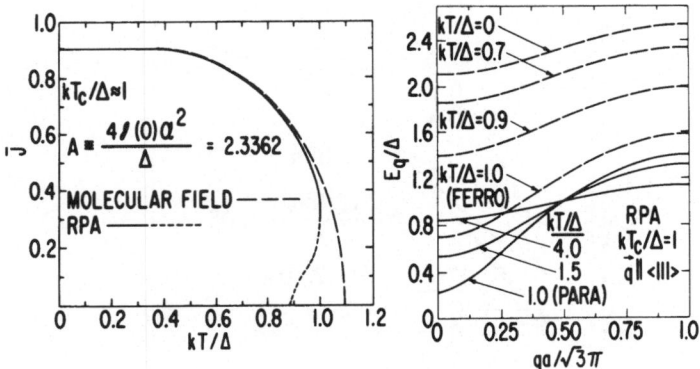

Figure 7 - (Left) Comparison of magnetization in RPA and
molecular field approximation for $A = 2.336$. (Right)
Thermal variation of excitation spectrum for two singlet
level model in RPA for $A = 2.336$. (After [12]).

persists, the size of discontinuity decreases; and the magnetization approaches the molecular field behaviour.

The approach to molecular field behaviour in the RPA for large
exchange is easily understood. First, we point out that there
are no spin waves in our induced moment system even if we allow
$A \to \infty$. This is because the two states of a single ion go to
$|1 \pm m\rangle$ with $m > \frac{1}{2}$ in our model as $A \to \infty$. (For example, for Pr^{3+}
in a hexagonal crystal field, the two lowest singlet states are
$\frac{1}{2}(|3\rangle - |-3\rangle)$ and $(1/\sqrt{2})(|3\rangle + |-3\rangle)$. As exchange increases from
zero, these states mix; and in the limit of large exchange, the
states become $|+3\rangle$ and $|-3\rangle$). The spin waves involve transitions
to $|m \pm 1\rangle$, and these are higher lying states not taken into ac-
count for the two singlet state system. Second as $A \to \infty$, there
are also no excitations of the type we have considered, since the
coupling vanishes between the two single ion states (i.e.
$\langle m|J_z|-m\rangle = 0$).

Basically, this first order transition occurs because the long
wavelength, 'soft', modes dropping in energy offer a catastroph-
ically effective channel of depopulation of the ground state.
*I want to emphasize that the theory yields a first order transi-
tion because it has been done self-consistently.*

It is much more difficult to do a self-consistent calculation
for the TSCA. One of the difficulties [12] is related to the
fact that the equations of motion techniques really amount to
perturbation theories; and for the TSCA in the magnetically ordered regime, the molecular field states do not offer a good
starting basis for such a perturbation theory. So one has to
perform an additional transformation. However, one can with full
self-consistency, for the paramagnetic régime calculate the critical value of A at $T = 0$ at which the $q = 0$ mode goes unstable.
This value of A is $A_{crit} = 1.18$ for the simple cubic lattice with
nearest neighbour exchange. This compares to $A = 1$ for the molecular field theory and $A = 1.04$ for the RPA. Now, as shown in

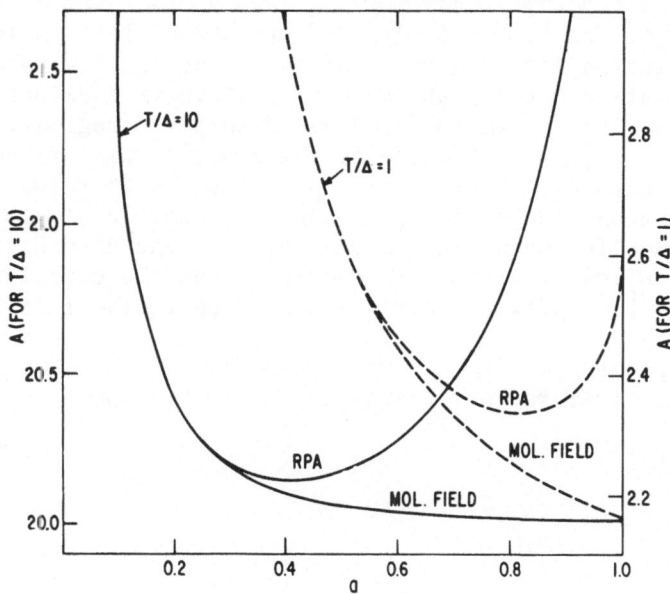

Figure 8 - Variation of A with a for fixed T/Δ in RPA and molecular field approximation. (After [12]).

figure 8, the first order transition in the RPA is associated with double valued behaviour of the A vs a curve. Another way of stating this criterion is that if $dA/da|_{a=1}$ is positive, there is a minimum in the A vs a curve, and hence there is a first order transition. Wang and I [12] examined the TSCA at $T = 0$ and found that $dA/d\tilde{a}|_{\tilde{a}=1}$ (the quantity corresponding to $dA/da|_{a=1}$ for the RPA) is negative, indicating that even at $T = 0$, the transition to ferromagnetic ordering with increasing A is first order. Thus the system has a first order transition for an A less than the value $A = 1.18$ at which the $k = 0$ mode would go soft if the transition were second rather than first order.

I have already mentioned the exact equivalence of the two singlet level problem to the Ising system in a transverse magnetic field. This allows us to discuss the relevance of calculations by Elliott and Wood [14] and by Pfeuty and Elliott [15] to the two singlet level problem. Translating to the nomenclature of the two singlet level problem, Elliott and Wood have studied the transition temperature T_C by a power series expansion of the susceptibility at high T up to fifth order in J/kT to all orders in Δ/kT. By assuming a second order transition and finding the temperature for divergence of the susceptibility, they find a critical value of $A \approx 1.16$ at $T = 0$ for the simple cubic lattice with nearest neighbour exchange. *I want to note and emphasize that the susceptibility for which Elliott and Wood obtained the divergence is that for the Ising transverse field problem. This* is $\langle S_z \rangle / \Delta$, *and is not the physical susceptibility for the two* *singlet level problem. That is, the second order transition be-* *ing assumed is for* $\langle S_z \rangle$ *not* $\langle J \rangle$. Pfeuty and Elliott in related work have studied the properties of the Ising model with a trans-verse field at zero temperature using perturbation expansions. This was done both in the ordered and disordered regions. In the ordered region, $\langle S_z \rangle$ is treated as the order parameter, and is expanded in a power series in Δ/J up to the fourth term. Pfeuty and Elliott then studied how $\langle S \rangle$ tended to zero as Δ goes to the critical value for magnetic disordering. In the disordered state at $T = 0$, the lowest excitation energy E_0 and the correlation function $G = \sum_m \langle S_{iz} S_{i+mz} \rangle$ were calculated up to the fifth term as a series expansion in J/Δ.

In the critical region, Pfeuty and Elliott assumed a power law behaviour,

$$\langle S_z \rangle \sim (\Delta_c - \Delta)^{\beta^*} , \qquad \qquad (21a)$$

$$G \sim (\Delta - \Delta_c)^{t^*} , \qquad \qquad (21b)$$

$$E_0 \sim (\Delta - \Delta_c)^s . \qquad \qquad (21c)$$

The critical values of Δ_c and the indices were estimated from the power series. Only a small number of terms in the series were obtained; nevertheless, the values of Δ_c obtained from the different expressions were very consistent.

For the simple cubic lattice with nearest neighbour exchange, the results are as follows. In the ordered régime, from the ex-pression in (21a), $A_c = 1.15$, $\beta^* = 0.43$. For G in (21b), in the disordered régime, $A_c = 1.16$ and $t^* = 0.535$. For E_0 in (21c) in the disordered régime, $A_c = 1.18$ and $s = 0.58$. *We note that this* *critical value of* A_c *for the mode softening is exactly the same* *as that found by Wang and myself* [12] *in the TSCA calculation.* Unfortunately, there has been no examination of the consequences of the critical behaviour for $\langle S_z \rangle$ on the corresponding behaviour of the physical magnetization $\langle J \rangle$ in the two singlet level prob-lem. Therefore we cannot at present say what information the Pfeuty and Elliott calculation yields as whether the magnetic or-dering transition for the two singlet level problem is first or second order.

I have reviewed the essential features of the theory for the magnetic exciton behaviour as it had developed in the period prior to neutron scattering experiments. The first neutron ex-periments to examine magnetic excitons in singlet ground state systems were those of Holden *et al.* [16] on TbSb and of Rainford and Houmann [17] on single crystal dhcp Pr. Holden *et al.* were able to fit the magnetic exciton dispersion curves at low T for

TbSb using the effective boson theory. They obtained exchange
and crystal field parameters consistent with those obtained by
Vogt and myself [5] from the macroscopic susceptibility measurements shown in figure 3. While the lowest energy exciton does
begin to shift to lower energy as temperature starts to increase, as
the temperature is raised further, there simply are no well defined excitons. Thus one cannot follow the exciton behaviour
through the Néel temperature. It has been suggested that the
disappearance of the excitons is associated with lifetime damping effects because the exchange is much larger than the threshold value for magnetic ordering.

Rainford and Houmann [17] have observed the magnetic exciton
spectrum in single crystal dhcp Pr which does not order magnetically. The behaviour is rather complicated, partially because
of the two types of crystal site present. Qualitatively there
is soft mode behaviour as temperature is lowered indicating an
approach toward magnetic ordering; however, the exchange is not
large enough to order the system even at zero temperature, and
consistent with this the soft mode does not go all the way to zero
energy.

The greatest amount of study and discussion of the magnetic
exciton behaviour has been for the Pr_3Tl system [18-20]. Here
the Pr sites occupy three out of the four sites on an fcc lattice, with the Tl occupying the fourth site. To a very good approximation one can treat the Pr ions as though they occupy an
fcc lattice. The ferromagnetic ordering temperature is 11.6°K,
while the crystal field splitting from the Γ_1 singlet ground
state to the Γ_4 triplet first excited state is approximately
77°K. Thus one has a case where the exchange is only slightly
greater than the critical value for magnetic ordering.

Using the value of crystal field splitting indicated by specific heat measurements [21] and estimating the exchange constant from the ordered moment [21] per Pr^{3+} ion, prior to the
performance of the neutron scattering experiments, as shown in
figure 9 I calculated the expected exciton spectrum [19] at low
temperature using the Bogoliubov type of boson approximation.
The samples used in the experiments were powder, so the observed
spectrum [18] corresponded to an average over crystallographic
directions. Taking the wave vector along $\langle 110 \rangle$ in the calculations gives a good approximation to such an average. Also the
experiments did not resolve the difference between longitudinal
and transverse excitons. The experimental results, also shown
in figure 9, gave excellent agreement with the prediction of the
boson theory. In fact, as shown in figure 10, the value I used
for the crystal field splitting was about 10% too low, and by
increasing this crystal field splitting Birgeneau obtained almost perfect agreement between theory and experiment.

Substituting La for Pr in Pr_3Tl serves to lower the exchange
while leaving the crystal field splitting approximately unchanged.
Substitution of about 7% La is sufficient [8] to reduce the exchange below the threshold value for ferromagnetism. The lower

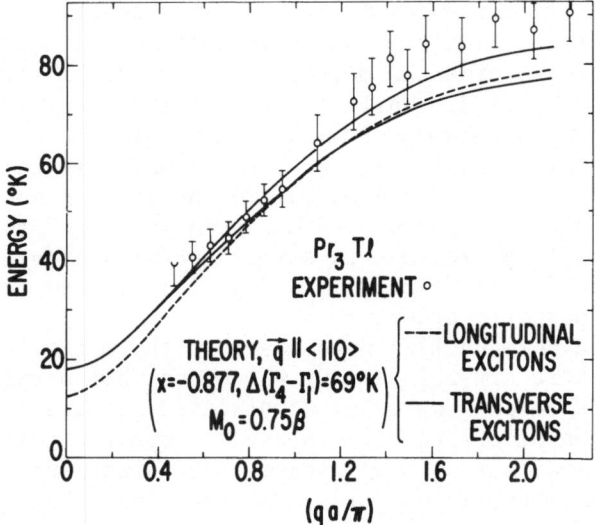

Figure 9 - Comparison of theory and experiment for magnetic exciton dispersion in Pr$_3$Tl. (After [19]).

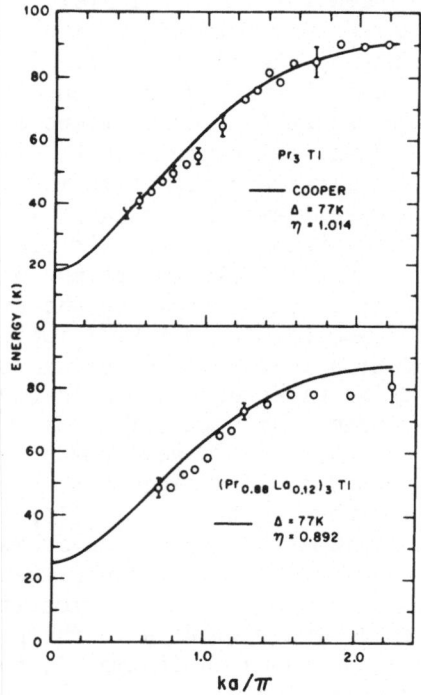

Figure 10 - Γ_1-Γ_4 spherically averaged dispersion relations in Pr Tl and (Pr$_{0.88}$La$_{0.12}$)$_3$Tl. The solid lines are the effective boson theory dispersion curves. (After [20]).

panel in figure 10, shows the exciton spectrum for a sample with
12% La substitution, so that this sample is on the paramagnetic
side of the critical value of exchange for ferromagnetic ordering. Birgeneau [20] fits this spectrum using the same parameters
as for pure Pr_3Tl, just decreasing the exchange by the amount of
the La substitution. The fact that the fit remains quite good
is further confirmation of the success of the boson theory in
treating the low temperature exciton spectrum.

Birgeneau *et al.* [18] found surprising behaviour for the temperature dependence of the exciton energies. In both the pure
Pr_3Tl and the sample with 12% La dilution, the excitation energies are essentially temperature independent up to quite high
temperatures, well above the Curie temperature for Pr_3Tl. The
excitations seem simply to decrease continuously in intensity.
Also there is no evidence of the mode softening anticipated from
the RPA studies of the two singlet level problem. However, in
that connection it is important to note that the fact that the
samples were powder prevented the observation of small wave-vector
excitons. If the temperature dependence of the mode energies
was confined to $\vec{k}$ vectors less than about 0.2 $Å^{-1}$, it would not
have been observed in the powder experiment. The RPA calculations indicate a distinctive linear variation of the exciton energies with $\vec{k}$ at low $\vec{k}$ as the $\vec{k} = 0$ mode becomes soft, and the
experiments show no evidence for approach to that linear behavi-
our. Thus these experiments seem to rule out the simple type of
soft mode behaviour predicted by the RPA, but they do not pre-
clude more complex soft mode dynamics.

Thus, about two years ago there were two, perhaps related,
questions remaining about the magnetic exciton behaviour. First
was the lack of temperature dependence of the exciton dispersion
curve. Second was the question of whether the $\vec{k} = 0$ mode completely softens, i.e. goes to zero energy, at the Curie temperature. In the past year or two, several groups have focussed
their efforts on these questions.

Birgeneau [20] suggested that one might understand the lack
of temperature dependence as coming from repulsion effects from
higher lying exciton modes which would inhibit the renormalization. Calculations by Holden and Buyers [22] support this picture, and the reason for the lack of temperature dependence at
larger wave numbers now seems clear. However, the nature of the
approximations involved in the various calculations [22-27] including excited crystal field states in addition to an only singlet (i.e. for the singlet-triplet model [23] or for complete crystal-field level schemes [22,24-27] is such as to leave the soft
mode question still open).

The calculation of Holden and Buyers [22] adopts the pseudo-
fermion formalism. As I pointed out earlier, this formalism suffers from the defect that it allows unphysical states corresponding to the simultaneous occupation of two or more different
single-ion states at the same site simultaneously. However,
Fulde and Peschel [24,25] used a technique owing to Abrikosov

[*28*] to eliminate such states. One replaces the state energies
ε_i by $\varepsilon_i + \lambda$. The elimination of the unphysical states is done
by taking the limit,

$$\lim_{\lambda \to \infty} \left\{ \frac{1}{C} \exp\left(\frac{\lambda}{T}\right) [\ldots] \right\} , \tag{22a}$$

with

$$C = \sum_{\substack{i \\ states}} \exp\left(- \frac{\varepsilon_i}{T}\right) , \tag{22b}$$

at the end of every calculation of thermal expectation values.
Thus λ plays the rôle of a chemical potential in freezing out
the unphysical states. Using a diagrammatic summation technique
Fulde and Pesche [*24,25*] find the dynamic susceptibility

$$\chi(q,\omega) = \frac{u(\omega)}{1 - J(\vec{q},\omega)u(\omega)} , \tag{23a}$$

$$u(\omega) = \sum_{i,j} \frac{\langle i|\vec{J}|j\rangle \cdot \langle j|\vec{J}|i\rangle (e^{-E_i/T} - e^{-E_j/T})}{\hbar\omega - (E_j - E_i)} . \tag{23b}$$

The excitation energies are then given by the poles of the dynamic susceptibility.

Because of the need to eliminate the unphysical states, the
pseudo-fermion theory of Fulde and Peschel suffers from a rather
severe restriction. In freezing out the unphysical states, the
linked cluster theorem is no longer valid. In order to avoid
this difficulty, one is restricted to calculating the susceptibility in molecular field theory. This means that the unenhanced
single ion susceptibility of (23b) is of necessity a molecular
field average. Thus the excitation spectrum found is of necessity effectively the result of a hybrid Random Phase Approxima-
tion-Molecular Field (RPA-MF) calculation, where one decouples
the equations of motion for appropriate generating operators by
replacing certain operators by thermal averages. However, these
thermal averages are the molecular field values and not truly
self-consistent values evaluated by summations over the excitation spectrum.

Holden and Buyers [*22*] have outlined a calculation deriving
the dynamic susceptibility by studying the equations of motion
for the retarded time Green's function,

$$G^z(g,\ell) = -i\langle [J_g{}^z(t), J^z(0)]\rangle \theta(t) \tag{24}$$

for the longitudinal excitations, and

$$G^-(g,\ell) = -i\langle [J_g^-(t), J_\ell^+(0)]\rangle \theta(t), \qquad (25a)$$

$$G^+(g,\ell) = -i\langle [J_g^+(t), J_\ell^+(0)]\rangle \theta(t), \qquad (25b)$$

for the transverse excitations.

Probably the simplest way to see how one can obtain the collective excitation energies is to look directly at the equations of motion for the appropriate generating operators. By decoupling, by using ground state expectation values or thermal averages respectively, one obtains the excitation spectrum at $T = 0$ or at finite temperatures. Thus one takes the fermion Hamiltonian of (6,7), and one finds the equations of motion for the Fourier transformed generating operators, $\sum_k d_{kn}^\dagger d_{-kn}$ by taking the commutator,

$$[\sum_k d_{kn}^\dagger d_{-km}, \mathcal{H}_0 + \mathcal{H}_1] \qquad (26)$$

This is done for all pairs of levels n and m. The decoupling is done in the RPA after taking the commutator by making the replacement,

$$d_{km}^\dagger d_{-kn} d_{k'm'}^\dagger d_{-k'n'} \rightarrow \delta_{mn} f_m d_{k'm'}^\dagger d_{-k'n'} \qquad (27)$$

$$+ \delta_{m'n'} f_{m'} d_{km}^\dagger d_{-kn},$$

where f_m and $f_{m'}$ are the Boltzmann factors for the molecular field energy values. By introducing the molecular field population factors, one automatically excludes the unphysical states. This gives the same excitation spectrum as obtained from the poles of the dynamic susceptibility.

Figure 11 shows the magnetic exciton spectrum for Pr$_3$Tl calculated by Holden and Buyers [22] compared to the experimental results. At $T = 0$, Holden and Buyers have repeated the pseudo-boson calculation I have already discussed. The calculation at 21.3°K, above the Curie temperature, has been done by the pseudo-fermion technique. In the paramagnetic régime a new mode appears, corresponding to an excitation from the Γ_4 first excited crystal-field state to the Γ_3 higher excited crystal-field state (see figure 1); and the strength of the mode depends on the population of the Γ_4 state. When the energy of the excitation from the ground state equals the energy of this new mode, signified by the arrow at the left, resonant scattering and mode-mode repulsion occurs as shown. This effect is absent at $T = 0$ because the population of the triplet Γ_4 state is zero. This effect is necessarily excluded in the singlet-singlet and singlet-triplet models since higher levels than the Γ_4 level are excluded.

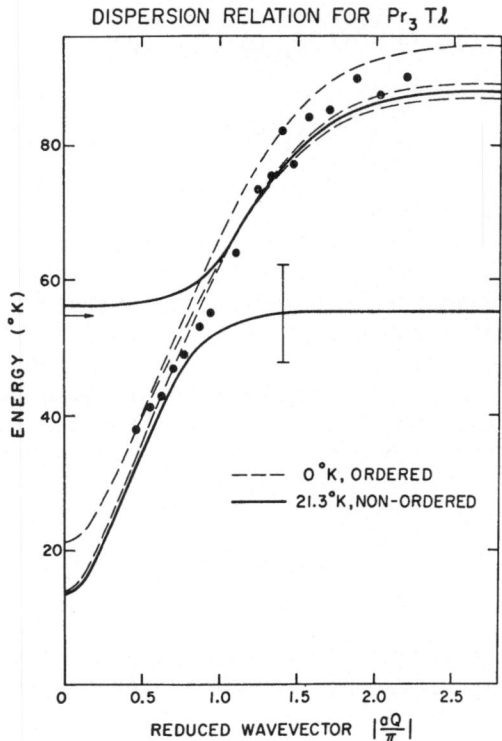

Figure 11 - Dispersion relation of magnetic excitons in
Pr$_3$Tl. Theory at 0°K is for effective boson calculation
and at 21.3°K for pseudo-fermion calculation (i.e. poles
of exchange enhanced dynamical susceptibility) as described in text. (After [22]).

The intensity variation with wave vector is such that one expects to observe the Γ_4 to Γ_3 exciton only near the crossover,
where one expects to observe two excitons at a given wave vector.
The observation of such mode splitting has also been predicted
by Peschel *et al.* [26]. This effect has now been observed by
Birgeneau and coworkers [29] in a Pr$_3$Tl alloy diluted by 7% La.
This is an alloy with exchange just at the critical value for
ferromagnetism at $T = 0$.

Figure 12 (from reference [22]) shows the temperature dependence of the mode frequency for a wave vector falling in the
crossover region. Here the solid curve gives the intensity
weighted average of the two modes in the crossover region for
the pseudo-fermion calculation of Holden and Buyers; while the
dashed curve shows the behaviour expected for the singlet-singlet model. The pseudo-fermion calculation including the effect
of all levels is in excellent agreement with experiment, and
satisfactorily explains the observed lack of temperature dependence in the spectrum of excitations from the ground state.

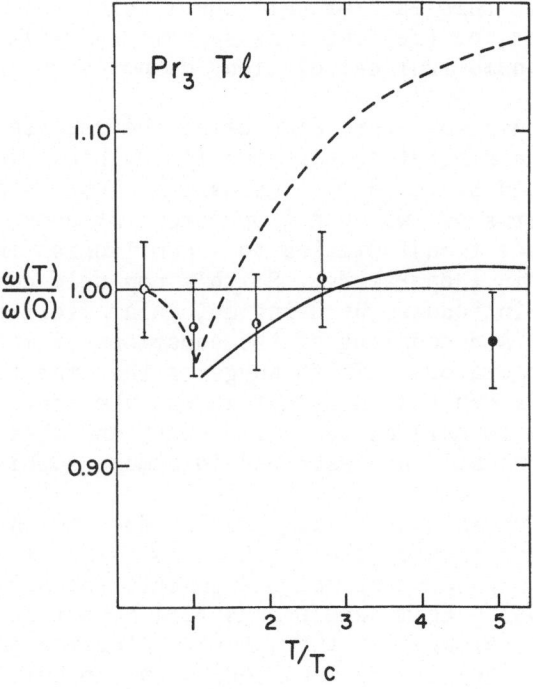

Figure 12 - Temperature dependence of the magnetic excitons in Pr_3Tl according to experiment [18,20], singlet-
singlet model (broken line), and pseudo-fermion (dynamical susceptibility) theory (solid line). (After [22]).

Indeed the renormalization is simply inhibited by the coupling
to higher states. Holden and Buyers have also performed calculations satisfactorily showing the lack of temperature dependence
for the magnetic exciton spectrum in $(Pr_{0.88}La_{0.12})_3Tl$ as experimentally observed.
    Finally, I would like to discuss the question of the $\vec{k} = 0$
mode softening at the Curie temperature. Several groups of workers [23,27,30] have performed calculations indicating that the
soft mode associated with magnetic ordering transition at T_c,
when there are degenerate, i.e. non-singlet, excited states, is
not the $\vec{k} = 0$ exciton corresponding to transitions from the singlet ground state to an excited crystal-field level; and that the
energy of this mode, while falling as T_c is approached, does not
go to zero energy. The calculations of references [27,30] are
of the pseudo-fermion type I have just discussed. They suffer
from the built in defect of a lack of self-consistency, i.e. the
expectation values used in the RPA decoupling are molecular field
expectation values; and this lack of self-consistency may be
crucial for the soft mode question.
    Smith's calculations [23], on the other hand, are for the
singlet-triplet problem. While the present calculation is not

self-consistent, this lack of self-consistency does not seem to
be a requirement for treating this system; and self-consistency
by some cyclic numerical calculations do not seem out of the
question.

Smith's calculations were done using the representation first
employed for the singlet-triplet problem by Pink [13], and also
used by Hsieh and Blume [31]. The matrix elements of $\vec{J}$ are represented in terms of two spin $\frac{1}{2}$ operators at each site. This
leads to an exact transformation to a Hamiltonian in terms of
these pseudo-spin $\frac{1}{2}$ operators. Smith's non-self-consistent approximation is introduced by using molecular field expectation
values in the RPA decoupling of the equations of motion for the
pseudo-spin $\frac{1}{2}$ operators. Smith suggests that the true soft mode
of the system is one that does not change the state of the system, that always occurs at zero frequency, and that has a finite
line width which could be described in a more sophisticated
theory.

My own feeling is that as long as one does not have a self-
consistent calculation of the $k = 0$ modes as T_C is approached,
or some rigorous general symmetry argument, the soft mode question remains open. Egami and Brooks [32] have recently proposed
a new method of calculation that provides for the best single-
site representation for the Hamiltonian in the spin wave approximation. The only discussion of this technique available as yet
is quite abbreviated; but this technique may prove useful in resolving the soft mode question. Actually the soft mode question
may very well be first resolved from the experimental side.
Once one obtains single crystals for a relevant material such as
Pr_3Tl, one can go in to $\vec{k} = 0$ and see what is happening.

In my opinion there are three outstanding theoretical problems
left open in the understanding of singlet ground state magnetism:

(1) Whether the transition is first or second order for the
    singlet-singlet problem;

(2) Whether for systems with excited crystal field levels
    in addition to a single singlet level, the soft mode of
    the system (for a second order ferromagnetic transition) is
    a $\vec{k} = 0$ longitudinal exciton corresponding to single ion ex-
    citations from the crystal field ground state to a crystal
    field excited level;

(3) The inclusion of life time effects in treating the ex-
    citon behaviour as exchange increases beyond the criti-
    cal values for magnetic ordering.

3. ANOMALOUS MAGNETIC PROPERTIES OF CERIUM MONOPNICTIDES

Now I would like to change subjects somewhat, and treat the
second topic I mentioned in my introduction. I would like to
give a brief progress report on our present understanding of the
magnetic properties of the cerium monopnictides, especially CeSb
and CeBi.

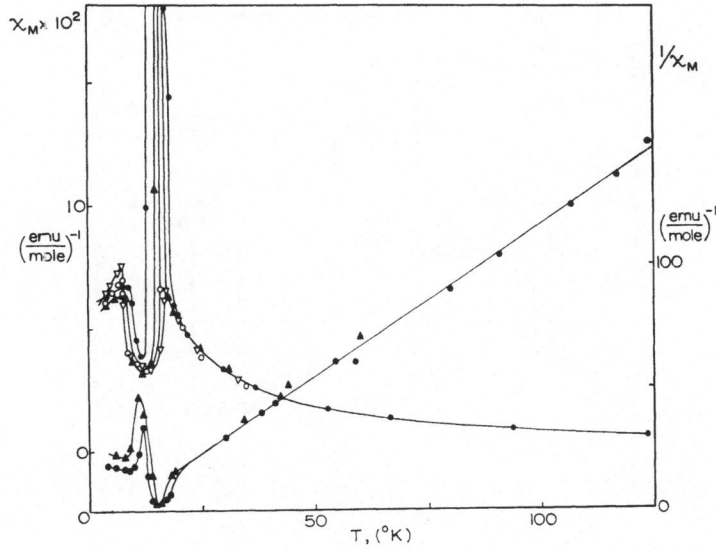

Figure 13 - Inverse susceptibility and susceptibility versus temperature for CeSb. (After [33]).

To begin, I will briefly review some of the experimental features that draw our attention to these materials. First is the susceptibility behaviour of these materials. As shown in figure 13, in a plot of inverse susceptibility versus temperature, instead of the usual antiferromagnetic behaviour showing a simple minimum at the Néel temperature, for [33-36] CeSb and CeBi for decreasing temperature there is a minimum, followed by a maximum, with an additional upturn at the lowest temperatures.

The magnetization curves at low temperature also show unusual behaviour. The magnetization is quite anisotropic and increases in sharply defined steps which show considerable hysteresis. For CeBi, as shown [37] in figure 14 the initial step is to half the saturation magnetization, while for [38] CeSb as shown in figure 15 it is to one third of the saturation magnetization.

Finally, and probably most striking of all, are the magnetic structures. For [39] CeBi, the initial ordering at 26°K is via a second order transition to a simple type 1 ordering with moments of alternate ⟨100⟩ planes antiparallel; and where the moments are aligned along the ⟨100⟩ direction. At approximately half the Néel temperature there is a first order transition to a type 1a structure where two planes with moment pointing up the ⟨100⟩ axis are followed by two planes with moments pointing down the ⟨100⟩ axis. (The fact that the first step in the magnetization curve for CeBi is to half the saturation value, which is close to the free ion moment of 2.14 u_B, is consistent with having an intermediate phase between the type 1a and type 1

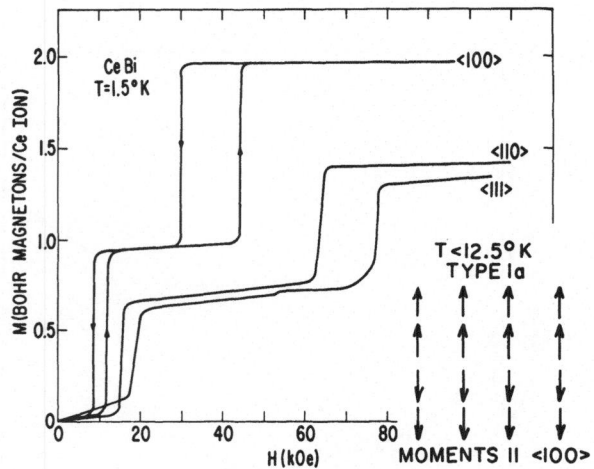

Figure 14 - Magnetization of CeBi at 1.5°K. (After [37]).

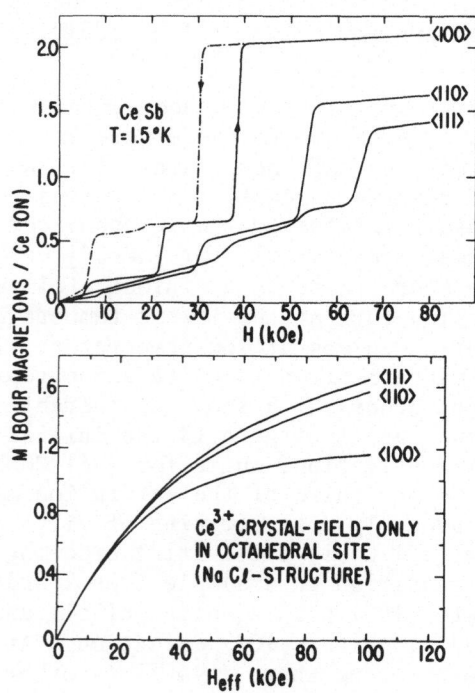

Figure 15 - (Upper) Magnetization of CeSb at 1.5°K (after
[38]). (Lower) Magnetization per Ce at 1.5°K for crystal-field-only theory with Γ_7 doublet ground state and
crystal-field splitting of 20°K.

structures as the field increases; where this structure has
three layers with moments pointing up the $\langle 100 \rangle$ axis followed
by one layer with moment pointing down).

The magnetic structures of CeSb are even more unusual [40].
The ordering at T_N is with a first order transition into a structure where the moments of successive $\langle 100 \rangle$ layers point along
$\langle 100 \rangle$, but vary sinusoidally in magnitude. With decreasing temperature there is a series of four transitions into other such
sinusoidal structures with differing periods.

With properties just described for motivation, I would now
like to discuss the model for magnetic behaviour that is developing for the cerium monopnictides. The key feature of this model
is the presence of a strongly ansotropic exchange interaction
that increases sharply on going form CeP toward CeBi, and a crystal field splitting that has exactly opposite behaviour. The
two kinds of interaction are comparable for CeSb, so it is not
surprising that this material is the one showing the most complex
behaviour.

Ce^{3+} has $J = 5/2$ in the ground state manifold. In a cubic
crystal field this ground state manifold splits into a Γ_7 doublet
and a Γ_8 quartet. For all the cerium monopnictides, the evidence
is that the crystal field ground state is the Γ_7 doublet. For
both CeAs and CeP the ordered moment per Ce at $T = 0$ is quite
close, (0.68 μ_B and 0.8 μ_B respectively [41,42]) to that expected for the Γ_7 doublet (0.71 μ_B), which is quite different from
that for the Γ_8 quartet (1.57 μ_B). Also for CeP, neutron scattering intensity measurements [43] somewhat favour Γ_7 as ground
state. Magnetic anisotropy measurements [35,37,44] for CeSb and
CeBi diluted by La and Y provide definitive evidence that the
ground state for Ce^{3+} in these compounds is the Γ_7 doublet.

Now I want to discuss the direct evidence for the presence of
the anisotropic interaction in CeSb and CeBi, and the inferential
evidence for its presence in CeP and CeAs. As shown in figure
15, the low temperature $\langle 100 \rangle$ easy direction for CeSb is in contrast to the $\langle 111 \rangle$ easy direction expected for crystal field anisotropy with a Γ_7 crystal field ground state. This led to Vogt
and myself [35] to think that the major source of anisotropy in
CeSb is an anisotropic interaction between the Ce ions. If this
is the case, then by diluting the cerium ions, one could decrease
the anisotropic Ce-Ce interaction, and change the easy direction
to the $\langle 111 \rangle$ direction characteristic of crystal field anisotropy.
As shown in figure 16, this was done by diluting the Ce with La
or Y; and in both cases, at sufficiently low Ce concentration the
easy direction became $\langle 111 \rangle$.

For CeBi, just as for CeSb, dilution of the Ce causes a change
in easy direction from $\langle 100 \rangle$ toward the $\langle 111 \rangle$ crystal-field easy
direction [37]. However, even at a cerium concentration of 10%,
in contrast to the dilation experiments for CeSb, the anisotropic
magnetization curves differ substantially from those expected for
crystal-field-only effects. This is consistent with the crystal-
field splitting being smaller, and the exchange interaction
larger in CeBi than in CeSb.

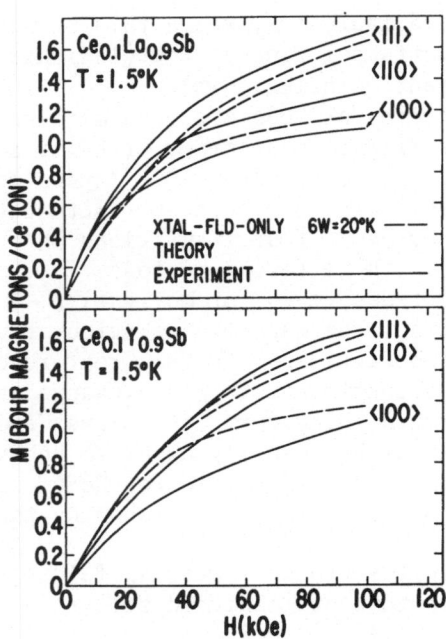

Figure 16 - Comparison of experimental magnetization per
Ce ion for $Ce_{0.1}Y_{0.9}Sb$ and $Ce_{0.1}La_{0.9}Sb$ at 1.5°K to crystal-field-only theory with splitting of 20°K. (After
[44]).

Both CeP and CeAs order in simple type 1 ordering at their
Néel temperatures, and have no further transitions with decreasing temperature. However, both these materials have ⟨100⟩ easy
directions [41,42] just as in CeSb and CeBi, even though the crystal-field anisotropy would give a ⟨111⟩ easy direction.

When we consider the Néel temperatures and crystal-field
splittings as found from bulk magnetic and neutron scattering ex-
periments† for the cerium monopnictides, tabulated immediately
below, we can envisage a universal model for this family of materials, featuring an anisotropic Ce-Ce interaction and a crystal-
field interaction changing drastically in magnitude in opposite
directions on going down the column, as indicated by the experi-

---

† These are summarized in Table II of reference [43]. Note that
the crystal field ground state for $Ce_{0.5}Y_{0.5}Sb$ is incorrectly
identified as Γ_8 as discussed in reference [44]. Recent measurements by H. Heer, A. Furrer, and O. Vogt ((1974). *Neutronen Streuung*, (Progress Report AF-SSP-74), (Eidgenössiches Institut für
Reaktorforschung, Wurenlingen), (January, 1974), p. 29) gives a crystal field splitting of 161°K for CeP, somewhat greater than the value
found earlier. These measurements agree with the earlier measurements in favouring the Γ_7 as the crystal-field ground state.

	T_N	Δ_{CF}
CeP	9.5°K	∿140°K
CeAs	7.8°K	∿140°K
CeSb	∿16°K	∿ 25°K
CeBi	25-26°K	∿ 10°K

mental behaviour of the Néel temperatures and crystal field splittings, respectively.

Next I would like to discuss how the unusual variety of susceptibility behaviours observed for these materials can be understood on this basis. The theory of Wang and myself [45] shows that the anomalous $1/\chi$ behaviour of CeSb and CeBi arises because, as shown in figure 17, there is a range of exchange field for which the splitting within the Γ_7 ground state doublet decreases with increasing field until the doublet levels cross. *I want to stress that this crossover phenomenon occurs only for molecular field along the $\langle 100 \rangle$ crystal field hard direction. It does not occur for molecular field along the $\langle 111 \rangle$ crystal field easy direction. Thus implicit in these calculations is the fact that*

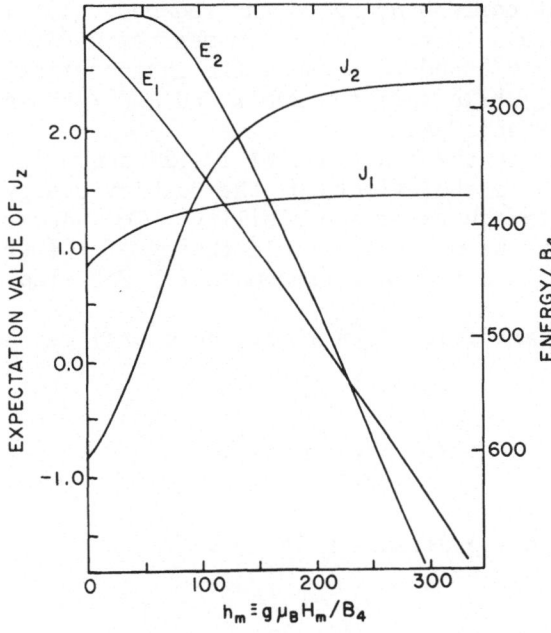

Figure 17 - Energy variation and corresponding change in magnetic moment with increasing molecular field for ground state crystal-field doublet of Ce^{3+}. (After [45]).

the exchange is anisotropic and causes the ordered magnetic moments to point along $\langle 100 \rangle$.

The calculations of Wang and myself are for a model with type 1 ordering along a $\langle 100 \rangle$ axis, where the relative magnitudes of T_N and Δ_{CF} vary through the complete range of interest. We feel that the essence of the susceptibility behaviour lies in the Γ_7 crossover phenomenon; and that the complex magnetic structures of CeSb and CeBi do not represent as essential complication in that regard.

As shown in figure 17, as the molecular field at a site increases from zero, and the wave functions of the Γ_8 excited state mix into the ground state Γ_7 doublet changing the energies to give the Γ_7 crossover, there is a corresponding drastic change in the magnetic moment of the ground state. It is straightforward procedure to calculate the transverse and longitudinal susceptibilities. This is fairly tricky numerically, because one has to take account of the behaviour shown in figure 17 in order to treat properly the self-consistency at varying temperatures.

Figure 18 shows the powder average inverse susceptibility $(\chi = (\chi_{\parallel}/3) + (2\chi_{\perp}/3))$ with this model for a fixed Néel temperature of 20°K for various values of the crystal field splitting Δ. For small Δ, $1/\chi$ behaves normally, as for a simple antiferromagnet. As Δ increases to 6°K and 20°K, the minimum in $1/\chi$ for decreasing temperature is followed by a maximum, and finally an upturn at the lowest temperatures. Finally, as Δ becomes quite large compared with T_N, one returns to normal behaviour, but with an ordered moment corresponding to the reduced value in the Γ_7 doublet, rather than to the free ion moment of Ce^{3+}. Thus the behaviour for $\Delta = 6$°K and 20°K shows the characteristic behaviour of CeBi and CeSb; while that for 200°K and 300°K shows that characteristic of CeP and CeAs.

The variations in the behaviour of $1/\chi(T)$ shown in figure 18 can all be traced to the effects of the doublet energy level crossing on the transverse susceptibility. To understand this, we review briefly the molecular field theory of the transverse susceptibility of a simple antiferromagnet. For simplicity we assume $S = \frac{1}{2}$.

In the molecular field, $H_m = \lambda \langle m \rangle$, the energy gap between the two levels is

$$\epsilon = g\mu_B H_m = g\mu_B \lambda \langle M \rangle, \qquad (28)$$

and the sublattice magnetization is

$$\langle M \rangle = \frac{1}{2} g\mu_B \frac{\left(1 - \exp\left(-\frac{\epsilon}{kT}\right)\right)}{\left(1 + \exp\left(-\frac{\epsilon}{kT}\right)\right)} \qquad (29)$$

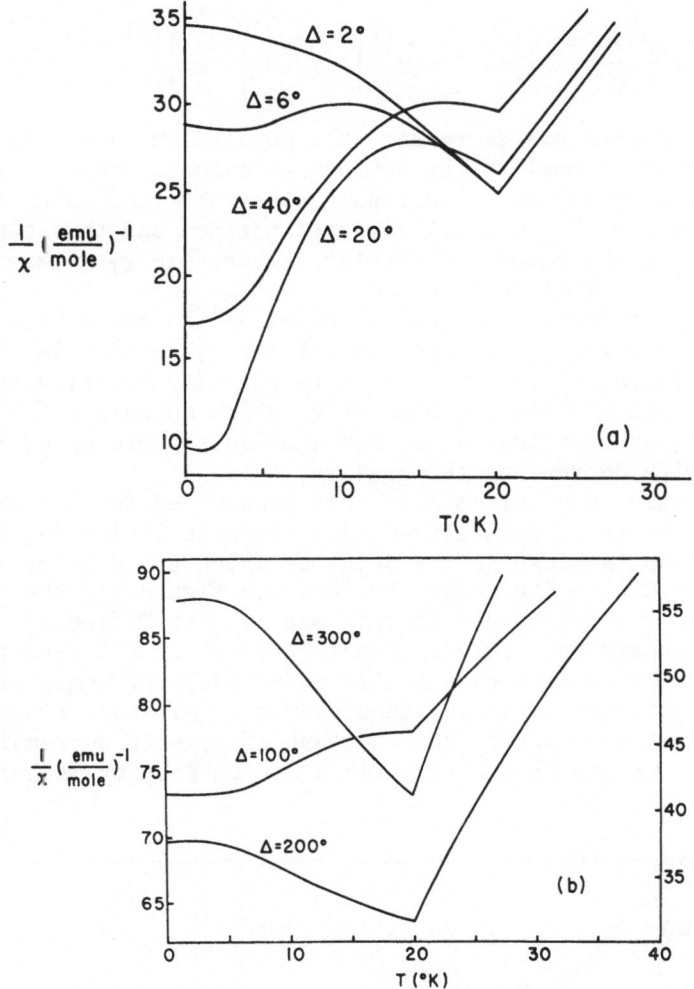

Figure 18 - Inverse susceptibility versus temperature
with fixed $T_N = 20°K$, $\theta = 0°K$, and varying crystal field
splitting between the Γ_7 doublet and Γ_8 quartet for Ce^{3+}.
(Upper) $1/\chi$ for Δ up to $40°K$. (Lower) $1/\chi$ for larger Δ.
(After [45]).

The transverse susceptibility is

$$\chi_\perp = \frac{\chi_\perp^0}{(1 - \bar{\lambda}\chi_\perp^0)} , \tag{30}$$

where $\bar{\lambda}$ is the paramagnetic molecular field constant [46] and

$$\chi_\perp^0 = 2g^2\mu_B^2 \left|\left\langle \tfrac{1}{2}\left|S^x\right|-\tfrac{1}{2}\right\rangle\right|^2 \left\{1 + \exp\left(-\frac{\varepsilon}{kT}\right)\right\} \left\{1 + \exp\left(-\frac{\varepsilon}{kT}\right)\right\}^{-1} = \text{(Contd)}$$

(Contd) $= \dfrac{g^2 \mu_B^{\,2}}{2\varepsilon}\left(1 - \exp\left(-\dfrac{\varepsilon}{kT}\right)\right)\left(1 + \exp\left(-\dfrac{\varepsilon}{kT}\right)\right)^{-1}.$ (31)

Now as temperature decreases, the population factor in (31), which is proportional to the sublattice magnetization, increases. However, the energy gap ε increases at exactly the same rate proportional to the sublattice magnetization, and thus cancels the change in the population factor. Therefore $\chi_\perp$ remains constant below the Néel temperature.

For Ce^{3+} in the cerium monopnictides on the other hand, because of the energy level crossing of the Γ_7 doublet in the molecular field, $\chi_\perp(T)$ can be a sharply rising function as decreases. It is this increase of $\chi_\perp$ which outweighs the decrease of $\chi_{||}$ that brings about the anomalous increase of the powder χ with decreasing temperature.

The anomaly in χ versus T is most pronounced for $\Delta \approx kT_N$. The self-consistent molecular field, shown in figure 19, for the $\Delta = 20°K$ case is close to the value at which the doublet levels cross ($h_m \approx 230$). Therefore, as shown in figure 20, the energy gap between the two levels is very small. For T from $5°K$ to $14°K$, the energy gap is less than $1.5°K$. It is the fact that the energy gap is both essentially independent of temperature and small compared to temperature over a significant range of temperature that leads to the increase of $\chi_\perp$ with decreasing temperature within that range. This can be seen by isolating

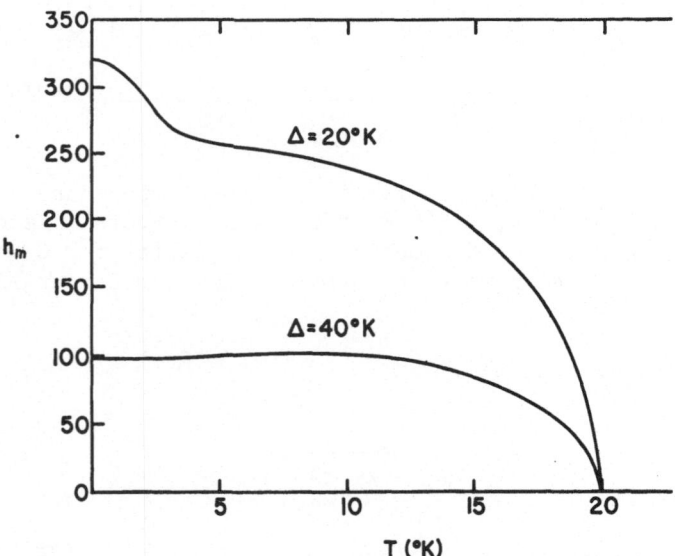

Figure 19 - Molecular field as a function of temperature for $T_N = 20°K$ with $\Delta = 20°K$ and $\Delta = 40°K$. (After [45]).

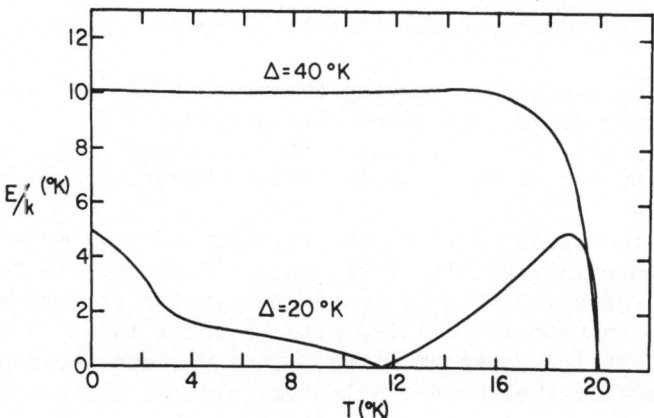

Figure 20 - Energy gap between the two lowest energy levels of Ce^{3+} as a function of temperature for $T_N = 20°K$
with $\Delta = 20°K$ and $\Delta = 40°K$. (After [45]).

the contribution of the doublet to the total transverse susceptibility. The contribution of the doublet to $\chi_\perp^0$ is given by

$$\chi^0_{12} = \left[\chi^0_1 + \chi^0_2 \exp\left(-\frac{\varepsilon}{kT}\right)\right]\left[1 + \exp\left(-\frac{\varepsilon}{kT}\right)\right]^{-1}, \qquad (32)$$

with

$$\chi^0_1 \equiv \frac{a}{\varepsilon} + \delta_1, \qquad (33a)$$

$$\chi^0_2 \equiv -\frac{a}{\varepsilon} + \delta_2, \qquad (33b)$$

$$\varepsilon \equiv E_2 - E_1, \qquad (34)$$

$$a \equiv 2g^2\mu_B^2|\langle\Psi_1|J_x|\Psi_z\rangle|^2, \qquad (35)$$

$$\delta_{1,2} \equiv 2g^2\mu_B^2 \sum_{\substack{n \\ (n\neq 1,2)}} \frac{|\langle\Psi_{1,2}|J_x|\Psi_n\rangle|^2}{(E_n - E_{1,2})}, \qquad (36)$$

where Ψ_1 and Ψ_2 are the two states of the doublet, and the ordered moment is along the z-axis.

At temperatures low compared to the splitting between the
doublet quartet, χ^0_{12} dominates the overall behaviour of $\chi_\perp$.
When ε is small compared with T and essentially constant,

$$\chi^0{}_{12} \simeq \frac{a}{2kT} + \frac{1}{2}(\delta_1 + \delta_2).$$

(37)

Here a, δ_1, δ_2 are slowly varying functions of T, so that, as shown in figure 21, $\chi_\perp$ increases with decreasing T.

So the crystal field plus anisotropic exchange model successfully explains the unusual susceptibility behaviour of the cerium monopnictides.

In order to examine further the validity of this model, as shown in figure 22, Landolt, Vogt, and I [37] examined the Néel temperature variation with Ce concentration for ternary alloys, where the Ce in CeSb was diluted with La and Y in a proportion to hold the lattice constant fixed. This was done to separate out the effect of the change in lattice size on the Ce-Ce interaction, from the simple effect of diluting the Ce. The Néel temperature drops much more steeply than linearly with decreasing Ce concentration.

As shown in figure 23, the change in strength of the Ce-Ce interaction with lattice constant variation can be studied by keeping the Ce content constant, and varying the ratio of La to Y content in ternary alloys. The Néel temperature measurements shown here indicate a smooth increase of interaction strength with decrease of Ce ion separation. At present, there is no definite evidence as to whether the anisotropic component of the Ce-Ce interaction changes at a different rate than the overall interaction as the lattice constant varies.

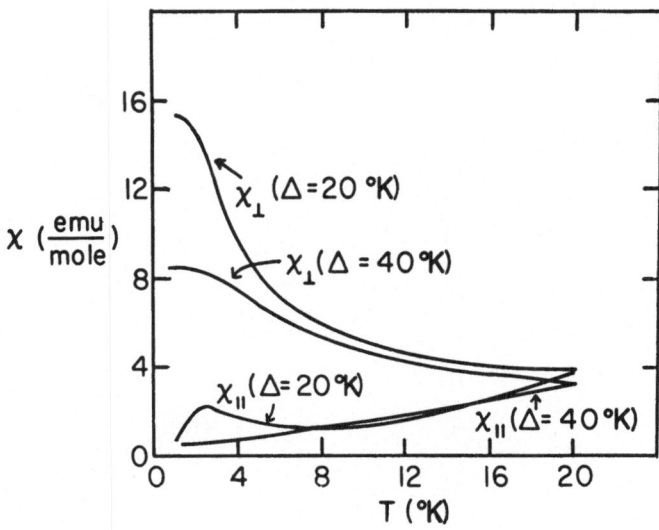

Figure 21 - Transverse and longitudinal susceptibilities for T_N = 20°K with Δ = 20°K and Δ = 40°K. (After [45]).

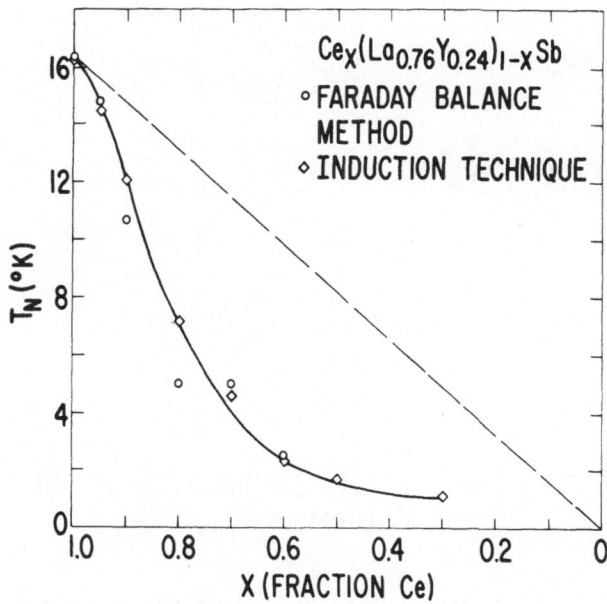

Figure 22 - Variation of Néel temperature with cerium concentration in $Ce_x(La_{0.76}Y_{0.24})_{1-x}Sb$ (alloys with lattice constant equal to that of CeSb). (After [37]).

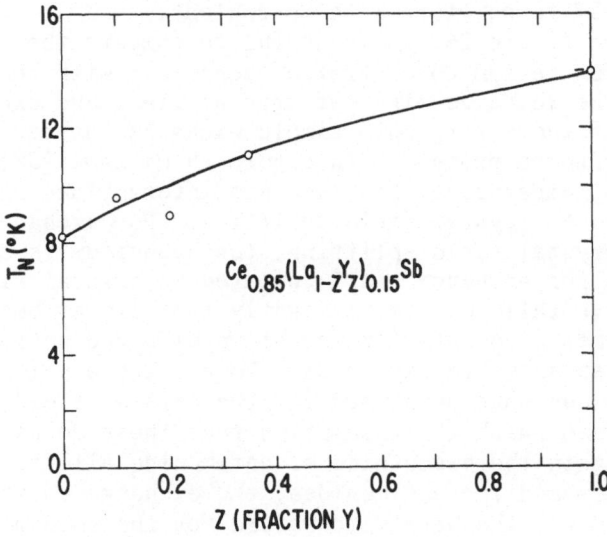

Figure 23 - Variation of Néel temperature with yttrium concentration in $Ce_{0.85}(La_{1-z}Y_z)_{0.15}Sb$ (alloys with varying lattice constant). (After [37]).

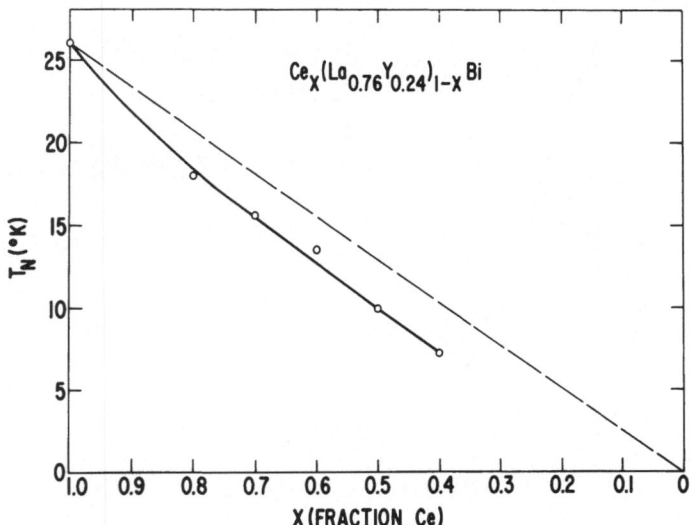

Figure 24 - Variation of Néel temperature with cerium con-
contration in $Ce_x(La_{0.76}Y_{0.24})_{1-x}Bi$ (alloys with lattice
constant equal to that of CeBi). (After [37]).

As shown in figure 24, in contrast to the behaviour for the
ternary antimonide alloy system with fixed lattice constant, the
variation of the Néel temperature for the bismuthide compounds
with lattice constant fixed at that of CeBi is close to a linear
scaling of Néel temperature with Ce content.

As shown in figure 25, it is useful to compare the Néel temperature versus cerium concentration behaviour with that expected
considering the relative sizes of crystal field and exchange interactions, if there were only simple exchange interactions and
magnetic structures present. In figure 25 we show [37] the variation of Néel temperature with exchange interactions where both
are normalized to crystal field splitting. For exchange large
compared to crystal field splitting, the behaviour is close to
linear; while for exchange small compared to crystal field splitting, the curve falls off significantly from linear behaviour.

The experimental points for the bismuthide and antimonide compounds are also shown in figure 25. To get these points, the experimental values have been used for the crystal field splitting
of CeBi and CeSb, with the assumption that these do not vary significantly within the bismuthide or antimonide alloys. For both
the bismuthides and the antimonides, the exchange constants have
been chosen to fit the Néel temperature for the pure material,
and have been scaled linearly with Ce content. The points for
the bismuthide system fall along the linear portion of the curve,
and the departure from the theoretical curve may not be significant. The points for the antimonide system fall along the non-
linear part of the theoretical curve, but they show significantly
greater departure from linear behaviour than the theoretical curve.

THEORY FOR TYPE I ORDERING AT T_N
WITH $\vec{J_i} \cdot \vec{J_j}$ ISOTROPIC EXCHANGE

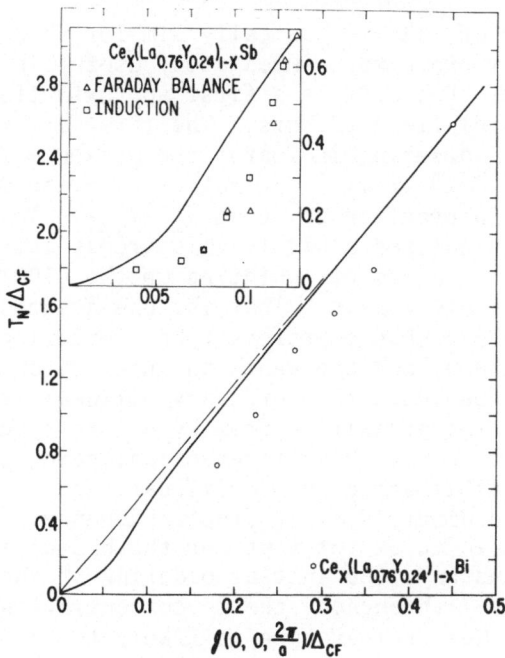

Figure 25 - Molecular-field theory variation of T_N/Δ_{CF} with $J(0,0,2\pi/a)/\Delta_{CF}$. Here $J(0,0,2\pi/a)$ is the appropriate combination of exchange constants, and the calculations are for type 1 ordering with isotropic bilinear exchange. The theory is compared with the experimental points for $Ce_x(La_{0.76}Y_{0.24})_{1-x}Bi$ and $Ce_x(La_{0.76}Y_{0.24})_{1-x}Sb$ obtained as described in the text. (After [37]).

In fact, as already discussed, the magnetic structures for CeSb and CeBi are not simple; and, as already stated, the Ce-Ce interaction is clearly not a simple isotropic exchange interaction. I have indicated the general nature of the model in terms of which we are seeking to understand the magnetic behaviour of the cerium monopnictides. The essential features of that model are the presence of crystal field splitting and of anisotropic exchange interaction. The effects of these two interactions, particularly as their relative sizes vary, are sufficiently distinctive for the behaviour of the properties already discussed to convince us that this model contains the essentials necessary for understanding the magnetic behaviour of the cerium monopnictides. What remains to be done? I think there are two main th things. First is to arrive at a form for the anisotropic interaction, that on a phenomenological basis allows us to understand the variety of magnetic structure behaviour and transitions in these materials. Second is to get some, at least semiquantitative, idea of the physical origins of this anisotropic interaction.

The behaviour of CeBi is basically simpler than that for CeSb, and will probably prove more immediately useful in achieving these objectives. For CeBi is a first approximation, one can neglect the crystal field effects. The first order transtion at 12.5°K to type *1a* ordering indicates the presence of an interaction going as $J_{iz}J_{jz}{}^3$ or $J_{iz}{}^3J_{jz}{}^3$ rather than, or in addition to, any quadrupolar interaction going as $J_{iz}{}^2J_{jz}{}^2$. The reason for this is that one requires a highly anisotropic interaction which distinguishes between moments pointing up the $\langle 100 \rangle$ axis as opposed to down the $\langle 100 \rangle$ axis. That is, one assumes that ordinary bilinear isotropic exchange prefers type *1* ordering as occurs the Néel temperature, and one wants an interaction having stronger temperature dependence than ordinary bilinear isotropic exchange, to drive the transition from type *1* to type *1a* with decreasing temperature. Such an interaction, going as $J_{iz}J_{jz}{}^3$ or $J_{iz}{}^3J_{jz}{}^3$ may possibly arise form modifications, due to Jahn-Teller effects, of the ordinary $\vec{J_i} \cdot \vec{J_j}$ isotropic exchange.

It is reasonable to expect that for the bismuthide alloys with fixed lattice constant, the initial ordering at the Néel temperature remains type *1* throughout the Ce concentration range studied. The value of the Néel temperature is likely to be determined by simple isotropic exchange scaling with cerium concentration, and one then expects the approximately linear behaviour for the variation of Néel temperature with cerium concentration that is experimentally observed. For the bismuthide alloys, the low temperature anisotropic magnetization curves should be useful in providing quantitative values for the anisotropic interaction constants.

For CeSb the situation is more complex, probably because exchange and crystal field effects are comparable in size. The crystal field energy is minimized when the moment per Ce is reduced to the value pertinent to the Γ_7 doublet state. This is 1/3 of the free ion moment. The exchange energy, on the other hand, is minimized when the moment per Ce ion has the free ion value. The sinusoidal structure presumably represents a compromise between exchange and crystal field effects, where the anisotropic interaction always holds the moments along $\langle 100 \rangle$. In such a picture, it is possible that a sinusoidal structure may be stable even at zero temperature.

The fact that exchange and crystal field effects are comparable, and that the magnetic structure at the Néel temperature is not a simple type *1* structure, probably plays a rôle in bringing about the highly non-linear variation of Néel temperature with concentration for the antimonide alloys with fixed lattice constant. In this connection, one of the effects of varying the cerium concentration in these alloys may be to change the magnetic structure at the Néel temperature.

To conclude, I will just make the obvious statement that we

hope to make these ideas more quantitative in a theory for the
cerium monopnictides, relating the magnetic structures found to
the anisotropic magnetic forces present.

It should be pointed out that susceptibility measurements in
CeBi at fields of 5,500 Oe (by Tsuchida and Wallace [34]) and
5,000 Oe (by Bartholin et al. [36]) on magnetically 'virgin'
material show a quite small change in susceptibility at the 12.5°K
first order transition from type 1 to type 1a ordering. This is
also true of measurements (M. Landolt, unpublished) by the induction technique with an AC field of approximately 5 Oe and either
zero DC biasing field or a DC field of 1,000 Oe. This is in contrast to measurements in fields of more than 5,000 Oe, or with no
attempt to prepare a magnetically 'virgin' sample. Then there is
a large, almost discontinuous, change in χ at a temperature close
to that of the type 1 to 1a transition. The Δ = 6°K curve in
figure 18 is consistent with the induction technique and 'virgin'
material measurements in CeBi.

For CeSb the situation is not so well defined or satisfactory.
Susceptibility measurements at 2,300 Oe by Tsuchida and Wallace
[33] and at 3,000 Oe by Vogt (unpublished) show the behaviour
described in the text, i.e. $1/\chi$ vs T for decreasing T has a minimum, followed by a maximum, with an additional upturn at the lowest temperatures. This is qualitatively consistent with the Δ =
20°K curve in figure 18. On the other hand, induction technique
measurements in fields of approximately 10 Oe (M. Landolt, unpublished) show a simple minimum in $1/\chi$ vs T. Induction measurements in a DC biasing field to test for the absence of domain effects are not available.

ACKNOWLEDGMENT

I should like to acknowledge with thanks the receipt of a
travel grant under the Special Foreign Currency Programme of the
National Science Foundation, which made it possible for me to
present these lectures.

REFERENCES

1. Bleaney, B. (1963). *Proc. Roy. Soc.*, **A276**, 19.
2. Cooper, B.R. and Vogt, O. (1971). *J. de Phys.*, **32**, C1-958.
3. Trammell, G.T. (1963). *Phys. Rev.*, **131**, 932.
4. Child, H.R., Wilkinson, M.K., Cable, J.W., Koehler, W.C. and
   Wollan, E.O. (1963). *Phys. Rev.*, **131**, 922.
5. Cooper, B.R. and Vogt, O. (1970). *Phys. Rev.*, **B1**, 1218.
6. Stutius, W. (1969). *Phys. Kond. Mat.*, **9**, 341.
7. Bucher, E., Andres, K., Maita, J.P. and Hull, G.W. Jr.,
   (1968). *Helv. Phys. Acta*, **41**, 723; Bucher, E., Chu. C.W.,
   Maita, J.P., Andres, K., Cooper, A.S., Buehler, E. and
   Nassau, K. (1969). *Phys. Rev. Lett.*, **22**, 1260.
8. Bucher, E., Maita, J.P. and Cooper, A.C. (1972). *Phys. Rev.*,

B6, 2709; Andres, K., Bucher, E., Darack, S. and Maita, J.P. (1972). *Phys. Rev.*, **B6**, 2716.

9. For a summary see: Cooper, B.R. (1972). CRC Critical Reviews, in *Solid State Sci.*, **3**, 83.

10. Trammell, G.T. (1960). *J. Appl. Phys.*, **31**, 362S.

11. Bozorth, R.M. and Van Vleck, J.H. (1960). *Phys. Rev.*, **118**, 1493.

12. Wang, Y.L. and Cooper, B.R. (1968). *Phys. Rev.*, **172**, 539; (1969). *Phys. Rev.*, **185**, 696.

13. Pink, D. (1968). *J. Phys.*, **C1**, 1246.

14. Elliott, .R.J. and Wood, C. (1971). *J. Phys.*, **C4**, 2359.

15. Pfeuty, P. and Elliott, R.J. (1971). *J. Phys.*, **C4**, 2370.

16. Holden, T.M., Svensson, E.C., Buyers, W.J.L. and Vogt, O. (1972). In *Neutron Inelastic Scattering 1972*, (International Atomic Energy Agency, Vienna), pp. 553-561.

17. Rainford, B.D. and Houmann, J.C.G. (1971). *Phys. Rev. Lett.*, **26**, 1254; Rainford, B.D. (1971). *AIP Conference Proceedings No.5, Magnetism and Magnetic Materials 1971*, (eds. Graham, C.D., Jr. and Rhyne, J.J.), p. 591.

18. Birgeneau, R.J., Als-Nielsen, J. and Bucher, E. (1971). *Phys. Rev. Lett.*, **27**, 1530; (1972). *Phys. Rev.*, **B6**, 2724.

19. Cooper, B.R. (1972). *Phys. Rev.*, **B6**, 2730.

20. Birgeneau, R.J. (1972). *AIP Conference Proceedings, No.10, Magnetism and Magnetic Materials 1972*, (eds. Graham, C.D., Jr. and Rhyne, J.J.), p. 1664.

21. Bucher, E., Chu, C.W., Maita, J.P., Andres, K., Cooper, A.S., Buehler, E. and Nassau, K. (1969). *Phys. Rev. Lett.*, **22**, 126.

22. Holden, T.M. and Buyers, W.J.L. (1974). *Phys. Rev.*, **B**, (to be published).

23. Smith, S.R.P. (1972). *J. Phys.*, **C5**, L157.

24. Fulde, P. and Peschel, I. (1971). *Z. Phys.*, **241**, 82.

25. Fulde, P. and Peschel, I. (1972). *Adv. Phys.*, **21**, 1.

26. Peschel, I., Klenin, M. and Fulde, P. (1972). *J. Phys.*, **C5**, L194.

27. Klenin, M. and Peschel, I. (1973). *Phys. Kond. Mat.*, **16**, 219.

28. Abrikosov, A.A. (1965). *Physica*, **2**, 5.

29. Birgeneau, R.J. (1974). (Private communication).

30. Buyers, W.J.L. and Holden, T.M. (1973). *19th Conference on Magnetism and Magnetic Materials, Boston, November, 1973*, (paper 7C-5).

31. Hsieh, Y.Y. and Blume, M. (1972). *Phys. Rev.*, **B6**, 2684.

32. Egami, T. and Brooks, M.S.S. (1973). *19th Conference on Magnetism and Magnetic Materials, Boston, November, 1973*, (paper 7C-4).

33. Tsuchida, T. and Wallace, W.E. (1965). *J. Chem. Phys.*, **43**, 2885.

34. Tsuchida, T. and Wallace, W.E. (1965). *J. Chem. Phys.*, **43**, 2087.

35. Cooper, B.R. and Vogt, O. (1971). *J. de Phys.*, **32**, C1-1026.

36. Bartholin, H., Jacobs, I.S., Cooper, B.R. and Vogt, O. (1972).

Bull. Am. Phys. Soc., **17**, 248, (paper AJ7).

37. Cooper, B.R., Landolt, M. and Vogt, O. (1973). In *Proceedings of the International Conference on Magnetism, Moscow, 1973*, (paper 27a-S4), (to appear).

38. Busch, G. and Vogt, O. (1967). *Phys. Lett.*, **25A**, 449.

39. Cable, J.W. and Koehler, W.C. (1972). *AIP Conference Proceedings No.5, Magnetism and Materials 1971*, p. 1381.

40. Lebech, B., Fischer, P. and Rainford, B.D. (1971). *Proceedings of the Durham Conference on Rare Earths and Actinides*, (The Insitute of Physics, London), p. 204.

41. Rainford, B., Turberfield, K.C., Busch, G. and Vogt, O. (1968). *J. Phys.*, **C1**, 679.

42. Schobinger, P., Fischer, P. and Vogt, O. (1972). *Neutronen Streuung* (Progress Report Af-SSP-57), (Eidgenössicher Institut für Reaktorforschung, Wurenlingen), (January), p. 12.

43. Furrer, A., Buhrer, W., Heer, H., Halg, W., Benes, J. and Vogt, O. (1972). *Neutron Inelastic Scattering 1972*, (International Atomic Energy Agency, Vienna), p. 563.

44. Cooper, B.R., Furrer, A., Buhrer, W. and Vogt, O. (1972). *Solid State Commun.*, **11**, 21.

45. Wang, Y.-L. and Cooper, B.R. (1970). *Phys. Rev.*, **B2**, 2607.

ELEMENTARY EXCITATIONS
OF HIGH DEGREE PAIR INTERACTIONS
IN RARE EARTH COMPOUNDS†

PETER M. LEVY

Department of Physics, New York University,
New York, New York 10003, USA

1. INTRODUCTION

Rare earth ions retain many of their free ion properties when they form solids. In particular, the total angular momentum $\vec{J} = \vec{L} + \vec{S}$ remains a relatively good quantum number. In a first approximation the effect of the crystal field surrounding the rare earth ion is to split the $(2J + 1)$-fold degeneracy of the ground manifold. If the surrounding has high point group symmetry, e.g., cubic O_h symmetry, the split levels will still be degenerate; the orbital angular momentum of the rare earth ion is not quenched. Therefore we expect that the effects of orbital angular momentum on the magneto-thermal properties of rare earth compounds are quite marked. In terms of a 'phenomenological' Hamiltonian describing the pair interactions between rare earths in a solid, we will have in addition to the conventional bilinear (dipole) interaction

$$\mathcal{H}^{(1)} = \sum_{i,j} \Gamma^{(1)}(i - j)\vec{J}_i \cdot \vec{J}_j = \sum_{i,j} \Gamma^{(1)}(i - j)O_i^{(1)} \cdot O_j^{(1)}, \quad (1)$$

higher degree pair interactions of the generic form

$$\mathcal{H} = \sum_{i,j} \sum_{\substack{\ell,m \\ \ell',m'}} \Gamma_{mm'}{}^{\ell\ell'}(i - j)O_m{}^\ell(J_i)O_{m'}{}^{\ell'}(J_j), \quad (2)$$

† Research supported in part by the National Science Foundation.

where $\ell, \ell' \leqslant 7$ for f electrons and $-\ell \leqslant m \leqslant \ell$. We will call interactions of the form equation (2) multipolar, e.g., quadrupolar
interactions, even when they are not coupled so as to transform
as electric multipole interactions.

There are several mechanisms which give rise to these higher
degree pair interactions, e.g., electric multipole interactions,
Jahn-Teller effects (virtual phonon coupling), multi-electron exchange (Schrodinger's idea) and the contribution of orbital
anisotropy to exchange interactions. All of these will contribute
to the coupling constants $\Gamma_{mm'}{}^{\ell\ell'}(i - j)$. The Jahn-Teller effects together with anharmonic phonons also produce temperature
dependent coupling constants.

Examples of real systems which contain these interactions are
the rare earth pnictides (RSb, RAs, RP where R = rare earth) and
monosulphides (RS), both of which have a NaCl structure; and the
compounds RCd, RZn, and RAg which have a CsCl structure. All
these compounds exhibit metallic behaviour. The point group symmetry at the rare earth site is cubic O_h, therefore there is no
quadrupole moment $\langle O_m{}^2 \rangle$ at high temperatures. Most of these compounds order antiferromagnetically at low temperatures typically
in the range 5-30°K. Structural transitions have also been observed to occur from the cubic to a tetragonal or trigonal phase.
These transitions usually occur simultaneously with the magnetic
transition; however, for those materials which do not order magnetically, one still does find structural transitions. Compounds
of gadolinium do not have the structural transitions (except possibly for small exchange striction effects) because there is no
orbital angular momentum associated with these Gd^{3+} ions. The
gadolinium ion has a half-filled shell configuration $^8S_{7/2}$.

In addition to the above compounds there are the rare earth
arsenates, phosphates, and vanadates with the zircon structure.
For these, one notices a structural phase transition at a temperature much higher than the magnetic transition, if there is one
at all! For example, in $DyVO_4$ the structural transition is at
13.8°K and the antiferromagnetic transition is at $\sim$3°K.

To ascertain the form of the Hamiltonian necessary to describe
both a structural and a magnetic phase transition, we will consider dysprosium antimonide (DySb). It is cubic for $T > T_N =$
9.5°K and at this temperature undergoes a strong first order
phase transition to a tetragonal phase in which the spins are
antiferromagnetically ordered. From an analysis of data on the
softening of the elastic constant $C_\theta = \frac{1}{2}(C_{11} - C_{12})$, magnetization and specific heat, we find that large quadrupole interac-
tions are present in this compound. The Hamiltonian used to describe DySb is

$$\mathcal{H} = \sum_{i,j} \Gamma_{ij}{}^{(T_1)} \vec{S}_i \cdot \vec{S}_j + \sum_{i,j} \Gamma_{ij}{}^{(E)} \sum_{\alpha=1,2} O_\alpha{}^E(i) O_\alpha{}^E(j)$$

$$+ \sum_{i,j} \Gamma_{ij}{}^{(T_2)} \sum_{\alpha=1,2,3} O_\alpha{}^{T_2}(i) O_\alpha{}^{T_2}(j), \qquad (3)$$

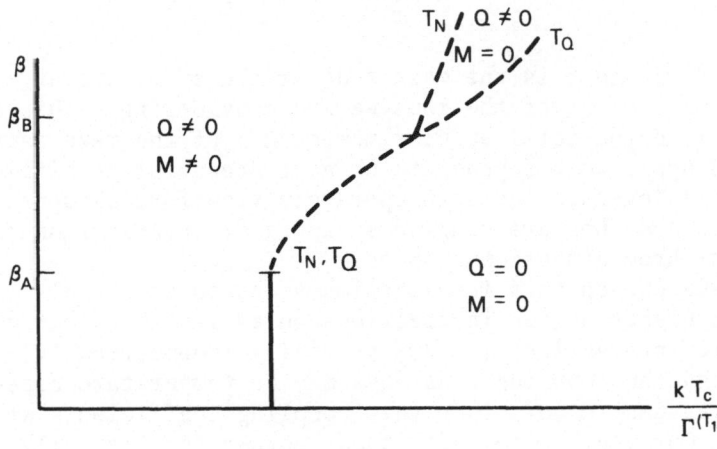

Figure 1

where E, T_1, T_2 are labels for the irreducible representations of the cubic group and the sums over the partners of the representations are denoted by α. We could also write $S_i \cdot S_j$ as $\sum O_\alpha^{T1} O_\alpha^{T1}$, and we have constructed the operators O^E, O^{T2} only from second rank angular momentum operators. We have not considered higher rank terms. We find that $\Gamma^{(E)} \gg \Gamma^{(T2)}$ for DySb although in general there is no reason for this.

If no higher degree coupling existed $\Gamma^{(E)} = \Gamma^{(T2)} = 0$, DySb would simply have a second order magnetic phase transition with no structural change except for magnetostriction. As a function of the ratio of quadrupolar to dipolar coupling $\beta \equiv \Gamma^{(E)}/\Gamma^{(T1)}$ the phase diagram for DySb is sketched in the following figure 1. For the coupling β less than a certain value β_A we have one *second* order phase transition, for $\beta > \beta_A$ a *first* order phase transition, therefore at β_A we have a tricritical point. For $\beta > \beta_B$ DySb would have two separate transitions. From all the data available DySb lies in between β_A and β_B. Therefore there is sizable quadrupole-like coupling in DySb, i.e., the quadrupolar coupling $\Gamma^{(E)}$ is of the same size as the dipolar coupling $\Gamma^{(T1)}$.

2. SYMMETRY OF THE HAMILTONIAN

From the preceding section we see that it is necessary to include higher degree pair interactions in the Hamiltonian describing rare earth compounds. We consider an *idealization* of the general interactions equations (2,3)

$$\mathcal{H} = \sum_\ell \sum_{i,j} \Gamma^{(\ell)}(i-j) O^\ell(S_i) \cdot O^\ell(S_j) = \qquad \text{(Contd)}$$

(Contd) $= \sum_{\ell} \sum_{i,j} \Gamma^{(\ell)}(i - j) \sum_{m} (-1)^{m} O_{m}^{\ell}(S_{i}) O_{-m}^{\ell}(S_{j}),$ (4)

where $\ell \leqslant 2S$ and S is the magnitude of the effective angular momentum or 'spin' of the ions we are considering. This S need not refer to the total angular momentum J of the rare earth ions. It could apply to a degenerate or near degenerate manifold of $n = 2S + 1$ levels. The spin operators transform as spherical harmonics and they are coupled so as to be invariant under rotations in three-dimensional space.

We have chosen this idealization so as to study only the effects of higher degree interactions on excitation spectra. More realistic interactions (3) are spatially (geometrically) anisotropic and the coupling constants may be temperature dependent. We will assume in our study that coupling coefficients are constants. Our goal is to understand whether the spatially isotropic Hamiltonians are dynamically anisotropic (see next section) and if so, how this anisotropy can be observed. Again it should be stressed that for real systems we usually have in addition to any dynamical anisotropy, geometrical or spatial anisotropy which does produce gaps in the excitation spectra at the zone centre $k = 0$.

The Hamiltonian (4) has been written in terms of operators O_{m}^{ℓ} which induce unimodular (unitary) transformations in a space of dimension $2S + 1$. They form the complete set of generators for an $SU(2S + 1)$ continuous group algebra. They also have been constructed so as to transform according to the irreducible representations (ℓ) of $O(3)$ or $SU(2)$ under three-dimensional rotations. The operators O_{m}^{ℓ} define a Lie algebra and therefore have well defined commutation properties. We will use these commutation properties to find excitation spectra of the Hamiltonian (4). Let us define operators in momentum space as

$$O_{m}^{\ell}(\vec{k}) \equiv \frac{1}{\sqrt{N}} \sum_{i} e^{-i\vec{k}\cdot\vec{R}_{i}} O_{m}^{\ell}(i).$$ (5)

In terms of this operator the Hamiltonian (4) is written as

$$\mathcal{H} = \sum_{\ell} \sum_{\vec{k}} \Gamma^{(\ell)}(\vec{k}) O^{\ell}(\vec{k}) \cdot O^{\ell}(-\vec{k}),$$ (6)

where

$$\Gamma^{(\ell)}(\vec{k}) \equiv \sum_{(i-j)} e^{i\vec{k}\cdot(\vec{R}_{i}-\vec{R}_{j})} \Gamma^{(\ell)}(i - j).$$

The operators $O_{m}^{\ell}(k)$ of $SU(2S + 1)$ satisfy the commutation relations,

$$[O_m{}^\ell(k), O_{m'}{}^{\ell'}(k')] = \frac{1}{\sqrt{N}} \sum_{\substack{\ell''m'' \\ (\ell+\ell'+\ell''=odd)}} C^{\ell\ell'\ell''}_{mm'm''} O_{m''}{}^{\ell''}(k+k') \quad (7)$$

where

$$C^{\ell\ell'\ell''}_{mm'm''} = (-)^{2S-m''+1} 2([\ell][\ell'][\ell''])^{\frac{1}{2}} \begin{Bmatrix} \ell & \ell' & \ell'' \\ S & S & S \end{Bmatrix} \begin{pmatrix} \ell & \ell' & \ell'' \\ mm' & -m'' \end{pmatrix}$$

and $[\ell] \equiv 2\ell + 1$, $\{::\}$ is a 6-j symbol and $(::)$ is a 3-j symbol. The commutator of one of the generators of the group with our Hamiltonian is:

$$[O_{m'}{}^{\ell'}(k'), \mathcal{H}] = \frac{2}{\sqrt{N}} \sum_{\substack{\ell''m'' \\ (\ell+\ell'+\ell''=odd)}} \sum_{\ell m} (-)^{m+1} C^{\ell\ell'\ell''}_{mm'm''}$$

$$\times \sum_k \Gamma^{(\ell)}(k) O_{-m}{}^\ell(k) O_{m''}{}^{\ell''}(k'-k). \quad (9)$$

In *general* these commutators of $O_m{}^\ell$ ($k = 0$) with $\mathcal{H}$ do not vanish, therefore our $\mathcal{H}$ is not an invariant of the group $SU(2S + 1)$, it does not have the maximum symmetry possible in this operator space. However, it may be an invariant of a subgroup of $SU(2S + 1)$. The way we constructed $\mathcal{H}$ it is at least always an invariant of $SU(2)$ or $O(3)$. We can readily show with the commutation relation (9) that:

$$[O_m{}^1(0), \mathcal{H}] = 0 \quad \text{or} \quad [S_q, \mathcal{H}] = 0,$$

and (10)

$$[O^1(0) \cdot O^1(0), \mathcal{H}] = 0 \quad \text{or} \quad [S^2, \mathcal{H}] = 0,$$

where

$$O_m{}^1(0) = \frac{1}{\sqrt{N}} \sum_i O_m{}^1(i), \qquad S_q \equiv \sum_i s_q(i)$$

and

$$S^2 = S \cdot S = S_x{}^2 + S_y{}^2 + S_z{}^2.$$

The $O_m{}^1(0)$ are the generators of $SU(2)$ and $O^1(0) \cdot O^1(0)$ is the bilinear invariant or Casimir operator of $SU(2)$. If one has a special relation amongst the coupling coefficients $\Gamma^{(\ell)}(k)$ so that $\mathcal{H}$ is an invariant of a higher subgroup of $SU(2S + 1)$ we say that $\mathcal{H}$ has *dynamical* symmetry. For example if our range functions $\Gamma^{(\ell)}(k)$ are identical for all the multipole interactions, i.e., $\Gamma^{(\ell)}(k) \equiv \Gamma(k)$ for all ℓ, we find

$$[O_m,^{\ell'}(0),\mathcal{H}] = 0, \tag{11}$$

for *all* $\ell' \leqslant 2S$ and $-\ell' \leqslant m' \leqslant \ell'$. This $\mathcal{H}$ is similar to the Casimir operator G (bilinear invariant) of $SU(2S + 1)$, i.e.,

$$G(SU(2S + 1)) = \sum_{\ell=1}^{2S} O^\ell \cdot O^\ell = \sum_{\ell=1}^{2S} \sum_{ij} O^\ell(i) \cdot O^\ell(j),$$

while

$$\mathcal{H} = \sum_{\ell=1}^{2S} \sum_{ij} \Gamma_{ij} O^\ell(i) \cdot O^\ell(j) \tag{12}$$

and from (11) we have $[G,\mathcal{H}] = 0$. Note that $\Gamma_{ii} = 0$. There are no interactions on the same site, therefore the term $O^\ell(i) \cdot O^\ell(i)$ in G is just a constant. Now, for the Casimir operators we have that the commutator of $O_m,^{\ell'}$ with G,

$$[O_m,^{\ell'}(k'),G] = 0,$$

vanishes for all k', not only $k' = 0$. By making the additional assumption $\Gamma_{ij} = \Gamma$ our $\mathcal{H}$ becomes identical to the Casimir operator (to within a constant); however, this is *not* an interesting model, because it is equivalent to the molecular field approximation and does not lead to any dispersion in the excitation spectra. Thus our special $\mathcal{H}$, has $SU(2S + 1)$ symmetry for $k = 0$ excitations; G has this symmetry for all k.

To label the eigenstates of a Hamiltonian one uses the eigenvalues of operators which commute with $\mathcal{H}$ and among themselves. The only operators which commute with the general Hamiltonian (4) are S^2 and S_q, $q = x,y,z$, i.e., $G(SU(2))$ and $O_m^1(0)$. As the O_m^1 do not commute, one chooses only one component and as $\mathcal{H}$ is rotationally invariant, we call it S_z. When $\mathcal{H}$ has some dynamical symmetry there will be more operators which commute with it, e.g., when all $\Gamma^{(\ell)}$ are equal all $O_m^\ell(0)$, $G(SU(2S + 1))$ and the Casimir operators of the sub-groups of $SU(2S + 1)$ commute with $\mathcal{H}$. These operators (rather a commuting subset thereof) provide additional labels to specify the $k = 0$ states of our system.

3. DYNAMICAL SYMMETRY

Whenever the Hamiltonian for a system has more symmetry than the *spatial* or *geometrical* symmetry of the true lattice dictates,

we say that the system has dynamical symmetry. This additional
symmetry comes about because of a special form of the pair interactions between particles or ions. For example, the hydrogen
atom

$$\mathcal{H} = \frac{p^2}{2m} - \frac{e^2}{r}$$

has not only $O(3)$ symmetry but also $O(4)$. The degeneracy n^2 of
all levels with the same principal quantum number n (different
ℓ,m) is explained by the $O(4)$ symmetry. With only $O(3)$ symmetry
there would be no reason to have levels with different angular
momentum ℓ degenerate. Similarly the three-dimensional isotropic harmonic oscillator

$$\mathcal{H} = \frac{p^2}{2m} + \tfrac{1}{2}kr^2$$

has $SU(3)$ symmetry. This can best be seen by writing the Hamiltonian as

$$\mathcal{H} = \sum_n \left(n + \frac{3}{2}\right)\hbar\omega$$

where $n = n_x + n_y + n_z$, and the oscillator eigenstates are
$|n_x\rangle|n_y\rangle|n_z\rangle$. The eigenfunctions written in spherical coordinate notation $\psi_{n\ell m}$ again have more degeneracy than one could
anticipate for a system with $O(3)$ symmetry, i.e., the even n
levels have all even values of orbital angular momentum ℓ degenerate, and the same for odd n and ℓ. This 'accidental' degeneracy is explained by the $SU(3)$ symmetry of the isotropic harmonic
oscillator and is thus a manifestation of the dynamical symmetry
of the system, therefore it is not an accidental symmetry.

In our case we consider only interactions with $SU(2)$ or $O(3)$
symmetry, even though our ions are placed on a three-dimensional
lattice and realistic interactions between ions are probably anisotropic. We can contemplate *additional* dynamical symmetry whenever the interaction constants $\Gamma^{(\ell)}(k)$ assume special relationships among themselves. In these lectures we consider the interactions have *at least* $SU(2)$ symmetry, and we will focus our attention on *additional* dynamical symmetry. We will be particularly interested in finding out how to observe additional dynamical
symmetry or considered conversely, whether the lack of maximum
dynamical symmetry, i.e., 'dynamical anisotropy', is observable.

Dynamical symmetry or anisotropy can be observed by noting
the degeneracies of $k = 0$ levels for our Hamiltonian (4), i.e.,
when special relations exist between the $\Gamma^{(\ell)}(k)$ we expect that
certain gaps between levels will disappear. There is particular
interest in degeneracies (or gaps as the case may be) in the
ground and low-lying states of a system, and therefore we focus

our attention on the Goldstone Theorem. This theorem states that for each symmetry operation of a Hamiltonian, i.e., $[O_m, l'(k = 0), \mathcal{H}] = 0$ there exists a gapless excitation mode, the Goldstone boson $O_m, l'(k = 0)$, which acts on the ground state and restores a *continuous* symmetry that has been broken by a phase transition. A system described by Hamiltonian (4) has at least *one* Goldstone boson (see equations (10)), when it has an ordered state which is not rotationally invariant. This boson restores the rotational invariance broken by the ordered phase. When additional dynamical symmetry is present, new Goldstone bosons may exist in low temperature excitation spectra of ordered systems. These bosons restore the additional symmetry broken by the ordering. Conversely, eyed from the position of a Hamiltonian like (12) with additional dynamical symmetry, the general Hamiltonian is dynamically *anisotropic* and instead of a gapless excitation mode, one may find gaps in the spectrum at $k = 0$. This corresponds to generators $O_m l(0)$ which are no longer symmetry transformations of $\mathcal{H}$. Therefore we can say that *dynamical anisotropy is a possible origin of gaps at $k = 0$ in some modes of the elementary excitation spectra of systems described by the Hamiltonian* (4). Now we will present an explicit example which illustrates the existence of a new Goldstone boson when extra symmetry exists and, conversely, which shows a gap in the excitation spectrum when this symmetry is not present.

4. ELEMENTARY EXCITATION SPECTRA

The work I am going to present has been done in close collaboration with Dr. Sou-Tong Chiu-Tsao. Section 4.2, in particular, is based on her recent research results. A more complete account of our work will be published elsewhere.

4.1 SPIN ONE

Let us consider the example of a linear chain of ferromagnetically coupled $S = 1$ ions with nearest neighbour dipole and quadrupole interactions of the form of equation (4), at $T = 0°K$,

$$\mathcal{H} = -\Gamma^{(1)}\left\{ \sum_i \vec{S}_i \cdot \vec{S}_{i+\sigma} + \beta \sum_i O_i^2 \cdot O_{i+\delta}^2 \right\} , \qquad (13)$$

where $\Gamma^{(1)} > 0$ and $\beta \equiv \Gamma^{(2)}/\Gamma^{(1)}$ is the ratio of quadrupole to dipole coupling. We will consider $\beta \leqslant 1$ for now, because for this case we know the *exact* ground state of the system, it is ferromagnetic, and we can calculate exactly the elementary excitations of the linear chain at $T = 0°K$. The ground state for $\beta \leqslant 1$ at $T = 0°K$ has all the spins pointing in one common direction, e.g., $M_S = 1$,

$$|g\rangle = |111.....111\rangle. \qquad (14)$$

For this ground state

$$\sum_i s_z(i) \equiv S_z(0) = NS = N.$$

As eigenstates of $\mathcal{H}$ are also eigenstates of $S_z(0)$, that is to say, $[S_z(0),\mathcal{H}] = 0$, we can classify excitations from the ground state to excited eigenstates by changes in S_z from its ground state value by m where $m \equiv N - S_z(0)$.

For $m = 1$ we find the excitation energy for the state,

$$|k\rangle \equiv \frac{1}{\sqrt{N}} \sum_i e^{-i k \cdot R_i} S^-(i)|g\rangle = S^-(k)|g\rangle, \qquad (15)$$

is given as

$$\varepsilon_k - \varepsilon_g = 2\Gamma^{(1)}(1 + \beta)(1 - \cos ka), \qquad -\frac{\pi}{a} \leqslant k \leqslant \frac{\pi}{a}, \qquad (16)$$

where ε_g is the ground state energy. As we increase the quadrupolar coupling the width of the energy band increases as $4\Gamma^{(1)}(1+\beta)$, as shown in figure 2 below. Each excitation $|k\rangle$ carries with it a change in S_z by -1 and in the quadrupole operator $O_0{}^2(k = 0)$ by -3, i.e.,

$$[S_z(k = 0), S^-(k)]|g\rangle = -S^-(k)|g\rangle$$

and $\qquad\qquad\qquad\qquad\qquad\qquad\qquad\qquad\qquad\qquad\qquad\qquad (17)$

$$[O_0{}^2(k = 0), S^-(k)]|g\rangle = -3S^-(k)|g\rangle.$$

From the above excitation spectra one cannot identify the presence of quadrupolar interaction, nor any additional symmetry

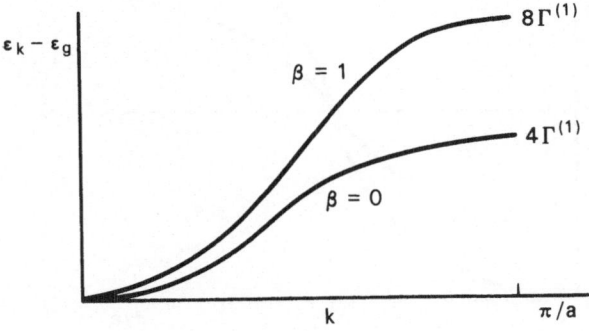

Figure 2

when $\beta = 1$. If $\Gamma^{(1)}$ is not fixed by some other experimental data one can easily explain the increased dispersion as caused by larger $\Gamma^{(1)}$.

For $m = 2$ we must consider two spin deviations $m = 1$ at *different* sites as well as both at the *same* site. In a lowest order of approximation of single-site decouplings, we need consider only single site $m = 2$ excitations. After presenting these results we will consider the complete problem of two $m = 1$ spin deviations at different sites interacting with $m = 2$ excitations at the same site. The solutions, as we will see, are rather complicated; it is for this reason that we first present the approximate but simpler results.

The excitation energy for an $m = 2$ mode,

$$|k\rangle_{m=2} \equiv \frac{1}{\sqrt{N}} \sum_i e^{-ik \cdot R_i} (S^-(i))^2 |g\rangle = O_{-2}{}^2(k)|g\rangle, \qquad (18)$$

where $O_{-2}{}^2(i) \sim (S^-(i))^2$ is given as

$$\varepsilon_k^{(2)} - \varepsilon_g = 4\Gamma^{(1)}(1 - \beta\cos ka), \qquad (19)$$

where $\varepsilon_k^{(2)}$ is the energy of the double spin deviation or quadrupole excitations. In figure 3 below we plot this energy for various values of β. Each excitation changes $S_z(k = 0)$ by 2, but leaves $O_0{}^2(0)$ unchanged, i.e., $[\mathcal{H}, O_{-2}{}^2(k)]|g\rangle = (\varepsilon_k^{(2)} - \varepsilon_g) O_{-2}{}^2(k)|g\rangle$, $[S_z(0), O_{-2}{}^2(k)]|g\rangle = -2O_{-2}{}^2(k)|g\rangle$, and $[O_0{}^2(0), O_{-2}{}^2(k)]|g\rangle = 0$.

We note that as the quadrupole interaction strength increases so does the dispersion of the quadrupole or $m = 2$ excitations.

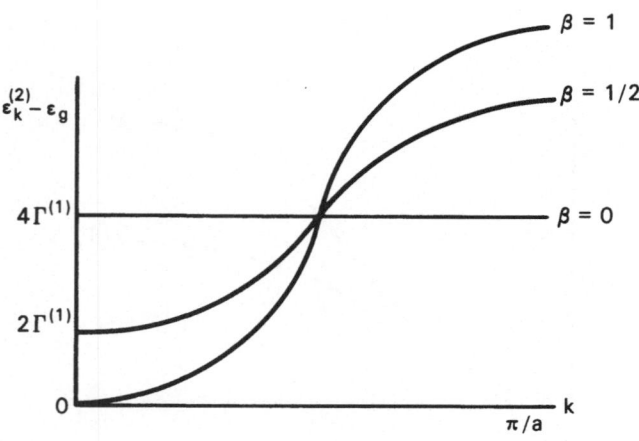

Figure 3

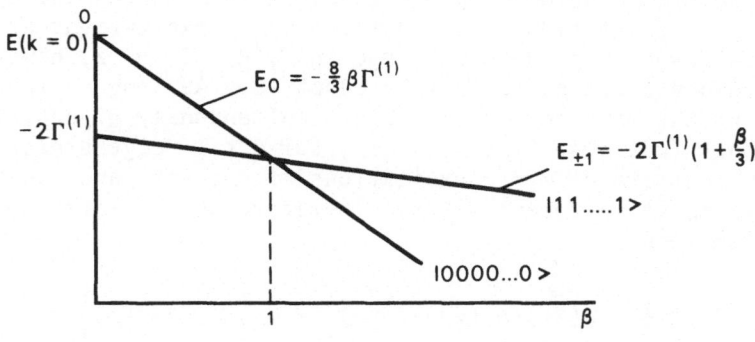

$$E_0 = -\tfrac{8}{3}\beta\Gamma^{(1)}$$

$$E_{\pm 1} = -2\Gamma^{(1)}\left(1 + \tfrac{\beta}{3}\right)$$

Figure 4

For $\beta = 1$ we note that the $\varepsilon_k^{(2)}$ spectrum is *gapless* at $k = 0$, and we have a new Goldstone boson. One way of looking at the 'softening' of the $k = 0$ mode as β increases is that it is the mode which drives the system at $T = 0°K$ from the ferromagnetic phase $M \equiv \langle S_z \rangle = 1$, $Q \equiv \langle O_0^2(0) \rangle = 1$ when $\beta < 1$ to the non-magnetic quadrupolar phase $M = 0$, $Q = -2$ when $\beta > 1$. As seen from figure 4 below, for $\beta < 1$ the ground state is $|1111...\rangle$, whilst for $\beta > 1$ it is $|00000\rangle$.
For $\beta = 1$ either ground state is valid, and we also note that excitation spectra for $m = 1$ and $m = 2$ are identical, see figures 2,3. We readily understand how this degeneracy arises as the single site excitations $|11...0...111\rangle$ and $|11...-1...11\rangle$ require the same energy for $\beta = 1$.

We can understand the *gap* in the $m = 2$ excitation spectra for the *isotropic $O(3)$ Hamiltonian* (13) as follows. For $\beta = 0$ the only term in $\mathcal{H}$ which can transfer excitations from one site to another is $S_i^+ S_{i+\delta}^-$; this can only change m by ± 1, *not* 2. Therefore an $m = 2$ excitation at one site is not propagated or transferred by a dipole interaction in a one-step process (we will presently consider two-step processes). It remains dispersionless because $m = 2$ is a single site excitation for $\beta = 0$. For $\beta \neq 0$ our $\mathcal{H}$ includes $O_i^2 \cdot O_{i+\delta}^2$ which contains the term $O_2^2(i)$ $O_{-2}^2(i + \delta) \sim (S_i^+)^2(S_{i+\delta}^-)^2$. This term is able to transfer an $m = 2$ excitation in one step and is responsible for the dispersion in the $\varepsilon_k^{(2)}$ spectra shown in figure 3, as well as the decrease in the $k = 0$ gap energy.

When $\Gamma^{(1)} = \Gamma^{(2)}$ or $\beta = 1$ we note that the gap disappears. This is readily explained by the fact that the Hamiltonian (13) with $\beta = 1$ is an invariant of $SU(3)$ and that then $[O_{\pm 2}^2(0), \mathcal{H}] = [(S^\pm)_{k=0}^2, \mathcal{H}] = 0$. This is the origin of the new Goldstone mode, the appearance of a new symmetry when $\beta = 1$. Conversely we can notice the lack of $SU(3)$ symmetry, i.e., dynamical anisotropy by the existence of a gap in the $\varepsilon_k^{(2)}$ spectra even when one still has isotropic or $O(3)$ invariant Hamiltonians.

4.2 TWO-SITE EXCITATIONS

In the preceding work on $m = 2$ excitations we considered only single site excitations. In fact two $m = 1$ excitations at different sites also produce a change in $S_z(k = 0)$ by 2, however, they produce a change in $O_0{}^2(k = 0)$ of -6. As long as the eigenstates of the Hamiltonian (13) are simultaneously eigenstates of $S_z(0)$ only (not also $O_0{}^2(0)$), i.e., for $\beta < 1$, eigenstates of $\mathcal{H}$ are *combinations* of $m = 2$ excitations at one site and two $m = 1$ excitations at different sites. Therefore we now introduce a trial function,

$$|m = 2\rangle \equiv \left\{ \sum_i \phi(ii)(S^-(i))^2 + \sum_{\substack{ij \\ i \neq j}} \phi(ij)S^-(i)S^-(j) \right\} |g\rangle, \quad (20)$$

and determine the amplitude function $\phi(i,j)$ and eigenvalues (energies) by making the states $|m = 2\rangle$ eigenstates of the Hamiltonian (13)

If we neglect spin wave interactions, the excitation energy of two free spin waves is

$$\varepsilon_{Kk}{}^{(2)} - \varepsilon_g = \varepsilon_{k_1} + \varepsilon_{k_2} - \varepsilon_g$$

$$= 4\Gamma^{(1)}(1 + \beta)[1 - \cos\tfrac{1}{2}Ka \cos ka], \quad (21)$$

where $K \equiv k_1 + k_2$ is the momentum of the centre of mass of two magnons and $k \equiv \tfrac{1}{2}(k_1 - k_2)$ is their relative momentum. The amplitude function for these states is

$$\phi_{k_1 k_2}(ij) = e^{-ik_1 \cdot r_i} e^{-ik_2 \cdot r_j} + e^{-ik_2 \cdot r_i} e^{-ik_1 \cdot r_j},$$

or

$$\phi_{Kk}(ij) = e^{-i\frac{1}{2}K \cdot (r_i + r_j)} 2\cos k \cdot R, \quad (22)$$

where $R \equiv r_i - r_j$.

The two-spin-wave energy spectrum $\varepsilon^{(2)}$ forms a continuum which is bounded from below by

$$\varepsilon_{K,0} - \varepsilon_g = 4\Gamma^{(1)}(1+\beta)[1 - \cos\tfrac{1}{2}Ka] \qquad \text{for } k = 0$$

and from above by

$$\varepsilon_{K,\pi/a} - \varepsilon_g = 4\Gamma^{(1)}(1 + \beta)[1 + \cos\tfrac{1}{2}Ka] \qquad \text{for } k = \frac{\pi}{a}$$

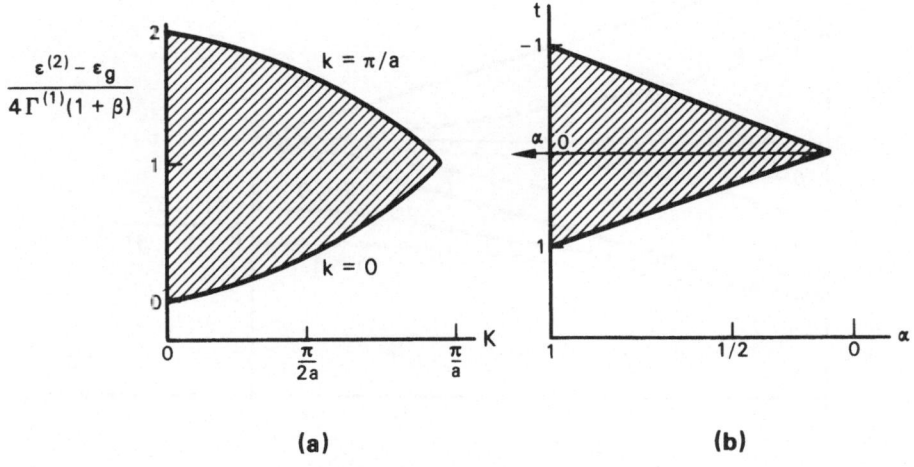

Figure 5

We show the two-spin-wave continuum in figure 5(a). If we define

$$t \equiv 1 - \frac{\varepsilon^{(2)} - \varepsilon_g}{4\Gamma^{(1)}(1 + \beta)} \quad \text{and} \quad \alpha \equiv \cos\tfrac{1}{2}Ka$$

the two-spin-wave continuum boundaries are straight lines as shown in figure 5(b).

When we consider the interactions between two spin waves we find that the energy levels in the continuum $\varepsilon_{Kk}^{(2)}$ are slightly shifted by the order of $1/N$, and the amplitude functions (22) remain about the same. However, three new solutions for each value of K do exist when the spin waves interact. These are solutions to the bound state problem. For $\beta < 1$ one root is real and corresponds to a true bound state and two roots, complex conjugates of one another, are resonant states in the continuum (see figure 6). The latter states have a finite lifetime which increases with β and decreases with α. For $\beta = 1$ the complex solutions become real and are just the solutions $\varepsilon_k^{(2)}$ for $\beta = 1$, see equations (18,19).

To better understand the different solutions let us look at their various amplitude functions. The continuum solutions (22) have a plane wave character with their amplitude modulated by $\cos k \cdot (r_i - r_j)$, see figure 7(a). The true bound state below the continuum has an exponential decaying amplitude function. For $\beta = 1$ this is given as

$$\phi_k(R) = e^{-q|R|/a}(1 - \delta_{RO}) \tag{23}$$

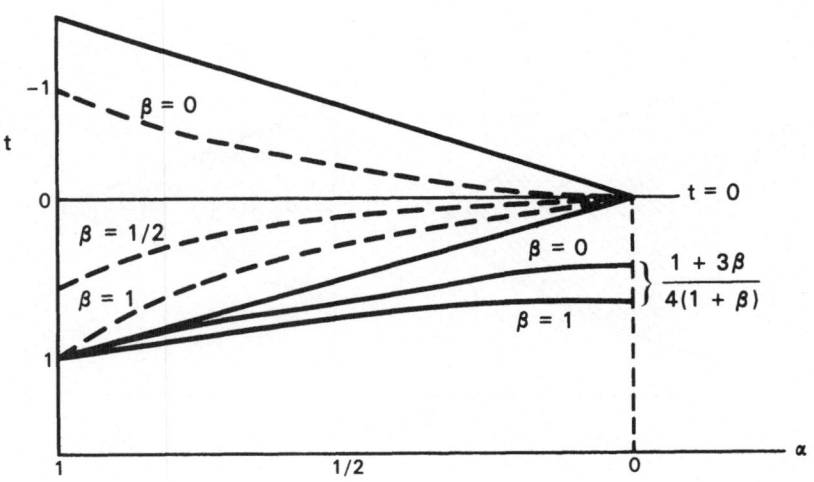

Figure 6

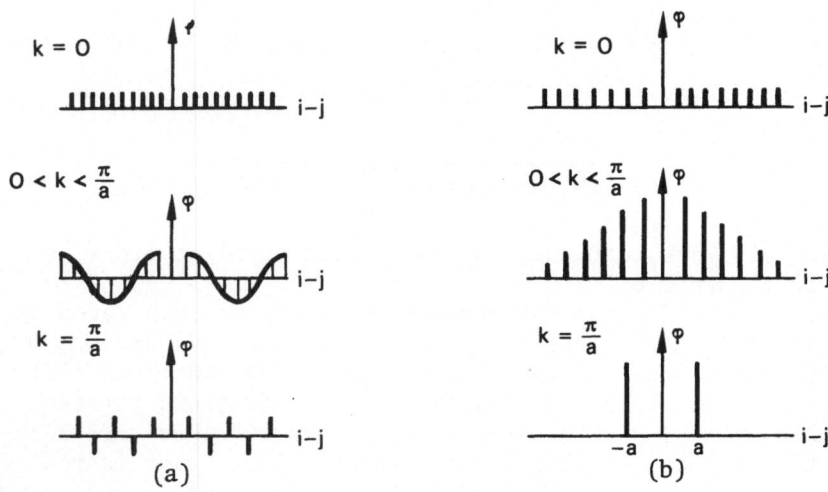

Figure 7

where $q = |\ln \cos\frac{1}{2}Ka|$. The probability of two spin waves being
at the same site, i.e., $R = 0$, is zero for $\beta = 1$. For $\beta < 1$
there is a finite probability. The function (23) is plotted in
figure 7(b). As K increases towards $K = \pi/a$, the state is more
strongly bound and the binding energy depresses this state below
the continuum accordingly. Also the binding energy increases as
β, therefore *quadrupolar coupling tends to bind neighbouring spin
deviations more strongly*.

Finally, the resonant or virtual bound states have an oscilla-
tory decaying amplitude function $\phi_K(R)$. The period of oscillation depends on K, i.e., the oscillations increase with K, and the

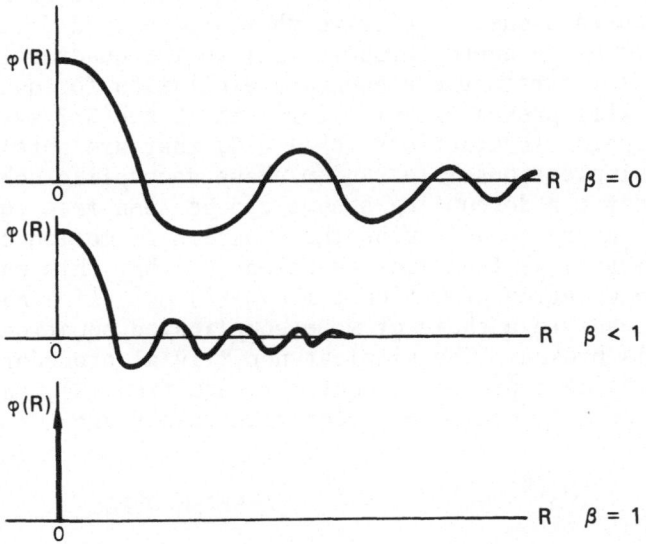

Figure 8

decay rate depends on β. In figure 8 we show some typical functions $\phi_K(R)$ for resonant states. Note that for these states there is finite probability that two $m = 1$ excitations are on the same site ($R = 0$). For β = 1, the state has infinite lifetime and represents a true $m = 2$ single site excitation wave $\phi_K(R) \sim \delta_{RO}$ or quadrupole excitation wave with energy $\varepsilon_k{}^{(2)}$ given by equation (19) with β = 1. It differs from the two-spin wave bound state in that it does not cause any change in the quadrupole moment of the solid, i.e., $\Delta Q = 0$, while the bound state has $\Delta Q = -6$. For β = 1 the quadrupole moment of the solid is a good quantum number and the two excitations $\Delta Q = 0$, and $\Delta Q = -6$ do not mix because they lead to states with different quantum numbers. Conversely for β ≠ 1 the quadrupole moment of the solid is *not* a good quantum number, i.e., $[Q,\mathcal{H}] \neq 0$, and the single site $m = 2$, $\Delta Q = 0$ excitations *mix* with the two site $m = 2$, $\Delta Q = -6$ excitations. This leads to the coupling of single site excitations with the two-spin wave continuum, and concomitantly to resonant or virtual bound states with finite lifetimes.

This completes our results on the $S = 1$ linear chain at $T = 0°K$ for β ≤ 1. We are at present extending these results to two- and three-dimensional systems.

4.3 QUADRUPOLE HAMILTONIAN

Systems with dominant quadrupolar coupling β > 1, are more complicated to treat because one does not know the exact ground state. We do know the approximate ground state; for $S = 1$ this

is a non-magnetic state with all ions in their $s_z = 0$ states.
The excited states $s_z = \pm 1$ are degenarate for pure quadrupolar
coupling and in a one ion picture they are removed from the
ground state by an energy proportional to the quadrupole moment
$\langle O_0{}^2 \rangle$. To illustrate the elementary excitation for quadrupolar
systems we will present results for spin -1 and 3/2 systems with
pure quadrupolar interactions, $\Gamma^{(1)} = 0$, that are obtained by
using the simplest possible random phase decoupling scheme.

The need for a decoupling scheme can be seen from equation
(9). When we try to determine the equation of motion for an operator $O_m{}'^{\ell'}(k')$ we find from equation (9) that this entails our
finding the equation of motion for $O_{-m}{}^{\ell}(k)O_m{}''^{\ell''}(k' - k)$. This
procedure leads to a chain of more complicated equations unless
the chain is broken. The simplest non-trivial procedure is to
symmetrically decouple the commutator, equation (9), and to assume that the only non-zero expectation values are

$$\langle O_m{}'^{\ell'}(k) \rangle = \sum_{\ell} \langle O_0{}^{\ell}(0) \rangle \delta_{\ell\ell'} \delta_{m'0} \delta_{k0}. \qquad (24)$$

The sum ℓ is over the non-zero order parameters and we consider
a one sublattice system whose ordering is homogeneous ($k = 0$)
and axial ($m' = 0$). The brackets $\langle \cdots \rangle$ denote a thermal expectation value taken by using the density matrix $e^{-\beta H_0}/Tr e^{-\beta H_0}$ where
H_0 is the molecular field Hamiltonian. The restriction to one
sublattice is possible as we shall consider only ferroquadrupolar
coupling. We could consider ordering with $m \neq 0$ and $k \neq 0$; this
does not complicate the problem too much.

Applying the above random phase approximation (RPA) to equation (9) we find

$$[O_m{}'^{\ell'}(k'), \mathcal{H}^{(\ell)}]_{RPA} = \sum_{\ell''} \alpha_{\ell'\ell''}{}^{(\ell)} O_m{}'^{\ell''}(k'). \qquad (25)$$

We have put a superscript ℓ on the Hamiltonian to denote that we
consider only one coupling; in our case of quadrupolar coupling
$\ell \equiv 2$. For states of angular momentum S this leads to $N \leqslant 2S$
coupled homogeneous equations. As the basis $O_m{}^{\ell}(k)$ is complex,
our matrices $\alpha_{\ell'\ell''}{}^{(\ell)}$ are *non-Hermitian* and are diagonalized by
a bi-orthogonal transformation. That is, if

$$\alpha_{\ell'\ell''}{}^{(\ell)} \equiv \mathcal{L}, \qquad (26)$$

we can diagonalize this matrix by the transformation

$$\vec{f}_m{}^* \cdot \mathcal{L} \cdot \vec{e}_n = \lambda_n \delta_{mn}, \qquad (27)$$

where $\vec{e}_n$ is a column vector of the coefficients of an eigenfunction of $\mathcal{L}$ with eigenvalue λ_n, and $\vec{f}_m$ the coefficients of an eigenfunction of the adjoint matrix $\mathcal{L}^+$,

$$\mathcal{L}e_n = \lambda_n e_n,$$

$$\mathcal{L}^+ f_m = \mu_m f_m,$$

and

$$\mu_m = \lambda_m^*. \tag{28}$$

The eigenfunction e_n and f_m^* have the property that they are orthogonal

$$\vec{f}_m^* \cdot \vec{e}_n = \delta_{mn}. \tag{29}$$

If we diagonalize $\mathcal{L}$ we find that equation (25) can be reduced to the form

$$[O^\lambda m(k'), \mathcal{H}^{(\ell)}]_{RPA} = e_{\lambda_m}(k')O^\lambda m(k'), \tag{30}$$

where $O^\lambda m(k')$ is an operator which produces an elementary excitation with $\Delta S_z(0) = m$ and energy $\varepsilon_k(m) - \varepsilon_g = \varepsilon_{\lambda m}(k)$. The $\varepsilon_{\lambda m}$ are eigenvalues of the matrix $\alpha_{\ell'\ell''}^{(\ell)}(m)$, i.e.,

$$F\alpha E = e, \tag{27'}$$

and the $O^\lambda m$ are the eigenvectors of the adjoint matrix α^+, i.e.,

$$O^\lambda m(k) = \sum_{\ell''} f_{\lambda_m, \ell''} O_m^{\ell''}(k). \tag{31}$$

We fix the normalization of our eigenvectors by requiring that they satisfy boson-like commutation rules.

$$[O^\lambda m(k), O^{\lambda' m'}(k')^+]_{RPA} = \pm \delta_{\lambda\lambda'} \delta_{mm'} \delta_{kk'}. \tag{32}$$

The $\pm$ sign depends on whether O^λ is a raising or lowering operator. The condition (32) is a rather severe approximation, somewhat similar to that used in the $S = \frac{1}{2}$ spin wave approximation,

$$[S^+, S^-] = 2S^z \xrightarrow{RPA} [S^+, S^-]_{RPA} = 2\langle S^z \rangle.$$

If we define

$$a\dagger \equiv \frac{S^-}{(2\langle S^z\rangle)^{\frac{1}{2}}} \ , \ a \equiv \frac{S^+}{(2\langle S^z\rangle)^{\frac{1}{2}}} \ ,$$

we find that the spin wave operators satisfy the relation (32),

$$[a,a^+] = 1.$$

Owing to the approximation made in using the commutation rule (32), our results are only reliable at temperatures for which excited states are negligibly populated. Now, having presented the formalism let us apply it to two examples: the elemetary excitations of $S = 1$ and 3/2 quadrupolar Hamiltonians

For $S = 1$ and $\ell = 2$ we find from equations (9,25) that for $m = 1$ excitations

$$\mathbf{\alpha}(2) = \begin{bmatrix} 0 & \Gamma(0) - \Gamma(k) \\ \Gamma(0) & 0 \end{bmatrix} \tag{33}$$

We note that the matrix is not hermitian. It is readily brought to diagonal form

$$e_{m=1} \sim \langle O_0{}^2\rangle \begin{bmatrix} (\Gamma_0(\Gamma_0 - \Gamma_k))^{\frac{1}{2}} & 0 \\ 0 & -(\Gamma_0(\Gamma_0 - \Gamma_k))^{\frac{1}{2}} \end{bmatrix}, \tag{33'}$$

and the eigenvectors (31) are

$$O^\lambda{}_1(k) = \frac{1}{(2K\gamma)^{\frac{1}{2}}} \begin{bmatrix} 1 & \gamma \\ 1 & -\gamma \end{bmatrix} \begin{bmatrix} O_1{}^1(k) \\ O_1{}^2(k) \end{bmatrix}, \tag{34}$$

where $\gamma \equiv (1 - \Gamma(k)/\Gamma(0))^{\frac{1}{2}}$ and $K \sim \langle O_0{}^2\rangle$. The dispersion relation (33') for the quadrupole excitations is *linear* for $k \to 0$, i.e., $\varepsilon \sim k$, in contrast to that for spin waves, equation (16), which is quadratic in k. Also, as $k \to 0$ the eigenvector (34) is

$$O^\lambda{}_1 \sim \frac{1}{\sqrt{\gamma}} O_1{}^1(k).$$

Its amplitude blows up as $k \to 0$; this is the behaviour indicative of a Goldstone mode which restores the broken symmetry of the Hamiltonian $\mathcal{H}(2)$ owing to the axial ordering $\langle O_0{}^2\rangle \neq 0$. At $k = 0$ the operator $O_1{}^2(0)$ does not enter; as it does not commute with the Hamiltonian, $[O_m{}^2(0),\mathcal{H}(2)] \neq 0$, it cannot be part of the $k = 0$ Goldstone mode. As the levels $m_S = \pm 1$ are degenerate for zero dipolar coupling, the energy of $m = 2$ excitations are zero.

Finally, let us consider $S = 3/2$ and again $\ell = 2$. The molecular field ground state is $m_S = \pm 3/2$ or $\pm \frac{1}{2}$. From equations (9, 25) we find that for $m = 1$ transitions the matrix α is

$$\boldsymbol{\alpha}^{(2)} \sim \langle 0_0{}^2 \rangle \begin{bmatrix} 0 & a(\Gamma_0 - \Gamma_k) & 0 \\ a\Gamma_0 & 0 & b\Gamma_0 \\ 0 & b(\Gamma_0 - \Gamma_k) & 0 \end{bmatrix}, \qquad (35)$$

which has the eigenvalues

$$e_{m=1} = 0, \ \pm K\left(1 - \frac{\Gamma_k}{\Gamma_0}\right)^{\frac{1}{2}} = 0, \ \pm K\gamma \qquad (36)$$

and eigenvectors

$$O^{\lambda}1(k) \sim \frac{1}{\sqrt{\gamma}} \begin{bmatrix} \xi & 0 & -\xi \\ \eta & \gamma/\sqrt{2} & \eta' \\ \eta & -\gamma/\sqrt{2} & \eta' \end{bmatrix} \begin{bmatrix} 0_1{}^1(k) \\ 0_1{}^2(k) \\ 0_1{}^3(k) \end{bmatrix}, \qquad (37)$$

where K, ξ, η and η' are constants. The $\varepsilon_1 = 0$ mode corresponds to transitions between degenerate states ($m_S = -\frac{1}{2}$ to $m_S = +\frac{1}{2}$). As $k \to 0$ we note again that the $m = 1$ excitations of a quadrupolar Hamiltonian have a linear dispersion, $\varepsilon \sim k$, and that the Goldstone mode

$$O^{\lambda}1 \sim \frac{1}{\sqrt{\gamma}}\left[0_1{}^1 + \frac{\eta'}{\eta}\, 0_1{}^3\right], \qquad (k \to 0)$$

does not contain the operator $0_1{}^2(0)$. It does contain $0_1{}^3(0)$, which is a rather special case. In general only the $0_m{}^1(0)$ commutes with the isotropic Hamiltonian $\mathcal{H}^{(\ell)}$; however, for $S = 3/2$ and $\ell = 2$ one finds from equation (9) that

$$[O_m{}^3(k = 0) \,\mathcal{H}^{(2)}] = 0.$$

Therefore the $0_m{}^3(0)$ are constants of the motion for this special case and can be part of a Goldstone mode.

For $m = 2$ excitations we find

$$\boldsymbol{\alpha} \sim \langle 0_0{}^2 \rangle \begin{bmatrix} 0 & 1 \\ \gamma^2 & 0 \end{bmatrix}, \qquad (38)$$

whose eigenvalues are

$$e_{m=2} = \pm K\gamma, \tag{39}$$

and whose eigenvectors are

$$o^{\lambda 2} \sim \frac{1}{\sqrt{\gamma}} \begin{bmatrix} \gamma & 1 \\ & \\ \gamma & -1 \end{bmatrix} \begin{bmatrix} O_2{}^2(k) \\ \\ O_2{}^3(k) \end{bmatrix}. \tag{40}$$

The energy for the $m = 2$ excitations (39) are the same as those for $m = 1$ (36) (except for the $\varepsilon_{m=1} = 0$ mode) because they represent excitations between states which have the same energies, i.e., $m = 1$ gives a transition $m_S = \frac{1}{2}$ to $m_S = 3/2$, while $m = 2$ gives a transition $m_S = -\frac{1}{2}$ to $m_S = 3/2$. As both $m_S = \pm\frac{1}{2}$ are degenerate the excitations have identical energies. As $k \to 0$ we find from equation (40) that

$$o^{\lambda 2} \sim \frac{1}{\sqrt{\gamma}} O_2{}^3.$$

This represents a second Goldstone mode in addition to $o^{\lambda 1}$. For $m = 3$, the excitation energy $\varepsilon_3 = 0$ represents transitions between degenerate states.

5. PERTURBATION THEORY

When the quadrupole or higher degree interactions are small they can be considered as perturbations on the energy levels and excitations found for pure dipolar interactions. However, there exist systems with rare earth ions for which sizeable higher degree interactions cannot be considered as perturbations. In such cases it may be useful to consider the Hamiltonian (12) with all $\Gamma^{(\ell)}(k)$ equal as the unperturbed Hamiltonian $\mathcal{H}_0$ and the *smaller* difference between the full $\mathcal{H}$ and $\mathcal{H}_0$ as the perturbation. The high symmetry of the Hamiltonian (12) makes it about as simple to calculate its eigenvalues and eigenfunctions as those of the dipolar Hamiltonian.

This idea has been applied by Elliott to the interaction betweens nucleons. The principal interaction is represented as that of a three-dimensional isotropic harmonic oscillator $\mathcal{H}_0 = (n + 3/2)\hbar\omega$. This $\mathcal{H}_0$ is invariant under the group of $SU(3)$ operations. Elliott considers, in addition to $\mathcal{H}_0$, a scalar quadrupole interaction as a perturbation, i.e.,

$$\mathcal{H} = \mathcal{H}_{\text{iso-osc}} + V \sum_m (-)^m Q_m Q_{-m}. \tag{41}$$

The isotropic oscillator $\mathcal{H}$ can be written as a linear combination of a dipole and quadrupole interaction,

$$\mathcal{H} = \alpha \sum_m (-)^m S_m S_{-m} + \beta \sum_m (-)^m Q_m Q_{-m}, \qquad (42)$$

where $\beta = \alpha + V$.

When the Hamiltonian is written in this form we see that whereas the quadrupole interaction in (41) is a *small* perturbation on the isotropic oscillator $\mathcal{H}_0$, this is not true for the quadrupole interaction of strength $\alpha + V$ in equation (42), where $\mathcal{H}_0$ is a dipole interaction. Therefore, in general, Hamiltonians with higher or dynamical symmetries often serve as appropriate 'unperturbed' Hamiltonians from which to consider that a Hamiltonian deviates.

6. SUMMARY

In general, real systems do not possess $SU(3)$ nor, in general, $SU(2S + 1)$ symmetry. We can only say that compared with an ideal system whose Hamiltonian is invariant under $SU(3)$ or $SU(2S + 1)$, real systems are dynamically anisotropic. The systems are not necessarily anisotropic in a geometrical sense (they might possibly still have $O(3)$ symmetry), but in the sense that real systems do not possess $SU(n)$ symmetry. This dynamical anisotropy may be observed in quadrupole ($m = 2$ single site) excitation spectra in that these excitations have finite lifetimes and that a gap exists for $k = 0$. When $\Gamma^{(1)}(k) = \Gamma^{(2)}(k)$ the gap disappears, a new Goldstone mode appears, and the lifetimes of the single site quadrupole excitations are infinite.

Compared to the behaviour of systems with only dipolar interactions, the presence of quadrupole interactions in real systems is manifested by: (1) increases in the lifetime of the virtual bound or resonant $m = 2$ states; (2) a lowering of the real part of the energy of these states, particularly near $k = 0$; and (3) an increase of the binding energy of the true bound states, which means that the pair of spin deviations comprising the bound states is more strongly bound together by quadrupole interactions.

We have considered the effects of mixing the two-spin wave continuum with single site $m = 2$ excitations only for a linear chain. However, we expect that the qualitative features of this mixing on the excitation spectra also apply to two- and three-dimensional systems. That is to say, we expect that quadrupole interactions in two- and three-dimensional systems manifest themselves by increasing the lifetimes of the virtual bound states, and by lowering the real part of the energy of these states. Systems with $SU(3)$ symmetry have two Goldstone modes. The one for $m = 1$ is present as long as the system has $SU(2)$ or $O(3)$ symmetry; the second one for $m = 2$ excitations only appears when $\Gamma^{(1)}(k) = \Gamma^{(2)}(k)$.

7. RÉSUMÉ OF THE BIBLIOGRAPHY

The effects of quadrupolar interactions on the true two spin waves bound states have been previously studied in references [1,2]. The elementary excitations of quadrupolar Hamiltonians have been independently studied by J. Sivardière [3]. The treatment of spin greater than one-half Hamiltonians is to be found in references [4-9]. The two-spin-wave bound state problem was first studied in references [10,11].

ACKNOWLEDGMENTS

I should like to acknowledge with thanks the support of the National Science Foundation which made it possible for me to present these lectures. I should also like to thank Dr. J. Sivardière for drawing my attention to this problem, and for the numerous discussions we had on this topic. Finally my thanks to Dr. Sou-Tung Chiu-Tsao for her permission to present her recent results and for a careful reading of this manuscript.

REFERENCES

1. Pink, D.A. and Tremblay, P. (1972). *Can. J. Phys.*, **50**, 1728.
2. Pink, D.A. and Ballard, R. (1974). *Can. J. Phys.*, **52**, 33.
3. Sivardière, J. (To be published).
4. Murao, T. and Matsubara, T. (1968). *J. Phys. Soc. Japan*, **25**, 352.
5. Raich, J.C. and Etters, R.D. (1968). *Phys. Rev.*, **168**, 425.
6. Haley, S.B. and Erdös, P. (1972). *Phys. Rev.*, **B5**, 1106.
7. Westwański, B. and Pawlikowski, A. (1973). *Phys. Lett.*, **A43**, 201.
8. Westwański, B. (1973). *Phys. Lett.*, **A44**, 27.
9. Westwański, B. (Preprint E4-7486, JINR).
10. Wortis, M. (1963). *Phys. Rev.*, **132**, 85.
11. Hanus, J. (1963). *Phys. Rev. Lett.*, **11**, 336.

MAGNETIC PROPERTIES OF
ACTINIDES AND THEIR COMPOUNDS†

DANIEL J. LAM

Argonne National Laboratory,
Argonne, Illinois 60439, *USA*

1. INTRODUCTION

The first studies of the magnetic properties of the actinides
concentrated on ionic and covalent compounds to obtain information concerning the actinide ionic charge. For the most part,
these studies were confined to magnetic susceptibility measurements at room temperature and higher. Later, when sufficient
information became available to indicate that the crystal-field
interaction was strong, the measurements were extended down to
liquid helium temperature. In the early 1950's, the properties
of uranium alloys were systematically studied by Professor Bates
and his co-workers in England to obtain information on the band
structure. About the same time, Prof. Trzebiatowski and his as-
sociates in Poland began a systematic examination of the magnetic properties of uranium intermetallic compounds, which is ac-
tively pursued today. Subsequently, groups at Harwell in England, at Fontenay in France, at Phillips Research Laboratory in
the Netherlands. and at the AEC National Laboratories in the
United States have studied the magnetic properties of neptunium,
plutonium, and americium compounds using magnetization, neutron
and X-ray diffraction, nuclear gamma-ray resonance (Mössbauer
effect), nuclear magnetic resonance, electrical-transport properties, and low-temperature specific-heat measurements.

The actinide metals and their compounds exhibit a wide variety of not well understood but interesting and unusual magnetic
properties. The complexity of the magnetic behaviour is mainly

† Work performed under the auspices of the U.S. Atomic Energy
Commission.

287

due to the relatively large spatial extension of the $5f$ wavefunctions. The $5f$ electrons, which are not well localized in
space, experience smaller Coulomb correlation interactions than
the $4f$ electrons in the rare earths, and a stronger hybridization with the $6d$ and $7s$-like band. The apparent degree of localization of the $5f$ electrons in the metals, alloys, and compounds depends on the interatomic spacing because of the dependence of the effective band width on the overlap of the neigh-
boring $5f$ orbitals. Greater overlap favours larger band widths
that, when larger than the Coulomb correlation energy, lead to
an itinerant electron magnetism. The band width is effectively
increased by the hybridization of the $5f$ bands with the wider
$6d$-$7s$-p bands. Smaller overlap leads, in turn, to greater importance for the Coulomb correlation and localization of the
$5f$'s.

2. REVIEW OF EXPERIMENTAL RESULTS

As one goes along the actinide series with increasing atomic
number z, one goes from uranium, which can be a superconductor
and is non-magnetic, through neptunium and plutonium, which are
non-superconducting and not magnetic, to a phase boundary at
americium, above which well-defined local moments exist. The
appearance of magnetism in the later portion of the actinide
series has been related by Jullien et $al.$ [1] to strong f-d hybridization in the actinide metals. Since Dr. Coqblin will discuss this problem, I will not elaborate further.

One of the fascinating features of actinide compounds is the
widespread occurrence of long-range magnetically ordered structure, although the component elements are non-magnetic in the
pure metallic state. Hill [2] noted that all the intermediate
compounds of uranium, neptunium, and plutonium can be roughly
separated into 'magnetic' and non-magnetic' groups, and critical
values of the interionic spacing exist that separates the two
groups. He showed that in uranium compounds the transition from
'non-magnetic' to 'magnetic' behaviour seems to occur over a
range of U-U spacings (3.4 to 3.6 Å). He also indicated a critical value for Np-Np spacing of 3.25 Å for Np compounds and a
Pu-Pu spacing of 3.4 Å for Pu compounds. Subsequently, it was
found [3] that the separation of magnetic and non-magnetic actinide compounds by the 'critical value' of actinide-actinide (An-An)
spacing only applied to compounds with 'non-magnetic' B partners.
Magnetic ordering exists in compounds with $3d$ transition elements as B partners, although the An-An separation is less than
the 'critical value'. Because of the possible magnetism of the
$3d$ elements, it is difficult to evaluate the magnetism of the
actinide ion for compounds in this group.

The strong dependence of the magnetic properties of compounds
on the An-An separation near the 'critical value' can best be
illustrated by the variation of magnetic behavior of NpX_2 cubic
Laves phases, where X is a non-magnetic element. Aldred et $al.$
[4] have studied the magnetic properties of a series of NpX_2

compounds (where X = Al, Os, Ir and Ru) by means of magnetization and nuclear gamma-ray-resonance (NGR) measurements, and
these properties show a systematic variation with Np-Np separation. Thus, $NpAl_2$ appears to be a good local-moment system with
T_C = 56°K and μ_{sat} = 1.6 μ_B/mole; $NpOs_2$ is weakly magnetic
(probably itinerant) with T_C = 7.5°K and μ_{sat} = 0.4 μ_B/mole;
$NpIr_2$ is weakly magnetic (antiferromagnetic) with T_N = 7.5°K and
μ_{sat} = 0.6 μ_B/mole; and $NpRu_2$ does not order magnetically but
shows evidence, in the resistivity data [5], of spin-fluctuation
effects. These results can be qualitatively understood in terms
of a band model in which the neptunium $5f$ electrons form a narrow band close to the Fermi level. In $NpAl_2$, the band width Δ
of this level is small, and the moment is essentially localized.
As the Np-Np separation decreases (as in $NpOs_2$ and $NpIr_2$), overlap of the $5f$ wave function and hybridization of the $5f$ states
with the spd conduction band causes an increase in Δ and a com-
mensurate de-localization of the $5f$ electrons. This, in turn,
results in a decrease in the amount of unpaired $5f$ spin, which
eventually causes the disappearance of local moments. In addition, Δ becomes so large that itinerant magnetism does not occur. Below an Np-Np separation of $\sim$3.23.Å ($NpRu_2$), magnetic
ordering from the neptunium alone is no longer possible. Isostructural compounds of UAl_2 and $PuAl_2$, where the $An-An$ separation is close to the critical value, show a strong localized
spin-fluctuation behaviour, as indicated by the results obtained
from magnetization, electrical resistivity, and nuclear magnetic
resonance measurements [6].

At $An-An$ spacings larger than 3.5 Å, such as in most of the
NaCl-type compounds, well-developed localized moment behaviour
is observed. The magnetic properties of NaCl-type compounds
have been studied extensively by many workers with various experimental techniques. Perhaps the most interesting feature of
the magnetic properties of the NaCl-type compounds of uranium
with group V elements are the multiple transitions that occur
in UP and UAs. Recently, these transitions have been the subject of theoretical studies [7,8]. In an effort to learn more
about these low-temperature transitions and about the exchange
processes in uranium monopnictides, a number of experiments on
the solid solutions of uranium monopnictides and uranium monochalcogenides have been performed, for example, UP-US [9]†,
IAs-US [10]† and UAs-USe [11]. The neutron-diffraction experiments on these systems showed the presence of long-range magnetic structures, in contrast to the simple type I antiferromagnetic structure of the uranium monopnictide binary compounds.
The appearance of long-range magnetic structure has been interpreted as an indication of the importance of the Rudeman-Kittel-
Kasuya-Yosida (RKKY) interaction [12]. Although the crystal-
field and exchange interactions are present, their relative influence on the magnetic ordering is still a subject of specula-
tion. Recent results obtained from isostructural binary neptun-

---

† See these references, for example.

ium monopnictides indicate that the magnetic structure can be quite complicated [13]. Neptunium phosphide has an incommensurate magnetic structure and, in addition, at the lowest temperatures the magnetic moments on all the metal atoms are not equivalent. NpAs becomes antiferromagnetic at T_N = 175°K. At ~160°K, the highest-temperature 4+, 4- structure transforms to the type I arrangement. The crystal lattice of NpAs becomes tetragonal at T_N, exhibits a first order transition to cubic symmetry at 142°K, and remains cubic at lower temperatures. This transition from tetragonal to cubic symmetry within the ordered regime is, as far as we know, unique. The complexity of magnetic ordering in NaCl-type actinide compounds not withstanding, the experimental results support a localized description of the magnetic moments. In this type of compound, the simple crystal-field model can be used to correlate different experimental results, provided the effects of strong spin-orbit and crystal-field interactions have been fully taken into account. The general similarity of the magnetic structure of NpP and NpAs to that of CeSb is particularly interesting. In cerium compounds the f band lies close to the Fermi level, and the difficulties associated with understanding their magnetic properties are, in many respects, similar to those encountered in the actinide series.

3. APPLICATION OF CRYSTAL-FIELD THEORY

In terms of the crystal-field model, the f electrons in the solid are not characterized individually by the crystal momentum vector K and the intrinsic spin but collectively by a correlated several-electron wave function with appropriate quantum numbers. The quantum numbers to be used depend on the relative strength of the interactions that must be included. The pertinent interactions are the Coulomb (H_C), spin-orbit (H_{so}), crystal-field (H_{crys}), and exchange (H_{ex}) interactions. Infrared spectroscopy results for solutions of trivalent actinides [12] indicate that the Coulomb interactions are ~10^5 cm^{-1}, and the spin-orbit constants are ~3×10^3 cm^{-1}. Magnetic and spectroscopic studies show that the crystal-field interactions are of the order of 10^3 cm^{-1}. The exchange interaction estimated from the magnetic ordering temperatures is ~10^2 cm^{-1}. In the presence of the Coulomb interactions only, the wave function is characterized by quantum numbers S and L, which are the total spin and the total orbit angular momentum of the f electrons, respectively. The spin-orbit interaction causes mixing of various SL configurations. When the strength of the spin-orbit interaction is comparable to the separation between these SL multiplets, the mixing becomes significant. The appropriate quantum is then J, the total angular momentum. The inclusion of the crystal field destroys the rotational symmetry of the ion, and thereby causes an admixture of different J multiplets. When the crystal-field interaction is comparable to the separation between these J multiplets, the only quantum number appropriate to characterize the several

electron wave function is Γ, the irreducible representation of
the point group symmetry operations. Therefore, in the case of
actinides, the combination of strong spin-orbit and crystal-
field interactions dictates the use of the intermediate coupling
scheme rather than the Russell-Saunders coupling scheme.

The correlated several electron crystal-field wave function
may be obtained in the following manner [14]. Russell-Saunders
basis functions are used to set up matrix elements for the combined Coulomb and spin-orbit Hamiltonian $H_0 = H_c + H_{so}$ of the
free ion. This is then diagonalized to obtain eigenfunctions
in the intermediate coupling scheme characterized by given J's.
The matrix elements involved may be expressed in terms of only
four variables, which differ from element to element in the
actinide series. The variables are the three Slater integrals
F_2, F_4 and F_6 and the spin-orbit integral ξ_{5f}. These four variables may be treated as parameters derived from spectroscopic
data. After diagonalizing H_0, we obtain the wave functions of
the free ion, which are characterized by J. These are linear
combinations of the various ν, S and L components that span the
manifold of the constant J of an f^n electron configuration and
are in the form

$$|\phi_r\rangle = |\alpha J M\rangle = \sum_{\nu S L} \langle f^n \nu S L J M | \alpha J M \rangle | f^n \nu S L J M \rangle. \tag{1}$$

Here ν represents seniority v and the two other quantum W and ξ
introduced by Racah [15].

To obtain the crystal-field wave functions, the complete set
of free-ion eigenfunctions should, in principle, be used to set
up the crystal-field matrix, which is then diagonalized. In
practice, this is not possible for an f^n configuration, where
$n > 2$, even with the present generation of computers. Since we
are interested in the magnetic properties of the f^n configuration, we only need to know the low-lying states accurately because the Boltzmann factor cuts off the effects of the higher
energy states. We can therefore truncate the crystal-field matrix, provided we allow the size of the truncated basis set to
vary and check the low-lying states for convergence. The problem is then reduced to solving the eigenvalue problem

$$(PHP)P|\bar{\psi}\rangle = EP|\bar{\psi}\rangle, \tag{2}$$

where $H = H_c + H_{so} + H_{crys}$, and

$$P = \sum_{r=1}^{n} |\phi_r\rangle\langle\phi_r|$$

is a projection operator that projects on the n low-lying states
of interest. The crystal-field states $|\Gamma\rangle$ finally obtained are
related to the Russell-Saunders basis set used to set up the

Coulomb and spin-orbit matrices by two successive unitary transformations. We have

$$|\Gamma\rangle = \sum_{\alpha JM} \langle \alpha JM | \Gamma \sum | \alpha JM\rangle$$

$$= \sum_{\alpha JM} \langle \alpha JM | \Gamma\rangle \sum_{\nu SL} \langle f^n \nu SLJM | \alpha JM\rangle | f^n \nu SLJM\rangle. \qquad (3)$$

The crystal-field potential may be expanded in terms of the
tensor operator U_Q^K introduced by Racah. Because of equation
(1), the crystal-field matrix elements $\langle \Gamma | H_{\text{crys}} | \Gamma'\rangle$ are linear
combinations of the basis matrix elements

$$\langle f^n \nu SLJM | H_{\text{crys}} | f^n \nu' S'L'J'M'\rangle.$$

In standard notation [16], the latter may be written in the form

$$\langle f^n \nu SLJM | H_{\text{crys}} | f^n \nu' S'L'J'M'\rangle$$

$$= \sum_{KQ} B_Q^K \langle f^n \nu SLJM | U_Q^K | f^n \nu' S'L'J'M'\rangle (\ell \| C^K \| \ell'),$$

where

$$(\ell \| C^K \| \ell') = (-1)^\ell \sqrt{(2\ell + 1)(2\ell' + 1)} \begin{bmatrix} \ell K \ell' \\ 000 \end{bmatrix}.$$

The B_Q^K's depend on crystal structure. The values of K and Q
are limited by the requirement that the crystal-field potential
be invariant under the symmetry operations of the relevant point
group. The matrix elements of U_Q^K may be further simplified by
using the Wigner-Eckart theorem [17]. We have

$$\langle f^n \nu SLJM | U_Q^K | f^n \nu' S'L'J'M'\rangle$$

$$= \delta_{SS'} (-1)^{K+S+L'+2J-M} [(2J + 1)(2J' + 1)]^{\frac{1}{2}}$$

$$\times \begin{bmatrix} J K J' \\ -M Q M' \end{bmatrix} \begin{Bmatrix} L & J & S \\ J' & L' & K \end{Bmatrix} (f^n \nu SL \| U^K \| f^n \nu' S'L').$$

The reduced matrix elements $(f^n \nu SL \| U^K \| f^n \nu' S'L')$ are defined as

$$(f^n \nu SL \| U^K \| f^n \nu' S'L') = \delta_{SS'} [(2L + 1)(2L + 1)]^2 \times \text{(Contd)}$$

(Contd) $\quad \times \; n \sum\limits_{\psi} \; (\psi\{|\bar{\psi})(\psi'\{|\bar{\psi})(-1)^{\bar{L}+K+3+L} \begin{bmatrix} 3 & L\bar{L} \\ L' & 3K \end{bmatrix}$,

where the $(\psi\{|\bar{\psi})$'s are the fractional percentage coefficients.
These reduced matrix elements have been tabulated by Nielson
and Koster [18] for all f^n configurations. The $3j$ and $6j$ coefficients can be generated from the standard programmes, and
the calculation of the crystal-field matrix elements therefore
involves only straightforward computational operations. Once
these matrix elements are determined, the eigenvalue problem of
the crystal-field potential may be solved. The crystal-field
wave functions thus obtained have been used to calculate experimentally observable physical quantities.

Comparision with experimental results from NaCl- and CaF_2-
type actinide compounds has been carried out in several cases,
and good quantitative agreements have been obtained [19]†. It
is emphasized that in obtaining crystal-field wave functions
for the interpretation of experimental results, the effects of
intermediate coupling and J-mixing should be taken fully into
account. A detailed evaluation of the successes and shortcom-
ings of the model, however, can only be made through systematic
comparision with experiments.

4. CONCLUSION

I have tried to present a brief summary of the magnetic properties of actinides and their compounds and the application of
the crystal-field model in actinide compounds. The experimental
evidence indicates that neither the simple band theory nor the
simple localized electron theory, as presently constituted, will
be able to describe completely the observed magnetic behaviour.
It is hoped that the present report serves both as an illustration of the challenge of actinide magnetism and also as a spur
to theoretical involvement in the subject.

ACKNOWLEDGMENT

The author wishes to express his appreciation to the National
Science Foundation for the travel grant and to many colleagues
for valuable discussions.

REFERENCES

1. Jullien, R., Galleani d'Agliano, E. and Coqblin, B. (1972).
Phys. Rev., **B6**, 2139.
2. Hill, H.H. (1970). In *Plutonium 1970*, (ed. Miner, W.N.),

---

† See this reference, for example.

(Metallurgical Society of AIME, New York), pp. 2-19.

3. Lam, D.J. and Aldred, A.T. (1974). In *The Actinides: Electronic Structure and Related Properties*, (eds. Freeman, A.J. Darby, J.B. Jr.), (Academic Press, New York).

4. Aldred, A.T., Dunlap, B.D., Lam, D.J. and Nowik, I. (1974). *Phys. Rev.*, **B10**, 1011.

5. Harvey, A.R. Electrical Resistivities of the Intermetallic Compounds $NpRu_2$ and $NpOs_2$, *Solid State Commun.*, (to be published).

6. Arko, A.J., Fradin, F.Y. and Brodsky, M.B. (1973). *Phys. Rev.*, **B8**, 4104.

7. Long, C. and Wang, Y.L. (1971). *Phys. Rev.*, **B3**, 1656.

8. Erdos, P. and Robinson, J.M. (1973). *Am. Inst. Phys. Conf. Proc.*, **10**, 1070.

9. Lander, G.H., Kuznietz, M. and Cox, D.E. (1969). *Phys. Rev.*, **188**, 963; Allbutt, M., Dell, R.M., Junkison, A.R. and Marples, J.A. (1970). *J. Inorg. Nucl. Chem.*, **32**, 2159; Trzebiatowski, W. and Palewski, T. (1969). *Phys. Stat. Sol.*, **34**, K51; Friedman, F. and Grunzweig-Genossar, J. (1971). *Phys. Rev.*, **B4**, 180.

10. Trzebiatowski, W. and Palewski, T. (1971). *Bull. Acad. Polon. Sci. Ser. Sci. Chim.*, **19**, 83; Leciejewicz, J., Murasik, A., Troc, R. and Palewski, T. (1971). *Phys. Stat. Sol.*, **46**, 391; Lander, G.H., Mueller, M.H. and Reddy, J.F. (1972). *Phys. Rev.*, **B6**, 1880.

11. Leciejewicz, J., Murasik, A., Palewski, T. and Troc, R. (1970). *Phys. Stat. Sol.*, **38**, K89; (1971). *Phys. Stat. Sol.*, **48**, 445; Palewski, T., Suski, W. and Mydlarz, T. (1970). *Int. J. Magn.*, **3**, 269.

12. Kuznietz, M. and Grunzweig-Genossar, J. (1970). *J. Appl. Phys.*, **41**, 906.

13. Aldred, A.T., Dunlap, B.D., Harvey, A.R., Lam, D.J., Lander, G.H. and Mueller, M.H. (1974). *Phys. Rev.*, **B9**, 3766.

14. Chan, S.-K. and Lam, D.J. (1970). In *Plutonium 1970*, (ed. Miner, W.N.), (The Metallurgical Society of AIME, New York), pp. 219-232, and reference [3].

15. Racah, G. (1949). *Phys. Rev.*, **76**, 1352.

16. Wybourne, B.G. (1965). *Spectroscopic Properties of Rare Earths*, (Wiley, New York).

17. Judd. B.R. (1963). *Operator Techniques in Atomic Spectroscopy*, (McGraw-Hill, New York).

18. Nielson, C.W. and Koster, G.F. (1963). *Spectroscopic Coefficients for the p^n, d^n, and f^n Configuration*, (MIT Press, Cambridge, Massachusetts).

19. Lam, D.J. and Fradin, F.Y. (1974). *Phys. Rev.*, **B9**, 238.

EXPLORATORY BAND STRUCTURE CALCULATIONS
FOR ACTINIDE COMPOUNDS†

H.L. DAVIS

*Solid State Division, Oak Ridge National Laboratory,
Oak Ridge, Tennessee 37830, USA*

1. INTRODUCTION AND GENERAL COMMENTS

The actinides are well known for forming compounds with many of the elements of groups IVB, VB, and VIB of the periodic table. The specific compounds we wish to bring to attention are the carbides, the pnictides (N, P, As, Sb, and Bi), and the chalcogenides (S, Se, and Te) of the actinide metals. Now, a wide variety of interesting physical properties are possessed by the various members of this class of solid state compounds; e.g., all members are of a refractory nature, some members are semiconducting whilst most are fair metallic conductors, and some members are paramagnetic at all investigated temperatures whilst others undergo transitions to ordered magnetic states. But detailed discussions of these compounds' properties is not our purpose, and since a sampling of their properties may be gained from an inspection of the literature [1¶,2§], we wish to note here only that a diversity of properties does exist, and that its existence has provided the motivation for our work. Rather, our purpose is to discuss some of the fundamentals necessary for the eventual theoretical understanding of the compounds' properties. In doing this, we will mainly present a

---

† Research sponsored by the U.S. Atomic Energy Commission under contract with Union Carbide Corporation.

¶ An excellent review of the properties of actinide compounds may be found in some of the chapters of this book.

§ This is a concise review of the properties of some uranium compounds.

review of some of our work [3-5] which has been concerned with
investigating the electronic structure of a subclass of these
compounds. The subclass of our interest has been those actinide
monocarbides, monopnictides, and monochalcogenides which possess
the NaCl type of structure. Throughout, we will denote a general compound in this NaCl-structured subclass as AX. Specific
examples of AX compounds are ThP, UC, UAs, NpSb, PuS, etc..

Any similarities and differences in properties of the AX compounds must be traceable to actual similarities and differences
in their electronic structures, since the solid state properties of any compound are connected with how the electrons in
the outer shells of the constituent atoms have redistributed
themselves to form the compound's electronic structure. So some
insight into the electronic structure of AX compounds may be
gained by inspecting free atom configurations. Free actinide
atoms have ground electronic configurations of the form
$5f^i6d^j7s^2$, with i and j both being $\geqslant 0$ and whose values depend
upon the actinide element under consideration. For example,
for Th $i = 0$ and $j = 2$, for U $i = 3$ and $j = 1$, and for Pu $i = 6$
and $j = 0$. The AX compounds' 'anions' are derived from free
atoms having configurations of the form p^k, with $k = 2$ for C,
or $k = 3$ for a pnictogen, or $k = 4$ for a chalcogen. Then as an
AX compound is formed, its bonding electronic states are derived from the free atom states present in the ground, and low
lying excited configurations of the free atoms. That is, the
free atom s, p, d and f states will convert to energy band
states, in the solid, which possess varying widths, overlaps,
and relative energy placements, depending upon explicit details
of wavefunctions, lattice parameters, etc.. But one of the
major difficulties in attempting to provide a theoretical description of these energy band states in AX compounds is the
presence of the $5f$ states in the ground, or low lying excited,
configurations of the free actinide atoms. To illustrate the
difficulties associated with the $5f$ electrons it is useful first
to consider AX compounds as if they did not possess $5f$ electrons, or, alternatively, the $5f$ electrons present may be considered as very highly localized. Under these assumed conditions we will refer to the compounds as being hypothetical AX
compounds.

Either the absence of, or very highly localized, $5f$ electrons
means $5f$ states would not be expected to contribute to any
bonding of an AX compound. So the bonding would take place as
if the AX compound were being formed from constituent atoms
having free atom configurations of the form $6d^j7s^2$ and p^k.
(The j value here would not necessarily be identical with the
value for a free actinide atom, but is meant to denote the
value for the configuration giving the major component in the
bonding electronic distribution). Clues to the type of electronic structure these hypothetical compounds would have can
be found by noticing that, generally speaking, they resemble a
a higher order transition metal compound. Clues can also be

found by investigating rare earth compounds which are known to have well localized $4f$ electrons, and hence bonding states being formed from $5d^j6s^2$ and p^k configurations. Thus, it should be of interest to study some band structure calculations which have been performed to provide theoretical descriptions of the one-electron states for transition metal and rare earth compounds.

Some examples of calculations performed on transition metal compounds are those by Ern and Switendick [6] for TiC and TiN, by Conklin and Silversmith [7] for TiC, by Switendick and Jones [8] for ScN, ScP, and ScAs, by Conklin and Simpson [9] for NbC, by Davis [10] for HfC and TaC, and by Mattheiss [11] for NbN. Some rare earth compound calculations are those by Cho [12,13] for some europium chalcogenides, and by Davis [4,14,15] for various rare earth compounds. From a study of these calculations it is possible to formulate a qualitative picture of what a band structure might be like for a hypothetical AX compound. The result of such an educated, but imaginary, process is illustrated by figure 1, where any explicit representation of $5f$ states has been omitted, owing to the present assumption of either the absence of, or the perfect localization of, the $5f$ electrons. Figure 1 is a plot of energy, $\varepsilon_{\vec{k}}$, vs wave number $\vec{k}$, for $\vec{k}$ along three major symmetry directions of the Brillouin zone (BZ). Since the NaCl lattice has fcc translational symmetry, the BZ corresponding to AX compounds is shown in figure 2.

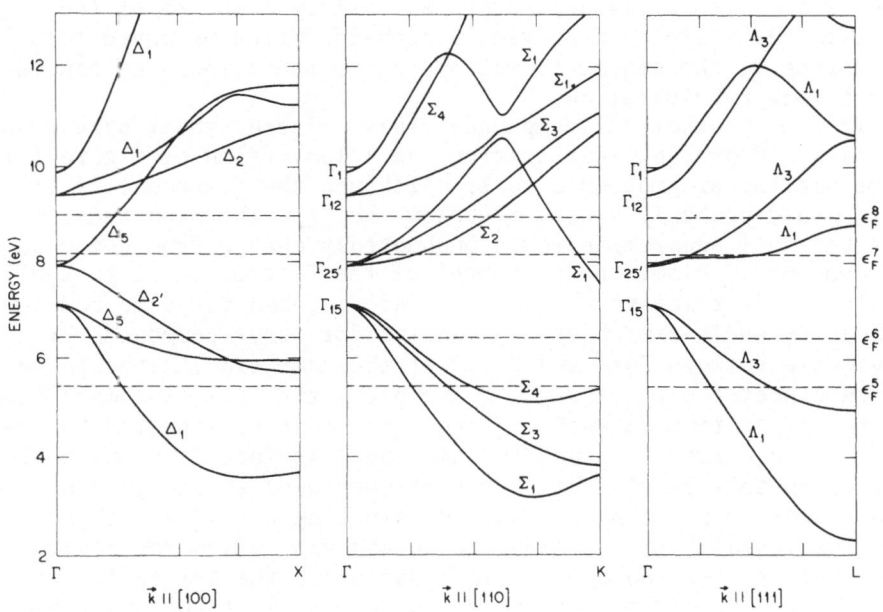

Figure 1 - Postulated bonding band structure for a hypothetical AX compound where $5f$ electrons are assumed to be either absent or highly localized.

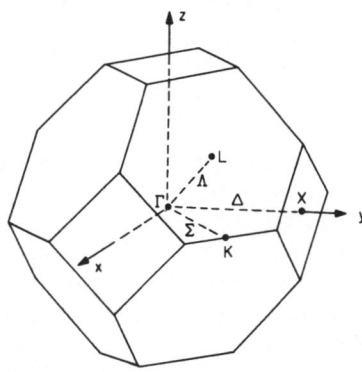

Figure 2 - Brillouin zone for the NaCl type of lattice.

Although figure 1 is being presented as an educated guess at the band structure of a hypothetical AX compound, let us consider some of its features relative to the $6d^j7s^2$ and p^k configurations of the free atoms. Consider the electron states at Γ. The Γ_{15} states are spatially triply degenerate and have as their basic component the p states of the 'anion' atom. The $\Gamma_{25'}$ states are also triply degenerate, but are primarily composed of the 'cation' $6d$ states, as are the Γ_{12} which are doubly degenerate, and Γ_1 is singly degenerate and mainly composed of the 'cation' $7s$ state. Of course, as consideration is moved to a $\vec{k}$ away from Γ, the states corresponding to the allowed $\varepsilon_{\vec{k}}$ can reflect more hybridization.

If the hypothetical compounds truly existed, their band structures would probably differ from the illustration of figure 1 in some smaller or greater detail. Although the placement of the $\Gamma_{25'}$ would probably always be below the Γ_{12}, the magnitude of the $\Gamma_{12}-\Gamma_{25'}$ separation would undoubtedly change from compound to compound. Also, the placement of the Γ_1 could, and probably would, shift relative to the $\Gamma_{25'}$ and Γ_{12}, and for some compounds Γ_1 would lie above Γ_{12}, whilst for other compounds Γ_1 would lie between $\Gamma_{25'}$ and Γ_{12}. At the same time, although we would expect the Γ_{15} always to lie below the $\Gamma_{25'}$, the magnitude of the separation between Γ_{15} and $\Gamma_{25'}$ would be expected to undergo change from compound to compound. In fact, the possibility definitely exists that for some compounds an energy gap would open up between the bands originating at Γ_{15} and those originating at $\Gamma_{25'}$. To imagine such a gap, allow the separation between Γ_{15} and $\Gamma_{25'}$ to increase until the $\Delta_{2'}$ tail, from $\Gamma_{25'}$ in the Γ to X direction, is raised until it no longer overlaps the bands originating from Γ_{15}. Thus, although one expects qualitative features to remain intact, the band structure for a specific hypothetical AX compound would not be expected to resemble that of figure 1 in all details.

It is important to note how band structural states are occupied by electrons. Since band theory is based on a one-particle type of model, its resulting states are to be occupied in accordance with the rules of Fermi statistics. This eventually means that each band can be occupied by two electrons from each primitive unit cell of the solid. For example, the lowest band of figure 1 is composed of all $\varepsilon_{\vec{k}}$ which continuously connect with the illustrated Δ_1, Σ_1 and Λ_1, and it is this total band which can hold two electrons from each primitive unit cell. Thus, if figure 1, for a hypothetical AX compound, was formed from atoms having configurations $6d^17s^2$ and p^2, the Fermi energy of this band structure would be the dotted line ε_F^5. ε_F^5 would also be the Fermi energy if the compound was formed from atoms having configurations $7s^2$ and p^3. If the atoms had configurations $6d^17s^2$ and p^3, or configurations $7s^2$ and p^4, or $6d^27s^2$ and p^2, the Fermi energy would be ε_F^6. Likewise, if the atoms had a total of seven or eight bonding electrons, the Fermi energy would be, respectively, ε_F^7 or ε_F^8.

It is interesting to note that certain features of figure 1 have a bearing on understanding the bonding mechanisms, and possible 'ionicity', not only of the AX compounds but also of transition metal and rare earth compounds. First, notice that the metal 's states', represented by the band originating at Γ_1, have been pushed to higher energies during the bonding process, so, to a good first order, no metal s states are occupied in the solid. Also, the 'p bands' originating at the Γ_{15} are at lower energies than the 'd bands' originating at the $\Gamma_{25'}$ and Γ_{12}. Thus, it might be concluded that a charge transfer to the nonmetal has taken place and produced ions whose charge magnitude is given by the value for a completely filled p shell. But this is not the true state of affairs, because the wavefunctions for the electrons which occupy these Γ_{15} 'p bands' must be considered. These wave functions would reflect diffuse character and have a reasonable density at the metal site in the compound. Consideration of this aspect in more detail would show that there is some metal to non-metal charge transfer, but its magnitude is reduced from the value expected for a completely filled p shell. Also, more detailed considerations would show that the metal to non-metal bonding of the 'p bands' has considerable covalent character. This band result is entirely consistent with the earlier semi-empirical conclusions drawn by Rundle and Bilz [33,34] for, e.g., transition metal carbides in order to explain their strong bonding.

Finally, it is to be noticed that the overall width of the 'd bands' in figure 1 is of the same magnitude as found for 'd bands' in the pure transition metals. Thus, the density of states as a function of energy for this band structure would not have peak values greatly exceeding peak values found for the pure transition metals. So it would be expected that low temperature electronic specific heat coefficients, γ, be of the same order

for both metallic transition metal compounds and the pure transition metals. Also, the γ's that would be estimated for the hypothetical AX compounds would be of the same order as found for
transition metals and their compounds. Although these qualitative expectations are fulfilled when comparing results for many
transition metal compounds with results for the pure metals, it
is a very important observation that the available experimental
γ's for, say, uranium compounds are inconsistent with this expectation [2]. That is, it is difficult to reconcile the observed γ values for the lighter actinide AX compounds with the
type of band structures expected for the present hypothetical
model. Thus, for the compounds of the lighter actinides, their
electronic structures cannot be adequately accounted for via
band structures of the type of figure 1. The probable reason
for such inadequacy is the complete neglect of any consideration
of $5f$ effects in the bonding of the AX compounds.

With the exception of Th, the free actinide elements do possess $5f$ electrons in their ground electronic configurations, and
even for Th low lying excited configurations contain $5f$ electrons. Thus, any attempt at elucidation of the electronic
structure of AX compounds must consider these $5f$ electrons. In
other words, if figure 1 is to represent all the important electronic states of an AX compound, what changes must be made so it
will properly indicate the presence of $5f$ electrons? In order
to consider such changes one must first answer a major question:
Are the $5f$ electrons in an AX compound best described by a localized model or by an itinerant model? Of course, these two models
are limiting extremes of theoretical utility, and the actual
circumstances in an AX compound must lie somewhere between the
extremes. But owing to the spatial contraction of the $5f$ shell
as the total number of $5f$ electrons is increased, the answer to
the above question may vary from AX compound to compound; e.g.,
the itinerant model may be best for US, but the localized model
best for AmSb.

Now, controversy exists when attempting to answer the above
question for AX compounds, but there is agreement as to the correct answer to a similar question for rare earth compounds.
For such rare earth systems, the constituent atoms would have
configurations $4f^i5d^j6s^2$ and p^k. Then, since it is universally
accepted that the correct model for the $4f$ electrons is the
localized one, the bonding electronic structures for these compounds would be formed from the $5d^j6s^2$ or (or $5d^{j+1}6s^2$) and p^k
configurations. If actual band structure calculations are performed for rare earth compounds, the results [4,12-15] look
qualitatively like figure 1, but, in addition, a narrow band of
f states is found. But experience with such calculation shows
it is difficult to place the $4f$ states relative to the other
states with any degree of reliability. This difficulty is owing
to the extreme sensitivity of the $4f$ states to minor variations
in the band theoretic potentials. Regardless, the results of
the calculations do produce various placements of the $4f$ states.

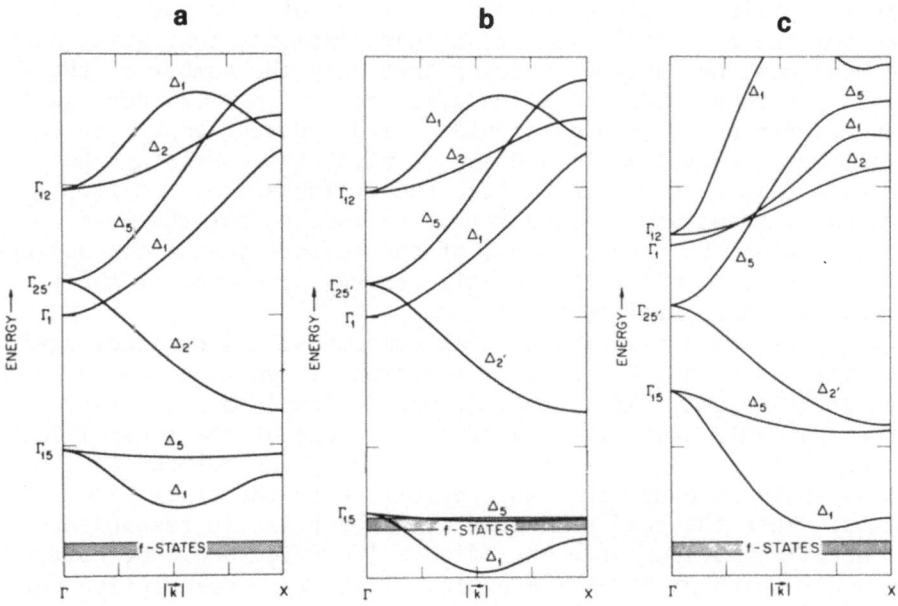

Figure 3 - Illustrative examples of calculated band
structures for rare earth compounds.

Three possible variations of $4f$ placement are illustrated by
figure 3, where only the band structures for the Γ to X $\vec{k}$-direc-
tion are plotted. Notice the $4f$ states of figure 3, or '$4f$ bands',
are quite narrow, show little dispersion, reflecting the fact
that the $4f$ electrons are localized, and, thus, are not really
band states. But the existence of band structures, including
the placement of the $4f$ states, of the types of figure 3 have
been qualitatively verified by photoemission experiments. Work
by Eastman and Kuznietz [16,17] indicates that EuS has an electronic structure like figure 3(b), GdS has one like figure 3(a),
and Gambino et al. [18] have shown GdSb's electronic structure
resembles figure 3(c). The point being that although the present state of calculations does not allow the $4f$ states to be
placed with confidence, the other features of the calculations
are more reliable, and by combining experimental and calculational results a reasonably reliable picture of the electronic
structure of a rare earth compound emerges.

As mentioned, the $4f$ states for rare earth compounds are
found to occupy a narrow range of energies both in experiments
and calculations. Thus, one should not consider these $4f$ states
as band states; i.e., the one-electron model fails for these
states. This might cause some conceptual difficulties when one
applies statistics to the electron occupation of the calculated
band structures, since the $4f$ states contain seven bands. One
might envisage occupying the states of, say, figure 3(a) or figure 3(c) according to the rules of Fermi statistics. Though

this is applied to the other bands, it is not, and must not be,
done for the f states. Rather, a 'correlational configurational'
argument must be invoked, meaning that only the number of the f
states are to be occupied which give the $4f^n$ ($n \leqslant 14$) configuration 'expected' for the particular solid. Having empty states
at a lower energy than filled states might be surprising, but
the same result occurs in Hartree-Fock calculations for free
rare earth atoms and in band structure results for the pure rare
earth metals. Such occurences, of course, are just a consequence
of attempting to mimic the relative energy placement of the f
states by a one-particle model.

We return to a discussion of AX compounds, and consider again
the basic question: Are the $5f$ electrons in an AX compound localized or itinerant? If localized, the AX compound's electronic
structure would probably be a similar to one of the types illustrated by figure 3. However, if itinerant, the compounds' $5f$
states would be band-like and hybridize with the other band
states. Then the band structure would bear little resemblance
to the illustrations of either figure 1 or figure 3. Of course,
the model which provides the better first order description for
an AX compound must be the one providing the better framework to
explain its properties. It is our belief that, at least for the
lighter actinide AX compounds, the band picture provides the required framework, and our remaining discussion will be an attempt
to explain reasons for this belief. To this end, we will present in section 2 a brief outline of the band model and some of
the calculational details useful in obtaining results from the
model. Then, in section 3 some of our calculational results for
actual AX compounds will be presented and discussed.

2. THE MODEL AND CALCULATIONAL DETAILS

The band structure for a solid is defined here as the allowed
solutions of the one-particle Schrödinger equation

$$[-\nabla^2 + V(\vec{r}) - E_{\vec{k}}]\psi_{\vec{k}}(\vec{r}) = 0, \tag{1}$$

where $V(\vec{r})$ is a periodic potential reflecting the translational
symmetry of the solid. Use of this equation implies a very
basic approximation has been made to obtain a workable theoretical model to describe the solid's electronic structure. Simply,
this approximation is that the solid's electronic structure is
considered as a collection of non-interacting Fermi particles
all moving in the same potential. So it is important to note
that use of equation (1) implies some neglect of N-body type of
interactions, e.g., electron-electron correlation. Thus, depending upon the solid or the type of electron in that solid, neglect of electron-electorn correlation could invalidate the use
of equation (1).

A rule useful in guiding qualitative judgments on the validity
of equation (1), with regard to its neglect of correlation which

might be important for 5f electrons, may be stated in terms of
bandwidths resulting when use is made of this equation blindly
to calculate electronic bands. If broad bands are obtained, the
electrons under consideration are probably more towards the
itinerant limit, and the validity of the use of equation (1)
would not be so much in question; however, if narrow bands are
obtained, the electrons should be considered as being more towards the localized limit, which probably invalidates the use of
equation (1). In particular, if equation (1) is to be semi-
quantitatively valid for an AX compounds' 5f electrons, its application should lead to reasonably broad '5 bands'; however, if
narrow '5 bands' are obtained, a cautionary flag should be raised.
But it is to be expected that equation (1) is a valid model for the
X's p and the A's 6d and 7s electrons. So application of equation
(1) to the study of an AX compound will always result in some useful information.

Besides the approximations inherent in a one-particle model,
other approximations are used when equation (1) is applied to
calculating the electronic structure for an AX compound. Since
it is not practical to obtain rigorously a Hartree-Fock solution
for the electrons in these compounds, nor is it certain that
this would give the best results, approximations are made in
picking the form of $V(\vec{r})$. The potentials used in our work were
obtained by invoking a common heuristic prescription [6,11,19]
which uses free atom charge densities. With this prescription
the potential about a given site is approximated by a spherical-
ly symmetric, muffin tin, potential $V(\vec{r}) \simeq V(r) = V_C(r) + V_{ex}(r)$.
Both the Coulomb part, $V_C(r)$, and the exchange part, $V_{ex}(r)$, of
the potential are approximated by a lattice superposition of
free atom quantities.

At a given lattice site, $V_C(r)$ is taken to be the spherically
averaged sum of two components. The first component is the
Coulomb potential owing to the free atom at the given site,
whilst the second component is the sum of the contributions owing
to the tails of free atom Coulomb potentials centred on other lat-
tices sites in the vicinity. Then, a Slater [20] type of approximation is used for the exchange potential: $V_{ex}(r) \simeq -6\alpha \times$
$[3\rho(r)/8\pi]^{1/3}$. Here, $\rho(r)$ is the spherically symmetric lattice
superposition of atomic charge densities, and obtained in a manner analogous to that used for $V_C(r)$. The 'parameter' α in the
expression for $V_{ex}(r)$ may be used to control the exchange contribution to the total potential, with $\alpha = 1$ being commonly
called the 'full Slater limit', and $\alpha = 2/3$ called the Dirac-
Gaspar-Kohn-Sham limit. The details of a calculated band structure will depend on the value of α used, so it is sometimes useful to vary α between the above limits. When this is done, it
is in the spirit of trying to obtain the most realistic model
Hamiltonian for the solid considered.

Once α is fixed, two spherically symmetric potentials may be
calculated, one centred about each of the two different lattice
sites in an AX compound. The radii for these two muffin tin

spheres are determined by requiring that the two site potentials
be equal at their point of contact. The remaining part of the
potential to be specified is that part located in the interstitial region between the spheres, which we assume is constant
throughout the interstitial region. This constant is taken to
be the common value of the two site potentials at sphere contact,
and its value is denoted as V_0. The band energies quoted herein
are relative to the V_0's for the respective compounds.

For almost all our calculations on AX compounds, to obtain
the band structure, $E_{\vec{k}}$ vs $\vec{k}$, which is associated with a given
potential, we have used the non-relativistic form of the Korringa-Kohn-Rostoker (KKR) method [21,22], as formulated by Segall
[23] and Treusch and Sandrock [24] for crystals having more than
one atom per unit cell. The few relativistic results presented
in section 3 were obtained by use of the Onodera and Okazaki
[25] formulation. Under the stated potential restrictions, and
when considering NaCl structured AX compounds, it is convenient
to express the wavefunctions in the KKR method as

$$
\psi_{\vec{k}}(r) = \begin{cases}
\sum_{\ell,m} C_{\ell m}{}^A R_\ell{}^A(r_A) Y_{\ell m}(\theta_A,\phi_A) & \text{if } \vec{r} \text{ inside an A} \\
& \text{muffin tin sphere,} \\[1em]
\sum_{\ell,m} C_{\ell m}{}^X R_\ell{}^X(r_X) Y_{\ell m}(\theta_X,\phi_X) & \text{if } \vec{r} \text{ inside a X} \\
& \text{muffin tin sphere,} \quad (2) \\[1em]
\sum_{j} F_j \exp[i\vec{r}\cdot(\vec{K}_j + \vec{k})] & \text{if } \vec{r} \text{ outside both} \\
& \text{muffin tin spheres,}
\end{cases}
$$

The $C_{\ell m}{}^A$, $C_{\ell m}{}^X$, and F_j are expansion coefficients depending on
the $(E,\vec{k})$ point, and determinable with KKR techniques. The $Y_{\ell m}$
are spherical harmonics, whilst the A and X subscripts and super-
scripts denote quantities relative to origins on those sites.
The $\vec{K}_j$ are reciprocal lattice vectors.

The radial parts, $R_\ell{}^A$ and $R_\ell{}^X$, depend only on the energy and
are calculable from radial equations valid inside the two muffin
tin spheres:

$$
\left[-\frac{1}{r^2} \frac{d}{dr}\left(r^2 \frac{d}{dr} \right) + \frac{\ell(\ell + 1)}{r^2} + V(r) - E \right] R_\ell(r) = 0. \quad (3)
$$

To obtain $R_\ell{}^A(r_A)$ from this equation one inserts for $V(r)$ the
quantity $V_A(r_A)$, which is the potential inside the A muffin tin
sphere, with a like procedure to obtain $R_\ell{}^X(r_X)$. It is convenient to normalize the radial parts of the wave function, so

$$
\int_0^{\tau^A} [R_\ell{}^A(r)]^2 r^2 dr = \int_0^{\tau^X} [R_\ell{}^X(r)]^2 r^2 dr = 1, \quad (4)
$$

for each ℓ value being considered, with τ^A and τ^X being the muffin tin sphere radii.

Use of the wave function expansion (2) in the KKR method leads to the determinant

$$\det \left| A_{\ell m, \ell' m'}{}^{i,i'} + \sqrt{E} \delta_{\ell \ell'} \delta_{mm'} \delta_{ii'} \cot\eta_\ell{}^i(E) \right|, \qquad (5)$$

whose zeroes implicitly generate the dispersion relations, $E_{\vec{k}}$ vs $\vec{k}$. In expression (5), i and i' = A or X. The order of this determinant is governed by the truncation value, ℓ_{max}, used for both of the (ℓ, m) sums in equation (2), with the order being $2(\ell_{max} + 1)^2$. Since it is well established that the KKR method is rapidly convergent as ℓ_{max} is increased, we have considered at most $\ell_{max} = 3$ in our work. Also, use of $\ell_{max} = 3$ is necessary to allow for 'f bands'.

The quantities $\cot\eta_\ell{}^i(E)$ in expression (5) are the cotangents of the phase shifts resulting from scattering an electron, of energy E, from one of the individual muffin tin potentials at sites i = A or X, and calculable from equations (3). The $A_{\ell m, \ell' m'}{}^{i,i'}$ are complicated functions of $\vec{k}$ and E, but these quantities are completely potential independent. Since explicit expressions for these quantities may be found elsewhere [25], their exact form need not be reproduced here; however, it is noted that they may be efficiently generated when using a modern computer [26]. Once the quantities $\cot\eta_\ell{}^i(E)$ and $A_{\ell m, \ell' m'}{}^{i,i'}$ are available, the band structure corresponding to a given potential directly follows by finding those $(E, \vec{k})$ points which cause the determinant (5) to vanish.

3. CALCULATIONAL RESULTS AND DISCUSSION

The results of performing KKR band structure calculations for some AX compounds will now be presented and briefly discussed. Although the results have all been reported elsewhere [3-5], they are reproduced here to illustrate general features obtained when actual calculations are performed. The results given below have been chosen, firstly, to illustrate some of the difficulties when attempting to perform predictive calculations, and, secondly, to demonstrate how even exploratory band calculations can contribute to the qualitative understanding of the electronic structure of AX compounds. The specific compounds to be considered here are listed in table I. Since more than one potential has been used for some compounds, table I also indicates the differences between potentials for the same compound. All potentials were obtained via the procedures described above using relativistic Hartree-Fock-Slater wave functions for the actinide atoms as calculated by Nestor *et al.* [27,28], while the non-actinide atomic wave functions were of the non-relativistic Hartree-Fock type as calculated by Fischer [29] for their ground configurations.

TABLE I

Potentials for AX Compound Band Structures

Potential	Config-uration	α	a (Å)	τ^A (Å)	$-V_0$ (Ry)
ThP	d^2s^2	1.0	5.82	1.57	1.51
PuN1	f^6s^2	1.0	4.91	1.39	1.81
PuN2	f^6s^2	0.9	4.91	1.39	1.70
NpSb1	$f^4d^1s^2$	1.0	6.25	1.58	1.35
NpSb2	f^4d^3	1.0	6.25	1.57	1.32
NpSb3	$f^4d^2s^1$	1.0	6.25	1.57	1.33
NpSb4	f^5s^2	1.0	6.25	1.57	1.34
NpSb5	$f^3d^2s^2$	1.0	6.25	1.60	1.35
US	$f^3d^1s^2$	1.0	5.49	1.47	1.64

The second column is the configuration used in the free atom actinide calculation, whilst α is the exchange parameter. The quantity a is the lattice parameter, τ^A is the muffin tin sphere radius for the actinide part of the potential, and V_0 is the value of the potential between the muffin tin spheres.

ThP

The band structure for the potential ThP is plotted along the major symmetry directions of the BZ in figure 4. Also in figure 4 is a histogram density of states based on sampling 152 $\vec{k}$-points in 1/48-th of the BZ (4000 points in the total BZ). It is interesting to consider similarities and differences between figure 4 and those for a hypothetical AX compound given by figure 1. The main differences between these two figures are caused by the 'f bands' in figure 4, which originate at $\Gamma_{2'}$, Γ_{25}, and the upper Γ_{15}. If these 'f bands' were removed from figure 4 it would be quite similar to figure 1. So the results of figure 4 may be pictured as being owing to overlaying 'f bands' onto the band structure of a higher order transition metal compound, with accompanied hybridization of the 'f bands' with the bands originally present. Notice the result of overlaying the $5f$ bands to give the band structure of figure 4 for ThP is quite different from that produced when $4f$ states are overlaid to produce the rare earth compound type of results of figure 3. Of course, the resulting differences are owing to the different f 'bandwidths', which are traceable to the fact thorium $5f$ wavefunctions (for

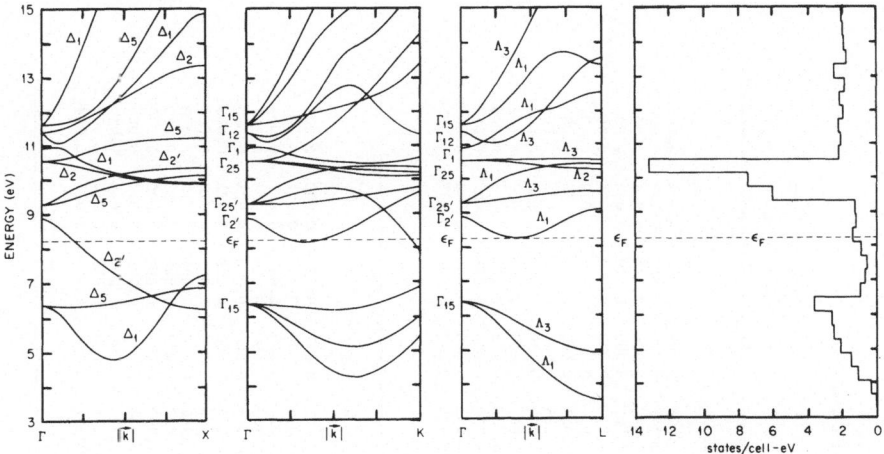

Figure 4 - Electronic band structure and density of
states for ThP.

excited configurations) are more spatially diffuse than rare
earth $4f$ wave functions. This implies the $5f$ states of figure
4 are more towards the itinerant limit, while the $4f$ states of
figure 3 are almost at the localized limit.

To provide a pedagogical illustration of how superposition of
'$5f$ bands' results in hybridization, the potential ThP has also
been used to do calculations where the $5f$ states were artifical-
ly supressed, and such results are in figure 5. The difference
between figures 4,5 has been the use of, respectively, $\ell_{max} = 3$
and 2, where ℓ_{max} defines the cutoff ℓ value of equations (2).
Compare these figures and envisage the $5f$ bands being laid on
figure 5 with the resulting hybridization producing figure 4.
Let us dwell further on the superposition-hybridization process.
First, it is noticed, as symmetry considerations dictate, no
change in energy placement or magnitude occurs between figures
5,4 for the $\Gamma_{25'}$, Γ_1, and Γ_{12}. However, the change from $\ell_{max} = 2$
to $\ell_{max} = 3$ does decrease the energy of the lower Γ_{15} by about
1 eV, whilst, of course, the upper Γ_{15} only exists when $\ell_{max} = 3$
is used. This energy shift is a consequence of the relatively
large overlap of the P's p-electrons onto the Th sites, and thus
allowing interaction with available $5f$ states at the Th sites,
even though these $5f$ states will remain unoccupied.

In table II are tabulated the wave function ℓ-character for
some $(E, \vec{k})$ points of the band structure for the potential ThP
when $\ell_{max} = 3$. The tabulated quantities are sums of the $C_{\ell m}$ of
equation (2),

$$D_\ell{}^{Th} = \sum_m |C_{\ell m}{}^{Th}|^2 \qquad \text{and} \qquad D_\ell{}^{P} = \sum_m |C_{\ell m}{}^{P}|^2, \qquad (6)$$

TABLE II

Wavefunction Character for the ThP Band Structure of Figure 4

$\vec{k} = (0.00, 0.01, 0.02)$

E (eV)	D_0^{Th}	D_1^{Th}	D_2^{Th}	D_3^{Th}	D_0^{P}	D_1^{P}	D_2^{P}	D_3^{P}
6.421	0.000	0.054	0.002	0.371	0.000	0.572	0.000	0.001
6.428	0.000	0.054	0.001	0.372	0.000	0.572	0.000	0.001
6.430	0.000	0.055	0.000	0.372	0.000	0.572	0.000	0.001
8.895	0.000	0.000	0.026	0.962	0.000	0.000	0.005	0.007
9.310	0.000	0.000	0.848	0.005	0.000	0.000	0.144	0.003
9.312	0.000	0.000	0.849	0.007	0.000	0.000	0.144	0.000
9.327	0.000	0.000	0.785	0.082	0.000	0.000	0.133	0.000
10.541	0.000	0.000	0.001	0.998	0.000	0.000	0.000	0.000
10.543	0.000	0.000	0.001	0.998	0.000	0.000	0.000	0.001
10.544	0.000	0.000	0.001	0.998	0.000	0.000	0.000	0.001
10.992	0.484	0.001	0.000	0.019	0.495	0.000	0.000	0.001
11.354	0.000	0.001	0.602	0.245	0.000	0.021	0.131	0.000
11.399	0.000	0.001	0.756	0.072	0.000	0.006	0.165	0.000
11.623	0.000	0.009	0.001	0.883	0.000	0.107	0.000	0.000
11.636	0.000	0.009	0.030	0.849	0.000	0.105	0.007	0.000
11.685	0.000	0.009	0.117	0.750	0.000	0.099	0.025	0.000

$\vec{k} = (0.40, 0.01, 0.02)$

E (eV)	D_0^{Th}	D_1^{Th}	D_2^{Th}	D_3^{Th}	D_0^{P}	D_1^{P}	D_2^{P}	D_3^{P}
4.898	0.066	0.028	0.168	0.104	0.004	0.630	0.000	0.000
6.455	0.000	0.104	0.053	0.277	0.000	0.564	0.001	0.001
6.460	0.000	0.105	0.052	0.277	0.000	0.564	0.001	0.001
7.489	0.000	0.000	0.597	0.288	0.000	0.003	0.105	0.007
9.785	0.000	0.007	0.128	0.829	0.000	0.001	0.033	0.002
9.788	0.000	0.000	0.126	0.831	0.000	0.001	0.033	0.002
10.124	0.000	0.000	0.098	0.886	0.000	0.000	0.015	0.001
10.196	0.000	0.000	0.040	0.948	0.000	0.000	0.010	0.002
10.294	0.005	0.000	0.082	0.859	0.004	0.025	0.024	0.001
10.989	0.000	0.008	0.067	0.897	0.000	0.021	0.005	0.002
10.997	0.000	0.008	0.067	0.896	0.000	0.022	0.005	0.002
12.103	0.178	0.017	0.176	0.399	0.192	0.017	0.019	0.002
12.148	0.000	0.000	0.698	0.152	0.000	0.001	0.149	0.000
12.522	0.000	0.000	0.334	0.504	0.000	0.123	0.037	0.002
12.538	0.001	0.001	0.343	0.491	0.000	0.123	0.404	0.002

$\vec{k} = (0.60, 0.61, 0.02)$

E (eV)	D_0^{Th}	D_1^{Th}	D_2^{Th}	D_3^{Th}	D_0^{P}	D_1^{P}	D_2^{P}	D_3^{P}
4.705	0.108	0.128	0.173	0.026	0.009	0.552	0.002	0.002
5.484	0.000	0.062	0.256	0.041	0.000	0.639	0.002	0.000
6.557	0.000	0.346	0.145	0.027	0.000	0.474	0.004	0.004
9.086	0.001	0.024	0.106	0.797	0.000	0.029	0.042	0.001
9.197	0.000	0.004	0.061	0.847	0.000	0.060	0.024	0.004
9.657	0.000	0.000	0.054	0.924	0.000	0.000	0.019	0.003
10.132	0.001	0.001	0.024	0.953	0.000	0.010	0.008	0.003
10.259	0.000	0.000	0.000	0.988	0.000	0.003	0.008	0.001
10.391	0.000	0.002	0.005	0.992	0.000	0.000	0.000	0.001
10.509	0.000	0.000	0.008	0.980	0.000	0.009	0.002	0.001
11.935	0.003	0.004	0.271	0.583	0.008	0.111	0.015	0.005
12.647	0.000	0.000	0.239	0.558	0.000	0.178	0.023	0.002
13.145	0.000	0.000	0.654	0.292	0.000	0.002	0.040	0.008
14.605	0.072	0.005	0.614	0.128	0.073	0.006	0.097	0.005

$\vec{k} = (0.40, 0.41, 0.42)$

E (eV)	D_0^{Th}	D_1^{Th}	D_2^{Th}	D_3^{Th}	D_0^{P}	D_1^{P}	D_2^{P}	D_3^{P}
3.836	0.169	0.011	0.210	0.023	0.013	0.570	0.000	0.004
5.116	0.000	0.003	0.322	0.016	0.000	0.657	0.001	0.001
5.137	0.000	0.006	0.320	0.017	0.000	0.655	0.001	0.001
8.943	0.002	0.073	0.039	0.840	0.000	0.009	0.036	0.001
9.631	0.000	0.006	0.019	0.951	0.000	0.000	0.024	0.000
9.639	0.000	0.006	0.015	0.956	0.000	0.000	0.023	0.000
10.309	0.000	0.001	0.007	0.979	0.000	0.001	0.012	0.000
10.310	0.000	0.005	0.006	0.980	0.000	0.001	0.012	0.000
10.428	0.000	0.000	0.003	0.987	0.002	0.000	0.003	0.000
10.569	0.000	0.000	0.000	1.000	0.006	0.000	0.000	0.000
12.361	0.012	0.008	0.297	0.464	0.006	0.193	0.014	0.006
13.059	0.000	0.003	0.768	0.132	0.000	0.055	0.026	0.016
13.134	0.000	0.003	0.784	0.114	0.000	0.058	0.024	0.017
13.632	0.005	0.200	0.105	0.397	0.047	0.189	0.053	0.004

The k's are expressed in units of $(2\pi/a)$.

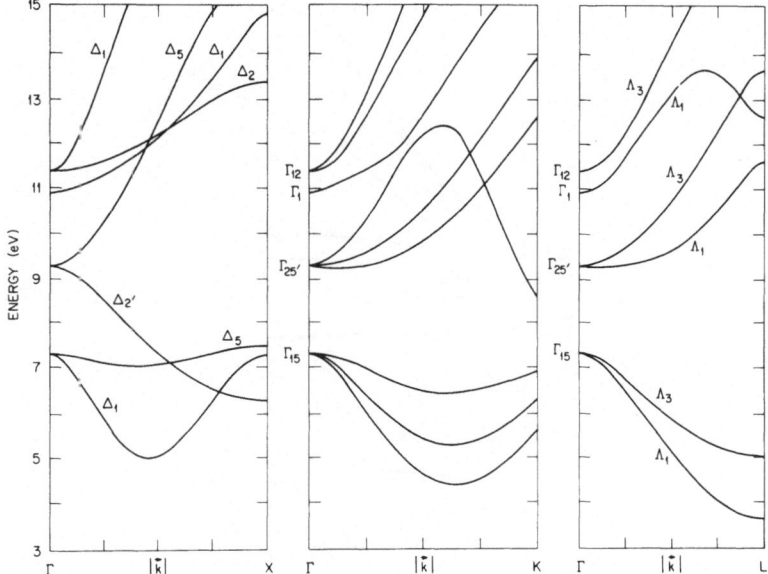

Figure 5 - Electronic band structure for ThP where the
'5f bands' have been arbitrarily removed.

with the normalization condition

$$\sum_{\ell} (D_{\ell}^{\text{TH}} + D_{\ell}^{\text{P}}) = 1. \tag{7}$$

In table II, we have considered $\vec{k}$ points displaced slightly from
symmetry points in order to simplify the programming. Such minor
displacements still allow direct comparison of table II with the
band structure displayed in figure 4. Undue elaboration upon
the results contained in table II is not necessary, although
they are important, since they illustrate the type of hybridza-
tion processes in many AX compounds containing lighter actinides.

The relativistic Γ to X band structure for the potential ThP
is plotted in figure 6. Non-relativistic results for the same
potential are given in figure 4. The results of figure 6 are
preliminary, since we have not performed the analysis to obtain
the symmetry designations of the bands and the smooth curves of
figure 6 were drawn by connecting energy values at 21 discrete
$\vec{k}$-points. But the results of figure 6 do indicate some small
spin-orbit splitting of the bands; e.g., the non-relativistic
lowest Δ_5 band of figure 4 is spin-orbit split, which is a re-
flection of the f-character in these 'p bands'. Such spin-orbit
splitting is also the cause of the 'f bands' at the X $\vec{k}$-point to
increase their width from 1.3 eV in figure 4 to 1.9 eV in figure
6. But, the overall width of all the Γ states of figure 6 is
4.9 eV, which is a slight decrease from the corresponding value
of 5.2 eV for figure 4. Also, the $\Gamma_{2'}$ to lower Γ_{15} splitting of
figure 4 undergoes a relativistic decrease from 2.5 to 2.0 eV.
These relativistic perturbations appear quite minor when making

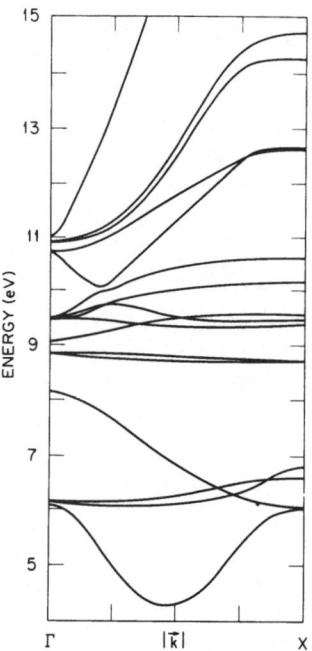

Figure 6 – Calculated relativistic band structure for
ThP along the Γ to X direction.

a qualitative perspective comparison of figures 4,6. That is,
in comparing the non-relativistic and relativistic calculations
no drastic energy reordering or large increases or decreases in
overall widths are observed, and other uncertainties exist in
all AX calculations which can make the non-relativistic-relativ-
istic differences seem quite moot.

PuN

An uncertainty in the calculations for AX compounds is the
choice of the parameter α in the exchange potential. To illustrate how this choice influences the calculated band structures,
displayed in figure 7 are results for PuN based on two choices
of α. As would be expected from any cursory examination of the
general literature of band theory, figure 7 does indicate sensitivity to the exchange potential. Although the underlying s, p
and d features of the band structure are not as sensitive to α
as the f parts, there is a tendency for the 'gap' between the
$\Delta_{2'}$ tail and the p bands, originating at the lower Γ_{15}, to close
α is decreased. This tendency is a systematic feature observed
in all our calculations on AX compounds and for similar rare
earth based compounds. Further decrease in α would cause the
$\Delta_{2'}$ tail to overlap the p states. Another universal feature
seen in figure 7 is the tendency of the 'f bands' to move up in
energy as α is decreased. Coupled with this movement is also a

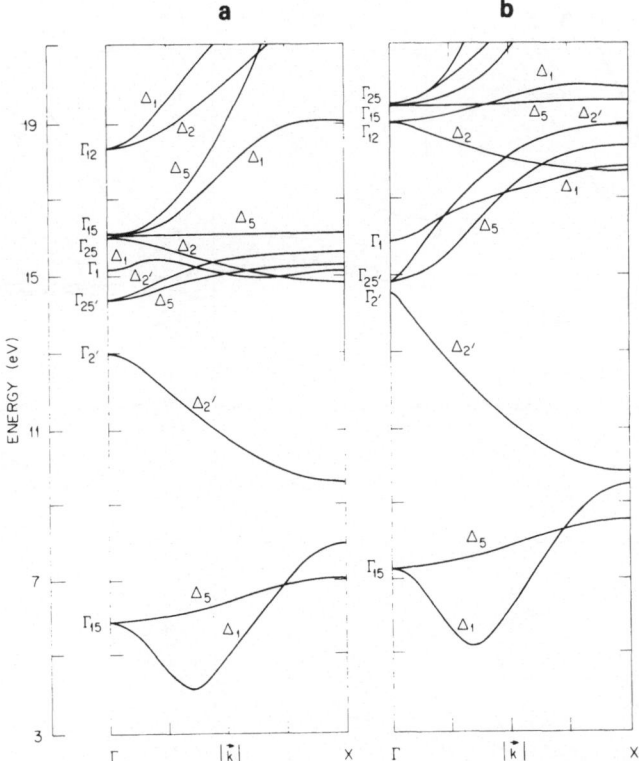

Figure 7 - Band structures for PuN corresponding to different exchange potentials. The results (a) are for the potential PuN1 with $\alpha = 1.0$, whilst (b) is for the potential PuN2 with $\alpha = 0.90$.

spatial expansion of the $5f$ orbitals, which allows more hybridization of the $5f$ bands with available states and increases their bandwidth. This feature directly bears on calculational attempts to determine whether the $5f$ electrons in an AX compound are best described via an itinerant or localized model.

NpSb

Another uncertain procedure used in obtaining the potential for a given AX compound is the choice of the configuration for the actinide atomic calculation. To illustrate effects which can occur when using potentials generated from different configurations for the same actinide element, a series of calculations for the compound NpSb are illustrated in figure 8. Perhaps the most striking result in figure 8 is the upward movement of the '$5f$ bands' as the number of f electrons is increased in the atomic calculation. That is, for NpSb5 ($f^3s^2d^2$) very narrow and almost unhybridized 'f bands' exist just below the 'p bands', whilst for NpSb4 (f^5s^2) the 'f bands' show more hybridization and occur in the 'd bands' well above the 'p bands'. The results

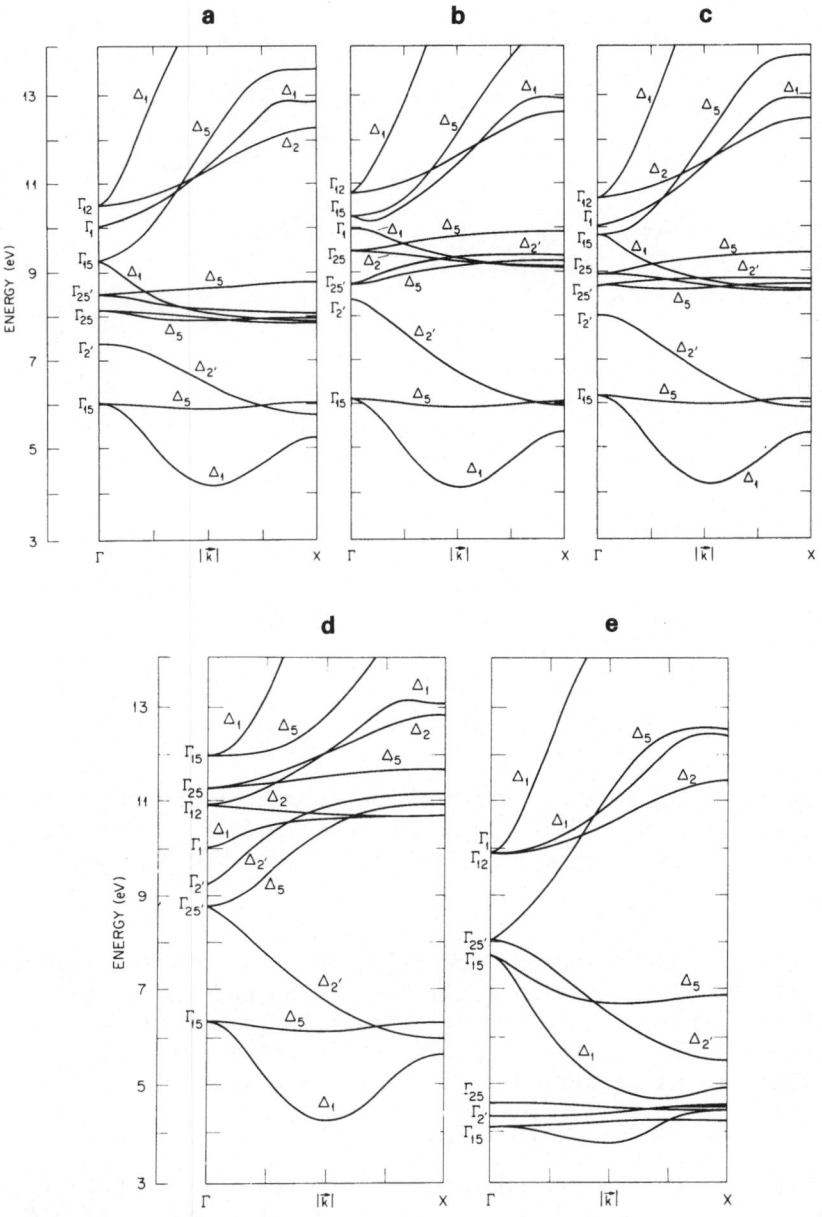

Figure 8 - Band structures for NpSb along the Γ to X
direction. These are for the various NpSb potentials of
table I and are: (a) NpSb1; (b) NpSb2; (c) NpSb3; (d)
NpSb4; and (e) NpSb5.

of figure 8 for the potentials NpSb1, 2, and 3 also are for different atomic configurations, although the number of f electrons
remain fixed. Although differences are seen to occur among these
results, they are minor compared with those observed as the

number of f electrons is varied. Not all the choices of configuration made in obtaining the results of figure 8 would be considered 'reasonable', and our purpose has only been to demonstrate that the results depend on choice of configuration.

US

The above result for PuN and NpSb have been included in order
to illustrate some of the difficulties associated with attempts
to perform band structure calculations for AX compounds. We
purposely have indicated these difficulties in order not to con-
vey the mistaken impression that the calculations are expected
to possess the highest degree of quantitative predictability,
but rather possess an exploratory, first attempt flavour. Thus,
it now might be asked, considering the difficulties discussed,
how can such calculations be used in furthering the quest toward
understanding the AX compounds' properties? We will partially
answer this question by using results calculated for US. These
results nicely demonstrate that, with proper care, band calculations for AX compounds can contain many semi-quantitatively
valid features.

In 1970 were reported [3] some non-relativistic KKR calculations for six UX compounds (X = C, N, P, As, S, and Se). These
calculations were based on the procedures discussed in section 2,
and resulted in U $5f$, U $6d$, and U $7s$ mutually hybridized bands
overlapping each other and falling within an energy range of
about 5-7 eV. Owing to the heavy hybridization of these bands,
it was difficult to extract values for $5f$ bandwidths for each of
those UX compounds. However, rough considerations indicated
their bandwidths had values $\sim$ 2-3 eV. It is interesting to compare these $5f$ bandwidths, which might appear unexpectedly large,
with the bandwidth obtained for a cubic crystallographic form of
metallic uranium. For bcc γ-U Kmetko and Hill [30] have calculated a $5f$ bandwidth of about 3.0 eV, a value which is also consistent with the calculations of Freeman and Koelling [31]. Obtaining $5f$ bandwidths for UX compounds about the same as that for
γ-U might appear to indicate an error, since the closest U-U
distance varies from 3.4 to 4.1 Å for the UX compounds, but is
only 3.0 Å for γ-U. So, owing to the increased U-U distances in
the compounds, one might expect their $5f$ bandwidths to be less
than that for γ-U. This would be the case if only $5f$-$5f$ overlap
and $5f$ hybridization with uranium $7s$ and $6d$ were the controlling
bandwidth factors, as they must be for γ-U. However, for the
UX compounds, overlap and hybridization of the $5f$ states with
anion p states could be a contributing factor to the $5f$ bandwidth, since the closest uranium-anion distance varies from 2.5
to 2.9 Å. Such circumstances are actually reflected in the UX
calculation results when the Bloch functions are analyzed [3].
Thus $5f$ bandwidths of $\sim$2-3 eV appear reasonable for UX compounds.
Also, $5f$ bandwidths of $\sim$2-3 eV indicates that the U $5f$ states in
the UX compounds reflect considerable itinerant behaviour.

One verification of the semi-quantitative validity of the

procedures used for the UX calculations has been the subsequent
photo-emission work of Eastman and Kuznietz [16,17] for the compound US. These authors reported photoemission results for US,
where, by varying the incident photon energies, they were able
to determine the placements and widths of occupied bands and
their approximate angular momentum character. Their resulting
energy distribution curves (EDC) for US are shown in figure 9.
The strong growth in the EDC's immediately below the zero of
energy (equal to the Fermi energy), as the incident photon energy is increased, indicates that these states predominantly
have higher angular momentum character than the states in the
region ∿3-7 eV below the Fermi energy. By an analysis of such
intensity changes Eastman and Kuznietz established that the
states ∿0-2 eV below the Fermi energy are predominantly 'd-f'
character, whilst those ∿3-7 eV below are predominantly 'p'
character. As indicated by these authors, their analysis of the
photo-emission data was in semi-quantitative agreement with the
published US band structure [3].

To expand the comparison between the photoemission data and
calculations the US calculations have been extended recently [5]
and are presented in figure 10. By comparing the results of
figures 9,10, it is seen that the calculated 'p bands', originating at the lower Γ_{15}, are slightly more below (∿7 eV) the Fermi
energy than from the photoemission data (∿5 eV). However, the
'p' bandwidths are in better agreement, with the calculations
and experiments both producing a value of ∿4-5 eV. Also, there
is no serious conflict between the calculated relative values of

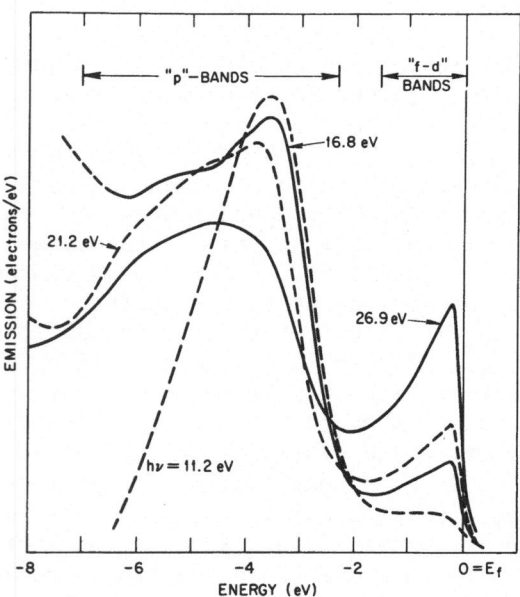

Figure 9 - Photoemission results for US as a function
of incident energy (after Eastman and Kuznietz [17]).

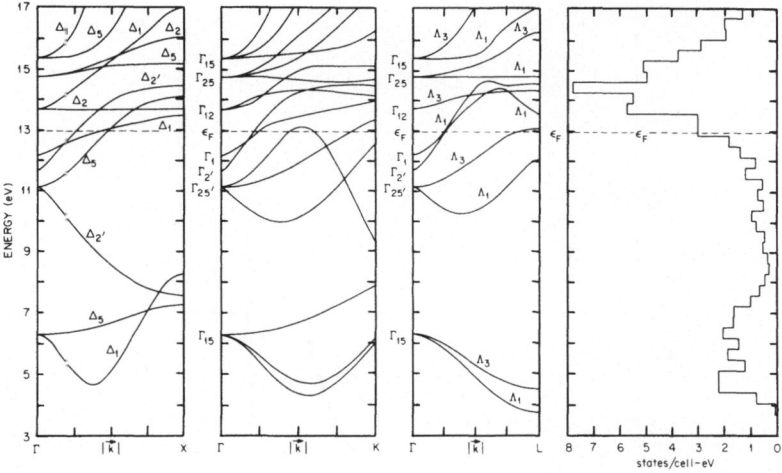

Figure 10 - Electronic band structure and density of states for US.

the density of states at the vicinity of the Fermi energy and at the centre of the 'p band' when compared to the relative photoemission intensities. That is, from the data of figure 9 for an incident photon energy of 26.9 eV one would expect, after allowance is made for relative emission strengths [16,17], that the density of states would be slightly higher in the vicinity of the Fermi energy compared to the peak in the 'p band' region. Just below the Fermi energy if US it is possible to make a more quantitative comparison between the photo-emission results and the calculated results. Since the curves of figure 9 for energies within ~ 2 eV of the Fermi energy do not significantly change their shape as the incident photon energy is increased above 16.8 eV, Eastman and Kuznietz reasoned that this constant shape should approximately resemble the shape of the density of states. By normalizing this shape to contain four electrons, since US has ten electrons in the outer shells of the constituent atoms and the 'p valence bands' would hold six, they obtained an approximation to the 'f-d conduction band' density of states, which is plotted in figure 11 as the smooth curve. The dotted portion of this curve was uncertain, although a necessary extrapolation to normalize the results. Also plotted in figure 11 is part of the calculated histogram density of states for US. By obtaining the smooth curve of figure 11 Eastman and Kuznietz estimated a value of $\simeq 2.5$ electrons/cell eV for US's density of states at the Fermi energy, whilst the calculated histogram value is $\simeq 3.0$. Also, it is relevant to compare these with the value of 9.9 which would be extracted from measured low temperature specific heat of US [32], by assuming no enhancement processes are acting in this compound. However, the above comparisons appear to imply that such enhancement

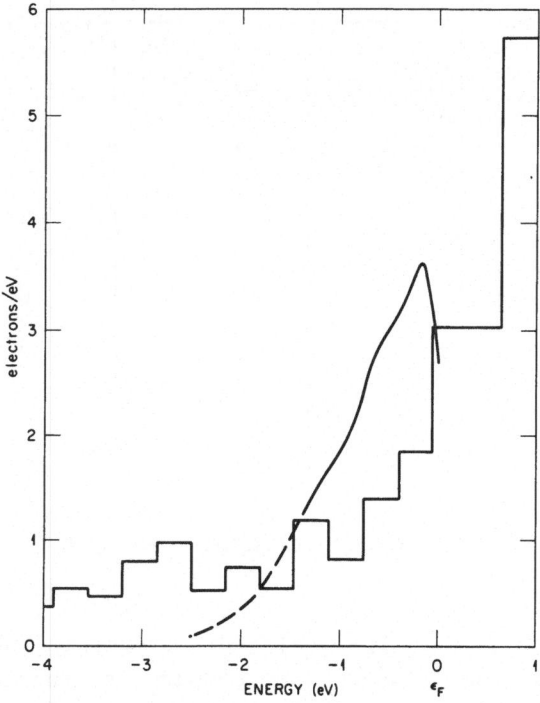

Figure 11 - Comparison of density of states directly below the Fermi energy as extracted from photoemission data (smooth curve) and calculated band structurally (histogram).

processes do contribute to the specific heat in US, and, hence, raises the possibility that the same could be true for other AX compounds.

It is to be noticed from figure 11 that both the smooth curve and the histogram reflect about the same decrease just below the Fermi energy, but an absolute comparison implies the smooth curve is higher. One explanation for this may be the normalization process used by Eastman and Kuznietz. That is, by normalizing their curve to contain four electrons, they implicitly assumed that the 'p band' and the 'd-f conduction band' do not overlap and that a region of zero density of states existed between these two bands. But it is seen from figure 10 that the calculations have these two major band regions overlapping. Thus, it appears quite plausible that these major bands overlap, which would imply that Eastman and Kuznietz's curve should have been normalized to contain less than four electrons. In fact, the calculated density of states has the result that the bands hold seven electrons below -2.85 eV (energy scale of figure 11). Thus, the calculations imply it would have been more reasonable to normalize the experimentally extracted curve to contain slightly less than three electrons. If this were done, considerable more agreement in absolute magnitude would be obtained between the experimentally derived curve and the histogram. Regardless, the photoemission work of Eastman and Kuznietz provides some semi-quantitative tests for the US calculations of figure 10, with the results of such testing strongly indicating that the calculations reasonably approximate experimental reality. At the same time, such indications for US imply that the calculations for other AX compounds might bear some reasonable approximation to their actual electronic structures.

REFERENCES

1. Freeman, A.J. and Darby, J.B., Jr. (eds.) (1974). *The Actinides: Electronic Structure and Other Related Properties*. (Academic Press, New York).

2. Kuznietz, M. (1971). In *Conference Digest No.3: Rare Earths and Actinides, Durham 1971*, (Institute of Physics, London), p. 162.

3. Davis, H.L. (1970). In *Plutonium 1970 and Other Actinides* (ed. Miner, W.N.), (The Metallurgical Society of A.I.M.E., New York), p. 209.

4. Davis, H.L. (1971). In *Conference Digest No.3: Rare Earths and Actinides, Durham 1971*, (Institute of Physics, London), p. 126.

5. Davis, H.L. (1974). *The Actinides: Electronic Structure and Other Related Properties, Vol.2*, (eds. Freeman, A.J. and Darby, J.B., Jr.), (Academic Press, New York), chapter 1.

6. Ern, V. and Switendick, A.C. (1965). *Phys. Rev.*, **137**, A1927.

7. Conklin, J.B. and Silversmith, D.J. (1968). *Int. J. Quant. Chem.*, **2S**, 243.

8. Switendick, A.C. and Jones, E.D. (1968). *Bull. Amer. Phys. Soc.*, **13**, 365.

9. Conklin, J.B. and Simpson, R.W. (1970). *Bull. Amer. Phys. Soc.*, **15**, 310.

10. Davis, H.L. (1971). *Bull. Amer. Phys. Soc.*, **16**, 637.

11. Mattheiss, L.F. (1972). *Phys. Rev.*, **B5**, 290.

12. Cho, S.J. (1967). *Phys. Rev.*, **157**, 632.

13. Cho, S.J. (1970). *Phys. Rev.*, **B1**, 4589.

14. Davis, H.L. (1971). *Proceedings of The Rare Earth Research Conference Blacksburg*, (ed. Field, P.E.), (USAEC CONF-711001), p. 3.

15. Davis, H.L. (1972). *Bull. Amer. Phys. Soc.*, **17**, 346.

16. Eastman, D.E. and Kuznietz, M. (1971). *Phys. Rev. Lett.*, **26**, 846.

17. Eastman, D.E. and Kuznietz, M. (1971). *J. Appl. Phys.*, **42**, 1396.

18. Gambino, R.J., Eastman, D.E., McGuire, T.R., Moruzzi, V.L. and Grobman, W.D. (1971). *J. Appl. Phys.*, **42**, 1468.

19. Mattheiss, L.F. (1964). *Phys. Rev.*, **133**, A1399.

20. Slater, J.C. (1951). *Phys. Rev.*, **81**, 385.

21. Korringa, J. (1947). *Physica*, **13**, 392.

22. Kohn, W. and Rostoker, N. (1954). *Phys. Rev.*, **94**, 1111.
23. Segall, B. (1957). *Phys. Rev.*, **105**, 108.
24. Treusch, J. and Sandrock, R. (1966). *Phys. Stat. Sol.*, **16**, 487.
25. Onodera, Y. and Okazaki, M. (1966). *J. Phys. Soc. Jap.*, **21**, 1273.
26. Davis, H.L. (1971). In *Computational Methods in Band Theory*, (eds. Marcus, P.M., Janak, J.F. and Williams, A.R.), (Plenum Press, New York), p. 183.
27. Nestor, C.W., Tucker, T.C., Carlson, T.A., Roberts, L.D., Malik, F.B. and Froese, C. (1966). (Oak Ridge National Laboratory Report 4027).
28. Nestor, C.W., Tucker, T.C., Carlson, T.A. and Malik, F.B. (1969). *Phys. Rev*, **178**, 998.
29. Fischer, C.F. (1969). *Comput. Phys. Commun.*, **1**, 151.
30. Kmetko, E.A. and Hill, H.H. (1970). In *Plutonium 1970 and Other Actinides*, (ed. Miner, W.N.), (The Metallurgical Society of A.I.M.E., New York), p. 233.
31. Freeman, A.J. and Koelling, D.D. (1972). *J. Phys. (Paris)*, **33**, C3-57.
32. Westrum, E.F., Walters, R.R., Flotow, H.E. and Osborne, D.W. (1968). *J. Chem. Phys.*, **48**, 155.
33. Rundle, R.E. (1948). *Acta Cryst.*, **1**, 180.
34. Bilz, H. (1958). *Z. Phys.*, **153**, 338.

MAGNETISM IN ACTINIDES

B. COQBLIN and R. JULLIEN

*Laboratoire de Physique des Solides†,
Université Paris-Sud, Centre d'Orsay,
91405 Orsay, France*

INTRODUCTION

Pure actinides and metallic actinide systems are very interesting from a theoretical point of view since they show a great variety of experimental magnetic properties. One can observe:

— Clear magnetism, as in pure curium and berkelium;

— Kondo effect, as in many transition based alloys with neptunium and plutonium impurities;

— Nearly magnetism with spin fluctuations as in plutonium, neptunium and in a great number of actinide compounds;

— Superconductivity, as in thorium or uranium under pressure.

In these lectures, we will be interested in some features of this experimental situation. A first part will be devoted to pure actinides, a second part to transition based alloys with actinide impurities and a third part to nearly magnetic actinide systems.

1. PURE ACTINIDE METALS

1.1 EXPERIMENTAL SITUATION

From an anlysis of the experimental data and especially from

---

† Laboratoire associé au CNRS.

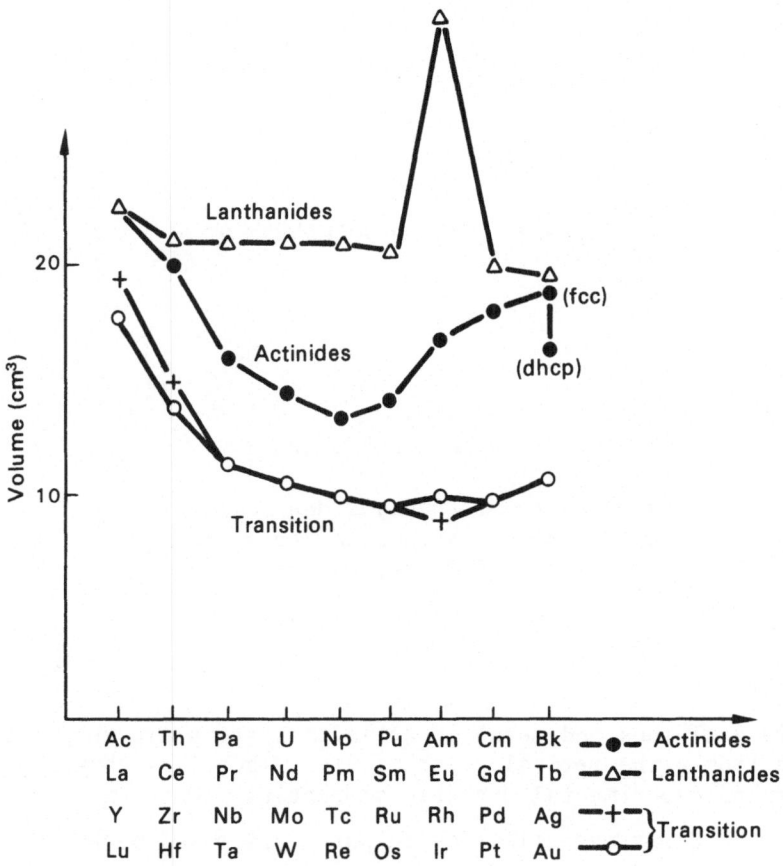

Figure 1 - Plot of the molar volume (in cm^3) for the actinide series compared with lanthanide, 4d and 5d series.

the plot of the atomic volume (figure 1) along the series, we can distinguish three groups in the actinide series [1]:

(i) Actinium and thorium are transition-like metals with a pronounced 6d character and a negligible 5f character. Actinium is very similar to lanthanum with a $7s^26d^1$ configuration. Thorium is superconducting, and the values of its magnetic susceptibility χ and its electronic specific heat constant suggest that it is tetravalent, with roughly the atomic configuration $7s^26d^2$;

(ii) Protoactinium, uranium, neptunium and plutonium have atomic volumes intermediate between those of the rare earths and those of the 4d and 5d transition metals. Their densities of states, estimated by the χ and γ values are too large to be attributed only to a simple 6d character, and the absence of magnetism shows that their 5f bands are certainly much broader than in rare earths. They are mixed $6d$-$5f$ metals with a hybridized d-f character. The two last ones, neptunium and plutonium are nearly magnetic, as is suggested by their high values of χ compared with those of γ, and by the large T^2 term observed in their resistivity curves at low temperatures;

(iii) Americium, curium and berkelium have atomic volumes similar to those of the corresponding rare earths and similar hexagonal structures, suggesting a localized $5f$ character with a valence close to three. Americium is nonmagnetic, curium and berkelium are magnetic with magnetic moments corresponding roughly to the $5f^7$ and $5f^8$ configurations respectively.

We would like to explain here the appearance of magnetism only near the middle of the series, for curium and berkelium, when the $5f$ shell is almost half filled (unlike for the rare earth series) by taking into account a large $6d$-$5f$ hybridization.

1.2 THE LOCALIZED APPROACH

In a first simple model [2], we have described a pure actinide metal as a collection of $6d$ and $5f$ impurity levels mixed with the $7s$ conduction band by the usual Friedel-Anderson resonant scattering mechanism [3]: this is equivalent to assuming a negligible overlap between localized states at different sites of the lattice. This assumption is not well checked in actinide metals, but it permits a first simple derivation of the appearance of magnetism, and the results will be justified *a posteriori* by the band approach of subsection 1.3. The Coulomb repulsion between d and f electrons is taken into account within the Hartree-Fock approximation, and all the different Coulomb and exchange integrals are assumed in the following to have the same value U. Finally, a one-body mixing term V_{df} between $5f$ and $6d$ states is introduced here to describe phenomenologically the d-f hybridization.

Thus the starting Hamiltonian is

$$H = \sum_{k,\sigma} \varepsilon_k c_{k\sigma}{}^+ c_{k\sigma} + E_{0d} \sum_\sigma c_{d\sigma}{}^+ c_{d\sigma} + E_{0f} \sum_\sigma c_{f\sigma}{}^+ c_{f\sigma}$$

$$+ \sum_{k,\sigma} (V_{kd} c_{k\sigma}{}^+ c_{k\sigma} + \text{h.c.}) + \sum_{k,\sigma} (V_{kf} c_{k\sigma}{}^+ c_{k\sigma} + \text{h.c.})$$

$$+ U(n_{d\uparrow} n_{d\downarrow} + n_{f\uparrow} n_{f\downarrow} + n_{d\uparrow} n_{f\downarrow} + n_{d\downarrow} n_{f\uparrow})$$

$$+ \sum_\sigma (V_{df} c_{d\sigma}{}^+ c_{f\sigma} + \text{h.c.}). \tag{1}$$

The usual notations have been adopted here.

We use standard Green function formalism within the Hartree-Fock self-consistency scheme. The diagonal elements of the Green function are given by

$$G_{ff}^{\sigma}(E) = \left[E - E_f^{\sigma} + i\Gamma_f - \frac{|V_{df}|^2}{E - E_d^{\sigma} + i\Gamma_d} \right]^{-1},$$

$$(2)$$

$$G_{dd}^{\sigma}(E) = \left[E - E_d^{\sigma} + i\Gamma_d - \frac{|V_{df}|^2}{E - E^{\sigma} + i\Gamma} \right]^{-1},$$

where Γ_f and Γ_d are the half widths of the two f and d virtual bound states without V_{df} interaction, and E_f^{σ}, E_d^{σ} their respective energies, shifted by the spin dependent Hartree-Fock field and given by:

$$E_d^{\sigma} = E_{0d} + U\langle n_{-\sigma}\rangle,$$

$$E_f^{\sigma} = E_{0f} + U\langle n_{-\sigma}\rangle,$$

where

$$n_{\sigma} = n_{d\sigma} + n_{f\sigma}. \qquad (4)$$

From (2), one can derive the total density of d- and f-states by using the relation:

$$\rho_{\sigma}(E) = -\frac{1}{\pi} \, \text{Im}(G_{ff}^{\sigma}(E) + G_{dd}^{\sigma}(E)), \qquad (5)$$

which, in the simple case $V_{df} = 0$, is equal to:

$$\rho_{\sigma}(E) = \frac{1}{\pi}\left[\frac{\Gamma_f}{(E - E_f^{\sigma})^2 + \Gamma_f^2} + \frac{\Gamma_d}{(E - E_d^{\sigma})^2 + \Gamma_d^2} \right]. \qquad (6)$$

This is an obvious simple addition of two Anderson-like virtual bound states [3]. In the general mixed case, $V_{df} \neq 0$, this density of states still comes out as a sum of two Lorentzians:

$$\rho_{\sigma}(E) = \frac{1}{\pi}\left[\frac{\Gamma_1}{(E - E_1^{\sigma})^2 + \Gamma_1^2} + \frac{\Gamma_2}{(E - E_2^{\sigma})^2 + \Gamma_2^2} \right]. \qquad (7)$$

but the centres E_1^{σ} and E_2^{σ} and the corresponding half widths Γ_1 and Γ_2 are now modified as a result of hybridization effects.

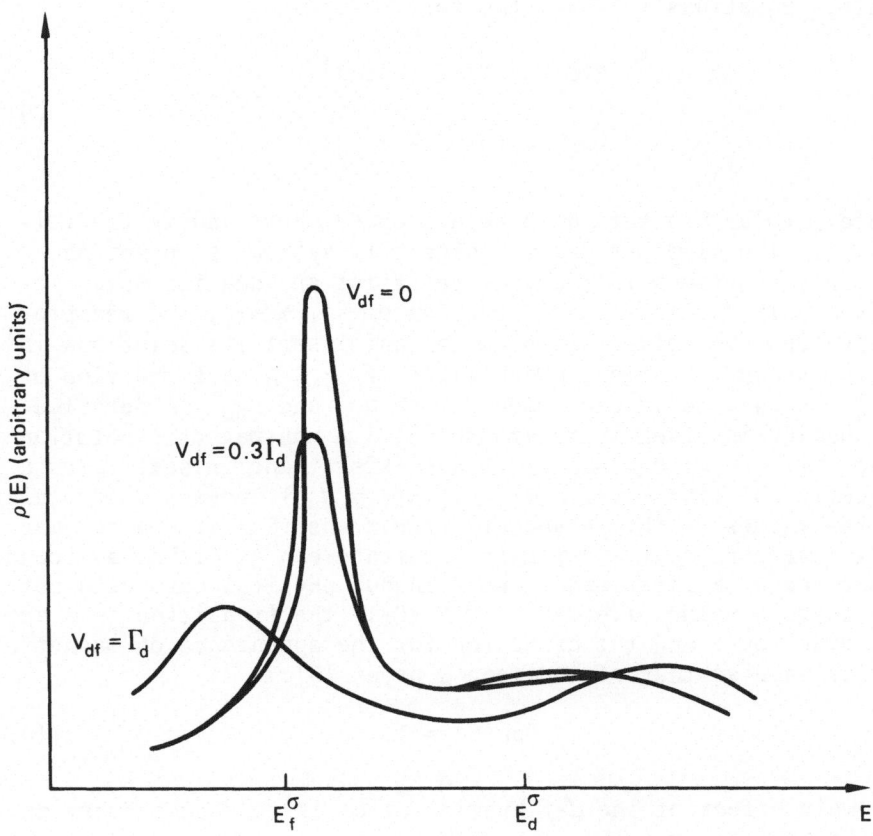

Figure 2 - Theoretical d and f extra density of states for a typical set of parameters $\Gamma_d = 10\Gamma_f$, $E_d^{(0)} - E_f^{(0)} = \Gamma_d$, and three different values of $V_{df} = 0$, 0,3 Γ_d, Γ_d, which show the effect of hybridization.

More precisely, as can be seen from figure 2, hybridization 'pushes away' from each other the centres and 'equalizes' the widths. In the physical case for actinides $\Gamma_f \ll \Gamma_d$, the smallest of the two widths, let us say Γ_1, is greatly increased with respect to Γ_f.

To compute $\langle n_{f\sigma} \rangle$ and $\langle n_{d\sigma} \rangle$, one has to solve two equations:

$$\langle n_{f\sigma} \rangle = -\frac{1}{\pi} \operatorname{Im}\int_{-\infty}^{0} G_{ff}^{\sigma}(E)dE,$$

$$\langle n_{d\sigma} \rangle = -\frac{1}{\pi} \operatorname{Im}\int_{-\infty}^{0} G_{dd}^{\sigma}(E)dE,$$

(8)

where the Fermi level has been taken as the zero on the energy
scale. Equations (8) have the form

$$n_{f\sigma} = F(n_{f-\sigma} + n_{d-\sigma}),$$

$$n_{d\sigma} = D(n_{f-\sigma} + n_{d-\sigma}), \tag{9}$$

where angular brackets have been dropped, here and in the following. The simplest way to solve this system is to sum and
subtract equations (9); so, we get first an equation which in-
volves only $n_{d\uparrow} + n_{f\uparrow}$ as a function of $n_{d\downarrow} + n_{f\downarrow}$, and recipro-
cally; one can solve this equation and insert its solutions in
to the second equation, which gives $n_{f\sigma} - n_{d\sigma}$ as a function of
$n_{f-\sigma} + n_{d-\sigma}$. So we can compute both $n_{f\sigma}$ and $n_{d\sigma}$. A detailed
discussion is given in reference [2]. A non-magnetic solution
(i.e. $n_{f\uparrow} = n_{f\downarrow}$; $n_{d\uparrow} = n_{d\downarrow}$) always exists, and in some cases a
magnetic solution (i.e. $n_{f\uparrow} \neq n_{f\downarrow}$; $n_{d\uparrow} \neq n_{d\downarrow}$) appears. According
to the values of the parameters, the transition from a non-mag-
netic state to a d- and f-magnetic state can be either a second
order one or a first order one. In our physical case of a not
too large U value (i.e. $\pi\Gamma_d \gtrsim U \gtrsim \pi\Gamma_f$), the transition is a second order one, and the criterion for the appearance of magnetism
is the same as that of Anderson's paper [3]:

$$U\rho_\sigma(0) = 1. \tag{10}$$

The main effect of the d-f hybridization is to reduce drastically the magnetic region. This fact can be easily understood, because, for $\Gamma_f < \Gamma_d$, the smallest of the two effective widths Γ_1
and Γ_2, which becomes equal to Γ_f for $V_{df} = 0$, increases with
V_{df}.

We now discuss the properties of the actinide series, within
our model, in the following way: we suppose that, whilst the $5f$
level shifts down with respect to the Fermi level along the series, the uperturbed $6d$ level E_{0d} stays at a fixed position. To
find a reasonable value for E_{0d} and for the Coulomb repulsion
parameter U, one can adjust the $6d$ occupation numbers in order
to fit the expected atomic values of 0.2 at the beginning and 0.1 at
the end of the series, when the f level is respectively empty or filled
up (i.e. $E_{0f} = \pm\infty$). The following choice:

$$\frac{E_{ad}}{\Gamma_d} = \frac{U}{\Gamma_d} = 1, \tag{11}$$

roughly satisfies this condition. Finally, we have taken the
ratio between the two widths Γ_d and Γ_f equal to 10; this gives,
for example, $\Gamma_f = 0.2$ eV and $\Gamma_d = 2$ eV, which are of the order
of magnitude suggested by band calculations for the widths of
non-hybridized $5f$ and $6d$ bands [4]. Changing slightly the para-

meters around the chosen values does not affect the physical results greatly.

In figure 3 we have plotted $n_{f\sigma}$ and $n_{d\sigma}$, the total number of d and f electrons, and the spin magnetic moment, both as functions of E_{0f}/U. We neglect completely here the orbital contribution to the magnetism, which is certainly important in actinides, and which can be described only by taking into account the orbital degeneracy in the Hamiltonian (1). Here we assume, phenomenologically, that the total number of d- and f-electrons is given by:

$$N = 5(n_{d\uparrow} + n_{d\downarrow}) + 7(n_{f\uparrow} + n_{f\downarrow}). \tag{12}$$

Similarly the total spin magnetic moment defined in the Hartree-Fock approximation as its projection along the Oz axis is given by:

$$M = 5(n_{d\uparrow} - n_{d\downarrow}) + 7(n_{f\uparrow} - n_{f\downarrow}). \tag{13}$$

We see from figure 3 that the reasonable parameters used here, and especially a large V_{df} parameter of order U, can explain how magnetism appears only after Am, and not after U, as would have been given by the simple $5f$ virtual bound state theory without hybridization. Therefore the d-f hybridization can explain the delay, experimentally observed, in the appearance of magnetism in the actinide series. However, the theoretical magnetic moments for Cm and Bk are smaller than the experimental ones. This discrepancy is connected with some of our approximations: the constancy of V_{df} and U along the series and the Lorentzian shape of the d virtual bound state.

1.3 THE BAND APPROACH

In the preceeding approach, V_{df} was introduced as a phenomenological parameter. In fact, a one-body mixing term such as $V_{df}c_{d\sigma}{}^{+}c_{f\sigma}$ cannot be found in this localized picture. The d-f hybridization, either in pure actinide metals or in transition based alloys with actinide impurities, comes from the overlap between d and f wavefunctions centred on neighbouring sites of the lattice. So we present now a more realistic band approach [5].

We start from d and f bands:

$$E_{df} = E_{0d} + \Gamma_d \alpha_d(\mathbf{k}),$$

$$E_{fk} = E_{0f} + \Gamma_f \alpha_f(\mathbf{k}), \tag{14}$$

and from the corresponding k-dependent d-f hybridization:

$$V_{dfk} = \Gamma_{df} \beta(\mathbf{k}). \tag{15}$$

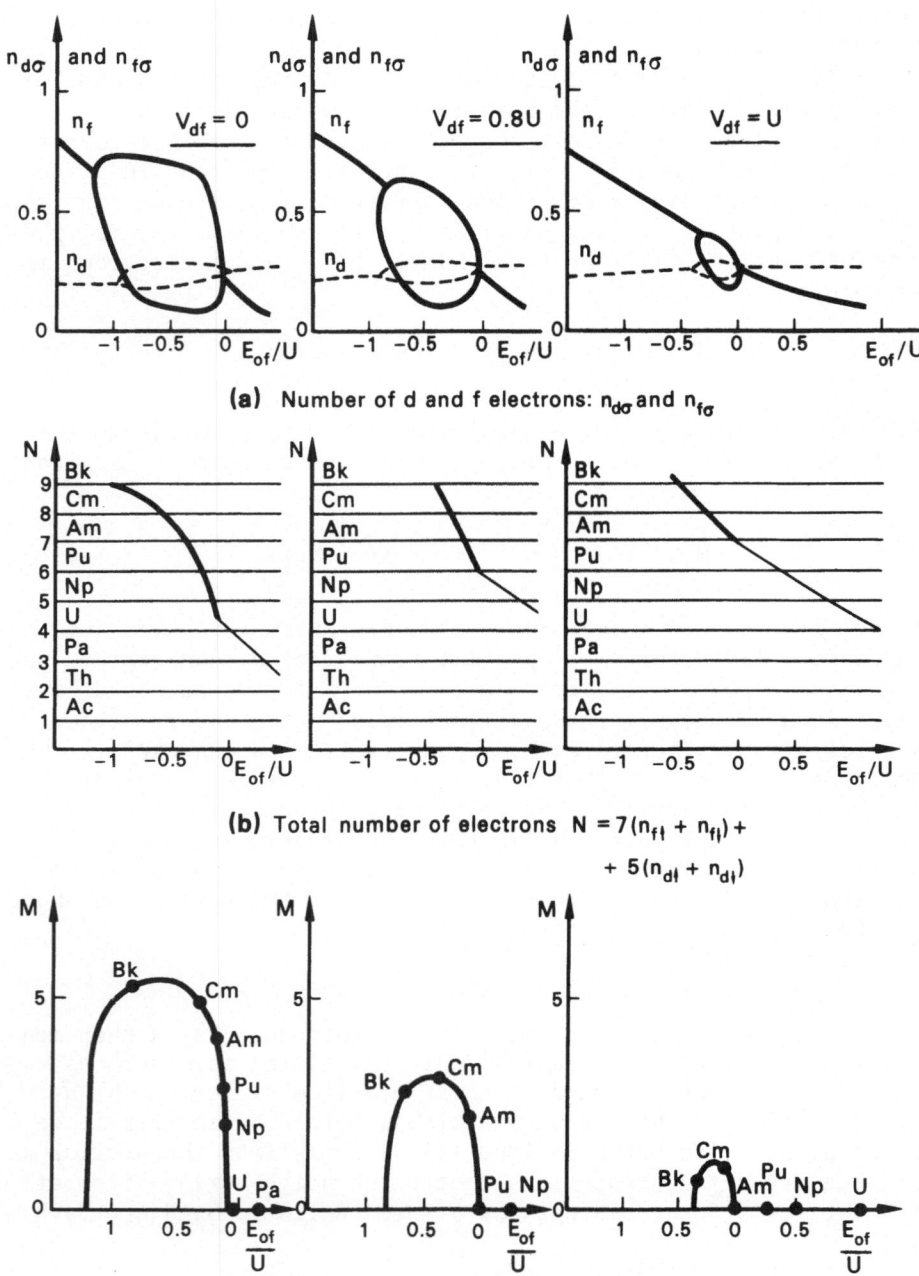

(a) Number of d and f electrons: $n_{d\sigma}$ and $n_{f\sigma}$

(b) Total number of electrons $N = 7(n_{f\uparrow} + n_{f\downarrow}) +$
$$+ 5(n_{d\uparrow} + n_{d\downarrow})$$

(c) Magnetic moment $M = 7(n_{f\uparrow} - n_{f\downarrow}) + 5(n_{d\uparrow} - n_{d\downarrow})$

Figure 3 - Plot of (a) $n_{d\sigma}$ and $n_{f\sigma}$; (b) the total number of d and f electrons; and (c) the magnetic moment M in Bohr magnetons vs E_{0f}/U for three different values ($V_{df} = 0$, $0.8\ U$, U). The choice of the other parameters and the identification of each actinide is explained in the text.

With the usual notation the Hamiltonian is given by

$$H = \sum_{k,\sigma} E_{dk} c_{dk\sigma}{}^{+} c_{dk\sigma} + \sum_{k,\sigma} E_{fk} c_{fk\sigma}{}^{+} c_{fk\sigma} + \sum_{k,\sigma} V_{dfk} c_{dk\sigma}{}^{+} c_{fk\sigma}$$

$$\tag{16}$$

$$+ \text{ h.c.} + U \sum_{i} (n_{di\uparrow} n_{di\downarrow} + n_{fi\uparrow} n_{fi\downarrow} + n_{di\uparrow} n_{fi\downarrow} + n_{fi\uparrow} n_{di\downarrow}),$$

where we consider only the d and f electrons and their two-body interactions by the U parameter as in the preceeding approach, but here we neglect the broad $7s$ band. We use the simplifying approximations:

APPROXIMATION (a): $\alpha_d(\mathbf{k}) \equiv \alpha_f(\mathbf{k})$, $\beta(\mathbf{k}) \equiv 1$;

APPROXIMATION (b): $\alpha_d(\mathbf{k}) \equiv \alpha_f(\mathbf{k})$, $(\beta(\mathbf{k}))^2 = 1 - (\alpha(\mathbf{k}))^2$.

Both approximations (a,b) yield the same shape for the d- and f-densities of states without hybridization which can be described by the same dimensionless function $f(x)$:

$$\rho_{d0}(E) = \frac{1}{\Gamma_d} f\left(\frac{E - E_{0d}}{\Gamma_d}\right),$$

$$\rho_{f0}(E) = \frac{1}{\Gamma_f} f\left(\frac{E - E_{0f}}{\Gamma_f}\right).$$

Approximation (a) corresponds to a constant d-f hybridization, whilst approximation (b) corresponds to a variable d-f hybridization which is a maximum at the band centres ($\alpha = 0$) and zero at the band edges ($\alpha = \pm 1$). Approximation (b) corresponds to a more realistic situation: for example, along a given k direction in a cubic crystal, the tight binding approach yields $\alpha_d(\mathbf{k}) = \alpha_f(\mathbf{k}) = \alpha_f(\mathbf{k}) = \cos(k_x a)$ and $\beta(\mathbf{k}) = \sin(k_x a)$.

We calculate the densities of states with hybridization by use of the Green function method within the Hartree-Fock approximation, as in subsection 1.2, and we express them in terms of the function $f(x)$ which describes the densities of states without hybridization. After some algebra, we find:

$$\rho_d{}^{\sigma}(E) = \{[E - E_{0f}{}^{\sigma} - \Gamma_f \alpha_-{}^{\sigma}(E)] f[\alpha_-{}^{\sigma}(E)]$$

$$- [E - E_{0f}{}^{\sigma} - \Gamma_f \alpha_+{}^{\sigma}(E)] f[\alpha_+{}^{\sigma}(E)]\}$$

$$\times \{[\Gamma_d(E - E_{0f}{}^{\sigma}) + \Gamma_f(E - E_{0d}{}^{\sigma})]^2 \tag{18}$$
$$\text{(Contd)}$$

$$- 4(\Gamma_d \Gamma_f + z\Gamma_{df}{}^2)[(E - E_{0d}{}^{\sigma})(E - E_{0f}{}^{\sigma}) - \Gamma_{df}{}^2]\}^{-\frac{1}{2}},$$

$$\rho f^\sigma(E) = \{[E - E_{0d}^\sigma - \Gamma_d \alpha_-^{-\sigma}(E)]f[\alpha_-^{-\sigma}(E)]$$

$$- [E - E_{0d}^\sigma - \Gamma_d \alpha_+^\sigma(E)]f[\alpha_+^\sigma(E)]\}$$

(Contd)
(18)

$$\times\{[\Gamma_d(E - E_{0f}^\sigma) + \Gamma_f(E - E_{0d}^\sigma)]^2$$

$$- 4(\Gamma_d\Gamma_f + z\Gamma_{df}^2)[(E - E_{0d}^\sigma)(E - E_{0f}^\sigma) - \Gamma_{df}^2]\}^{-\frac{1}{2}},$$

where

$$\alpha_\pm^\sigma(E) = \frac{1}{2(\Gamma_d\Gamma_f + z\Gamma_{df}^2)} \{\Gamma_d(E - E_{0f}^\sigma) + \Gamma_f(E - E_{0d}^\sigma)\}$$

$$\pm \{[\Gamma_d(E - E_{0f}^\sigma) + \Gamma_f(E - E_{0d}^\sigma)]^2$$

(19)

$$- 4(\Gamma_d\Gamma_f + z\Gamma_{df}^2)[(E - E_{0d})(E - E_{0f}) - \Gamma_{df}^2]\}^{-\frac{1}{2}},$$

$$E_{0d} = E_{0d} + U(n_{d-\sigma} + n_{f-\sigma}),$$

(20)

$$E_{0f} = E_{0f} + U(n_{f-\sigma} + n_{f-\sigma}).$$

$z = 0$ corresponds to approximation (a) and $z = 1$ corresponds to approximation (b).

The effect of the hybridization can be seen in figure 4, where we have plotted the total density of d and f states versus E for $E_{0d} - E_{0f} = 0.35 \, \Gamma_d$, $\Gamma_d = 10 \, \Gamma_f$ and for three values of Γ_{df}: $\Gamma_{df} = 0$ (———), $\Gamma_{df} = 0.2 \, \Gamma_d$ (----), $\Gamma_{df} = 0.4 \, \Gamma_d$ (····). The left part corresponds to approximation (a) and the right part to approximation (b). The top corresponds to 'square' bands:

$$f(x) = 0.5 \quad \text{for} \quad |x| < 1,$$

$$f(x) = 0 \quad \text{for} \quad |x| > 1,$$

(21)

and the bottom to 'parabolic' bands:

$$f(x) = \frac{3}{4}(1 - x^2) \quad \text{for} \quad |x| < 1,$$

$$f(x) = 0 \quad \text{for} \quad |x| > 1.$$

(22)

The same broadening effect as in figure 2 is observed when the hybridization parameter Γ_{df} increases.

Then the self-consistent equations giving $n_{d\sigma}$ and $n_{f\sigma}$ as a

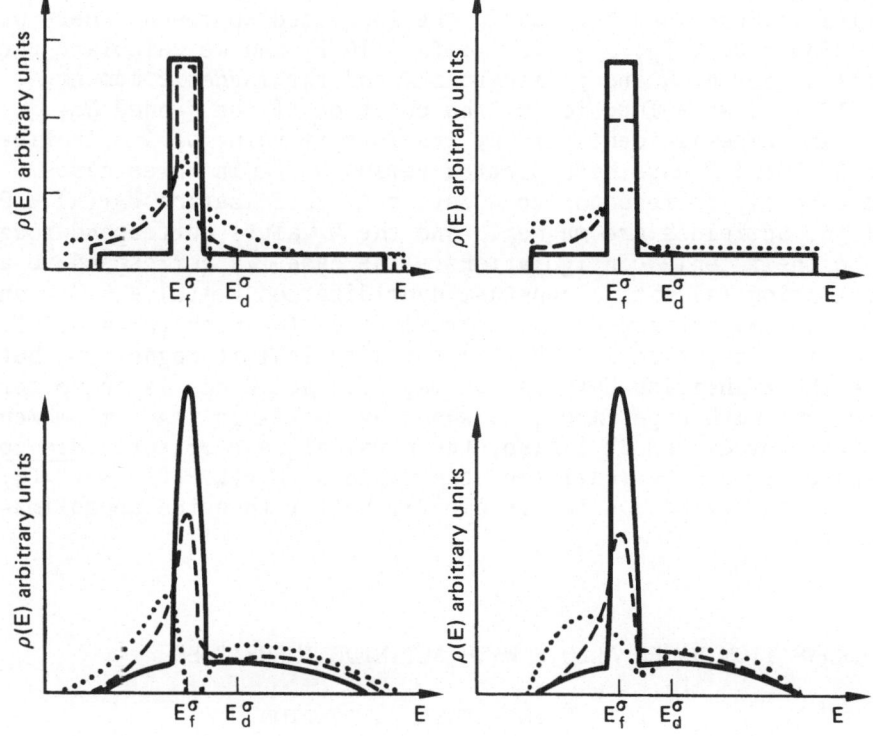

Figure 4 - Total density of d- and f-states. Details are given in the text.

function of $n_{d-\sigma}$ and $n_{f-\sigma}$ are obtained by:

$$n_{d\sigma} = \int_{-\infty}^{0} \rho_d^{\sigma}(E)\,dE,$$

$$n_{f\sigma} = \int_{-\infty}^{0} \rho_f^{\sigma}(E)\,dE. \tag{23}$$

This set of self-consistent equations also takes the form (9) which can be solved in the same way.

If we take a Lorentzian for $f(x)$ within approximation (a) we come back exactly to the same results as in the virtual bound states model. So the localized approach is in fact a reasonable approximation to describe pure actinides, and, moreover, the assumption of taking a phenomenological d-f hybridization in this approach is now perfectly justified.

Here, we can take more realistic band shapes without tails and a physical variable d-f hybridization. The results do not depend very much on the precise form of the bands, and we have taken square bands here with $f(x)$ given by (21). The actinide series is described here as in the localized approach. Here we take E_{0d} = 0.35 Γ_d, U = 0.5 Γ_d, Γ_d = 10 Γ_f and we calculate the total number of d and f electrons N and the magnetic moment M, by (12,13), as a function of the position of the f-band E_{0f}. Each actinide is identified by its integer value of N. In figure 5, N and M have been plotted versus E_{0f}/U in three cases. The case (1) corresponds to a zero d-f hybridization $V_{df}(\mathbf{k}) \equiv 0$, all the actinides are magnetic and the M values correspond roughly to their ionic configurations. The case (2) correspond to approximation (a) with a constant hybridization $V_{df}(\mathbf{k}) \equiv 0.7\ U$ and the case (3) corresponds to approximation (b) with Γ_{df} = 0.7 U.

    As in subsection 1.2, $V_{df}(\mathbf{k})$ tends to inhibit magnetism, but here the transition between non-magnetic actinides is in better agreement with experiment, as shown by the large magnetic moments ($\sim 6\mu_B$) for Cm and Bk. Also, the physical case of actinides corresponds to a more realistic mean value of $V_{df}(\mathbf{k})$ ($\Gamma_{df} \sim \sqrt{\Gamma_d \Gamma_f}$) and the approximation (b) is clearly better than the approximation (a).

2. TRANSITION BASED ALLOYS WITH ACTINIDE IMPURITIES

2.1 EXPERIMENTAL SITUATION

    Np and Pu, which are non-magnetic as pure metals, become magnetic when they are diluted in many transition hosts such as La, Th, Y, Pd,... . A review of this experimental situation can be found in reference [6]. An illustrative example is the case of lanthanum with actinide impurities, which has been measured by Hill *et al.* [7] and is plotted in figure 6. This experiment is very interesting because it covers the whole first half-series up to americium. The depression of T_c is small for thorium, uranium and americium, indicating that they are not magnetic, whilst it is very large for neptunium and plutonium, showing a magnetic behaviour. If we assume that the depression of T_c is proportional to the square of the magnetization, we can deduce that the magnetic moment of *La*-Np is roughly 50% larger than that of *La*-Pu. These results are checked by the presence of a resistivity minimum in *La*-Pu alloys [8] and by the magnetic susceptibility measurements which give a 0.75 μ_B value for the magnetic moment of *La*-Pu [9] and 1.7 μ_B value for that of *Y*-Np [9] in a similar host.

2.2 CASE OF FERROMAGNETIC HOSTS:
ANALYSIS OF HYPERFINE FIELDS ON ACTINIDE IMPURITIES

    In this section, we would like to show that, from an analysis of the hyperfine fields of actinides diluted in iron or nickel,

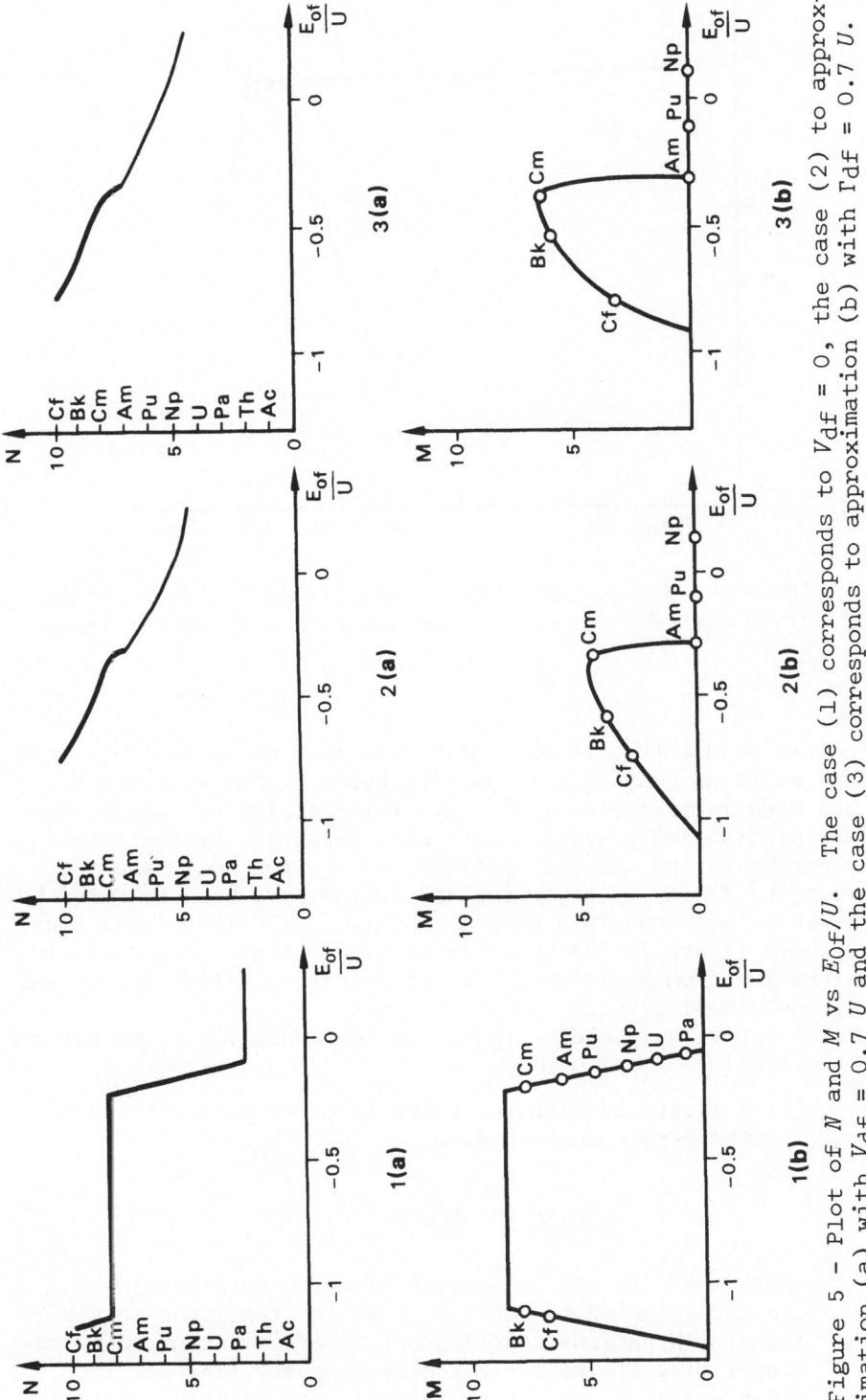

Figure 5 – Plot of N and M vs E_{0f}/U. The case (1) corresponds to $V_{df} = 0$, the case (2) to approximation (a) with $V_{df} = 0.7\ U$ and the case (3) corresponds to approximation (b) with $\Gamma_{df} = 0.7\ U$.

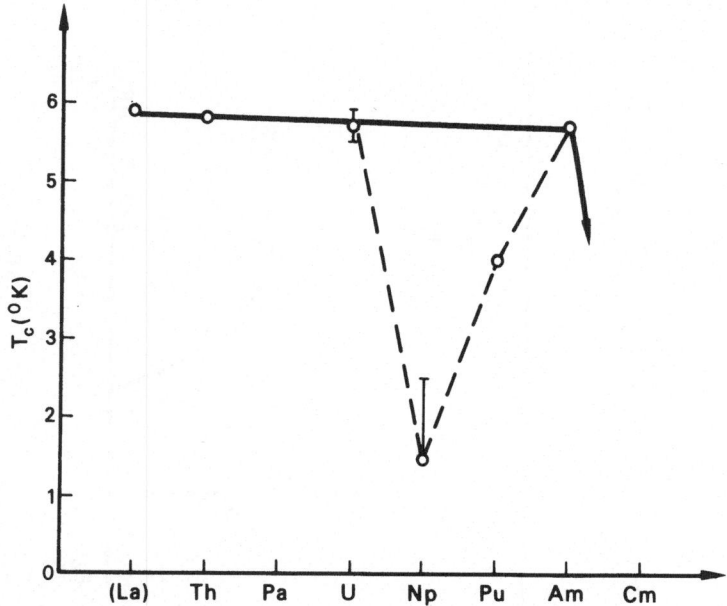

Figure 6 - Superconducting temperature of lanthanum-based alloys with 0.5 at. % actinide impurities (from reference [7]).

one can deduce that the magnetic situation of Np and Pu is the same as in non-magnetic transition hosts. This analysis has been made in reference [10]. Hyperfine fields of radium, thorium, uranium and plutonium have been measured, by the technique of perturbed angular correlations, in an iron host, and the hyperfine fields of neptunium and curium have been measured by the same technique in a nickel host [11,12]. Their values are given in figure 7: the hyperfine field is negative and slowly decreasing from Ra to U, whilst it becomes positive for Np and Pu, and almost zero for Cm.

The hyperfine field on the actinide impurity has several contributions:

(i) A first contribution comes from the polarization of the s-p conduction band:

$$h_{\text{cep}} = A(Z)m_{\text{s}}^{(0)}, \qquad (24)$$

where $m_{\text{s}}^{(0)}$ is the conduction electron polarization and can be estimated to be -0.2 $_B$ as in transition metals. The $A(Z)$ hyperfine coupling parameter is estimated by extrapolating its values given by Campbell [13] for the other elements, and by taking the same relative variation along the actinide series as in the rare earth series.

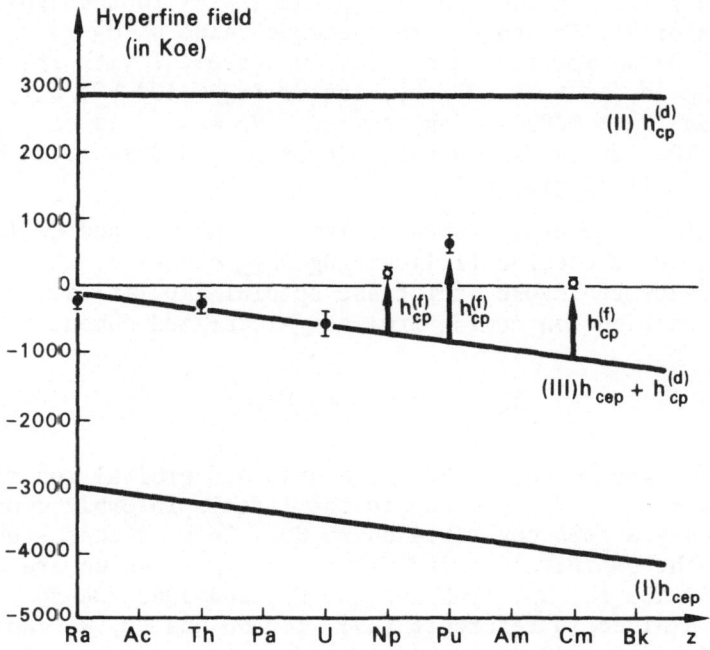

Figure 7 - Hyperfine field of actinides in ferromagnetic hosts (from references [12]) and its different contributions.

Therefore the first contribution is given (in kilo Oersteds) by curve (I) of figure 7:

$$h_{cep} = - 3000 - 120(Z - Z_0),\tag{25}$$

where Z_0 is the atomic number of radium.

(ii) A second contribution comes from the d-like electrons:

$$h_{cep}^{(d)} = - \alpha_d m_d^{(0)}.\tag{26}$$

For evaluating the d magnetic moment $m_d^{(0)}$ on the impurity site, we extend here the well known results of $3d$ impurities in ferromagnetic hosts [14]. If we do not take into account the contribution of the $5f$ electrons to the screening, the difference of charge ΔZ between iron and the actinide impurity varies from -5 to -2 from Ra to U, and then remains smaller than -2 for the following elements. Since $\Delta Z < -2$, $m_d^{(0)}$ remains negative and roughly constant of order $-1\mu_B$. Since α_d remains constant in each d series, $h_{cp}^{(d)}$ has to be constant. The constancy of $h_{cp}^{(d)}$ is checked by the experimental curve which is roughly parallel to the curve (I) up to Uranium, and verifies the assumption of only two contributions h_{cep} and $h_{cp}^{(d)}$ to the hyperfine field of Ra, Th and U. The deduced value of α_d is of order 3000 kOe/μ_B for the $6d$ series, in agreement with its value in the $3d$ ($\alpha_d \sim 50$ kOe/μ_B), the $4d$ ($\alpha_d \sim 400$ kOe/μ_B), and the $5d$ ($\alpha_d \sim 1000$ kOe/μ_B) series. So $h_{cp}^{(d)}$ is of order 3000 kOe for the whole actinide series, and is given by curve (II) of figure 7.

(iii) Since the experimental values for Np, Pu and Cm lie above the curve (III) giving $h_{cep} + h_{cp}^{(d)}$, it is necessary to invoke, for these special impurities, an extra contribution coming from an f-localized moment $m_f^{(0)}$:

$$h_{cp}^{(f)} = -\alpha_f m_f^{(0)}, \tag{27}$$

The f magnetic moment has both spin and orbital contributions in Np and Pu, owing to the large spin-orbit coupling, and only a spin contribution in Cm. This is consistent with the positive deduced values of $h_{cp}^{(f)}$ which are roughly 750 kOe for Np, 1500 kOe for Pu, and 1000 kOe for Cm. As explained in reference [10], for Np and Pu, we can take α_f of order 1500-2000 kOe/μ_B in Fe host, and three to five times smaller (as for rare earths) in Ni host. On the contrary, α_f has been taken much smaller in the case of Cm, which has only a spin contribution to the f magnetic moment.

The analysis of the hyperfine field data in a ferromagnetic host is, therefore, consistent with a spin and orbital f magnetic moment of order $2\mu_B$ for Np and $1\mu_B$ for Pu as is also observed in non-magnetic transition hosts.

2.3 THEORETICAL DISCUSSION

So, as described in subsections 2.1,2, one observes an increase of magnetism when one goes from pure actinide metals to actinides diluted in transition hosts. This physical situation can be theoretically accounted for by considering approaches similar to those described in subsections 1.2,3, and by just reducing the d-f hybridization parameter from the situation of pure actinides to the situation of actinide impurities. But in order to explain why Am remains non-magnetic either as pure metal or as impurity in La, it is necessary to include the spin-orbit coupling for the $5f$ electrons. The spin-orbit coupling split the $5f$ level into two $j = 5/2$ and $j = 7/2$ levels, and, in this approach, americium is non-magnetic because it corresponds to an almost completely filled up $j = 5/2$ level.

A theoretical model taking into account a $6d$ and two $5f$ virtual bound states split by a large spin-orbit coupling has been developed in reference [6]. With a reasonable set of parameters, such as that used in subsections 1.2,3, and by including a spin-orbit splitting of order 1.5 eV for the $5f$ electrons, this model can account for the small magnetic moments of order 1 to 2 μ_B

observed for Np and Pu impurities, just by reducing the d-f hybridization parameters from a situation corresponding to pure metals where Np and Pu are not magnetic.

The starting assumptions of the model are:

(i) We consider $6d$ and $5f$ levels on each independent impurity;

(ii) Only the $5f$ level is split by the spin-orbit coupling into two levels f_1 ($j = 5/2+$ and f_2 ($j = 7/2$) separated by the spin-orbit splitting λ_{so}. Therefore, we start from three localized levels: $6d$, $5f_1$ ($j = 5/2$), $5f_2$ ($j = 7/2$) at energies E_{0d}, E_{0f}, E_{0f} with $E_{0f_2} - E_{0f_1} = \lambda_{so}$;

(iii) We neglect the orbital degeneracy of each level, and in the case of the two f levels, we treat them as two-fold degenerate levels like those of a spin $\frac{1}{2}$ with usual spin components $\sigma = \uparrow$ and $\sigma = \downarrow$;

(iv) All the levels are treated within the resonant scattering mechanism, giving three virtual bound states with half widths Γ_d, Γ_{f_1}, Γ_{f_2};

(v) We introduce the d-f hybridization by a phenomenological one-body Hamiltonian:

$$\sum_\sigma (V_{df_1}c_{d\sigma}{}^+c_{f_1\sigma} + V_{df_2}c_{d\sigma}{}^+c_{f_2\sigma}) + \text{h.c.,} \qquad (27)$$

where the parameters V_{df_1} and V_{df_2} are assumed to be constant. Let us describe here only a first approach, corresponding to the case of an infinite spin-orbit coupling which describes relatively well the first half-series, and gives the same qualitative results as those obtained in the general case described in reference [6].

Within the approximation of an infinite spin-orbit splitting, the study of the first half-series is described by the simple Hamiltonian which describes only the f_1 level interacting with the d level:

$$H = \sum_{k,\sigma} \varepsilon_k c_{k\sigma}{}^+c_{k\sigma} + E_{0d} \sum_\sigma c_{d\sigma}{}^+c_{d\sigma} + E_{0f_1} \sum_\sigma c_{f_1\sigma}{}^+c_{f_2\sigma} \qquad (29)$$

$$+ \sum_{k,\sigma} (V_{kd}c_{k\sigma}{}^+c_{d\sigma} + \text{h.c.}) + \sum_{k,\sigma} (V_{kf_1}c_{k\sigma}{}^+c_{f_1\sigma} + \text{h.c.})$$

$$+ U(n_{d\uparrow} + n_{f_1\uparrow})(n_{d\downarrow} + n_{f_1\downarrow}) + \sum_\sigma (V_{df_1}c_{d\sigma}{}^+c_{f_1\sigma} + \text{h.c.}).$$

This Hamiltonian is similar to Hamiltonian (1) and can be solved in the same way.

In order to describe the first half-series, we have chosen the different parameters as follows:

(i) E_{0d} is taken equal to Γ_d, in order to fit approximately the total number of d electrons on the impurity site;

(ii) Γ_d and Γ_{f_1} are chosen, as in the case of pure actinide metals, by comparison with band calculations; here we have taken $\Gamma_d = 20\Gamma_{f_1} \sim 2$ eV;

(iii) U is chosen equal to Γ_d, $U = \Gamma_d \sim 2$ eV.

With that choice of parameters, we have calculated the total number of d and f_1 electrons:

$$N = 5(n_{d\uparrow} + n_{d\downarrow}) + 3(n_{f_1\uparrow} + n_{f_1\downarrow}). \tag{30}$$

and we have plotted it versus E_{0f_1}/U for four values of the V_{df_1} parameter in figure 8. In this figure we have also plotted the magnetic moment:

$$M = 5(n_{d\uparrow} - n_{d\downarrow}) + 3(n_{f_1\uparrow} - n_{f_1\downarrow}). \tag{31}$$

We see that, for a large d-f hybridization, magnetism disappears completely in the whole first half-series (case (d)), and, if this hybridization is reduced by 30%, we obtain (case (c)) a situation where Np and Pu become magnetic (with a larger magnetic moment for Np) and where U is very close to being magnetic. If we reduce more and more this hybridization, U can become magnetic (and even Pa in the case of a zero d-f hybridization).

The case (d) of a large d-f hybridization can account for the situation of pure actinide metals which are non-magnetic up to Am, as has already been described in subsections 1.2,3 without spin-orbit coupling. But the interesting result is that, if we reduce the d-f hybridization, Np and Pu become magnetic, as is experimentally observed in transition based alloys with actinide impurities.

So, including spin-orbit coupling into the localized picture of subsection 1.2 gives a good description of both pure actinide metals and actinide impurities in transition hosts. We could have introduced the spin-orbit coupling into the band approach of subsection 1.3, but the mathematical treatment would be more difficult, while certainly the results would be physically the same as those obtained here.

It remains to explain why the d-f hybridization is smaller when actinides are diluted in transition hosts like lanthanum. Since the d-f hybridization comes in both cases from the overlap between two neighbouring $6d$ and $5f$ wavefunctions, it is well described in the band approach of subsection 1.3, and it is given at the Fermi level by the formula:

$$V_{df} = \Gamma_{df}\beta(\mathbf{k}_F). \tag{32}$$

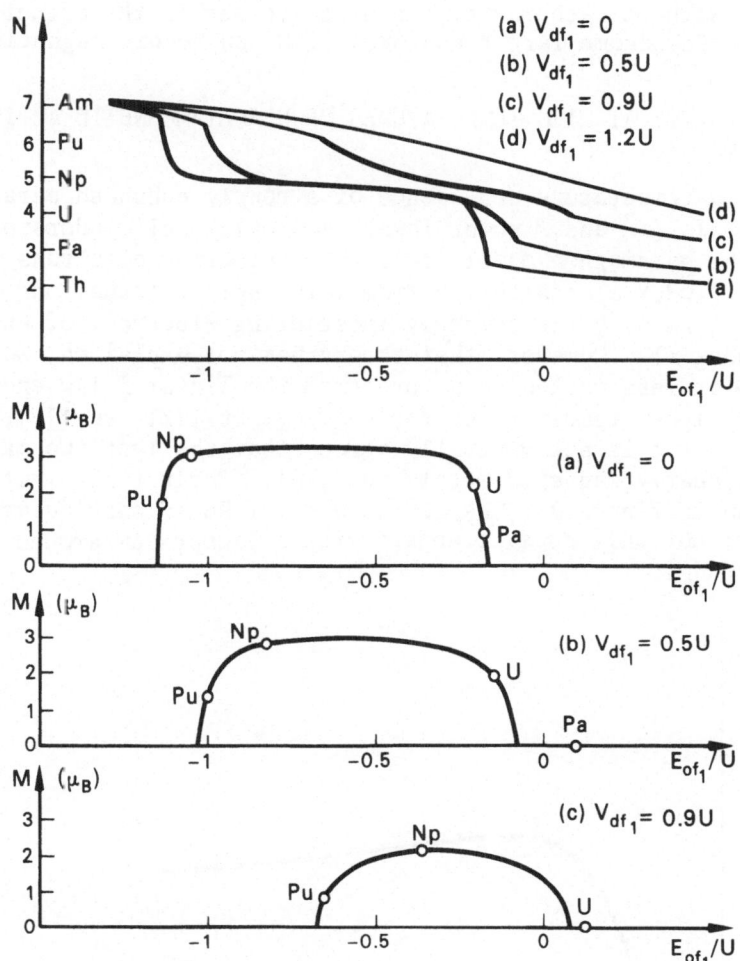

Figure 8 - Plot of N and M vs E_{0f_1}/U in the case of an infinite spin-orbit splitting, for $E_{0d} = \Gamma_d = 20\Gamma_{f_1} = U$ and for four values of V_{df_1}: $V_{df_1} = 0$ (case (a)); $V_{df_1} = 0.5\ U$ (case (b)); $V_{df_1} = 0.9\ U$ (case (c)); $V_{df_1} = 1.2\ U$ (case (d)). In each case the thick part of the curve giving N corresponds to magnetic solutions.

The formula (32) can connect the importance of the hybridization to the d-character of the matrix: Γ_{df} increases when the d-electrons are less localized and $\beta(k_F)$ increases when the number of d-electrons increases. For example, the number of d-electrons is one in lanthanum, whilst it is two or more in actinides, so that one can obtain a smaller V_{df} in lanthanum than in pure metals. This kind of analysis permits also a qualitative comparison between the magnetic properties of Np and Pu in different hosts, as has already been done in reference [5]. Finally, we suggest studying magnetic properties of actinide impurities

in hosts without d-character, in order to see if the magnetic
of Np and Pu become larger and even if U can become magnetic.

3. RESISTIVITY AT HIGH TEMPERATURES OF NEARLY MAGNETIC ACTINIDE SYSTEMS

The low temperature dependence of strongly enhanced paramagnetic metals (T^2 and T power laws) is usually well understood
within the paramagnon model [15]: the conduction electrons of
one band (index c) scattering from large spin fluctuations (paramagnons) formed by the strongly interacting electrons of another
band (index i). However, that theory has not explained, so far,
the more or less marked departure from the linear T law encountered at higher temperatures for Pd [16], Pt [17], Pu [18], Np
[16],... . It is now generally agreed that the last two materials are nearly magnetic metals too [19]. Their resistivities
are given in figure 9. The resistivity of Pu is particularly
striking: not only does it depart from a linear law around 50°K

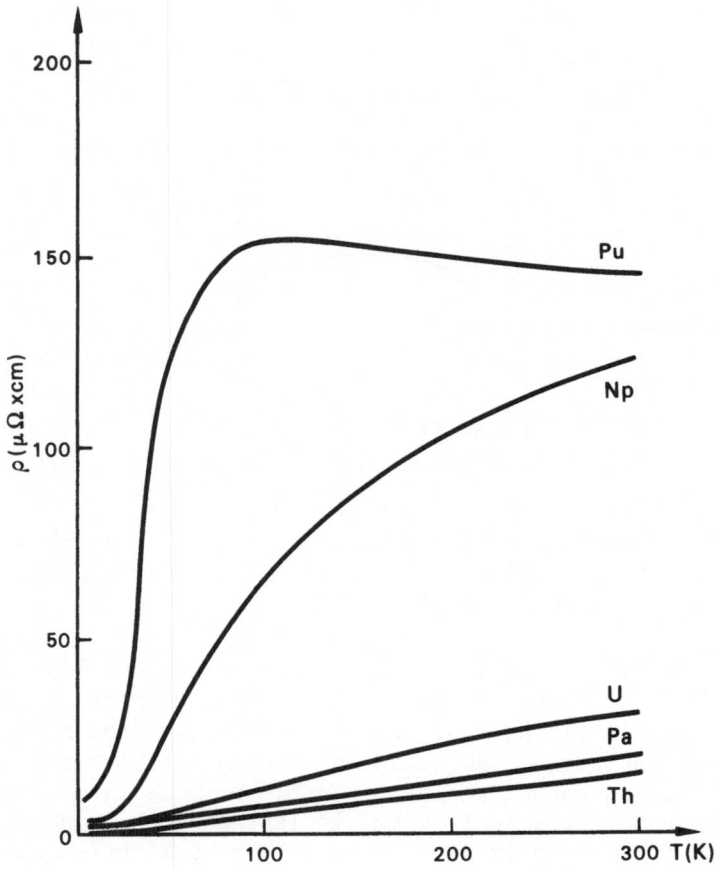

Figure 9 - Resistivities of actinides vs temperature.

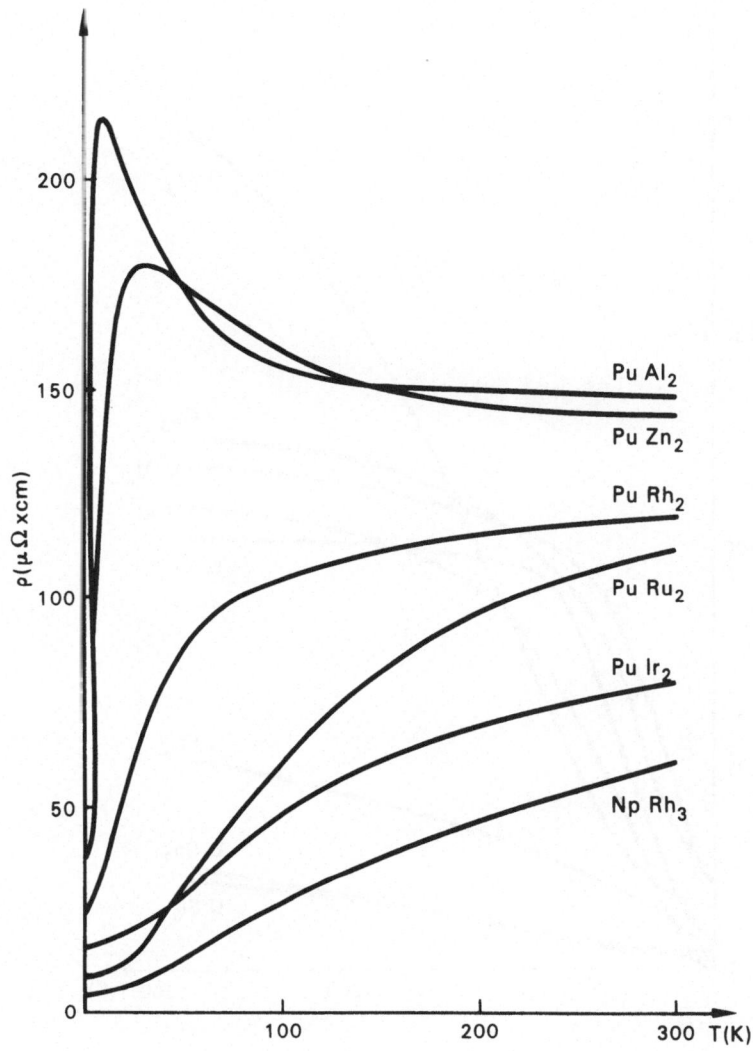

Figure 10 - Resitivities of PuAl$_2$, PuZn$_2$, PuRh$_2$, PuRU$_2$, PuIr$_2$, NpRh$_2$ vs temperature.

but it reaches a maximum of the order of 160 μΩ cm at about 100°K, and then decreases weakly. Such behaviour has been ob- served for all stabilized phases of Pu [18], and therefore is an intrinsic property which was not clearly understood so far. The resistivities of nearly magnetic actinide compounds such as some uranium compounds [20] or plutonium and neptunium compounds [21] are shown in figures 10,11, and present the same kinds of high temperature behaviours.

We have proposed [22] a crude but very simple way of account- ing for the general features of these resistivities as well at

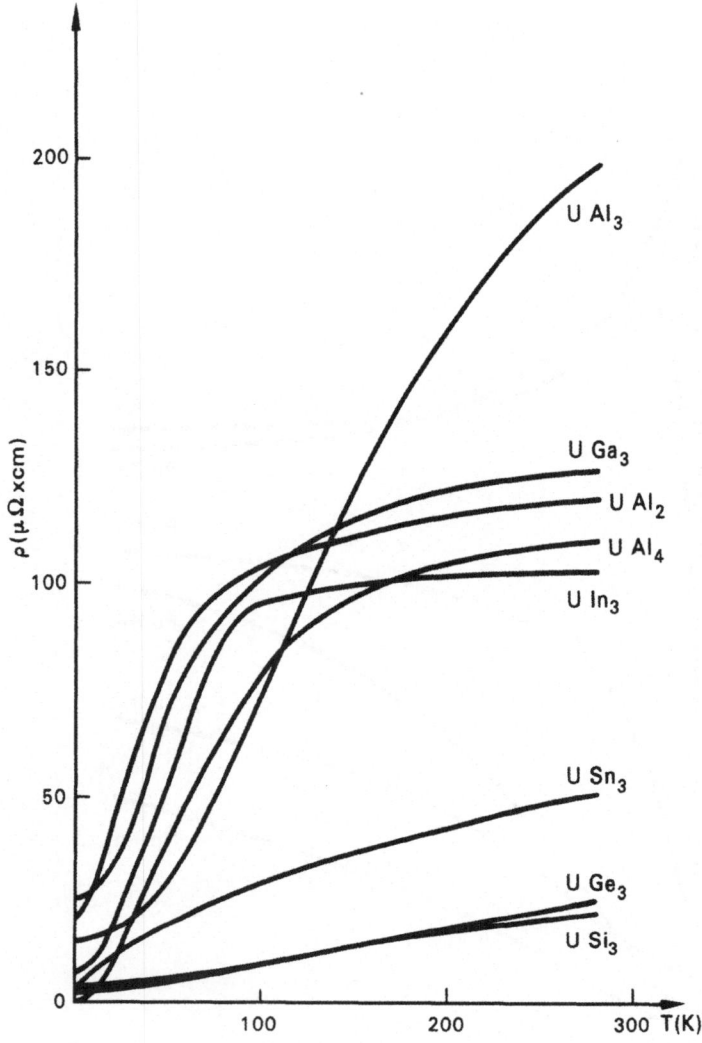

Figure 11 - Resistivities of UAl_2, UAl_3, UAl_4, UGa_3, UIn_3, USn_3, UGe_3, USi_3 vs temperature.

high temperatures as at low temperatures; more particularly, we can explain the observed maxima in Pu, $PuAl_2$, $PuZn_2$ and compute quite reasonable fits for the resistivities of Np and Pu. We show that, if it were not for phonon contributions, the electron-paramagnon resistivity would be expected to saturate at high temperatures at a value independent of the Stoner enhancement, and directly comparable with the high temperature spin-disorder scattering [23] of magnetic metals, with a similar T^{-1} correction for decreasing T. We only sketch our calculation here and present the fits with experiments.

We follow th Kaiser-Doniach [15] notation for the formal expression of the electron-paramagnon resistivity:

$$\rho = \frac{\rho_0}{T} \int_0^{2k_{F_c}} \frac{q^3}{k_{F_c}^4} \, dq \int_0^\infty 2 \text{Im} \chi(q,\omega,T) \, \frac{\omega d\omega}{(e^{\omega/T} - 1)(1 - e^{-\omega/T})} \, , \qquad (33)$$

with

$$\rho_0 = [\tfrac{1}{4} J N(E_{F_c})]^2 \times \frac{m_c}{n_c e^2 \tau_{F_c}} \times \frac{\nu}{n_c} \, . \qquad (34)$$

J is the coupling constant per unit cell between electrons of the two bands, ν is the number of atoms per unit volume, and $\hbar \tau_{F_c}^{-1} = \hbar^2 k_F^2 / 2m_c$. k_F, $N(E_F) = m k_F / \hbar^2 \pi^2 \nu$, m and n (with indices c or i) are respectively the Fermi momentum (defined at $T = 0$), the density of states per atom at the Fermi level, the effective mass, and the number of electrons per units volume. $\chi(q,\omega,T)$ is the paramagnon propagator with momentum q and energy ω, whose usual expression is given, in the random phase approximation (RPA), by

$$\chi(q,\omega,T) = \frac{\chi^0(q,\omega,T)}{1 - I \chi^0(q,\omega,T)} \, , \qquad (35)$$

$\chi^0(q,\omega,T)$ being the dynamical susceptibility of the i electrons in absence of the interaction I; $[1 - I\chi^0(0,0,0)]^{-1} \equiv (1 - I)^{-1} \equiv S$ is the Stoner enhancement factor:

$$\chi^0(q,\omega,T) = \sum_{\vec{k}} \frac{f(E_k,T) - f(E_{k+q},T)}{E_{\vec{k}+\vec{q}} - E_{\vec{k}} - \omega} \, , \qquad (36)$$

$f(E_{\vec{k}},T)$ being the Fermi distribution, depending both on T directly and through the chemical potential.

The new step introduced here is that we take into account, over the whole temperature range, the T-dependence of $f(E_k,T)$ in χ^0, whilst usual theories [15], confined to low temperatures, did not need to do so. Thus we use the fact that the static Stoner susceptibility decreases at high temperature until it reaches a Curie law behaviour. We point out that this Curie law behaviour is directly responsible for the saturation of ρ which we find at high T, where

$$\rho_{T \to \infty} \simeq \rho_0 \int_0^{2k_{F_c}} \frac{q^3}{k_{F_c}^4} \, dq \left\{ T \int_0^\infty 2 \, \frac{\text{Im}\chi(q,\omega,T)}{\omega} \, d\omega \right.$$

$$\left. - \frac{1}{12T} \int_0^\infty 2\omega \text{Im}\chi(q,\omega,T) d\omega \right\} \, . \qquad (37)$$

The two terms in the brackets are directly given by the Kramers-Kronig relations and the f sum rule; hence:

$$\rho_{T \to \infty} \simeq \rho_\infty \left\{ 1 + \frac{2}{3} \frac{T_{Fi}}{T} (\bar{I} - \frac{4}{3} \xi^2) \right\} , \tag{38}$$

with

$$\rho_\infty = \frac{2\pi\rho_0 n_i}{\nu} = \frac{\pi}{8} [JN(E_{F_C})]^2 \frac{m_c}{n_c e^2 \tau_{F_C}} \frac{n_i}{n_c} , \tag{39}$$

$$\xi = \frac{k_{F_C}}{k_{F_i}} , \tag{40}$$

where T_{Fi} is the Fermi temperature in the i band at $0°K$. At low temperatures, we obtain again the usual T^2 and T power laws [15], slightly modified as a result of the full q-, ω- and T-dependence of χ^0. In order to handle such complicated formulae, we have performed a numerical computation (on and IBM 370) to get the full T-depdendence of ρ.

Note that formula (38) is identical with the spin-disorder resistivity of ferromagnetic metals at high T. This is a reasonable result: at very high T, the product $I\chi^0$ in formula (35) is negligible compared with unity; therefore the i electrons can be regarded as not interacting anymore amongst themselves, and scattering independently the c electrons. They act then like the independent paramagnetic ions in the spin-disorder problem. at lower T, the approximate formula (38) indicates that, before reaching the low temperature T and T^2 power laws, one either gets a maximum or does not, depending upon the sign of $I - 4\xi^2/3$. In the spin-disorder problem, too, one has deviations from the saturation in T^{-1}.

At this stage, we must comment upon the roughness of our model:

(a) We have supposed I to be temperature independent, which is actually no true. I is probably weaker at higher T. However, since I appears as a phenomenological interaction in the paramagnon theory, it is difficult to know its exact temperature dependence. A weaker I would modify the T^{-1} quantitative contribution. However, the saturation ρ_∞ independent of I must be valid, since at very high T there is no longer any interaction between the i electrons which act as a classical gas, and it is thus reasonable to recover the spin-disorder high T resistivity;

(b) We have implicitly assumed the same form of the free energy at high and low T, leading to the same diagrammatic expansion of the RPA series (35) with (36). It is doubtful that particle-particle interactions negligible at low T are also negligible at high T. Furthermore, the temperature dependence of the chemical potential of the i band obtained here in the usual way is not obviously the correct one: in particular it is not obvious that is should remain insensitive to I.

(c) We have neglected the temperature dependence of the conduction electron Fermi energy, which is assumed to be much larger than T_{F_i} and all usual temperatures. Also we have neglected band structure effects and calculated χ^0 in the usual way [15] for a parabolic band model.

However, this simple minded method accounts satisfactorily for the observed resistivity data. We present in figure 12 the results of the numerical computation of ρ/ρ_∞ versus T/T_{F_i} for different values of S and $\xi = k_{F_c}/k_{F_i}$. These parameters have opposite rôles: the curves exhibit a more pronounced maximum for larger S or smaller ξ; for $\xi = 1$ and $\xi = 2$ there is no maximum for any value of S. So, we explain all the experimental

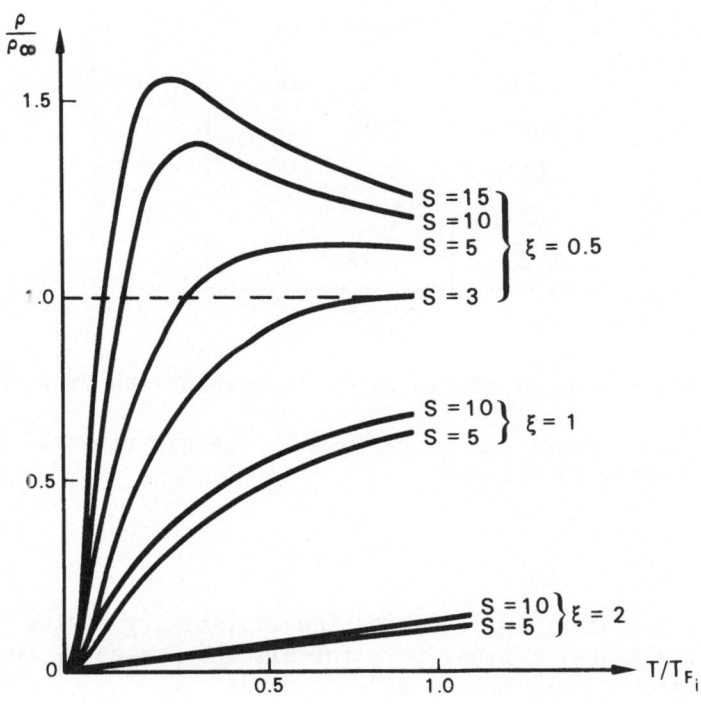

Figure 12 - Computed variation of ρ/ρ_∞ vs T/T_{F_i} for different values of S and $\xi = k_{F_c}/k_{F_i}$.

B. COQBLIN AND R. JULLIEN

TABLE I

	A_{exp} ($\mu\Omega \times cm/K^2$)	T_d (K)
Pt	1.3×10^{-5}	400
Pd	3×10^{-5}	300
Np	2×10^{-3}	80
$NpRh_3$	5×10^{-3}	90
UAl_3	5.3×10^{-3}	150
$PuIr_2$		75
$PuRu_2$		75
Pu	$\sim \times 10^{-2}$	40
UGa_3	1.75×10^{-2}	60
UAl_4	2×10^{-2}	60
UIn_3	2.5×10^{-2}	70
USn_3	2.5×10^{-2}	40
$PuRh_2$	0.127	30
UAl_2	~ 0.2	30
$PuAl_2$	1	10
$PuZn_2$	1.75	10

resistivity curves of nearly magnetic metals with or without a maximum.

The table I gives the value A_{exp} of the experimental T^2 term of the resistivity defined by

$$\rho = A_{exp}T^2, \qquad (41)$$

and the temperature T_d, at which the resistivity begins to depart from a linear T behaviour, for all the presently known nearly magnetic pure metals and compounds.

To apply our model to these systems, the interacting band i is the d band for Pd and Pt, whilst it is the f band for actinide systems. In the case of actinide systems, we consider that the i band is the very narrow $5f$ band and the c band is formed by the $7s$ and $6d$ bands, and we neglect here the d-f hybridization.

First, two interesting comments can be drawn from table I:

(a) There is a connection between A_{exp} and T_d; the value
    of T_d decreases when the A_{exp} coefficient increases,
or when the material is closer to becoming magnetic. This
results agrees with the model for which the saturation at
high temperatures is clearly linked to the importance of
the paramagnons yielding the T^2 law;

(b) The temperature T_d is larger by an order of magnitude
    for Pd and Pt than for the actinides, whilst the A_{exp}
coefficient of Pd is smaller by at least two orders of magnitude than the A_{exp} coefficients of actinides. Since the
Fermi energy T_{F_i} is larger by an order of magnitude for d

bands than for $5f$ bands, this result agrees with the theory
where T_d varies as T_{F_i} and A_{exp} as $(1/T_{F_i})^2$. It results

also that the resistivity is smaller for d metals than for
f metals.

Next, we present the fits of the resistivities for neptunium
and plutonium in figures 13,14. We have chosen $S = 10$ in the

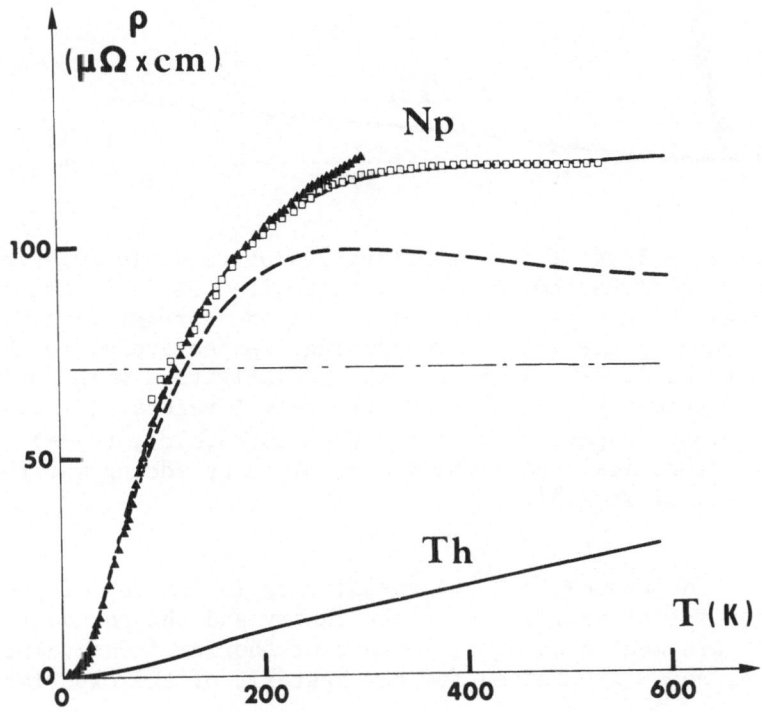

Figure 13 - Fit of the resistivity of Np; (▲,□) experimental data of Np according to Meaden [18]; (----) theoretical resistivity with $S = 10$, $\xi = 0.5$, $\rho_\infty = 71.5$ µΩcm,
$T_{f_i} = 750°$K; (——) total theoretical resistivity of Th
[18].

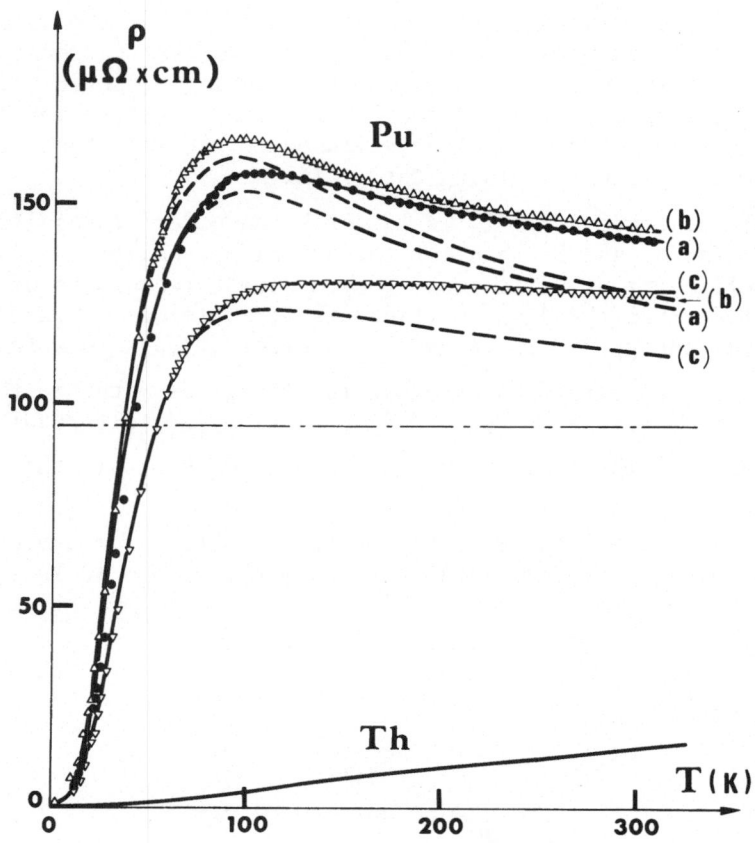

Figure 14 - Fits of the resistivities of randomly oriented
(a) sample of Pu and of monocrystals with (010) direction
parallel (b) and perpendicular (c) to the current ($\bullet$,$\triangle$,$\triangledown$)
experimental data [24] corresponding, respectively, to the
cases (a,b,c); (----) theoretical resistivities with $S = 10$,
$\rho_\infty = 95$ $\mu\Omega$cm, $T_{F_i} = 280°$K and different ξ values: $\xi = 0.4$,
0.37, 0.53, respectively, for the cases (a,b,c); (———)
total theoretical resistivities obtained by adding the resistivity of Th [18].

two cases. We assume that the resistivity is the sum of the
paramagnon resistivity given by the theory and the phonon resistivity assumed to be equal to that of thorium. The values
of T_{F_i}, ρ_∞ and ξ are chosen for the best fit of the experimental
curves.

Figure 13 shows the experimental resistivities of neptunium
and thorium up to 550°K, as well as the theoretical plot for Np.
The resistivity of Np is almost constant from 300 to 500°K, so
that the paramagnon resistivity has a maximum at roughly 300°K.
The parameter ξ is chosen here equal to 0.5, in order to have

such a maximum. We have taken here $S = 10$, $T_{F_i} = 750°K$, $\xi = 0.5$ and $\rho_\infty = 71.5$ $\mu\Omega cm$.

Figure 14 shows the experimental resistivities of plutonium either (a) with a maximum for a polycrystal, or (b) with a more pronounced maximum for a monocrystal with the current along the (010) axis, or (c) with a plateau from 120°K to 300 °K for a monocrystal along the (100) axis [24]. In the cases (a,b), the total resistivities exhibit a maximum, so that the ξ parameter has to be small; in the case (c), ξ is a little larger to give a maximum only for the paramagnon resistivity. So, the theoretical fits have been obtained for $S = 10$, $T_{F_i} = 280°K$, $\rho_\infty = 95$ $\mu\Omega cm$, and three different ξ values $\xi = 0.4$ for the case (a), $\xi = 0.37$ for the case (b), and $\xi = 0.53$ for the case (c). The choice of different ξ values for fitting the highly anisotropic resistivity of α-Pu single crystals can be qualitatively understood, because the Fermi surface of plutonium is certainly complicated and anisotropic; however, we cannot compute this effect quantitatively.

The ρ_∞ limits have practically the same value of Np and Pu, which is consistent with similar conduction bands and number of $5f$ electrons; but in order to obtain a pronounced maximum for Pu, we must choose different T_{F_i} and ξ values without any clear justification in the parabolic band model.

Thus, very good fits have been obtained for Np and Pu. According to figure 12, it would be possible to fit the data on $PuAl_2$ and $PuZn_2$ by, for example, taking roughly the same ρ_∞ and T_{F_i} as for Pu, but much smaller ξ and larger S than in Pu, in order to obtain their more pronounced maxima. The other curves in figures 10,11 can be fitted with reasonable parameters also.

Finally, the high-temperature decrease of the resistivity of plutonium is less pronounced when pressure is applied up to 13 kbar [25]. Figure 12 indicates that this variation can be described by a decrease of S for increasing pressure, as a result of the weaker f character in actinides, as well as in rare earths under pressure. It is worth while noting that the change in the density of states amplified by the effect of S in a nearly magnetic metal can be experimentally observed, whilst the simple change in the density of states in a normal metal would have been too small to be seen: this fact is a strong argument for the present paramagnon model.

However, the value of T_{F_i} used for plutonium is very small — of the order of room temperature — and, furthermore, the temperature variation of the Stoner susceptibility for the i band must be very large to account for the observed behaviour of ρ. The magnetic susceptibilities of $NpRh_3$, UAl_4, USn_3, $PuRh_2$, UAl_2, $PuZn_2$ have a Curie-Weiss temperature dependence at high temperatures, and deviate from this at low temperatures, as well as for Pd. These results on the magnetic susceptibility are in good agreement with the theoretical model and consistent with the results on the resistivity. But on the contrary, the total observed magnetic susceptibility of Np, Pu, UAl$_3$, PuIr$_2$, PuRu$_2$, UIn$_3$ is almost temperature independent up to room temperature, although the theoretical Stoner susceptibility of the i band has to decrease rapidly with temperature to account for the observed behaviour of ρ. So, there are spin-fluctuation systems which present a temperature-independent susceptibility. This point is not clearly understood here. But our present model for actinides is greatly over-simplified, because the bands are taken to be parabolic, the $5f$ band is assumed to be extremely narrow, and the d-f hybridization is neglected. Moreover, the total observed susceptibility contains a large $6d$ contribution in addition to the temperature dependent $5f$ contribution. For all these reasons, the relation between the susceptibility of our i band and the total observed susceptibility is not obvious.

Another explanation [21,26] has been proposed to explain the different behaviours of the magnetic susceptibilities: the spin-fluctuation systems which experimentally exhibit a temperature dependent susceptibility are of the nearly ferromagnetic type, like those studied here, while those which exhibit a temperature independent susceptibility would be of the nearly antiferromagnetic type.

To conclude, we suggest some experiments as a further check of our model:

(a) The resistivity of Pd and dilute Pd-Ni alloys has been recently measured under pressure up to 4.5 kbar, but only at temperatures below 12°K [27]; such measurements should be extended to higher temperatures and pressures for Pd and Pd alloys;

(b) High temperature measurements of the resistivities would be very interesting also in exchange enhanced compounds like Ni$_3$Ga and Ni$_3$Al [28];

(c) The actinide metals and compounds are very promising as spin-fluctuation systems. The resistivity and the magnetic susceptibility under pressure of the actinide metals and compounds listed in table I would be obviously very interesting. The resistivity of americium is worth measuring, since the magnetic susceptibility of americium is larger than that of neptunium and plutonium.

REFERENCES

1. Friedel, J. (1958). (Report No.766, C.E.A.), (unpublished).
2. Jullien, R., Galleani d'Agliano, E. and Coqblin, B. (1972). *Phys. Rev.*, **B6**, 2139.
3. Friedel, J. (1958). *Nuovo Cim.*, **52**, 287; Anderson, P.W. (1961). *Phys. Rev.*, **124**, 41.
4. Kmetko, E.A. and Hill, H.H. (1970). *Proceedings of the Metallurgical Society*, (A.I.M.E., Santa Fe, New Mexico). p. 58; Koelling, D.D., Freeman, A.J. and Arbman, G.O. (1970).

Proceedings of the Metallurgical Society, (A.I.M.E., Santa Fe, New Mexico), p. 194.

5. Jullien, R. and Coqblin, B. (1973). *Phys. Rev.*, **B8**, 5263.

6. Jullien, R., Galleani d'Agliano, E. and Coqblin, B. (1973). *J. Low Temp. Phys.*, **10**, 685.

7. Hill, H.H., Lindsay, J.D.G., White, R.W., Asprey, L.B., Strueling, V.O. and Mathias, B.T. (1971). *Physica*, **55**, 615.

8. Hill, H.H., Elliot, R.O. and Miner, W.N. (1969). *Colloque du C.N.R.S. No.180, Les Elements de terres rares, Paris-Grenoble, 1969*, (C.N.R.S., Paris-Grenoble), p. 541.

9. Gatesoupe, J.P. and de Novion, C.H. (1971). *Proceedings of the Durham Conference on Rare Earths and Actinides*, (Institute of Physics, London), p. 84.

10. Jullien, R., Gomes, A.A. and Coqblin, B. (1972). *Phys. Rev. Lett.*, **29**, 482.

11. Koster, T.A. and Shirley, D.A. (1971). In *Hyperfine Interactions in Excited Nuclei*, (eds. Goldring, R. and Kalish, R.), (Gordon and Breach, New York).

12. Ansaldo, E.J. and Grodzins, L. (1969). *Phys. Lett.*, **B30**, 538; see also references cited in reference [11].

13. Campbell, I.A. (1969). *J. Phys.*, **C2**, 1339.

14. Campbell, I.A. and Gomes, A.A. (1967). *Proc. Phys. Soc.*, **91**, 319; (1968). *Solid State Commun.*, **6**, 395.

15. Mills, D.L. and Lederer, P. (1966). *J. Phys. Chem. Solids*, **27**, 1805; Schindler, A.J. and Rice, M.J. (1967). *Phys. Rev.*, **164**, 759; Kaiser, A.B. and Doniach, S. (1970). *Int. J. Magn.*, **1**, 11.

16. White, G.K. and Woods, S.B. (1959). *Phil. Trans. Roy. Soc.*, **A251**, 273; Schindler, A.I., Smith, R.J. and Salkovitz, E.I. (1956). *J. Phys. Chem. Solids*, **1**, 39; (1957). (Report No. 4974, Naval Research Laboratory), (unpublished).

17. See, for example: (1966). *E.T.J. Bulletin*, No.1181, (Rosemont Engineering Company, Minneapolis, Minnesota), (unpublished).

18. See, for example: Meaden, G.T. (1966). *Electrical Resistance of Metals*, (Heywood Books, London), and references cited therein.

19. Arko, A.J., Brodsky, M.B. and Nellis, W.J. (1972). *Phys. Rev.*, **B5**, 4564; Doniach, S. (1972). In *A.I.P. Conference Proceedings No.5: Magnetism and Magnetic Materials, 1971*, (eds. Graham, C.D., Jr. and Rhyne, J.J.), (American Institute of Physics, New York).

20. Buschow, K.H.J. and Van Daal, H.J. (1973). *A.I.P. Conference Proceedings No.10: Magnetism and Magnetic Materials*, (ed. Douglass, D.H.), (American Institute of Physics, New York), p. 1464.

21. Arko, A.J., Brodsky, M.B. and Nellis, W.J. (1972). *Phys. Rev.*, **B5**, 4564; (1973). *Phys. Rev.*, **B7**, 4137; (to be published).

22. Jullien, R., Béal-Monod, M.T. and Coqblin, B. (1974). *Phys. Rev. Lett.*, **30**, 1057; *Phys. Rev.*, **B9**, 1441.

23. de Gennes, P.G. and Friedel, J. (1958). *J. Phys. Chem. Solids*, **4**, 71.

24. Olsen, C.E. and Elliot , R.O. (1965). *Phys. Rev.*, **A139**, 437.

25. Mortimer, M.J. (1972). (Report No. 7030, A.E.R.E.), (unpublished).

26. Friedel, J. (Private Communication).

27. Beyerlein, R.A. and Lazarus, D. (1973). *Phys. Rev.*, **B7**, 511.

28. Fluitman, J.H.J., Boem, R., de Chatel, P.F., Schinkel, C.J., Tilanus, J.L.L. and de Vries, B.R. (1973). *J. Phys.*, **F3**, 109.

THERMODYNAMIC PROPERTIES OF
THE ONE-DIMENSIONAL HUBBARD MODEL

U. BRANDT

Institute of Physics, University of Dortmund,
46 Dortmund, West Germany

In the last few years a lot of theoretical work has been done in order to illuminate the properties of the Hubbard Model. This model is one of the simplest which might be able to show the physically important mechanisms in the behaviour of highly correlated electron systems. The motivation for discussion of this model in one dimension is two-fold: first, there exists a series of pseudo-one-dimensional substances, which are suspected of being described, at least quantitatively, by this model (e.g. certain TCNQ-compounds). On the other hand, one may be able to apply well established approximation schemes to the one-dimensional case, and compare them with (approximate or exact) results, which in one dimension can be obtained in other ways.

We discuss in the following the properties of a Hubbard chain, containing N sites, and if necessary we go to the thermodynamic limit. The Hubbard Hamiltonian is given by

$$\mathcal{H} = \mathcal{T} + U\mathcal{H}_0,$$

where $\mathcal{T}$ and $\mathcal{H}_0$ are given by

$$\mathcal{T} = -t \sum_{\substack{i=1 \\ \sigma}}^{N-1} c_{i,\sigma}^+ c_{i+1,\sigma} + c_{i+1,\sigma}^+ c_{i,\sigma},$$

$$\mathcal{H}_0 = \sum_i n_{i\uparrow} n_{i\downarrow};$$

(1)

351

here $c_{i,\sigma}$ (resp. $c_{i,\sigma}{}^{+}$) are the fermion operators of electrons of spin σ in the Wannier state at site i, and $n_{i,\sigma}$ are the corresponding occupation number operators $c_{i,\sigma}{}^{+}c_{i,\sigma}$. We obtain our results by a perturbative method. We regard $\mathcal{J}$ as the perturbation, and within quantum mechanical perturbation theory we obtain the many particle energies. In first order perturbation theory the calculation can be performed exactly, and we can sum the partition function even in the presence of an external magnetic field. The second order perturbation theory can be obtained only approximately, and we can give the thermodynamic only in the case where the number of electrons (in this band) is equal to the number of sites.

The eigenstates of $\mathcal{H}_0$ are trivially found, the eigenvalues are $n = 0,1,2,..$ (the number of doubly occupied sites), and the first order correction is obtained by diagonalizing the operator $P_n \mathcal{J} P_n$ within the eigenspace of $\mathcal{H}_0$ for the eigenvalue n. P_n is the projection operator on this eigenspace. This problem is the same as has been discussed recently by Klein [1], and we shall use a characterization of the states similar to that used in that paper. Although in this reference the eigenvalues obtained are correct (in the case of vanishing magnetic field), the results for the thermodynamic properties are spurious because of the use of obscure mathematics.

We classify our eigenstates by three vectors $\{\lambda_i\}$, $\{\mu_\ell\}$, $\{\sigma_m\}$. $\lambda_i = 0$ if the i-th site is singly occupied, and $\lambda_i = 1$ if the i-th site is empty or doubly occupied. $\mu_\ell = 0$ if the ℓ-th non-singly occupied site is empty, and $\mu_\ell = 1$, if the ℓ-th non-singly occupied site is doubly occupied. $\sigma_m = +1$, if the m-th singly occupied site has an electron of spin $\uparrow$, and $\sigma_m = -1$, if the spin is $\downarrow$.

For example, for six sites the state

$$c_{2\uparrow}{}^{+}c_{4\uparrow}{}^{+}c_{4\downarrow}{}^{+}c_{6\downarrow}{}^{+}|\text{vacuum}\rangle$$

is characterized by

$$\{1,0,1,1,1,0\}, \quad \{0,0,1,0\}, \quad \{+1,-1\},$$

and if one chooses a convention for the phase factor, one can also reconstruct the state from these three vectors. Now we see that the operator $P_n \mathcal{J} P_n$ can move the holes (empty sites) and doubly occupied sites, but it does not affect the $\{\mu_\ell\}$ and $\{\sigma_m\}$.

Thus we can map such a state on a system of spinless fermions —each hole and each doubly occupied site corresponds to such a spinless fermion—and by choosing an appropriate phase factor the operator $P_n \mathcal{J} P_n$ has the same matrix elements as the operator $\mathcal{J}' = t \sum a_i{}^{+}a_{i+1} + a_{i+1}{}^{+}a_i$ in this spinless fermion system. The a_i (resp. $a_i{}^{+}$) are the annihilation (resp. creation) operators of these spinless fermions. Therefore the eigenvalues are that of a system of p non-interacting spinless fermions with the Hamiltonian, where the number of fermions $p = n +$ number of holes =

$N + 2n - N_e$ for a given number, N_e, of electrons. Thus we obtain for the canonical partition function for N_e electrons

$$Q_{N_e} = \sum_n e^{-\beta nU} \sum_{\substack{\sigma_1=\pm 1 \\ \sigma_{N_e-2n}=\pm 1}} \exp\beta h \sum_m \sigma_m \sum_{\substack{\mu_1=0,1 \\ \mu_{N-N_e+2n}=0,1 \\ \sum\mu_\ell=n}} Q^0_p, \qquad (2)$$

where Q^0_p is the canonical partition function for p $(=N + 2n - N_e)$ spinless fermions for N sites; the h-dependent exponential occurs in the case of non-vanishing magnetic fields. The number of doubly occupied sites, n, is restricted by

$$0 \leqslant 2n \leqslant N_e \qquad \text{for} \quad N_e \leqslant N$$

or

$$2(N_e - N) \leqslant 2n \leqslant N_e \qquad \text{for} \quad N_e \geqslant N.$$

The sum over the spins yields the factor $(2\cosh\beta h)^{N_e-2n}$, and the sum over the μ_ℓ yields the binomial factor

$$\binom{N - N_e + 2n}{n} .$$

Thus we obtain the grand canonical partition function

$$z = \sum_{N_e=0}^{2N} Q_{N_e} e^{\beta\mu N_e}$$

$$= \sum_{N_e} e^{\beta\mu N_e} \sum_n e^{-\beta nU} (2\cosh\beta h)^{N_e-2n} \binom{N - N_e + 2n}{n} Q^0_{N+2n-N_e}. \qquad (3)$$

Instead of summing over the number of electrons and number of doubly occupied sites, we may sum over the number of singly occupied sites and doubly occupied sites, i.e. we write $N_e = n_s + 2n$, and we have the restrictions $0 \leqslant n_s \leqslant N$ and $0 \leqslant n \leqslant N - n_s$, and because of the particle-hole symmetry in $\mathcal{I}'$ we have $Q^0_{N-n_s} = Q^0_{n_s}$, and we end up with

$$z = \sum_{n_s=0}^{N} \exp\beta\mu n_s (2\cosh\beta h)^{n_s}\{1 + \exp(2\beta\mu - \beta U)\}^{N-n_s} Q^0_{n_s}$$

$$= \{1 + \exp(2\beta\mu - \beta U)\}^N z^0(\mu'), \qquad (4)$$

where z^0 is the grand canonical partition function for free, spinless fermions, and μ' is given by

$$\mu' = \mu + \frac{1}{\beta} \ln \left[\frac{2\cosh\beta h}{1 + \exp(2\beta\mu - \beta U)} \right] . \tag{5}$$

Therefore the free energy per site is given by

$$f = - T\ln\{1 + \exp(2\beta\mu - \beta U)\} + f^0(\mu'), \tag{6}$$

where f^0 is the free energy per site of the spinless fermions, i.e.

$$f^0(\mu') = - T\int_{-\pi}^{\pi} \frac{dk}{2\pi} \ln\{1 + \exp\beta(\mu' + 2t\cos k)\} \tag{7}$$

in the thermodynamic limit.

Because we know the many particle energies and eigenstates, we may also calculate the one-particle Green's function G. $G_{R,R'}(\tau)$ is given by (for $\tau > 0$):

$$G^\sigma_{R,R'}(\tau) = - \frac{1}{z} \sum_{i,j} e^{-\beta E_i} e^{E_i \tau} e^{-E_j \tau} \langle i|c_{k,\sigma}|j\rangle \langle j|c_{k',\sigma}^+|i\rangle, \tag{8}$$

where $|i\rangle$ and $|j\rangle$ are eigenstates of the many particle system, and E_i and E_j their energies (the chemical potential is now included in these energies). The calculation can be performed easily under the following restrictions:

(1) We regard the case $\mu = \frac{1}{2}U$; this corresponds to one electron per site;

(2) In the sum we neglect all terms which are smaller at least by a factor $\exp(-\frac{1}{2}\beta U)$ for *all* τ compared to terms taken into account;

(3) For the eigenstates $|i\rangle$ and $|j\rangle$ we regard only the zeroth order approximation.

The approximations (2,3) can be justified for large U because the neglected contributions tend to zero for large U uniformly, i.e. after Fourier transformation uniformly for all frequencies.

From approximation (2) we see that we have two different types of contributions for the sum in equation (8):

(a) $|j\rangle$ is a state with all sites singly occupied, and

(b) $|i\rangle$ is a state with all sites singly occupied.

Now we see that the matrix elements in equation (8) are the same as for the corresponding spinless fermion system, if in the case of contribution (a) in the state $|j\rangle$ all spins between the sites R and R' (endpoints included) are equal and equal to σ, and in the case (b) if all spins in the state $|i\rangle$ between the sites R and R' (endpoints included) are equal and equal to $-\sigma$; otherwise the product of the matrix elements is zero. Thus the

Fourier transform $G(\omega)$ of the Green's function can be expressed with the help of the free particle Green's function $G^0(\omega)$ as follows:

$$G^\sigma{}_{R,R'}(\omega) = x^{|R-R'|+1} G^0{}_{R,R'}(\omega + \tfrac{1}{2}U + \sigma h)$$

$$+ x'^{|R-R'|+1} G^0{}_{R,R'}(\omega - \tfrac{1}{2}U + \sigma h), \qquad (9)$$

where $x = \exp(\beta\sigma h)/2\cosh(\beta h)$, and $x' = 1 - x$; the pre-factors can be interpreted as the probability of finding all spins between R and R' (including these two points) to be equal σ (or $-\sigma$ resp.). It is more convenient to discuss the Green's function (or the spectral weight function) in k-space; by Fourier transformation we obtain for the spectral weight function $A^\sigma(k,\omega)$ the following convolution integral, which can be performed easily:

$$A^\sigma(k,\omega)$$

$$= \int_{-\pi}^{\pi} \frac{dk'}{2\pi} \left\{ \delta(\omega + \tfrac{1}{2}U + \sigma h + 2t\cos(k - k')) \frac{x(1 - x^2)}{1 + x^2 - 2x\cos k'} \right.$$

$$\left. + \delta(\omega - \tfrac{1}{2}U + \sigma h + 2t\cos(k - k')) \frac{x'(1 - x'^2)}{1 + x'^2 - 2x'\cos k'} \right\}. \qquad (10)$$

A plot of $A^\sigma(k,\omega)$ as a function of ω for several k-vectors is given in figures 1,2. $A^\uparrow$ is plotted in the positive direction, $A^\downarrow$ in the negative one. The origin is the point $\omega = -\tfrac{1}{2}U$, t has been chosen to be unity, $h = \tfrac{1}{2}$, and the plot is given for two induced magnetizations $m = x - x'$: $m = 0.9$ and $m = 0.5$. At $\omega \approx \tfrac{1}{2}U$ one obtains the same plot upside down. The essential points are the following: the spectral weight function is non-zero in two ω-intervals of the same width as the band width of the non-interacting system, the centres of these intervals have separation U. The spectral weight function is not at all a sum of two δ-functions; this is true only in the limit $m \to 1$ or $t \to 0$, thus a two-pole approximation is acceptable only if one is interested in characteristic energies large compared to t.

The first order calculation for the eigenvalues is not always sufficient. The eigenvalues are still highly degenerate, because the energies do not depend on the $\{\mu\}$ and $\{\sigma\}$ vector (for zero magnetic field). Thus the thermodynamic behaviour is described correctly only if the temperature is large compared with the actual characteristic splitting energy of these states (i.e. t^2/U), or if the magnetic field is large compared to this quantity. Therefore we shall perform second order perturbation theory, i.e. we have to diagonalize the operator

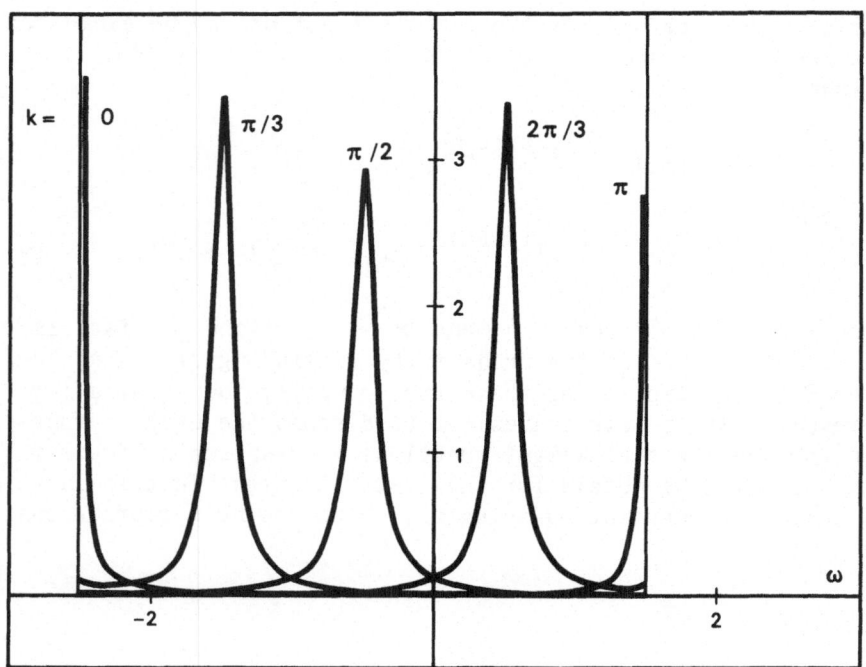

Figure 1 - The spectral weight function $A(k,\omega)$ is plotted for spin up electrons as a function ω for several k-vectors. The relative magnetization has been taken to be 0.9. The origin of the axis is $\omega = -\frac{1}{2}U$, the energy unit is t. The applied magnetic field is 0.5.

$$- \frac{1}{U} P_n \mathcal{J} P_{n+1} \mathcal{J} P_n + \frac{1}{U} P_n \mathcal{J} P_{n-1} \mathcal{J} P_n \tag{11}$$

within the space of still degenerate eigenstates. In general this problem cannot be solved, although we can give an approximate solution for states where the number of doubly occupied sites and the number of holes is small compared with the number of sites, thus our approximation is acceptable for states where the number of electrons is equal to the number of sites, for small temperatures. In this case the second term of the operator of equation (11) can be neglected, and the first term acts only on the vector $\{\sigma\}$ within this approximation, and it has essentially the form of an antiferromagnetically coupled Heisenberg chain:

$$\mathcal{J}'' = - \frac{t^2}{U} \lambda_{\{k\}} \sum_{m=1}^{N-2n-1} \frac{1}{2}(1 - S_i{}^z S_{i+1}{}^z)$$

$$\times (2 - 2S_i{}^+ S_{i+1}{}^- - 2S_i{}^- S_{i+1}{}^+), \tag{12}$$

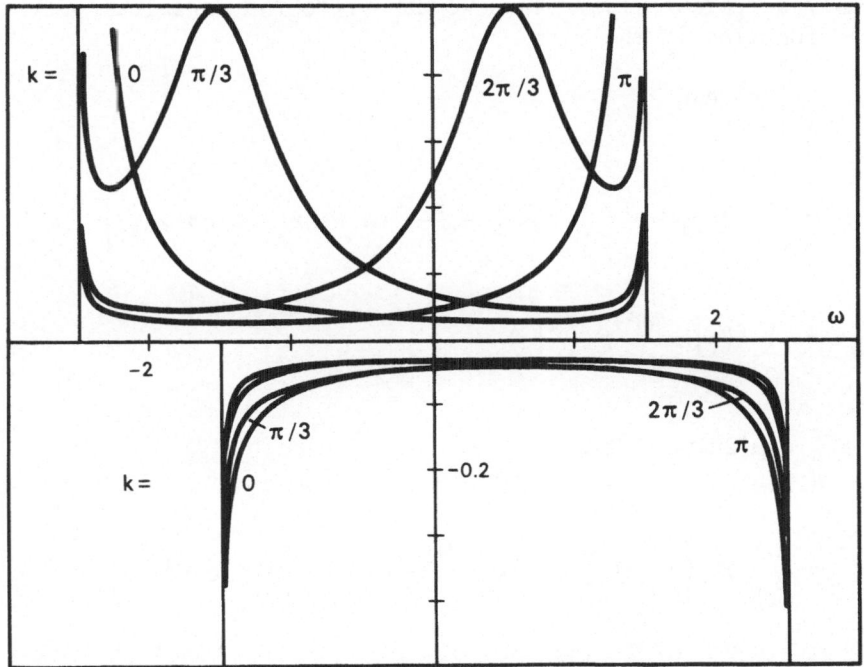

Figure 2 - The spectral weight function $A(k,\omega)$ is plotted as a function of ω for several k-vectors. For spin up A is plotted in the positive direction, for spin down A is plotted in the negative direction. The relative magnetization has been taken to be 0.5; the other parameters are the same as in figure 1.

with

$$\lambda = 1 - \frac{2n}{N} + \frac{1}{N} \sum_i \cos 2k_i,$$

(up to first order in n/N). This expression can be evaluated:

$$\mathcal{J}" \quad - \frac{t^2}{u}(N - 2n - 1)\lambda + \mathcal{H}_{N-2n}^{HS}(J'),\tag{13}$$

where $\mathcal{H}_{N-2n}^{HS}$ is the Hamiltonian of a $(N - 2n)$-site Heisenberg chain, coupled with the coupling constant J':

$$J' = -\frac{2t^2}{U}\left(1 - \frac{2n}{N} + \frac{1}{N} \sum_i \cos 2k_i\right) = J^0\lambda.\tag{14}$$

A magnetic field can also be incorporated in $\mathcal{H}^{HS}$. The factor $2n$ occurs because the number of holes is equal to the number of doubly occupied sites. Thus we can write the canonical partition function in the form

$$Q_N = \sum_n \binom{2n}{n} e^{\beta nU}$$

$$\times \sum_{\{k\}} \exp\left\{ -2\beta t \sum_i \cos k_i + \frac{\beta t^2}{U} (N - 4n + \sum_i \cos 2k_i) \right\}$$

$$\times \sum_{\{\sigma\}} \langle \sigma | e^{-\beta \mathcal{H}^{HS}(J')} | \sigma \rangle$$

$$= \sum_n \exp(2n \ln 2 - \beta nU)$$

$$\times \sum_{\{k\}} \exp\left\{ -2\beta t \sum_i \cos k_i + \frac{\beta t^2}{U} (N - 4n + \sum_i \cos 2k_i) \right\}$$

$$\times \exp\{ -\beta(N - 2n) f^{HS}(J') \}. \tag{15}$$

The sum over $\{k\}$ is the sum over all combinations of $2n$ k-values. This equality holds up to factors of order $\exp\{o(N)\}$, and $f^{HS}(J')$ is the free energy per site of a Heisenberg chain with the coupling constant J'. $f^{HS}(J')$ may be expanded up to terms first order in n/N:

$$f^{HS}(J') = f^{HS}(J^0) + \frac{1}{N} h_{int}^{HS}(\sum_i \cos 2k_i - 2n), \tag{16}$$

where h_{int}^{HS} is the expectation value of the interaction energy per site for the coupling J^0. Again we take into account in the exponent only first order terms in n/N and obtain

$$Q_N = \exp\left\{ \frac{\beta N t^2}{U} - \beta N f^{HS}(J^0) \right\}$$

$$\times \sum_n \exp\left\{ 2n \ln - \beta nU + 2\beta n h_{int}^{HS} - \frac{4n \beta t^2}{U} + 2n \beta f^{HS}(J^0) \right\}$$

$$\times \sum_{\{k\}} \exp\left\{ -2\beta t \sum_i \cos k_i + \frac{\beta t^2}{U} \sum_i \cos 2k_i \right.$$

$$\left. - \beta h_{int}^{HS} \sum_i \cos 2k_i \right\}. \tag{17}$$

The sum over $\{k\}$ is just the canonical partition function Q_{2n}^0 for $2n$ spinless, non-interacting fermions with the energy dispersion

$$\varepsilon_k = 2t\cos k + \left[h_{\text{int}}^{\text{HS}} - \frac{t^2}{U} \right] \cos 2k. \tag{18}$$

Therefore we can express the partition function Q_N in the thermodynamic limit by the thermodynamic potentials of the Heisenberg chain and the grand canonical partition function z_0 of a non-interacting spinless fermion system with a temperature and magnetic field dependent energy dispersion ε_k:

$$Q_N = \exp\left[\frac{\beta t^2}{U} N - \beta N f^{\text{HS}}(\mathcal{J}^0) \right] + z^0(\mu', \varepsilon_k), \tag{19}$$

where ε_k is given by equation (18) and

$$\mu' = \frac{1}{\beta} \ln 2 - \tfrac{1}{2}U + h_{\text{int}}^{\text{HS}} + f^{\text{HS}} - \frac{2t^2}{U}. \tag{20}$$

Again this equality holds up to factors of order $\exp(o(N))$. Therefore the free energy per site,

$$\frac{F}{N} = - \frac{T}{N} \ln Q_N,$$

is given in the thermodynamic limit by

$$\frac{F}{N} = - \frac{t^2}{U} + f^{\text{HS}} + f^0(\mu', \varepsilon_k), \tag{21}$$

where the grand canonical free energy f^0 of the spinless fermion system reads:

$$f^0 = - T \int_{-\pi}^{\pi} \frac{dk}{2\pi} \ln[1 + \exp\beta(\mu' - \varepsilon_k)]. \tag{22}$$

The second order energy correction has been calculated for small densities of non-singly occupied sites, thus this expression is applicable for low temperatures — of course we always assume $U \gg t$ for allowing for perturbation theory. But at higher temperatures, where the density of non-singly occupied sites is no longer small, the second order correction gives no essential contribution at all, thus equation (21) can be used for all temperatures. The thermodynamic quantities now can be obtained by taking the derivatives of F/N with respect to the corresponding parameters such as temperature or magnetic field.

The entropy per site $S/N = - \partial(F/N)/\partial T$ for zero magnetic field is given by

$$\frac{S}{N} = s^{HS} + s^0 + n^0(\ln 2 - s^{HS} - c^{HS}) - c^{HS}\int_{-\pi}^{\pi} \frac{dk}{2\pi} n_k^0 \cos 2k. \qquad (23)$$

Here s^{HS} is the Heisenberg chain entropy per site, s^0 is the entropy of the spinless fermion system with the given chemical potential μ' and the energy dispersion of equation (18). c^{HS} is the specific heat of the Heisenberg chain (at zero magnetic field). n_k^0 is the mean occupation number in the fermi system, i.e.

$$n_k^0 = \{\exp\beta(\varepsilon_k - \mu') + 1\}^{-1},$$

and

$$n^0 = \int_{-\pi}^{\pi} \frac{dk}{2\pi} n_k^0$$

is the mean density of the fermions; this quantity is approximately twice the number of doubly occupied sites. The last two terms in equation (23) occur because of the temperature dependence of μ' and ε_k, but this contribution is almost negligible.

By taking the second derivative with respect to h we obtain the magnetic susceptibility (at zero field):

$$\chi = \chi^{HS} + n^0\left\{\frac{\partial}{\partial T}(T\chi^{HS}) - \chi^{HS}\right\} - \frac{\partial}{\partial T}(T\chi^{HS})\int_{-\pi}^{\pi} \frac{dk}{2\pi} n_\mu^0 \cos 2k, \qquad (24)$$

Here χ^{HS} is the susceptibility of the Heisenberg chain, and we have used the identity:

$$\left.\frac{\partial^2 h_{int}^{HS}}{\partial h^2}\right|_{h=0} = \left.\frac{\partial}{\partial T} T\chi^{HS}\right|_{h=0} \qquad (25)$$

Because the thermodynamics of the Heisenberg chain are not known analytically we have used the numerical results of Bonner and Fisher [2] for temperatures smaller than $2|J^0| = 4t^2/U$, and for higher temperatures we have used the first few terms in the high temperature expansions for the thermodynamic quantities of of the Heisenberg chain. It should be mentioned that this does not correspond to a complete high temperature expansion as has been performed by Beni *et al*. [3], who took the quantity t/T to be small, while our high temperature expansion is restricted to the Heisenberg chain, and we take $2t^2/UT$ to be small.

The results for $U/t = 12$ are plotted in figures 3-5. Here

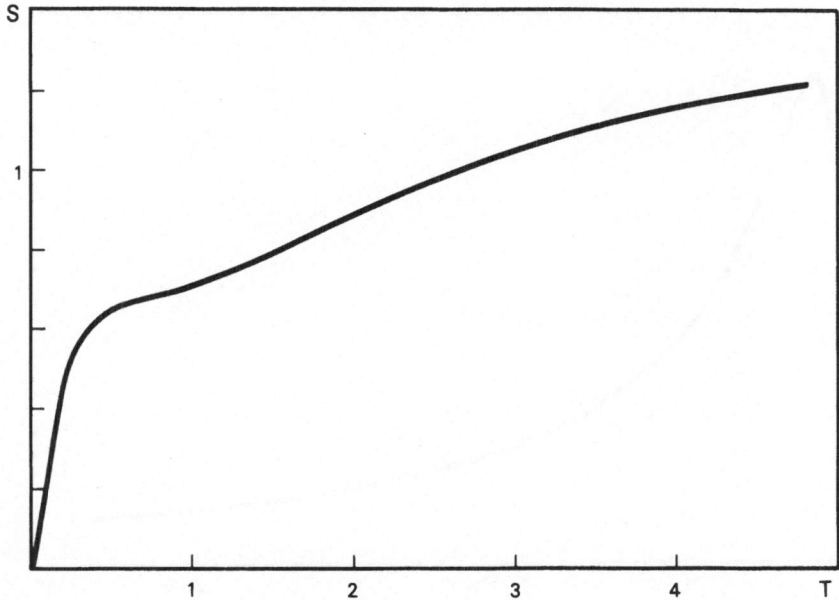

Figure 3 - The entropy per site has been plotted as a function of the temperature. As the energy unit we have used the transfer constant t. U has been taken to be 12.

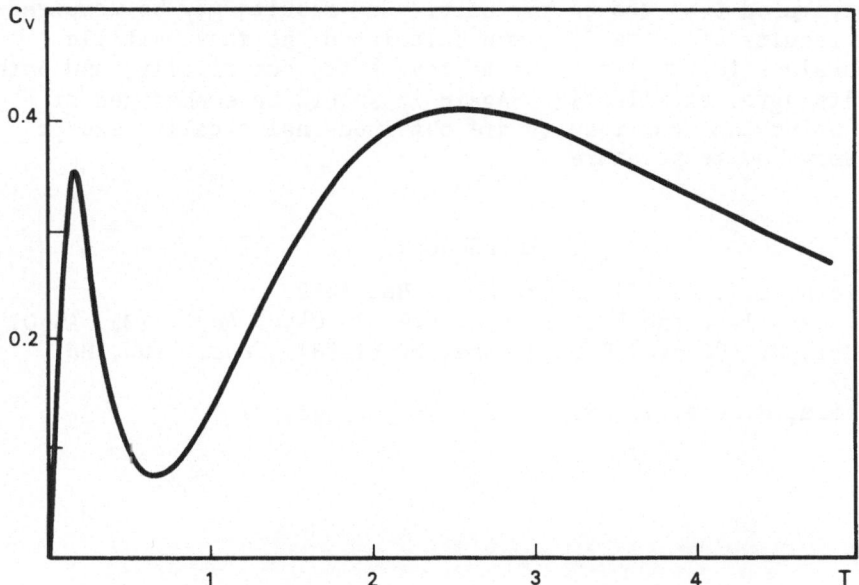

Figure 4 - The specific heat at constant volume is plotted as a function of the temperature. Again the temperature is measured in units of t, and $U = 12t$.

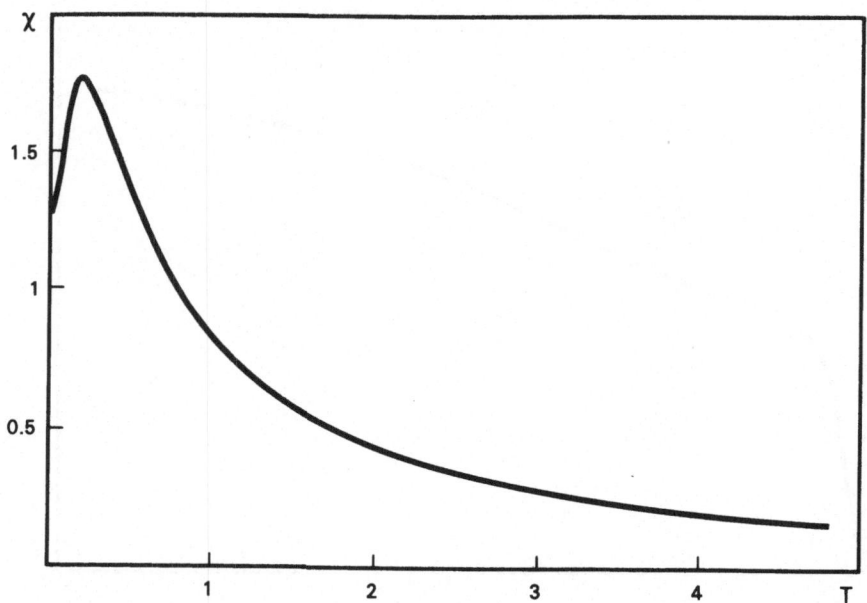

Figure 5 – The magnetic susceptibility is plotted as a
function of the temperature. Parameters and energy units
are the same as in figures 3,4.

we have used t as the energy unit. Our results may be compared
with results of Shiba [4], who calculated the many particle
eigenvalues for chains of up to five sites numerically, and both
results agree excellently. Again it should be emphasized at
this point that our results are obtained analytically, except
for very low temperatures.

REFERENCES

1. Klein, D.J. (1973). *Phys. Rev.*, **B8**, 3452.
2. Bonner, J.C. and Fisher, M.E. (1964). *Phys. Rev.*, **135**, A640.
3. Beni, G., Pincus, P. and Hone, D. (1973). *Phys. Rev.*, **B8**, 3389.
4. Shiba, H. (1972). *Prog. Theor. Phys.*, **48**, 2171.

WEAK SELF-CONSISTENT APPROXIMATION SCHEME

HAJO LESCHKE

*Institut für Physik, University of Dortmund,
West Germany*

1. INTRODUCTION

Even the simpler models, discussed in the theory of magnetism, exhibit the full complexity of the general quantum statistical many body problem. It is for this reason that the more
or less powerful many body techniques and their corresponding
approximations are applied to the theory of magnetism.

The simplest non-trivial approximation in magnetism is the
'molecular field' approximation (MFA), being essentially equivalent to the Hartree-Fock-Approximation (HFA) widely used by
many body theorists. The HFA belongs to the general class of
the so called self-consistent approximations. However, the
same expression 'self-consistent' is often used in quite different connections, thereby creating a certain confusion in the
literature. Therefore, the purpose of this lecture is two-fold:

   (a) First, I want to explain—using Green's function lan-
       guage —which approximations for many body systems I
   think one should call 'self-consistent', or even better,
   'thermodynamically (self-) consistent'. A sufficient condition for an approximation to be self-consistent is its
   so called φ-derivability, a notion first introduced by
   Kadanoff and Baym [1,2], and generalized recently by Brandt
   et al. [3,4], mainly in order to guarantee the conservation
   laws in transport theory;

   (b) Secondly, I want to describe an idea I worked out in
       collaboration with Brandt [5]. We called it the 'weak
   self-consistent' approximation scheme. In this scheme we
   propose to weaken the requirement of self-consistency.

As a consequence, one gets as equilibrium condition a system of non-linear algebraic equations, which seems mathematically more convenient than Dyson's non-linear integral equation of the usual (strong) self-consistent approximation scheme. This yields the possibility of constructing new types of approximations, which may go beyond the HFA.

2. DEFINITIONS AND THE PROBLEM OF SELF-CONSISTENCY

Most of the interesting physics of a *many-Fermion-system* described by a Hamiltonian $\mathcal{H}$ not explicitly time dependent is contained in the following functional:

$$W(U) := - \ln \, \mathrm{tr}\{e^{-\beta(\mathcal{H}-\mu\mathcal{N})}T_\tau S(U)\} \tag{1}$$

Here T_τ means the time ordering operation, β is the inverse temperature, μ is the chemical potential, $\mathcal{N}$ is the total particle number (operator), and

$$S(U) := \exp\{-\int d1d1'U(1,1')\psi^+(1)\psi(1')\} \tag{2}$$

is the *S-matrix* corresponding to an arbitrary *external field* $U(1,1')$, which may be non-local in space and (at least formally) in time. The symbol 1 is shorthand for the space, spin and (imaginary) time variables $(\vec{r}_1,\sigma_1,\tau_1)$, and $\psi(1)$ is the corresponding Fermion field operator in the Heisenberg picture with respect to $\mathcal{H} - \mu\mathcal{N}$. An integral $\int d1$ means

$$\sum_{\sigma_1} \int d^3r1 \int_0^\beta d\tau_1.$$

The importance of W is due to the fact that W itself and its first and second (functional) derivatives with respect to U have well known physical meanings.

In this lecture we only regard (static) equilibrium situations for the sake of simplicity. Therefore, we do not write down explicitly the second derivative, which is essentially the most general (isothermal) *linear response function*. The first derivative is just (a generalization of) the (one-particle Matsubara) *Green's-function* of the system [6]

$$G(1,1'|U) := \frac{\delta W}{\delta U(1,1')} = - \frac{\langle T_\tau S(U)\psi(1)\psi^+(1')\rangle}{T_\tau S(U)} \, . \tag{3}$$

Here $\langle \ \rangle$ means the grand canonical average with respect to $\mathcal{H}$. At the special 'point' of a *time-independent external field*

$$U(1,1') = U_{\sigma_1\sigma_1'},(\vec{r}_1,\vec{r}_1')\delta(\tau_1 - \tau_1'), \tag{4}$$

one can interpret $\beta^{-1}W(U)$ as the *thermodynamical potential* and $G(U)$ as the (usual) Greenian corresponding to the Hamiltonian $\mathcal{H} + \mathcal{A}$, where

$$\mathcal{A} := \beta^{-1}\int d1 d1' U(1,1')\psi^+(1)\psi(1'). \tag{5}$$

In order to keep the formulae transparent, we will frequently use a *matrix notation* for any two-point function $X = X(1,1')$ thereby doing away with hordes of indices. For instance, we define an inverse matrix X^{-1} of X by

$$\int d2 X^{-1}(1,2)X(2,1') = \delta(1,1')$$

$$:= \delta_{\sigma_1 \sigma_1'}\delta(\vec{r}_1 - \vec{r}_1')\delta(\tau_1 - \tau_1'), \tag{6}$$

and we write

$$\text{tr}X := \int d1 X(1,1^+) \tag{7}$$

for the trace of X, etc., where 1^+ refers to a time argument being infinitesimally greater than τ_1. Furthermore, we omit all indices in functional derivatives, e.g., we simply write equation (3) as

$$\frac{dW}{dU} = G(U). \tag{8}$$

A special version of this relation is obtained, if one substitutes U by λU — λ being a parameter — and differentiates with respect to λ

$$\frac{dW}{d\lambda} = \text{tr}G(\lambda U)U. \tag{9}$$

This relation has an obvious physical interpretation on the case where U is time independent:

> The expectation value of the one-particle quantity $\mathcal{A}$ (in the grand canonical ensemble corresponding to the Hamiltonian $\mathcal{H} + \lambda\mathcal{A}$) is the same, whether it is calculated as the derivative $\beta^{-1}dW/d\lambda$ of the thermodynamic potential or as $\beta^{-1}\text{tr}GU$.

Let us call the formal extension of this property to any (even time dependent) one-particle quantity (thermodynamical) *self-consistency* [1-4]. Of course, in the exact theory this is a trivial matter, because it is nothing else than the definition (8) of the Greenian. However, as there is no real hope of ex-

plicitly calculating the exact functional W at sufficiently many
'points' U, it is important to investigate what can be said about
approximations. The basic problem is to find an approximate
thermodynamic potential corresponding to a given approximate
Greenian.

To take an example, consider a system in a static external
field λU with fixed U. Suppose one has found an approximate
Greenian $\tilde{G}(\mu,\lambda)$ (e.g. by a certain decoupling procedure in the
equation of motion) depending of course on the chemical potential μ and on the parameter λ (and on the temperature and possibly on several other parameters not explicitly mentioned).

From G one may calculate the mean particle number

$$\langle \mathcal{N} \rangle = \beta^{-1} \mathrm{tr}\tilde{G}(\mu,\lambda). \tag{10}$$

Comparing this with the thermodynamic relation

$$\langle \mathcal{N} \rangle = - \beta^{-1} \frac{\partial W}{\partial \mu}, \tag{11}$$

it seems natural to define an approximate potential $\tilde{W}$ corresponding to $\tilde{G}$ by

$$\tilde{W}(\mu,\lambda) := - \int_{-\infty}^{\mu} d\mu' \mathrm{tr}\tilde{G}(\mu',\lambda), \tag{12}$$

where it is assumed, that $\tilde{W}(\mu = -\infty,\lambda) = 0$. In order to calculate the average value of $\mathcal{A}$ one should be allowed to use either
of the following two expressions:

$$\beta^{-1} \frac{\partial \tilde{W}}{\partial \lambda} \quad \text{or} \quad \beta^{-1} \mathrm{tr}\tilde{G}U. \tag{13}$$

However, in general these expressions do not coincide! The
mathematical reason for this thermodynamic inconsistency is that
the approximate $\tilde{G}(\mu,\lambda)$ in general fails to fulfil the *'condition
of complete integrability'*,

$$\frac{\partial}{\partial \lambda} \mathrm{tr}\tilde{G}(\mu,\lambda) = - \frac{\partial}{\partial \mu} \mathrm{tr}\tilde{G}(\mu,\lambda)U, \tag{14}$$

being necessary (and in non-pathological examples also sufficient) for the existence of an approximate thermodynamic potential corresponding to $\tilde{G}$.

3. ϕ-DERIVABLE APPROXIMATIONS AND STATIONARITY PRINCIPLE

A simple way to guarantee the existence of an approximate
thermodynamic potential corresponding to an approximate Greenian $\tilde{G}(U)$, is to do something more than is really necessary,

from a physical point of view, namely to allow only for those approximations in which even the whole $\tilde{G}(U)$ has a generating functional $\tilde{W}(U)$ in the same sense (8) as in the exact theory.

In this way, one gets (thermodynamically) self-consistent approximations completely defined by a certain functional $\tilde{W}(U)$. In order to get more insight into the structure of the possible $\tilde{W}(U)$'s, let us first return to the exact theory.

From now on we assume as usual that the Hamiltonian $\mathcal{H}$ consists of two parts; one corresponding to a one-particle operator H_0, and one corresponding to a two-particle interaction V. Therefore, it is convenient to introduce besides $G = G(U)$ the *'free' Greenian* G_0 corresponding to $V = 0$ and $U = 0$. According to Baym [2], the exact $W = W(U)$ has the following *(Baym-) representation*

$$W = \phi - \mathrm{tr}\Sigma G + \ln G. \tag{15}$$

Here ϕ is the generating functional of the *self-energy* (functional),

$$\Sigma := \frac{\partial \phi}{\partial G} , \tag{16}$$

which depends on $G = G(U)$ (and V but not explicitly on U), $\ln G$ is by definition a functional obeying the differential equation

$$\frac{\mathrm{d}}{\mathrm{d}G} \ln G = G^{-1}, \tag{17}$$

and (the 'equilibrium') G has to fulfil Dyson's equation

$$G^{-1} = G_0^{-1} - u - \Sigma(G). \tag{18}$$

Because in the representation (15) W depends on U only via G, one may verify from (15-18) the *self-consistency relation* (8) without using additional information about ϕ. This is the reason why any approximation to W (and thereby to G), which is characterized by an 'appropriate' approximation $\tilde{\phi}$ to ϕ, is automatically self-consistent as long as $\partial\tilde{\phi}/\partial G$ is used as approximation to Σ. These approximations are called by Baym [2] ϕ-*derivable*, because they are precisely those approximations in which the self-energy is derivable from a functional $\tilde{\phi} = \tilde{\phi}(G,V)$. In what follows we will only consider such approximations. Thus we may drop the tilde '~' for approximate quantities. Of course, all the results are true for the exact theory too, but in practice the exact ϕ is not known explicitly. As an example we write down diagrammatically [6] the most simple ϕ. It corresponds to the *Hartree-Fock Approximation* (HFA):

$$\phi_{HFA} = \frac{1}{2} \; \text{(diagram)} \; + \; \frac{1}{2} \; \text{(diagram)} \tag{19}$$

To summarize the above, I want to stress the remark of Baym [2] "that the requirement that an approximation be ϕ-derivable is a useful criterion to apply in terminating the infinite hierarchy of coupled Green's function equations".

So far, we have considered W as a functional of the variable U getting the corresponding fixed 'equilibrium' $G = G(U)$ by differentiation. For practical calculations, however, it is often convenient to look upon W in a slightly different way. Let us define the following *auxiliary functional*

$$\bar{W}(U,G) := \phi(G) - \text{tr}(G_0^{-1} - U - G^{-1})G + \ln G. \tag{20}$$

Here, both U and G are considered as independent variable quantities! The significance of $\bar{W}$ is due to the following:

Stationarity Principle:

(a) The equilibrium Greenian $G_{eq}(U)$ (which has been written till now simply as $G(U)$) may be determined as a stationary 'point' of $\bar{W}$ for fixed U, because the stationarity condition,

$$\left(\frac{\partial \bar{W}}{\partial G}\right)_U = 0, \tag{21}$$

is equivalent to Dyson's equation (18);

(b) At the stationary point W and $\bar{W}$ are identical, more precisely,

$$\bar{W}(U,G_{eq}(U)) = W(U), \tag{22}$$

for all U.

4. WEAK SELF-CONSISTENCY

With the help of the stationarity principle it can easily be shown that the self-consistency relation (8) is essentially equivalent to the equilibrium condition given by Dyson's equation (18). In order to do so, let us assume $G_{eq}^{app}(U)$ to be an approximate stationary point of $\bar{W}$, i.e. an approximate solution of Dyson's equation. Then

$$W^{app}(U) := \bar{W}(U, G_{eq}^{app}(U)), \tag{23}$$

is the corresponding approximate potential. Taking the derivative with respect to U, one gets

$$\frac{dW^{app}}{dU} = \frac{\partial \bar{W}}{\partial G}\bigg|_{G_{eq}^{app}} \frac{dG_{eq}^{app}}{dU} + \frac{\partial \bar{W}}{\partial U}\bigg|_{G_{eq}^{app}}. \tag{24}$$

According to (20) the second term on the right hand side of (24) is equal to G_{eq}^{app}. Hence one gets the self-consistency relation (8) if and (disregarding possible pathological counterexamples) only if $\partial \bar{W}/\partial G|_{G_{eq}^{app}}$ vanishes. In other words: If one wants to have self-consistency exactly, one has to solve Dyson's equation exactly!

By the way, at this stage one clearly sees that our definition of self-consistency essentially coincides with the more or less precise notion of *'self-consistency'* used in *mean* (or molecular) *field theories*. The reason for this is simply that one may look upon the self-energy $\Sigma = \Sigma(G)$ as the most general (internal) *'mean field'* with Dyson's equation as its corresponding 'self-consistency equation' [7].

Unfortunately, Dyson's equation is in most non-trivial approximations a highly non-linear integral equation, and in practice it seems nearly impossible to find an exact solution of it. Of course, one can expect self-consistency to hold at least more and more precisely as Dyson's equation is solved with increasing accuracy [2]. However, even numerical calculations on Dyson's equation are extremely hard to do especially if one wants to go beyond the HFA [8,9]. Instead of improving the methods for solving Dyson's equation there is another more physical way [5] of overcoming the stringent implications of the requirement of (strong) self-consistency. The idea is to weaken the self-consistency condition and, with it, Dyson's equation. The starting point for such a weakening procedure is as simple as it is physical. Instead of requiring self-consistency for any (even nonphysical) one-particle quantity, we impose it for the *physically relevant quantities* only, thereby losing nothing of the physically relevant information contained in the Greenian but gaining — as we hope — a lot of mathematical convenience. Of course it depends on the actual system under consideration which quantities one should regard as physically relevant. This is as in phenomenological thermodynamics, which by itself does not give the 'thermodynamic' variables. Instead, one must guess them or know them from experiment.

Now let us derive the formal consequences and the corresponding equations of weak self-consistency.

Let us try to calculate (at least approximately) the stationary point of $\bar{W}$ by means of a *parametric variational ansatz*, which is due to (18) conveniently chosen in the form

$$G^{-1} = G_0^{-1} - U - \hat{\Sigma}(\alpha,\beta). \tag{25}$$

Here the 'matrix' $\hat{\Sigma}$ depends on a set α of variational parameters α_n, and eventually explicitly on the (inverse) temperature β. Inserting the ansatz (25) into the functional (20) and requiring stationarity with respect to the α's yields the following *weak equilibrium condition*

$$0 \stackrel{!}{=} \left(\frac{d\bar{W}}{d\alpha_n}\right)_U = \text{tr}\{(\Sigma(G) - \hat{\Sigma})G \frac{\partial\hat{\Sigma}}{\partial\alpha_n} G\}. \tag{26}$$

This is a system of non-linear algebraic equations for the equilibrium, $\alpha_{eq} = \alpha_{eq}(U)$, and it is in general a weaker condition than the usual (strong) equilibrium condition given by Dyson's equation

$$\Sigma(G) = \hat{\Sigma}. \tag{27}$$

Hence, one may find a solution of (26) which is not a solution of (27). As I mentioned above, such a solution cannot be (strongly) self-consistent, but, as I want to point out, it may be nevertheless weakly self-consistent, if one impose some formal restrictions to the possible ansatz for $\hat{\Sigma}$.

Moreover there is at least one physical restriction, in fact one should require that for static fields U the equilibrium 'value' $\hat{\Sigma}_{eq} := \hat{\Sigma}(\alpha_{eq}(U))$ of $\hat{\Sigma}$ depends only on the difference of the two time variables. Here, we may assume this property to hold even for the ansatz $\hat{\Sigma}$ itself, because we do not regard time dependent fields at all.

Now we can give a precise formulation of weak self-consistency. The *key argument* goes as follows. We are interested only in certain physically relevant (one-particle) quantities instead of the whole Greenian. Therefore we restrict U to be a linear combination,

$$U = \sum_j \lambda_j U_j, \tag{28}$$

of fields U_j which couple to the relevant quantities. The observable expectation value of the relevant quantites may be defined as $d\bar{W}/d\lambda_j$ or as $\text{tr}G_{eq}U_j$. For that very reason we require the *weak self-consistency relation*

$$\frac{d\bar{W}}{d\lambda_j} \stackrel{!}{=} \text{tr}G_{eq}U_j, \tag{29}$$

where G_{eq} corresponds via (25) to $\hat{\Sigma}_{eq}$. Equation (29) is equivalent to

$$0 \overset{!}{=} \frac{dW}{d\lambda_j} - \frac{\partial \bar{W}}{\partial \lambda_j} = \text{tr} \left. \frac{\partial \bar{W}}{\partial G} \right|_{G_{eq}} \frac{dG_{eq}}{d\lambda_j}$$

$$= \text{tr}(\Sigma(G_{eq}) - \hat{\Sigma}_{eq}) G_{eq} \frac{dG_{eq}^{-1}}{d\lambda_j} G_{eq}$$

$$= - \text{tr}(\Sigma(G_{eq}) - \hat{\Sigma}_{eq}) G_{eq} U_j G_{eq}$$

$$- \sum_n \frac{d\alpha_n^{eq}}{d\lambda_j} \text{tr}(\Sigma(G_{eq}) - \hat{\Sigma}_{eq}) G_{eq} \left. \frac{\partial \hat{\Sigma}}{\partial \alpha_n} \right|_{\alpha_{eq}} G_{eq}. \qquad (30)$$

The second term vanishes because of the weak equilibrium condition (26). A sufficient (and presumably also necessary) condition to make the first term vanish too, is to allow in (25) only for those $\hat{\Sigma}$'s which fulfil the following *representability condition*

$$U_j = \sum_n \xi_{jn} \frac{\partial \hat{\Sigma}}{\alpha \partial n} \qquad \text{for all } j, \qquad (31)$$

with appropriate numbers ξ_{jn} depending on the (equilibrium) α's.

The condition (31) shows that the variational ansatz for $\hat{\Sigma}$ has to be in close connection with the physical problem and with the one-particle quantities regarded as physically relevant.

As an *example*, let us consider a physical system where the z-component ζ^z of the total spin is a relevant quantity coupled to a homogeneous external magnetic field h. In this case one has to use a variational ansatz, which fulfils

$$U_{-h}(1,1') = s_{\sigma_1 \sigma_1'}{}^z \delta(\vec{r}_1 - \vec{r}_1') \delta(\tau_1 - \tau_1')$$

$$= \sum_n \xi_n \frac{\partial \hat{\Sigma}(\vec{r}_1 \tau_1, \vec{r}_1' \tau_1'; \alpha, \beta)}{\partial \alpha_n} . \qquad (32)$$

Here $s_{\sigma\sigma'}{}^z$ is the corresponding Pauli Matrix. Now we obtain

$$- \beta \langle \zeta^z \rangle \equiv \left(\frac{dW}{dh} \right)_{\beta, \mu} = \frac{\partial \bar{W}}{\partial h} + \text{tr} \left. \frac{\partial \bar{W}}{\partial G} \right|_{G_{eq}} \frac{dG_{eq}}{dh} . \qquad (33)$$

The second term on the right hand side vanishes, owing to (26) and to (32). The first term becomes, more explicitly,

$$\frac{\partial \bar{W}}{\partial h} = - \text{tr} G U_{-h} = - \beta \sum_{\sigma\sigma'} s_{\sigma\sigma'}{}^z \int d^3 r G_{eq}{}^{\sigma'\sigma}(\vec{r}\tau, \vec{r}\tau^+), \qquad (34)$$

which is the desired result.

In conclusion, I want to make two remarks concerning the weak self-consistent approximation scheme:

(a) Also the (internal) energy may be calculated in the usual way,

$$\langle \mathcal{H} - \mu N \rangle \equiv \frac{dW}{d\beta} = \beta^{-1} \text{tr}(H_0 + \tfrac{1}{2}\Sigma(G_{eq}))G_{eq},$$

if one imposes the following additional restrictions for the possible approximations (for the details see [5]!):

(i) H_0 has a representation the same as in equation (31);

(ii) The temperature dependence of the ansatz $\hat{\Sigma}$ is of the form

$$\hat{\Sigma}(\tau,\tau';\alpha,\beta) =: \beta^{-2}\hat{\sigma}(\beta^{-1}\tau,\beta^{-1}\tau';\alpha);$$

(iii) The functional ϕ has a scaling property which is fulfilled by the exact ϕ and also by any diagrammatic approximation. It may be formulated as,

$$\varphi(\lambda g, \lambda^{-2}\beta) = \varphi(g,\beta),$$

where $g(\tau,\tau') := G(\beta\tau,\beta\tau')$, and

$$\varphi(g,\beta) := \phi(G,\beta).$$

(b) The whole procedure works also in the presence of external time dependent fields. Therefore this approximation scheme is, in principle, also applicable to transport phenomena.

ACKNOWLEDGEMENT

The author is very much indebted to Professor Dr. U. Brandt, without whose help the idea of weak self-consistency would not have been created.

REFERENCES

1. Baym, G. and Kadanoff, L.P. (1961). *Phys. Rev.*, **124**, 287.
2. Baym, G. (1962). *Phys. Rev.*, **127**, 1391.
3. Brandt, U., Pesch, W. and Tewordt, L. (1970). *Z. Phys.*, **238**, 121.
4. Brandt, U., Lustfeld, H., Pesch, W. and Tewordt, L. (1971). *J. Low Temp.*, **4**, 79.
5. Brandt, U. and Leschke, H. (1973). *Z. Phys.*, **260**, 147.

6. Abrikosov, A.A., Gor'kov, L.P. and Dzyaloshinski, I.Ye. (1965). *Quantum Field Theoretical Methods in Statistical Physics*, (Pergamon Press, Oxford).
7. Mattuck, R.D. and Johansson, B. (1968). *Adv. Phys.*, **17**, 509.
8. Brandt, U. (1971). *Z. Phys.*, **244**, 217.
9. Edwards, D.M. and Hertz, J.A. (1973). *J. Phys.*, **F3**, 2174, 2191.

ON THE CONNECTION BETWEEN EQUATION OF MOTION TECHNIQUE AND PERTURBATION THEORY APPROACH TO THE EVALUATION OF DOUBLE TIME GREEN'S FUNCTION

C.R. NATOLI

C.N.R.S. Grenoble, France†

and

J. RANNINGER

Institut Laue-Langevin, Grenoble, France

In this lecture of a methodological character we establish a connection between two widely used methods for the evaluation of double time Green's functions: namely the equation of motion technique for Zubarev's Green's functions, and perturbation method for Matsubara's Green's functions. We shall show that the equation of motion technique, when supplemented with an appropriate decoupling procedure is equivalent to summing a certain class of diagrams in the sense of perturbation theory.

The so called Matsubara [1], or temperature dependent, Green's function is defined by:

$$G_{AB}(t,t') = - i\langle T[A(t)B(t')]\rangle$$

$$= - iZ^{-1}\text{Tr}\{T[A(t)B(t')]e^{-\beta(H-\mu N)}\}, \qquad (1)$$

$$\beta = \frac{1}{k_B T}, \qquad Z = \text{Tr}\{e^{-\beta(H-\mu N)}\},$$

where t and t' are purely imaginary and lie in the interval $(0, -i\beta)$ $(-\beta \leqslant \text{Im}(t - t') \leqslant 0)$, and the time-ordering operator is defined as:

---

† Permanent address: C.N.E.N., Frascati, Roma, Italy

$$T[A(t)B(t')] = A(t)B(t') \qquad \text{if } it > it',$$

$$T[A(t)B(t')] = \pm B(t')A(t) \qquad \text{if } it' > it. \tag{2}$$

Here the upper sign applies when either of both A and B have Bose character, the lower sign when both A and B have Fermi character (an operator has Fermi character if it involves the product of an odd number of creation and annihilation operators for Fermi particles, otherwise it is of Bose character). Moreover, also for complex times, we have

$$A(t) = e^{iHt} A e^{-iHt}. \tag{3}$$

This means that the usual equation of motion is still valid

$$i \frac{\partial}{\partial t} G_{AB}(t,t') = \delta(t - t')\langle [A,B]_{\mp} \rangle + G_{[A,B],B}(t,t'), \tag{4}$$

where the time differentiation is now carried out along the imaginary time axis provided one interprets the δ-function for the imaginary times t, t' according to

$$\delta(t - t') = i\delta(it - it'), \qquad (-\beta \leqslant -it, -it' < 0), \tag{5a}$$

or alternatively

$$\int_0^{-i\beta} \delta(t)dt = 1, \qquad \delta(t) = 0 \quad \text{unless} \quad t = 0. \tag{5b}$$

The boundary condition with which equation (4) is to be supplement follows directly from the definition (1) and is given by

$$G_{AB}(0,t') = \pm\, e^{n\beta\mu} G_{AB}(-i\beta,t'), \tag{6a}$$

where n is the number of particles annihilated by the operator A. ($[N,A] = -nA$ where N is the particle number operator). Similarly one derives

$$G_{AB}(t,0) = \pm\, e^{-n\beta\mu} G_{AB}(t, -i\beta). \tag{6b}$$

Equations (6a,b) together with the fact that $G_{AB}(t,t')$ is a function of $t - t'$, that is to say,

$$G_{AB}(t,t') = G_{AB}(t - t') = G_{AB}(\tau), \tag{7}$$

(where, if t and t' lie in the interval $(0, -i\beta)$ τ must lie in the interval $(-i\beta, i\beta)$) leads us to conclude that the function

$$\hat{G}_{AB}(\tau) = e^{in\mu\tau}G_{AB}(\tau) \tag{8}$$

is periodic on $(i\beta, -i\beta)$ with period $2i\beta$, hence

$$\hat{G}_{AB}(i\beta) = \hat{G}_{AB}(-i\beta). \tag{9}$$

It follows at once that $\hat{G}_{AB}(\tau)$ can be expanded in a Fourier series on $(-i\beta, i\beta)$ according to

$$\hat{G}_{AB}(\tau) = \frac{1}{-i\beta} \sum_{\nu} A_{\nu} e^{-(\pi\nu/(-i\beta))\tau}, \tag{10}$$

where the sum over ν runs over all positive and negative integers $(\nu = 0, \pm1, \pm2, \dots)$. However, from the relation,

$$\hat{G}_{AB}(i\beta) = \pm \hat{G}_{AB}(0), \tag{11}$$

easily derivable from above, we can conclude at once that

$$A_{\nu}[1 \mp (-1)^{\nu}] = 0 \tag{12}$$

so that in the Bose case the A_{ν} are non-vanishing only when ν is even, and in the Fermi case the A_{ν} are non-vanishing only when ν is odd

$$G_{AB}(\tau) = \frac{1}{-i\beta} \sum_{\nu} A_{\nu} e^{-iz_{\nu}\tau}, \qquad z_{\nu} = -\frac{\pi\nu}{i\beta} + n\mu, \tag{13}$$

and can determine the A_{ν}'s by inverting this Fourier series

$$A_{\nu} = \int_{0}^{-i\beta} G_{AB}(\tau)e^{iz_{\nu}\tau}d\tau, \tag{14}$$

since the functions $e^{-iz_{\nu}\tau}$ are orthogonal in $(0, -i\beta)$ when the ν's are all even or all odd and constitute a complete set. The spectral representation for $G_{AB}(\tau)$ allows us to derive from (14) the following expression for A_{ν}

$$A_{\nu} = \int_{-\infty}^{+\infty} \frac{d\omega}{2\pi} \frac{\rho_{AB}(\omega)}{z_{\nu} - \omega}, \tag{15}$$

where $\rho_{AB}(\omega)$ is the function

$$\rho_{AB}(\omega) = 2\pi Z^{-1} \sum_{\ell m} \langle \ell|A|m \rangle \langle m|B|\ell \rangle e^{-\beta(E_{\ell} - \mu N_{\ell})}$$

$$\times(1 \mp e^{-\beta(\omega-\mu n)})\delta(\omega + E_\ell - E_m). \qquad (16)$$

One recognized that the function

$$G(z) = \int_{-\infty}^{+\infty} \frac{d\omega}{\pi} \frac{\rho_{AB}(\omega)}{z - \omega} \qquad (17)$$

is analytic throughout the whole z-plane apart from a cut along those portions of the real axis on which $\rho_{AB}(\omega) \neq 0$, and that the Fourier transforms $G_{AB}^{(r/a)}(\omega)$ of the retarded and advanced Zubarev Green's functions given by

$$G_{AB}^{(r/a)}(\omega) = \lim_{\varepsilon \to 0^+} -i \int_{-\infty}^{+\infty} dt e^{i(\omega \mp i\varepsilon)t} \langle [A(t),B(0)]_\mp \rangle \theta(\pm t) \qquad (18)$$

are equal to $G_{AB}(z)$ taken at the points $z = \omega \mp i\varepsilon$. This establishes the well known connection [2] between temperature dependent or Matsubara Green's functions and advanced and retarded Zubarev Green's functions. Indeed, knowing the coefficients,

$$A_\nu = G(z_\nu), \qquad (19)$$

in the series expansion (13) for $G_{AB}(\tau)$ at the points z_ν of a set of points having an accumulation point in the region of analyticity of $G(z)$ suffices to determine this latter function. To summarize, one may say that the time-ordered Green's function technique consists of solving the equation of motion (4) on the strip $(0, -i\beta)$ of the imaginary time axis subject to the boundary condition (6a), expanding the solution in a Fourier series according to (13), obtaining $G_{AB}(z)$ from the coefficients $A_\nu = G_{AB}(z_\nu)$ by analytic continuation, finally deriving $\rho_{AB}(\omega)$ from the relation

$$\rho_{AB}(\omega) = i \lim_{\varepsilon \to 0^+} [G_{AB}(\omega + i\varepsilon) - G_{AB}(\omega - i\varepsilon)]. \qquad (20)$$

If one substitutes the expansion (13) which can be rewritten as

$$G_{AB}(\tau) = \frac{1}{-i\beta} \sum_\nu G_{AB}(z_\nu)e^{-iz_\nu\tau} \qquad (13a)$$

into the equation of motion (4) and integrates over t from 0 to $-i\beta$ one obtains

$$z_\nu G_{AB}(z_\nu) = \langle [A,B]_\mp \rangle + G_{[A,H]B}(z_\nu). \qquad (21)$$

This equation is now equivalent to the equation of motion (4) and the boundary condition (6a) in the sense that if one can find a solution of (21) and the Fourier series (13a) for this solution converges, then the function given by (13a) satisfies both (4) and (6a).

The equation (21) is to be compared with analogous equation for the function (17)

$$zG_{AB}(z) = \langle [A,B]_{\mp} \rangle + G_{[A,H]B}(z), \qquad (z \to \omega \pm i\varepsilon), \qquad (22)$$

in the equation of motion approach for advanced and retarded Green's function. The advantage of the previous formulations of the problem lies in the fact that there exists for the calculation of the Fourier component $G_{AB}(z_\nu)$ a diagramatic technique [2] which enables us to link the results obtained by the two methods. The connection is best illustrated on a specific example. Suppose we have to evaluate the quantity $G_{\psi(x)\psi^+(x')}(\tau)$ where $\psi(x)$ is the Fermion field relative to an Hamiltonian of the type

$$H = \int \psi^+(x) \left[-\frac{\nabla^2}{2m} + V(x) \right] \psi(x) dx$$

$$+ \tfrac{1}{2} \int\int v(x - x') \psi^+(x) \psi^+(x') \psi(x') \psi(x) dx dx'. \qquad (23)$$

To write the equations of motion for this quantity we shall need the commutators

$$[\psi(x),H] = \left[-\frac{\nabla^2}{2m} + V(x) \right] \psi(x) + \int v(x - t) \psi^+(t) \psi(t) \psi(x) dt,$$

$$[\psi^+(x),H] = - [\psi(x),H]^+ \qquad (24a)$$

$$= - \left[-\frac{\nabla^2}{2m} + V(x) \right] \psi^+(x) - \int v(x - t) \psi^+(x) \psi^+(t) \psi(t) dt,$$

together with the mixed double commutators

$$[[\psi^+(x_1),H]_-\psi^+(x_2)]_+ = - v(x_2 - x_1) \psi^+(x_1) \psi^+(x_2),$$

$$[\psi(x_1)[\psi(x_2),H]_-]_+ = v(x_2 - x_1) \psi(x_1) \psi(x_2), \qquad (24b)$$

where as above $[A,B]_+ = AB + BA$, and $[A,B]_- = AB - BA$.

Writing the ν-th Fourier component of $G_{AB}(\tau)$ as $\langle\langle A/B \rangle\rangle_{z_\nu}$, where in the following we shall drop the index z_ν, we obtain for the first equation

$$\left\{ z_\nu - \left[-\frac{\nabla_x^2}{2m} + V(x) \right] \right\} \langle\!\langle \psi(x) | \psi^+(x') \rangle\!\rangle$$

$$= \delta(x - x') + \int dt v(x - t) \langle\!\langle \psi^+(t)\psi(t)\psi(x) | \psi^+(x') \rangle\!\rangle. \quad (25)$$

The next equation is for the quantity $\langle\!\langle \psi^+(y)\psi(x_1)\psi(x_2) | \psi^+(x') \rangle\!\rangle$ and reads

$$\left\{ z_\nu + \left[-\frac{\nabla_y^2}{2m} + V(y) \right] - \sum_{i=1}^{2} \left(-\frac{\nabla_{x_i}^2}{2m} + V(x_i) \right) - v(x_2 - x_1) \right\}$$

$$\times \langle\!\langle \psi^+(y)\psi(x_1)\psi(x_2) | \psi^+(x_1) \rangle\!\rangle$$

$$= \langle [\psi^+(y)\psi(x_1)\psi(x_2), \psi^+(x')] \rangle$$

$$+ \int dt \left\{ -v(y - t) + \sum_{i=1}^{2} v(x_i - t) \right\}$$

$$\times \langle\!\langle \psi^+(y)\psi^+(t)\psi(t)\psi(x_1)\psi(x_2) | \psi^+(x') \rangle\!\rangle, \quad (26)$$

where we have put the operators ψ and ψ^+ in normal order in the higher order Green's function using the mixed double commutator in (24b). In doing so we have projected out of the spectral representation of

$$\langle\!\langle \psi^+(y)\psi^+(t)\psi(t)\psi(x_1)\psi(x_2) | \psi^+(x') \rangle\!\rangle$$

all the contributions coming from the two particle states which are taken into account by the term $V(x_2 - x_1)$ in the LHS of (26). Following the same procedure of putting in normal order the operators in the higher order Green's functions by using the equations (24b), for the equation of motion for

$$\langle\!\langle \prod_{j=1}^{n} \psi^+(y_j) \prod_{i=1}^{n+1} \psi(x_i) | \psi^+(x') \rangle\!\rangle$$

we find

$$\left\{ z_\nu + \sum_{j=1}^{n} \left[-\frac{\nabla_{y_j}^2}{2m} + V(y_j) \right] + \sum_{j<j'}^{1,n} v(y_{j'} - y_j) \right.$$

$$- \sum_{i=1}^{n+1} \left(-\frac{\nabla_{x_i}^2}{2m} + V(x_i) \right) - \sum_{i<i'}^{1,n+1} v(x_{i'} - x_i) \Bigg\}$$

$$\times \left\langle\!\!\left\langle \prod_{j=1}^{n} \psi^+(y_j) \prod_{i=1}^{n+1} \psi(x_i) \big| \psi^+(x') \right\rangle\!\!\right\rangle$$

$$= \left\langle \left[\prod_{j=1}^{n} \psi^+(y_j) \prod_{i=1}^{n+1} \psi(x_i), \psi^+(x') \right]_+ \right\rangle$$

$$+ \int dt \Bigg\{ - \sum_{j=1}^{n} v(y_j - t) + \sum_{i=1}^{n+1} v(x_i - t) \Bigg\}$$

$$\times \left\langle\!\!\left\langle \prod_{j=1}^{n} \psi^+(y_j)\psi^+(t)\psi(t) \prod_{i=1}^{n+1} \psi(x_i) \big| \psi^+(x') \right\rangle\!\!\right\rangle. \qquad (27)$$

If we were to solve the Green's function problem in the Hilbert sub-space of $n + 1$ particles, the last Green's function in (27) would be zero and we could achieve a closed solution for $\langle\!\langle \psi(x) | \psi^+(x') \rangle\!\rangle$, provided we knew the solution of the Schrödinger equation for a number of particles ranging from 1 to $n + 1$. In any practical evaluation we are forced to truncate the chain of equations. The way to do this in the present scheme is immediate. Namely we define the many time Green's function

$$G_{\psi_1 \dots \psi_n \psi_1^+ \dots \psi_n^+}(t_1 \dots t_n, t_1' \dots t_n')$$

$$= - iZ^{-1} \text{Tr}\{T[\psi(x_1 t_1) \dots \psi(x_n t_n)\psi^+(y_1 t_1') \dots \psi^+(y_n t_n')]$$

$$\times e^{-\beta(H-\mu N)}\}, \qquad (28)$$

for $t_1 \dots t_n, t_1' \dots t_n'$ all lying on the interval $(0, -i\beta)$ of the imaginary axis where the time-ordering operator T is defined by

$$T[A_1(t_1)A_2(t_2) \dots A_m(t_m)]$$

$$= \pm A_{i_1}(t_{i_1})A_{i_2}(t_{i_2}) \dots A_{i_m}(t_{i_m}), \qquad (29)$$

where $i_1, i_2, \dots, i_m$ are chosen so that $it_{i_1} > it_{i_2} > \dots > it_{i_m}$ and the $\pm$ sign is taken according to whether an even (upper sign) or odd (lower sign) number of interchanges of Fermi operators has to be made in permuting $A_1 A_2 \dots A_m$ into $A_{i_1} A_{i_2} \dots A_{i_m}$. A perturbation theory is known to exist for the calculation of the

quantity in (28), and a decoupling rule arises from taking all
the diagrams which decompose, at least, into two separate parts
which are not connected to each other by any line, and discard
all those which do not decompose in such a way (these latter
give rise to a complete vertex part). After Fourier transforming, the result is simply the cumulant approximation for

$$\left\langle\!\!\left\langle \prod_{j=1}^{n} \psi^+(y_j) \prod_{i=1}^{n+1} \psi(x_i) \middle| \psi^+(x') \right\rangle\!\!\right\rangle$$

given by

$$\left\langle\!\!\left\langle \prod_{j=1}^{n} \psi^+(y_j) \prod_{i=1}^{n+1} \psi(x_i) \middle| \psi^+(x') \right\rangle\!\!\right\rangle$$

$$\simeq \sum_{partitions} c(m) \left\langle\!\!\left\langle \prod_{j=1}^{n_1} \psi^+(y_j) \prod_{i=1}^{n_1+1} \psi(x_i) \middle| \psi^+(x') \right\rangle\!\!\right\rangle$$

$$\times \left\langle \prod_{j=1}^{n_2} \psi^+(y_j) \prod_{i=1}^{n_2} \psi(x_i) \right\rangle \dots \left\langle \prod_{j=1}^{n_m} \psi^+(y_j) \prod_{i=1}^{n_m} \psi(x_i) \right\rangle, \quad (30)$$

$$(n_1 + n_2 + \dots + n_m = n),$$

where the sum is over all partitions of $\psi^+(y_1)\dots\psi^+(y_n)$
$\times\psi(x_1)\dots\psi(x_{n+1})\psi^+(x')$ apart from those in which $\psi^+(x')$ stands
alone and $c(m)$ is given by

$$c(m) = (-1)^m (m - 1)!,$$

m being the number of factors appearing in (30).

A self-consistent solution of the chain thus truncated is in-
terpretable in terms of diagrams by solving the chain by an iteration procedure starting from the last equation. Since the
free part of any Green's function,

$$\left\langle\!\!\left\langle \prod_{j=1}^{n} \psi^+(y_j) \prod_{i=1}^{n+1} \psi(x_i) \middle| \psi^+(x') \right\rangle\!\!\right\rangle,$$

is expressible in terms of averages $\langle \psi^+\psi \rangle$ and the one particle
Green's function $\langle\!\langle \psi(x) | \psi^+(x') \rangle\!\rangle$; the resulting expansion for
the latter must be a partial series of the total perturbation
series. This expansion contains averages of the type

$$\left\langle \prod_{j=1}^{m} \psi^+(y_j) \prod_{i=1}^{m} \psi(x_i) \right\rangle$$

which must be evaluated in terms of the approximate expression
for the corresponding Green's function obtained at the same order of the iteration. This partial series consists of all the
diagrams in which at most $n + 1$ particles are interacting simultaneously if the decoupling is made at the stage (27). This
kind of approach is applicable whenever the physical situation
is best described by clusters of n particles interacting between
themselves in the presence of the mean field due to the other
particles in the system, or when the repeated scattering among
the n particles is predominant over other types of interaction
(the case of low density, for example). This is, for instance,
the case for the two magnon light or neutron excitations in insulating antiferromagnets. The same kind of procedure can be
followed in the case of density fluctuations when studying

$$\langle\langle \psi^+(x_1)\psi(x_2) | \psi^+(x_1')\psi(x_2') \rangle\rangle$$

or in the case of excitations of two particles when studying

$$\langle\langle \psi(x_1)\psi(x_2) | \psi^+(x_1')\psi^+(x_2') \rangle\rangle .$$

Finally, we want to point out that the evaluation of Green's
functions relative to spin operators can follow the same patterns
as above. Indeed, a perturbation expansion for spin Green's functions which is based on the Matsubara technique and takes into
account in a controlled way the effect of the kinematical interactions has been given by Vaks, Larkin and Pikin [3].

REFERENCES

1. Matsubara, T. (1955). *Prog. Theor. Phys.*, **14**, 351.
2. Abrikosov, A.A., Gorkov, Z.P. and Dzyaloshinski, I.E. (1963).
   Methods of Quantum Field Theory in Statistical Physics,
   (Prentice-Hall, Inc.).
3. Vaks, V.G., Larkin A.I. and Pikin S.A. (1968). *Sov. Phys.
   JETP*, **26**, 188.

FIRST ORDER GREEN FUNCTION THEORY
OF A HEISENBERG FERROMAGNET

A. KÜHNEL

Sektion Physik der Karl-Marx-Universität,
Leipzig, DDR

1. INTRODUCTION

Green functions (GF) containing spin operators have their own peculiarities owing to the commutation relations of spin operators, which are neither Bose nor Fermi operators. In a perturbation theory the resulting series become even more difficult than in the Bose or Fermi case. In particular, Dyson's equation is not sufficient for summing all the contributions appearing in the perturbation series. After some contradicting opinions the present answer to this problem is the following.

A first order GF theory using spin operators yields exactly the spin wave energy (SWE) obtained by Dyson in the first Born approximation, but in the case of spin $\frac{1}{2}$ the resulting low temperature magnetization contains a term proportional to T^3 but no term T^4, in contrast with the result of Dyson.

The comparison of results has to be performed at two levels: for the microscopic spin wave energy and for the thermodynamic quantities as magnetization or heat capacity. In spite of the agreement of the resulting SWE obtained by Dyson [1], by GF theory with ideal spin wave operators (Bose operators) [2,3,4] and by GF theories with spin operators [5-11], the low temperature magnetization results obtained by ideal spin operator and by spin operator GF theories differ from each other. This difference is caused by the shape of GF. Spin operator theories yield the GF in the form

$$G = \frac{\sigma}{i\omega - \varepsilon(\mathbf{k})} \, , \tag{1}$$

and one gets the relative magnetization per lattice site as

$$\sigma = 1 - 2\bar{n} = 1 + 2G_{ii}(-0) = 1 - 2\sigma\phi,$$

$$\sigma = \frac{1}{1 + 2\sigma\phi} = 1 - 2\phi + 4\phi^2 - \dots , \tag{2}$$

where

$$\phi = \sum_{\mathbf{k}} (e^{\epsilon(\mathbf{k})/T} - 1)^{-1}.$$

However, Bose operator theories give

$$G = \frac{1}{i\omega - \epsilon(\mathbf{k})} \qquad \text{and} \qquad \sigma = 1 - 2\phi. \tag{3}$$

The appearance of $\sigma(T)$ in the numerator of the GF is the origin of the different results for the low temperature magnetization. This factor has been a matter of some dispute in the past, but one may consider the above statement now as well established. There are two papers of Ortenburger [19] and of Rudoy and Tserkovnikov [10] in which the term T^3 in the magnetization has been removed and a term T^4 has appeared in second order theory. Rudoy and Tserkovnikov find that a term $-4\phi^2$ results in the expansion of σ in second order exactly cancelling the unpleasant term in equation (2). However, the SWE now gets an imaginary part, and the exact agreement with ideal spin wave theory is destroyed, even if the damping of the spin waves were to be very small.

2. DIAGRAM TECHNIQUE FOR ARBITRARY SPIN AND FOR SPIN $\frac{1}{2}$

We deal with the Heisenberg model

$$H = - \mu_B \mathcal{H} \sum_f S_f{}^z - \sum_{f,g} J_{f,g} \vec{S}_f \cdot \vec{S}_g, \tag{4}$$

or for spin $\frac{1}{2}$,

$$H = (\mu_B \mathcal{H} + J(0)) \sum_f b_f{}^+ b_f - \sum_{f,g} J_{f,g}(b_f{}^+ b_g + b_f{}^+ b_f b_g{}^+ b_g), \tag{5}$$

in the usual notations. The commutation relations of the spin operators are

$$[S_f{}^\pm, S_g{}^z] = \mp \delta_{f,g} S_g{}^\pm,$$

$$[S_f{}^+, S_g{}^-] = 2\delta_{fg} S_g{}^z, \qquad S_f{}^{\pm(2s+1)} = 0, \tag{6}$$

or in the case of spin $\frac{1}{2}$ using Pauli operators $b = S^+$, $b = S^-$:

$$[b_f, b_g^+] = \delta_{fg}(1 - 2b_f^+ b_f), \qquad b_f^z = b_f^{+z} = 0;$$

$$S_f^z = \frac{1}{2} - b_f^+ b_f. \tag{7}$$

We want to calculate the thermodynamic Green function defined as

$$G_{\ell m}(\tau_\ell - \tau_m) = - \frac{\langle T\{S_\ell^+(\tau_\ell)S_m^-(\tau_m)\gamma(T^{-1})\}\rangle_0}{\langle \gamma(T^{-1})\rangle_0}, \tag{8}$$

where the spin operators are to be taken in the interaction representation. $\gamma(T^{-1})$ is the usual S-operator, the expansion of which gives the perturbation series.

It can be proved [12,6] that Wick's theorem holds for spin operators in the following form

$$\langle T(S_1^{\alpha_1} S_2^{\alpha_2} \ldots) \rangle_0$$

$$= \frac{1}{2\langle S^z \rangle_0} \{G^0_{12}(\tau_1 - \tau_2)\langle T([S_1^{\alpha_1}, S_2^{\alpha_2}]S_3^{\alpha_3} \ldots)\rangle_0$$

$$+ G^0_{13}(\tau_1 - \tau_3)\langle T(S_2^{\alpha_2}[S_1^{\alpha_1}, S_3^{\alpha_3}] \ldots)\rangle_0 + \ldots \}. \tag{9}$$

As it stands, relation (9) is valid for $S_1^{\alpha_1} = S_1^+$, but a corresponding relation may be proved for $S_1^{\alpha_1} = S_1^-$. The zeroth order GF is defined as

$$G^0_{\ell m}(\tau_\ell - \tau_m) = \begin{cases} -\langle S_\ell^+ S_m^- \rangle_0 = \dfrac{2\langle S^z \rangle_0}{1 - e^{-\epsilon_0/T}} \delta_{\ell m} e^{-\tau\epsilon_0}, & \tau > 0, \\[4mm] -\langle S_m^- S_\ell^+ \rangle_0 = -\dfrac{2\langle S^z \rangle_0}{1 - e^{\epsilon_0/T}} \delta_{\ell m} e^{-\tau\epsilon_0}, & \tau < 0, \end{cases} \tag{10}$$

where $\tau = \tau_\ell - \tau_m$. Note that the GF in (9) are multiplied by $1/(2\langle S^z \rangle_0)$.

Now we introduce the following diagrammatic representation in the case of arbitrary spin. A solid line represents a GF G^0. A point stands for the transversal interaction $-\int d\tau \sum\limits_{f,g} J_{fg}$, a wavy line is the symbol for the longitudinal interaction $2\int d\tau \sum\limits_{f,g} J_{fg}$, the ends of a wavy line belong to different lattice points f and g. A point is connected always with two solid lines, getting an additional factor $1/(2\langle S^z\rangle_0)$, with a circle getting an additional factor $\langle S^z\rangle_0$, or with a broken line. A broken line represents $b'\delta_{fg}$, where b' is the derivative of $b = \langle S^z\rangle_0$. The connection of a broken line and a solid line gets a factor $1/\langle S^z\rangle_0$. The triangle means a factor $1/(2\langle S^z\rangle_0)$, and if it is situated with one angle on a solid line it gets an additional factor $1/\langle S^z\rangle_0$. A broken line can be introduced between all parts of diagrams not belonging to the same lattice site.

The diagrammatic representation introduced above is in general equivalent to the diagrams of Izyumov and Kassan-Ogly [12], and in the case of spin $\frac{1}{2}$ it is identical with the diagrams proposed by the author [6]. Using these diagrams we get the results of figure 1 for the first and second order terms in the perturbation series; terms coming from $\langle T(S_1{}^z S_f{}^z S_g{}^z S_r{}^z S_s{}^z)\rangle_{oc}$ are not drawn explicitly because they all together appear in the expansion of $\langle S^z\rangle_0$ (see below). In the case of spin $\frac{1}{2}$ we may reduce the broken line to two solid lines running forth and back. This

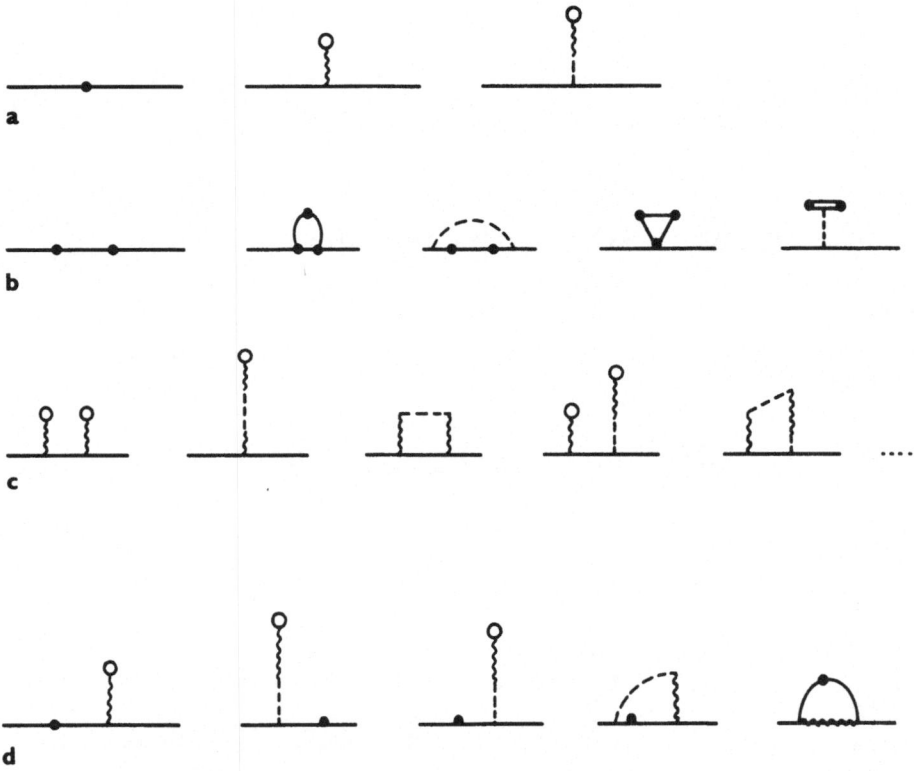

Figure 1 - (a) First order contributions; (b) twice transversal; (c) twice longitudinal; and (d) transversal-longitudinal second order contributions.

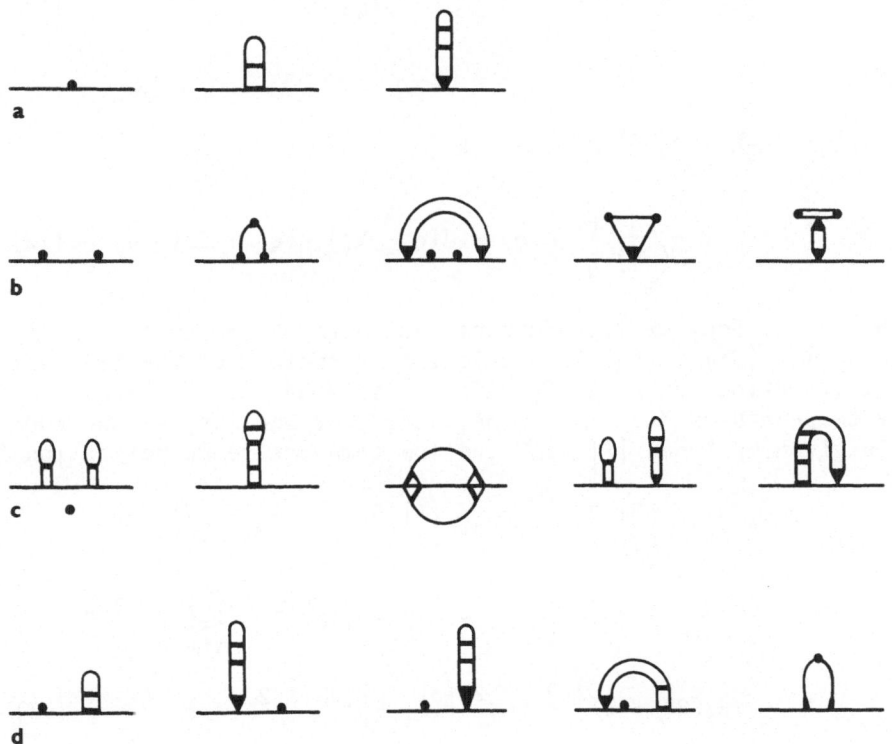

Figure 2 - Diagrams corresponding to figure 1 in the case of spin $\frac{1}{2}$.

is possible owing to the relations $(S_f{}^z)^2 = \frac{1}{4}$ and $S_f{}^z = \frac{1}{2} - b_f{}^+ b_f$, and we get $b' = \bar{n}(1 - \bar{n})$. The corresponding diagrams are shown in figure 2 (for details see [6]). The square in figure 2 is identical with the wavy interaction line in figure 1.

The diagrams consisting of one solid line connected only by a broken line (figure 1) or only by a (figure 2) to the remaining part of the diagram cannot be summed with the help of Dyson's equation. However, if we expand $\langle S^z \rangle$ into a perturbation series we find just those parts of the diagrams mentioned which are connected with a solid line; amongst others we find also the term not represented by diagrams in figure 1. In this way we find out that the diagrams which cannot be included in Dyson's self energy part are summed by replacing $\langle S^z \rangle_0$ in the numerator of the unperturbed GF by $\langle S^z \rangle$. In the case $S > \frac{1}{2}$ we find the replacement $\langle S^z S^z \rangle_0 \to \langle S^z S^z \rangle$ as the summation of another class of diagrams. The remaining diagrams may be summed with help of Dyson's equation, and in first order in $J(\mathbf{k})$ we obtain

$$G(\mathbf{k},\omega) = \frac{2\langle S^z\rangle}{i\omega_n - \varepsilon(\mathbf{k})} , \tag{11}$$

where

$$\varepsilon(\mathbf{k}) = \mu_B \mathcal{H} + 2\langle S^z\rangle [J(0) - J(\mathbf{k})]$$

$$+ \frac{1}{N\langle S^z\rangle} \sum_{\mathbf{q}} [J(\mathbf{q}) - J(\mathbf{q} - \mathbf{k})][n(\mathbf{q}) + 2K(\mathbf{q})]. \tag{12}$$

$K(\mathbf{q})$ is the Fourier transform of the correlation function $K_{\ell m}^{zz} = \langle S_\ell^z S_m^z\rangle - \langle S^z\rangle^2$; $n(\mathbf{q})$ is the Fourier transform of the transversal correlation function $\langle S_\ell^- S_m^+\rangle$ and is equal to $2\langle S^z\rangle\phi(\mathbf{q})$. At low temperatures $K(\mathbf{q})$ may be neglected, or one may use the sum rule given by Kenan [7], and the low temperature magnetization is

$$\langle S^z\rangle = S - \zeta\left(\frac{3}{2}\right)t^{3/2} - \frac{3}{4}\pi\nu\zeta\left(\frac{5}{2}\right)t^{5/2} - \pi^2\omega\nu^2\zeta\left(\frac{7}{2}\right)t^{7/2}$$

$$- \frac{3}{2s}\pi\nu\zeta\left(\frac{3}{2}\right)\zeta\left(\frac{5}{2}\right)t^4 + (2S + 1)\zeta^{(2S+1)}\left(\frac{3}{2}\right)2S+3/2$$

$$+ (2S + 1)^2 \frac{3}{4}\pi\nu\zeta^{2S}\left(\frac{3}{2}\right)\zeta\left(\frac{5}{2}\right)t^{3S+5/2} + \ldots . \tag{13}$$

The results (12,13) for the spin wave energy and for the low temperature magnetization, respectively, have been discussed in the introduction.

3. COMPARISON OF DIFFERENT DIAGRAMMATIC APPROACHES

Recently we were able to show [13,14] that in the case of $S = \frac{1}{2}$ the perturbation series are identical if they are calculated for the GF (8) according to our approach, with the help of Spencer's drone-fermion representation [16], or using the diagrams of Izyumov and Kassan-Ogly [12] (after the correction of some misprints).

As far as one is concerned with the summation of diagrams, the problem becomes more difficult. We could show [15] that unjustified neglections are made by Lewis and Stinchcombe [17] in order to obtain complete agreement with Dyson's results. Spencer [16] neglects all diagrams containing a broken line (parallel lines running back and forth in figure 2) and finds full agreement with Dyson's results, too. Izyumov and Kassan-Ogly divide their diagrams into single tails, double tails, etc., without having an order principle for this subdivision. We think that our summation is well justified and unique, and, furthermore, it yields results consistent with those of non-perturbative methods.

4. CORRESPONDENCE WITH OTHER METHODS

In this last point we mention briefly that the generalized Wick theorem (9) may be used as a guide for a decoupling scheme. One is led to propose the decoupling ([18,8])

$$\langle\langle b_f^+ b_f b_f{}' , b_g^+ \rangle\rangle \rightarrow \bar{n}\langle\langle b_f{}' , b_g^+ \rangle\rangle + \frac{1}{\sigma} \langle b_f^+ b_f{}' \rangle\langle\langle b_f , b_g^+ \rangle\rangle$$

$$- \frac{2}{\sigma} \langle b_f^+ b_f{}' \rangle\langle b_f b_f{}'^+ \rangle\langle\langle b_f , b_g^+ \rangle\rangle , \qquad (14)$$

and the result of the first order decoupling is just the GF (11). The last term of (14) contains the splitting of K^{zz}.

The same result was obtained by a more complicated decoupling by Mubayi and Lange and by Kenan [7]. Rudoy and Tserkovnikov found the same first order using a time dependent formalism [10], and Plakida found them using a differentiation technique.

REFERENCES

1. Dyson, F.J. (1956). *Phys. Rev.*, **102**, 1217, 1230.
2. Tahir-Kheli, R.A. and ter Haar, D. (1962). *Phys. Rev.*, **127**, 95.
3. Szaniecki, J. (1961). *Acta Phys. Polon.*, **20**, 983, 995.
4. Kühnel, A. (1969). *Wiss. Z. Univ. Jena, Math.-Naturwiss. Reihe*, **18**, 165.
5. Vaks, V.G., Larkin, A.I. and Pikin, S.A. (1967). *Zh. Eksp. Teor. Fiz.*, **53**, 281.
6. Kühnel, A. (1969). *J. Phys.*, **C2**, 704, 711; Haberlandt, H. and Kühnel, A. (1973). *Phys. Stat. Sol.*, **60**, No.2.
7. Mubayi, V. and Lange, R.V. (1969). *Phys. Rev.*, **178**, 882; Kenan, R.P. (1970). *Phys. Rev.*, **B1**, 3205.
8. Kühnel, A. and Trimper, S. *Acta Phys. Polon.*, (submitted).
9. Oguchi, T. (1970). *Prog. Theor. Phys.*, **44**, 1548.
10. Rudoy, Yu.G. and Tserkovnikov, Yu.A. (1973). *Teor. Mat. Fiz.*, **14**, 102.
11. Plakida, N.M. (1973). *Phys. Lett.*, **43A**, 481.
12. Izyumov, Yu.A. and Kassan-Ogly, F.A. (1970). *Fiz. Met. Metalloved.*, **30**, 225.
13. Kühnel, A. (1973). *Phys. Stat. Sol. (b)*, **55**, 559.
14. Kühnel, A. and Trimper, S. (1973). *Acta Phys. Polon.*, **A44**, 493.
15. Kühnel, A. and Trimper, S. (1973). *Phys. Stat. Sol.*, **60**, K15.
16. Spencer, H.J. (1967). *Phys. Rev.*, **167**, 430, 434.
17. Lewis, W.W. and Stinchcombe, R.B. (1967). *Proc. Phys. Soc.*, **92**, 1002, 1010.
18. Kühnel, A. and Trimper, S. (1971). *Wiss. Z. Univ. Leipzig, Math.-Naturwiss. Reihe*, **20**, 303.
19. Ortenburger, I. (1967). *Phys. Rev.*, **136**, A1374.

SPIN-ONE LATTICE-GAS MODEL

JEAN SIVARDIERE

Départment de Recherche Fondamentale,
Centre d'Etudes Nucléaires,
BP 85, 38041 Grenoble Cedex, France

1. THE SPIN HALF ISING MODEL

Many cooperative systems are characterized by a *single order parameter* and a second order phase transition at a *critical point*. Their thermodynamical behaviour can be simulated by a spin half Ising model with positive dipolar interactions. For instance [1]:

— magnetic ordering in a uniaxial or isotropic system;

— phase separation in a binary mixture;

— condensation of a simple fluid (Lee and Yang);

— freezing of a liquid (Lennard-Jones and Devonshire);

— suprafluid ordering (London, Tisza [2]);

— magnetic ordering in a two-singlet system (Bleaney, Cooper);

— ferroelectric ordering in KDP (De Gennes);

— ordering in molecular crystals: NH_4Cl, $NaNO_2$, ...

Many of these systems are *two-state* systems and accordingly, the order parameter $M = \langle S_i{}^z \rangle$ of the model has only two possible phases, positive and negative. In other cases (isotropic magnetic system, liquid helium) the phase of the order parameter can be varied continuously; however, many qualitative properties of the these systems are still correctly reproduced. In particular the analogy between a simple fluid, a magnet, a binary mixture, ... together with the analogy between critical opalescence and critical neutron or X-ray scattering, support the theory of Tisza [2]: the Andrews critical point, as well as the other λ-points, is the limit of stability of a homogeneous high temperature phase.

2. THE SPIN ONE ISING MODEL

Other physical systems are characterized by *two order parameters* η_1 and η_2. Their main interests are: the existence of *critical lines* and possibly a *tricritical point* [3]; the interplay between two ordering processes.

Some of these systems can be represented by a spin half Ising model with negative dipolar interactions: for instance a two-sublattice antiferromagnet (η_1 and η_2 are the sublattice and net magnetizations): an ordered alloy (η_1 is the structural order parameter and η_2 is the concentration). Many other *three state* systems, which are considered below, can be represented by a spin one Ising model with positive dipolar, quadrupolar and crossed dipolar-quadrupolar interactions. This model offers two kinematically coupled order parameters: $M = \langle S_i^z \rangle$ and $Q = \langle (S_i^z)^2 \rangle$. We consider successively:

— the applicability of the spin one Ising model to the description of three-state systems;

— the general form of the Hamiltonian and some particular cases;

— the equations of state in the molecular field approximation and the construction of the phase diagram;

— some results concerning simple, binary and ternary fluids.

3. APPLICABILITY OF THE SPIN ONE ISING MODEL

We first consider systems with a *variable pressure*. The state $|0\rangle_i$ is used to represent a vacancy in the cell or at the site i. The states $|\pm1\rangle_i$ represent a cell i occupied by a molecule with an internal degree of freedom which can give rise to a cooperative ordering: magnetic, suprafluid, orientional, configurational or chemical, generally described by a spin half Ising model.

$$S_i^z = +1: \quad \uparrow \quad He4^+ \quad \bigcirc \quad \boxed{ | \times} \quad A$$

$$S_i^z = -1: \quad \downarrow \quad He4^- \quad \ominus \quad \boxed{\times | } \quad B$$

Consequently the model applies to the following systems:

— paramagnetic gas [4] (magnetic ordering);

- pure He^4 [5] (suprafluid ordering);

- nematic fluid [5] (nematic ordering);

- simple fluid [5] (solidification);

- binary mixture [6] (phase separation).

The model describes the interplay between condensation (or adsorption) and the above ordering in the condensed phase, or conversely the influence of pressure on this ordering. Q represents the density, M the magnetization, orientation, suprafluid or crystalline order parameter, or deviation from equiconcentration.

We consider now systems *kept at a fixed pressure*. The state $|0\rangle_i$ is now used to represent an atom or molecule B in the cell i. The states $|\pm1\rangle_i$ are used as above, so that the following systems can be represented:

- magnetic alloy [4] (A magnetic, B non-magnetic);

- He^3-He^4 mixtures [7];

- nematic mixtures [5] (A elongated, B spherical);

- binary liquid [5] (A small, B big);

- ternary liquid or solid mixture [6].

The spin one model describes the interplay between a phase separation and some other cooperative ordering, or conversely the influence of dilution on this ordering.

4. THE SPIN ONE ISING HAMILTONIAN

We consider first the general case of a ternary mixture ABC [8]. The composition is given by:

$$x_A = \tfrac{1}{2}(Q + M),$$

$$x_B = 1 - Q, \tag{1}$$

$$x_C = \tfrac{1}{2}(Q - M).$$

Let us define operators $\varpi_i{}^A$ ($\varpi_i{}^B, \varpi_i{}^C$) which are equal to one if the cell i is occupied by an A (B,C) molecule, and zero otherwise:

$$\varpi_i{}^A = \tfrac{1}{2}(Q_i + S_i{}^z),$$

$$\varpi_i{}^B = 1 - Q_i, \tag{2}$$

$$\varpi_i{}^C = \tfrac{1}{2}(Q_i - S_i{}^z).$$

Introducing the six pair interaction parameters J_{AA}, J_{AB}, ...
and the chemical potentials μ_A, μ_B, μ_C we get the Hamiltonian of
the system:

$$\mathcal{H} = - \sum_{i>j} [J_{AA}\varpi_i^A\varpi_j^A + \ldots + J_{AB}(\varpi_i^A\varpi_j^B + \varpi_i^B\varpi_j^A) + \ldots]$$

$$- \mu_A \sum_i \varpi_i^A - \mu_B \sum_i \varpi_i^B - \mu_C \sum_i \varpi_i^C. \qquad (3)$$

$S_i^Z Q_i = S_i^Z$, whence a Hamiltonian characterized only by three
interaction parameters K, L and J, and two fields D and H:

$$\mathcal{H} = - \sum_{i>j} [KQ_iQ_j + L(S_i^ZQ_j + Q_iS_j^Z) + JS_i^ZS_j^Z]$$

$$- D \sum_i Q_i - H \sum_i S_i^Z, \qquad (4)$$

with:

$$K = \tfrac{1}{4}(J_{AA} + J_{CC} + 2J_{AC}) + (J_{BB} - J_{BC} - J_{BC}),$$

$$L = \tfrac{1}{4}(J_{AA} - J_{CC}) + \tfrac{1}{2}(J_{BC} - J_{BA}),$$

$$J = \tfrac{1}{4}(J_{AA} + J_{CC} - 2J_{AC}), \qquad (5)$$

$$D = \tfrac{1}{2}(\mu_A + \mu_C - 2\mu_B) + (J_{BA} + J_{BC} - 2J_{BB}),$$

$$H = \tfrac{1}{2}(\mu_A - \mu_C) + (J_{BA} - J_{BC}).$$

The crossed dipolar-quadupolar interaction L, which is not found
in the Blume-Emery-Griffiths model [7], is in general different
from zero, even if B is neutral with respect to A and C ($J_{BA} =
J_{BC}$) or if B is a vacancy (binary fluid AC). L is related to
the difference between A and C; J describes the tendency towards
phase separation in an AC mixture; K the tendency towards AC-B
phase separation in the ABC mixture (or towards condensation in
an AC gas).

PARTICULAR CASE: $L = 0$

Suppose now that the state $|0\rangle$ represents either a neutral
atom or a vacancy so that $J_{BA} = J_{BC}$, and that A and C are a given
molecule in different spin states A^+ and A^-, or orientations, or
positions so that by symmetry $J_{AA} = J_{CC}$. Then $L = 0$, and:

$$K = \bar{J}_{AA} + J_{BB} - 2J_{AB},$$

$$D = \mu_A - \mu_B + 2(J_{AB} - J_{BB}),$$

$$H = \tfrac{1}{2}(\mu_{A^+} - \mu_{A^-}), \tag{6}$$

$$\mu_A = \tfrac{1}{2}(\mu_{A^+} + \mu_{A^-}),$$

$$4\bar{J}_{AA} = J_{A^+A^+} + J_{A^-A^-} + 2J_{A^+A^-}.$$

$\bar{J}_{AA}$ is the mean interaction between A molecules, and μ_A the mean chemical potential of A. $L = 0$ for the following systems:

- paramagnetic gas; magnetic alloy;
- pure He^4; He^3-He^4 mixtures;
- nematic fluid; nematic alloy;
- simple fluid; binary liquid.

If, moreover, $J = 0$, condensation or phase separation are driven only by molecular interactions. If, on the contrary, $K = 0$, they are triggered by magnetic, suprafluid, nematic or positional ordering.

5. EQUATIONS OF STATE

The Hamiltonian (4) can be treated in the molecular field approximation. M and Q are then given by:

$$ZQ = 2\exp\beta\{D + 2KQ + 2LM\}\cosh\beta(H + 2JM + 2LQ), \tag{7}$$

$$M = Q\tanh\beta(H + 2JM + 2LQ), \tag{8}$$

with

$$Z = 1 + 2\exp\beta\{D + 2KQ + 2LM\}\cosh\beta(H + 2JM + 2LQ). \tag{9}$$

Suppose the model describes a system with a variable pressure P, Q is the density. In order to calculate P, we use the relation: $P = -\phi$, where ϕ is the free energy of the grand ensemble, whence the equations of the isotherms:

$$P = -\frac{1}{\beta}\log(1 - Q) - KQ^2 - JM^2 - 2LMQ, \tag{10}$$

M being given by (8) as a function of Q. One or two Maxwell plateaux can be constructed numerically on each isotherm (10). However, they have a physical meaning only for $L = 0$: it is well known that in binary mixtures AC ($L \neq 0$), the composition and

pressure of the two phases at equilibrium vary during the isothermal condensation, so that (10) can represent only homogeneous
states.

More generally, the phase diagram can be constructed only by
numerical methods. The parameters J, K, L, D and H are chosen:
at a given temperature T, there is in general only one solution
(M,Q) of (7) and (8) of lowest free energy. T is then varied until two solutions (M_1,Q_1) and (M_2,Q_2) with the same lowest free
energy are found (first order transition): these solutions represent two distinct phases at equilibrium. D and H are then varied
in such a way that a first order transition is found at the same
temperature T, so that an isothermal equilibrium line is constructed. The equation of critical lines can be found using the
Landau theory of second order phase transitions. If M is very
small ($H = 0$, $L = 0$), we get from (8).

$$Q = \frac{1}{2\beta J} , \tag{11}$$

whence from (7):

$$D = T\log \frac{T}{2(2J - T)} . \tag{12}$$

6. APPLICATION TO SIMPLE, BINARY, AND TERNARY FLUIDS

SIMPLE FLUIDS [5]

We must choose $L = 0$, and $H = 0$, since in general no physical
field induces solidification. Equations (8,10) give the iso-
therms. $M = 0$ is always a solution (liquid or gas) but if $v = Q^{-1} < v_B = 2J/T$, (8) has also a solution $M \neq 0$. There is a
branching point B on the isotherm and for $v < v_B$, only the low
pressure branch ($M \neq 0$) may represent stable states. If the
slope of this branch in B is positive, a solidification plateau
can be constructed; if not, solidification is second order (see
figure 1). Figures 2,3 illustrate the various possible aspects
of the phase diagram when the ratio J/K is changed.

The model indicates the possibility of critical melting, which
contradicts a well known result due to Landau [9]. This is owing
to the fact that the order parameter M has an orientational symmetry and cannot properly represent a three-dimensional density
wave. However, solidification at high pressures is driven mainly
by orientational interactions and leads to a very small volume
discontinuity [10], so that the model is not unrealistic.

The phase diagrams of figures 2,3 apply to other systems: He^3-
He^4 mixtures, magnetic alloys, paramagnetic gas.

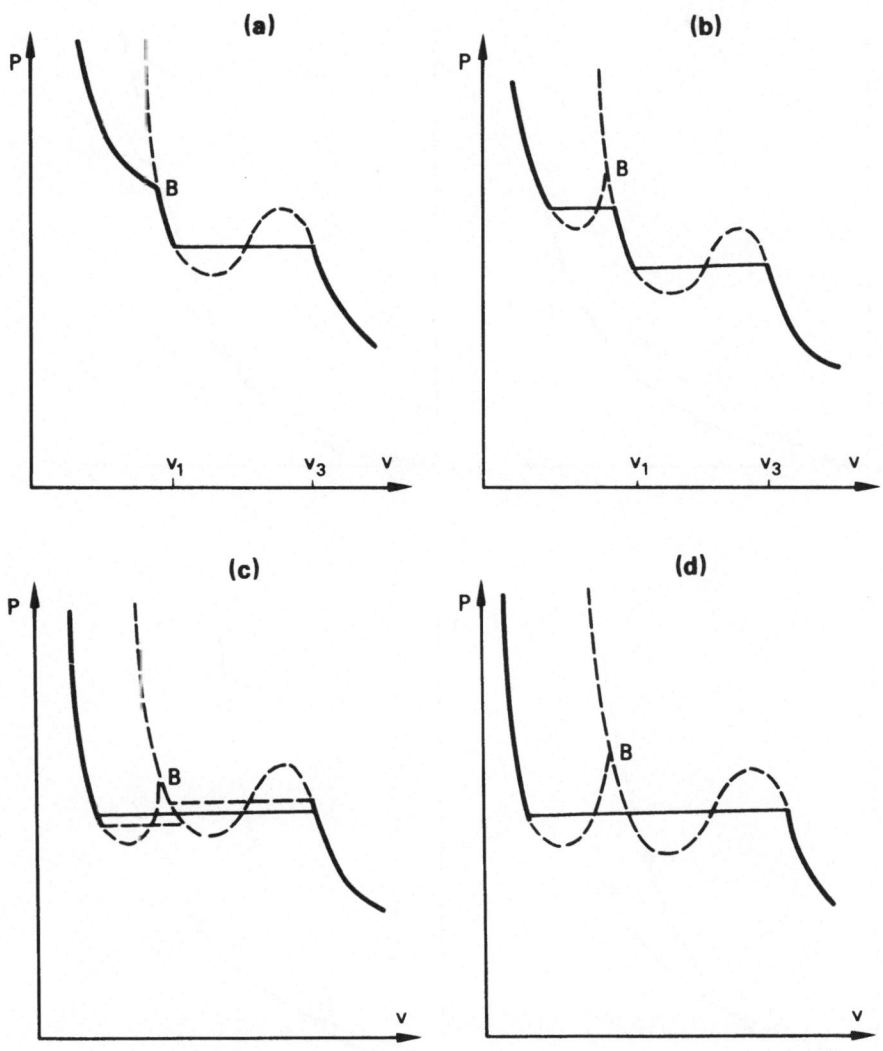

Figure 1 - Isotherms of a simple fluid. (a) Condensation
followed by second order solidification; (b) condensation
followed by first order solidification; (c) solid-gas
equilibrium; (d) triple point.

BINARY FLUIDS [6]

The two chemical species are now called A and B. In general
$L \neq 0$ ($J_{AA} \neq J_{BB}$). If the interactions J_{AA} and J_{BB} are kept constant, the properties of the mixture depend on $J_{AB} = J$.

$$J = 0$$

The solution is ideal (figure 4).

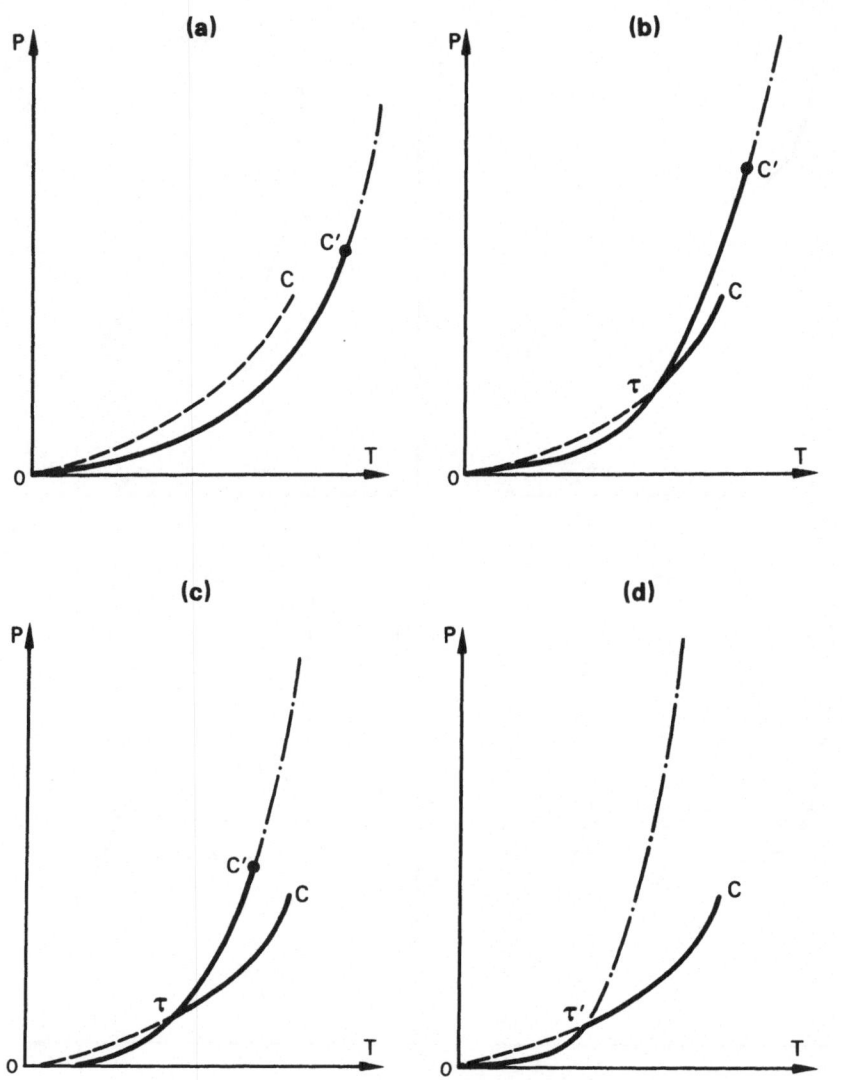

Figure 2 - (P, T) phase diagram of a simple fluid $K = 1$,
$T_C = 0.5$.   (a) $J > 0.365$, no liquid-gas equilibrium curve;
(b) $0.365 > J > 0.33$, $T_{C'} > T_C$; (c) $0.33 > J > 0.26$,
$T_{C'} < T_C$; (d) $J < 0.26$, no tricritical point.

$$-0.5 < J < 0$$

The solution exhibits negative deviation from ideal behaviour;
no azeotrope or maximum critical temperature are found (figure 5).

$$J < -0.5$$

Negative azeotropy is found (figure 6); the maximum critical

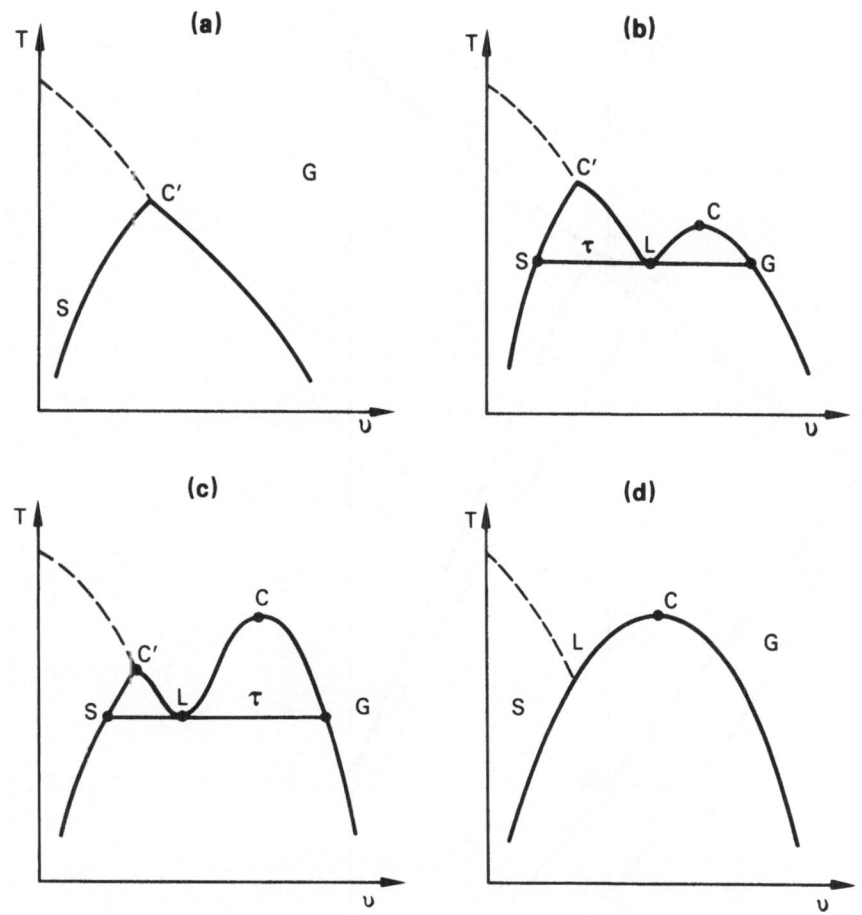

Figure 3 - (T,v) phase diagram of a simple fluid. If the
system is an AB mixture, the following situations are described: (a) phase separation driven by internal ordering
of A (He^3-He^4) mixtures); (d) phase separation driven by
AB molecular interactions.

point is also the critical azeotrope.

$$0.5 > J > 0$$

Positive deviation from ideal behaviour and phase separation
in the liquid phase ($J > 0$) are found; the phase diagram exhibits a liquid-gas critical line, a liquid-liquid critical line
and a liquid-liquid-gas triple line.

$$J > 0.5$$

Positive azeotropy is found and the minimum critical point is
also the critical azeotrope (figure 7). Liquid immiscibility is
also found at lower temperatures (figure 8). The phase diagram
exhibits two critical lines, an azeotropic line and a triple
line [11].

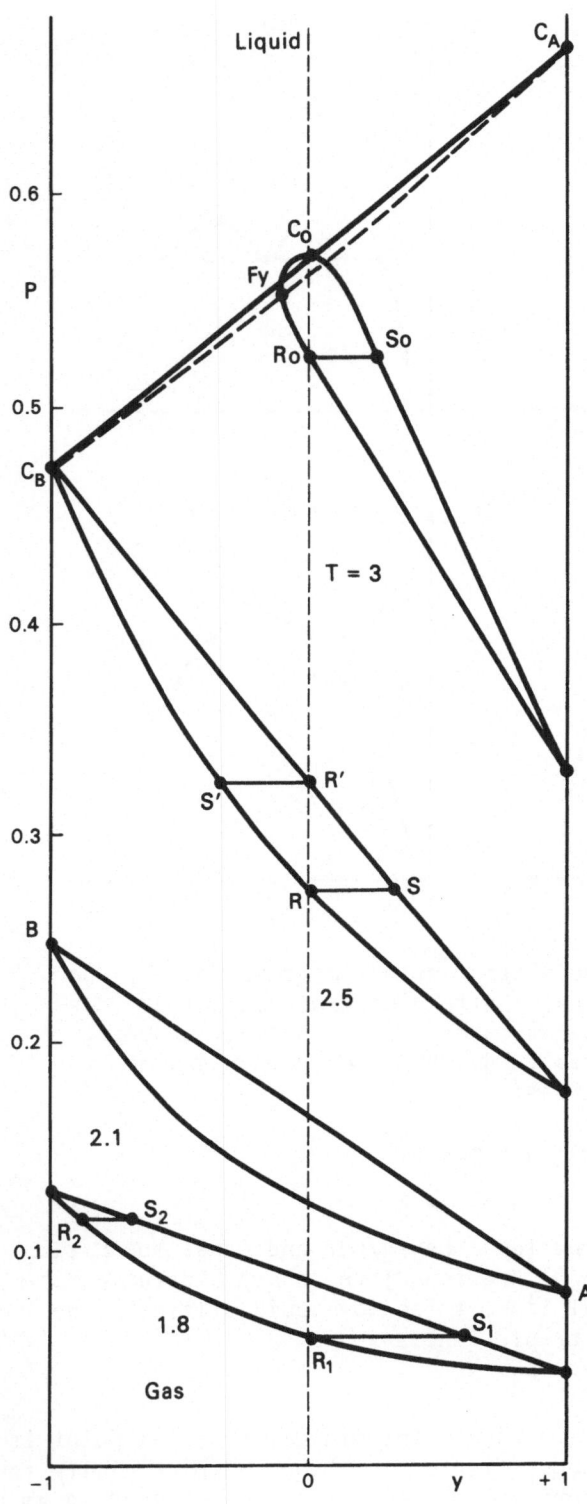

Figure 4 - Ideal AB
mixture (J_{AA} = 7,
J_{BB} = 5). R and S
are two phases at
equilibrium, $C_A C_O C_B$
is the critical
line, $C_A F_Y C_B''$ the
cricondentherm line.

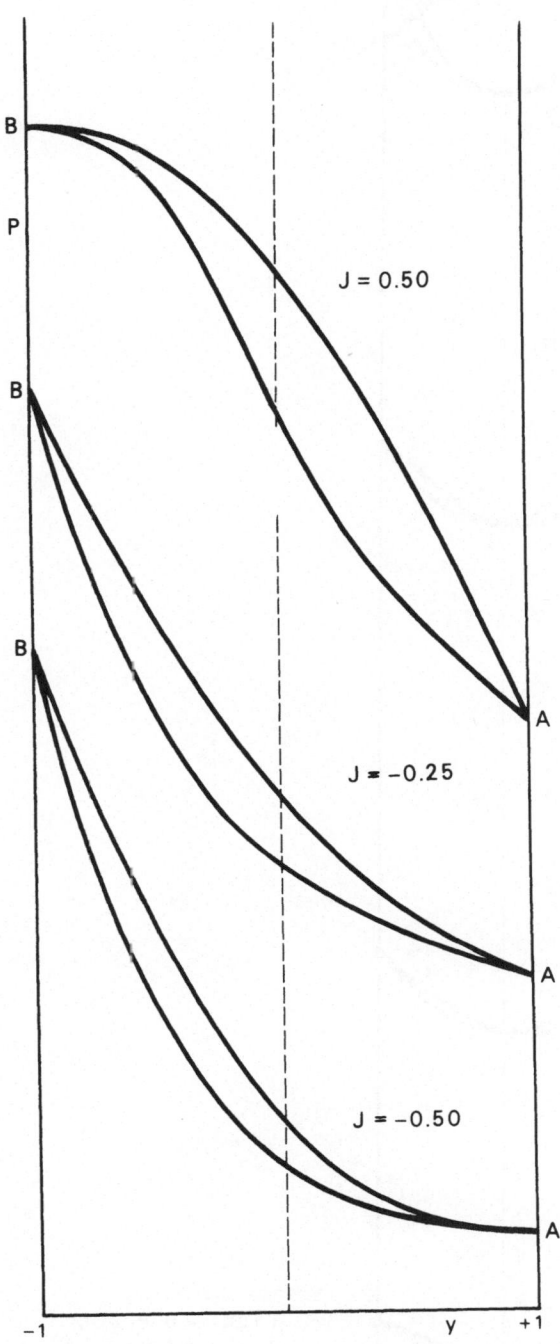

Figure 5 - Non-ideal AB mixture: positive or negative deviation from ideal behaviour without azeotropy.

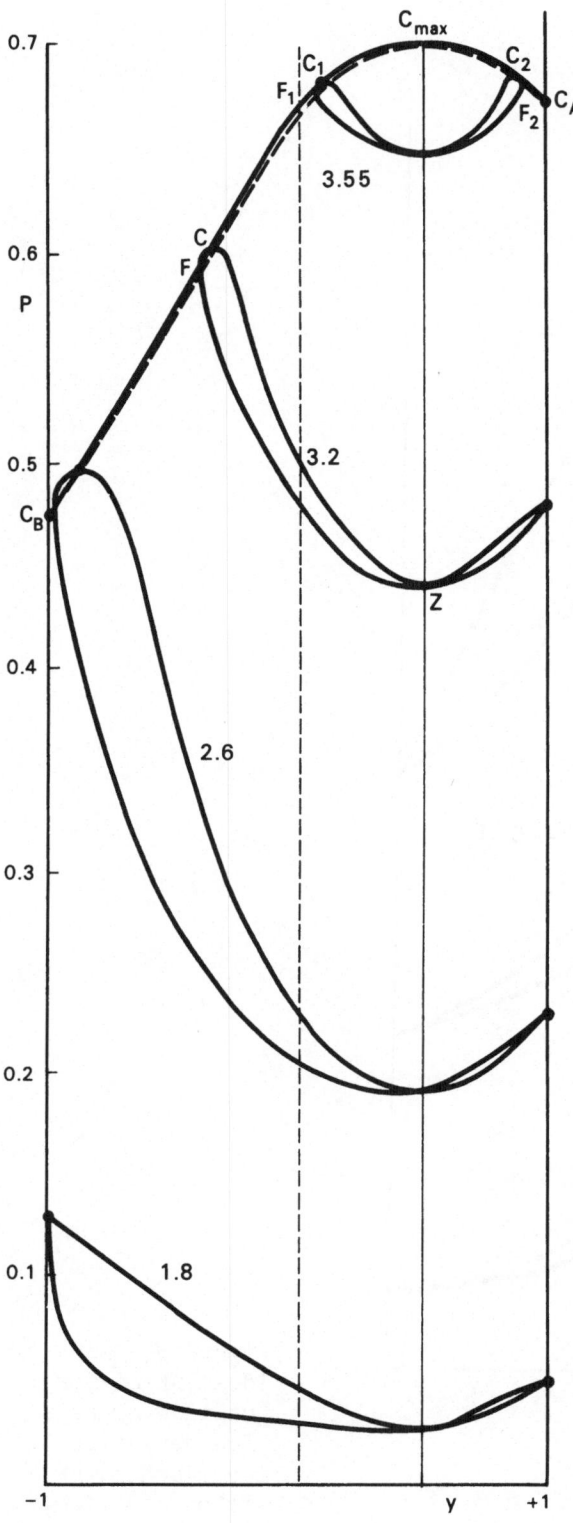

Figure 6 - Non-ideal mixture ($J = -1$) exhibiting negative azeotropy and a maximum critical point.

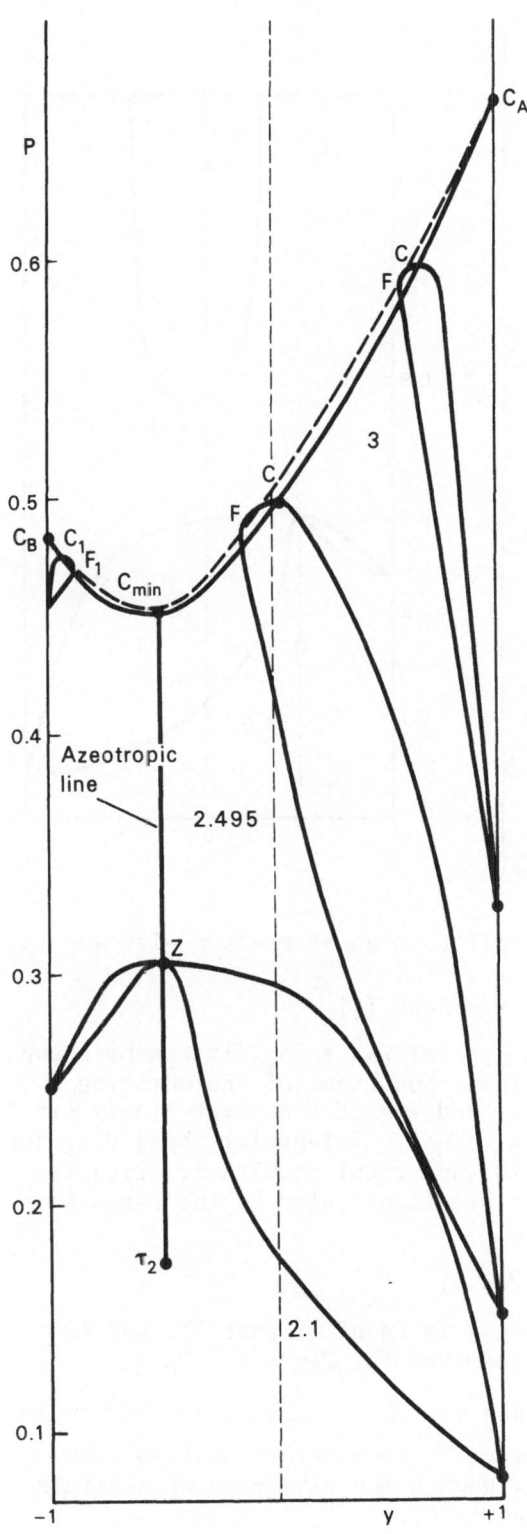

Figure 7 - Non-ideal mixture (J = +1) exhibiting positive azeotropy and a minimum critical point.

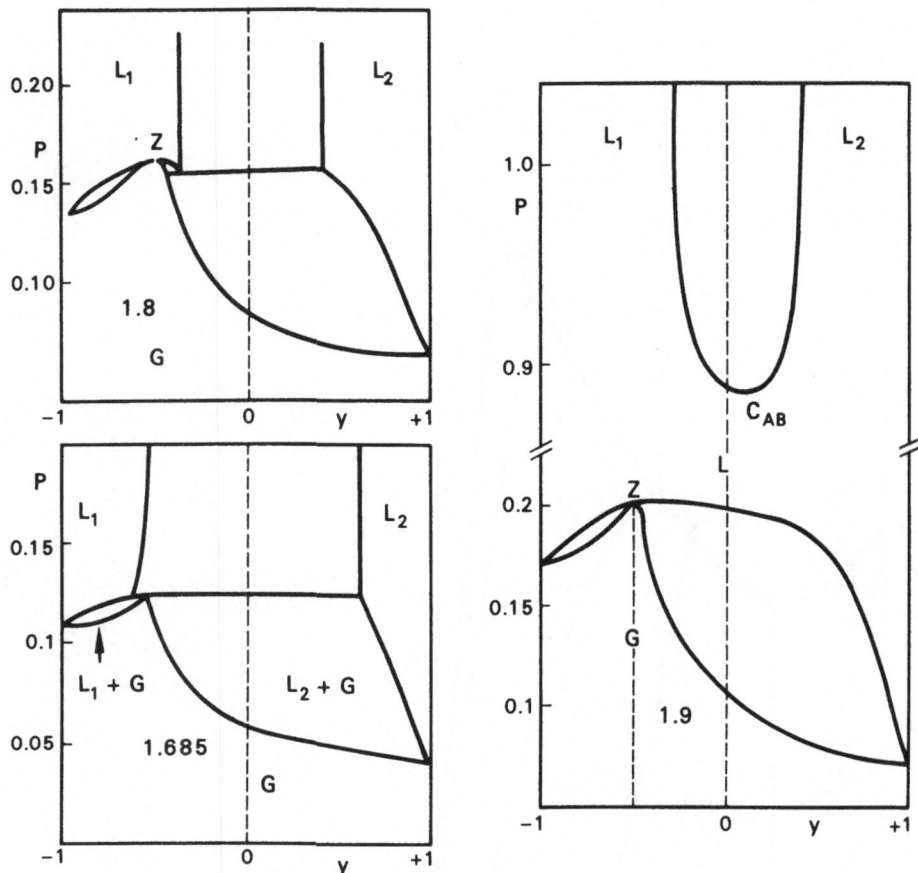

Figure 8 - Liquid immiscibility in an AB mixture with $J = +1$.

TERNARY MIXTURES [6]

In general $L \neq 0$ since in general $T_{AB} \neq T_{BC}$ (it can be shown easily that K, L and J are linear functions of the unmixing critical temperatures T_{AB}, T_{BC} and T_{AC} of the three binary mixtures AB, BC and AC: $\frac{1}{2}L = T_{AB} - T_{BC}$). Triangular phase diagrams are easily constructed from the numerical results relative to binary mixtures. The results are illustrated by the cases $J = -1$ and $J = +1$.

$$J = -1$$

A ternary critical point P_{max} is found (figure 9), and two isothermal plait points are observed for $T_{AB} < T < T_{max}$.

$$J = +1$$

The plait point curve presents a temperature minimum. AC phase separation is possible, whence the existence of a triple line at low temperatures (figure 10).

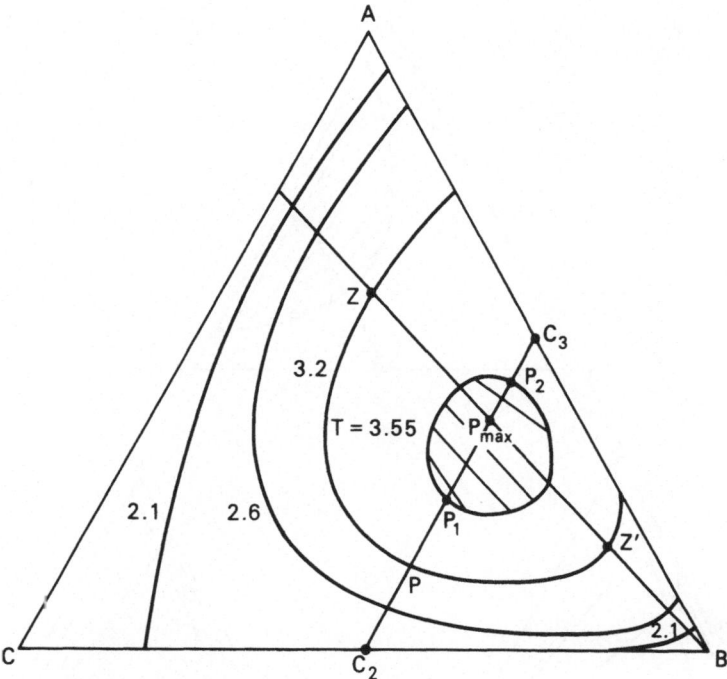

Figure 9 - Ternary mixture with $K = 7$, $L = 0.5$ and $J = -1$: P_{max} is a ternary critical point.

TRICRITICAL POINTS IN BINARY AND TERNARY MIXTURES [6]

No tricritical point can be found in a boundary mixture, or in a ternary mixture kept at a fixed pressure, as a consequence of the phase rule [12,13]. Accidental tricritical points can however, be found for special values of the interactions [14], for instance if $L = 0$ as suggested by the results of Blume, Emery and Griffiths [7].

The property $L = 0$ corresponds, in fact, to a special symmetry of the system: $\mathcal{H}$ is invariant if A and B are interchanged (binary mixture) or if two of the three components are interchanged (ternary mixture). Consequently, M and $-M$ being simultaneous solutions of (7,8) for $H = 0$, the coefficient of the third order term of the Landau free energy expansion is zero: A tricritical point may be found since it corresponds to the realization of only two conditions: $a = c = 0$.

Figures 11,12 illustrate the possibility of a tricritical point in binary and ternary mixtures. Critical points found at $M \neq 0$ generate critical lines which cross with the critical line $M = 0$ at the tricritical point, in agreement with the general scheme of Griffiths [3].

In a ternary mixture described by $L = 0$, $K = 3J$, the Hamil-

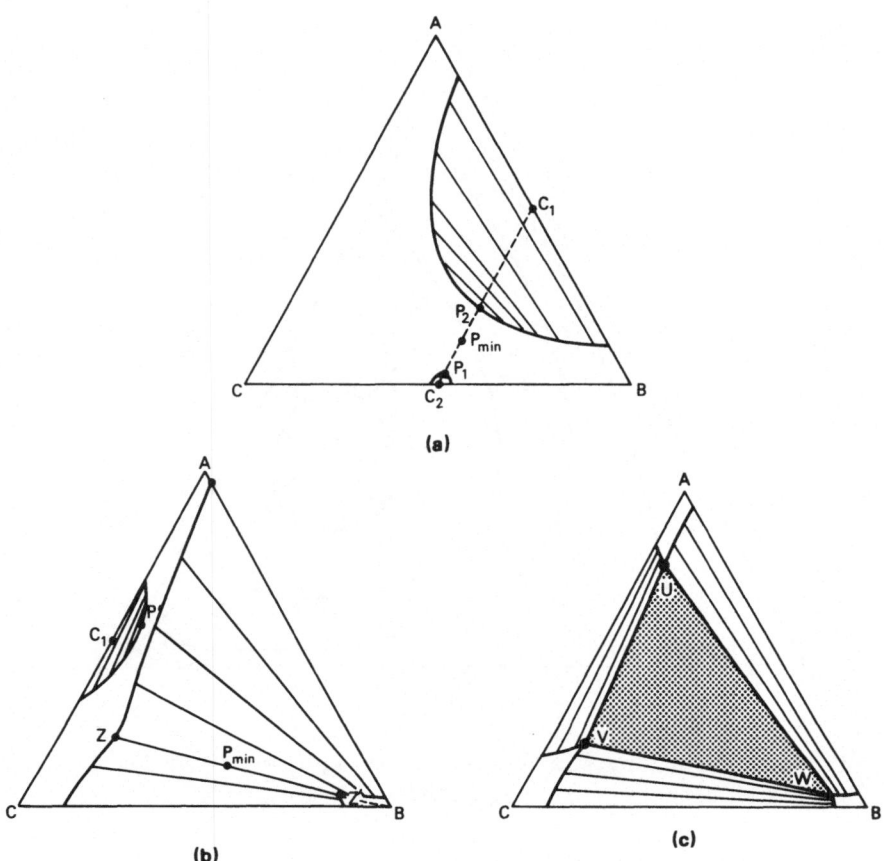

(a)

(b)   **(c)**

Figure 10 - Ternary mixture with $K = 5$, $L = 0.5$ and $J = +1$.
(a) $T = 2.49$, two plait points P_1 and P_2 are found;
(b) $T = 1.9$, AC phase separation is found;
(c) $T = 1.685$, triple point (the stippling denotes the
three-phase region).

tonian is invariant with respect to any interchange of two chemical species. This situation has been described by Straley and
Fisher [14] using the three state Potts model. The Hamiltonian
(4) with $L = D = H = 0$ and $K = 3J$ can be written in the form:

$$\mathcal{H} = - \sum_{i>j} K(3Q_i Q_j + P_i P_j),\qquad(13)$$

since the operator $P_i = (S_i^x)^2 - (S_i^y)^2$ has the same matrix
elements in the basis $|0\rangle$, $(|1\rangle + |-1\rangle)/\sqrt{2}$, $(|1\rangle - |-1\rangle)/\sqrt{2}$ as
S_i^z in the basis $|0\rangle$, $|1\rangle$, $|-1\rangle$. $\mathcal{H}$ which is equivalent to the
Potts model, has a Γ_3 cubic symmetry and three equivalent
ground states (quadrupoles perpendicular to x, y or z). This
explains the symmetry of $\mathcal{H}$ discovered for $K = 3J$ by Kim and
Joseph [15].

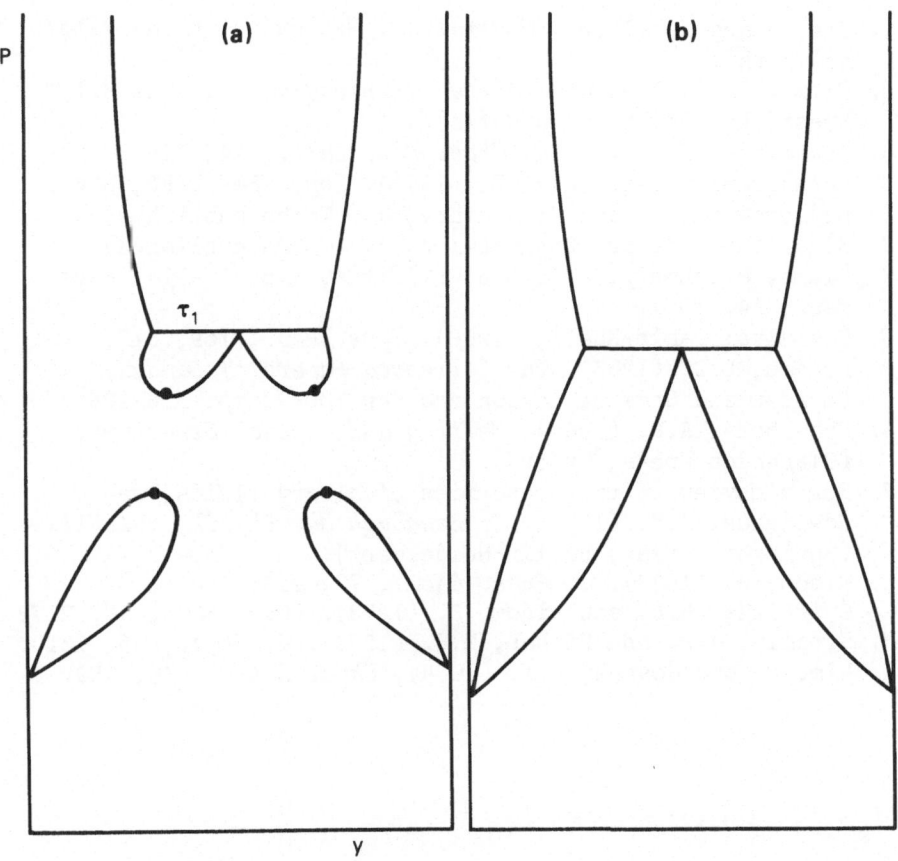

Figure 11 - Tricritical point in a binary mixture with $L =$ 0. (a) The four critical points are found for $H \neq 0$; (b) The critical points disappear at a lower temperature.

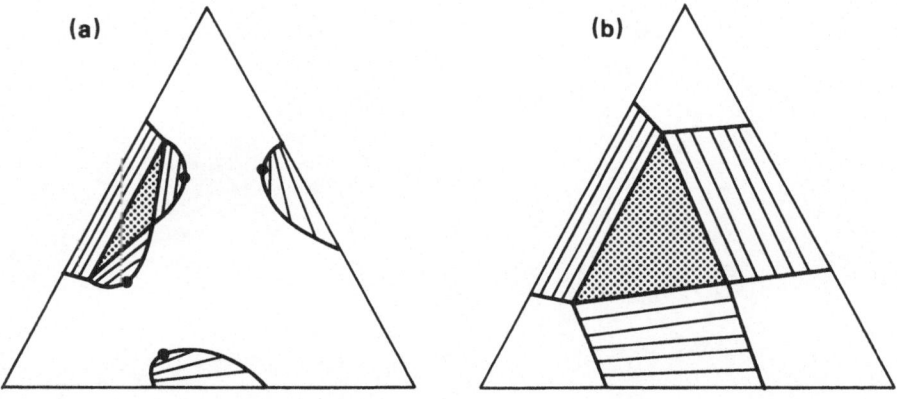

Figure 12 - Tricritical point in a ternary mixture with $L = 0$.

REFERENCES

1. See: Huang, K. (1963). *Statistical Mechanics*, (John Wiley, New York).
2. Tisza, L. (1966). *Generalized Thermodynamics*, (The M.I.T. Press, Boston, Massachusetts).
3. Griffiths, R.B. (1970). *Phys. Rev. Lett.*, **24**, 715.
4. Bernasconi, J. and Rys, F. (1971). *Phys. Rev.*, **B4**, 3045.
5. Lajzerowicz, J. and Sivardière, J. (To be published).
6. Sivardiere, J. and Lajzerowicz, J. (To be published).
7. Blume, M., Emery, V.J. and Griffiths, R.B. (1971). *Phys. Rev.*, **A4**, 1071.
8. See also: Tahir-Kheli, (1968). *Phys. Rev.*, **169**, 517.
9. Landau, L.D. (1965). *The Collected Papers of Landau*, (Gordon and Breach, London and New York), pp. 193-225.
10. Ubbelhode, A.R. (1965). *Melting and Crystal Structure*, (Clarendon Press, Oxford).
11. For a review of the properties of binary fluids, see: Rowlinson, J.S. (1966). In *Handbuch der Physik, Vol.XXI*, (Springer-Verlag, Berlin-Heidelberg).
12. Widom, B. (1973). *J. Phys. Chem.*, **77**, 2196. Griffiths, R.B. and Widom, B. (1973). *Phys. Rev.*, **A8**, 2173.
13. Straley, J.P. and Fisher, M.E. (1973). *J. Phys.*, **A6**, 1310.
14. Kim, D. and Joseph, R.I. (1974). *Phys. Lett.*, **A46**, 359.

DIPOLAR AND QUADRUPOLAR ORDERING
IN MAGNETIC CRYSTALS

JEAN SIVARDIERE

Département de Recherche Fondamentale,
Centre d'Etudes Nucléaires,
BP 85, 38041 Grenoble Cedex, France

1. GEOMETRICAL DESCRIPTION OF DIPOLAR AND QUADRUPOLAR ORDERING IN MAGNETIC CRYSTALS

1.1 INTRODUCTION: DIPOLAR AND QUADRUPOLAR ORDERING IN MOLECULAR CRYSTALS

Consider an array of interacting electric dipoles, for instance polar molecules such as HCl. Bilinear interactions between these dipoles may induce a dipolar ordering below a transition temperature T_d: each dipole i then has a mean orientation along some direction $+z_i$. Neglecting the possibility of non-collinearity and sub-lattice structure, suppose that the ordering is ferroelectric. The order parameter of the transition is the polarization $\langle \cos\theta \rangle$, θ being the angle between a dipole and the polarization axis $+z$. Similarly, consider an array of interacting electric quadrupoles, for instance centro-symmetric linear molecules such as H_2, having no dipolar moment. Interactions between these quadrupoles may induce some kind of ordering below a transition temperature T_q. In particular, if all quadrupoles are aligned by positive interactions, the ordering is called ferroquadrupolar. The order parameter is then the orientation $\langle \cos^2\theta - 1/3 \rangle$, θ being the angle between the quadrupoles and the orientation axis z.

Consider, finally, an array of interacting molecules having dipolar and quadrupolar moments. This situation has been described by Krieger and James [1] in the case of positive interactions. If the quadrupole-quadrupole interaction K is much larger than the dipole-dipole interaction J, a first transition from a para-electric state to a ferro-quadrupolar state is observed

411

as temperature is lowered below T_q; in this state, $\langle \cos^2\theta - 1/3 \rangle \neq 0$
and $\langle \cos\theta \rangle = 0$: the molecules have a mean orientation, but the
dipolar moments are directed at random. Dipole-dipole interac-
tions may, however, induce a second transition from the quadrupolar state to a ferroelectric state below a second transition
temperature T_d. If, on the contrary, the interaction J is larger than the interaction K, a single transition from the para-
electric state to the ferroelectric state is observed at T_{dq};
ferroelectric ordering implies quadrupolar ordering, since
$\langle \cos\theta \rangle \neq 0$ implies $\langle \cos^2\theta - 1/3 \rangle \neq 0$ (kinematical coupling).
For instance, two separate phase transitions are observed in HBr
and HI, whereas a single one is observed in HCl [2].

Multipolar ordering in molecular crystals is also observed,
for instance octupolar ordering in NH_4Cl and CH_4, which contain
regular tetrahedral ions or molecules. Nematic ordering in
liquid crystals made of rod-like molecules is a special kind of
ferro-quadrupolar ordering [3]. Suppose, however, that the molecules are elongated and flat (rectangular plates). Two different
quadrupolar ordered phases are possible, as discussed by Freiser
[4]: a bi-axial phase in which both the longest axis and the flat
faces of the molecules tend to be parallel ($\langle \cos^2\theta - 1/3 \rangle \neq 0$
and $\langle \cos 2\phi \rangle \neq 0$, θ being the angle between the longest axis and
the nematic axis, and ϕ defining the orientation of the flat
face); and a uniaxial phase in which only the longest axis of
the molecules tend to be parallel ($\langle \cos^2\theta - 1/3 \rangle \neq 0$ and
$\langle \cos 2\phi \rangle = 0$).

1.2 DIPOLAR AND QUADRUPOLAR ORDERING
IN MAGNETIC CRYSTALS

In most magnetic crystals, only bilinear interactions between
magnetic moments need to be considered. They may induce ordinary magnetic ordering, which we call here dipolar ordering. In
the simplest case, the ordering is ferromagnetic and the order
parameter is the magnetization. However in some crystals such
as MnAs [5], UO_2 [6], UP [7], DySb [8], ... bi-quadratic interactions cannot be neglected. They may even be larger than the dipolar ones if the magnetic ions are Jahn-Teller ions, and lead
to a separate phase transition. Consequently the possibility of
purely quadrupolar ordering in magnetic crystals must be investigated, and in this section we discuss mainly the geometrical
aspects of this problem.

1.2.1 Origin of the Dipolar and Quadrupolar Interactions

Bilinear and biquadratic interactions will be called dipolar
and quadrupolar respectively, in order to suggest that they produce dipolar or quadrupolar ordering. Dipolar interactions may
be magnetic dipole-dipole, or isotropic or anisotropic bilinear
exchange interactions. Quadrupolar interactions may be electric
quadrupole-quadrupole interactions [8], or biquadratic exchange
interactions due to the presence of the orbital moment [9,10].
They may arise as well from bilinear exchange projected into the

low lying levels of a multiplet, or from virtual phonon exchange between ions via the quadrupole-lattice coupling (Jahn-Teller coupling) [9,10]. The rôle of super-exchange interactions has been considered too [11]. In the following, these interactions will be introduced as phenomenological parameters, their exact nature is unimportant as far as thermodynamics is concerned.

Quadrupolar interactions between magnetic quadrupoles can be compared to the interactions among rigid molecular quadrupoles. In molecular and liquid crystals however, the existence of local quadrupoles is owing to strong intramolecular forces which fix the shape of the molecules. The much weaker intermolecular forces depend on the relative orientation of the molecules. In magnetic crystals, on the contrary, the existence of the local quadrupoles (i.e. the asphericity of the spin density) *and* their alignment arise in general together from the interionic interactions.

1.2.2 Dipolar and Quadrupolar Order Parameters

We consider only ferro-ordering of dipoles or quadrupoles. The three dipolar order parameters are the three components of the magnetization (a vector):

$$M_x = \langle S^x \rangle ,$$

$$M_y = \langle S^y \rangle , \tag{1}$$

$$M_z = \langle S^z \rangle .$$

They transform in the D_1 representation of the rotation group, and only M_z is different from zero if z is the magnetization axis. Similarly five quadrupolar order parameters must be introduced [12] in order to describe non-magnetic states in which the spin density is not spherical but ellipsoidal, and can be represented by a second order tensor. They can be expressed as simple linear combinations of the traceless spin harmonics of order two, and transform in the D_2 representation of the rotation group:

$$Q = \langle (S^z)^2 - \frac{S(S+1)}{3} \rangle \sim Y_2^0 ,$$

$$P = \langle (S^x)^2 - (S^y)^2 \rangle \sim Y_2^2 + Y_2^{-2} ,$$

$$P' \text{ or } P_{xy} = \langle S^x S^y + S^y S^x \rangle \sim Y_2^2 - Y_2^{-2} , \tag{2}$$

$$P_{xz} = \langle S^x S^z + S^z S^x \rangle \sim Y_2^1 + Y_2^{-1} ,$$

$$P_{yz} = \langle S^y S^z + S^z S^y \rangle \sim Y_2^1 - Y_2^{-1} ,$$

Only P and Q are different from zero if x, y and z are the principal axis of the quadrupole, and then

$$\left\langle (S^x)^2 \right\rangle = \frac{S(S+1)}{3} - \frac{Q}{2} + \frac{P}{2} \, ,$$

$$\left\langle (S^y)^2 \right\rangle = \frac{S(S+1)}{3} - \frac{Q}{2} - \frac{P}{2} \, , \qquad (3)$$

$$\left\langle (S^z)^2 \right\rangle = \frac{S(S+1)}{3} + Q .$$

$M = 0$, $P = 0$, $Q \neq 0$ represents an axial non-magnetic spin density, a situation found also in nuclear oriented states. See table I for a comparison with molecular and liquid crystals. According to (3), $(Q,0)$, $(-Q/2,3Q/2)$ and $(-Q/2,-3Q/2)$ describe equivalent axial quadrupoles, elongated along z, x or y. Dipolar and quadrupolar parameters are not independent: for instance, $M_z \neq 0$ or $P \neq 0$ implies $Q \neq 0$ (kinematical coupling). Multipolar order parameters can be defined too [13]: for instance the octupolar parameter $\left\langle S^x S^y S^z \right\rangle$ describes a tetrahedral spin density.

TABLE I

Comparison Between Molecular (or Liquid) Crystals and Magnetic Crystals

Type of Ordering	Molecular or Liquid Crystals	Magnetic Crystals	Schematic Picture
Dipolar	$\langle \cos\theta \rangle \neq 0$	$M_z \neq 0$	
Axial Quadrupolar	$\langle \cos^2\theta \rangle \neq 0$ $\langle \cos\theta \rangle \neq 0$, $\langle \cos 2\phi \rangle = 0$	$Q \neq 0$ $M_z = 0$, $P = 0$	
Biaxial Quadrupolar	$\langle \cos^2\theta \rangle \neq 0$ $\langle \cos 2\phi \rangle \neq 0$ $\langle \cos\theta \rangle = 0$	$Q \neq 0$ $P \neq 0$ $M_z = 0$	

1.2.3. Examples of Dipolar and Quadrupolar Phase Transitions

1.2.3.1 *Transition-Metal-Ion Compounds*

Cooperative distortions are frequently observed in substances containing ions with orbital degeneracy, for instance:

> — spinels with cations, notably Mn^{3+} and Cu^{2+}, in a doubly degenerate state E at the octahedral sites (Mn_3O_4, $ZnMn_2O_4$, $MgMn_2O_4$, $CuFe_2O_4$) [14];

> — spinels with cations in a triply degenerate state T_1 or T_2 at the tetrahedral sites ($NiCr_2O_4$ [15], $CuCr_2O_4$);

> — perovskites with octahedrally coordinate cations in an E state ($LaMnO_3$, $KCuF_3$, $KCrF_3$) [11,16].

Most spinels exhibit below T_q a cubic to tetragonal ferro-quadrupolar distortion described by the parameter Q, whereas in perovskites an orbital super-structure is observed. In these compounds magnetic ordering is found below a second critical temperature T_d; orbital ordering induces a strong anisotropy of the magnetic properties. In other crystals ($RbFeF_3$, FeO, CoO, $MgCr_2O_4$), a simple transition is observed from an undistorted paramagnetic state to a distorted magnetic state at T_{qd}.

1.2.3.2 *Rare Earth Compounds*

Various cooperative Jahn-Teller and magnetic phase transitions have been discovered recently in the tetragonal rare earth vanadates, arsenates and phosphates [17]. In some cases, two transitions at T_q and T_d have been found. When $T > T_q$, only the parameter Q is different from zero, because of the tetragonal symmetry. In $DyVO_4$ and $DyAsO_4$, a distortion along [100] or [010] is observed below T_q: the order parameter is P', and the two orthorhombic domains differ only in the sign of P'. The distortion is followed by magnetic ordering along [110] ($P' > 0$) or [1$\bar{1}$0] ($P' < 0$). Similarly, in $TbVO_4$, $TbAsO_4$, $TmVO_4$ and $TmAsO_4$, a distortion is observed below T_q along [100] or [010]: the order parameter is P. Magnetic ordering may be found along [100] ($P > 0$) or [010] ($P < 0$) below $T_d < T_q$. Finally, in $TbPO_4$ and $TmAsO_4$, a monoclinic distortion is likely to be observed [18]: the order parameters are P and P_{xz} or P_{yz}, and four monoclinic domains are found ($P \gtrless 0$, P_{xz} or $P_{yz} \neq 0$). Below T_d, the moments order off the c-axis in the a-b or b-c plane. See figure 1 for a comparison between $DyVO_4$, $TbVO_4$ and $TbPO_4$.

In TmCd [19] a cubic $\rightarrow$ tetragonal first order distortion is found. $PrAlO_3$ [20] exhibits three successive structural transitions. Finally in UO_2 [6] and DySb [8] there is a single first order transition from the paramagnetic undistorted state to a distorted magnetic state.

1.2.3. Classification of Quadrupolar Structures

It is interesting to predict and classify all the ordered

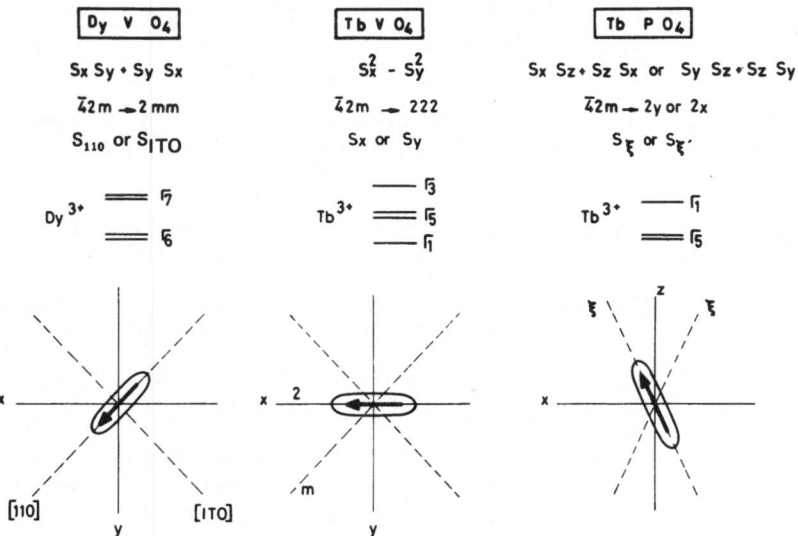

Figure 1 - Comparison between TbVO$_4$, DyVO$_4$ and TbPO$_4$. For each compound we give the quadrupolar order parameter, the corresponding symmetry reduction, the dipolar order para- meter, an illustrative picture and the low lying crystal- line field levels above T_q.

quadrupolar structures or arrangements in a crystal. Two meth- ods have been proposed and can be shown to be equivalent, as in the analogous case of the classification of magnetic structures: the Opechowski and Guccione method [21], and the Landau method of group representations [22]. In the Landau method, which is most adapted to the study of phase transitions, normal quadru- polar modes are defined, and classified according to the rep- resentations of the high temperature space group. In the fol- lowing we shall consider only ferro-distortive transitions: only some representations of the point group G need to be considered to describe the deformation of the unit cell, namely the ir- reducible components of $[V^2]$, V being the vector representation.

Table II gives the classification of deformations for $G = \bar{4}2m(D_{2d})$. If the representation Γ_α which describes the de- formation is unidimensional, only 'antiphase' domains are found (e.g. $P \gtrless 0$ in TbVO$_4$). If the dimension of Γ_α is larger than one, domains may be found which are analogous to the S-domains in magnetic structures (e.g. P_{xz} or P_{yz} in TbPO$_4$). Table II in- dicates also the elastic constants which go to zero at the tran- sition point if the transition is second order.

According to Landau, the transition is permitted to be second order if a single representation $\Gamma_\alpha \neq \Gamma_1$ is involved just below T_q; if $[\Gamma_\alpha^3]$ does not contain Γ_1 (no third order term in the free energy expansion), and if $\{\Gamma_\alpha^2\}$ has no components in common with V (stability condition). Since $[\Gamma_3^3] = \Gamma_3$; $[\Gamma_4^3] = \Gamma_4$;

TABLE II

Dipolar and Quadrupolar Order Parameters,
Deformations and Corresponding Elastic Constants
for the Group G = $\bar{4}2m$
(Examples are given in the last column)

Γ_1 (A_1)		$S_x{}^2 + S_y{}^2$; $\quad S_z{}^2$	$U_{xx} + U_{yy}$; $\quad U_{zz}$		
Γ_2 (A_2)	S_z				DyPO$_4$
Γ_3 (B_1)		$S_x{}^2 - S_y{}^2$	$U_{xx} - U_{yy}$	$C_{11} - C_{12}$	TbVO$_4$
Γ_4 (B_2)		$S_xS_y + S_yS_x$	U_{xy}	C_{66}	DyVO$_4$
Γ_5 (E)	$\begin{bmatrix} S_x \\ S_y \end{bmatrix}$	$\begin{bmatrix} S_yS_z + S_zS_x \\ -S_zS_x - S_xS_z \end{bmatrix}$	$\begin{bmatrix} U_{yz} \\ -U_{zx} \end{bmatrix}$	C_{44}	TbPO$_4$

$[\Gamma_5{}^3] = 2\Gamma_5$; $\{\Gamma_5{}^2\} = \Gamma_4$ and $V = \Gamma_2 + \Gamma_5$, the last two Landau conditions are satisfied for all Γ_α. Table II shows also that the quantities $P.(P_{xz}{}^2 - P_{yz}{}^2)$ and $P'.P_{xz}.P_{yz}$ are invariant; this means that ordering in P_{xz} or P_{yz} implies ordering in P, and ordering in $P_{xz} = P_{yz}$ implies ordering in P' (i.e. monoclinic distortion implies orthorhombic distortion). P or P' is a secondary order parameter (similarly a transition in S_x would induce an orthorhombic distortion). These third order terms are negligible just below T_q, so that the first Landau condition is satisfied too. Finally the quantities $(P_{xz}{}^2 - P_{yz}{}^2)^2$ and $(P_{xz}.P_{yz})^2$ are invariant and are equal to $p^4\cos^2 2\theta$ and $p^4\sin^2 2\theta$ respectively with $p_{xz} = p\cos\theta$ and $P_{yz} = p\sin\theta$; they are minimum for $\theta = 0, \frac{1}{2}\pi$ so that if the transition is second order the distortion is necessarily monoclinic and not triclinic.

Table III gives the classification for $G = 23$ and 432. Since $[\Gamma_3{}^3] = \Gamma_1 + \Gamma_2 + \Gamma_3$, the free energy expansion contains a third order term $Q^3 - QP$ whence in general a first order transition:

$$\phi_3 = \phi_0 + a(T)(3Q^2 + P^2)$$

$$+ \; b(T)(Q^3 - QP^2) + c(T)(3Q^2 + P^2)^2 + \ldots . \qquad (4)$$

With

$$\sqrt{3}Q = q\cos\theta,$$
$$P = q\sin\theta, \qquad\qquad (5)$$

TABLE III

Classification for the Groups G = 432 and 23

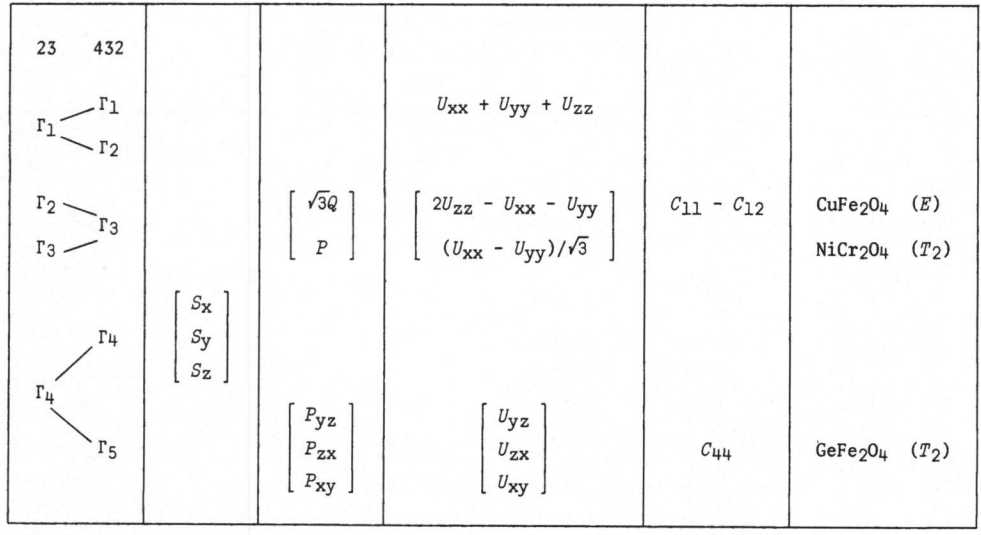

one gets

$$\phi_3 = \phi_0 + a(T)q^2 + b(T)q^3\cos3\theta + c(T)q^4 + \dots ,$$

(6)

$$\frac{\partial\phi_3}{\partial\theta} = 0 \quad \text{for} \quad \theta = 0 \quad \text{or} \quad \pm\frac{2\pi}{3} \quad (P = 0 \quad \text{or} \quad P = \pm 3Q),$$

which corresponds to tetragonal distortions along z, x or y.
The first order character of the cubic → tetragonal or orthorhombic (432 → 422 or 23 → 222) distortion is in agreement with a theorem by Landau (the order of the group is reduced by a factor 3).
We consider finally a transition in which the order parameter is
some linear combination of P_{xy}, P_{yz} and P_{zx}. Since $[\Gamma_5^3] = \Gamma_1 + \Gamma_4 + 2\Gamma_5$, the free energy expansion contains the third order invariant $P_{xy} \cdot P_{yz} \cdot P_{zx}$. Since P_{xy}, P_{yz} and P_{zx} transform in Γ_3, we
may write

$$P_{xy} = p\cos\theta,$$

$$P_{yz} = p\sin\theta\sin\phi,$$ (7)

$$P_{zx} = p\sin\theta\cos\phi,$$

whence

$$P_{xy}^2 + P_{yz}^2 + P_{zx}^2 = 1,$$

$$\left.\begin{array}{c} \\ \\ \end{array}\right\} \quad (8)$$

$$P_{xy} \cdot P_{yz} \cdot P_{zx} = p^3 \sin^2\theta\cos\theta\sin\phi\cos\phi,$$

which is minimum for: $\phi = \frac{1}{4}\pi + n\frac{1}{2}\pi$ and $\cos\theta = 1/\sqrt{3}$, or $P_{xy} = \pm P_{yz} = \pm P_{zx}$, which describes a trigonal distortion along a [111] direction.

2. DIPOLAR AND QUADRUPOLAR PHASE TRANSITIONS IN $S = 1$ AND $3/2$ MODELS

2.1 THE ISING HAMILTONIAN AND THE MOLECULAR FIELD TREATMENT

We consider the simplest, Ising-like Hamiltonian involving dipolar and quadrupolar interactions between $S = 1$ or $3/2$ ions on a lattice:

$$\mathcal{H} = - \sum_{i,j} J_{ij}S_i^z S_j^z - \sum_{i,j} K_{ij}Q_iQ_j - H\sum_i S_i^z - D\sum_i Q_i. \quad (9)$$

The uniaxial operators S_i^z and $Q_i = (S_i^z)^2 - S(S+1)/3$ are traceless and each interaction is counted twice. Within the molecular field approximation (MFA), we shall investigate the possibility of ordering in the independent parameters $M = \langle S_i^z \rangle$ and $Q = \langle Q_i \rangle$, and discuss the phase diagram of the system as a function of the ratio J/K, where J and K are the Fourier transforms of the interaction J_{ij} and K_{ij}; for instance, $J = J(0) = \sum_j J_{ij}$.

We consider first the case of $S = 1$ ions, derive self-consistent MFA equations for M and Q and discuss the phase diagram, which is contrasted then with the phase diagram for $S = 3/2$ ions. We are mainly interested in the possibility of successive transitions. Hamiltonians with a different symmetry will be considered later.

In order to derive self-consistent equations in M and Q, we use the variational principle [23] for the free energy F. If ρ is any approximate density matrix for the system:

$$F(\rho_{exact}) \leqslant F(\rho) = \mathrm{tr}\rho\mathcal{H} + \frac{1}{\beta} \mathrm{tr}\rho\log\rho. \quad (10)$$

Let us take $\rho = \rho_0$, the molecular field density matrix:

$$\rho_0 = \frac{e^{-\beta\mathcal{H}_0}}{\mathrm{tr}e^{-\beta\mathcal{H}_0}} = \frac{e^{-\beta\mathcal{H}_0}}{z_0}, \quad (11)$$

whence

$$F(\rho_{exact}) \leqslant F(\rho_0) = \phi, \tag{12}$$

with

$$\phi = tr\rho_0 \mathcal{H} + \frac{1}{\beta} tr\rho(-\beta\mathcal{H}_0 - tre^{-\beta\mathcal{H}_0})$$

$$= -\frac{1}{\beta} logZ_0 + tr\rho(\mathcal{H} - \mathcal{H}_0)$$

$$= -\frac{1}{\beta} logZ_0 + \langle\mathcal{H}\rangle_0 - \langle\mathcal{H}_0\rangle_0. \tag{13}$$

We have

$$\langle\mathcal{H}\rangle_0 = -NJM^2 - NKQ^2 - NHM - NDQ, \tag{14}$$

and

$$\mathcal{H}_0 = -N(h + H)S^z - N(d + D)\left[(S^z)^2 - \frac{S(S+1)}{3}\right], \tag{15}$$

h and d being the dipolar and quadrupolar molecular fields respectively, whence:

$$\langle\mathcal{H}_0\rangle_0 = -N(h + H)M - N(d + D)Q, \tag{16}$$

h and d are determined by the conditions

$$\frac{\partial\phi}{\partial h} = 0, \tag{17}$$

and

$$\frac{\partial\phi}{\partial d} = 0, \tag{18}$$

which give

$$h = 2JM, \tag{19}$$

$$d = 2DQ, \tag{20}$$

and

$$\phi = -\frac{1}{\beta} logZ_0 + JM^2 + KQ^2. \tag{21}$$

The order parameters M and Q are defined by

$$M = \frac{1}{\beta Z_0} \frac{\partial Z_0}{\partial h} ,$$ (22)

and

$$Q = \frac{1}{\beta Z_0} \frac{\partial Z_0}{\partial d} .$$ (23)

The values of h and d (equations (19,20)) are reported in the above equations (22,23), whence two self-consistent equations in M and Q which are solved numerically. If several solutions are found at a given temperature, only the solution which leads to the lowest free energy is kept. Three types of solutions must be considered:

$Q = 0$, $M = 0$: disordered state;

$Q \neq 0$, $M = 0$: paramagnetic quadrupolar state;

$Q \neq 0$, $M \neq 0$: ferromagnetic state.

2.2 PHASE DIAGRAMS

2.2.1 $S = 1$ Model [24-26]

M and $Q' = Q + 2/3 = \langle S_z^2 \rangle$ are solutions of the two self-consistent equations:

$$Q' = \frac{2e^{\beta(D+2KQ')}\cosh\beta(H + 2JM)}{1 + 2e^{\beta(D+2KQ')}\cosh\beta(H + 2JM)} ,$$ (24)

and

$$\frac{M}{Q'} = \tanh\beta(H + 2JM).$$ (25)

The phase diagram in the $(T/K, J/K)$ plane is represented in figure 2 for $D = H = 0$ ($K = 1$). τ is a triple point, C' a tricritical point (successive phase transitions are found only if $D \neq 0$ [12]).

When J/K is large, a single second order transition is found. When the ratio K/J, i.e. the importance of biquadratic interactions, is increased, the transition becomes first order, as in MnAs or DySb. When a single transition in M and Q is found, one may say that M is the principal order parameter since it has a lower symmetry than Q. The tricritical point C' is located using the Landau free energy expansion. Three critical lines meet in C': the line of second order dipolar ordering at $H = 0$ and two critical lines at $H > 0$ and $H < 0$ (when the transition at $H = 0$

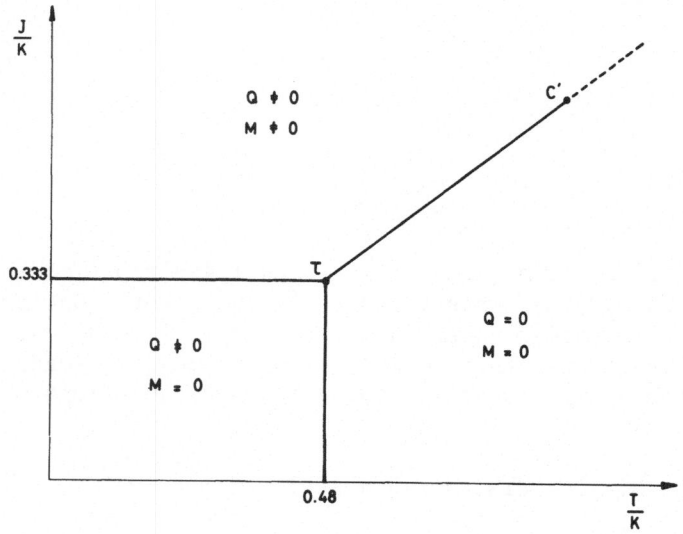

Figure 2 - Phase diagram of an $S = 1$ Ising system with
dipolar and quadrupolar interactions. The dashed line
is a second order transition line.

is first order, the jump of M and Q at the transition point is
reduced by the application of H till a critical point is obtained).

When $J/K < 1/3$, there is a single first order transition between the paramagnetic quadrupolar state and the disordered
state. The thermal variation of Q and of the single ion levels
is given in figure 3. In the ordered state, the quadrupoles are
oblate ('pancakes') whereas for $S > 1$ they are prolate ('cigars').
Two remarks explain why the transition is first order: the transition involves no symmetry change; Q and $-Q$ describe different
states, whence a third order term in the free energy expansion
(whereas M and $-M$ describe equivalent domains in the $S = \frac{1}{2}$ Ising
Model).

2.2.2 $S = 3/2$ Model [27]

The self-consistent equations in M and Q are given in reference [27]. As for $S = 1$, the dipolar transition is first order
when biquadratic exchange is important. However, even if $D = 0$,
two successive phase transitions are found when $J \ll K$ (figure 4:
$D = H = 0$). τ is a triple point, C' and C'' are tricritical
points. Figure 5 illustrates the case of two successive second
order transitions. The phase diagram, found also for $S > 3/2$,
is typical of systems with two kinematically coupled order parameters and competing interactions.

The quadrupolar transition is second order, whereas for any
other value of S it is first order. This accidental situation
is due to the fact that the coefficient of the third order term

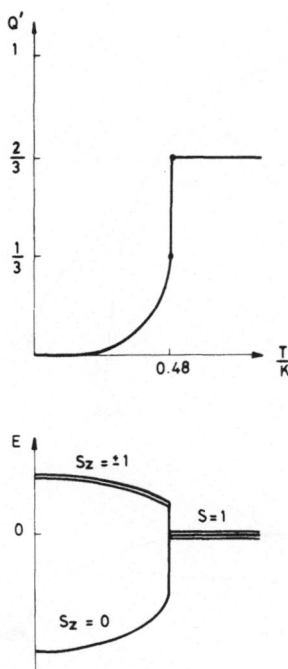

Figure 3 - Quadrupolar transition in an $S = 1$ Ising system.

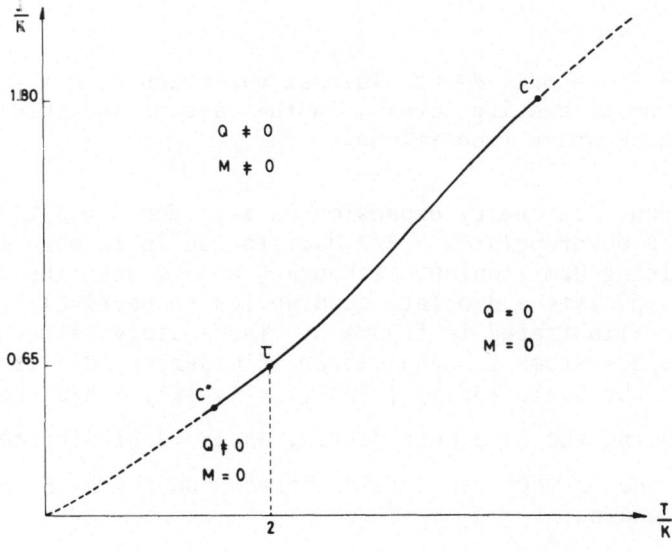

Figure 4 - Phase diagram of an $S = 3/2$ Ising system with
dipolar and quadrupolar interactions.

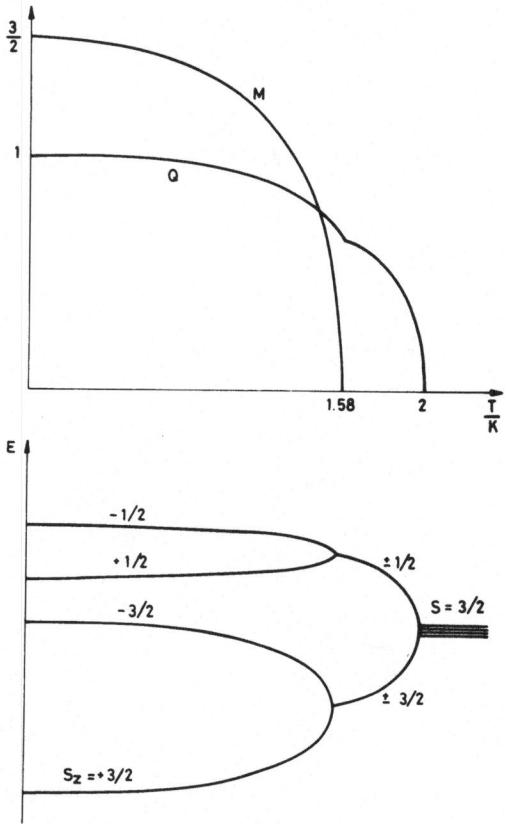

Figure 5 - $S = 3/2$, $J \ll K$: Thermal variation of Q and M, and of the single ion levels, in the case of two successive second order transitions.

of the Landau free energy expansion is zero for $S = 3/2$. More generally, a quadrupolar $S = 3/2$ Hamiltonian is isomorphic to the $S = \frac{1}{2}$ Ising Hamiltonian, although Q and $-Q$ describe different states (prolate and oblate quadrupoles respectively). These remarks are illustrated in figure 6. The analogy between the $S = \frac{1}{2}$ and $3/2$ systems is even closer if ordering in P is considered: in the basis $|3/2\rangle$, $|-3/2\rangle$, $|-\frac{1}{2}\rangle$, $|\frac{1}{2}\rangle$, P has the form $\sqrt{3}\begin{bmatrix} 01 \\ 10 \end{bmatrix}$, 1 being the 2×2 unit matrix, which is similar to S_x for $S = \frac{1}{2}$. To the $\pm x$ magnetic domains correspond the x ($P > 0$) and y ($P < 0$) domains.

2.2.3 More General Hamiltonians

A quadrupolar Hamiltonian involving uniaxial and biaxial interactions has been considered in reference [12]. When the single ion anisotropy D is different from zero, successive transitions in Q and P are found. Hamiltonians with other symmetries

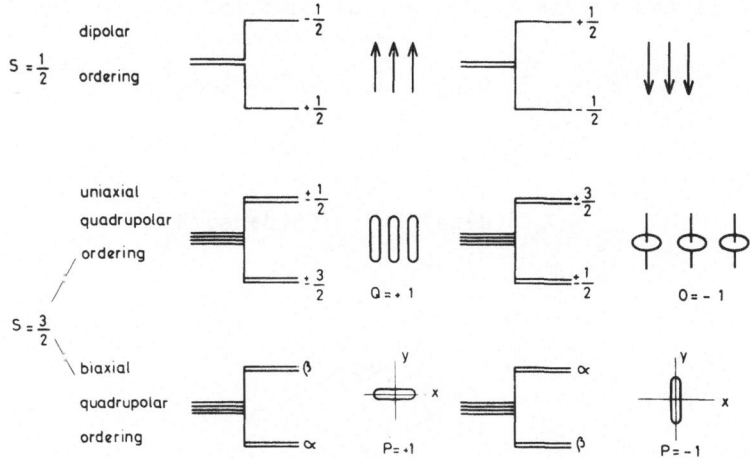

Figure 6 - Comparison between the $S = \frac{1}{2}$ dipolar and $S =$ 3/2 quadrupolar Ising models.

and $S = 1$ have been studied in reference [28]; quadrupole re-
orientations are found in some cases. Negative quadrupolar and
dipolar interactions may induce anti-quadrupolar ordering and
ferromagnetism. Finally, if multipolar interactions are introduced ($\ell = 3$: octupolar; $\ell = 4$: hexadecapolar; $\ell < 2S$), successive phase transitions are found too [13].

2.2.4 Spin-Lattice Coupling:
Calculation of the Elastic Constants

Quadrupolar ordering will in general induce a distortion of
the lattice. We consider for instance ordering in Q for an $S = 1$
uniaxial system. We add to $\mathcal{H}$ a coupling $-\eta \sum_i eQ_i$ between the

electronic states and the lattice strain e having the same symmetry as Q, and the elastic strain energy $\frac{1}{2}c_0 e^2$. c_0 is an elastic constant in the disordered state. We want to calculate the
elastic constant c in the ordered state [29]; c plays the same
rôle as the susceptibility in a magnetic system and is related
to the fluctuations in Q. Using the variational method, we calculate $\Phi(d,e)$. The independent molecular field parameters d and
e are determined by the conditions

$$\frac{\partial \phi}{\partial d} = 0, \tag{26}$$

and

$$\mathcal{E} = \frac{\partial \phi}{\partial e} = 0, \tag{27}$$

where $\mathcal{S}$ is the stress at mechanical equilibrium. We get

$$\phi = -\frac{1}{\beta} \log Z_0 + dQ - KQ^2 + \tfrac{1}{2}c_0 e^2, \tag{28}$$

with

$$Z_0 = 2e^{\beta(d+\eta e)/3} + e^{-2\beta(d+\eta e)/3}, \tag{29}$$

whence

$$d = 2KQ, \tag{30}$$

and

$$\mathcal{S} = c_0 e - \eta Q = 0. \tag{31}$$

Q is then given by

$$Q = \frac{2}{3} \frac{e^{\beta(2KQ+\eta e)} - 1}{2e^{\beta(2KQ+\eta e)} + 1}, \tag{32}$$

where $e = e_0$, the equilibrium strain

$$e_0 = \frac{\eta}{c_0} Q, \tag{33}$$

so that the effective interaction K' is given by

$$K' = K + \frac{\eta^2}{2c_0}. \tag{34}$$

Finally, the elastic constant is

$$c = \frac{\partial \mathcal{S}}{\partial e} = \frac{\partial^2 \phi}{\partial e^2}, \tag{35}$$

or

$$c = c_0 - \eta \frac{\partial Q}{\partial e}$$

$$= c_0 - \frac{\eta^2 Q^2}{kT}. \tag{36}$$

3. PHENOMENOLOGICAL DESCRIPTION OF PHASE TRANSITIONS IN REAL SYSTEMS

3.1 CUBIC TRANSITION METAL-ION COMPOUNDS

We consider tetragonal distortions induced by the Jahn-Teller effect in cubic materials. The more general purely quadrupolar Hamiltonian is

$$\mathcal{H} = \mathcal{H}_3 + \mathcal{H}_5, \tag{37}$$

with

$$\mathcal{H}_3 = - \sum_{i,j} K_{ij}(\sqrt{3}Q_i\sqrt{3}Q_j + P_iP_j), \tag{38}$$

and

$$\mathcal{H}_5 = - \sum_{i,j} K_{ij}{}'(P_i{}^{xy}P_j{}^{xy} + P_i{}^{yz}P_j{}^{yz} + P_i{}^{zx}P_j{}^{zx}). \tag{39}$$

$\mathcal{H}_3$ has a cubic Γ_3-symmetry and $\mathcal{H}_5$ a cubic Γ_5 symmetry. ($\mathcal{H}$ is isotropic if $K_{ij}{}' = K_{ij}$). We take $K_{ij}{}' = 0$ since we are looking for tetragonal distortions. (If $P \neq 0$ or $\pm 3Q$, the symmetry is orthorhombic). Two situations must be distinguished according to the value of the orbital degeneracy of the magnetic ions.

3.1.1 Doublet (Γ_3) Systems

The doublet is described by an effective-spin $\frac{1}{2}$ [30]. Since $[\Gamma_3{}^2] = \Gamma_1 + \Gamma_3$, $\sqrt{3}Q$ and P have non-zero matrix elements within the doublet. $\sqrt{3}Q$ is traceless and diagonal since it separates the orbitals d_{z^2} and $d_{x^2-y^2}$, and is represented by S^z. P mixes the two orbitals and is represented by S^x. Hence

$$\mathcal{H}_3 = - \sum_{i,j} J_{ij}(S_i{}^zS_j{}^z + S_i{}^xS_j{}^x). \tag{40}$$

$\mathcal{H}_3$ is isotropic in the x-z plane of the spin space and is similar to the X-Y model. Ordering takes place in any direction and is second order.

Consequently the distortion is second order, and in general orthorhombic since the angle θ ($\tan\theta = \langle S^x\rangle/\langle S^z\rangle = P/\sqrt{3}Q$, is not determined, although according to Landau it is in general first order and tetragonal. First order tetragonal distortions ($\theta = 0$, $\pm 2\pi/3$) are stabilized once anharmonic Jahn-Teller couplings are introduced in the form of a $\cos3\theta$ anisotropy [30]. This is also the case if excited crystalline field (as in TmCd [19]) or vibronic [31] levels are considered.

The above degeneracy, which is similar to the degeneracy found in $S = 3/2$ quadrupolar systems (section 2, second order transition in Q), is reflected in the excitation spectrum which satisfies the Goldstone theorem [32]: the continuous broken symmetry is not the cubic symmetry but the rotational symmetry in the effective-spin space.

3.1.2 Triplet (Γ_4 or Γ_5) Systems

The electronic levels are described using an effective-spin one formalism [30]. The Hamiltonian [38] is then similar to [9] with $H = D = 0$ and $J_{ij} = 3K_{ij}$. Indeed the operator P has the same form in the basis $(|1\rangle + |-1\rangle)/\sqrt{2}$, $(|1\rangle - |-1\rangle)/\sqrt{2}$, $|0\rangle$ or $|0\rangle_y$, $|0\rangle_x$, $|0\rangle_z$ (the index denotes the quantization axis) than the operator S^z in the basis $|1\rangle$, $|-1\rangle$, $|0\rangle$. From figure 2 we see immediately that a first order transition is found, leading to one of the three equivalent ground states:

$$(Q,P) = \left[-\frac{2}{3},0\right] \; ; \quad \left[\frac{1}{3},1\right] \; ; \quad \left[\frac{1}{3},-1\right] \; .$$

The distortion is first order and tetragonal along z, x, or y, in agreement with the Landau theory [33]. Contrary to the case of Γ_3 ions, one gets three distinct potential minima even if higher order Jahn-Teller and anharmonic terms are neglected.

Remark: In the above description of the Jahn-Teller distortion in doublet or triplet systems, there is no local distortion above the transition temperature: the appearance of local distortions and their alignment are simultaneous (displacive transition). A basically different model has been considered by Wojtowicz [34]. Large local distortions due to the static Jahn-Teller effect exist at high temperature and are oriented at random along x, y or z. Interactions between these distortions induce a first order (order-disorder) transition to a state in which the local distortions are aligned in one particular direction. This description [14,34] is classical since it does not depend on the type of orbital degeneracy, and is related to quadrupolar ordering in molecular crystals, ordering in a three well ferroelectric model, and the three state Potts model [35] (in fact both Wojtowicz and Potts models can be represented by an effective-spin one quadrupolar model). Intermediate situations have been described by Englman and Halperin [36] and Thomas and Müller [31].

3.2 RARE EARTH COMPOUNDS: $TmVO_4$, $DyVO_4$, $TbVO_4$ AND $TbPO_4$

3.2.1 Crystalline Field Properties

The decomposition of the $^7F_6$ (Tb^{3+}) or $^6H_{15/2}$ (Dy^{3+}) multiplets by the crystalline field leads in these compounds to the existence of several low lying levels well separated from the excited states and forming a 'band'. In some cases, as discus-

sed by Trammel [37], the occurrence of such a band made of n
levels is owing to the fact that the moments may have only a
discreet number n of equivalent orientations. The wave functions of the band levels describe symmetry-adapted tunnelling
states between these equivalent orientations:

DyVO$_4$ The band is made of two doublets Γ_6 and Γ_7. According to Pytte and Stevens [38], the corresponding four
allowed orientations of the moments are $\pm[110]$ and $\pm[1\bar{1}0]$;

TbVO$_4$ The band is made of a doublet Γ_5 and two singlets Γ_1
and Γ_3. The four allowed orientations of the moments
are $\pm[100]$ and $\pm[010]$;

TbPO$_4$ The band is made of a doublet Γ_5 and a singlet Γ_1.
It can be interpreted as the lowest part of a Trammel
band corresponding to the existence of eight equivalent
orientations of the moments: $[\pm h, 0, \pm \ell]$ and $[0, \pm h, \pm \ell]$ [18];

TmVO$_4$ The situation is as in TbPO$_4$, except that only the
doublet Γ_5 has to be considered at low temperature.

In the case of DyVO$_4$, TbPO$_4$ and TmVO$_4$, effective spins
$S' = 3/2$, 1 and $\frac{1}{2}$ respectively can be introduced to describe the
low lying crystalline levels, since the splitting of the band
can be simulated by a uniaxial anisotropy energy of the form
$-D \sum_i Q_i$. In the case of TbVO$_4$, the splitting of the band cannot

be reproduced in such a simple way, and either two effective
spins $\frac{1}{2}$ per site must be introduced, or, a simple procedure, a
four-dimensional spin Hamiltonian must be constructed directly
using symmetry considerations.

3.2.2 Symmetry-Adapted Hamiltonian

The distortion observed in TmVO$_4$, DyVO$_4$, TbVO$_4$, and TbPO$_4$
can be considered as a Jahn-Teller induced phase transition
which lifts the degeneracy of the crystalline field band and can
be followed by an ordinary magnetic transition. These two transitions may also be considered as an order-disorder process involving successive reductions of the number of possible equivalent orientations of the moments [38]. The phenomenological description which follows is, however, independent of these physical interpretations, since only phenomenological symmetry-adapted interactions are introduced in order to describe the coupled
spin-phonon system.

TmVO$_4$ Since $[\Gamma_5^2] = \Gamma_1 + \Gamma_3 + \Gamma_4$, the Γ_5 degeneracy is lifted by an orthorhombic distortion. The operators P
and P' may be represented by S_x and S_y respectively, whence
an anisotropic X-Y Hamiltonian:

$$\mathcal{H} = - \sum_{i,j} J_{ij}(S_i^x S_j^x + \eta S_i^y S_j^y) - H_z \sum_i S_i^z. \qquad (41)$$

The Zeeman energy represents the influence of a magnetic field along z, which hinders the distortion. Distortion is second order, and occurs along [100] ($\eta < 1$) or [110] ($\eta > 1$). If $\eta = 0$, the system is Ising-like and the picture of Finch *et al.* [14] applies quite well. The transition is second order since P and $-P$ describe equivalent domains.

DyVO$_4$ and TbVO$_4$ The following simplified Hamiltonian has been used [17] to reproduce the observed successive phase transitions:

$$\mathcal{H} = \mathcal{H}_c - \sum_{i,j} J_{ij} S_i^x S_j^x - \sum_{i,j} K_{ij} P_i P_j. \qquad (42)$$

$\mathcal{H}_c$ is a single ion operator describing the splitting of the Trammel band, x is the direction of the distortion in the basal plane. The form of the 4×4 operators S^x and P is determined by symmetry arguments. The phase diagram in the ($J/K, T/K$) plane is similar to that of figure 3. For $K \gg J$, a second order distortion to an orthorhombic state at T_q is followed by a second order magnetic ordering along x at T_d (the sign of J_{ij} is unimportant as far as $H_x = 0$). A field along z stabilizes the quadratic phase, whereas a field along x induces the distortion at $T > T_q$.

Whereas TmVO$_4$ undergoes a true Jahn-Teller transition, DyVO$_4$ undergoes a pseudo-JT transition since the two doublets Γ_6 and Γ_7 are already separated above T_q.

The Pytte and Stevens model provides a simple interpretation of the above results: at $T > T_q$, the magnetic moment tunnels between the four equivalent orientations in the a-b plane described by the crystal field band. For $T_d < T < T_q$ only two opposite orientations are allowed (quadrupolar ordering). Finally, for $T < T_d$, a single orientation is selected, whence magnetic ordering.

TbPO$_4$ The main problem is to understand why magnetic ordering occurs off the z axis. A first possibility is the existence of a single first order transition in M_{zx}, but it must be discarded since the specific heat shows two λ-anomalies at low temperatures. A second possibility is the existence of a second order magnetic transition in M_z followed by a second order monoclinic distortion driving a canting of the moments off the c-axis. This is in agreement with the Landau theory since M_x and M_z transform in two different representations Γ_α of the groups different from Γ_1 and such that $[\Gamma_\alpha^3]$ does not contain Γ_1. But the first transition is purely crystallographic so that the only possibility is a monoclinic or orthorhombic distortion followed by a magnetic transition. The distortion is in fact monoclinic, otherwise magnetic ordering would be first order, or take place either along x or along z. Whence the following Hamiltonian with $S = 1$:

$$\mathcal{H} = \mathcal{H}_c - \sum_{i,j} J_{ij}(S_i{}^x S_j{}^x + S_i{}^z S_j{}^z) - \sum_{i,j} K_{ij} P_i{}^{xz} P_j{}^{xz}. \quad (43)$$

For $K \gg J$, a second order monoclinic distortion induces a strong monoclinic magnetic anisotropy and is followed by a second order magnetic ordering along the easy direction θ off the c-axis in the a-c plane ($\theta \simeq 40°$). As shown by crystal field calculations, the eight equivalent easy directions pre-exist in the paramagnetic state [18].

It is likely that the distortion observed in TmAsO$_4$ is also monoclinic, although the angle θ might be very weak since the energy separation between the Γ_5 and Γ_1 levels is very large. Indeed recent optical birefringence measurements [39] have revealed only orthorhombic symmetry below T_q, as in TmVO$_4$ in which only the doublet has to be considered.

4. CONCLUSION

We have shown that the introduction of quadrupolar order parameters provides a simple description of structural transitions induced by magnetic interactions. Phenomenological Hamiltonians adapted to the symmetry of particular crystals have been treated in the molecular field approximation and observed transitions have been reproduced qualitatively, in particular the successive transitions found in DyVO$_4$, TbVO$_4$ and TbPO$_4$. The molecular field approximation is correct if the range of the pair interactions is large, for instance in the case of a coupling between the spin system and acoustical phonons as in TbVO$_4$ (coupling to optical phonons leads to short range inter-ionic interactions).

Of course, as mentioned in subsection 1.2, the quadrupolar interactions may arise from different physical mechanisms. Their origin does not affect the qualitative thermodynamical properties of the system we have considered; it could be determined from the experimental study of elementary excitations. Some theoretical [40,42] and experimental [43] work on elementary excitations in quadrupolar systems has been performed recently.

One of the main points of interest of the above systems with dipolar and quadrupolar interactions is that they are characterized by two order parameters (M and Q for instance) and two competing interactions, so that tricritical points are expected, and indeed found in the theoretical phase diagrams. In fact similar spin models have been used to simulate the thermodynamical behaviour of various systems, such as He^3-He^4 mixtures [44], which exhibit, or might exhibit, tricritical points.

REFERENCES

1. Krieger, T.J. and James, H.M. (1954). *J. Chem. Phys.*, **22**, 796.

2. Kobayashi, K.K., Hanamura, E. and Shishido, F. (1969). *Phys. Lett.*, **28A**, 454.

3. De Gennes, P.G. (1969). *Phys. Lett.*, **30A**, 454.

4. Freiser, M.J. (1971). *Mol. Liq. Cryst.*, **14**, 165.

5. Bean, C.P. and Rodbell, D.S. (1962). *Phys. Rev.*, **126**, 104.

6. Allen, S.J. (1968). *Phys. Rev.*, **166**, 530; **167**, 492.

7. Long, C. and Wang, Y.L. (1971). *Phys. Rev.*, **B3**, 1656.

8. Moran, T.J., Thomas, R.L., Levy, P.M. and Chen, H.H. (1973). *Phys. Rev.*, **B7**, 3238.

9. Elliott, R.J. and Thorpe, M.F. (1968). *J. Appl. Phys.*, **39**, 802.

10. Levy, P.M. (1964). *Phys. Rev.*, **135**, A155.

11. Khomski, D.I. and Kugel, K.I. (1972). *JETP Lett.*, **15**, 629.

12. Sivardière, J., Berker, A.N. and Wortis, M. (1973). *Phys. Rev.*, **B7**, 343.

13. Sivardière, J. (1973). *J. Phys. Chem. Solids*, **34**, 267.

14. Finch, G.I., Sinha, A.P.B. and Sinha, K.P. (1957). *Proc. Roy. Soc.*, **A242**, 28.

15. Kino, Y., Lüthi, B. and Mullen, M.E. (1972). *J. Phys. Soc. Jap.*, **33**, 687.

16. Novak, P. (1970). *J. Phys. Chem. Solids*, **31**, 125.

17. See: Sivardière, J. (1972). *Phys. Rev.*, **B6**, 4284, and references therein.

18. Sivardière, J. (1973). *Phys. Rev.*, **B8**, 2004.

19. Lüthi, B., Mullen, M.E., Andres, K., Bucher, E. and Maita, J.P. (1973). *Phys. Rev.*, **B8**, 2639.

20. Harley, R.T., Hayes, W., Perry, A.M. and Smith, S.R.P. (1973). *J. Phys.*, **C6**, 2382.

21. Felsteiner, J., Litvin, D.B. and Zak, J. (1971). *Phys. Rev.*, **B3**, 2706.

22. Sivardière, J. (1972). *Phys. Rev.*, **B5**, 2094.

23. See: Falk, H. (1970). *Am. J. Phys.*, **38**, 858.

24. Blume, M. and Hsieh, Y.Y. (1969). *J. Appl. Phys.*, **40**, 1249.

25. Chen, H.H. and Levy, P.M. (1971). *Phys. Rev. Lett.*, **27**, 1383.

26. Nauciel-Bloch, M., Sarma, G. and Castets, A. (1972). *Phys. Rev.*, **B5**, 4603.

27. Sivardière, J. and Blume, M. (1972). *Phys. Rev.*, **B5**, 1126.

28. Chen, H.H. and Levy, P.M. (1973). *Phys. Rev.*, **B7**, 4267.

29. Elliott, R.J., Harley, R.T., Hayes, W. and Smith, S.R.P. (1972). *Proc. Phys. Soc.*, **A328**, 217.

30. Kanamori, J. (1960). *J. Appl. Phys.*, **31**, 145.

31. Thomas, H. and Müller, K.A. (1972). *Phys. Rev. Lett.*, **28**, 820.

32. Sarfatt, J. and Stoneham, A.M. (1967). *Proc. Phys. Soc.*, **91**, 214.

33. See a similar calculation applied to DySb in: Ray, D.K. and Young, A.P. *J. Phys.*, **C**, (to be published).

34. Wojtowicz, P.J. (1959). *Phys. Rev.*, **116**, 32.

35. See: Straley, J.P. and Fischer, M.E. (1973). *J. Phys.*, **A6**, 1310.

36. Englman, R. and Halperin, B. (1970). *Phys. Rev.*, **B2**, 75.

37. Trammel, G.T. (1963). *Phys. Rev.*, **131**, 932.

38. Pytte, E. and Stevens, K.W.H. (1971). *Phys. Rev. Lett.*, **27**, 862.

39. Becker, P.J. (private communication).

40. Novak, P. (1970). *Czech. J. Phys.*, **B20**, 196.

41. Halperin, B. and Englman, R. (1970). *Solid State Commun.*, **8**, 1555.

42. Levy, P.M. (to be published); Sivardière, J. (to be published).

43. Kjems, J.K., Shirane, G., Birgeneau, R.J. and van Uitert, L.G. (1973). *Phys. Rev. Lett.*, **31**, 1300.

44. Blume, M., Emery, V.J. and Griffiths, R.B. (1971). *Phys. Rev.*, **A4**, 1071.

ONE AND TWO MAGNON LIGHT SCATTERING
IN INSULATING ANTIFERROMAGNETS
AND ITS POSSIBLE RELEVANCE FOR THE INVESTIGATION
OF THE SHORT RANGE INTERATOMIC CORRELATIONS

C.R. NATOLI†

C.N.E.N., Frascati, Roma, Italy

and

J. RANNINGER

Institut Laue-Langevin, Grenoble, France

INTRODUCTION

Although the possibility of light scattering by magnons was considered theoretically as early as 1959 [1], it was only in 1966 that the first observation of the phenomenon was reported in FeF_2 [2]. Since then an intense experimental activity has followed in a variety of antiferromagnetic systems, ranging from rutiles (MnF_2, CoF_2, NiF_2) to perovskites ($RbMnF_3$, $KNiF_3$) and rock salt structures (NiO).

One of the most unusual features observed in these light scattering experiments was the observation of a two magnon scattering process with a cross section of the same order, or bigger, than the one found in one magnon Raman scattering. Several theoretical models were proposed to explain the experimental results on two magnon scattering, and were tested in respect of the polarization selection rules (Raman tensor symmetry), the intensity, the line shape, the temperature and magnetic field dependence of the scattered light. The conclusions are drawn in a paper by Fleury and Loudon [3] where it is shown that the predictions of the spin-orbit interaction mechanism are con-

---

† Current address: C.N.R.S., Grenoble, France

firmed in the case of one magnon scattering, whereas another
mechanism has to be invoked for the explanation of the two magnon scattering, the so called excited state exchange interaction.

While there has been little theoretical interest in one magnon light scattering after the proposed theoretical mechanism
[3] was confirmed experimentally, the two magnon process has,
on the contrary, quickly attracted the attention of the theoreticians owing to another unexpected feature: the Raman spectrum at $T = 0$ did not look like the density of the state curve
$\sum_{k} \delta(\omega - 2\omega_k)$ relative to two magnons of opposite moments since
the light carries practically no moment (we suppose $\omega_k = \omega_{-k}$).
This curve is sharply peaked at twice the zone boundary energy
of a single magnon (ω_B) with a width which is very small compared to the observed width of the Raman spectrum; moreover,
the centre of the observed peak, defined as that energy transfer for which the scattered intensity is a maximum, is shifted
toward lower energies compared to $2\omega_B$. This was explained by
Elliott and Thorpe [4] who argued that because of the nearly
zero momentum transfer in light scattering and the sharply
peaked one-magnon density of states at the zone edge, light
scattering predominantly consists of flipping two spins on adjacent sites. Such two spins are strongly coupled via the exchange interaction leading to a resonance state of two zone
boundary magnons of opposite wave vectors. The energy required
for creating two magnons should therefore not be $2\omega_B$, but this
energy minus the exchange energy. The observed width is in
this way connected with the lifetime of this bound state which
lies in the middle of the two magnon band. Herewith a quantitative explanation for the shift and the width of the Raman
peak at zero temperature was achieved. It remained to explain
the features of the spectrum as the temperature was raised,
namely the decreasing amplitude of the peak, its broadening and
shifting to lower energies. Figures 1,2 taken from Fleury [5],
show a typical two magnon spectrum for temperatures below and
above the Néel temperature T_N (the specific case refers to
NiF_2) and the temperature variation of the shift and the width
of the peak. As one can see, the peak is persistent well beyond T_N, in keeping with Elliott and Thorpe's argument that
light probes the short range order of the antiferromagnetic system.

The quantitative description of this state of affairs made
the subject of number of papers [6-12], most of them concerned
only with the description of the two magnon Raman process in
the ordered region. It was found that a repeated scattering of
the two magnon in a collisionless régime was sufficient to account for the temperature variation of the position of the peak
below T_N, whereas the decreasing amplitude and the broadening
remained unexplained. Although the feeling that the single magnon damping was important for the explanation of these features

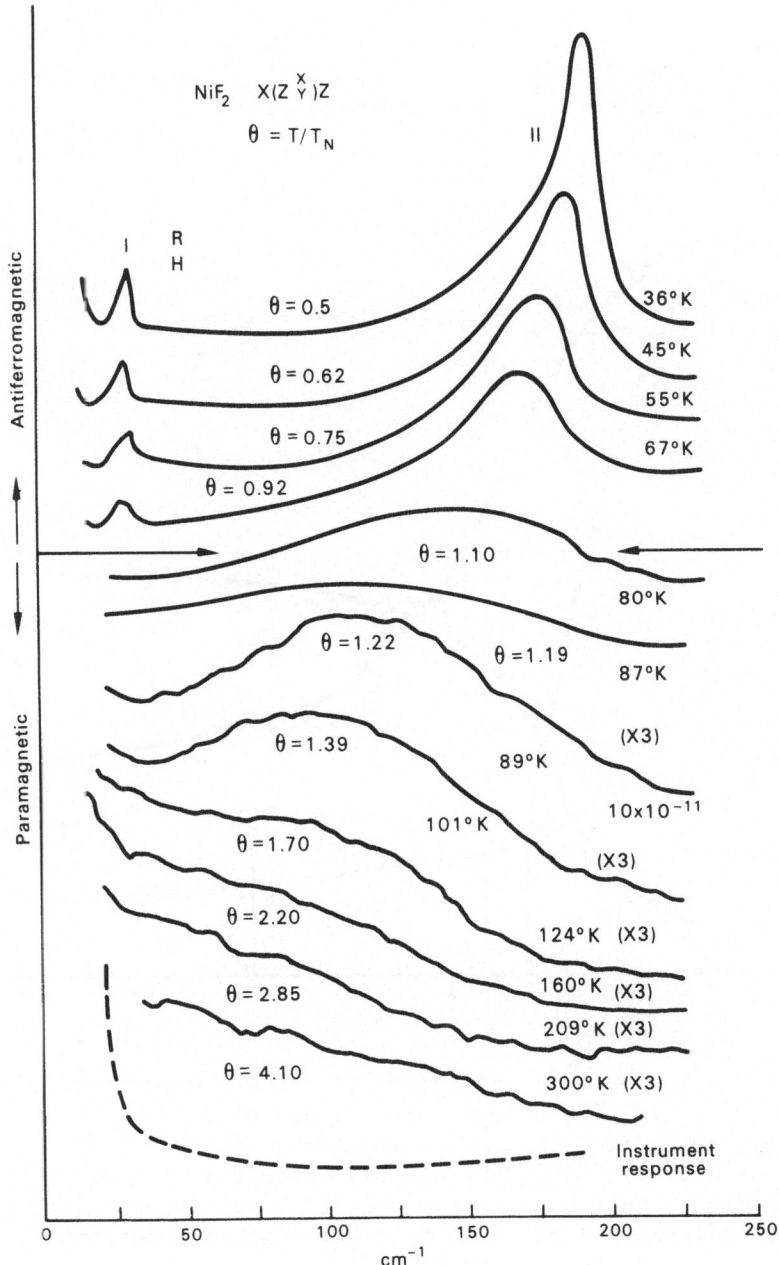

Figure 1 - Spectrum of magnetic light scattering in anti-ferromagnetic and paramagnetic NiF_2 (T_N = 73°K). Peak I is owing to a single zone-centre magnon and disappears with the disappearance of long range order at T_N. Peak II is owing to pairs of zone boundary magnons and clearly persists well above 89°K. The instrument response curve is appropriate to these more sensitive traces.

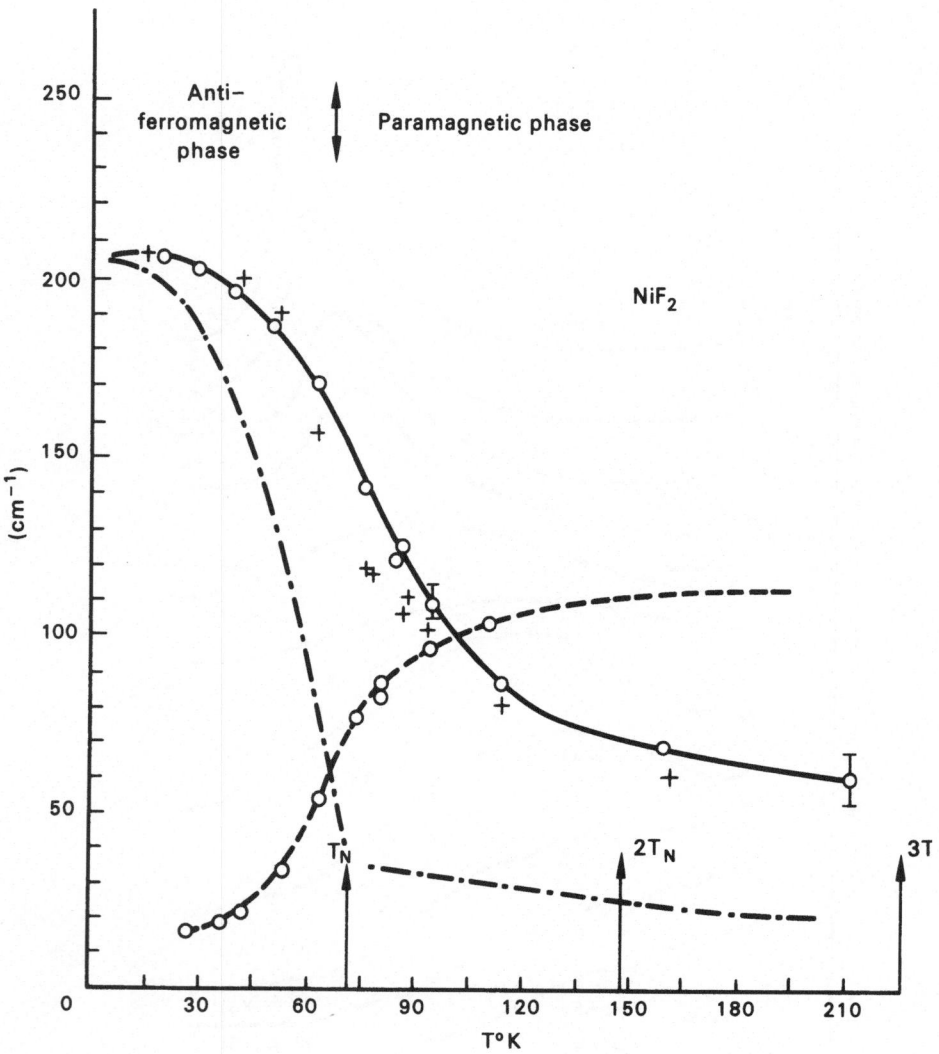

Figure 2 - Temperature dependence of various aspects of
peak II. Solid line: observed frequency of peak II.
Crosses: normalized integrated intensity of scattered
light in peak II (arbitrary units). Dashed line: observed
full width at half maximum for peak II. Dot-dashed line:
nearest-neighbour equal-time spin correlation function
for S = 1 calculated in reference [7].

was widespread, it was not until a reliable calculation of this
latter quantity was available [12-13] that a good quantitative
description of the spectrum was achieved, at least up to 0.8
T_N. Unfortunately an accurate calculation of the zone boundary
magnon damping above 0.8 T_N is still lacking so that up to now
we do not have a quantitative interpretation of the spectrum

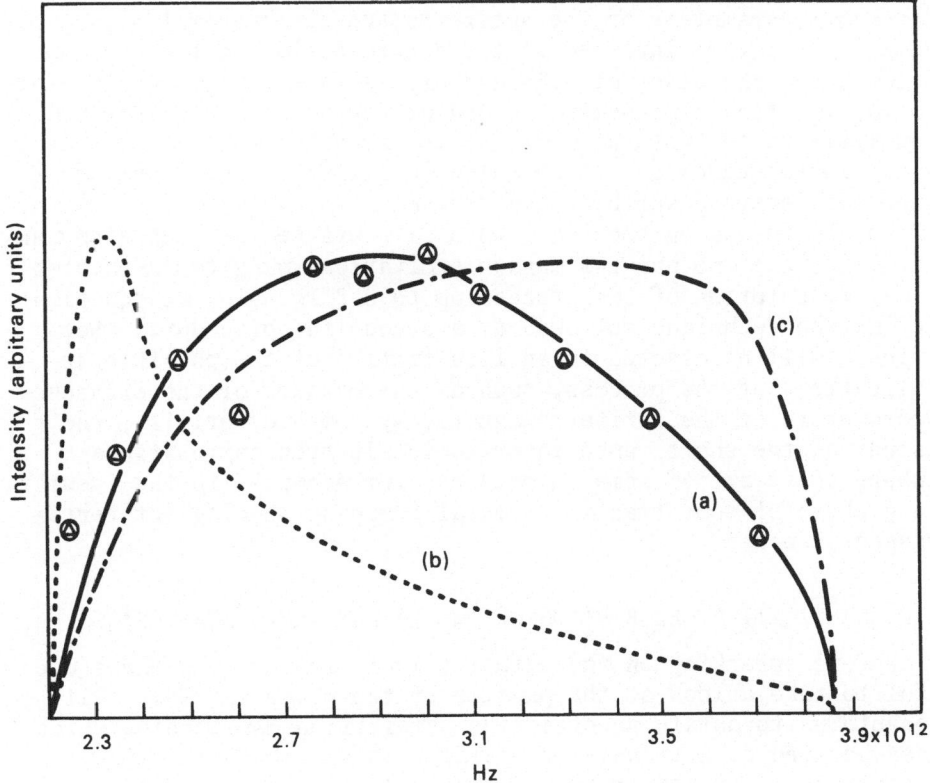

Figure 3 - Theoretical and experimental neutron scattering cross sections for CoF_2 at $Q = (2\pi/a)(1,0,0)$: (a)
using our Hamiltonian ($\alpha = 0.62$) (full curve); (b) using
Elliott and Thorpe's Hamiltonian ($\alpha = 0$) (dotted curve);
(c) neglecting every interaction between magnons (dot-
dashed curve). The circles represent the experimental
points by Cowley et al. [21]. (Taken from reference [22]).

around and above T_N; moreover, in this temperature region spin
density fluctuations also play a rôle further complicating any
quantitative description. Only the position of the Raman peak
in the paramagnetic region can be given a quantitative description in terms of the second moment of the one particle Green's
functions $\langle\langle S_{kl(2)}{}^+; S_{kl(2)}{}^- \rangle\rangle$ where $S_{kl(2)}{}^+$ is the Fourier
transform of the spin operator on sublattice 1 (2). Further-
more, it is to be noticed that although we have been speaking
for sake of brevity of zone boundary magnons excited above T_N,
the true nature of these excitations is not amenable to that
of a set of interacting boson-like particles. Hence the description of the two magnon Raman process in this temperature
region with all its related questions is, as far as we are
aware, still an open problem.

   The description of the phenomenon at $T = 0$ is rather accurate, and one can think of exploiting this process for measuring the exchange parameters of the antiferromagnetic system [14]. Since the errors involved in the determination of the peak position are of the order of 0.5% and with a good instrumental resolution function the observed width of the peak can be ascribed entirely to the natural width of the resonant two magnon state, this method turns out to be quite sensitive to the magnitude of the parameters describing the system.

In the following section 1 we shall outline the mechanism controlling the one and two magnon excitations and give the theoretical description of the process up to $0.8\ T_N$ using an equation of motion technique followed by a decoupling procedure. Section 2 will be devoted to an illustration of the possible applications of the process, such as the measure of the exchange parameters of the antiferromagnetic system and, briefly, the study of the short range interatomic electron correlations where these are of some importance, for example, in the insulating phase of some transition metal compound showing antiferromagnetic order.

1. QUANTITATIVE DESCRIPTION OF TWO MAGNON RAMAN SCATTERING

Before embarking on the actual calculation it is perhaps useful to take a look at the physics of the process. This will enable us to obtain an effective interaction Hamiltonian which can be used as a transition operator in a Green's function technique. For comparison we shall also briefly discuss the one magnon process. In this latter case one assumes that one has an $L = 0$ ionic ground state (or a quenched orbital angular momentum) with spin S split by the exchange field into $(2S + 1)$ states and an excited $L = 1$ state, itself split into three components by the spin orbit interaction (other values for L can be treated similarly). Since the magnon is a linear combination of the spin excitations of the individual ions from, say, their $S^z = S$ to their $S^z = S - 1$ state, one needs to evaluate the Raman transition probability for a simple magnetic ion. The Raman transition in this latter case can proceed by a pair of allowable electric dipole transitions via the $L = 1$ excited state. In the excited $L = 1$ state the spin orbit coupling can intervene to flip the spin before the second transition to the $L = 0$, $S^z = S - 1$ state takes place. Since the matrix element of the transition connects the $S^z = S$ ground state to the $S^z = S - 1$ state, it can be written as an operator linear in S_i^-, i being the site of the ion. Taking into account the anti-Stokes scattering and summing over all the magnetic ions in the system, the result is conveniently written in the form of a spin operator Hamiltonian

$$H_{eff} = -2i\Gamma \sum_{ij} [(E_1{}^x E_2{}^z - E_1{}^z E_2{}^x)(S_i{}^x + S_j{}^x)$$

$$+ (E_1{}^z E_2{}^y - E_1{}^y E_2{}^z)(S_i{}^y + S_j{}^y)], \quad (1.1)$$

where the electric field vectors $\vec{E}_1$ and $\vec{E}_2$ refer to the exciting and scattered radiation, i and j label the two sublattices of the antiferromagnetic system (we shall assume this is the case) and Γ is given by

$$\Gamma = \frac{e^2\lambda^2}{2^{3/2}} \left\langle L = 0, L^z = 0 \left| z \right| L = 1, L^z = 0 \right\rangle$$

$$\times \left\langle L = 1, L^z = 1, \left| x + iy \right| L = 0, L^z = 0 \right\rangle$$

$$\times \left\{ \frac{1}{(E_0 - \hbar\omega_{k_1})^2} - \frac{1}{(E_0 + \hbar\omega_{k_2})^2} \right\} , \qquad (1.2)$$

where λ is the excited state spin-orbit coupling as already evaluated and incorporated in the matrix element $\left\langle L = 1, L^z = 0 \right|$ $\left| \lambda L^- S^+ \right| L = 1, L^z = +1 \rangle$.

Moreover E_0 is the energy gap between the $L = 0$ and the $L = 1$ states, $\hbar\omega_{k_1}$, $\hbar\omega_{k_2}$ the incident and scattered photon energies and we have neglected in the energy denominators the magnon energy $\hbar\omega_0$ corresponding to the gap between the $L = 0$, $S^z = S-1$ and the $L = 0$, $S^z = S$ states. In equation (1.1) the electric field vectors are to be evaluated at the appropriate site of the ions and in second quantization formalism are given

$$\vec{E}_1(i) = -\frac{1}{c} \frac{d\vec{A}(\vec{r}_i)}{dt} ,$$

$$\qquad (1.3)$$

$$\vec{A}(\vec{r}_i) = c \sum_k \frac{\vec{\varepsilon}_k e^{i k \cdot r_i}}{(V\omega_k \eta_k^2)^{\frac{1}{2}}} (b_k + b_{-k}^+),$$

where η_k is the refractive index at the frequency ω_k of the radiation, k and ε_k are the wave and polarization vectors of the radiation, V is the crystal volume and b_k^+ and b_k are photon creation and destruction operators.

Inserting (1.3) in (1.1), one obtains for the spin part

$$H_{eff} = \sum_k (S_{k1}^+ + S_{k2}^+ + S_{k1}^- + S_{k2}^-)\delta(k_1 - k_2 - k), \quad (1.4)$$

where $S_{k\ell}^+$ is the Fourier transform of the spin operator on sublattice ℓ ($= 1,2$) and k_1, k_2 are the incoming and the outgoing wave vectors. Since for light $|k_1| \sim |k_2|$ is very small compared with the wave vectors ending on the Brillouin zone, the only magnons excited are those with $k \simeq 0$, i.e. the centre of the Brillouin zone. This fact is often referred to as the uniform mode antiferromagnetic resonance excited by light.

The two-magnon process is not a combination of two one-magnon processes, since in this case the intensity would be λ times smaller than in the case of the one magnon excitation. From magnetic field dependence and polarization selection rules Fleury and Loudon [3] find that the two magnon excited state involves a simultaneous excitation of one magnon on each of the sublattices. The two ions involved are coupled via electronic excitations in an exchange scattering mechanism. Let us call these two representative ions i and j, one on each sublattice and, for simplicity, suppose that in the ground state ion i has an electron at $\mathbf{r}_i$ with $S^z = \frac{1}{2}$ accomodated in an orbital ϕ_i, while ion j has an electron at $\mathbf{r_j}$ with $S^z = -\frac{1}{2}$ in an orbital ϕ_j. The interaction of the electrons with the electromagnetic field and with each other is described by the Hamiltonian

$$H = - e(\mathbf{E}(i)\cdot\mathbf{r}_i + \mathbf{E}(j)\cdot\mathbf{r}_j) + \frac{e^2}{r_{ij}} \; ,$$

where $r_{ij} = |\mathbf{r}_i - \mathbf{r}_j|$. A representative matrix element for the Raman process accompanied by simultaneous change in the spin components of ions i and j is

$$M_{ij} = \sum_{\mu\nu} \langle \phi_{j\uparrow}\phi_{i\downarrow} | e\mathbf{E}_2(j)\mathbf{r}_j | \phi_{\nu\uparrow}\phi_{i\downarrow}\rangle\langle\phi_{\nu\uparrow}\phi_{i\downarrow}| \frac{e^2}{r_{ij}} |\phi_{j\downarrow}\phi_{\mu\uparrow}\rangle$$

$$(1.5)$$

$$\times\langle\phi_{j\downarrow}\phi_{\mu\uparrow}|e\mathbf{E}_1(i)\cdot\mathbf{r}_i|\phi_{j\downarrow}\phi_{i\uparrow}\rangle\Big[(E_\nu + \hbar\omega_k - \hbar\omega_{k_1})(E_\mu - \hbar\omega_{k_1})\Big]^{-1} ,$$

where $\hbar\omega_k$ is the magnon energy and with obvious meaning of the other symbols. The process in equation (1.5) can be described as the excitation of the electron on site i to an excited state by the incoming photon without change of S^z, followed by the de-excitation of the same electron to the ground orbital state with connected excitation of the electron on site j to a state ν and simultaneous flipping of the spin of the two electrons involved, in turn followed by de-excitation of the electron on site j with emission of the outgoing photon. The matrix element (1.5) can be written in operator form as

$$\mathrm{M}_{ij} = M_{ij}S_i^- S_j^+. \qquad (1.6)$$

To obtain the effective interaction Hamiltonian we must sum over all pairs of ions i and j on opposite sublattices. Since M_{ij} includes an exchange integral between the pairs of ions decreasing exponentially with the inter-ionic distance, one can restrict the summation to nearest neighbours on opposite sublattices. In forming the spin Hamiltonian the relative phases of the contributions of the z neighbours of a given spin i can be adjusted by introducing a vector σ^{ij} defined as

$$\sigma_{ij}{}^\alpha = \text{sign}(r_j - r_i)_\alpha, \qquad \alpha = x,y,z.$$

Thus the $\sigma_{ij}{}^\alpha$ are equal to either +1 or -1 and merely determine the relative signs of the terms in the i,j summation.

To form H_{eff} we multiply the spin operator $S_i{}^-S_j{}^+ + S_i{}^+S_j{}^-$ by those combinations of E_1, E_2 and $\boldsymbol{\sigma}^{ij}$ which belong to the identical representation $\Gamma_1{}^+$ of the antiferromagnetic space group, since the combination $S_i{}^-S_j{}^+ + S_i{}^+S_j{}^-$ already transforms according to $\Gamma_1{}^+$. So we get, in the special case of rutiles, for example:

$$H_{eff} = \sum_{ij} \{M_{xx}(E_1{}^xE_2{}^x + E_1{}^yE_2{}^y) + M_{zz}E_1{}^zE_2{}^z$$

$$+ M_{xy}(E_1{}^xE_2{}^y + E_1{}^yE_2{}^x)\sigma_{ij}{}^x\sigma_{ij}{}^y$$

$$+ M_{xz}[(E_1{}^yE_2{}^z + E_1{}^zE_2{}^y)\sigma_{ij}{}^y\sigma_{ij}{}^z$$

$$+ (E_1{}^xE_2{}^z + E_1{}^zE_2{}^x)\sigma_{ij}{}^x\sigma_{ij}{}^z]$$

$$+ \bar{M}_{xz}[(E_1{}^yE_2{}^z - E_1{}^zE_2{}^y)\sigma_{ij}{}^y\sigma_{ij}{}^z$$

$$- (E_1{}^zE_2{}^x - E_1{}^xE_2{}^z)\sigma_{ij}{}^x\sigma_{ij}{}^z]\}(S_i{}^xS_j{}^x + S_i{}^yS_j{}^y), \quad (1.7)$$

where M_{xx}, M_{zz}, M_{xy}, M_{xz} and $\bar{M}_{xz}$ are coupling constants given by (1.5) with the electric fields removed and appropriate components of $\mathbf{r}_i$ and $\mathbf{r}_j$ substituted. Experimentally the Raman tensor turns out to be symmetric, so that $\bar{M}_{xz} = 0$.

By insertion of (1.3) into (1.7) we obtain for the spin part of the transition operator,

$$H_{eff} \propto \sum_{\alpha\alpha'} \sum_{ij} e^{-i\mathbf{k}(\mathbf{r}_i-\mathbf{r}_j)}\sigma_{ij}{}^\alpha\sigma_{ij}{}^{\alpha'}(S_{k2}{}^+S_{k1}{}^- + S_{k1}{}^+S_{k2}{}^-)M_{\alpha\alpha'}, \quad (1.8)$$

putting again the light momentum transfer equal to zero. Hence the Raman scattering cross section is obtained as being proportional to the imaginary part of a weighted average of the Zubarev Greens function $\langle\!\langle S_{k+q1}{}^+S_{k2}{}^- + S_{k+q2}{}^+S_{k1}{}^-;B\rangle\!\rangle$ for the case $q = 0$ and $B = S_{k1}{}^+S_{k2}{}^- + S_{k2}{}^+S_{k1}{}^-$, where $S_1{}^+$, $S_2{}^-$ $(S_1{}^-,S_2{}^+)$ create (destroy) magnons on different sublattices. If the weight is such as to damp the effect of the creation of pairs of boundary magnons, which give a strong contribution to the

density of states, then any interaction effect is likely to be
hidden; otherwise a pronounced effect is to be expected. Remembering that zone boundary magnons virtually flip spins on
adjacent sites, the statement just made confirms what was stated
in the introduction regarding the coupling of two adjacent spin
deviation via the exchange interaction. Another process depending on the same type of correlation function is the longitudinal neutron scattering in antiferromagnets. This is given in
terms of

$$\langle\!\langle S_{q1}{}^{z}\pm S_{q2}{}^{z};S_{-q1}{}^{z}\pm S_{-q2}{}^{z}\rangle\!\rangle \, , \qquad (1.9)$$

which can be shown to be equal to

$$\frac{1}{\omega}\frac{1}{2\sqrt{N}}\sum_{\mathbf{k}}(J_{\mathbf{k}+\mathbf{q}} + J_{\mathbf{k}})$$

$$\times\langle\!\langle S_{\mathbf{k}+q2}{}^{+}S_{\mathbf{k}1}{}^{-} \pm S_{\mathbf{k}+q1}{}^{+}S_{\mathbf{k}2}{}^{-};S_{-q1}{}^{z} \pm S_{-q2}{}^{z}\rangle\!\rangle \qquad (1.10)$$

using the first equation of motion in Fourier space with Hamiltonian

$$H = -\sum_{ij} J_{ij}\mathbf{S}_{i}\cdot\mathbf{S}_{j}$$

$$= -\sum_{\mathbf{k}} J_{\mathbf{k}}[S_{\mathbf{k}1}{}^{z}S_{-\mathbf{k}2}{}^{z} + \tfrac{1}{2}(S_{\mathbf{k}1}{}^{+}S_{\mathbf{k}2}{}^{-} + S_{\mathbf{k}2}{}^{+}S_{\mathbf{k}1}{}^{-})]. \qquad (1.11)$$

Equation (1.10) means that the longitudinal part of the scattering cross section in antiferromagnets gives information not
only on the spin density fluctuations in the system but also
on its two magnon excitation properties. However, by inspection one can easily see that the weighting factor in (1.10)
never enhances pairs of zone boundary magnons, so that strong
interaction effects are not to be expected in the longitudinal neutron
scattering cross section (figure 3). This fact is to be contrasted
with light scattering, where the weighting factors are given by
given by

$$\sum_{ij} e^{-i\mathbf{k}(\mathbf{r}_i-\mathbf{r}_j)}\sigma_{ij}{}^{\alpha}\sigma_{ij}{}^{\alpha'} = \phi_{\alpha\alpha'}(\mathbf{k}). \qquad (1.12)$$

If $\alpha = \alpha'$ ($\Gamma_1{}^+$ type of scattering) the weighting factor is just

$$\gamma_{\mathbf{k}} = \frac{1}{z}\sum_{\delta} e^{-i\mathbf{k}\cdot\boldsymbol{\delta}},$$

and the zone boundary magnons are depressed, contrary to $\alpha \neq \alpha'$ (Γ_5^+ type of scattering) $\phi_{\alpha\alpha'}(k)$ where they are strongly weighted (the functions $\phi_{\alpha\alpha'}(k)$ are given in the next section).

Explicit calculation of the Raman scattering cross section bears this out. The technique used is an equation of motion approach for Green's functions followed by a decoupling procedure which can be put into correspondence with the diagrammatic perturbation theory [11,12]. Although one can use spin operators directly, it turns out to be more practical for the sake of presentation to do a Dyson-Maleev transformation (we shall follow closely reference [12]). Then the Hamiltonian (1.11) takes the form

$$H = \text{constant} + \sum_{\mathbf{k}} (\Omega_{\mathbf{k}}(\alpha_{\mathbf{k}}^+\alpha_{\mathbf{k}} + \beta_{\mathbf{k}}^+\beta_{\mathbf{k}})) \qquad (1.13)$$

$$- \frac{Jz}{N} \sum_{qq',pp'} \delta(q + p - q' - p') \times (I_{qq',pp'}^{\alpha\alpha}\alpha_q^+\alpha_p^+\alpha_{q'}\alpha_{p'}$$

$$+ I_{qq',pp'}^{\beta\beta}\beta_{q'}^+\beta_{p'}^+\beta_q\beta_p + I_{qq',pp'}^{\alpha\beta}\alpha_q^+\alpha_q\beta_{p'}^+\beta_p)$$

where $\alpha_{\mathbf{k}}, \beta_{\mathbf{k}}$ ($\alpha_{\mathbf{k}}^+, \beta_{\mathbf{k}}^+$) are magnon destruction (creation) operators, $\Omega_{\mathbf{k}} = JSz(1 - \gamma_{\mathbf{k}}^2)^{\frac{1}{2}}$ is the magnon frequency, and z the number of nearest neighbours. The quantities

$$I_{\mathbf{c}q',pp'}^{\alpha\alpha} = I_{qq',pp'}^{\beta\beta} \qquad \text{and} \qquad I_{qq',pp'}^{\alpha\beta}$$

are some functions of the quantities

$$u_{\mathbf{k}} = \left(\frac{1}{2(1 - \gamma_{\mathbf{k}}^2)^{\frac{1}{2}}} + \frac{1}{2}\right)^{\frac{1}{2}},$$

$$\qquad (1.14)$$

$$v_{\mathbf{k}} = - \left(\frac{1}{2(1 - \gamma_{\mathbf{k}}^2)^{\frac{1}{2}}} - \frac{1}{2}\right)^{\frac{1}{2}},$$

which need not be specified for our purposes.

A first order decoupling of the equation of motion for the one-particle propagator $\langle\langle \alpha_{\mathbf{k}_1}; \alpha_{\mathbf{k}}^+ \rangle\rangle$ gives for the Hartree-Fock temperature dependent magnon frequencies

$$\bar{\Omega}_{\mathbf{k}} = \alpha(T)\Omega_{\mathbf{k}}, \qquad (1.15)$$

where the Hartree-Fock renormalization factor $\alpha(T)$ is determined by the implicit equation

$$\alpha(T) = 1 - \frac{1}{J_z S^2} \frac{1}{N} \sum_{\mathbf{q}} \Omega_{\mathbf{q}} (e^{\alpha(T)\Omega_{\mathbf{q}}\beta} - 1)^{-1}$$

$$= 1 - \frac{1}{J_z S^2} \frac{1}{N} \sum_{\mathbf{q}} \Omega_{\mathbf{q}} \langle \alpha_{\mathbf{q}}^+ \alpha_{\mathbf{q}} \rangle. \qquad (1.16)$$

Equation (1.15) seems to describe quite well the experimental data up to 0.8 T_N (see figure 4 in the case of RbMnF$_3$). When terms containing four Bose operators are treated via

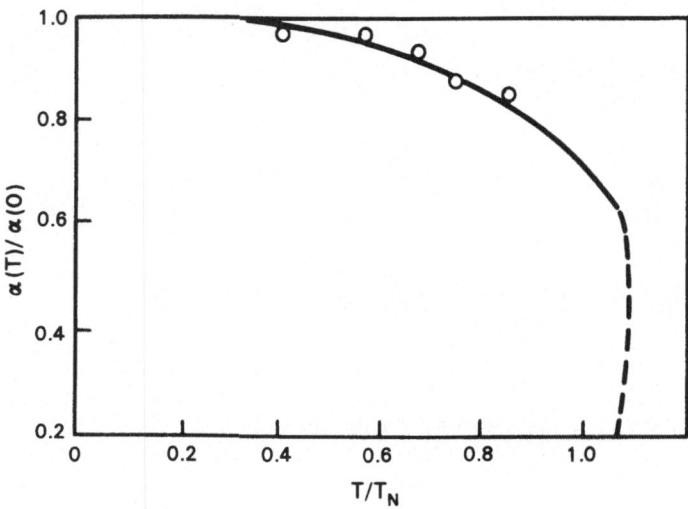

Figure 4 – Spin wave renormalization factor for RbMnF$_3$. The experimental points are from Saunderson *et al.* [23]. (Taken from reference [13]).

Hartree-Fock approximation, the transition operator (1.8) in the α's and β's representation takes the form

$$M_{\alpha\alpha'} = \alpha(T) S \sum_{\mathbf{k}} \phi_{\alpha\alpha'}(\mathbf{k})(u_{\mathbf{k}}^2 + v_{\mathbf{k}}^2)(\alpha_{\mathbf{k}}\beta_{\mathbf{k}} + \alpha_{\mathbf{k}}^+\beta_{\mathbf{k}}^+), \quad (1.17)$$

and the scattering cross section is given by

$$K^{\alpha\alpha'}(\omega) = (1 - e^{-\beta\omega})^{-1} \lim_{\varepsilon \to 0^+} \text{Im} \langle\!\langle M;M \rangle\!\rangle_{E=\omega+i\varepsilon}. \qquad (1.18)$$

For Stokes scattering ($\omega > 0$) we need only study the function $\langle\!\langle \alpha_{\mathbf{k}}\beta_{\mathbf{k}}; \alpha_{\mathbf{k}'}^+\beta_{\mathbf{k}'}^+ \rangle\!\rangle$ whose equation of motion reads

(Contd)

$$(E - 2\Omega_{\mathbf{k}})\langle\!\langle \alpha_{\mathbf{k}}\beta_{\mathbf{k}}; \;\rangle\!\rangle = \frac{1}{2\pi} (2\langle \alpha_{\mathbf{k}}^+ \alpha_{\mathbf{k}} \rangle + 1)\delta(\mathbf{k} - \mathbf{k}') - \frac{J_z}{N} \sum_{\mathbf{p}} I_{\mathbf{k}\mathbf{p},\mathbf{p}\mathbf{k}}^{\alpha\beta}$$

$$\times \langle\!\langle \alpha_{\mathbf{p}}\beta_{\mathbf{p}}; \;\rangle\!\rangle -$$

$$- \frac{Jz}{N} \sum_{q'pp'} \delta(k + p - q' - p')\{(I^{\alpha\alpha}_{kq',pp'} + I^{\alpha\alpha}_{pq',kp'})$$

$$\times \langle\!\langle \alpha_p^+ \alpha_{p'} \alpha_{q'} \beta_k; \,\rangle\!\rangle$$

$$+ (I^{\beta\beta}_{q'k,p'p} + I^{\beta\beta}_{q'p,p'k})\langle\!\langle \alpha_k \beta_p^+ \beta_{p'} \beta_{q'}; \,\rangle\!\rangle + I^{\alpha\beta}_{kq',pp'}$$

$$\times \langle\!\langle \alpha_{q'} \beta_{p'}^+ \beta_p \beta_k; \,\rangle\!\rangle$$

$$+ I^{\alpha\beta}_{pp',q'k}\langle\!\langle \alpha_{p'}^+ \alpha_p \alpha_k \beta_{q'}; \,\rangle\!\rangle\}. \tag{1.19}$$

The Bethe-Salpeter ladder approximation of references [6] in a collisionless regime for the individual particles is obtained with decoupling (as an example)

$$\langle\!\langle \alpha_p^+ \alpha_{p'} \alpha_{q'} \beta_k; \,\rangle\!\rangle$$

$$\simeq \langle \alpha_p^+ \alpha_{p'} \rangle \delta_{pp'} \langle\!\langle \alpha_{q'} \beta_k; \,\rangle\!\rangle + \langle \alpha_p^+ \alpha_{q'} \rangle \delta_{pq'} \langle\!\langle \alpha_{p'} \beta_k; \,\rangle\!\rangle.$$

The result is

$$(E - 2\bar{\Omega}_k)\langle\!\langle \alpha_k \beta_k; \,\rangle\!\rangle = \frac{1}{2\pi}(2n_k + 1)\delta(k - k') \tag{1.20}$$

$$- \frac{Jz}{N}(2n_k + 1)\sum_p I^{\alpha\beta}_{kp,pk}\langle\!\langle \alpha_p \beta_p; \,\rangle\!\rangle,$$

where $n_k = \langle \alpha_k^+ \alpha_k \rangle = \langle \beta_k^+ \beta_k \rangle$ is to be determined self-consistently form (1.16). From (1.20) an expression for $K^{\alpha\alpha'}(\omega > 0)$ can be derived in the form

$$K^{\alpha\alpha'}(\omega) = - \frac{\alpha^2(T)S^2}{1 - e^{-\beta\omega}} \tag{1.21}$$

$$\times \frac{1}{2\pi} \, \text{Im} \, \frac{L_2 + \frac{1}{2}J(L_1 L_1 - L_0 L_2)}{1 - \frac{1}{2}J(L_0 + L_2) - \frac{1}{4}J^2(L_1 L_1 - L_0 L_2)} \, ,$$

where

$$L_m^{\alpha\alpha'}(E) = - \frac{1}{N} \sum_k \phi_{\alpha\alpha'}(k)(u_k^2 + v_k^2)^m \frac{2n_k + 1}{E - 2\bar{\Omega}_k} \, , \quad (m = 0,1,2). \tag{1.22}$$

If the weighting function $\phi_{\alpha\alpha'}(k)$ is such as to depress zone boundary magnons (Γ^+ symmetry), the enhancement denominator in (1.21) is nearly one, so that $K(\omega) \propto \text{Im} L_2$, as in a non-interacting picture. However for those experimental geometries for which $\phi_{\alpha\alpha'}(k)$ strongly weights ZB magnons (Γ_5^+ symmetry in rutiles, Γ_3^+ symmetry in perovskites) a resonance condition is reached in the denominator. Moreover, owing to the sharp peaking of the magnon density of states in such a case, a negligible error is made if we take $u_k^2 + v_k^2 \sim 1$. In this case $L_0 \sim L_1 \sim L_2$ and equation (1.21) reduces to

$$K^{\alpha\alpha'}(\omega) = - n(-\omega)\alpha^2(T)S^2 \frac{1}{2\pi} \text{Im}\left(\frac{L_0(E)}{1 - JL_0(E)}\right)$$

$$= \frac{\alpha^2(T)S^2}{1 - e^{-\beta\omega}} \frac{\text{Im} L_0^{\alpha\alpha'}(\omega)}{[1 - RL_0^{\alpha\alpha'}(\omega)]^2 + (\text{Im} L_0^{\alpha\alpha'}(\omega))^2} . \quad (1.21')$$

The calculated spectrum (1.21') agrees at $T = 0$ with experiments and also reproduces the variation of the Stokes peak with temperature as far as $T \sim 0.8\, T_N$. Unfortunately it gives a width which is either constant or decreasing with increasing temperature. Some higher order effect must then be taken into account. This means considering the equation of motion for the higher order Green's functions in (1.19). Decoupling in the next stage according to the usual rule of taking all the possible pairs on the LHS of the semicolon and substituting the result in (1.19), we obtain

$$(E - 2\Omega_k - 2\Sigma_k(E))\langle\!\langle \alpha_k \beta_k; \;\rangle\!\rangle =$$

$$\frac{1}{2\pi}(2n_k + 1)\delta_{kk'} - \frac{Jz}{N}(2n_k + 1)\sum_p I_{kp,pk}^{\alpha\beta}\langle\!\langle \alpha_p \beta_p; \;\rangle\!\rangle$$

$$+ 2\left(\frac{Jz}{N}\right)^2 \sum_{q'pp'} \delta(k + p - p' - q')\left[(I_{kq',pp'}^{\alpha\alpha} + I_{pp',kq'}^{\alpha\alpha}\right.$$

$$\left. + I_{kp',pq'}^{\alpha\alpha} + I_{pq',kp'}^{\alpha\alpha}\right)$$

$$\times I_{p'p,q'k}^{\alpha\beta}\left(\frac{n_p(n_{p'} + n_k + 1) - n_{p'}n_k}{E + \bar\Omega_p - \bar\Omega_{p'} - \bar\Omega_{q'} - \bar\Omega_k} + \frac{n_{p'}(n_p + n_k + 1) - n_p n_k}{E + \bar\Omega_{p'} - \bar\Omega_p - \bar\Omega_{q'} - \bar\Omega_k}\right)$$

$$\times\langle\!\langle \alpha_{q'}\beta_{q'}; \;\rangle\!\rangle + \quad\quad (1.23)$$
$$\text{(Contd)}$$

(Contd)

$$+ I^{\alpha\beta}_{kq',pp'} I^{\alpha\beta}_{q'p,p'k} \frac{n_{p'}(n_{q'} + n_k + 1) - n_k n_{q'}}{E + \bar{\Omega}_{p'} - \bar{\Omega}_p - \bar{\Omega}_{q'} - \bar{\Omega}_k} \langle\langle \alpha_p \beta_p; \rangle\rangle \Big],$$

where the second order contribution to the magnon self-energy is given by

$$\Sigma_k(E) =$$

$$\left(\frac{Jz}{N}\right)^2 \sum_{q'pp'} \delta(k + p - p' - q') \left[\frac{n_p(n_{p'} + n_{q'} + 1) - n_{p'} n_{q'}}{E + \bar{\Omega}_p - \bar{\Omega}_{p'} - \bar{\Omega}_{q'} - \bar{\Omega}_k} \right.$$

$$\times (I^{\alpha\alpha}_{kq',pp'} + I^{\alpha\alpha}_{pq',kp'})(I^{\alpha\alpha}_{p'k,q'p} + I^{\alpha\alpha}_{q'p,p'k} + I^{\alpha\alpha}_{q'k,p'p}$$

$$+ I^{\alpha\alpha}_{p'p,q'k})$$

$$+ \frac{n_{p'}(n_p + n_{q'} + 1) - n_p n_{q'}}{E + \bar{\Omega}_{p'} - \bar{\Omega}_p - \bar{\Omega}_{q'} - \bar{\Omega}_k} I^{\alpha\beta}_{kq',pp'} I^{\alpha\beta}_{q'k,p'p} \Big]. \tag{1.24}$$

For studying the dynamical behaviour of high frequency magnons, for which we shall try to justify that the low density Boson gas approximation holds up to such elevated temperatures as 0.8 T_N [13], expressions (1.23,24) can be considerably simplified. First of all, the one-magnon properties such as the temperature dependent magnon frequency renormalization factor $\alpha(T)$ and the sublattice magnetization per site, if evaluated self-consistently in a Hartree-Fock scheme, are in excellent agreement with the available experimental data up to ~ 0.8 T_N (see $\alpha(T)$ for RbMnF$_3$ in figure 4). This is an indication that the particle picture is a sensible one up to such temperatures and gives good results for the average properties of the spin system in this régime. Furthermore the value of the sublattice magnetization at $T = 0.8$ T_N (around 0.75 of the maximum at $T = 0$) serves as an estimate for the density of the excited magnons (25% of the total), certainly not an exceedingly high value. Of these, the zone boundary magnons constitute a minority, owing to the low values of the Bose occupation number, compared with the low frequency ones. Moreover, even if one could argue that the particle picture is not entirely appropriate for the low frequency magnons at $T \approx 0.8$ T_N, the high frequency ones are short range excitations of the spin system which persist well above T_N [15]. In this perspective it is not unreasonable to evaluate the high frequency magnon damping up to 0.8 T_N by the golden rule, i.e. the expression in (1.24). One objection could still be the neglect of the renormalization of the magnon-magnon interaction by the spin fluctuations in a simple golden rule formula. This effect would show up in higher order decoupling. However, from (1.23) we can see what this correction looks like. Namely the last terms in square brackets is just this type of correction to the bare potential between magnons given by $I_{kp,pk}{}^{\alpha\beta}$. If evaluated in the case of a flat dispersion (Ising model), appropriate to high frequency magnons, it turns out to be more than a factor $1/z$ smaller than $I_{kp,pk}{}^{\alpha\beta}$. Hence as a first approximation we can neglect the last term in (1.23) and the result for $k(\omega)$ turns out to be the same as (1.21), but with a modified $L_m(E)$,

$$L_m{}^{\alpha\alpha'}(E) = - \frac{1}{N} \sum_k \phi_{\alpha\alpha'}(k)(u_k{}^2 + v_k{}^2)^m$$

$$\times \frac{2n_k + 1}{E - 2\bar{\Omega}_k - 2\Sigma_k(E)} . \qquad (1.22')$$

A rather accurate estimate of the zone boundary magnons [13],

$$\Gamma(T) = \frac{3\pi}{z^2} \frac{1}{S^2} \frac{\Omega_{k_B}{}^2}{\bar{\Omega}_{k_B}} \left[\sinh^2\left(\frac{1}{2} \frac{\Omega_{k_B}(T)}{kT_N} \frac{T_N}{T} \right) \right]^{-1} , \qquad (1.25)$$

gives the results shown in table I, in good agreement within 20%, with existing experimental data. 20% is just the estimated order of magnitude of the neglected corrections. Regarding two magnon Raman spectra, due to the sharp peaking of the density of states near the boundary of the Brillouin zone, $\Sigma_k(\omega)$ can be put equal to $\Sigma_{k_B}(E)$ where k_B is a boundary wave vector. The spectra calculated are shown in figures 5,6 for RbMnF$_3$ and KNiF$_3$, whilst figures 7,8,9 compare the peak's width and position with the experimental results [12].

Fitting the two magnon resonance line shape in terms of the adjustable parameter of the magnon width, one obtains this quantity within an accuracy of 20%, which compares favourably with its determination by neutron scattering.

2. DETERMINATION OF EXCHANGE CONSTANTS AND ANISOTROPY FIELDS BY ONE AND TWO-MAGNON RAMAN SCATTERING

Studies of insulating magnetic systems are generally based on the Heisenberg Hamiltonian

$$H = \sum_i H_s{}^i + \sum_j H_s{}^j + \sum_{ij} J_{ij} S_i \cdot S_j$$

$$+ \sum_{ii'} J_{ii'} S_i \cdot S_{i'} + \sum_{jj'} J_{jj'} S_j \cdot S_{j'}$$

$$H_{si}{}^i = - D S_{iz}{}^2 + E(S_{ix}{}^2 - S_{iy}{}^2),$$

$$H_{si}{}^j = - D S_{jz}{}^2 - E(S_{jx}{}^2 - S_{jy}{}^2),$$

$$(2.1)$$

TABLE I

Comparison Between the Existing Experimental Data and the Theoretical Values Obtained from Equation (2.21)

	RbMnF$_3$						KniF$_3$		CoF$_2$	
$\tau = \dfrac{T}{T_N}$	0.25	0.38	0.50	0.60	0.70	0.80	0.70	0.85	0.77	0.95
$\Gamma_{exp}{}^{-1}$	1.5×10^{-3}	1×10^{-2}	2.5×10^{-2}	4.5×10^{-2}	8.5×10^{-2}	1.4×10^{-1}	5×10^{-2}	1×10^{-1}	1.3×10^{-1}	3×10^{-1}
$\Gamma_{theor}{}^{-1}$	1.2×10^{-3}	8×10^{-3}	2×10^{-2}	4×10^{-2}	8×10^{-2}	1.2×10^{-2}	5×10^{-2}	9.5×10^{-2}	1.3×10^{-1}	3.8×10^{-1}

Theoretical values are evaluated using the formula $\Gamma'(T) = (3\pi/z^2)S^{-2}[\omega_B(0)/\omega_B(T)]^2\{(T_N/T)\sinh^2[\tfrac{1}{2}\omega_B(T)/k_BT_N]\}^{-1}$. The data for RbMnF$_3$ and CoF$_2$ are taken from neutron scattering experiments, using the formula $\Gamma_{obs}{}^2 = \Gamma_{inst}{}^2 + \Gamma_{true}{}^2$, with obvious meaning of the symbols (Γ_{true} is called Γ_{exp} in the table). Γ'_{exp} for KniF$_3$ has been taken from Davies et al. [6], who used these values to fit the two-magnon Raman peak at $\tau = 0.69$ and $\tau = 0.86$.

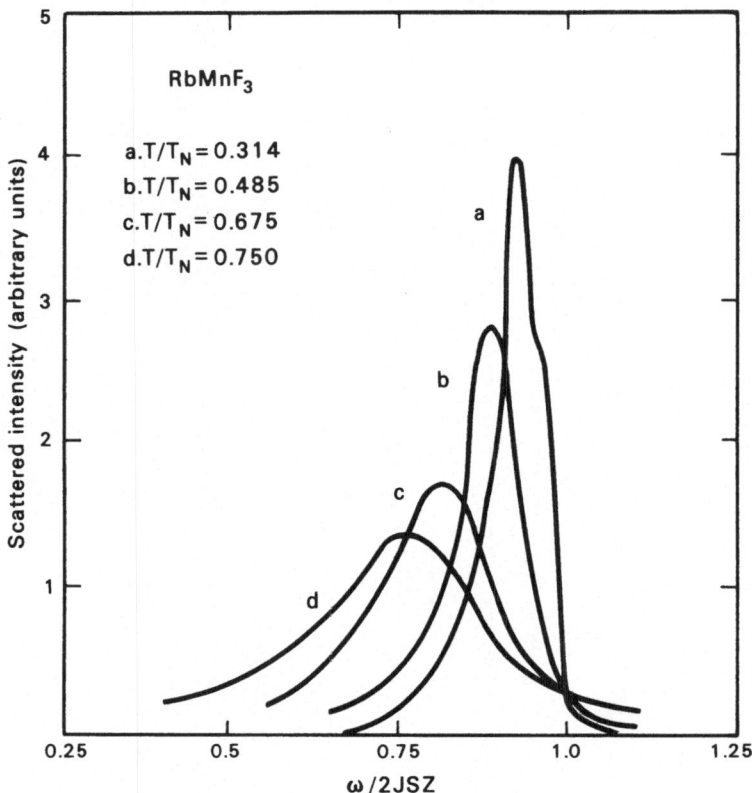

Figure 5 - Some computed two-magnon spectra in RbMnF$_3$
versus relative frequency at various temperatures.

which describes extremely well the static and dynamical behaviour of those systems. We shall restrict ourselves in this discussion to rutile structure and have chosen $H_{si}{}^i$ and $H_{si}{}^j$ accordingly. It must be remembered that this Hamiltonian represents in most cases an effective Hamiltonian with the exchange constants J and single ion anisotropy parameters E, D being chosen in such a way that the eigenvalues and eigenstates of this Hamiltonian reproduce the first few excited levels of the electronic structure of the magnetic ions, together with their crystal field and exchange splitting.

Using infrared absorption, magnetic susceptibility sublattice magnetization, and EPR measurements, the electronic structure of such systems can be determined, which in turn yields the parameters entering the Heisenberg Hamiltonian (2.1).

Neutron scattering techniques provide a direct method of determining the exchange constants without needing to know anything about the underlying electronic structure. This method is essentially based on fitting the experimentally observed magnon dispersion curve to the theoretically obtained one.

After many years of experimental [16-18] and theoretical

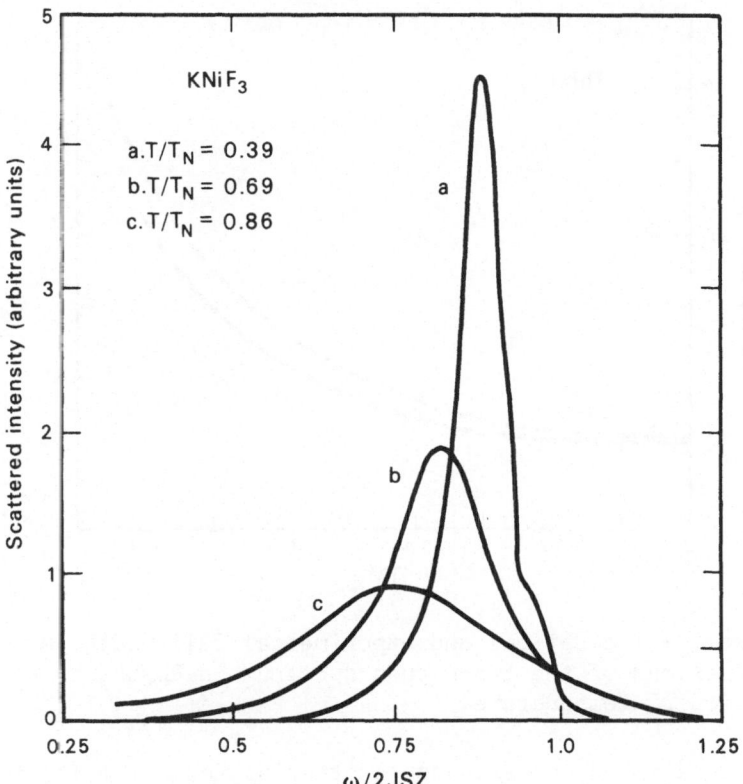

Figure 6 - Some computed two-magnon spectra in KNiF₃ versus relative frequency at various temperatures.

[4,6,8,9,11,14] work on the two-magnon resonance state in Heisenberg antiferromagnets, and because of the excellent agreement reached between experiment and theory, the two-magnon Raman scattering should prove to be an extremely accurate and inexpensive tool for determining exchange constants, anisotropy field parameters, and even magnon lifetimes [11] in those systems.

In the following, we shall outline how from the experimental data on one and two-magnon Raman scattering experiments one can derive these parameters, and shall, for the case of CoF_2, compare them with those obtained by other techniques.

There are three coincidences which make two-magnon Raman scattering experimentally interesting:

    (i) its large scattering amplitude;

   (ii) the sharp resonance (typically of the order of 10 cm^{-1}); and

  (iii) its strong dependence on the geometry of the scattering process.

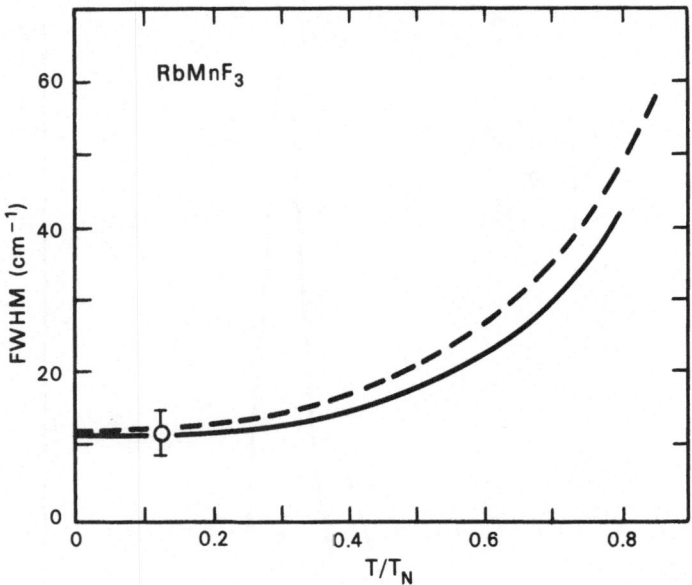

Figure 7 - Theoretical and experimental full widths at
half maximum of the two-magnon spectrum in RbMnF$_3$ as a
function of temperature.

On the basis of the exchange scattering mechanism [3] for two
magnon Raman scattering, the cross section at temperature $T = 0$
is found to be proportional to (1.21), valid also for the
slightly more general Hamiltonian (2.1) provided $\Omega_\mathbf{k}$ is given by

$$\Omega_\mathbf{k}^2 = S\left\{\left[J_2(0) - J_1(0) - J_3(0) + 2D\left(1 - \frac{1}{2S}\right) + J_1(\mathbf{k}) + J_3(\mathbf{k})\right.\right.$$

$$\left.+ 2E\left(1 - \frac{1}{2S}\right)^{\frac{1}{2}}\right]$$

$$\times\left[J_2(0) - J_1(0) - J_3(0) + 2D\left(1 - \frac{1}{2S}\right) + J_1(\mathbf{k}) + J_3(\mathbf{k})\right.$$

$$\left.\left. - 2E\left(1 - \frac{1}{2S}\right)^{\frac{1}{2}}\right] - J_2^2(\mathbf{k})\right\} , \qquad (2.2)$$

$$J_1(\mathbf{k}) = 2J_1\cos k_3 c,$$

$$J_3(\mathbf{k}) = 2J_3(\cos k_1 a + \cos k_2 a),$$

$$J_2(\mathbf{k}) = 8J_2\cos\tfrac{1}{2}k_1 a\cos\tfrac{1}{2}k_2 a\cos\tfrac{1}{2}k_3 c. \qquad (2.3)$$

J_2 denotes the exchange coupling between two sublattices, and

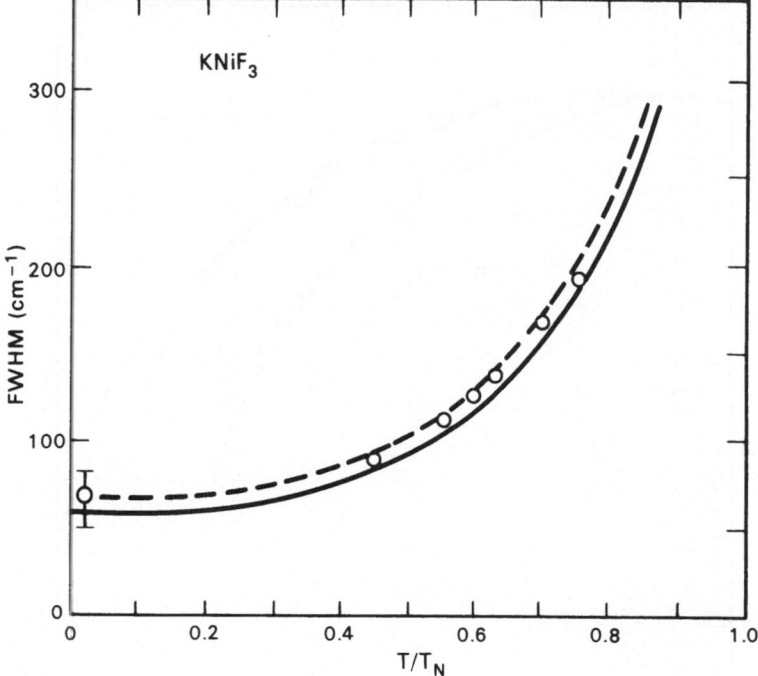

Figure 8 - Theoretical and experimental full widths at
half-maximum of the two-magnon spectrum in KNiF$_3$ as a
function of temperature.

J_1 and J_3 the exchange coupling between spins belonging to the
same sublattice along the c axis and in the basal plane respectively (figure 10). Equation (1.21), together with (2.2),
represents the two magnon Raman scattering cross-section.

For the present discussion of the two magnon resonance we
shall restrict ourselves to the zero temperature limit of the
two magnon Raman scattering cross section. Here we are only
interested in the physical implications of the two-magnon resonance, and therefore discard all the detailed algebra which
would be necessary for a quantitative analysis [11].

As already explained, the basic physical mechanism leading
to a two-magnon resonance state lies in the fact that less energy is required to excite simultaneously two spin deviations
on adjacent sites belonging to different sublattices than on
sites further apart. The reduction in energy is due to the attractive antiferromagnetic coupling between the spins. Exciting two spin deviations by Raman scattering means creating two
spin waves propagating in opposite direction due to the nearly
zero momentum transfer of the light. The high density of states
for zone boundary magnons implies that the two magnons created
by the Raman scattering process are predominantly magnons of
large wave vectors in the neighbourhood of the Brillouin zone.

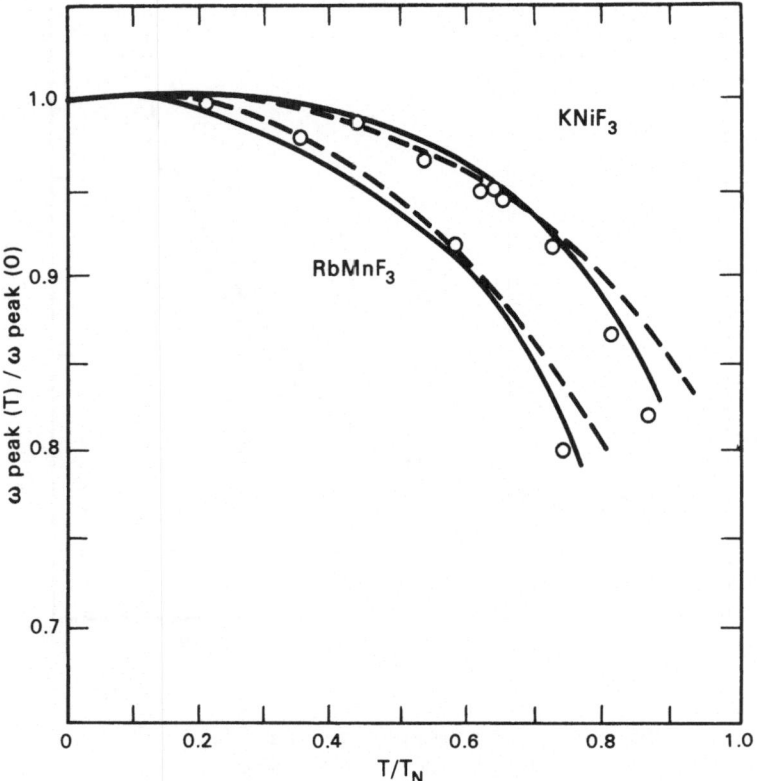

Figure 9 - Theoretical and experimental peak positions of two-magnon spectra as a function of temperature in KNiF$_3$ and RbMnF$_3$. Here and in the following figures, the full curves have been obtained by a direct numerical evaluation of equation (36), whereas the dashed curves refer to the simplified model mentioned at the end of section 5. The experimental data for KNiF$_3$ and RbMnF$_3$ are taken from references [2,13] respectively.

Yet a second and, for light scattering, rather important effect comes into play, and that is the geometric form factor (proportional to $\phi_{\alpha\alpha'}(k)$) entering the cross section. This form factor is a rapidly varying function of k and singles out certain restricted regions of the Brillouin zone contributing to the cross section.

As an example we indicate below some for Raman scattering representative geometries for the case of rutile structures:

$$\overset{\alpha\alpha'}{\phi_{\Gamma_5^+}}(k) = \sin\tfrac{1}{2}k_1 a \sin\tfrac{1}{2}k_2 a \cos\tfrac{1}{2}k_3 c, \quad xy \text{ scattering}$$

$$\phi_{\Gamma_5^+}(k) = \sin\tfrac{1}{2}k_1 a \cos\tfrac{1}{2}k_2 a \sin\tfrac{1}{2}k_3 c, \quad xz \text{ scattering} \qquad \begin{array}{c}(2.4)\\(\text{Contd})\end{array}$$

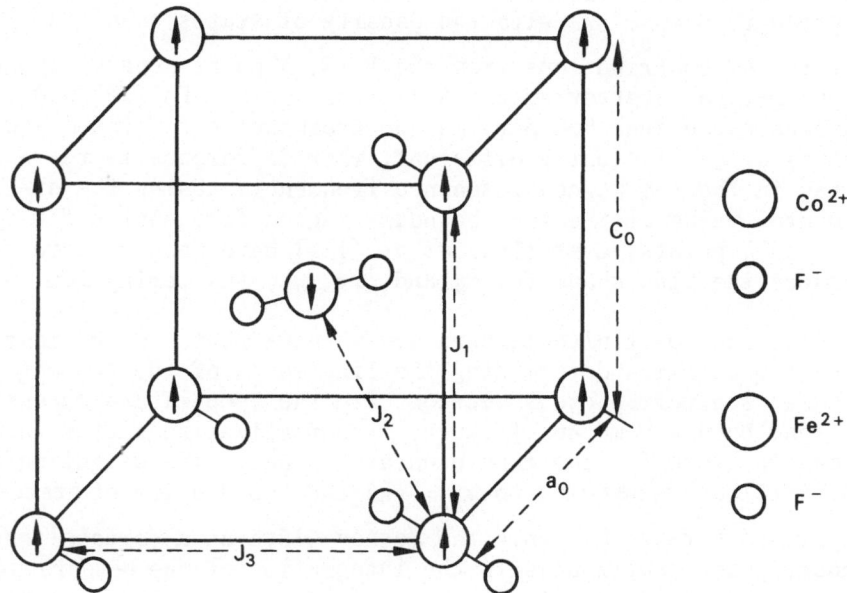

Figure 10 - Unitary magnetic cell for the rutile structure showing the spin direction and the location of the three nearest types of neighbours.

(Contd)   (2.4)

$$\phi_{\Gamma_5^+}^{\alpha\alpha'}(\mathbf{k}) = \cos\tfrac{1}{2}k_1a\sin\tfrac{1}{2}k_2a\sin\tfrac{1}{2}k_3c, \qquad yz \text{ scattering}$$

$$\phi_{\Gamma_1}(\mathbf{k}) = \cos\tfrac{1}{2}k_1a\cos\tfrac{1}{2}k_2a\cos\tfrac{1}{2}k_3c, \quad xx, \ yy, \ zz \text{ scattering}$$

where $\alpha\alpha'$ refer to the polarization of the incoming and outgoing photons.

From the form of the expressions (2.4) the weighting of certain regions of the Brillouin zone is now apparent. In particular we notice that in the xy, (xz,yz) scattering geometry certain magnons in the neighbourhood of the M, (R,U) point are singled out, whilst for xx, yy, zz scattering the centre of the Brillouin zone is strongly weighted. This has important consequences for the position of the two-magnon resonance frequency and the line shape of the Raman scattering cross section. While in the Γ_5^+ geometry, the weighting in the Brillouin zone owing to $\phi_{\Gamma_5^+}^2(\mathbf{k})$ coincides with the weighting caused by the density of states maximum, this is not the case for the Γ_1^+ geometry. Whereas for the Γ_5^+ geometry a sharp resonance behaviour is observed, nothing but a rather broad peak is found for Γ_1^+ scattering [5]. Moreover, in the Γ_5^+ scattering geometry, the regions in the Brillouin zone contributing to the

scattering cross section are sufficiently restricted owing to
the product of $\phi_{\Gamma_5^+}^2(\mathbf{k})$ with the density of states so as to

identify the contributions from the M (R,U) point separately in
the xy (xz, yz) scattering cross section. For CoF_2 [18] and
FeF_2, where the zone boundary magnon frequencies for the M and
R points are sufficiently different, this difference is reflected in the two magnon resonance frequency. Using the experimental value of the zone boundary magnon frequencies for
the M and R points, Castellani et al. [14] have been able to
reproduce the line shape for xy and (xz, yz) scattering for
CoF_2.

For the case of finite temperature we have that the two most
relevant quantities determining the line shape of the two magnon Raman scattering cross sections are the temperature dependent zone boundary magnon frequency renormalization factor and
the magnon width for the zone boundary magnon. The weighting
in the Brillouin zone owing to $\phi_{\Gamma_5^+}^2(\mathbf{k})$ and the density of states

is again so strong that only the magnon width at a certain zone
boundary point contributes in the integration of the temperature
dependent modified Green's function (1.22'). In turn, fitting
the theoretical expression to the experimentally obtained two-
magnon Raman line shape, one can determine the zone boundary
magnon widths with an accuracy comparable to its determination
by neutron techniques [15].

Having discussed the excellent agreement between theory and
experiment for the two magnon resonance one might ask the question of whether, by fitting the theoretical two magnon Raman
scattering cross section to the experimental results, one could
not obtain numerical values for the various exchange constants
and anisotropy parameters.

Expanding the ω dependent form factor in the scattering cross
section around the resonance frequency, we find for the Γ_5^+
geometry $(\alpha \neq \alpha')$, to which we shall restrict ourselves from
now on,

$$K^{\alpha\alpha'}(\omega, T = 0) \propto$$

$$\frac{G''_{\alpha\alpha'}(\omega_R^{\alpha\alpha'})\left[\frac{\partial G'_{\alpha\alpha'}(\omega)}{\partial\omega}\right]_{\omega=\omega_R^{\alpha\alpha'}}^{-1}}{(\omega - \omega_R^{\alpha\alpha'})^2 + \left[\frac{G''_{\alpha\alpha'}(\omega_R^{\alpha\alpha'})}{\left[\frac{\partial G'_{\alpha\alpha'}(\omega)}{\partial\omega}\right]_{\omega=\omega_R^{\alpha\alpha'}}}\right]^2} \cdot \frac{1}{\left[\frac{\partial G'_{\alpha\alpha'}(\omega)}{\partial\omega}\right]_{\omega=\omega_R^{\alpha\alpha'}}}, \quad (2.5)$$

with

$$1 - G'_{\alpha\alpha'}(\omega_R^{\alpha\alpha'}) = 0, \qquad (2.6)$$

putting for brevity

$$J_2 \lim_{\varepsilon \to 0^+} \mathrm{Re} L_0{}^{\alpha\alpha'}(E) = G'{}_{\alpha\alpha'}(\omega), \quad J_2 \lim_{\varepsilon \to 0^+} \mathrm{Im} L_0{}^{\alpha\alpha'}(E) = G''{}_{\alpha\alpha'}(\omega),$$

which is of Lorentzian form and has the width

$$\frac{G''{}_{\alpha\alpha'}}{\left[\dfrac{\partial G'{}_{\alpha\alpha'}(\omega)}{\partial \omega}\right]_{\omega=\omega_R{}^{\alpha\alpha'}}} . \qquad (2.7)$$

For simplicity let us suppose $J_3 = 0$. In this case we have four relations determining the resonance frequency and width for the xy and xz geometry (the yx geometry is indistinguishable from the xz one because of x-y symmetry for rutiles).

Explicit analytic evaluation of $L_0{}^{\alpha\alpha'}(E)$ leads to the following functional dependencies on J_1, J_2, E and $\omega_B = S[J_2(0) - J_1(0) + 2D(1 - (1/2S))]$ [13],

$$G'{}_{\alpha\alpha'}(\omega) = \frac{J_2(0)}{\omega_B} f_{\alpha\alpha'} \left[\left[1 - \frac{2J_1(0)}{\omega_B} - \frac{\bar{E}^2}{\omega_B{}^2} - \frac{\omega^2}{\omega_B{}^2}\right]\left(\frac{J_2(0)}{\omega_B}\right)^{-2}\right]$$

$$= \frac{\omega_B}{J_2(0)} f_{\alpha\alpha'}(\omega), \qquad (2.8)$$

$$G'{}_{\alpha\alpha'}(\omega) = \frac{J_2(0)}{\omega_B} g_{\alpha\alpha'} \left[\left[1 - \frac{2J_1(0)}{\omega_B} - \frac{\bar{E}^2}{\omega_B{}^2} - \frac{\omega^2}{\omega_B{}^2}\right]\left(\frac{J_2(0)}{\omega_B}\right)^{-2}\right]$$

$$= \frac{\omega_B}{J_2(0)} g_{\alpha\alpha'}(\omega),$$

$$\bar{E} = 2E\left(1 - \frac{1}{2S}\right)^{\frac{1}{2}}.$$

This functional dependence is obtained provided $J_1/\omega_B \ll 1$, which is true for most physical cases of interest. If this condition is not satisfied the functional dependence is changed, although the following consideration remains unaltered.

The relations which will determine J_1, J_2, E and D are now given by

$$1 - \frac{J_2(0)}{\omega_B} f_{xy}(\omega_R{}^{xy}) = 0, \qquad \frac{g_{xy}(\omega_R{}^{xx})}{f'_{xy}(\omega_R{}^{xy})} = \frac{\Gamma_{xy}}{\omega_B}, \qquad (2.9)$$
$$\text{(Contd)}$$

$$1 - \frac{J_2(0)}{\omega_B} f_{xz}(\omega_R{}^{xz}) = 0, \qquad \frac{g_{xz}(\omega_R{}^{xz})}{f'_{xz}(\omega_R{}^{xz})} = \frac{\Gamma_{xz}}{\omega_B}, \qquad \begin{matrix}\text{(Contd)}\\ \text{(2.9)}\end{matrix}$$

which in principle can determine I_1, J_2, E and D. In practice, it turns out that it is possible to get a greater accuracy by taking advantage of the knowledge of the uniform antiferromagnetic resonance

$$\Omega^2(0) = \left[J_2(0) + 2D\left(1 - \frac{1}{2S}\right) \right]^2 - \bar{E}^2 - J_2{}^2(0). \qquad (2.10)$$

In such a case, the extra relation (2.10) serves as a check on the accuracy of the parameters. For CoF_2, J_1, J_2, E and D have been determined [13] by this procedure within an accuracy of 1%. Using these parameters the following values for the zone boundary magnon frequencies were derived,

$$2\omega(R) = 128.5 \text{ cm}^{-1}, \qquad 2\omega(M) = 123.5 \text{ cm}^{-1},$$

which compare well with the one obtained from neutron scattering [15]

$$2\omega(R) = 130 \text{ cm}^{-1}, \qquad 2\omega(M) = 125 \text{ cm}^{-1}.$$

Before concluding these notes we should like to discuss briefly the possible relevance of the two-magnon Raman scattering for the investigation of the short range inter-atomic electron correlation. The two-magnon Raman scattering mechanism discussed above applies strictly to insulating antiferromagnets where the magnetic electrons can be considered localized, at least on the time scale of the Raman scattering process.

In this connection it is particularly interesting that two-magnon Raman scattering has been observed in NiO [19] exhibiting the same features as in the rutiles and perovskites Heisenberg antiferromagnets. Since NiO belongs to the family of the transition metal compounds for which Hubbard's ideas on narrow band systems seem to apply, the investigation of the local character of the magnetic moments in these systems by two-magnon Raman excitation should throw some light on the inter-atomic electron correlations. By local moments we mean moments in the sense of the Heisenberg picture. In this connection two-magnon Raman scattering could provide an indication of the variation of the amount of locality with the ratio Δ/U, Δ being the bandwidth and U the Coulomb inter-electronic repulsion. It would be interesting to see what the two-magnon Raman scattering looks like, if it exists at all, in the insulating phase of V_2O_3. V_2O_3 is a candidate for a Hubbard system with a gap of 0.1 ev. Moreover, it has recently been suggested [20] that there might be two types of antiferromagnetic order as a func-

tion of Δ/U, for which NiS and FeS are representative examples
(figure 11). In NiS the antiferromagnetic ground state seems

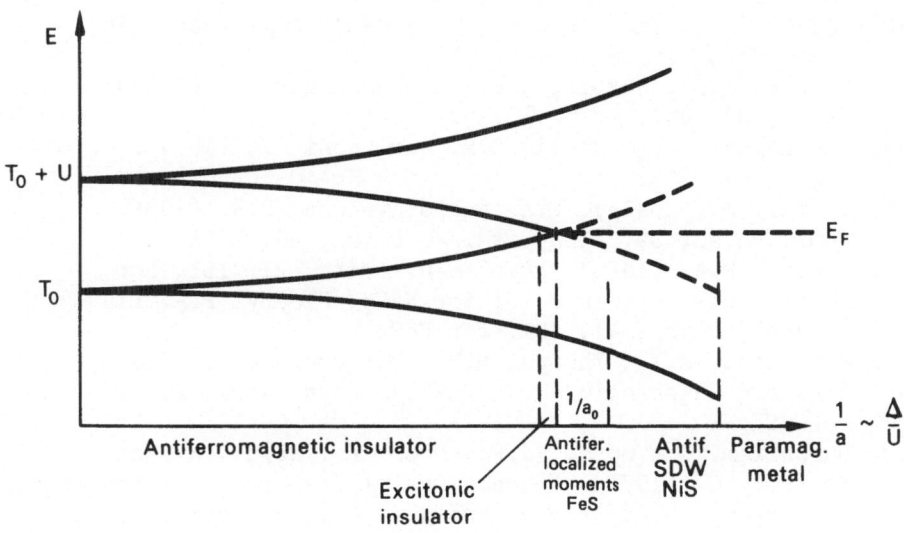

Figure 11

to have an itinerant character resembling the Cr spin density
wave ground state. No evidence of the two-magnon Raman process
could be found, consistently with the fact that itinerancy
means delocalization (these concepts could be put on a quantitative basis). On the contrary, FeS seems to have an antiferromagnetic ground state with some local character, and there
the two-magnon Raman scattering should be observable.

REFERENCES

1. Bass, F.G. and Kaganov, M.I. (1959). *Zh. Eksp. Teor. Fiz.*,
   37, 1390. [(1960). *Soviet Phys.*, *JETP*, **10**, 986].
2. Fleury, P.A., Porto, S.P., Chesman, L.E. and Guggenheim,
   H.J. (1966). *Phys. Rev. Lett.*, **17**, 84.
3. Fleury, P.A. and Loudon, R. (1968). *Phys. Rev.*, **166**, 514.
4. Elliott, R.J. and Thorpe, M.F. (1969). *J. Phys.*, **C2**, 1630.
5. Fleury, P.A. (1969). *Phys. Rev.*, **180**, 591. For a review,
   see also: Fleury, P.A. (1970). *Int. J. Magn.*, **1**, 75.
6. Davies, R.W., Chinn, S.R. and Zeiger, H.J. (1971). *Phys.
   Rev.*, **B4**, 992, 4017.
7. Kawasaki, T.K. (1970). *J. Phys. Soc. Jap.*, **28**, 1144.
8. Šolyom, J.S. (1971). *Z. Phys.*, **243**, 382; (1971). In *Light
   Scattering in Solids*, (ed. Balkanski, M.), (Flammarion,
   Paris), pp. 165-169.
9. Cottam, M.G. (1972). *Solid State Commun.*, **10**, 99; *J. Phys.*,
   C5, 1461.

10. Balucani, V., Barocchi, F. and Tognetti, V. (1972). *Phys. Lett.*, **A40**, 339.

11. Natoli, C.R. and Ranninger, J. (1972). *Phys. Lett.*, **A39**, 105; (1973). *J. Phys.*, **C6**, 345.

12. Balucani, U. and Tognetti, V. (1973). *Phys. Rev.*, **B8**, 4247; *Solid State Commun.*, **13**, 1811.

13. Bohnen, K.P., Natoli, C.R. and Ranninger, J. (1974). *J. Phys.*, **C7**, 947.

14. Castellani, C., Natoli, C.R. and Leoni, F. (1974). *J. Phys.*, **C7**, 1353.

15. Martel, P., Cowley, R.A. and Stevenson, R.W. (1968). *J. Appl. Phys.*, **39**, 1116; *Can. J. Phys.*, **46**, 1355.

16. Fleury, P.A. (1968). *Phys. Rev. Lett.*, **21**, 151, for $RbMnF_3$; see reference [5] for NiF_2; Fleury, P.A. (1970). *J. Appl. Phys.*, **41**, 881, for FeF_2.

17. Hutchings, M.T., Thorpe, M.F., Birgeman, R.J., Fleury, P.A. and Guggenheim, H.J. (1970). *Phys. Rev.*, **B2**, 1362, for NiF_2.

18. Dugantier, C., Gosso, J.P., Moch, P., Moyal, R. and Parisot, G. (1971). *Proceedings of The Second International Conference on Light Scattering in Solids, Paris, 1970*, (ed. Balkanski, M.), (Flammarion, Paris).

19. Dietz, R.E., Parisot, G.I. and Meixner, A.E. (1971). *Phys. Rev.*, **B4**, 2302.

20. Coey, J.H.D., Brusetti, R., Kallel, A., Schweizer, J., and Fuess, H. (1974). *Phys. Rev. Lett.*, **32**, 1257.

21. Cowley, R.A., Buyers, W.J.L., Martel, P. and Stevenson, R.W.H. (1969). *Phys. Rev. Lett.*, **23**, 86.

22. Natoli, C.R. and Ranninger, J. (1973). *J. Phys.*, **C6**, 370.

23. Saunderson, D.M., Windsor, C.G., Briggs, G.A., Evans, M.T. and Hutchinson, E. (1972). *I.A.E.A. Symposium on Neutron Inelastic Scattering, Grenoble*, (International Atomic Energy Authority, Vienna).

TWO-FLUID HYDRODYNAMIC DESCRIPTION
OF MAGNETIC CRYSTALS†

CHARLES P. ENZ

Département de Physique Théorique,
Université de Genève,
1211 Geneva 4, Switzerland

INTRODUCTION

The rôle of the exchange interaction in the transition to an ordered state of the spins in a magnetic crystal was first discussed by Heisenberg, Frenkel and Dorfman [1] in 1928. Two years later Bloch [2] showed that this coupling gives rise to wave excitations of the spin orientation. These spin waves, or magnons, were subsequently derived by Landau and Lifshitz [3] in the framework of a purely phenomenological description of magnetism. But it was only the hydrodynamic description by Halperin and Hohenberg [4] that established, for the planar and isotropic ferro- and antiferromagnets, the close analogy with the phenomenological description of superfluid Helium.

Halperin and Hohenberg [4] obtained for the isotropic (Heisenberg) ferromagnet the well known quadratic magnon spectrum $\omega = Dq^2$ originally derived, in the collisionless case and for small q, by Bloch [2]. But for the planar ferromagnet in which the preferred direction of magnetization is restricted to an 'easy plane' they found a sound-like spectrum, $\omega = c_1 q$, for small q. In the collisionless case such a linear dispersion relation had been discovered earlier for antiferromagnets by Hulthén [5]. Halperin and Hohenberg [4] also rederived this Hulthén mode for the planar antiferromagnet in which the sublattice magnetization is in an 'easy plane' and for the isotropic (Heisenberg antiferromagnet, They also obtained a new prediction for the magnon dampings. A deriva-

---

† Part 4 of the review article published in (1974). *Rev. Mod. Phys.*, **46**, 705.

tion of these results from the point of view of irreversible
thermodynamics was subsequently given by Enz [6].

In analogy with dielectric crystals, the two fluids in a magnetic crystal have to be identified as the classical magnetization field, $\vec{M}(\vec{r},t)$, and the fluid of the thermal magnons. Indeed, $\vec{M}$ describes the dynamics of the condensed phase of the
ordered spins. It is also the order parameter of the ferromagnet, whilst in the antiferromagnet the sublattice (staggered)
magnetization $\vec{N}$ plays this rôle. Whilst the magnetization satisfies a local conservation law analogous to the continuity
equation, this is not the case for the mass density of the magnon fluid, since magnon number is not conserved. This difference with superfluids justifies the omission of the 'first fluid'
of thermal magnons in the treatment of Halperin and Hohenberg [4].

The hydrodynamics of the magnon fluid was first considered in
a note by Gulayev [7] (see also reference [8]) in which the name
'second spin waves' was introduced for the excitations of this
fluid. Subsequently Reiter [9] derived a Boltzmann equation for
the magnon fluid of a Heisenberg ferromagnet. A generalized
form of this Boltzmann equation was used by Michel and Schwabl
[10] to derive coupled hydrodynamic equations for the two fluids,
in close analogy with the treatment of dielectric crystals by
Götze and Michel [11].

This analogy extends, in particular, to the introduction of
a magnon drift (quasi-momentum), which is essential for the fluid properties of the magnons and hence for the existence of second sound or second spin waves, called second magnons in reference [10]. In subsequent papers Michel and Schwabl [12] have
derived this hydrodynamics from microscopic equations of motion.
As was shown by Enz [13] the addition of the local balance equation for the magnon momentum to the hydrodynamic equations of
Halperin and Hohenberg [4,6] then also leads to second sound.

The existence of second sound depends on Landau's criterion for
superfluidity. This means that a Bloch mode, $\omega = Dq^2$, does not give
rise to second sound, whilst a Hulthén mode $\omega = c_1 q$, does. In insulating magnetic crystals, however, the long-range dipolar interaction
between the spins gives rise to a gap of the Bloch mode analogous
to the plasmon, and also transforms a Hulthén mode into a Bloch
mode with gap. In fact, up to date neutron scattering which
measures directly the magnon spectrum, has, to our knowledge,
not revealed any case of a purely sound-like dispersion relation. The search for second sound in insulating magnetic crystals is therefore a quite realistic problem. The quantitative
conditions for the realizability of a second magnon have been
analyzed by Michel and Schwabl [14], Forney [15] and Forney and
Jäckle [16], the conclusions being quite encouraging.

In this lecture we follow essentially the approach of reference [13] except for two generalizations. The first is a description of the response measured by neutron scattering. The second generalization concerns the cases of the axial and isotropic ferro- and antiferromagnets which are treated in much greater detail, including, in particular, the effects of the dipole-dipole interaction and a comparison with the phenomenological approach of Kittel [17,18]. This generalization allows a direct comparison with earlier results for the magnon spectra [19-22] but, in addition, also yields expressions for the damping of all the magnon modes.

Thus we arrive at a general and unified hydrodynamic description of magnetic crystals, within the classification into planar, isotropic and axial systems. This classification, of course, is not exhaustive, as can be seen from more specialized reviews [23,24]. It is also evident that lattice displacements introduce variations in the coupling functions of the exchange interaction giving rise to magnon-phonon interactions.

1. HYDRODYNAMICS OF THE PLANAR FERROMAGNET

A planar ferromagnet is described by a Hamiltonian

$$\mathcal{H} = \mathcal{H}_0 - m\mu \sum_i S_i{}^z,$$

$$\mathcal{H}_0 = - \sum_{ij} \left\{ J_{ij}{}^z S_i{}^z S_j{}^z + J_{ij}{}^\perp (S_i{}^x S_j{}^x + S_i{}^y S_i{}^y) \right\}, \qquad (1)$$

with coupling constants $J_{ij}{}^\perp > J_{ij}{}^z > 0$, for the dominant pairs of neighbours i,j. $m\mu$ is the microscopic magnetic field seen by the spins. (Magnetic fields are measured in energy units).

The crucial feature of a planar ferromagnetic state is the existence of a non-vanishing average of the spin raising operator $S_i{}^+ = S_i{}^x + iS_i{}^y$. The ferromagnetic average has the property that it is taken with a density matrix which commutes with $\mathcal{H}$ but not with $\sum_i S_i{}^z$. Since $S_i{}^+$ increases the value of $S_i{}^z$ by one, and since $[\sum_i S_i{}^z, \mathcal{H}_0] = 0$, the Heisenberg representation,

$$S_i{}^+(t) = \exp(i\mathcal{H}_0 t) S_i{}^+ \exp(-i\mathcal{H}_0 t),$$

together with (1) imply the existence of a non-vanishing order parameter $(1/v)\langle S_i^+\rangle$ with the time evolution

$$\frac{1}{v}\langle S_i^+(t)\rangle = \frac{1}{v} e^{+im\mu t}\langle S_i^+\rangle = M_\perp e^{+i\phi(t)} = M_x + iM_y. \qquad (2)$$

Here v is the volume of the unit cell, $M_\perp$ the perpendicular magnetization and $\phi(t)$ the precession angle. From (2) we have

$$\phi(t) = \phi(0) + m\mu t. \qquad (3)$$

The magnetization energy per unit mass is

$$\epsilon_S = (\rho V)^{-1}\langle \mathcal{H}_0\rangle. \qquad (4)$$

The index s refers to the identification of the magnetization with the 'superfluid' phase. ρ is the mass density of the crystal, which here is a constant, and V the volume of the crystal. Since ϵ_S and the parallel magnetization,

$$M_z = \frac{1}{v}\langle S_i^z\rangle, \qquad (5)$$

are conserved quantities, $M_\perp$ must be a function of ϵ_S and M_z. As pointed out in reference [4] this supposes the existence of a relaxation mechanism which brings $M_\perp$ to its value $M_\perp(\epsilon_S M_z)$ in a microscopic time.

In thermal equilibrium at a temperature T below the Curie point T_c, $\epsilon_S = \epsilon_S(M_z,T)$ so that

$$M_\perp = M_\perp(M_z,T) \qquad (6)$$

and the thermodynamic field is, for $M_z \ll M_\perp$,

$$h_z(M_z,T) = \left(\frac{\partial(\rho\epsilon_S)}{\partial M_z}\right)_{S_S} = \chi^{-1}(T)M_z, \qquad (7)$$

where $s_S = s_S(M_z,T)$ is the magnetization entropy per unit mass and χ is the isothermal longitudinal susceptibility. Thermal equilibrium also means that the precession is zero and that the external field H_z equals the thermodynamic field (7), that is,

$$\dot{\phi}_{eq} = m\mu_{eq} = 0, \qquad (H_z)_{eq} = h_z. \qquad (8)$$

For $\mu \neq 0$, equation (3) tells us that

$$\dot{\phi} = m\mu. \qquad (9)$$

On the other hand, for an external field H_z so small that $M_z = 0$ is maintained, we see from (1) that we must have $m\mu = H_z$ (this is a metastable state!), hence

$$\mu = \frac{1}{m} (H_z - h_z). \tag{10}$$

μ plays the rôle of a chemical potential per unit mass, where m is the atomic mass.

Going over to local variables $\varepsilon_S(r,t)$, etc., defined as averages of density operators (see reference [4]), we may again define a velocity field

$$\vec{v}_S = - \frac{1}{m} \nabla\phi, \tag{11}$$

which is exactly analogous to the superfluid velocity. According to (9) $\vec{v}_S$ satisfies the equation of motion

$$\dot{\vec{v}}_S = - \nabla (\mu + \mu'), \tag{12}$$

where μ' is the dissipative part of μ.

Equations (12,10) are the same, respectively, as equations (2.15,20) of reference [4] with the identification of $\vec{v}/m$ with our $\vec{v}_S$, and of ψ/m with our μ.

In view of (7), and because $\vec{v}_S$ is an independent thermodynamic variable, we have the thermodynamic relation

$$d(\rho\varepsilon_S) = h_z dM_z + \rho_S \vec{v}_S \cdot d\vec{v}_S + T d(\rho s_S). \tag{13}$$

The coefficient ρ_S, which for simplicity we treat as a scalar, must be positive. It has the meaning of a stiffness constant [4] (see below) and the dimension of a mass density and is the analogue of the superfluid mass density.

Local conservation of parallel magnetization is expressed by

$$m\dot{M}_z + \nabla \cdot (\vec{j}_S + \vec{j}_S') = 0, \tag{14}$$

where the dissipative part $\vec{j}_S'$ describes spin diffusion. Apart from a factor m equation (14) is the same as equation (2.16) of reference [4].

In analogy with the case of phonons in dielectric crystals, the thermal magnons are described by a momentum density $\vec{j}_M$ and a local drift $\vec{v}_M$ which define the magnon mass density ρ_M through

$$\vec{j}_M = \rho_M \vec{v}_M. \tag{15}$$

Again we treat ρ_M as a scalar for the sake of simplicity. The local magnon momentum balance equation is

$$\dot{\vec{j}}_M + \nabla(\pi_M + \pi_M') = -\frac{1}{\tau_J} \vec{j}_M. \tag{16}$$

Here,

$$\pi_M = -\rho f_M 1 + \rho_M \vec{v}_M \otimes \vec{v}_M \tag{17}$$

is the momentum flux, and f_M the magnon free energy per unit mass which satisfies the thermodynamic relation

$$d(\rho f_M) = -\vec{j}_M \cdot d\vec{v}_M - \rho s_M dT. \tag{18}$$

s_M is the magnon entropy per unit mass and π_M' the dissipative part of π_M.

The magnon energy density is given by

$$\rho \varepsilon_M = \vec{v}_M \cdot \vec{j}_M + \rho(f_M + T s_M). \tag{19}$$

The total energy is the sum of the magnetization energy, the magnon energy and the Zeeman energy of an arbitrary external field $\vec{H}$,

$$d(\rho \varepsilon) = d(\rho \varepsilon_S) + d(\rho \varepsilon_M) - \vec{H} \cdot d\vec{M}. \tag{20}$$

Since according to (2),

$$\vec{M} = (M_\perp \cos\phi, M_\perp \sin\phi, M_z), \tag{21}$$

we can choose $\vec{H} = (0, H_y, H_z)$. Making use of (13,19,18), equation (20) can then be written as

$$d(\rho \varepsilon) = (h_z - H_z)dM_z - H_y dM_y$$
$$+ \rho_s \vec{V}_s \cdot d\vec{V}_s + \vec{v}_M \cdot d\vec{j}_M + T d(\rho s), \tag{22}$$

where $s = s_S + s_M$.

The equilibrium condition for the total free energy at constant T and $\vec{j}_M$ is

$$\delta \int (\rho \varepsilon - T \rho s) d^3 r = 0, \qquad \delta T = 0, \qquad \delta \vec{j}_M = 0. \tag{23}$$

Remembering (21,6) and making use of (11), and of a partial integration to eliminate $\vec{v}_s$, the conditions (23) with (22) yield

$$\int \left\{ \left[h_z - H_z - H_y \left(\frac{\partial M_\perp}{\partial M_z} \right)_T \sin\phi \right] \delta M_z \right.$$

$$\left. + \left[-H_y M_\perp \cos\phi - \frac{\rho_s}{m^2} \nabla^2 \phi \right] \delta\phi \right\} d^3 r = 0.$$

Since δM and $\delta\phi$ are independent variations, we find for the Fourier components, for small H_y,

$$\phi = \frac{m^2 M_\perp}{q^2 \rho_s} H_y + O(H_y^3), \tag{24}$$

and

$$h_z = H_z + \frac{m^2 M_\perp}{q^2 \rho_s} \left(\frac{\partial M_\perp}{\partial M_z} \right)_T H_y^2 + O(H_y^4).$$

In equilibrium, $H_y = 0$, so that $\phi_{eq} = 0$, and we recover (8). Inserting (24) into (21) we also obtain

$$M_x = M_\perp + O(H_y^2), \tag{25}$$

$$M_y = \chi_\perp H_y + O(H_y^3),$$

where

$$\chi_\perp = \frac{m^2 M_\perp^2}{q^2 \rho_s} \tag{26}$$

is the isothermal transverse susceptibility (see equation (2.30) of reference [4]).

From (21,24,11) it also follows that for $H_y = 0$,

$$\nabla M_y = -m M_\perp \vec{\vartheta}_s, \tag{27}$$

so that

$$\rho_s \vec{\vartheta}_s \cdot d\vec{\vartheta}_s = \frac{\rho_s}{m^2 M_\perp^2} \nabla M_y \cdot d\nabla M_y \tag{28}$$

is recognized as an exchange energy contribution [17,23] to (13). It is of the same form as the elastic energy contribution,

$$\sum_{ijk\ell} C_{ij,k\ell} \nabla_i u_j d \nabla_k u_\ell,$$

Thus we see that $\rho_S/(m^2 M_\perp^2)$ plays the same rôle as the 'elastic stiffness constants' $C_{ij,k\ell}$ [17].

We now convert the energy conservation,

$$(\rho\varepsilon)^{\cdot} + \nabla \cdot (\vec{J}_\varepsilon + \vec{J}_\varepsilon') = 0, \tag{29}$$

into the entropy balance,

$$(\rho s)^{\cdot} + \nabla \cdot (\vec{J}_S + \vec{J}_S') = \sigma, \tag{30}$$

thus obtaining expressions for the magnetization flux $\vec{j}_S$, the energy flux $\vec{J}_\varepsilon$ and the entropy flux $\vec{J}_S$. The dissipative parts μ', π_M', $\vec{j}_S'$, $\vec{J}_\varepsilon'$ and $\vec{J}_S'$ will be constructed afterwards.

From (22) we obtain with (10),

$$T(\rho s)^{\cdot} = (\rho\varepsilon)^{\cdot} + m\mu\dot{M}_z + H_y\dot{M}_y - \rho_S\vec{v}_S \cdot \dot{\vec{v}}_S - \vec{v}_M \cdot \dot{\vec{j}}_M, \tag{31}$$

which apart from the magnon drift term, is the same, for $\vec{H} = 0$ as equation (2.31) of reference [4]. In order to compensate the term $H_y\dot{M}_y$ in (31), the equation of motion (14) has to be modified in the presence of the transverse field H_y. This is done by writing

$$\dot{M}_z + \frac{1}{m} \nabla \cdot (j_S + j_S') = - M_x H_y,$$

$$\dot{M}_x - \gamma M_x \dot{M}_z = - M_y(H_z - h_z), \tag{32}$$

$$\dot{M}_y - \gamma M_y \dot{M}_z = M_x(H_z - h_z),$$

where the last two equations follow from (21,9) with

$$\gamma = \frac{1}{M_\perp} \left(\frac{\partial M_\perp}{\partial M_z} \right)_T \ .$$

Since according to (26) $\chi_\perp \gg \chi$ for small q, the right hand side of (32) is just the precession term $\vec{M} \times (\vec{h} - \vec{H})$ with $\vec{h} = (0,0,h_z)$ which brings (32) into the form of the Bloch [17] or Landau-Lifshitz equation [23].

Substituting for the right hand side of (31) successively from the equations of motion (29), the first equation (32), (12)

and (16), making use, in the last equation, of (15,17,18) we
find

$$(\rho s)^{\cdot} = - \nabla [\frac{1}{T} (\vec{J}_{\epsilon} + \vec{J}_{\epsilon}' + \mu(\vec{j}_S + \vec{j}_S') - \mu' \rho_S \vec{v}_S - \pi_M' \vec{v}_M)]$$

$$+ [\vec{J}_{\epsilon} + \vec{J}_{\epsilon}' + \mu(\vec{j}_S + \vec{j}_S') - \mu' \rho_S \vec{v}_S - \pi_M' \vec{v}_M - T \rho s_M \vec{v}_M] \cdot \nabla \frac{1}{T}$$

$$+ \frac{1}{T} (\vec{j}_S + \vec{j}_S' + \rho_S \vec{v}_S) \nabla \mu - \frac{1}{T} [\mu' \nabla \cdot (\rho_S \vec{v}_S) + (\pi_M' \nabla) \cdot \vec{v}_M]$$

$$+ \frac{\rho_M}{T \tau_J} \vec{v}_M^{\,2} .$$

Since the non-dissipative terms on the right hand side must add
to a divergence we find

$$\vec{j}_S = - \rho_S \vec{v}_S, \tag{33}$$

and

$$\vec{J}_{\epsilon} = \mu \rho_S \vec{v}_S + T \rho s_M \vec{v}_M. \tag{34}$$

(Compare equations (2.22,35) and (2.21,36) of reference [4]).
Comparison with (30) then yields the entropy flux,

$$\vec{J}_S = \rho s_M \vec{v}_M, \tag{35}$$

its dissipative part,

$$\vec{J}_S' = \frac{1}{T} (\vec{J}_{\epsilon}' + \mu \vec{j}_S' - \mu' \rho_S \vec{v}_S - \pi_M' \vec{v}_M), \tag{36}$$

and the entropy production density,

$$\sigma = T \vec{J}_S' \nabla \frac{1}{T} + \frac{1}{T} [\vec{j}_S' \nabla \mu + \mu' \nabla \cdot \vec{j}_S + (\pi_M' \nabla) \cdot \vec{v}_M] + \frac{\rho_M}{T \tau_M} \vec{v}_M^{\,2}. \tag{37}$$

(Note that the partition of dissipation terms into $\vec{J}_S'$ and σ is
dictated by the positiveness of σ. In reference [13] this condition was overlooked, without serious consequences, however).
    For small perturbations, and for $H_y = 0$, the dissipative
parts $\vec{j}_S'$, μ', π_S' and $\vec{J}_S'$ must be linear functions of the gra-
dients of the coefficients in equation (22), that is, of $-m \nabla \mu$,
$\nabla \otimes (\rho_S \vec{v}_S)$, $\nabla \otimes \vec{v}_M$ and ∇T. Irreversibility requires that under
time reversal they transform with opposite sign to their respec-
tive non-dissipative parts. Since $\vec{J}_S$ and π_M transform with positive sign, μ, $\vec{v}_M$ and $\vec{J}_S$ (and also $M_\perp$!) with negative sign, we

find, with (33),

$$\vec{J}_S' = + m^2 \xi \nabla \mu,$$

$$\mu' = + (\zeta \nabla) \cdot \vec{J}_S,$$

$$\pi_{Mij}' = - \sum_{k\ell} \gamma_{ij,k\ell} \nabla_k v_{M\ell}, \tag{38}$$

$$\vec{J}_S' = - \frac{\kappa}{T} \nabla T.$$

Here we have neglected cross-terms coupling magnetic and thermal variables, that is, respectively, terms proportional to $\nabla \otimes \vec{v}_M$, ∇T, $\nabla \mu$ and $\nabla \otimes (\rho_S \vec{v}_S)$ in $\vec{J}_S'$, μ', π_M' and $\vec{J}_S'$. (Note that the last equation (38) differs from equation (30) of reference [13] which, through equation (27), led to additional damping terms). ξ is the spin diffusion tensor, ζ a second viscosity tensor analogous to ζ_3 in superfluid Helium, $\gamma_{ij,k\ell}$ is the magnon viscosity tensor, and κ the magnon heat conductivity tensor. With (38), equation (37) becomes

$$T\sigma = \nabla T \cdot \left[\frac{\kappa}{T} \nabla T \right] + m^2 \nabla \mu \cdot (\xi \nabla \mu) + \nabla \cdot (\rho_S \vec{v}_S)(\zeta \mu) \cdot (\rho_S \vec{v}_S)$$

$$+ \sum_{ijk\ell} \nabla_i v_{Mj} \gamma_{ij,k\ell} \nabla_k v_{M\ell} + \frac{\rho_M}{\tau_J} \vec{v}_M^2. \tag{39}$$

2. HYDRODYNAMIC MODES OF THE PLANAR FERROMAGNET

The hydrodynamic equations of motion (12,14,16,30) are, in linear approximation with $H_y = 0$, inserting from (10,15,17,18, 33,35,38),

$$\dot{\vec{v}}_S - \frac{1}{m} \nabla h_z - \rho_S \nabla((\zeta \nabla) \cdot v_S) = - \frac{1}{m} \nabla H_z, \tag{40}$$

$$\dot{M}_z - \frac{\rho_S}{m} \nabla \cdot \vec{v}_S - (\xi \nabla) \cdot \nabla h_z = - (\xi \nabla) \cdot \nabla H_z, \tag{41}$$

$$\dot{v}_{Mi} + \frac{\rho}{\rho_M} s_M \nabla_i T - \frac{1}{\rho_M} \sum_{jk\ell} \gamma_{ij,k\ell} \nabla_j \nabla_k v_{M\ell} + \frac{1}{\tau_J} v_{Mi} = 0, \tag{42}$$

$$\dot{s} + s_M \nabla \cdot \vec{v}_M - \frac{1}{\rho T} (\kappa \nabla) \cdot \nabla T = 0. \tag{43}$$

Choosing as independent variables δT, δM_z, v_S and v_M, we have

$$\delta s = \frac{c_M}{T}\,\delta T - \frac{\beta}{\rho}\,\delta M_z,$$

$$\delta h_z = \beta\delta T + \frac{1}{\chi}\,\delta M_z, \tag{44}$$

where c_M is the specific heat at constant magnetization per unit mass, χ is defined by equation (7), and

$$\beta = \left(\frac{\partial h_z}{\partial T}\right)_{M_z} = -\rho\left(\frac{\partial s}{\partial M_z}\right)_T \tag{45}$$

is the magnetic tension coefficient. Here the second equality follows from (22).

Going over to plane waves and introducing

$$(\lambda_N^2(\vec{q},\omega))_{ij} = \frac{\tau_J}{\rho_M}\sum_{m,n}\gamma_{im,nj}(\vec{q},\omega)\hat{q}_m\hat{q}_n, \tag{46}$$

where $\hat{q} = \vec{q}/|q|$, equations (40-43) become, after insertion of (44),

$$\omega\vec{v}_s + i\vec{q}\rho_s\vec{q}\cdot\vec{v}_s + \vec{q}\,\frac{1}{m\chi}\,\delta M_z + \vec{q}\,\frac{\beta}{m}\,\delta T = \vec{q}\,\frac{1}{m}\,\delta H_z, \tag{47}$$

$$\left[\omega + iq^2\,\frac{\xi_\ell}{\chi}\right]\delta M_z + \frac{\rho_s}{m}\,\vec{q}\cdot\vec{v}_s + iq^2\beta\xi_\ell\delta T = iq^2\xi_\ell\delta H_z, \tag{48}$$

$$\left[-i\omega + \frac{1}{\tau_J} + \frac{1}{\tau_J}\,\lambda_N^2 q^2\right]\vec{v}_M + i\vec{q}\,\frac{\rho}{\rho_M}\,s_M\delta T = 0. \tag{49}$$

and

$$\left[\omega + iq^2\,\frac{\kappa_\ell}{\rho c_M}\right]\delta T - \frac{Ts_M}{c_M}\,\vec{q}\cdot\vec{v}_M - \omega\,\frac{T\beta}{\rho c_M}\,\delta M_z = 0. \tag{50}$$

Here we have: $\xi_\ell = (\hat{q}\cdot\xi\hat{q})$ and $\kappa_\ell = (\hat{q}\cdot\kappa\hat{q})$. For $\delta H_z = 0$, equations (48,49,50) are the same, respectively, as equations (4.14b,c,a,) of reference [10] (see the criticism concerning reference [6] added under reference [6] below) if we put $\vec{v}_s = 0$ in equation (48), if the last term in (50) is substituted from (48), and if in reference [10] the external temperature and field variations and the coupling coefficient of the magnon drift to the magnetization are put equal to zero ($\theta = 0$, $h = 0$ and $c_{ij}^l = 0$). The latter could have been included by adding a term $-\rho_M c^l\chi dh_z$ to equation (18) and the term $+\rho_M c^l\chi dh_z$ to equation (19) which, however, would lead to an additional coupling between magnetic and thermal variables and to an additional term in (33), namely $\vec{J}_S = -\rho_S v_S + c^l \chi \rho_M v_M$. Note that in reference [10] the equivalent of equation (47) is absent. The resulting hydrodynamic modes therefore are modified in the general case (see below).

To solve the system (47-50) we first express the velocities from (47,49)

$$\vec{v}_S = -\frac{\vec{q}}{\omega m}\left\{\frac{1}{\chi}\,\delta M_z + \beta \delta T - \delta H_z\right\}\left\{1 + i\,\frac{q^2}{\omega}\,\rho_S \zeta_\ell\right\}^{-1},\qquad (51)$$

and

$$\vec{v}_M = -\frac{\phi}{T\rho s_M}\,i\vec{q}\delta T,\qquad (52)$$

where

$$\phi(\vec{q},\omega) = \frac{T\rho^2 s_M^2}{\rho_M}\left(-i\omega + \frac{1}{\tau_J} + \frac{1}{\tau_J}\,\lambda_N^2 q^2\right)^{-1}.\qquad (53)$$

is the heat convectivity tensor and ζ_ℓ is the longitudinal projection of ζ.

We first examine the special case of complete decoupling of magnetic variables $\vec{v}_S$, δM_z and the thermal variables $\vec{v}_M$, δT. This decoupling is obtained under the assumption

$$\beta = 0.\qquad (54)$$

There now exists a purely magnetic mode ($\vec{v}_M = 0$, $\delta T = 0$) and a purely thermal mode ($\vec{v}_S = 0$, $\delta M_z = 0$). Indeed, with (54) and $\vec{v}_M = 0$, $\delta T = 0$, equations (49,52) are identically satisfied, and (48,51) yield, for $\delta H_z = 0$, force-free damped isothermal spin waves,

$$\frac{\omega^2}{q^2} = c_1^{\,2}(1 + i\,\frac{q^2}{\omega}\,\rho_S \zeta_\ell)^{-1} - i\omega\,\frac{\xi_\ell}{\chi}\,,\qquad (55)$$

with the *longitudinal isothermal magnon velocity*

$$c_1 = \left(\frac{\rho_S}{m^2 \chi}\right)^{\frac{1}{2}}.\qquad (56)$$

This shows that ρ_S has the character of a stiffness constant. Assuming all the imaginary parts to be small, so that they need be retained only linearly, equation (55) simplifies to

$$\omega = c_1 q - \frac{i}{2}\left[\rho_s \zeta_\ell + \frac{\xi_\ell}{\chi}\right] q^2, \tag{57}$$

which is the same as equations (2.48,50) of reference [4]. Halperin and Hohenberg [4] were the first to predict such a *sound-like* magnon mode in the hydrodynamic domain.

The thermal mode is obtained with (54) and $\vec{v}_s = 0$, $\delta M_z = 0$, in which case equations (48,51) are identically satisfied.

Equations (50,52) take the form

$$- i\omega + \frac{1}{\rho c_M} [\phi_\ell(\vec{q},\omega) + \kappa_\ell(\vec{q},\omega)]q^2 = 0. \tag{58}$$

There exists a domain of second spin waves given by the conditions

$$\frac{1}{\tau_J} \ll \omega \ll \frac{1}{\tau_{eq}}, \qquad \omega\tau_J \gg q^2 \lambda_{Ni}^2. \tag{59}$$

The first is the 'window condition', typical for crystals, τ_{eq} being the relaxation time responsible for thermal equilibrium of the magnon fluid. In the domain (59) we have from (53)

$$\phi_\ell(\vec{q},\omega) = \frac{i}{\omega} \frac{T\rho^2 s_M^2}{\rho_M}\left(1 - i \frac{1 + (\lambda_N^2)\ell q^2}{\omega\tau_J}\right), \tag{60}$$

and, inserting into (58),

$$\frac{\omega^2}{q^2} = c_2^2\left(1 - i \frac{1 + (\lambda_N^2)\ell q^2}{\omega\tau_J} - i\omega\tau_2\right). \tag{61}$$

Here,

$$c_2 = \left(\frac{\rho T s_M^2}{\rho_M c_M}\right)^{\frac{1}{2}}, \tag{62}$$

is the second magnon velocity, which has exactly the same form as the second sound velocity in the phonon case, and

$$\tau_2(\vec{q},\omega) = \frac{1}{\rho c_M c_2^2} \kappa_\ell(\vec{q},\omega) \tag{63}$$

is the thermal conduction relaxation time, which, for a propagating second magnon, must satisfy the additional condition

$$\omega\tau_2 \ll 1. \tag{64}$$

Since, with the decoupling assumption (54), equation (47) does not influence the thermal mode, our second magnon is the same as that derived in reference [10] for $c^1 = 0$. As shown in reference [10], $c^1 \neq 0$ modifies c_2 to become the adiabatic second magnon velocity.

The window condition of equation (59) has been analyzed by Michel and Schwabl [10] for the cases of the axial antiferromagnet MnF_2 and the planar antiferromagnet K_2NiF_4. Their conclusions are favourable for the existence of a second magnon at temperatures of the order of 30°K for MnF_2 and of 90°K for K_2NiF_4 (see section 4 below). Detailed calculations of the window condition for the insulating ferromagnet EuS have been carried out by Forney [15] and Forney and Jäckle [16] who find an open window below approximately 2°K. This calculation is based on a Heisenberg plus a dipole-dipole coupling, the first giving rise to four-magnon processes, the second to three-magnon processes.

It is interesting to note that the conditions $\omega\tau_J \ll q^2\lambda_{Ni}{}^2$, $q^2\lambda_{Ni}{}^2 \gg 1$, which define the domain of Poiseuille flow in the phonon case, are also realisable for magnetic crystals of high magnon viscosity. To our knowledge this effect has never been looked for yet.

Heat diffusion by magnons is obtained in the domain

$$\omega\tau_J \ll 1, \qquad q^2\lambda_{Ni}{}^2 \ll 1, \tag{65}$$

where (50) reduces to

$$- i\omega + D_T{}'q^2 = 0, \tag{66}$$

with

$$D_T{}' = \lim_{\omega\to 0}\lim_{q\to 0} \frac{1}{\rho c_M} [\phi_\ell(\vec{q},\omega) + \kappa_\ell(\vec{q},\omega)]. \tag{67}$$

Neglecting here the convective contribution ϕ_ℓ, we recover equation (2.49) of reference [4].

We now discuss the solutions of the complete equations (48, 50) and (51,52), that is, including an external field variation δH_z and dropping the decoupling assumption (54). This discussion is of interest in view of the possibility, mentioned above, of detecting a second-magnon peak, in addition to the first-magnon peak, in a neutron scattering experiment.

Insertion of (51) into (48) yields, with (56),

$$\left\{\frac{\omega^2}{q^2} - c_1{}^2\left[\left(1 + i\frac{q^2}{\omega}D_1\right)^{-1} - i\omega\tau_1\right]\right\}\delta M_z$$

$$- c_1{}^2\beta\chi\left[\left(1 + i\frac{q^2}{\omega}D_1\right)^{-1} - i\omega\tau_1\right]\delta T$$

$$= - c_1{}^2\chi\left[\left(1 + i\frac{q^2}{\omega}D_1\right)^{-1} - i\omega\tau_1\right]\delta H_z, \qquad (68)$$

where we have introduced the definitions

$$D_1 = \rho_s\zeta_\ell, \qquad \frac{\xi_\ell}{\chi} = c_1{}^2\tau_1. \qquad (69)$$

Insertion of (52) into (50) yields

$$c_1{}^2\beta\chi\left[-i\omega + \frac{1}{\rho c_M}(\phi_\ell + \kappa_\ell)q^2\right]\delta T = -i\omega c_1{}^2\theta\delta M_z \qquad (70)$$

where

$$c_1{}^2\theta = \frac{\rho_s T\beta^2}{m^2\rho c_M}. \qquad (71)$$

There are three domains of ϕ, equation (53). In the domain

$$\omega\tau_J \ll q^2\lambda_{Ni}{}^2, \qquad q^2\lambda_{Ni}{}^2 \gg 1, \qquad (72)$$

we can write

$$\frac{1}{\rho c_M}\phi_\ell q^2 = \tau_N{}^{-1}. \qquad (73)$$

Inserting this into (68) and eliminating δT from (68,70), retaining only lowest powers in q and ω, but separately in real and imaginary parts, we obtain the *correlation function*

$$\chi(q,\omega) \equiv \frac{\delta M_z}{\delta H_z}$$

$$= -\chi[\omega^2 - c_1{}^2(1 - i\omega\tau_1')q^2]^{-1}c_1{}^2q^2, \qquad (74)$$

where

$$\tau_1' = \tau_1 + \theta\tau_N + \frac{D_1}{c_1{}^2} \quad . \tag{75}$$

Equation (74) shows that in the domain (72) δM_z propagates with the *isothermal* magnon velocity c_1.

In the domain

$$\omega\tau_J \ll 1, \qquad q^2\lambda_{Ni}{}^2 \ll 1, \tag{76}$$

insertion of (67) into (70) and elimination of δT from (68,70) gives for the correlation function defined in equation (74),

$$\chi(q,\omega) = \chi\Big\{1 - i\omega\theta(-i\omega + D_T{}'q^2)^{-1}$$

$$- i\omega(c_1 q)^{-2}[\tau_1 + (-i\omega + D_1 q^2)^{-1}]^{-1}\Big\}^{-1} . \tag{77}$$

In the case $D_T{}'q^2 \ll \omega$, $D_1 q^2 \ll \omega$, $\omega\tau_1 \ll 1$, we find

$$\chi(q,\omega) = -\chi[\omega^2 - \tilde{c}_1{}^2(1 - i\omega\tilde{\tau}_1)q^2]^{-1}c_1{}^2 q^2, \tag{77a}$$

where

$$\tilde{c}_1{}^2 = c_1{}^2(1 + \theta), \tag{78}$$

and

$$\tilde{\tau}_1 = \tau_1 + \frac{\left[D_1 + \dfrac{\theta}{1 + \theta} D_T{}'\right]}{\tilde{c}_1{}^2} \quad . \tag{79}$$

$\tilde{c}_1$ is the *longitudinal adiabatic first magnon velocity* which is connected with the adiabatic longitudinal susceptibility $\tilde{\chi}$ by the analogue of equation (56),

$$\tilde{\chi} = \frac{\rho_s}{m^2\tilde{c}_1{}^2} = \chi(1 + \theta)^{-1}. \tag{80}$$

In the opposite case, $\omega \ll D_T{}'q^2$, $\omega \ll D_1 q^2$, one finds from equation (77),

$$\chi(q,\omega) = \chi\left\{1 - i\omega\left[\frac{\theta}{D_T{}'q^2} + \frac{D_1}{c_1{}^2(1 + \tau_1 D_1 q^2)}\right]\right\} \qquad \text{(Contd)}$$

(Contd)

$$+ \omega^2 \left[\frac{\theta}{D_T'^2 q^4} - \frac{1}{c_1^2 q^2 (1 + \tau_1 D_1 q^2)^2} \right] + O(\omega^3) \Big\}^{-1} . \tag{77b}$$

This expression shows that in the limit $q \to 0$ the heat diffusion pole in equation (77) dominates the response, and gives rise to a Rayleigh peak.

In the second-magnon domain, (59), insertion of (60,62,63) into (68,70) yields for the correlation function of equation (74), after elimination of δT and neglecting damping terms proportional to q^2/ω,

$$\chi(q,\omega) =$$

$$- \chi G(q,\omega) \cdot (1 - i\omega\tau_1) \left[\omega^2 - c_2^2 \left(1 - \frac{i}{\omega\tau_J} - i\omega\tau_2 \right) q^2 \right] c_1^2 q^2, \tag{81}$$

where

$$G^{-1}(q,\omega) = [\omega^2 - c_1^2 (1 - i\omega\tau_1) q^2]$$

$$\cdot \left[\omega^2 - c_2^2 \left(1 - \frac{i}{\omega\tau_J} - i\omega\tau_2 \right) q^2 \right]$$

$$- c_1^2 \theta q^2 \omega^2 (1 - i\omega\tau_1). \tag{82}$$

Note that the isothermal limit yields correctly

$$\chi = \lim_{q\to 0} \lim_{\omega\to 0} \chi(q,\omega),$$

in agreement with equation (3.9a) of reference [4].
G^{-1} can be factorized in the form

$$G^{-1}(q,\omega) = \left[\omega^2 - c_I^2 \left(1 - i\omega\tau_I - \frac{i}{\omega\tau_{JI}} \right) q^2 \right]$$

$$\cdot \left[\omega^2 - c_{II}^2 \left(1 - i\omega\tau_{II} - \frac{i}{\omega\tau_{JII}} \right) q^2 \right]$$

$$\equiv A_I A_{II}, \tag{83}$$

valid to first order in the damping terms.

Hence

$$\chi(q,\omega) = -\chi\left[\frac{1-R}{A_I} + \frac{R}{A_{II}}\right]c_1^2 q^2, \tag{84}$$

with

$$R = r\left[1 - i\omega\tau_r + \frac{i}{\omega\tau_{Jr}}\right], \tag{85}$$

and, to first order in θ,

$$r \cong c_1^2 c_2^2 (c_1^2 - c_2^2)^{-1}\theta. \tag{86}$$

As in the case of dielectric crystals, $\beta \propto c_M \propto T^3$ and $s_M \propto T^3$ at low temperatures, so that, according to (71), $\theta \propto T^4$. Since the window condition restricts the occurrence of the second magnon to low temperatures, the second-magnon peak will be much smaller than the first-magnon peak in a magnetic excitation by inelastic neutron scattering.

The integrated intensity can be obtained from (84). However, the damping terms $i/\omega\tau_{JI}$, $i/\omega\tau_{JII}$ and $i/\omega\tau_{Jr}$ make the integration complicated; we refrain from giving the details.

3. THE AXIAL AND ISOTROPIC FERROMAGNETS

The *axial* ferromagnet is characterized by $J_{ij}^z > J_{ij}^\perp > 0$ for the dominant pairs of neighbours i,j. This has the effect that without an external field the magnetization aligns in the easy z-direction, and $M_\perp = 0$. Thus the precession angle is not a dynamic variable, and the analogy with superfluid Helium is lost. But a two-fluid description in terms of magnetic and thermal variables still exists.

The relation (7) is invalid in this case, since $M_z \neq 0$ but $h_z = 0$. On the other hand, for $|M_i| \ll M_z$, $i = x,y$, $\chi_\perp$ is not given by equation (26) but is finite, and

$$M_i = \chi_\perp h_i, \qquad i = x,y. \tag{87}$$

Choosing the external field as $\vec{H} = (H_x, H_y, 0)$, we have equations of motion analogous to (32),

$$\dot{M}_z + \frac{1}{m}\nabla\vec{j}_{sz} = -(\vec{M}\times m\vec{\mu})_z = -M_x H_y + M_y H_x \cong 0,$$

$$\dot{M}_x + \frac{1}{m}\nabla(\vec{j}_{sx} + \vec{j}_{sx}') = -(\vec{M}\times m\vec{\mu})_x, \tag{88}$$

$$\dot{M}_y + \frac{1}{m}\nabla(\vec{j}_{sy} + \vec{j}_{sy}') = -(\vec{M}\times m\vec{\mu})_y,$$

where

$$\mu_i = \frac{1}{m} (H_i - h_i), \qquad i = x,y, \qquad \mu_z = 0. \tag{89}$$

The energy can be written in a form analogous to equations (13, 20,28),

$$d(\rho\varepsilon) = \vec{h}\cdot d\vec{M} - \vec{H}\cdot d\vec{M} + \sum_{i=x,y} \frac{\rho_s^i}{m^2 M_z^2} \nabla M_i\cdot d\nabla M$$

$$+ \vec{v}_M\cdot d\vec{j}_M + T d(\rho s), \tag{90}$$

where $\rho_s^x = \rho_s^y = \rho_s < \rho_s^z$. The second and third terms are, respectively, the Zeeman and exchange energy contribution [17, 23].

From that equilibrium condition (23), supplemented by the further requirement that

$$- \vec{j}_{sz} \propto \nabla M_z = 0 \tag{91}$$

(note that with this assignment the first equation (88) gives rise to a diffusive mode), we obtain, with (90), for the Fourier components, after a partial integration,

$$H_i = h_i + \frac{\rho_s q^2}{m^2 M_z^2} M_i, \qquad i = x,y. \tag{92}$$

Thus, according to (87), $H_i = h_i$ in the limit $q \to 0$.

Writing in analogy with (27),

$$\vec{v}_{sx} = \frac{1}{mM_z} \nabla M_x,$$

$$\tag{93}$$

$$\vec{v}_{sy} = - \frac{1}{mM_z} \nabla M_y,$$

and with (89,91) the energy expression (90) takes the form

$$d(\rho\varepsilon) = - m\vec{u}\cdot d\vec{M} + \rho_s \sum_{i=x,y} \vec{v}_{si}\cdot d\vec{v}_{si} + \vec{v}_M\cdot d\vec{j}_M + T d(\rho s). \tag{94}$$

$\vec{v}_{sx}$ and $\vec{v}_{sy}$ are here introduced mainly to exhibit the analogy with the planar case, which is even more evident from their equations of motion. In fact, from (88,93) we obtain

$$\dot{\vec{v}}_{sx} = + \nabla(\mu_y + \mu_y') - \frac{1}{m^2 M_z} \nabla\nabla\cdot\vec{j}_{sx}, \qquad \text{(Contd)}$$

(Contd) $\dot{\vec{v}}_{sy} = + \nabla(\mu_x + \mu_x') + \dfrac{1}{m^2 M_z} \nabla\nabla\cdot\vec{j}_{sy},$ (95)

where

$$\mu_x' = + \frac{1}{m^2 M_z} \nabla\cdot\vec{j}_{sy}',$$

$$\mu_y' = - \frac{1}{m^2 M_z} \nabla\cdot\vec{j}_{sx}',$$

(96)

Equations (95) are of the form (12). The analogue of (33) can also be established by working out the transition from (29) to (30).

From (94) we have

$$T(\rho s)^{\cdot} = (\rho\varepsilon)^{\cdot} + m\vec{\mu}\cdot\dot{\vec{M}} - \rho_s \sum_{i=x,y} \vec{v}_{si}\cdot\dot{\vec{v}}_{si} - \vec{v}_M\cdot\dot{\vec{j}}_M. \qquad (97)$$

Inserting from (29,88,95,16) we obtain with (15,17,18)

$$(\rho s)^{\cdot} = - \nabla\cdot\left\{\frac{1}{T}\left[\vec{j}_\varepsilon + \vec{j}_\varepsilon' + \sum_i \mu_i(\vec{j}_{si} + \vec{j}_{si}') + \rho_s(\mu_y'\vec{v}_{sx} + \mu_x'\vec{v}_{sy})\right.\right.$$

$$\left.\left. - \frac{\rho_s}{m^2 M_z} (\vec{v}_{sx}\nabla\cdot\vec{j}_{sx} - \vec{v}_{sy}\nabla\cdot\vec{j}_{sy}) - \pi_M'\vec{v}_M\right]\right\}$$

$$+ \left[\vec{j}_\varepsilon + \vec{j}_\varepsilon' + \sum_i \mu_i(\vec{j}_{si} + \vec{j}_{si}') + \rho_s(\mu_y'\vec{v}_{sx} + \mu_x'\vec{v}_{sy})\right.$$

$$\left. - \frac{\rho_s}{m^2 M_z} (\vec{v}_{sx}\nabla\cdot\vec{j}_{sx} - \vec{v}_{sy}\nabla\cdot\vec{j}_{sy}) - \pi_M'\vec{v}_M - T\rho s M\vec{v}_M\right]\cdot\nabla \frac{1}{T}$$

$$+ \frac{1}{T} (\vec{j}_{sx} + \vec{j}_{sx}' - \rho_s\vec{v}_{sy})\cdot\nabla\mu_x + \frac{1}{T} (\vec{j}_{sy} + \vec{j}_{sy}' - \rho_s\vec{v}_{sx})\cdot\nabla\mu_y$$

$$- \frac{1}{Tm^2 M_z}\left[\nabla\cdot(\rho_s\vec{v}_{sx})\nabla\cdot\vec{j}_{sx} - \nabla\cdot(\rho_s\vec{v}_{sy})\nabla\cdot\vec{j}_{sy}\right]$$

$$+ \frac{1}{T}\left[\mu_y'\nabla\cdot(\rho_s\vec{v}_{sx}) + \mu_x'\nabla\cdot(\rho_s\vec{v}_{sy}) - (\pi_M'\nabla)\cdot\vec{v}_M\right] + \frac{\rho M}{T\tau_J} \vec{v}_M^2.$$

The form (30) then implies that

$$\vec{j}_{sx} = + \rho_s \vec{v}_{sy} = - \frac{\rho_s}{mM_z} \nabla M_y,$$

$$\vec{j}_{sy} = + \rho_s \vec{v}_{sx} = - \frac{\rho_s}{mM_z} \nabla M_x, \tag{98}$$

where we have used (93), and

$$\vec{J}_\varepsilon = - \sum_{i=x,y} \mu_i \vec{j}_{si} - \frac{1}{m^2 M_z} \nabla \times (\vec{j}_{sx} \times \vec{j}_{sy}) + T\rho_s s_M \vec{v}_M. \tag{99}$$

This leads again to equation (35) for $\vec{J}_s$ and to

$$\vec{J}_s{}' = \frac{1}{T}\left[\vec{J}_\varepsilon{}' + \sum_{i=x,y} (\mu_i \vec{j}_{si}{}' + \mu_i{}' \vec{j}_{si}) - \pi_M{}' \vec{v}_M\right] \tag{100}$$

and

$$\sigma = T\vec{J}_s{}' \cdot \nabla \frac{1}{T} + \frac{1}{T} \sum_{i=x,y} (\vec{j}_{si}{}'\nabla\mu_i + \mu_i{}'\nabla\cdot\vec{j}_{si})$$

$$- \frac{1}{T} (\pi_M{}'\nabla)v_M + \frac{\rho_M}{T\tau_J} v_M{}^2. \tag{101}$$

Note that addition of an independent dissipative part of the form of the second equation (38) to $\vec{\mu}$ on the right hand side of equations (88) would, according to equations (30,97), violate the positive definite character of σ. Equations (98) are the same as equations (7.3a,b,8) of reference [4].

The form of the dissipative terms $\pi_M{}'$ and $\vec{J}_s{}'$ is again given by equation (38). For $\vec{j}_{si}{}'$ rotational symmetry in the (x,y)-plane implies the form

$$\vec{j}_{si}{}' = - m^z(\xi\nabla)\nabla_i \sum_{j=x,y} \nabla_j\mu_j, \qquad i = x,y, \tag{102}$$

with (102,96) the second term of $T\sigma$, equation (101), becomes, making use of (87,89,98) and assuming $\nabla H_i = 0$,

$$\sum_{j=x,y} (\vec{j}_{si}{}'\cdot\nabla\mu_i + \mu_i{}'\nabla\cdot\vec{j}_{si}) =$$

$$- \sum_{i,j=x,y} \nabla_i\left\{(\nabla h_i)\cdot(\xi\nabla)\nabla_j h_j + \frac{\rho_s\chi_\perp}{m^2 M_z{}^2} (\nabla^2 h_i)(\xi\nabla)\cdot\nabla\nabla_j h_j\right\}$$

(Contd)

(Contd)

$$+ \sum_{i,j=x,y} \left\{ (\nabla\nabla_i h_i)\cdot(\xi\nabla)\nabla_j h_j + \frac{\rho s \chi_\perp}{m^2 M_z^2} (\nabla^2\nabla_i h_i)(\xi\nabla)\cdot\nabla\nabla_j h_j \right\}$$

Here the first term is a divergence and can be absorbed into $\nabla\cdot\vec{J}_s{}'$, and the second term is a positive quadratic form.

The hydrodynamic modes are obtained from the equations of motion (88) combined with (42,43). Equations (44) are modified as follows

$$\delta s = \frac{c_M}{T} \delta T - \sum_{i=x,y} \frac{\beta_i}{\rho} \delta M_i,$$

$$\delta h_i = \beta_i \delta T + \frac{1}{\chi_\perp} \delta M_i, \qquad i = x,y, \tag{103}$$

where

$$\beta_i = \left(\frac{\partial h_i}{\partial T}\right)_{\vec{M}} = -\rho\left(\frac{\partial s}{\partial M_i}\right)_T, \qquad i = x,y \tag{104}$$

Excluding precessional motions the perpendicular fields and the magnetization vary around zero averages, $H_i = \delta H_i$, $h_i = \delta h_i$, $M_i = \delta M_i$. Then equations (88), combined with (98,102,89) and the second equation (103), become

$$\dot{M}_x + \frac{M_z}{\chi_\perp}\delta M_y - \frac{\rho s}{m^2 M_z}\nabla^2 M_y + \frac{1}{\chi_\perp}(\xi\nabla)\cdot\nabla\nabla_x \sum_{i=x,y}\nabla_i M_i + \beta_y\delta T$$

$$+ (\xi\nabla)\cdot\nabla\nabla_x \sum_{i=x,y}\beta_i\nabla_i T$$

$$= + M_z\delta H_y + (\xi\nabla)\cdot\nabla\nabla_x \sum_{i=x,y}\nabla_i H_i,$$

$$\dot{M}_y - \frac{M_z}{\chi_\perp}M_x + \frac{\rho s}{m^2 M_z}\nabla^2 M_x + \frac{1}{\chi_\perp}(\xi\nabla)\cdot\nabla\nabla_y \sum_{i=x,y}\nabla_i M_i - \beta_x\delta T \tag{105}$$

$$+ (\xi\nabla)\cdot\nabla\nabla_y \sum_{i=x,y}\beta_i\nabla_i T$$

$$= - M_z\delta H_x + (\xi\nabla)\cdot\nabla\nabla_y \sum_{i=x,y}\nabla_i H_i.$$

Eliminating here δT with the aid of equations (42,43) and the first equation (103) gives us the response of the system to an external field variation δH_i as obtained in an inelastic neutron scattering experiment. Rather than treating this general case which gives rise to complicated expressions we treat here only decoupled free spin waves. That is, we assume

$$\beta_i = 0, \qquad \delta H_i = 0, \qquad i = x,y, \qquad (106)$$

In plane wave representation equations (105) then lead to

$$\left(\omega + i\,\frac{\xi_\ell}{\chi_\perp}\,q^2 q_x^{\,2}\right)\delta M_x = -\,i\left(H_e + Dq^2 + \frac{\xi_\ell}{\chi_\perp}\,q^2 q_x q_y\right)\delta M_y,$$

$$\left(\omega + i\,\frac{\xi_\ell}{\chi_\perp}\,q^2 q_y^{\,2}\right)\delta M_y = i\left(H_e + Dq^2 - \frac{\xi_\ell}{\chi_\perp}\,q^2 q_x q_y\right)\delta M_x, \qquad (107)$$

where we have introduced the definitions

$$H_e = \frac{M_z}{\chi_\perp}, \qquad D = \frac{\rho_s}{m^2 M_z}. \qquad (108)$$

Equations (107) describe spin waves in the (x,y)-plane with the dispersion relation

$$\omega = H_e + Dq^2 - \frac{i}{2}\,\frac{\xi_\ell}{\chi_\perp}\,q^4 \sin^2\theta_{\vec q} + O(q^8) \qquad (109)$$

where $\cos\theta_{\vec q} = q_z/q$.

Such a quadratic dispersion relation is typical for axial and for isotropic ferromagnets (see below) [17,23]. The gap H_e is important for the existence of a magnon drift and hence for the existence of a second magnon. Indeed, without a gap, the magnon distribution function,

$$n_q = \left\{\exp\left(\frac{(\omega_q - \vec v_M \vec q)}{k_B T}\right) - 1\right\}^{-1},$$

is negative for sufficiently small q with $\vec v_M \cdot \vec q > 0$. With a dispersion ω_q given by (109), however, the magnon drift exists for $|\vec v_M| < 2(H_e D)^{\frac{1}{2}}$ [15,16]. The gap H_e arose here from the precessional (Bloch) terms in the equations of motion (88). As we shall see, in an ideal isotropic ferromagnet this is no longer the case.

Before discussing the isotropic case we wish to generalize the above equations by taking into account the effect of the dipole-dipole interaction between the spins.

Phenomenologically (see section 30 of reference [23]), this

effect is described by a *dipolar field* $\vec{H}_d$ produced by the smeared out spin distribution $\vec{M}$. The discreteness of the spins can
be taken into account by the field produced in a small sphere
centred at $\vec{r}$ by the magnetic moment in the sphere.

$$\vec{H}_d{}'(\vec{r}) = (\int_{sphere} d^3r'\vec{M}(\vec{r}')\cdot\nabla)\nabla\frac{1}{|\vec{r}-\vec{r}'|} = -\frac{4\pi}{3}\vec{M}(\vec{r}).$$

The sum must satisfy [18]

$$\nabla\cdot(\vec{H}_d + \vec{H}_d{}' + 4\pi\vec{M}) = 0, \qquad \nabla\times(\vec{H}_d + \vec{H}_d{}') = 0,$$

which is equivalent to

$$\nabla^2(\vec{H}_d + \vec{H}_d{}') = -4\pi\nabla(\nabla\cdot\vec{M}).$$

Hence

$$\vec{H}_d = \frac{4\pi}{3}\vec{M} - 4\pi(\nabla^2)^{-1}\nabla(\nabla\cdot\vec{M}) - \alpha\vec{M}_0, \tag{110a}$$

where the last term is an integration constant which depends on the
shape of the sample (see section 16 of reference [23]). Physically $\vec{M}_0 = (0,0,M_z)$ and α is the *demagnetizing factor*.
   Then equations (88) are generalized by substituting for equations (89) (see equation (30.5) of reference [23]),

$$m\dot{\vec{\mu}} = \vec{H} - \vec{h} + \vec{H}_d + \vec{H}_A, \tag{110}$$

where $\vec{H}_A = (0,0,H_A)$ is the *anisotropy field* which is due to
higher order perturbation effects of the dipole-dipole (and
higher pole) interactions (see sections 22-25 of reference [23]).
Note that insertion of this modified form (110) of $m\dot{\vec{\mu}}$ into equations (88) does not give rise to modifications of equation (97).
Hence the connection (98) is still valid. We therefore find, in
plane wave representation with $H_i = 0$, $M_i = \delta M_i$ ($i = x,y$), the
following generalization of equations (107),

$$\left[\omega + i\frac{\xi_\ell}{\chi_\perp}q^2q_x^2 + 4\pi i M_z\hat{q}_x\hat{q}_y\right]\delta M_x$$

$$= -i\left[H_e + H_A + Dq^2 + \frac{\xi_\ell}{\chi_\perp}q^2q_xq_y - \alpha M_z + 4\pi M_z\hat{q}_y^2\right]\delta M_y,$$

$$\tag{107a}$$

$$\left[\omega + i\frac{\xi_\ell}{\chi_\perp}q^2q_y^2 - 4\pi i M_z q_xq_y\right]\delta M_y = \qquad\text{(Contd)}$$

(Contd)

$$= i\left[H_e + H_A + Dq^2 - \frac{\xi_\ell}{\chi_\perp} q^2 q_x q_y - \alpha M_z + 4\pi M_z \hat{q}_x^2\right]\delta M_x, \qquad (107a)$$

where $\hat{q} = \vec{q}/q$. Apart from the damping terms, equations (107a) are the same as equations (30.7) of reference [23]. They lead to the secular equation

$$\omega^2\left[1 + \frac{i}{\omega}\frac{\xi_\ell}{\chi_\perp} q^4 \sin^2\theta_{\hat{q}}\right] = \omega^2(\vec{q}),$$

where

$$\omega^2(\vec{q}) = (H_e + H_A - \alpha M_z + Dq^2)$$

$$\cdot (H_e + H_A - \alpha M_z + 4\pi M_z \sin^2\theta_{\hat{q}} + Dq^2), \qquad (111)$$

which has the solution

$$\omega = \omega(\vec{q}) - \frac{i}{2}\frac{\xi_\ell}{\chi_\perp} q^4 \sin^2\theta_{\hat{q}} + O(q^8). \qquad (112)$$

The dispersion relation (111) was first derived by Clogston, Suhl, Walker and Anderson [19] (see also reference [18]). The merit of our hydrodynamic derivation is that, in addition to the q-dependence of $\omega(\vec{q})$, it also determines the damping of the spin wave.

   In the *isotropic*, or Heisenberg, ferromagnet, $J_{ij}^\perp = J_{ij}^z$ and $\rho_s^x = \rho_s^y = \rho_s^z = \rho_s$ in equation (90). Assuming that in equilibrium without external field the magnetization points in the z-direction, we have the same situation as with the axial ferromagnet, except that $\vec{h} = 0$. This follows from the fact that in the planar case $\vec{h} = (0,0,h_z)$, in the axial case $\vec{h} = (h_x, h_y, 0)$ and that the isotropic case is the limit between the two. $\vec{h} = 0$ has the effect that equation (92) now becomes, in analogy with (25,26),

$$M_i = \chi_\perp H_i, \qquad i = x, y, \qquad (113)$$

with

$$\chi_\perp = \frac{m^2 M_z^2}{\rho_s q^2}. \qquad (114)$$

Since $\vec{h} = 0$, the second equation (103) implies that $\beta_i = 0$ and $1/\chi_\perp = 0$. With this modification the equation of motion (105)

also hold in the isotropic case. However, $1/\chi_\perp = 0$ eliminates
the damping of δM_x and δM_y. Now, since M_i and μ_i transform with
the same sign under time reversal, we can add a term to (102) in
which μ_i is replaced by M_i. Whilst in the axial case this did
not affect the form of $\vec{J}_{si}'$, it now gives rise to the leading
term,

$$\vec{J}_{si}' = + 2m(\eta\nabla)\nabla_i \sum_{j=x,y} \nabla_j M_j, \qquad i = x,y. \qquad (115)$$

With this expression, the dispersion relation (109) is modified
as follows,

$$\omega = Dq^2 - i n_\ell q^4 \sin^2\theta_{\vec{q}} + O(q^6), \qquad (116)$$

where D is given by the second equation (108). Here the real
part is the same as equation (7.9) of reference [4]. Equation
(116) is the well known dispersion relation of the isotropic
ferromagnet [17,23] first derived, for spin $\frac{1}{2}$, by Bloch [2].
The phenomenological form (108) of D was first given by Landau
and Lifshitz in their pioneering work on the macroscopic description of the magnetic crystals [3]. The dispersion relation
(116) has been observed by neutron scattering in Fe, Co, Ni and
their alloys [25].

Even in an ideally isotropic ferromagnet a gap arises through
dipole-dipole interaction [15,16]. In our phenomenological approach this effect is taken care of by substituting equations
(110,110a) for (89). This means that the dispersion relation
(116) is replaced by equations (111,112) with $H_e = 0$ which, to
lowest order, is again of the form (109). Hence the second magnon should also occur in an ideally isotropic ferromagnet due to
dipole-dipole integration [15,16].

4. HYDRODYNAMICS OF THE ANTIFERROMAGNETS

The antiferromagnet [24] is characterized by the existence of
two sublattices A and B such that the coupling constants J_{ij}^z
and $J_{ij}^\perp$ are positive or negative if the positions i and j are
on the same or on different sublattices, respectively. Defining
numbers [4]

$$\eta_i = \begin{cases} +1, & \text{if } i \text{ on } A, \\ -1, & \text{if } i \text{ on } B, \end{cases} \qquad (117)$$

the operators $\eta_i S_i^x$, $\eta_i S_i^y$, S_i^z satisfy the same commutation relations as S_i^x, S_i^y, S_i^z and $\eta_i \eta_i J_{ij}^\perp$ is ferromagnetic. Hence
the Hamiltonian (1) written in these variables describes ferro-
magnetism in the (x,y)-plane. Defining the sublattice magnetization by

$$\vec{M}_A = \frac{1}{v} \langle \vec{S}_i \rangle, \quad i \text{ on } A,$$

$$\vec{M}_B = \frac{1}{v} \langle \vec{S}_i \rangle, \quad i \text{ on } B,$$

(118)

the total magnetization is

$$\vec{M} = \frac{1}{v} \langle \sum_i \vec{S}_i \rangle = \vec{M}_A + \vec{M}_B, \tag{119}$$

and

$$\vec{N} = \frac{1}{v} \langle \sum_i n_i \vec{S}_i \rangle = M_A - M_B \tag{120}$$

is the staggered magnetization. This means for the Fourier components that $\vec{N}(\vec{q}) = M(\vec{q}_R - \vec{q})$, where $\vec{q}_R$ is the vector to the $(1,1,1)$-corner of the Brillouin zone. Hence, near $q = 0$, N_x and N_y obey the same equations of motion as M_x and M_y in the ferromagnetic case. From (119,120) we conclude that in thermal equilibrium

$$\vec{M} \cdot \vec{N} = \vec{M}_A^2 - \vec{M}_B^2 = 0, \tag{121}$$

since the two sublattices are equivalent except for reflections along symmetry or field directions. This is still true for hydrodynamic variations for which $\vec{q}$ is small compared with the border vectors of the Brillouin zone.

In the *planar* antiferromagnet $\vec{N}$ is in the easy (x,y)-plane, and

$$N_x + i N_y = N_\perp e^{i\phi} \tag{122}$$

is equivalent to equation (2), the motion being given by equations (9,10). M_z, equation (5), is again conserved together with ε_s, equation (4). Thus equation (7) again holds.

On the other hand, N_z is not conserved but is supposed to relax to zero in a microscopic time. The orthogonality (121) together with (122) then implies that

$$\vec{M} = (-M_\perp \sin\phi, M_\perp \cos\phi, M_z),$$

$$\vec{N} = (N_\perp \cos\phi, N_\perp \sin\phi, 0).$$

(123)

Here $M_\perp$ and M_z are small compared to $N_\perp$, and $N_\perp$ is supposed to relax to a value

$$N_\perp = N_\perp(M_z, T) \tag{124}$$

in a microscopic time.

Defining a velocity $\vec{v}_S$ by equation (11), the energy expression (22) is again valid and the equilibrium condition (23) yields

$$\int\left\{(h_z - H_z)\delta M_z + \left[-H_y M_\perp \sin\phi - \frac{\rho_S}{m^2}\nabla^2\phi\right]\delta\phi\right\}d^3r = 0.$$

Hence, for $H_y \neq 0$,

$$h_z = H_z, \qquad \phi = 0, \tag{125}$$

and

$$M_\perp = M_y = \chi_\perp H_y, \tag{126}$$

where the transverse susceptibility $\chi_\perp$ is now finite for $q \to 0$.

From (11,125) we find, in analogy with (27),

$$\vec{v}_S = -\frac{1}{mN_\perp}\nabla N_y, \tag{127}$$

so that the exchange energy contribution is now supplied by the staggered field,

$$\rho_S \vec{v}_S \cdot d\vec{v}_S = \frac{\rho_S}{m^2 N_\perp^2}\nabla N_y \cdot d\nabla N_y. \tag{128}$$

Thus we see that the dynamic of the planar antiferromagnet is exactly the same as that of the planar ferromagnet described in section 2, if there $M_\perp$ is replaced by $N_\perp$. In particular, in the decoupling approximation (54) the spin-wave mode (57) with velocity (56) and the thermal mode (58) are recovered. The existence of a sound-like spin-wave dispersion in antiferromagnets was discovered by Hulthén [5].

In the *axial* antiferromagnet $\vec{N}$ is in the easy z-direction. As noted above, the hydrodynamic (small q) motion of N_x and N_y is the same as the hydrodynamic motion of M_x and M_y in the ferromagnet. But here we need in addition the hydrodynamic motion of M_x and M_y.

Because of the condition (121) the motions of $\vec{M}$ and $\vec{N}$ are not independent but satisfy, at least for small q,

$$\vec{N}\cdot\dot{\vec{M}} + \vec{M}\cdot\dot{\vec{N}} = 0. \tag{129}$$

Leaving out for the moment gradient terms, these equations of motion then are

$$\dot{\vec{M}} = - \vec{M} \times m\vec{\mu} \tag{130}$$

and

$$\dot{\vec{N}} = - \vec{N} \times m\vec{\mu}. \tag{131}$$

As in the axial ferromagnet, $h_z = 0$, and (87) is valid. We choose again $\vec{H} = (H_x, H_y, 0)$, so that $m\vec{\mu}$ is given by equations (89). Then the initial conditions are

$$\vec{M}_0 = (M_x, M_y, 0),$$
$$\vec{N}_0 = (0, 0, N_z), \tag{132}$$

since

$$M_z = \chi H_z \tag{133}$$

with finite longitudinal susceptibility χ. Equations (129,132) together ensure that the orthogonality condition (121) is satisfied at all times.

Completing equation (130) by divergence terms, in analogy with equations (88), we have, in view of (132,89),

$$\dot{M}_z = - m(M_x\mu_y - M_y\mu_x) \cong 0,$$

$$\dot{M}_x + \frac{1}{m} \nabla \cdot (\vec{j}_{sx} + \vec{j}_{sx}') = mM_z\mu_x \cong 0, \tag{134}$$

$$\dot{M}_y + \frac{1}{m} \nabla \cdot (j_{sy} + j_{sy}') = - mM_z\mu_y \cong 0.$$

Similarly we complete equation (131) by divergence terms,

$$\dot{N}_z = - m(N_x\mu_y - N_y\mu_x) \cong 0,$$

$$\dot{N}_x + \frac{1}{m} \nabla \cdot (\vec{\ell}_{sx} + \vec{\ell}_{sx}') = mN_z(\mu_y + \mu_x'), \tag{134a}$$

$$\dot{N}_y + \frac{1}{m} \nabla (\vec{\ell}_{sy} + \vec{\ell}_{sy}') = - mN_z(\mu_x + \mu_x').$$

Here $\vec{\ell}_{si}$ and $\vec{\ell}_{si}'$ ($i = x, y$) are new currents and their dissipative parts, and μ_x', μ_y' are new dissipative parts of μ_x and μ_y. The latter were not allowed on the right hand side of equations (134, 88) because they would have violated the positive definite character of the entropy production density σ (see the remark after equation (101)). Since obviously the energy differential equation (94) does not contain a term $-m\vec{\mu} \cdot d\vec{N}$ these dissipative parts μ_x' and μ_y' in equations (134a) do not contribute to (97) and hence do not influence σ. They are in fact important, since they give the leading contribution to the damping (see below).

Writing the energy differential again in the form (94), $\vec{v}_{\text{sx}}$ and $\vec{v}_{\text{sy}}$ must now be related to N_{x} and N_{y}, respectively, since the dominant exchange energy obviously comes from the staggered field, as in equation (128), Writing in analogy with (93)

$$\vec{v}_{\text{sx}} = + \frac{1}{m N_z} \nabla N_{\text{x}},$$

$$\vec{v}_{\text{sy}} = - \frac{1}{m N_z} \nabla N_{\text{y}},$$

(135)

we deduce from equations (134a) that, in analogy with equations (95),

$$\dot{\vec{v}}_{\text{sx}} = + \nabla(\mu_{\text{y}} + \tilde{\mu}_{\text{y}}') - \frac{1}{m^2 N_z} \nabla \nabla \cdot \vec{\ell}_{\text{sx}},$$

$$\dot{\vec{v}}_{\text{sy}} = + \nabla(\mu_{\text{x}} + \tilde{\mu}_{\text{x}}') + \frac{1}{m^2 N_z} \nabla \nabla \cdot \vec{\ell}_{\text{sy}},$$

(136)

Here

$$\tilde{\mu}_{\text{x}}' = \mu_{\text{x}}' + \frac{1}{m^2 N_z} \nabla \cdot \vec{\ell}_{\text{sy}}',$$

$$\tilde{\mu}_{\text{y}}' = \mu_{\text{y}}' - \frac{1}{m^2 N_z} \nabla \cdot \vec{\ell}_{\text{sx}}',$$

(136a)

in analogy with equations (96).

Inserting equations (134,136) into (97), the entropy balance equation (30) follows again if we put

$$\vec{j}_{\text{sx}} = + \rho_s \vec{v}_{\text{sy}} = - \frac{\rho_s}{m N_z} \nabla N_{\text{y}},$$

$$\vec{j}_{\text{sy}} = + \rho_s \vec{v}_{\text{sx}} = + \frac{\rho_s}{m N_z} \nabla N_{\text{x}},$$

(137)

and

$$\nabla \cdot \vec{\ell}_{\text{sx}} \nabla \cdot \vec{j}_{\text{sy}} - \nabla \cdot \vec{\ell}_{\text{sy}} \nabla \cdot \vec{j}_{\text{sx}} = 0. \qquad (137a)$$

Then the energy current (99) is obtained, but with

$$- (m^2 M_z)^{-1} \nabla \times (\vec{j}_{sx} \times \vec{j}_{sy})$$

replaced by

$$- (m^2 N_z)^{-1} \cdot (\vec{j}_{sx} \nabla \cdot \vec{\ell}_{sy} - \vec{j}_{sy} \nabla \vec{\ell}_{sx}),$$

whilst equations (35) for $\vec{j}_s$, (100) for $\vec{j}_s{}'$, and (101) for σ are unchanged.

The new currents $\vec{\ell}_{sx}$ and $\vec{\ell}_{sy}$ are not determined by this procedure. This is because we have omitted an exchange term

$$\rho_s{}' \sum_{i=x,y} \vec{w}_{si} \cdot d\vec{w}_{si}$$

in the energy differential (94) where, in analogy with (135),

$$w_{sx} = (mN_z)^{-1} \nabla M_x,$$

$$w_{sy} = - (mN_z)^{-1} \nabla M_y,$$

(there could also be cross terms between $\vec{v}_{si}$ and $\vec{w}_{sj}$). For the sake of simplicity we leave out such additional exchange terms and put

$$\vec{\ell}_{si} = 0, \qquad \vec{\ell}_{si}{}' = 0, \qquad i = x,y. \tag{138}$$

This choice is compatible with the condition (129), which, after insertion of equations (134,134a) becomes

$$\sum_{i=x,y} N_i \nabla \cdot (\vec{j}_{si} + \vec{j}_{si}{}') = \sum_{i=x,y} M_i \nabla \cdot (\vec{\ell}_{si} + \vec{\ell}_{si}{}').$$

In fact, the initial conditions (132) immediately lead to (138).

In analogy with the second equation (38) we write

$$\mu_i{}' = + (\zeta \nabla) \cdot \vec{j}_{si}, \qquad i = x,y. \tag{139}$$

This is indeed compatible with rotational symmetry in the (x,y)-plane. On the other hand, $\vec{j}_{sx}{}'$ and $\vec{j}_{sy}{}'$ are given by equations (102). Excluding precessional motions, the perpendicular fields and magnetizations vary again around zero averages, $H_i = \delta H_i$, $h_i = \delta h_i$, $M_i = \delta M_i$, $N_i = \delta N_i$, whilst $M_z \cong 0$, $N_z \cong$ constant. Then equations (134,134a) become in plane wave representation, inserting from (89,102,103,137-9)

$$(- i\omega + \sigma_{xx})\delta M_x + \sigma_{xy}\delta M_y + \Omega \delta N_y = \sum_{i=x,y} \sigma_{xi}(\delta H_i - \beta_i \delta T),$$

$$\sigma_{xy}\delta M_x + (- i\omega + \sigma_{yy})\delta M_y - \Omega \delta N_x = \sum_{i=x,y} \sigma_{yi}(\delta H_i - \beta_i \delta T),$$

$$(140)$$

$$2H_e\delta M_y + (- i\omega + \lambda)\delta N_x = N_z(\delta H_y - \beta_y \delta T),$$

$$- 2H_e\delta M_x + (- i\omega + \lambda)\delta N_y = - N_z(\delta H_x - \beta_x \delta T).$$

Here we have introduced the abreviations

$$H_e = \frac{N_z}{2\chi_\perp} , \tag{141}$$

$$\lambda = \rho_s\zeta_\ell q^2, \tag{142}$$

and

$$\Omega = \frac{\rho_s}{m^2 N_z} q^2, \qquad \sigma_{ij} = \frac{\xi_\ell}{\chi_\perp} q^4 \hat{q}_i \hat{q}_j. \tag{142a}$$

Setting $\delta H_i = 0$ and $\beta_i = 0$, the solution is given by the following secular equation,

$$[\omega^2 - c_1^2 q^2 + i\omega(\lambda + \tfrac{1}{2}\sigma_\perp)]^2 + O(q^6\omega) = 0, \tag{143}$$

where, in analogy with (56),

$$c_1 = \left(\frac{\rho_s}{m^2 \chi_\perp}\right)^{\frac{1}{2}} \tag{144}$$

is the *transverse isothermal magnon velocity*, and $\sigma_\perp = \sigma_{xx} + \sigma_{yy}$. The solution of (143) is, to lowest order in q, inserting (142),

$$\omega = c_1 q - \tfrac{1}{2} i \rho_s \zeta_\ell q^2. \tag{145}$$

It is again of the form (57), valid also for the planar antiferromagnet.

As in the case of the axial ferromagnet, we now generalize the above equations by taking into account the effect of the dipole-dipole interaction. Leaving out gradient terms for the moment, this is done by replacing equations (130,131) by

$$\dot{\vec{M}} = - \vec{M} \times m\vec{\mu} - \vec{N} \times \vec{H}_A, \qquad (146)$$

and

$$\dot{\vec{N}} = - \vec{N} \times m\vec{\mu} - \vec{M} \times \vec{H}_A. \qquad (147)$$

Here, in distinction to equation (110),

$$m\vec{\mu} = \vec{H} - \vec{h} + \vec{H}_d, \qquad (148)$$

since the anisotropy field, $\vec{H}_A = (0,0,H_A)$, couples differently. In analogy with the equation (110a) the dipolar field is given by

$$\vec{H}_d = \frac{4\pi}{3} \vec{M} - 4\pi(\nabla^2)^{-1}\nabla(\nabla \cdot M) - \alpha\vec{N}_0, \qquad (148a)$$

where α is the demagnetizing factor and $\vec{N}_0 = (0,0,N_z)$ (see (132)). Equations (146,147) satisfy the condition (129). Together with equations (148,148a) they are equivalent to equations (17) of Anda [20] if we identify our quantities N_z, αN_z, and $N_z\vec{h}$, respectively, with Anda's $2m_0$, $2N_z\chi_\| B_0$, and $2H_e\vec{M}$.

Combination of equations (146,147) with (134,134a) leads to the general equations valid, as before, for the situation without precession ($H_z = 0$, $H_i = \delta H_i$, $h_i = \delta h_i$, $M_z \cong 0$, $M_i = \delta M_i$, $N_z \cong$ constant, $N_i = \delta N_i$)

$$\dot{M}_x + \frac{1}{m}\nabla \cdot (\vec{j}_{sx} + \vec{j}_{sx}') = - m\mu_z\delta M_y - H_A\delta N_y,$$
$$\dot{M}_y + \frac{1}{m}\nabla \cdot (\vec{j}_{sy} + \vec{j}_{sy}') = m\mu_z\delta M_x + H_A\delta N_x, \qquad (149)$$

and

$$\dot{N}_x + \frac{1}{m}\nabla \cdot (\vec{\ell}_{sx} + \vec{\ell}_{sx}') = mN_z(\mu_y + \mu_y') - m\mu_z\delta N_y - H_A\delta M_y,$$
$$\qquad (149a)$$
$$\dot{N}_y + \frac{1}{m}\nabla \cdot (\vec{\ell}_{sy} + \vec{\ell}_{sy}') = - mN_z(\mu_x + \mu_x') + m\mu_z\delta N_x + H_A\delta M_x,$$

In plane wave representation and with the choice (138), this leads to the following generalization of equations (140)

$$(i\omega + \tilde{\sigma}_{xx})\delta M_x - (\alpha N_z - \tilde{\sigma}_{xy})\delta M_y + \tilde{\Omega}\delta N_y = \sum_{i=x,y} \tilde{\sigma}_{xi}(\delta H_i - \beta_i\delta T),$$
$$\qquad (150)$$
$$(\alpha N_z + \tilde{\sigma}_{xy})\delta M_x + (- i\omega + \tilde{\sigma}_{yy})\delta M_y - \tilde{\Omega}\delta N_x = \sum_{i=x,y} \tilde{\sigma}_{yi}(\delta H_i - \beta_i\delta T),$$

(Contd)

(Contd)

$$2B_{xy}\delta M_x + (A + 2B_{yy})\delta M_y + (-i\omega + \lambda)\delta N_x - \alpha N_z \delta N_y \qquad (150)$$

$$= N_z(\delta H_y - \beta_y \delta T),$$

$$- (A + 2B_{xx})\delta M_x - 2B_{xy}\delta M_y + \alpha N_z \delta N_x + (-i\omega + \lambda)\delta N_y$$

$$= - N_z(\delta H_x - \beta_x \delta T),$$

Here we have introduced the abbreviations

$$A = 2H_e + H_A - \frac{4\pi}{3} N_z, \qquad B_{ij} = 2\pi N_z \hat{q}_i \hat{q}_j, \qquad (151)$$

and

$$\tilde{\Omega} = H_A + Dq^2, \qquad \tilde{\sigma}_{ij} = \frac{\xi_\ell}{\chi_\perp}\left[1 + \frac{8\pi}{3}\chi_\perp\right]q^4 \hat{q}_i \hat{q}_j, \qquad (151a)$$

which are consequences of equations (148,148a,141). The definitions (151a) are generalizations of (142a) and

$$D = \frac{\rho_s}{m^2 N_z}. \qquad (152)$$

Equations (152,141) are the analogue of equations (108) of the axial ferromagnet. H_e is the effective exchange field (see section 39 of reference [23]).

We consider again only the homogenous equations (150), setting $\delta H_i = 0$, $\beta_i = 0$. The associated secular equation is found, after considerable algebra, to be exactly given by

$$(\omega^2 - \omega_+^2)(\omega^2 - \omega_-^2) + i\omega^3(2\lambda + \tilde{\sigma}_\perp) - \omega^2\lambda(\lambda + 2\tilde{\sigma}_\perp)$$

$$- i\omega\{2\tilde{\Omega}(A + B_\perp)\lambda + \tilde{\Omega}A\tilde{\sigma}_\perp + (\alpha N_z)^2(2\lambda + \tilde{\sigma}_\perp) + \lambda^2\tilde{\sigma}_\perp\}$$

$$+ \tilde{\Omega}\lambda\tilde{\sigma}_\perp + (\alpha N_z)^2\lambda^2 = 0, \qquad (153)$$

where

$$B_\perp = B_{xx} + B_{yy}, \qquad \tilde{\sigma}_\perp = \tilde{\sigma}_{xx} + \tilde{\sigma}_{yy},$$

and

$$\omega_{\pm}{}^2 = \tilde{\Omega}(A + B_{\perp}) + (\alpha N_z)^2 \pm \{\tilde{\Omega}^2 B_{\perp}{}^2 + 4\tilde{\Omega}(\alpha N_z)^2(A + B_{\perp})\}^{\frac{1}{2}}. \qquad (154)$$

For $q = 0$, equation (154) is essentially the same as equation (23) of reference [20] (see also reference [21]); for $\alpha = 0$ it simplifies to

$$\omega_+{}^2 = \tilde{\Omega}(A + 2B_{\perp}) = (H_A + Dq^2)(2H_e + H_A - \frac{4\pi}{3} N_z + 4\pi N_z \sin^2\theta_{\hat{q}}),$$
$$(154a)$$
$$\omega_-{}^2 = \tilde{\Omega}A = (H_A + Dq^2)(2H_e + H_A - \frac{4\pi}{3} N_z),$$

where $\cos\theta_{\hat{q}} = \hat{q}_y$. To order q^2 this result is the same as equations (13a,b) of Brooks Harris [22]. Apart from the local field correction $-(4\pi/3)N_z$, formulas (154a) were already given in reference [21].

However, the merit of our hydrodynamic treatment is not only to be able to reproduce the quoted results, but, even more, to yield the damping terms. Indeed, the solutions of the secular equation (153) are, up to order q^4,

$$\omega_1 = \omega_+ + \frac{\omega_+{}^2 - (\alpha N_z)^2}{2\omega_+(\omega_+{}^2 - \omega_-{}^2)} \lambda^2 - \frac{i}{2}\left\{\lambda + \left(\frac{1}{2} + \frac{\tilde{\Omega}B_{\perp}}{\omega_+{}^2 - \omega_-{}^2}\right)\tilde{\sigma}_+\right\},$$
$$(155)$$
$$\omega_2 = \omega_- - \frac{\omega_-{}^2 - (\alpha N_z)^2}{2\omega_-(\omega_+{}^2 - \omega_-{}^2)} \lambda^2 - \frac{i}{2}\left\{\lambda - \left(\frac{1}{2} - \frac{\tilde{\Omega}B_{\perp}}{\omega_+{}^2 - \omega_-{}^2}\right)\tilde{\sigma}_{\perp}\right\}.$$

For $\alpha = 0$ these expressions reduce to

$$\omega_1 = (\tilde{\Omega}(A + 2B_{\perp}))^{\frac{1}{2}}\left(1 + \frac{\lambda^2}{4\tilde{\Omega}B_{\perp}}\right) - \frac{i}{2}(\lambda + \tilde{\sigma}_{\perp}),$$
$$(155a)$$
$$\omega_2 = (\tilde{\Omega}A)^{\frac{1}{2}}\left(1 - \frac{\lambda^2}{4\tilde{\Omega}B_{\perp}}\right) - \frac{i}{2}\lambda,$$

valid for $\tilde{\Omega}B_{\perp} \neq 0$. If, in addition, $H_A = 0$, we find, to order q^2, inserting (151,151a) and making use of (142,144,152),

$$\omega_1 = c_1 q\left[1 - (1 - 3\sin^2\theta_{\hat{q}}) \frac{4\pi}{3} \chi_{\perp}\right]^{\frac{1}{2}} - \frac{i}{2} \rho_s \zeta \ell q^2,$$
$$(155b)$$
$$\omega_2 = c_1 q\left[1 - \frac{4\pi}{3} \chi_{\perp}\right]^{\frac{1}{2}} - \frac{i}{2} \rho_s \zeta \ell q^2.$$

We notice that away from the easy axis the dipolar field correction splits the mode (145).

We also see that a sound-like Hulthén mode [5] is a general feature both of planar and, for $H_A = 0$, of axial antiferromagnets and hence, in particular, of *isotropic* antiferromagnets (see reference [4], p. 910, and reference [23], p. 107). On the other hand, neutron scattering has revealed that the quadratic dispersion relation with gap as described by equations (155, 155a) is valid not only for the *axial* antiferromagnets MnF_2 [26] and FeF_2 [27], but also for the *planar* antiferromagnets K_2NiF_4 [28] and K_2MnF_4 [29]. This shows that the dipole-dipole interaction acts similarly in the latter cases.

Thermodynamically there is, of course, an important difference between the quadratic dispersion laws (109,112,116,155, 155a) on the one hand, and the linear laws (57,145,155b) on the other. Indeed, if the gap is small compared to $k_B T$, quadratic dispersion gives rise to the well known $T^{3/2}$-dependence of the magnon specific heat, c_M, instead of the T^3-dependence valid for linear dispersions (for $k_B T$ small compared to the gap ω_0, the specific heat behaves as $e^{-\omega_0/k_B T}$). This $T^{3/2}$-dependence has the important consequence that the quantity θ defined in equation (71) varies as $T^{5/2}$, and not as T^4, at low temperatures. This slower decrease makes the ratio r between the second and the first magnon peaks, as defined in equation (86), more favorable. It is therefore advantageous to look for the second magnon peak in magnetic excitation experiments on systems with quadratic dispersion law with gap (because of drift, see above), that is for axial ferro- and antiferromagnets [14,15,16].

REFERENCES

1. Heisenberg, W. (1928). *Z.Phys.*, **49**, 619; Frenkel, J. (1928). *Z. Phys.*, **49**, 31; Dorfman, J. and Jaanus, R. (1929). *Z. Phys.*, **54**, 277.
2. Bloch, F. (1930). *Z. Phys.*, **61**, 206.
3. Landau, L.D. and Lifshitz, E.M. (1935). *Phys. Z. Sowjetunion*, **8**, 153.
4. Halperin. B.I. and Hohenberg, P.C. (1969). *Phys. Rev.*, **188**, 898.
5. Hulthén, L. (1936). *Proc. Roy. Acad. Sci. Amsterdam*, **39**, 190.
6. Enz, C.P. (1971). *Phys. Kondens. Mater.*, **12**, 262. This paper contains a confusion between magnon drift and 'superfluid' velocity which is corrected as follows: Put $\tau = \infty$; discard the comparison of equations (31-33,46) with equations (4.14a-c) and (4.20a,b), respectively, of reference [10]. In addition the condition for the planar case should read $J^z_{ij} < J_{ij}^{\perp}$. See also: Michel, K.H. and Schwabl, F. (1971). *Phys. Kondens. Mater.*, **14**, 78.
7. Gulayev, Yu.V. (1965). *Zh. Eksp. Teor. Fiz. Pism. Red.*, **2**, 3; [(1965). *Sov. Phys. JETP Lett.*, **2**, 1].
8. Dingle, R.B. (1952). *Proc. Phys. Soc.*, **A65**, 1044; Gurzhi, R.N. (1965). *Fiz. Tverd. Tela*, **7**, 3515; [(1966). *Sov. Phys. Solid State*, **7**, 2838].
9. Reiter, G.F. (1968). *Phys. Rev.*, **175**, 631.

10. Michel, K.H. and Schwabl, F. (1969). *Solid State Commun.*, **7**, 1781; (1970). *Phys. Kondens. Mater.*, **11**, 144.

11. Götze, W. and Michel, K.H. (1967). *Phys. Rev.*, **156**, 963.

12. Schwabl, F. and Michel, K.H. (1970). *Phys. Rev.*, **B2**, 189; Michel, K.H. and Schwabl, F. (1970). *Z.Phys.*, **238**, 264; **240**, 354.

13. Enz, C.P. (1972). In *Irreversibility in The Many Body Problem*, (eds. Biel, J. and Rae, J.), (Plenum Press, London and New York), p. 387.

14. Michel, K.H. and Schwabl, F. (1971). *Phys. Rev. Lett.*, **26**, 1568.

15. Forney, J.J. (1972). (Doctoral Thesis), (Geneva University), (unpublished).

16. Forney, J.J. and Jäckle, J. (1973). *Phys. Kondens. Mater.*, **16**, 147.

17. Kittel, C. (1963). *Quantum Theory of Solids*, (Wiley, New York), chapter 4.

18. Herring, C. and Kittel, C. (1951). *Phys. Rev.*, **81**, 869.

19. Clogston, A.M., Suhl, H., Walker, L.R. and Anderson, P.W. (1956). *J. Phys. Chem. Solids*, **1**, 129.

20. Anda, E. (1973). *J. Phys. Chem. Solids*, **34**, 1597.

21. Loudon, R. and Pincus, P. (1963). *Phys. Rev.*, **132**, 673.

22. Brooks Harris, A. (1966). *Phys. Rev.*, **143**, 353.

23. Keffer, F. (1966). *Handbuch der Physik, Vol. XVIII/2*, (ed. Flügge, S.), (Springer-Verlag, Berlin, Heidelberg), p. 1.

24. Nagamiya, T., Yosida, K. and Kubo, R. (1955). *Adv. Phys.*, **4**, 1.

25. Shirane, G., Minkiewicz, V.J. and Nathans, R. (1968). *J. Appl. Phys.*, **39**, 383.

26. Okazaki, A., Turberfield, K.C. and Stevenson, R.W.H. (1964). *Phys. Lett.*, **8**, 9; (1965). *Proc. Phys. Soc.*, **85**, 743.

27. Guggenheim, H.J., Hutchings, M.T. and Rainford, B.D. (1968). *J. Appl. Phys.*, **39**, 1120.

28. Skalyo, J., Jr., Shirane, G., Birgeneau, R.J. and Guggenheim, H.J. (1969). *Phys. Rev. Lett.*, **23**, 1394.

29. Birgeneau, R.J., Guggenheim, H.J. and Shirane, G. (1973). *Phys. Rev.* **B8**, 304.

IMPURITY FERROMAGNETISM
IN NON-MAGNETIC SEMICONDUCTORS

A.A. ABRIKOSOV and Yu.V. KOPAYEV

*Landau Institute for Theoretical Physics,
Academy of Sciences of the USSR, Moscow*

1. INTRODUCTION

Recent experiments demonstrated some phenomena in non-magnetic semiconductors with non-magnetic impurities which cannot be explained other than as an appearance of ferromagnetism. These phenomena were observed in InSb with an addition of tellurium and also $Sn_xPb_{1-x}Te$ complexes with an excess of tellurium, and look very much alike.

The first of these substances was investigated by Professor Brandt's group at Moscow University [1]. They studies the Shubnikov-de Haas oscillations with the aim of developing a systematic method of defining the free path length. They found that in the dependence of $\partial\rho/\partial H$ on H the usual oscillations are modulated by beats (figure 1). The reversal of the field direction changed the form of the beats. Particularly the positions of the knots (a) and (b) moved to other values of H. Since usually only one to two knots were observed and the amplitude of Shubnikov-de Haas oscillations increased strongly with the magnetic field, it was impossible to prove rigorously that only the phase of the beats changed after the reversal of H, but this appears very probable.

The essential feature of the phenomenon was the fact that the beats were observed only within a rather narrow range of impurity concentration, $0.8-1.1 \times 10^{18}$ cm^{-3}. Outside of this interval the beats disappeared. The temperature limit was approximately 30°K. There are some data about the dependence on the field direction relative to the crystal axes and other details, but to

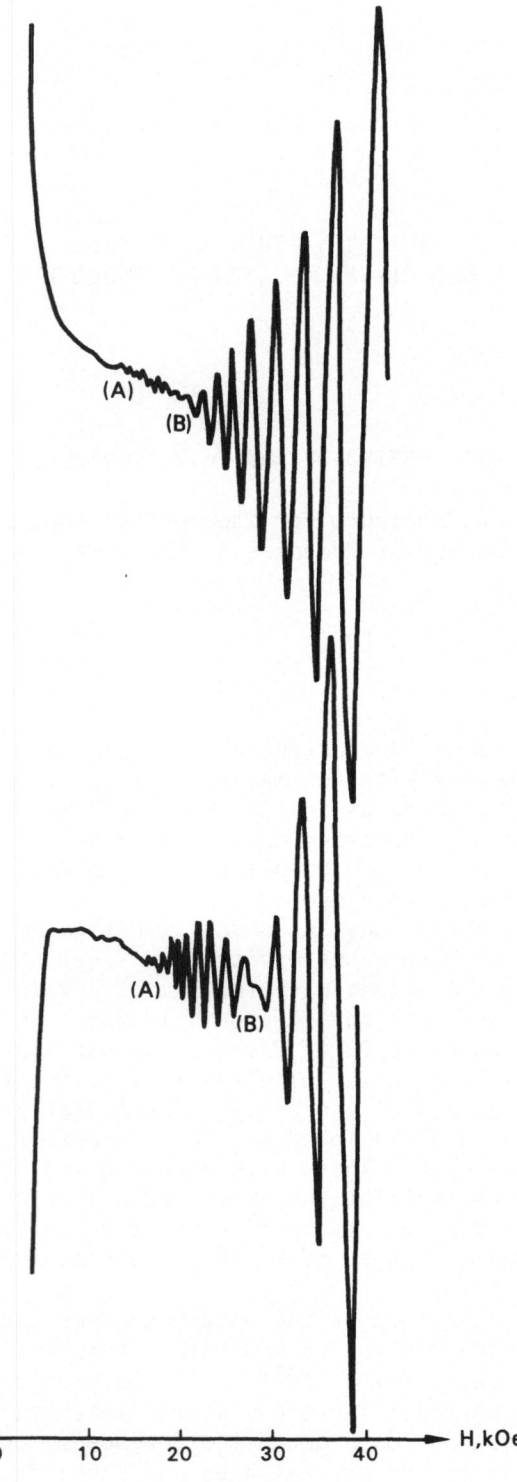

Figure 1

these I shall refer later. A short account on these investigations was published in 1973.

The other class of substance, i.e. $Sn_xPb_{1-x}Te$ with different concentrations of excess tellurium, are being investigated by Dr. Shotov's group at the Moscow Lebedev Physical Institute. The results will be published soon. In the first experiments they studied the magnetoresistance, and it was found that $\rho(H) - \rho(-H) \neq 0$. The effect was observed in a much wider concentration range than in InSb, namely 10^{17}-10^{19} cm^{-3}, and up to room temperatures. Quite recently they studied the quantum oscillations of the resistance at helium temperatures, and they found the same phenomena as in InSb.

So in both cases the properties change with the field commutation, i.e. with the change $H \rightarrow -H$. This is a definite reason for expecting ferromagnetism.

However, for the sake of completeness it should be mentioned that the beats in the quantum oscillations had already been observed in n-type HgSe (Whitesett [2]) and GaSb (Seiler and Becker [3]). These crystals have the same zincblende structure as InSb. A theory was proposed (Laura Roth et $al.$, [4]) which explained the beats in the following way. The zincblende structure has no inversion centre. Therefore the double degeneracy of the energy levels is lifted by the spin-orbit interaction. In the conduction band the spectrum follows essentially a square law $\varepsilon = p^2/2m^*$. But if we take cubic terms in p into account, an anisotropy appears and the degeneracy is lifted. Therefore in sufficiently doped samples two close Fermi surfaces appear. Since the Shubnikov-de Haas oscillation period is determined by the area of the central section of the Fermi surface, two close periods can exist and accordingly beats in oscillations appear.

Evidently the relative difference of these areas will increase with the Fermi momentum, i.e. with the carrier concentration. This contradicts the experimental fact that at high concentrations the effect disappears. Further on, according to the theory of Roth, the beats must not change with the field commutation, and this also contradicts experiment. Finally, in substances of the second type, $Sn_xPb_{1-x}Te$, the beats were observed at $x < 35\%$, where the crystal has an NaCl lattice which has an inversion centre.

All the observed features are explained naturally by the appearance of ferromagnetism. But neither the host substance nor the impurities are magnetic, i.e. they do not contain transitional element atoms. Therefore before considering the consequences of ferromagnetism one must understand how it can arise. Two theories have been proposed. The theory of A.A. Abrikosov (published in 1973) [5] explains, at least qualitatively, the situation in InSb. This approach we have called 'resonance ferromagnetism'. The theory of Yu.V. Kopayev and B.A. Volkov [6] explains the appearance of ferromagnetism in substances of the other type. Its authors have called it 'band ferromagnetism'. The reason for such names will become clear from the forthcoming.

Why were two different approaches necessary? The phenomena
look very much alike. But there are some differences. In InSb
the ferromagnetic phenomena are observed in a small concentration range with a relative width 0.3. In $Sn_xPb_{1-x}Te$ the concentration range is also limited, but it varies by two orders of
magnitude (10^{17}-10^{19} cm^{-3}). On the other hand, in these substances a band inversion takes place and phase transitions are
observed which can be of excitonic origin. There is nothing of
this kind in InSb. All this is of principal significance for
the proposed explanations.

2. 'RESONANCE FERROMAGNETISM'

The idea of the explanation of ferromagnetism in InSb with
tellurium impurity is based on the assumption that a part of the
tellurium atoms can occupy in the InSb lattice some special positions which act as localisation centres for electrons. The corresponding energy level falls inside the conduction band. In
principle, in these cases the level should acquire a finite width.
We suppose this width to be small (the rôle of a finite width
will be discussed later). This assumption means that the band
states are almost orthogonal to the localized states. We don't
know what these states are, but their existence was recently
proved by the independent experiments of Dr. Philipchenko in the
Lenigrad Ioffe Physical Institute. His measurements of the infrared absorption in InSb doped with Te demonstrate the existence
of narrow resonance energy levels within the conduction band,
one of them being exactly at the proper position.

Now imagine that the electrons in such localized states have
an 'exchange' interaction with the conduction electrons. Actually it must not be really a true exchange interaction. There can
be something like an Anderson mechanism, and therefore such a
spin dependent interaction can be rather strong and anisotropic.
The position of the Fermi boundary in a degenerate semiconductor
depends on the impurity concentration. If the Fermi boundary is
far from the localized level then this level will be either completely filled, if the boundary is high enough, or empty, if it
is low. In both cases the localized states have no net spin.
However, in the case where the Fermi boundary is close to the
localized level the situation may change. Let us imagine for
simplicity that the level corresponds to an S-state and the Fermi
boundary is below the level. If we put a single electron on
every such site we increase the energy. But if we imagine that
all the spins of the localized electrons are arranged parallel,
and owing to the 'exchange' interaction polarize the conduction
electrons, then the total energy can be decreased. The same will
take place if the Fermi boundary slightly exceeds the localized
level. We must draw your attention to the fact that the number
of tellurium atoms creating such sites must be smaller than their
total number, since otherwise the Fermi boundary would be unable
to cross the localized level. This means that such atoms have

very special positions indeed. This was already clear from the
narrowness of the localized level.

Evidently such an idea explains why ferromagnetism appears
only in a narrow impurity concentration range. Since the coincidence of the Fermi boundary with the localized level is demanded, we have called this phenomena 'resonance ferromagnetism'.

In order to imagine what the consequences of this idea are we
considered the simplest model with localized s-sites, an isotropic conduction band, and an exchange interaction $J\vec{\sigma}\vec{s}$. The Hamiltonian is taken in the form

$$\hat{H} - N\mu = \int \psi_\sigma^+(\vec{r})\hat{H}_0\psi_\sigma(\vec{r})d^3\vec{r} + \sum_i \varepsilon a_{is}^+ a_{is}$$

$$- J\sum_i a_{is}^+\vec{s}a_{is}\psi_\sigma^+(\vec{r}_i)\vec{\sigma}\psi_\sigma(\vec{r}_i), \qquad (1)$$

where $\hat{H}_0 = -\nabla^2/2m^* - \mu$, ε is the localized level energy with
respect to the chemical potential μ. About the influence of the
Coulomb interaction at one centre, and of the finite width, we
shall speak later on.

We apply the self-consistent field approximation. That means
that every localized spin is influenced by an 'effective' field
created by the conduction electrons

$$\vec{Q} = J\langle\vec{\sigma}\rangle. \qquad (2)$$

The average spin of an impurity centre is therefore

$$\langle S\rangle = n_F(\varepsilon - Q) - n_F(\varepsilon + Q) = \frac{1}{2}\left[\tanh\left(\frac{\varepsilon + Q}{2T}\right) - \tanh\left(\frac{\varepsilon - Q}{2T}\right)\right]. \qquad (3)$$

But the conduction electrons are influenced by the effective
field created by the impurity centres,

$$\vec{H} = JN_m\langle\vec{s}\rangle, \qquad (4)$$

where N_m is the number of centres in unit volume (we call them
'magnetic centres'). In this field the conduction electrons
develop an average spin equal to

$$\langle\vec{\sigma}\rangle = \chi\vec{H} = 2\nu\vec{H}. \qquad (5)$$

Where $\nu = p_0 m^*/2\pi^2$ is the state density.

From the formulae (2-5) we can get an equation for $\langle\sigma\rangle$

$$\langle\sigma\rangle = \nu JN_m\left[\tanh\left(\frac{Q + \varepsilon}{2T}\right) + \tanh\left(\frac{Q - \varepsilon}{2T}\right)\right]. \qquad (6)$$

It should be mentioned that this equation is invariant under change of sign of ε and change of sign J. The first means a symmetry to the Fermi boundary lying below or above the localized level, and the second means an independence of the magnitude of $\langle \vec{\sigma} \rangle$ on the sign of the 'exchange' interaction. Introducing new variables $u = Q/T$, $x = |\varepsilon|/T$, we get from (6)

$$u = \frac{2\alpha x \sinh u}{\cosh u + \cosh x} , \tag{7}$$

where

$$\alpha = \frac{\nu J^2 N_m}{|\varepsilon|} . \tag{8}$$

We shall need also the expression for the free energy. In the self-consistent field approximation we get from (1)

$$\frac{\partial \Omega}{\partial J} = \frac{\partial \langle H - N_\mu \rangle}{\partial J} = - N_m \langle \vec{s} \rangle \langle \vec{\sigma} \rangle = - \frac{\langle \vec{\sigma} \rangle^2}{2\nu J} . \tag{9}$$

If we write the equation (7) as

$$u = \alpha x f(u), \tag{10}$$

then we get from (9), after some transformations,

$$\delta \Omega = \tfrac{1}{2} N_m T \left[\tfrac{1}{2} u f(u) - \int_0^u f(u_1) du_1 \right] . \tag{11}$$

Substituting $f(u)$ from (7) we obtain:

$$\delta \Omega = \tfrac{1}{2} N_m T \left[\frac{u^2}{2\alpha x} - 2 \ln \left(\frac{2\alpha x \sinh u}{u(\cosh x + 1)} \right) \right] . \tag{12}$$

The equation (7) evidently has a trivial solution $u = 0$. But it can also have non-trivial solutions. Which of them can be true is defined by the condition $\delta \Omega \leqslant 0$.

We shall give the results in the form of graphs. The solutions of the equation (7) are given on figure 2. Every curve corresponds to a definite α. We would remind the reader that $u = Q/T$, where Q is proportional to the magnetization and $X = |\varepsilon|/T$. To every $\alpha > 0.5$ corresponds one or two curves. If $\alpha \rightarrow 0.5$ the whole curve moves to infinity. In the case where there are two curves for a given α, only the left one is stable. This is the region of large α, i.e. small $|\varepsilon| = |\varepsilon_0 - \mu|$, where the true position of the localized level is. The point where $u \rightarrow 0$ is the Curie point. It is defined according to (7) by the condition

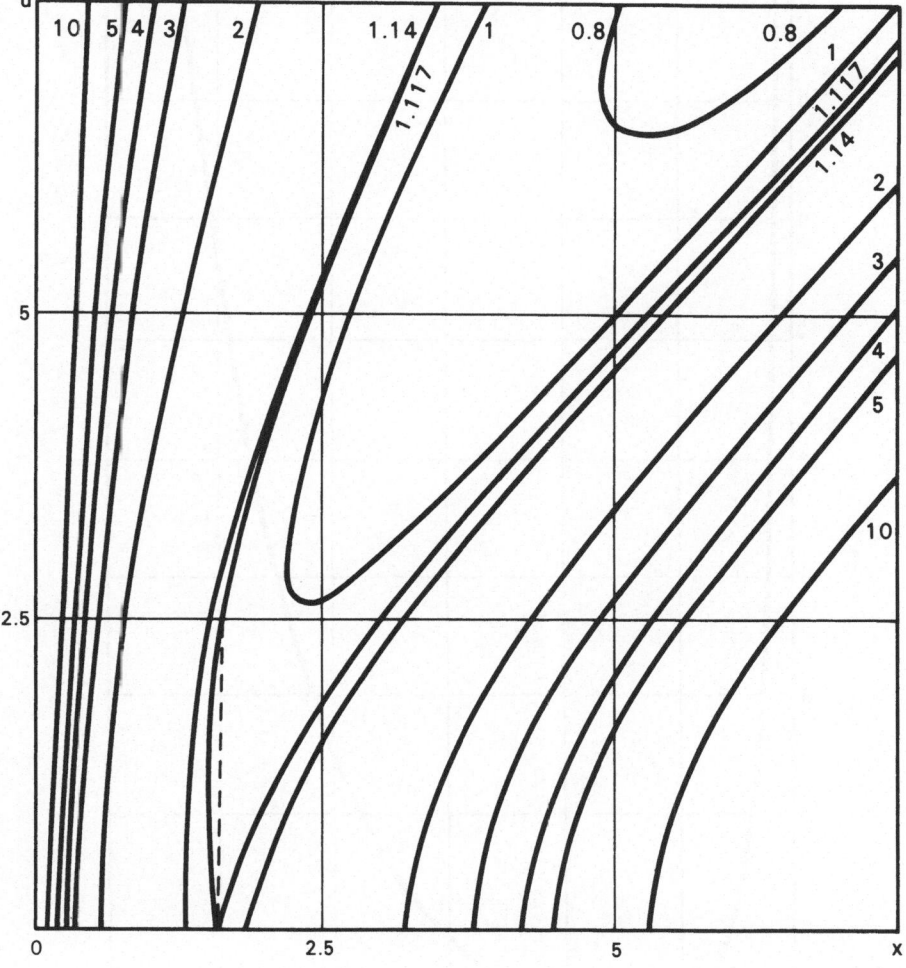

Figure 2

$$1 + \cosh x = 2\alpha x. \tag{13}$$

The corresponding dependence $\alpha(x)$ is drawn on figure 3. Equation (13) has solutions only for $\alpha > 1.117$. Even for larger α, namely $\alpha < 1.14$ the ferromagnetic transition becomes of the first order. This situation remains until $\alpha > 1$. If α is smaller we have $\delta\Omega > 0$ on the whole curve $u(x)$.

So the boundaries of the ferromagnetic phase are defined by the condition $\alpha = 1$. Substituting the value of α and taking into account that $|\varepsilon_0 - \mu| \ll \varepsilon_0$ we get

$$|\varepsilon_0 - \mu| < \frac{2m^{*3}\varepsilon_0{}^2 J^2}{(3\pi^2\hbar^6 q)}, \tag{14}$$

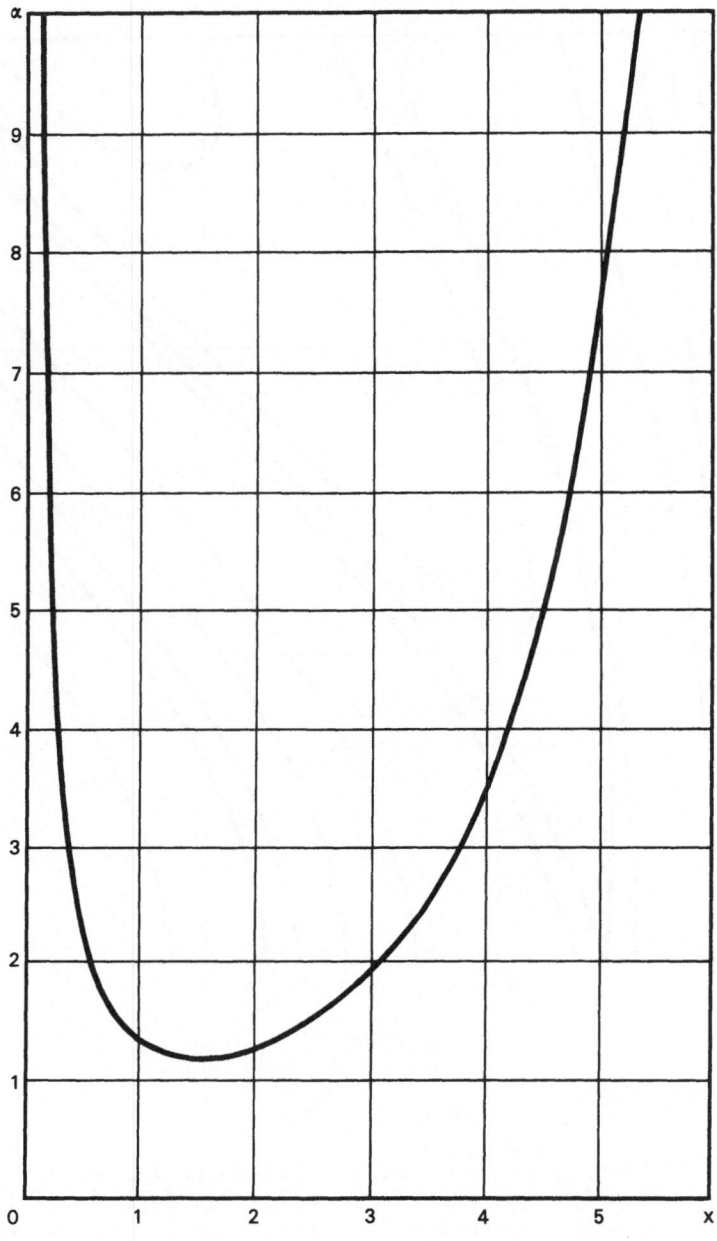

Figure 3

where $q = N_e/N_m$ is the relation of the concentration of conduction electrons to the concentration of 'magnetic' centres.
Actually it may happen that the effective interaction J also de-
pends on $|\varepsilon_0 - \mu|$ as it is in the Anderson model. Therefore the
condition (14) must not be taken too seriously. It is only a
demonstration of the concentration limitation.

In the region of big α, i.e. small $|\varepsilon_0 - \mu|$, we get

$$T_c = \nu J^2 N_m \qquad (15)$$

For smaller α it can be got from the left hand part of the curve on figure 3 or from the condition $\delta\Omega = 0$ (for $1.14 > \alpha > 1$).

Let us consider the influence of the spin polarisation on the quantum oscillations. For electrons with different spin orientations the Fermi moments are defined by the conditions

$$\frac{p_+^2}{2m^*} = \mu + H, \qquad \frac{p_-^2}{2m^*} = \mu - H.$$

The difference of the central cross section areas of the Fermi surface is

$$S_+ - S_- = \pi(p_+^2 - p_-^2) = 4\pi m^* H = \frac{2\pi m^*}{\nu} \langle \sigma \rangle. \qquad (16)$$

At $T = 0$ we get

$$S_+ - S_- = 4\pi m^* J N_m. \qquad (17)$$

This difference in the extremal areas leads to the beats in the quantum oscillations.

In order to explain the change of the picture with the commutation of the field we must take into account that, apart from the diamagnetic quantum levels, there exists the dependence of the electron energy on the spin orientation with respect to the external field: $-g\mu_B \vec{\sigma}\vec{H}$, where g is the gyromagnetic ratio, which in InSb is approximately 50. Let the sample consist of a single domain. The spins are polarized by the internal interaction and the (+) direction prevails. It is clear that by switching on the field $\vec{H}$ in the (+) direction we decrease the energy of these electrons and increase the energy of the electrons with spin (-) having a smaller Fermi sphere. Changing $\vec{H}$ to the opposite direction we increase the energy of the electrons with the big Fermi sphere and decrease the energy of the small group. This is the origin of the asymmetry.

For a single domain sample the Lifshitz-Kosevich formula for the quantum oscillations can be written as

$$\sin\left(\frac{cS_+}{\hbar eH} \pm \frac{\pi g m^*}{2m}\right) + \sin\left(\frac{cS_-}{\hbar eH} \mp \frac{\pi g m^*}{2m}\right),$$

where m is the mass of the free electron. The upper signs are taken for $\vec{H}$ in the (+) direction and the lower for $\vec{H}$ in the (-) direction. Assuming $S_+ - S_-$ to be small, introducing $S = \frac{1}{2}(S_+ + S_-)$ we get

$$2\sin\left[\frac{cS}{\hbar eH}\right]\cos\left[\frac{c(S_+ - S_-)}{2\hbar eH} \pm \frac{\pi g m^*}{2m}\right] .\tag{18}$$

One sees from here that the phase of the beats is changed if $H \to -H$.

If the sample consists of many domains and the concentration of (+) domains is c_d, then instead of the second factor in (18) the beats are described by the formula

$$[\cos^2 b + (1 - 2c_d)\sin^2 b]^{\frac{1}{2}}\cos[a + \arctan(2c_d - 1)\tan b],$$

$$\tag{19}$$

$$a = \frac{c(S_+ - S_-)}{2\hbar eH} , \quad b = \frac{\pi g m^*}{2m} .$$

The change of the sign of H means $b \to -b$. The amplitude of the beats stays invariant, but the phase is unchanged. The effect will disappear if $c_d = \frac{1}{2}$, i.e. if the total volumes occupied by different domains are the same, and if $\vec{H} \perp \langle \vec{\sigma} \rangle$. The latter is possible only for a uniaxial magnetization. Although InSb is a cubic crystal and should not have a uniaxial magnetization, nevertheless, the real samples which the Brandt group had at its disposal were strongly anisotropic. They were grown along the [110] direction and the 'commutation effect' disappeared for a whole plane of the directions of $\vec{H}$ (110).

Of course all this reasoning is valid only if the applied field H is small enough not to change the domain structure. Evidently a greater field would itself define 'the (+) direction' and the commutation effect would disappear (although the beats will remain). Experimentally the commutation effect remained up to H = 70kOe and this is rather unusual. This can be explained by the assumption that the original interation of spins which we wrote in a simple 'exchange' form is in fact strongly anisotropic. Therefore the turning of the magnetisation is counteracted not by a weak magnetic anisotropy but by the main interaction. To overcome it, much greater fields are demanded ($\sim$300 kOe, since $T_c \sim 30°K$).

The motion of the domain walls can be stopped by the non-uniformity of the real samples. The magnetism has a 'resonance' character and so it must be very sensitive to the non-uniformity of the electron concentration. Therefore one can imagine even an array of single domain ferromagnetic regions imbedded in a non-magnetic medium. This, of course, can considerably stiffen the sample towards domain wall motion.

Of course it would be interesting to see the spontaneous magnetisation directly. It must be equal to

$$\vec{M} = \mu_B(\tfrac{1}{2}g_i N_m\langle s \rangle + \tfrac{1}{2}g\langle \vec{\sigma} \rangle).$$

At T = 0 this gives

$$M = \mu_B N_m (\tfrac{1}{2} g_i + \nu J g).$$

(20)

Since the concentrations N_e are of the order of 10^{18} cm^{-3}, $N_m/N_e = 1/q < 1$, and further on, where the sample is not a single domain, it will not be easy to trace experimentally this spontaneous magnetisation. The Brandt group, however, is trying to do it.

To finish our discussion of this model we wish to speak about the influence of the Coulomb interaction and of the finite width of the level. If the Coulomb interaction is described as in the Hubbard model, or in the Anderson model, by a term

$$H_c = u \sum_i n_{i+} n_{i-}$$

(21)

in the Hamiltonian, where $n_{i\pm} = a_{i\pm}{}^+ a_{i\pm}$, with $u > 0$, it will evidently favour one electron at the localized level. Calculations show that in the case where $\nu J^2 N_m \ll u$ the range of ferromagnetism is defined by the condition

$$u > \mu - \varepsilon_0 > 0,$$

(22)

i.e. it will again be limited, although it will be wider than before. The Curie temperature will remain the same as previously obtained (15). Experimentally, for InSb $T_0 \sim 30°$K, and the width of the area is of the order of 30 meV $\sim 300°$K.

Now, if we take into account the transition between the localized levels and the conduction band by means of a term

$$H_{trans} = V \sum_i [\psi_\sigma{}^+(\vec{r}_i) a_{i\sigma} + a_{i\sigma}{}^+ \psi_\sigma(\vec{r}_i)]$$

(23)

in the Hamiltonian, then the level ε_0 acquires a finite width $\Gamma = \pi V^2 \nu$. In the calculations Γ competes with T and ε, and somewhere at $\nu J^2 N_m \approx T_c < \Gamma$ the ferromagnetism becomes impossible. So from experimental data we must assume $\Gamma \ll 3 \times 10^{-3}$ eV.

3. 'BAND FERROMAGNETISM'

The substances $Sn_x Pb_{1-x}Te$ at small tin concentrations have a lattice of the NaCl type and the band structure of the form of figure 4(a) close to the L-point in momentum space. With increase of tin concentration the spectrum is changed from figure 4(a) through a gapless spectrum, figure 4(b), to an 'inverted' one, figure 4(c). The situation of figure 4(b) corresponds to x = 35%. Experimentally, it appears that with x > 35% there is a low temperature phase having a distorted lattice of the bismuth type which undergoes, at a certain temperature, a first order transition to the cubic phase. The theoretical studies of

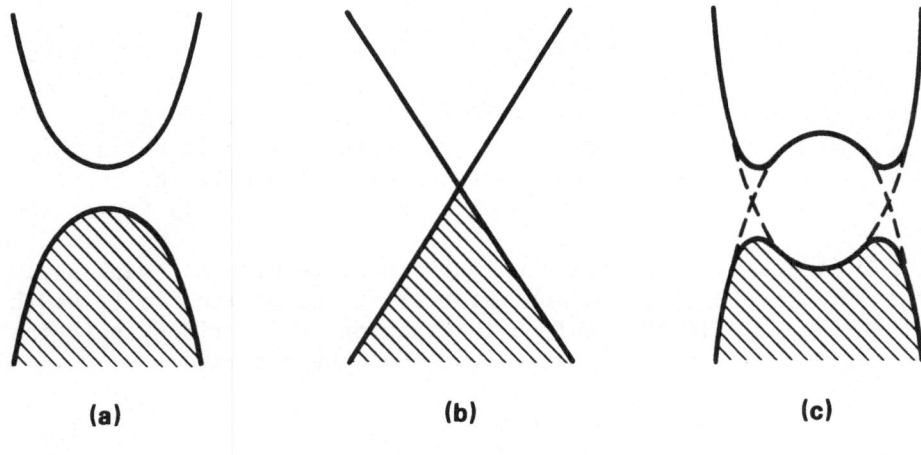

(a)   (b)   (c)

Figure 4

the energy spectrum and of the crystal structure of such substances have been done by Volkov and Kopayev [7] at the Moscow
Lebedev Physical Institute, also by Gor'kov and Gordiunin at the
Landau Institute for Theoretical Physics on the basis of the
idea of an excitonic phase transition. Some phenomena have been
understood, but as a whole this work is far from being finished.

At the same time, as we have said already, Shotov's group at
the Lebedev Institute has observed ferromagnetic phenomena. For
their explanation Volkov and Kopayev considered a simple model
having a possibility of an excitonic phase transition. Since
the model is over-simplified, its predictions can be only qualitatively compared with the experimental results.

The attempt to explain the ferromagnetism of a metal by the
interaction of the electrons in the conduction band starts as
early as the works of Bloch [8] and Stoner [9]. The comparison
of the energy of a paramagnetic and a ferromagnetic state of an
electron Fermi gas with Coulomb interaction shows that the ferromagnetic state becomes favourable if

$$p_F < \frac{5me^2}{(2^{1/3} + 1)2\pi\hbar} \, . \tag{24}$$

Hence ferromagnetism can appear only if the dimensionless interaction constant $e^2/\hbar v_F > 1$. But in this case the simple estimate itself becomes very doubtful, since the theory does not contain a small parameter, and an exact calculation of the energies
of different phases becomes impossible.

The theory of Volkov and Kopayev considers the simple Keldish-
Kopayev model with two intersecting isotropic bands which can
be divided in momentum space by a wave vector $\vec{q}$. This, however,
is inessential for the question of ferromagnetism, and so we
shall assume both extrema to be at the same point (figure 4(c)).

For simplicity the spectra of electrons and holes are assumed
to be equal and are written in the form

$$\varepsilon_{1,2} = -\,\delta\mu \pm \left[\frac{p^2}{2m^*} - \varepsilon_F\right].$$ (25)

These are the energies with respect to the Fermi level, 1 refers
to electrons, 2 refers to holes. In the presence of impurities
the numbers of carriers can be different, and this is described
by the introduction of $\delta\mu$. With equal numbers of electrons and
holes $\delta\mu = 0$.

The Hamiltonian is taken to be

$$\begin{aligned}
\hat{H} - N_\mu &= \sum_{\alpha,\sigma} \int \psi_{\alpha\sigma}{}^+ \hat{H}_{0\alpha}\psi_{\alpha\sigma}\, d^3\vec{r} \\
&+ \tfrac{1}{2}V_B \sum_{\alpha\sigma} \int \psi_{\alpha\sigma}{}^+\psi_{\alpha-\sigma}{}^+\psi_{\alpha-\sigma}\psi_{\alpha\sigma}\, d^3r \\
&+ V \sum_{\sigma\sigma'} \int \psi_{1\sigma}{}^+\psi_{2\sigma'}{}^+\psi_{2\sigma'}\psi_{1\sigma}\, d^3\vec{r} \\
&+ \tfrac{1}{2}[V_0 \sum_\sigma \int \psi_{1\sigma}{}^+\psi_{1-\sigma}{}^+\psi_{2-\sigma}\psi_{2\sigma}\, d^3\vec{r} + \text{c.c.}].
\end{aligned}$$ (26)

The index $\alpha = 1,2$. The interaction terms have the following
origin: V_B is the intra-band interaction, and V the inter-band
interaction of the density-density type; both can have Coulombic
origin. The term V_0 corresponds to the transition of two electrons
from band 2 to band 1. This term can be the result of a hybridisation
of the terms 2 and 1 (for the real case of the vicinity of an
L-point in PbTe the corresponding matrix element is in fact finite) or arise as a second order effect from the electron-phonon
interaction

$$g \sum a_{1\sigma}{}^+ a_{2\sigma}(b_q + b_{-q}{}^+) + \text{c.c.}.$$ (27)

In this case $V_0 = -4g^2\hbar\omega 0$. We assume $V > 0$, $V_B > 0$; V_0 may have
either sign but must not be too large. The case $T = 0$ is considered.

It is convenient to take the z-axis as a reference quantisation axis for the spin and at the same time to suppose that the
spontaneous magnetisation appears in the x-direction. In this
case

$$M_x = \sum_{\alpha,\sigma} \langle \psi_{\sigma\alpha}{}^+\sigma_x\psi_{\sigma\alpha}\rangle =$$ (Contd)

(Contd) $= -i \sum_{\sigma,\alpha} \int \frac{d^3\vec{p}}{(2\pi)^3} \frac{d\omega}{2\pi} G_{-\sigma\sigma}{}^{\alpha\alpha}(p,\omega)e^{i\omega\tau}$,   (28)

$$\tau \to +0,$$

(in an isotropic medium $\vec{M}$ can have an arbitrary direction).
One sees that for the appearance of M_x the anomalous Green function

$$G_{\sigma,\sigma}{}^{\alpha\alpha} = -i\langle T\psi_{\alpha,-\sigma}(x)\psi_{\alpha,\sigma}{}^{+}(x')\rangle \qquad (29)$$

must exist. The finite value of such an anomalous average means the instability of the system and the reconstruction of its ground state.

Apart from this instability the Hamiltonian (26) also permits instabilities in other channels, which are described by the anomalous Green functions

$$G_{\sigma\sigma}{}^{21} = -i\langle T\psi_{2\sigma}(x)\psi_{1\sigma}{}^{+}(x')\rangle, \qquad (30)$$

$$G_{-\sigma\sigma}{}^{21} = -i\langle T\psi_{2-\sigma}(x)\psi_{1\sigma}{}^{+}(x')\rangle, \qquad (31)$$

The first corresponds to a singlet inter-band pairing and the second one to a triplet inter-band pairing. The functions (29, 30,31) will be presented graphically as

$$\underset{1}{\overset{-\sigma}{\bullet}}\!\!\longrightarrow\!\!\underset{1}{\overset{\sigma}{\bullet}} \qquad \underset{2}{\overset{\sigma}{\bullet}}\!\!\longrightarrow\!\!\underset{1}{\overset{\sigma}{\bullet}} \qquad \underset{2}{\overset{-\sigma}{\bullet}}\!\!\longrightarrow\!\!\underset{1}{\overset{\sigma}{\bullet}}$$

In the Bloch-Stoner model only the inter-band interaction was taken into account (the term V_B in the Hamiltonian) and owing to it a $G_{\sigma\sigma}{}^{11} \neq 0$ should appear. In a self-consistent field approximation applied to the Hamiltonian (26) with the V_B term only, one can obtain a condition similar to (1), i.e. the constant must be large enough. But in this case the self-consistent field approximation is not justified.

However, in the model under consideration, owing to the interactions V and V_0 in the singlet and triplet inter-band channels, logarithmic singularities appear, and therefore, even with arbitrarily small interaction constants, the quasi-averages (30,31) can exist. If both are finite, then a Bloch-Stoner anomalous function (29) is also obliged to appear. Indeed one can write a diagrammatic equation, following immediately below, which proves our statement.

So the function $G_{\sigma-\sigma}{}^{11}$ can exist even in the case $V_B = 0$. In general, the constants V and V_B are of the same order of magnitude. In the expressions corresponding to (32) the elements

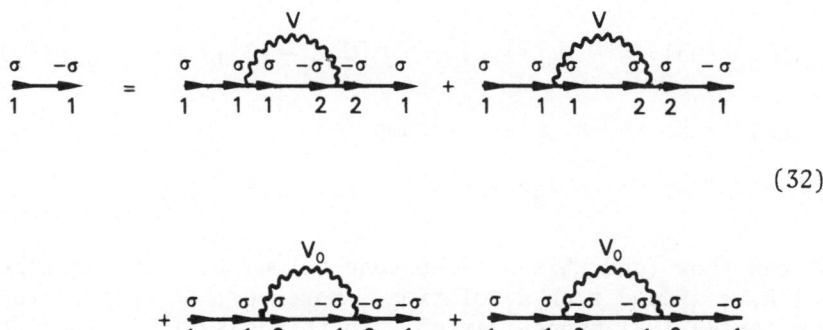

$$(32)$$

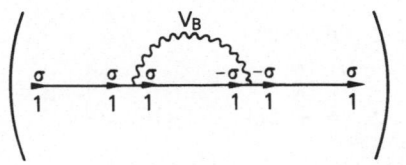 contain a logarithmic integral. However the corresponding term coming from the interaction does not contain an integral

$$\left(\begin{array}{c} \text{[diagram]} \end{array} \right).$$

Therefore if one assume the interaction constants to be small (e.g. owing to a large dielectric constant) V_B can be disregarded.

Equation (32) and corresponding equations for other G-functions can be written in the following matrix form.

$$
\begin{bmatrix}
\omega - \varepsilon_1 & 0 & \Delta_s & \Delta_t \\
0 & \omega - \varepsilon_1 & \tilde{\Delta}_t & \tilde{\Delta}_s \\
\Delta_s^* & \tilde{\Delta}_t^* & \omega - \varepsilon_2 & 0 \\
\Delta_t^* & \tilde{\Delta}_s^* & 0 & \omega - \varepsilon_2
\end{bmatrix}
\begin{bmatrix}
G_{\sigma\sigma}{}^{11} \\
G_{-\sigma\sigma}{}^{11} \\
G_{\sigma\sigma}{}^{21} \\
G_{-\sigma\sigma}{}^{21}
\end{bmatrix}
=
\begin{bmatrix}
1 \\
0 \\
0 \\
0
\end{bmatrix}, \quad (33)
$$

where

$$\Delta_s = - i \, [V G_{\sigma\sigma}{}^{12}(0) - V_0 G_{-\sigma-\sigma}{}^{21}(0)],$$

$$(34)$$

$$\Delta_t = - \ [V G_{\sigma-\sigma}{}^{12}(0) + V_0 G_{\sigma-\sigma}{}^{21}(0)].$$

The equation (33) is in fact two systems for $\sigma = (+)$ and $\sigma = (-)$, and $\tilde{\Delta}_s$, for example, means $\Delta_s((+) \not\gtrless (-))$.

From the definitions (30,31) it follows that

$$[G_{\sigma\sigma}12(0)]^* = - G_{\sigma\sigma}21(0), \qquad [G_{\sigma-\sigma}12(0)]^* = - G_{-\sigma\sigma}21(0).$$

We shall make the simplest assumption:

$$\Delta_s = \Delta_s^* = \tilde{\Delta}_s, \qquad \Delta_t = \Delta_t^* = \tilde{\Delta}_t.$$

One can show (and this will be done below) that the equations (33) have indeed such a solution. Moreover, it appears to produce the maximal magnetization. In this case it follows from (34) that V_0 must be real (V is seen to be real from (26)), and

$$G_{\sigma\sigma}12(0) = G_{\sigma\sigma}21(0) = G_{-\sigma-\sigma}12(0) = G_{-\sigma-\sigma}21(0),$$

$$G_{\sigma-\sigma}12(0) = G_{-\sigma\sigma}21(0) = G_{-\sigma\sigma}12(0) = G_{\sigma-\sigma}21(0).$$

The equations for Δ become:

$$\Delta_s = - i(V - V_0)G_{\sigma\sigma}21(0),$$

$$(35)$$

$$\Delta_t = - i(V + V_0)G_{-\sigma\sigma}21(0).$$

Equating the determinant of the system (33) to zero we get the energy spectrum

$$\omega = \tfrac{1}{2}(\varepsilon_1 + \varepsilon_2) \pm \{[\tfrac{1}{2}(\varepsilon_1 - \varepsilon_2)]^2 + (\Delta_s \pm \Delta_t)^2\}. \qquad (36)$$

Substituting the values of ε_1 and ε_2 (25) we obtain:

$$\omega = - \delta\mu \pm (\xi^2 + (\Delta_3 \pm \Delta_t)^2)^{\tfrac{1}{2}}. \qquad (37)$$

where $\xi = (p^2/2m^*) - \varepsilon_F$. From this formula it follows that the spectrum consists not of two, but of four branches, i.e. the spin degeneracy is lifted (figure 5). This fact leads to the existence of a favourable spin direction amongst the electrons, i.e. to a spontaneous magnetic moment. One can see also that if $\Delta_t = 0$ or $\Delta_s = 0$ this splitting disappears.

From the equations (33) we define $G_{\sigma\sigma}21$ and $G_{-\sigma\sigma}21$:

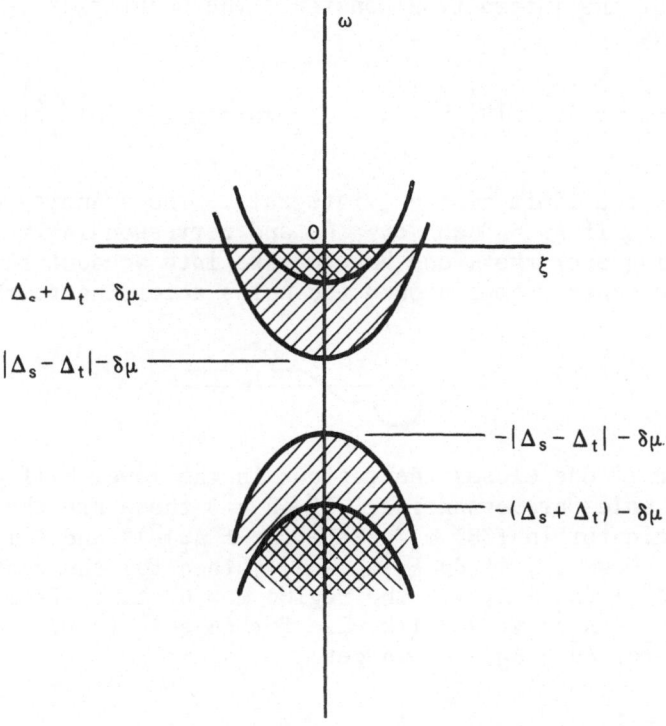

Figure 5

$G_{\sigma\sigma}{}^{21} =$

$$\frac{\Delta_s[\Delta_s{}^2 - \Delta_t{}^2 - (\omega - \varepsilon_1)(\omega - \varepsilon_2)]}{[(\omega + \delta\mu)^2 - \xi^2 - (\Delta_s + \Delta_t)^2][(\omega + \delta\mu)^2 - \xi^2 - (\Delta_s - \Delta_t)^2]} ,$$

$G_{-\sigma\sigma}{}^{21} =$

$$(38)$$

$$\frac{\Delta_t[\Delta_t{}^2 - \Delta_s{}^2 - (\omega - \varepsilon_1)(\omega - \varepsilon_2)]}{[(\omega + \delta\mu)^2 - \xi^2 - (\Delta_s + \Delta_t)^2][(\omega + \delta\mu)^2 - \xi^2 - (\Delta_s - \Delta_t)^2]} .$$

According to (35) Δ_s and Δ_t are expressed in terms of

$$G(0) = \int G(\omega, \vec{p})d\omega \, \frac{d^3\vec{p}}{(2\pi)^3} = \nu \int d\xi \, \frac{d\omega}{2\pi} \, G(\omega, \xi),$$

where $\nu = p_0 m^*/2\pi^2$ is the state density.

Inserting the formulae (38) into (35) it is convenient to use instead of the interaction constants the quantities $\Delta_s{}^0$ and $\Delta_t{}^0$ defined as

$$\frac{1}{2\nu(V - V_0)} = \ln\left(\frac{2\bar{\omega}}{\Delta_s{}^0}\right) , \qquad \frac{1}{2\nu(V + V_0)} = \ln\left(\frac{2\bar{\omega}}{\Delta_t{}^0}\right) . \qquad (39)$$

Here $\bar{\omega}$ is the limit of the ξ-integrals. The quantity $\Delta_s{}^0$ is the value of Δ_s if $\Delta_t = 0$ and $\delta\mu = 0$, and correspondingly for $\Delta_t{}^0$. Integrating over the ω one has to take into account that the integration contour goes around the poles according to the rule

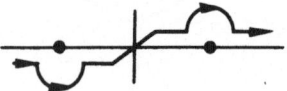

Therefore if one closes the contour in the lower half plane the positive poles are essential. If $\delta\mu = 0$ these are the roots of the denominator in (38) $\omega = (\xi^2 + (\Delta_s \pm \Delta_t)^2)^{\frac{1}{2}}$ and ξ may be arbitrary. However, if $\delta\mu > |\Delta_s - \Delta_t|$, then for the root $\omega = -\delta\mu + (\xi^2 + (\Delta_s - \Delta_t)^2)^{\frac{1}{2}}$ the region $\omega > 0$ starts from $\xi = (\delta\mu^2 - (\Delta_s - \Delta_t)^2)^{\frac{1}{2}}$, and likewise for $\delta\mu > \Delta_s + \Delta_t$.

Hence for $\delta\mu > \Delta_s + \Delta_t$ we get

$$\ln\left(\frac{\bar{\omega}}{\Delta_s{}^0}\right) = \frac{1}{2}\left(1 + \frac{\Delta_t}{\Delta_s}\right)\ln\left(\frac{\bar{\omega}}{(\delta\mu^2 - (\Delta_s + \Delta_t)^2)^{\frac{1}{2}} + \delta\mu}\right)$$

$$+ \frac{1}{2}\left(1 - \frac{\Delta_t}{\Delta_s}\right)\ln\left(\frac{\bar{\omega}}{(\delta\mu^2 - (\Delta_s - \Delta_t)^2)^{\frac{1}{2}} + \delta\mu}\right)$$

$$\ln\left(\frac{\bar{\omega}}{\Delta_t{}^0}\right) = \frac{1}{2}\left(1 + \frac{\Delta_s}{\Delta_t}\right)\ln\left(\frac{\bar{\omega}}{(\delta\mu^2 - (\Delta_s + \Delta_t)^2)^{\frac{1}{2}} + \delta\mu}\right) \qquad (40)$$

$$+ \frac{1}{2}\left(1 - \frac{\Delta_s}{\Delta_t}\right)\ln\left(\frac{\bar{\omega}}{(\delta\mu^2 - (\Delta_s - \Delta_t)^2)^{\frac{1}{2}} + \delta\mu}\right)$$

If $\delta\mu < (\Delta_s + \Delta_t)$ or $\delta\mu < |\Delta_s - \Delta_t|$ then in the corresponding terms the logs have the form $\ln(\bar{\omega}/|\Delta_s \pm \Delta_t|)$.

One must add also the charge conservation condition which defines $\delta\mu$. If the impurities supplied N electrons to the conduction band, then

$$N = - i \sum_{\alpha\sigma} \int [G_{\sigma\sigma}{}^{\alpha\alpha}(\omega,\vec{p},\mu) - G_{\sigma\sigma}{}^{\alpha\alpha}(\omega,\vec{p},0)]e^{i\omega\tau}\frac{d^3\vec{p}d\omega}{(2\pi)^4} . \qquad (41)$$

$$\tau \rightarrow +0$$

But one can act more simply, knowing the energy spectrum (37). If $\delta\mu > |\Delta_s - \Delta_t|$ then the electrons appear in one of the upper bands, and if $\delta\mu > \Delta_s + \Delta_t$, then they appear in both upper bands (figure 5). The corresponding Fermi moments are equal to

$$p_F - p_0 = \frac{1}{v} ((\delta\mu)^2 - (\Delta_s \pm \Delta_t)^2)^{\frac{1}{2}}.$$

Hence for the total excess number of electrons we can write:

$$N = 4\nu n,$$

$$n = \frac{1}{2}\{((\delta\mu)^2 - (\Delta_s + \Delta_t)^2)^{\frac{1}{2}}\theta[\delta\mu - (\Delta_s + \Delta_t)]$$

$$+ ((\delta\mu)^2 - (\Delta_s - \Delta_t)^2)^{\frac{1}{2}}\theta[\delta\mu - |\Delta_s - \Delta_t|]\}. \tag{42}$$

Now consider the equations (40). As already mentioned, the existence of the spontaneous magnetization demands simultaneously $\Delta_s \neq 0$, $\Delta_t \neq 0$. Therefore the boundaries of the region can be defined from the conditions $\Delta_s \to 0$ or $\Delta_t \to 0$. Consider first the case, $\delta\mu = 0$. Let us search the boundary $\Delta_s = 0$. The equation (40b) gives $\Delta_t = \Delta_t^0$. In the equation (40a) one has to develop the right hand side by $z = \Delta_s/\Delta_t$. The result is ln $(\Delta_s^0/\Delta_t^0) = 1$. So in this case

$$\Delta_t = \Delta_t^0 = \frac{\Delta_s^0}{e}. \tag{43}$$

One gets a similar expression for the other boundary. The region of the coexistence of both Δ's is presented in figure 6.

The strange feature of this graph is the fact that at the boundary $\Delta = 0$, $\Delta_s^0 > \Delta_t^0$, i.e. the singlet interaction constant,

$$V_s = V - V_0,$$

is larger than the triplet constant

$$V_t = V + V_0.$$

This already raises a suspicion that the result obtained is unphysical. Such a conclusion is confirmed by the free energy calculation.

From the Hamiltonian (26) with $V_B = 0$ we get

$$\frac{\partial\langle \hat{H} - N_\mu \rangle}{\partial V} = \sum_\sigma [G_{\sigma\sigma}^{12}(0)G_{\sigma\sigma}^{21}(0) + G_{\sigma-\sigma}^{12}(0)G_{-\sigma\sigma}^{21}(0)],$$

(Contd)

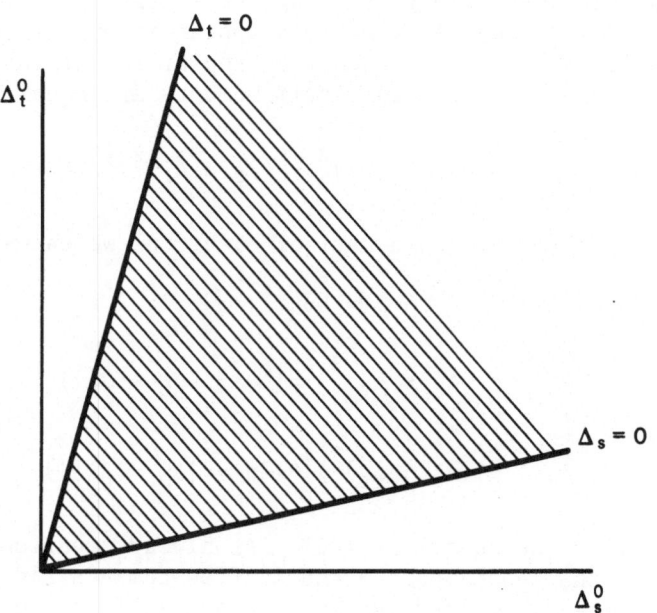

Figure 6

(Contd)

$$\frac{\partial \langle \hat{H} - N_\mu \rangle}{\partial V_0} = - G_{--}{}^{21}G_{++}{}^{21} - G_{--}{}^{12}G_{++}{}^{12}$$

$$+ G_{-+}{}^{21}G_{+-}{}^{21} + G_{-+}{}^{12}G_{+-}{}^{12}.$$

Introducing V_s and V_t we get

$$\frac{\partial \langle \hat{H} - N_\mu \rangle}{\partial V_t} = - 2 \frac{\Delta_s{}^2}{V_s{}^2} \; ; \qquad \frac{\partial \langle \hat{H} - N_\mu \rangle}{\partial V_t} = - 2 \frac{\Delta_t{}^2}{V_t{}^2} \; .$$

Accordingly

$$\delta\Omega = \Omega_{\text{ferr}} - \Omega_{\text{norm}} = - 2 \int_C \left[\frac{\Delta_s{}^2}{V_s{}^2} \, dV + \frac{\Delta_t{}^2}{V_t{}^2} \, dV_t \right] . \qquad (44)$$

The integration must be taken along a contour starting from the point $\Delta_s{}^0 = 0$, $\Delta_t{}^0 = 0$ and lying entirely within the region of coexistence of Δ_s and Δ_t.

The equations (40) with $\delta\mu = 0$ can be transformed to the form

$$\ln(z_0) = \left[\frac{1 - z^2}{2z}\right]\ln\left[\frac{1 + z}{|1 - z|}\right] \, , \tag{45}$$

$$\ln\left(\frac{\Delta_t{}^0}{\Delta_t}\right) = \ln|1 - z^2| + z\ln\left[\frac{1 + z}{|1 - z|}\right] \, , \tag{46}$$

where $z = \Delta_s/\Delta_t$, $z = \Delta_s{}^0/\Delta_t{}^0$. Let us take the integral along a contour z = constant. From (45) it follows that in this case z_0 = constant. By definition of the Δ^0's we have

$$\ln\left[\frac{2\bar{\omega}}{\Delta_s{}^0}\right] = \frac{1}{2\nu V_s} \, ; \qquad \ln\left(\frac{2\bar{\omega}}{\Delta_t{}^0}\right) = \ln\left(\frac{2\bar{\omega}z_0}{\Delta_s{}^0}\right) = \frac{1}{2\nu V_t} \, .$$

Hence

$$\frac{d\Delta_s{}^0}{\Delta_s{}^0} = \frac{dV_s}{2\nu V_t{}^2} = \frac{dV_t}{2\nu V_t{}^2} \, .$$

But according to (46) $\ln(\Delta_t{}^0/\Delta_t) = \ln(\Delta_s{}^0/z_0\Delta_t)$ = constant and therefore

$$\frac{d\Delta_s{}^0}{\Delta_s{}^0} = \frac{d\Delta_t}{\Delta_t} \, .$$

It follows that

$$\delta\Omega = -4\nu\int(\Delta_s{}^2 + \Delta_t{}^2)\frac{d\Delta_t}{\Delta_t}$$

$$= -4\nu(1 + z^2)\int\Delta_t d\Delta_t \tag{47}$$

$$= -2\nu(1 + z^2)\Delta_t{}^2 = -2\nu(\Delta_s{}^2 + \Delta_t{}^2) \, .$$

If $\Delta_s = 0$, then $\delta\Omega = \delta\Omega_t = -2\nu(\Delta_t{}^0)$, and similarly for $\Delta_t = 0$. From (47), by means of (45) and (46) one can obtain:

$$\delta\Omega = f(z^{-1})\delta\Omega_s = f(z)\delta\Omega_t,$$

where

$$\ln f(z) = \ln\left(\frac{1 + z^2}{|1 - z^2|}\right) - z\ln\left(\frac{1 + z}{|1 - z|}\right) \, . \tag{48}$$

The function $f(z)$ is monotonically decreasing with z from $f(0) = 1$ to $f(\infty) = 1/e^2$. Hence the free energy with $\Delta_t \neq 0$, $\Delta_s \neq 0$ is higher than in the case where, e.g., only $\Delta_s \neq 0$ and $\Delta_t = 0$. Therefore the ferromagnetic state with $\delta\mu = 0$ cannot exist. The reason for this is clear. Until $\delta\mu = 0$ the upper bands on figure 5 are empty and the lower ones are filled. Therefore the substance is non-magnetic all the same (the same number of electrons with both directions of the spin) and there are no reasons why the bands should split.

Now let $\delta\mu \neq 0$. Let us find the boundary $\Delta_s = 0$ from equations (40). From (40b) follows

$$(\delta\mu^2 - \Delta_t^2)^{\frac{1}{2}} + \delta\mu = \Delta_t^0.$$

Using the charge conservation condition (42) according to which, in the case $\Delta_s = 0$,

$$h = (\delta\mu^2 - \Delta_t^2)^{\frac{1}{2}} \tag{49}$$

we get

$$\Delta_t = (\Delta_t^0(\Delta_t^0 - 2n))^{\frac{1}{2}}. \tag{50}$$

Developing the right hand side of equation (40a) in terms of z at $z \to 0$, and using the condition (49), we find:

$$\ln\left(\frac{\bar{\omega}}{\Delta_s^0}\right) = \ln\left(\frac{\omega}{n + (n^2 + \Delta_t^2)^{\frac{1}{2}}}\right) + \frac{\Delta_t^2}{n(n + (n^2 + \Delta_t^2)^{\frac{1}{2}})} \, .$$

But according to the preceding formulas $n + (n^2 + \Delta_t^2)^{\frac{1}{2}} = \Delta_t^0$, and hence[†]

$$\ln\left(\frac{\Delta_s^0}{\Delta_t^0}\right) = -\frac{\Delta_t^2}{n\Delta_t^0} = -\frac{\Delta_t^0 - 2n}{n} \, . \tag{51}$$

The boundary curves in the Δ_s^0-Δ_t^0 plane acquire the form on figure 7. Now everything is alright. The decrease of Δ_s^0 results in a decrease of Δ_s.

The calculation of the free energy was done for the bisector $\Delta_s^0 = \Delta_t^0$, and it produced a value smaller than in the case of

---

[†] One can note that the formula (51) with $n = 0$ does not coincide with the formula for the case $\delta\mu = 0$. This is owing to the fact that at $T = 0$ even an infinitesimal amount of electrons in the conduction band makes $\delta\mu$ jump to the bottom of the band, i.e. in our case to $\delta\mu = \Delta_t^0$.

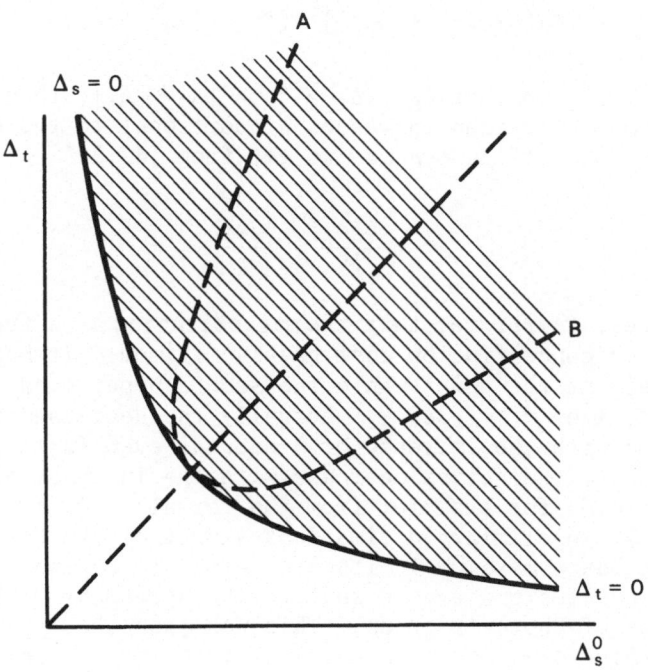

$\Delta_s = 0$

Δ_t

A

B

$\Delta_t = 0$

Δ_s^0

Figure 7

only one of the Δ's being finite. It is not excluded, however,
that if we move from the bisector to one of the boundaries at
figure 7 the free energy changes non-monotonically. This would
mean the existence of a line of first order phase transitions.
Such possible lines are denoted on figure 7 as A and B. In this
case the limits $\Delta_s = 0$ and $\Delta_t = 0$ are the absolute limits of
stability of the ferromagnetic phase. This question is now under investigation.

Let us, finally, find the spontaneous magnetization. Accord-
ing to (28) we must find $G_{-\sigma\sigma}11$ and $G_{-\sigma\sigma}22$. $G_{-\sigma\sigma}11$ can be obtained from (33) and $G_{-\sigma\sigma}22$ by means of $1 \gtrless 2$. After that we can
find the magnetization from (28). However, in fact one can act
more easily, namely in the same way that we obtained the charge
conservation condition. The magnetization is defined by the excess of electrons in one of the upper bands over the other[†].
Hence we get

$$U = 4\nu m,$$

$$m = \tfrac{1}{2}\{[\delta\mu^2 - (\Delta_s - \Delta_t)^2]^{\tfrac{1}{2}}\theta(\delta\mu - |\Delta_s - \Delta_t|) - \qquad (52)$$

(Contd)

† This reasoning does not permit the definition of the sign of
M. The G-function calculation shows that if Δ_s and $\Delta_t > 0$, then
$M < 0$. Hence, the band starting at $\Delta_s + \Delta_t$ corresponds to the
spin (+).

(Contd) $- [\delta\mu^2 - (\Delta_s + \Delta_t)^2]^{\frac{1}{2}}\theta(\delta\mu - (\Delta_s + \Delta_t)).$ (52)

If n is small, so that $\Delta_s + \Delta_t > \delta\mu > |\Delta_s - \Delta_t|$, then $m = n$, i.e. the magnetization increases proportional to the carrier concentration. If $\delta\mu \gg \Delta_s, \Delta_t$ we get from (42,52)

$$m = \frac{\Delta_s \Delta_t}{\delta\mu} = \frac{\Delta_s \Delta_t}{n}.$$

One sees, therefore, that the magnetization as a function of the carrier concentration must have a maximum. It happens approximately at the point where the second upper band starts to be filled, i.e. $n \sim (\Delta_s \Delta_t)^{\frac{1}{2}}$. Actually the decrease of the magnetization with the increase of n happens even faster, since the Δ_s and Δ_t themselves decrease with the increase of n. This can be traced, for example, from the formula (50) for Δ_t at $\Delta_s = 0$. The preliminary estimations show that the Curie temperature also has a maximum in its concentration dependence.

These conclusions are in qualitative agreement with the observations of Shotov's group. This was just the type of behaviour of $\rho(H) - \rho(-H)$ as a function of impurity concentration. Contrary to the 'resonance ferromagnetism', there are no reasons in this case for the concentration interval to be very narrow and for the Curie temperature to be very small. As we have already said, the effect was observed in fact with $N = 10^{17}-10^{19}$ cm^{-3} and T up to room temperatures.

The influence of the 'band ferromagnetism' on the quantum oscillations is essentially the same as that of the 'resonance ferromagnetism', i.e. one expects beats and the 'commutation effect', which were in fact observed. In the model considered the picture should be a little bit more complicated than in the InSb case. The carriers are concentrated close to p_0, i.e. they occupy one spherical layer (if $\Delta_t + \Delta_s > \delta\mu > |\Delta_s - \Delta_t|$) or two such layers (if $\delta\mu > \Delta_t + \Delta_s$). In the first case there are two close periods and in the second case there are four. In the two-period case the corresponding Fermi moments are

$$p_F = p_0 \pm \frac{1}{\nu} [\delta\mu^2 - (\Delta_s - \Delta_t)^2]^{\frac{1}{2}}.$$

and the difference of the areas of the central sections of the Fermi surfaces is

$$S_1 - S_2 = 4\pi m^* [\delta\mu^2 - (\Delta_s - \Delta_t)^2]^{\frac{1}{2}}.$$ (53)

Theoretical studies of 'band ferromagnetism' are being now continued. The details of the phase transitions are defined and the behaviour at finite temperature is being studied.

The authors also have in mind to study more realistic models which could account for the properties of real substances which are investigated experimentally. The same also applies to the theory of 'resonance ferromagnetism'.

REFERENCES

1. Andrianov, D.G., Brandt, N.B., Yon, E.P., Fistul', V.I. and Cudinov, C.M. (1973). *Pism. Zh. Eksp. Teor. Fiz.*, **17**, 9; [*Sov. Phys. JETP Lett.*, **17**, 356].

2. Whiteself, C.R. (1965). *Phys. Rev.*, **138**, 829.

3. Seiler, D.G. and Becker, N.M. (1967). *Phys. Lett.*, **A26**, 96.

4. Roth, L.M., Groves, S.M. and Wyatt, P.W. (1967). *Phys. Rev. Lett.*, **19**, 576.

5. Abrikosov, A.A. (1973). *Zh. Eksp. Teor. Fiz.*, **65**, 814; [(1974). *Sov. Phys. JETP*, **38**, 403].

6. Volkov, B.A. and Kopayev, Yu.V. (1974). *Pism. Zh. Eksp. Teor. Fiz.*, **19**, 168; [*Sov. Phys. JETP Lett.*, **19**, 104].

7. Volkov, B.A. and Kopayev, Yu.V. (1973). *Zh. Eksp. Teor. Fiz.*, **64**, 2184; [*Sov. Phys. JETP*, **37**, 1103].

8. Bloch, F. (1929). *Z. Phys.*, **57**, 545.

9. Stoner, E.C. (1938). *Proc. Roy. Soc.*, **A165**, 327.

PHONON HYDRODYNAMICS IN SOLIDS

A. THELLUNG

*Institut für Theoretische Physik,
Universität, Zurich, Switzerland*

1. INTRODUCTION

In this lecture we shall deal with hydrodynamic motion of phonons in solids, i.e. with situations where the phonon gas can flow or carry out density oscillations like an ordinary fluid.

What has this to do with magnetism? Nothing directly, but the methods applied here to phonons can also be used for magnons, and, in a very similar manner, hydrodynamic equations for the magnon gas can be derived [1,2].

Hydrodynamic motion of phonons may occur at low temperatures provided the two following conditions are fulfilled:

(a) One deals with a very pure, 'perfect', dielectric crystal. Scattering of phonons by impurities and other lattice imperfections can be neglected;

(b) The temperature is so low that Umklapp processes (U-processes) can be neglected, but it is still high enough so that many normal processes (N-processes) occur in order to establish local thermal equilibrium.

Since the mean free path for U-processes, ℓ_u, and the mean free path for N-processes, ℓ_N, have the following temperature dependences [3],

$$\ell_u \propto e^{\frac{1}{2}\theta/T}, \qquad \ell_N \propto T^{-5}, \tag{1}$$

(θ = Debye temperature) one has

$$\ell_N \ll \ell_u \tag{2}$$

at sufficiently low temperatures. Then the phonon gas can acquire a macroscopic drift velocity and hydrodynamic motion of the phonons is possible. The two main examples are *Poiseuille flow* of the phonon gas in a dielectric cylinder to whose ends a temperature difference is applied, and *second sound*, a kind of almost undamped temperature wave due to oscillations of the phonon density. If, in the case of Poiseuille flow, the diameter of the cylinder is denoted by d, and in the case of second sound the wave length of the temperature wave by λ, the condition for these phenomena to occur is

$$\ell_N \ll d, \lambda \ll \ell_\mu. \tag{3}$$

According to (1,2) this can be fulfilled if one has a large enough sample (of dimensions $\sim d, \lambda$) of a 'perfect' crystal.

Hydrodynamic equations for the phonon gas in solids were first derived theoretically, and the two effects, Poiseuille flow and second sound, predicted [4,5]. Later, Poiseuille flow and second sound were detected experimentally, first in solid ^{4}He [6,7], but afterwards also in ^{3}He, NaF and Bi.

2. PHONONS IN EQUILIBRIUM

Let us start by summarizing a few facts about phonons. We characterize a phonon by a wave vector k and a polarization index s; its circular frequency is denoted by $\omega(k,s)$. Instead, it is also customary to use the quantities

$$\mathbf{p} = \hbar \mathbf{k}, \qquad s; \qquad \varepsilon(\mathbf{p},s) = \hbar\omega(\mathbf{k},s). \tag{4}$$

p is called the momentum or quasi-momentum. No dynamical signification is attached to it; its only physical meaning comes from its proportionality to the wave vector. $\varepsilon(\mathbf{p},s)$ is the energy of the phonon. At low temperatures, when only acoustic phonons of low energies are excited, it is sufficient to take a linear dispersion law

$$\omega(\mathbf{k},s) = c_s k, \qquad \varepsilon(\mathbf{p},s) = c_s p, \tag{5}$$

where c_s is the velocity of sound of polarization s. In general, c_s also depends on the direction of k. However, we shall limit ourselves to the case where c_s only depends on s, i.e. to a Debye model with three branches.

Owing to the anharmonic terms in the potential energy of the lattice, the phonons interact with each other. This leads to collision processes (scattering, annihilation and creation of phonons). In those processes one has conservation of energy (the primed symbols refer to the quantities after the collision),

$$\sum \varepsilon = \sum \varepsilon', \tag{6}$$

and 'quasi-conservation' of quasi-momentum,

$$\sum p = \sum p' + \hbar K, \tag{7}$$

where K is a vector of the reciprocal lattice. If K = 0 one
speaks of N-*processes*; the momentum is conserved. If K ≠ 0 one
has U-*processes*; the momentum is not conserved. As can be seen
from (1), the U-processes die out exponentially at low temperatures and can be neglected if (3) is fulfilled. If there are *no
other momentum destroying processes* (like scattering by impurities or other lattice imperfections), one has *two* integrals of
the motion, the total energy and the total momentum of the phonon
gas,

$$E = \sum_{p,s} n(p,s)\varepsilon(p,s) = \text{constant}, \tag{8}$$

$$P = \sum_{p,s} n(p,s)p = \text{constant}. \tag{9}$$

Here $n(p,s)$ is the number of phonons (p,s). As a consequence
the mean occupation numbers in thermal equilibrium are

$$\bar{n}(p,s) = \frac{1}{e^{\beta[\varepsilon(p,s)-u\cdot p]} - 1} . \tag{10}$$

(This follows by standard methods from the general canonical
distribution for this case,

$$\rho = Ce^{-\beta(E-u\cdot P)}). \tag{11}$$

u is a constant vector, meaning the *drift velocity* of the *phonon
gas*. For u ≠ 0 the phonon distribution is anisotropic, the total momentum P and the energy current do not vanish. Actually,
in *full thermal equilibrium* u ≠ 0 is only possible for infinite
perfect crystals or for perfect crystals with periodic boundary
conditions. For an actual finite crystal the phonon momentum is
not conserved at the boundaries and u must vanish in full thermal equilibrium. But in *non-equilibrium situations* the distribution (10) is approximately valid. However, u and β are now
space- and time-dependent, corresponding to *local thermal equilibrium*, and u vanishes at the boundaries. The situation is the
same as for an ordinary gas: at the boundaries the drift velocity
u = 0, but several mean free paths away (where the particles do
not feel the presence of the surface) u ≠ 0 is possible.

In the following we shall deal with such non-equilibrium situations.

3. THE PEIERLS EQUATION. HYDRODYNAMIC EQUATIONS

In order to describe non-equilibrium phenomena, we start from the Peierls equation, which is the analogue for phonons of the Boltzmann equation for a gas of molecules. It reads

$$\frac{\partial F}{\partial t} + \mathbf{v} \cdot \frac{\partial F}{\partial \mathbf{r}} = \Delta_{coll}[F].$$
(12)

$F(\mathbf{p},s,\mathbf{r},t)$ is the phonon distribution function such that the number of phonon wave packets of mean momentum in d^3p around $\mathbf{p}$ and of position in d^3r around $\mathbf{r}$, with polarization s, at time t, is given by $h^{-3}F(\mathbf{p},s,\mathbf{r},t)d^3pd^3r$; $\mathbf{v}$ is the group velocity

$$\mathbf{v}(\mathbf{p},s) = \frac{\partial \varepsilon(\mathbf{p},s)}{\partial \mathbf{p}} = c_s \frac{\mathbf{p}}{p},$$
(13)

where (5) has been used. $\Delta_{coll}[F]$ is the change of F per unit time owing to collisions. No force term $\dot{\mathbf{p}}\partial F/\partial \mathbf{p}$ has been included on the left hand side of (12). $\dot{\mathbf{p}} \neq 0$ might be caused by density variations on the lattice. Such a force term leads to a coupling between the motion of the phonon gas and elastic deformations of the crystal. As a result one obtains two-fluid equations [8]. However, this coupling has a small influence on the effects we are interested in. (The same is true in liquid He II at low temperatures as soon as ρ_n, the normal density, is small compared to ρ_s, the superfluid density). Derivations of the Peierls equation by means of various (equilibrium and non-equilibrium) Green's function techniques have been given by many authors. It is also possible to obtain hydrodynamic equations directly by means of such techniques (including linear response theory and correlation function methods). A list of references can be found in [9].

We now assume that in the collisions not only the energy but also the momentum of the phonons is conserved. If we multiply (12) by the conserved quantities $\varepsilon(\mathbf{p},s)$ or p_i, and sum over the momenta and polarizations, the right hand side vanishes and the following macroscopic conservation laws are obtained

$$\frac{\partial E}{\partial t} + \frac{\partial q_i}{\partial r_i} = 0,$$
(14)

$$\frac{\partial P_i}{\partial t} + \frac{\partial P_{ij}}{\partial r_j} = 0,$$
(15)

where

$$
\left\{
\begin{array}{c}
E(\mathbf{r},t) \\[2mm]
q_i(\mathbf{r},t) \\[2mm]
P_i(\mathbf{r},t) \\[2mm]
P_{ij}(\mathbf{r},t)
\end{array}
\right\}
= \sum_s \frac{1}{h^3}\!\int\! d^3 p\, F(\mathbf{p},s,\mathbf{r},t) \cdot
\left\{
\begin{array}{c}
\varepsilon \\[2mm]
\varepsilon v_i \\[2mm]
p_i \\[2mm]
p_i v_j
\end{array}
\right\}
\qquad (16)
$$

are the densities of energy, energy current, momentum and momentum flux. In contrast to the case of molecules, one has no
conservation of particle number. (14,15) constitute *hydrodynamic equations* provided that the densities (16) can be expressed
in terms of the hydrodynamic variables $\beta(\mathbf{r},t)$ and $\mathbf{u}(\mathbf{r},t)$. In
order to achieve this, one has to solve the Peierls equation,
which is a difficult task. In the next section we shall use a
simple approximation, the mean free time theory, which, however,
gives a lot of insight into the physical situation. A more sophisticated method of solving the Peierls equation will be given
in section 7.

4. MEAN FREE TIME APPROXIMATION

The distribution function corresponding to local equilibrium
would be (see (10))

$$
F_{LE}(\mathbf{p},s,\mathbf{r},t) = \frac{1}{e^{\beta(\mathbf{r},t)[\varepsilon(\mathbf{p},s)-\mathbf{u}(\mathbf{r},t)\cdot\mathbf{p}]} - 1} \,. \qquad (17)
$$

In the mean free time theory one assumes F to be given by

$$
F(\mathbf{p},s,\mathbf{r},t) \approx F_{LE}(\mathbf{p},s,\mathbf{r} - \mathbf{v}\tau,t - \tau)
$$

$$
\approx F_{LE}(\mathbf{p},s,\mathbf{r},t) - \tau\left(\frac{\partial}{\partial t} + \mathbf{v}\,\frac{\partial}{\partial \mathbf{r}}\right)F_{LE}(\mathbf{p},s,\mathbf{r},t). \qquad (18)
$$

$\tau(\mathbf{p},s)$ is the mean free time for a phonon $(\mathbf{p},s)$, i.e. the mean
time between two collisions. The physical assumption underly-
ing (18) is that a local equilibrium has been established at the
place and time where the phonon suffered its last collision.
(18) is expected to be a good approximation if the time variation of F is slow over a mean free time τ and, similarly, the
spatial variation of F is small over a mean free path τv. We
linearize F_{LE} in (18) with respect to $\mathbf{u}(\mathbf{r},t)$ and the deviation
of $\beta(r,t)$ from its equilibrium value β_0, i.e.,

$$
F_{LE}(\mathbf{p},s,\mathbf{r},t) \approx N(\varepsilon(\mathbf{p},s)) \qquad (19)
$$

$$
+ \frac{\partial N(\varepsilon(\mathbf{p},s))}{\partial \varepsilon(\mathbf{p},s)}\,\beta_0\left[\varepsilon(\mathbf{p},s)\,\frac{\beta(\mathbf{r},t) - \beta_0}{\beta_0} - \mathbf{p}\cdot\mathbf{u}(\mathbf{r},t)\right],
$$

where

$$N(\varepsilon) = \frac{1}{e^{\beta_0 \varepsilon} - 1} \ . \tag{20}$$

We take the spectrum (5) and assume an isotropic mean free time $\tau(|\mathbf{k}|,s)$. Then the densities (16) can be evaluated and the hydrodynamic equations (14,15) for $\beta(\mathbf{r},t)$ and $u(\mathbf{r},t)$ are obtained. The terms $\propto \tau$ are considered to be small corrections. If, for a moment, we put them equal to zero, (14,15) become

$$\frac{\partial}{\partial t} \frac{\beta}{\beta_0} = \frac{1}{3} \, \boldsymbol{\nabla} \cdot \mathbf{u}, \tag{21}$$

$$\frac{\partial}{\partial t} \, \mathbf{u} = 3c_{II}^{2} \, \boldsymbol{\nabla} \, \frac{\beta}{\beta_0} \ , \tag{22}$$

where

$$c_{II}^{2} = \frac{1}{3} \frac{\sum\limits_{s} c_{s}^{-3}}{\sum\limits_{s} c_{s}^{-5}} \ . \tag{23}$$

(21,22) constitute the hydrodynamic equations of lowest order in τ and can be used to eliminate the time derivatives of β and $\mathbf{u}$ occurring in the terms $\propto \tau$ of the densities (16), which then read

$$E = \varepsilon_0 - 4\varepsilon_0 \, \frac{\beta - \beta_0}{\beta_0} \ ,$$

$$q_i = \frac{4}{3} \, \varepsilon_0 u_i + \frac{1}{3} \, \varepsilon_0 \, \frac{\nabla_i \beta}{\beta_0} \sum\limits_{s} (c_s^{2} - 3c_{II}^{2}) \sigma_s,$$

$$\tag{24}$$

$$P_i = \frac{4}{9c_{II}^{2}} \, \varepsilon_0 u_i + \frac{1}{3} \, \varepsilon_0 \, \frac{\nabla_i \beta}{\beta_0} \sum\limits_{s} \left(1 - \frac{3c_{II}^{2}}{c_s^{2}}\right) \sigma_s,$$

$$P_{ij} = \frac{1}{3} \, \delta_{ij} E + \frac{1}{15} \, \varepsilon_0 \left[\frac{2}{3} \, \delta_{ij} \, \boldsymbol{\nabla} \cdot \mathbf{u} - \nabla_i u_j - \nabla_j u_i\right] \sum\limits_{s} \sigma_s,$$

Here

$$\varepsilon_0 = \frac{\pi^2}{30 \beta_0^{4} \hbar^3} \sum\limits_{s} c_s^{-3} \tag{25}$$

is the thermal energy density in full equilibrium and

$$\sigma_s = c_s^2 \frac{1}{\varepsilon_0 \hbar 3} \int d^3p \; p^2 \left(- \frac{\partial N(\varepsilon(\mathbf{p},s))}{\partial \varepsilon(\mathbf{p},s)} \right) \tau(|\mathbf{p}|,s). \tag{26}$$

The conservation laws (14,15) then yield

$$\frac{\dot{\beta}}{\beta_0} = \frac{1}{3} \, \mathbf{\nabla} \cdot \mathbf{u} + \frac{1}{3} \frac{\nabla^2 \beta}{4\beta_0} \sum_s \sigma_s(c_s^2 - 3c_{II}^2), \tag{27}$$

$$\dot{u}_i = 3c_{II}^2 \frac{\nabla_i \beta}{\beta_0} - 3c_{II}^2 \frac{\nabla_i \beta}{4\beta_0} \sum_s \sigma_s \left(1 - \frac{3c_{II}^2}{c_s^2} \right)$$

$$+ \frac{3c_{II}^2}{20} \left(\frac{1}{3} \nabla_i \mathbf{\nabla} \cdot \mathbf{u} + \nabla^2 u_i \right) \sum_s \sigma_s. \tag{28}$$

(27,28) are the hydrodynamic equations including terms up to first order in τ.

5. SECOND SOUND

If in (27,28) we again neglect the terms $\propto \tau$, that is, if we take (21,22), we can eliminate u and obtain

$$\left(\frac{\partial^2}{\partial t^2} + c_{II}^2 \nabla^2 \right) \beta(\mathbf{r},t) = 0. \tag{29}$$

(29) means that undamped temperature waves with phase velocity c_{II} can propagate. Such waves are called second sound. The terms $\propto \tau$ in (27,28) lead to dissipative effects, i.e. to damping of second sound. We shall not pursue this matter any further, since in section 7 we shall calculate the damping of second sound by means of more rigorous methods, without a mean free time approximation.

6. POISEUILLE FLOW

In this section we consider stationary heat flow in a long cylinder of radius R to whose ends a temperature difference is applied. With vanishing time derivatives, (27,28) show that $\mathbf{\nabla} \cdot \mathbf{u}$, being $\propto \tau^2$, has to be set equal to zero,

$$\mathbf{\nabla} \cdot \mathbf{u} = 0, \tag{30}$$

whilst up to terms linear in τ,

$$\frac{\nabla_i \beta}{\beta_0} = - \frac{\nabla^2 u_i}{20} \sum_s \sigma_s. \tag{31}$$

We assume $\nabla\beta$ and u to have only axial components, i.e. in the z-direction. A solution of (30,31) is then given by

$$\nabla_z \beta = \text{constant}, \tag{32}$$

$$u_z(r) = - \frac{1}{\tau_p} (r^2 - R^2) \frac{\nabla_z \beta}{\beta_0} , \tag{33}$$

where $r = (x^2 + y^2)^{\frac{1}{2}}$ and where the boundary condition $u_z(r = R) = 0$ has been used. The value of the relaxation time τ_p is given by

$$\tau_p = \frac{1}{5} \sum_s \sigma_s. \tag{34}$$

The drift velocity (33) has a parabolic shape as a function of r; it corresponds to Poiseuille flow [4,5].

Substituting (32,33) into the expression for the energy current density q_i (24), we obtain,

$$q_z(r) = - \frac{4}{3} \frac{\varepsilon_0}{\tau_p} (r^2 - R^2) \frac{\nabla_z \beta}{\beta_0}$$

$$+ \frac{1}{3} \varepsilon_0 \frac{\nabla_z \beta}{\beta_0} \sum_s (c_s^2 - 3c_{II}^2)\sigma_s. \tag{35}$$

The second term is of relative order τ^2 compared to the first one and should be dropped, in order to be consistent with the fact that only corrections up to first order in τ have been considered. The total heat flow Q is obtained by integrating (35) over the cross section of the cylinder:

$$Q = \frac{2\pi}{3} \frac{\varepsilon_0}{\tau_p} R^4 \frac{\nabla_z \beta}{\beta_0} . \tag{36}$$

We *formally* define a thermal conductivity κ by

$$\bar{q}_z = \frac{Q}{\pi R^2} = - \kappa \nabla_z T. \tag{37}$$

Comparision with (36) shows $((\nabla_z \beta)/\beta_0 = - (\nabla_z T)/T_0)$:

$$\kappa = \frac{2}{3} \frac{\varepsilon_0}{\tau_p} R^2 \frac{1}{T_0} . \tag{38}$$

The energy density ε_0 (25) is proportional to T_0^4. The mean free time $\tau(p,s)$ of a phonon (p,s) can be calculated by perturbation theory with the result $\tau(p,s) \propto p^{-m}T^{m-5}$, with $m = 1,2,3$ or 4 [10]. An inspection of (26) shows that $\sigma_s \propto T^{-1}$, irrespective of m, and $\tau_p(34) \propto T^{-5}$. As a consequence, in the *Poiseuille flow régime*

$$\kappa \propto R^2 T^8 . \tag{39}$$

According to (33) the variation of u, and hence of the distribution function F_{LE}, is large over a distance R. Thus the approximation (18) and our results hold only if the average mean free path of the thermal phonons satisfies

$$\bar{\ell} = c\bar{\tau} \ll R. \tag{40}$$

Furthermore, the neglect of U-processes requires

$$R \ll \ell_u. \tag{41}$$

At sufficiently low temperatures $\bar{\ell} \propto T^{-5}$ will exceed R, the phonons will behave like a Knudsen gas, and heat transport is determined by boundary scattering of the phonons. This case has been dealt with by Casimir [11]. The thermal conductivity as defined in (37) behaves like

$$\kappa \propto RT^3 \qquad \text{for } \bar{\ell} \gg R. \tag{42}$$

For higher temperatures, where $\ell_u \lesssim R$, κ will be determined by U-*processes* with a temperature dependence for $T \ll \theta$ [3]

$$\kappa \propto e^{\frac{1}{2}\theta/T} \qquad \text{for } \ell_u \lesssim R. \tag{43}$$

In this case κ does not depend on R.

The general temperature behaviour of κ is shown in figure 1.

7. SOLUTION OF THE PEIERLS EQUATION BY MEANS OF PERTURBATION THEORY. DAMPING OF SECOND SOUND

In this section we shall develop a method of solving the linearized Peierls equation if the rate of U-processes as well as hydrodynamic frequencies are small [13]. No mean free time approximation will be made and the phonon spectrum (5) with three branches will be taken.

Introducing the following ansatz for the distribution function

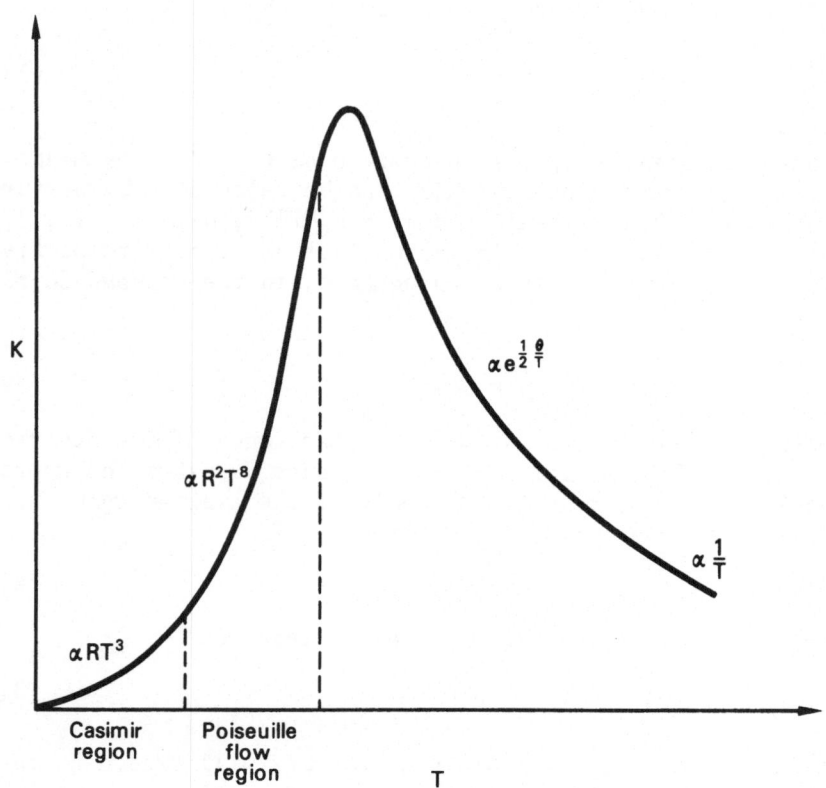

Figure 1

$$F(\mathbf{k},s,\mathbf{r},t) = N + N(N + 1)G(\mathbf{k},s,\mathbf{r},t), \tag{44}$$

where

$$N = \frac{1}{e^{\beta\varepsilon(\mathbf{k},s)} - 1} , \tag{45}$$

into the Peierls equation and keeping only the terms of first order in G, one obtains the *linearized Peierls equation*. Since we want to study second sound, we make a plane wave ansatz for the space and time dependence of G,

$$G(\mathbf{k},s,\mathbf{r},t) = g(\mathbf{k},s)e^{-iqx+\Omega t}. \tag{46}$$

We assume q real, then Ω will become complex. The linearized Peierls equation then reads

$$\Omega g = c_N[g] + \lambda c_U[g] + iqv_xg. \tag{47}$$

Here c_U is the linearized collision operator for U-processes

[3,12,13]:

$$c_U[g] = \frac{1}{N} \sum_{k \neq 0} \sum_{s's''} \int d^3k' \int d^3k'' \omega\omega'\omega''$$

$$\times [P_{ss's''}(k,k,'-k'')\delta(k + k' - k'' - K)$$

$$\times \delta(\omega + \omega' - \omega'')(N'+1)N''(g''-g-g')$$

$$+ \tfrac{1}{2}P_{ss's''}(k,-k,'-k'')\delta(k - k' - k'' - K)$$

$$\times \delta(\omega - \omega' - \omega'')N'N''(g''-g-g')], \qquad (48)$$

where ' and " refer to (k',s') and (k'',s'') and $P_{ss's''}$ are related to third order expansion coefficients of the potential energy of the crystal. c_N, the linearized collision operator for N-processes, is obtained from (48) by putting $K = 0$. In order to sort out the different order contributions of c_U in the following perturbation calculation we have multiplied c_U in (47) by a factor λ which at the end of the calculation will be put equal to one. (47) is an eigenvalue equation for eigenvalue Ω and eigenvector g. We consider a Hilbert space with a scalar product defined as

$$(g,f) = (f,g)^*$$

$$= \frac{V}{(2\pi)^3} \sum_s \int d^3k N(N+1)g^*(\mathbf{k},s)f(\mathbf{k},s), \qquad (49)$$

(V = volume of the crystal). Then c_U and c_N have the properties of being Hermitian and negative [12,13],

$$(g,c_{U \atop N}[f]) = (f,c_{U \atop N}[g])^* \qquad (50)$$

$$(g,c_{U \atop N}[g]) \leqslant 0. \qquad (51)$$

We are going to solve (47) by means of perturbation theory, where $|c_U|$ and the term $\propto q$ are considered to be small compared to c_N. The physical meaning is that the rate of U-processes is small compared to the rate of N-processes and that the wave length of second sound is large compared to the mean free path for N-processes.

We write the perturbation expansion of Ω and g as

$$\Omega_\alpha = \Omega_\alpha(0) + \Omega_\alpha(1) + \Omega_\alpha(2) + \dots , \qquad (52)$$

$$g_\alpha = g_\alpha(0) + g_\alpha(1) + g_\alpha(2) + \dots , \qquad (53)$$

where $\Omega_\alpha{}^{(n)}$, $g_\alpha{}^{(n)}$ are the terms of n-th order in the small parameters λ and q. α labels the eigenvectors and eigenvalues. At the moment nothing is assumed about the relative size of $|\lambda c_U[g]|$ and $|iqv_xg|$. Substituting (52,53) into (47), we obtain for the three lowest orders of the perturbation calculation

$$\Omega_\alpha{}^{(0)}g_\alpha{}^{(0)} = c_N[g_\alpha{}^{(0)}], \qquad (54)$$

$$\Omega_\alpha{}^{(1)}g_\alpha{}^{(0)} + \Omega_\alpha{}^{(0)}g_\alpha{}^{(1)}$$

$$= c_N[g_\alpha{}^{(1)}] + \lambda c_U[g_\alpha{}^{(0)}] + iqv_xg_\alpha{}^{(0)}, \qquad (55)$$

$$\Omega_\alpha{}^{(2)}g_\alpha{}^{(0)} + \Omega_\alpha{}^{(1)}g_\alpha{}^{(1)} + \Omega_\alpha{}^{(0)}g_\alpha{}^{(2)}$$

$$= c_N[g_\alpha{}^{(2)}] + \lambda c_U[g_\alpha{}^{(1)}] + iqv_xg_\alpha{}^{(1)}. \qquad (56)$$

ZERO ORDER

We are interested in solutions belonging to

$$\Omega_\alpha{}^{(0)} = 0. \qquad (57)$$

As a consequence (see (54)), we have to solve

$$c_N[g_\alpha{}^{(0)}] = 0. \qquad (58)$$

There are four linearly independent solutions, namely the four summation invariants

$$\{\psi_1, \psi_2, \psi_3, \psi_4\} = \{a_1k_x, a_2k_y, a_3k_z, a_4\omega\}, \qquad (59)$$

where the a_i are chosen in such a way that

$$(\psi_\alpha, \psi_\beta) = \delta_{\alpha\beta}. \qquad (60)$$

The $g_\alpha{}^{(0)}$, belonging to a degenerate eigenvalue, must be taken as linear combinations

$$g_\alpha{}^{(0)} = \sum_{\beta=1}^{4} t_{\alpha\beta}\psi_\beta. \qquad (61)$$

We require the $g_\alpha{}^{(0)}$ to be orthonormal, too,

$$(g_{\alpha'}{}^{(0)}, g_{\alpha}{}^{(0)}) = \delta_{\alpha'\alpha}. \tag{62}$$

Then $t_{\alpha\beta}$ is a unitary (orthogonal) matrix,

$$\sum_{\beta} t_{\alpha'\beta}{}^* t_{\alpha\beta} = \delta_{\alpha'\alpha}, \tag{63}$$

to be determined later.

FIRST ORDER

We form the scalar product of ψ_β with the first order equation (55). Taking into account (57), the property (50) of c_N, and (61), we obtain

$$\Omega_\alpha{}^{(1)} t_{\alpha\beta} = \sum_{\gamma} \{\lambda t_{\alpha\gamma}(\psi_\beta, c_U[\psi_\gamma]) + iq t_{\alpha\gamma}(\psi_\beta, v_x \psi_\gamma)\}. \tag{64}$$

For α fixed, we have a system of four linear equations for the $t_{\alpha\beta}$ and the corresponding secular determinant has to vanish. The matrix elements of v_x can be calculated. The only ones not equal to zero are

$$(\psi_1, v_x \psi_4) = (\psi_4, v_x \psi_1) = c_{II}, \tag{65}$$

With c_{II} given by (23). The only non-vanishing matrix elements of c_U are

$$(\psi_\beta, c_U[\psi_\beta]) \overset{\text{def}}{=\!=} -u < 0, \qquad \beta = 1,2,3. \tag{66}$$

The secular equation becomes

$$\begin{vmatrix} -\lambda u - \Omega^{(1)} & 0 & 0 & iq c_{II} \\ 0 & -\lambda u - \Omega^{(1)} & 0 & 0 \\ 0 & 0 & -\lambda u - \Omega^{(1)} & 0 \\ iq c_{II} & 0 & 0 & -\Omega^{(1)} \end{vmatrix} = 0, \tag{67}$$

giving the eigenvalues

$$(\Omega_1{}^{(1)}, \Omega_2{}^{(1)}, \Omega_3{}^{(1)}, \Omega_4{}^{(1)})$$

$$= (-\tfrac{1}{2}\lambda u - (\tfrac{1}{4}\lambda^2 u^2 - q^2 c_{II}^2)^{\frac{1}{2}}, -\lambda u, .. \text{ (Contd)} \tag{68}$$

(Contd) $\ldots, -\lambda u, -\frac{1}{2}\lambda u + (\frac{1}{4}\lambda^2 u^2 - q^2 c_{II}^2)^{\frac{1}{2}}).$ (68)

The $t_{\alpha\beta}$ and $g_\alpha^{(0)}$ can be calculated [13]. We are not giving the results here.

In the discussion of (68) one has to distinguish three cases:

(a) $|q c_{II}| \approx |\lambda u|$: The $\Omega_\alpha^{(1)}$ are either real negative, or complex with $|\mathrm{Im}\Omega_\alpha^{(1)}| \lesssim |\mathrm{Re}\Omega_\alpha^{(1)}|$. The corresponding solution is aperiodic in time or a strongly damped oscillation;

(b) $|q c_{II}| \ll |\lambda u|$: One has an exponential decay with time. The first three eigenvalues are $\Omega_1^{(1)} \approx \Omega_2^{(1)} = \Omega_3^{(1)} = -\lambda u$. But

$$\Omega_4^{(1)} = -\frac{q^2 c_{II}^2}{\lambda u}$$ (69)

is much smaller. $g_4^{(0)}$ turns out to be $\approx \psi_4$. This is a case of ordinary heat conduction. Comparison with the phenomenological heat conduction equation

$$c_v \dot{T} - \kappa \nabla^2 T = 0,$$ (70)

(c_v = specific heat, κ = thermal conductivity) and its solutions of the form

$$T = T_0 + \mathrm{constant} e^{-iqx - (\kappa q^2/c_v)t}$$ (71)

shows that

$$\kappa = -\frac{c_{II}^2 c_v}{\lambda(\psi_1, c_U[\psi_1])} .$$ (72)

This is the so called Ziman limit of thermal conductivity [14], valid if $|\lambda c_U| \ll |c_N|$.

(c) $|\lambda u| \ll |q c_{II}|$: Then

$$(\Omega_1^{(1)}, \ldots, \Omega_4^{(1)})$$

$$\approx (iq c_{II} - \frac{1}{2}\lambda u, -\lambda u, -\lambda u, -iq c_{II} - \frac{1}{2}\lambda u).$$ (73)

The eigenvalues $\Omega_1^{(1)}$ and $\Omega_4^{(1)}$ correspond to second sound with phase velocity c_{II} and damping owing to U-processes. An estimate of the integral (66) for u gives [13]

$$u \propto T^{-5} e^{-\frac{1}{2}\theta/T}. \tag{74}$$

SECOND ORDER

$\Omega_\alpha(2)$ can be estimated by means of a variational principle [12,13]. We only give the main results for case (c), $|\lambda u| \ll |qc_{II}|$. The N-processes lead to a damping,

$$\Omega_{N\alpha}(2) = - aT^{-5}, \tag{75}$$

whereas the U-processes (to first order in $|\lambda u|/|qc_{II}|$) yield purely imaginary terms of the form

$$\Omega_{U\alpha}(2) = ibT^{-5} e^{-\frac{1}{2}\theta/T}. \tag{76}$$

For $\alpha = 1,4$ they give a shift of the velocity of second sound. For $\alpha = 2,3$ they vanish.

ACKNOWLEDGMENT

I should like to thank Dr. H. Beck and Dr. P.F. Meier for their valuable help during the preparation of these lecture notes.

REFERENCES

1. Halperin, B.I. and Hohenberg, P.C. (1969). *Phys. Rev.*, **188**, 898.
2. Michel, K.H. and Schwabl, F. (1969). *Solid State Commun.*, **7**, 1781; (1970). *Phys. Cond. Matt.*, **11**, 144.
3. Peierls, R.E. (1955). *Quantum Theory of Solids*, (Clarendon Press, Oxford).
4. Sussmann, J.A. and Thellung, A. (1963). *Proc. Phys. Soc.*, **81**, 1122.
5. Gurzhi, R.N. (1964). *Sov. Phys., JETP*, **19**, 490; (1966). *Sov. Phys., Solid State*, **7**, 2838.
6. Mezhov-Deglin, L.P. (1966). *Sov. Phys., JETP*, **22**, 47; (1967). *Sov. Phys., JETP*, **25**, 568.
7. Ackerman, C.C., Bertman, B., Fairbank, H.A. and Guyer, R.A. (1966). *Phys. Rev. Lett.*, **16**, 789.
8. Götze, W. and Michel, K.H. (1967). *Phys. Rev.*, **156**, 963.
9. Beck, H., Meier, P.F. and Thellung, A. (1974). 'Phonon Hydrodynamics in Solids', (review article), *Phys. Stat. Sol.*, (a)**24**, 11.
10. Herring, C. (1954). *Phys. Rev.*, **95**, 954.
11. Casimir, H.B.G. (1938). *Physica*, **5**, 495.
12. Weiss, K. (1968). *Phys. Cond. Matt.*, **7**, 221.
13. Thellung, A. and Weiss, K. (1969). *Phys. Cond. Matt.*, **9**, 300.

14. Guyer, R.A. and Krumhansl, J.A. (1966). *Phys. Rev.*, **148**, 766.

THE SPIN-PHONON INTERACTION
IN ANHARMONIC FERROMAGNETIC CRYSTALS

N.M. PLAKIDA

Joint Institute for Nuclear Research,
Laboratory of Theoretical Physics, Dubna, USSR

and

H. KONWENT

University of Wrocław,
Institute of Chemistry, Wrocław, Poland

INTRODUCTION

The coupling between magnetic and lattice degrees of freedom
in magnetic crystals is taken into account in the framework of
the Heisenberg model by the simple assumption that the exchange
integral depends on the instantaneous positions of atoms changing as a result of their thermal vibrations. In the standard
approach, the exchange integral is expanded in a power series in
terms of the displacement of atoms from the equilibrium positions,
and only the first and the second terms of the expansion are retained. The first term of the expansion leads to the Heisenberg
Hamiltonian describing the spin interaction in the rigid lattice,
and the second one describes the interaction between the spins
(magnetic moments) and the lattice vibrations (the spin-phonon
interaction). The lattice vibrations are assumed to be harmonic.
This is the harmonic model. It was extensively developed in numerous papers [1-13] and used to explain the properties of ferromagnetic crystals. The assumptions made in the harmonic model
restrict its application to the cases where the displacements
of the atoms from equilibrium positions are small relative to
inter-atomic spacing.

Recently, a self-consistent theory of the spin-phonon inter-action was proposed by S.V. Tyablikov [14] and developed by the present authors [15-21]. In this approach the anharmonicity of the crystal lattice vibrations and all terms in the expansion of the exchange integral in a power series in terms of the displace-ment of atoms from equilibrium positions are taken into account. These terms may play an important rôle in some circumstances. In papers [15-21] the new approach has been used to investigate, in a self-consistent manner, the lattice and magnetic excita-tions in an anharmonic ferromagnetic crystal.

The aim of these lectures is to present, in some detail, the theory of the spin-phonon interaction developed in the papers [15-21]. Note that a similar approach was considered in [22-24].

1. THE HAMILTONIAN AND THE EQUATION OF STATE

Let us consider a single-domain isotropic ferromagnetic crys-tal consisting of N identical atoms in the volume V. The total Hamiltonian describing the lattice vibrations and the exchange interaction of the atomic spin in the Heisenberg model may be written in the form

$$\mathcal{H} = H_\ell + H_s + H_1, \tag{1.1}$$

$$H_\ell = \sum_f \frac{\vec{p}_f^{\,2}}{2M} + \mathcal{V}(\ .. \ \vec{R}_f \ .. \ ), \tag{1.1a}$$

$$H_s = - h \sum_f S_f^z - \frac{1}{2} \sum_{f,g} J(\vec{R}_f - \vec{R}_g)\vec{S}_f \cdot \vec{S}_g, \tag{1.1b}$$

$$H_1 = - \sum_f \vec{F}_f \cdot \vec{R}_f, \tag{1.1c}$$

where p_f^α and R_f^α are the α-Cartesian components of the momentum and the coordinate of an atom with spin $\vec{S}_f$ and the mass M_f in the lattice site f. It is assumed that the potential energy of the inter-atomic interactions $\mathcal{V}(\ .. \ \vec{R}_f \ .. \ )$ and the exchange in-tegral $J(\vec{R}_f - \vec{R}_g)$ in the Heisenberg model, in adiabatic approxi-mation, depend only on instantaneous values of coordinates of atoms $\vec{R}_f$. $h = \mu H^z$ is the Zeeman energy of the magnetic moment μ in the external magnetic field H^z. The action of the external forces F_f^α, introduced for convenience to describe the lattice deformations, is determined by the Hamiltonian H_1.

In an anharmonic crystal the equilibrium positions of atoms, determining the crystal lattice, depend on temperature. For this reason it is convenient to introduce the definition:

$$R_f^\alpha = \langle R_f^\alpha \rangle + u_f^\alpha = f^\alpha + u_f^\alpha, \tag{1.2}$$

where $u_f{}^\alpha$ is the time-dependent displacement of an atom from the equilibrium position f. The expectation values $\langle \dots \rangle$ are calculated with the equilibrium density matrix of the crystal with Hamiltonian (1.1):

$$\langle A \rangle = \frac{\mathrm{Sp}\{e^{-\mathcal{H}/\theta}A\}}{\mathrm{Sp}\{e^{-\mathcal{H}/\theta}\}} \ , \qquad \theta = kT. \qquad (1.3)$$

Equilibrium positions of the atoms can also be determined from the condition that at equilibrium the average force acting on an atom is equal to zero. Let us consider the averaged equation of motion for the momentum operator in the Heisenberg picture $p_f{}^\alpha(t) = e^{i\mathcal{H}t}p_f{}^\alpha e^{-i\mathcal{H}t}$

$$i\frac{\partial}{\partial t}\langle ip_f{}^\alpha(t)\rangle = \langle [ip_f{}^\alpha,\mathcal{H}]\rangle = \langle \frac{\partial}{\partial R_f{}^\alpha}(H_\ell + H_s)\rangle - F_f{}^\alpha = 0,$$

$$(1.4)$$

$$\hbar = 1.$$

In the case of homogenous deformation, the condition (1.4) can be written in a more convenient form, if we consider the work performed by the external forces during the infinitesimal displacement of the equilibrium positions of atoms $\delta f^\alpha = \sum_\beta u_{\alpha\beta}f^\beta$:

$$-\delta H_1 = \sum_f \vec{F}_f \cdot \vec{\delta f} = \sum_{\alpha,\beta,f} F_f{}^\alpha u_{\alpha\beta}f^\beta = V\sum_{\alpha,\beta}\sigma_{\alpha\beta}u_{\alpha\beta},$$

where $\sigma_{\alpha\beta}$ is the stress tensor and $u_{\alpha\beta}$ is the corresponding deformation. Taking into account (1.4), we get the following equation of state:

$$\sigma_{\alpha\beta} = \frac{1}{V}\sum_f F_f{}^\alpha f^\beta = \frac{1}{V}\sum_f \langle \frac{\partial}{\partial R_f{}^\alpha}(\mathcal{V} + H_s)\rangle f^\beta. \qquad (1.5)$$

For isotropic external pressure P this equation takes the form

$$P = -\frac{1}{3}\sum_\alpha \sigma_{\alpha\alpha} = -\frac{1}{3V}\sum_{f,\alpha} f^\alpha \langle \frac{\partial}{\partial R_f{}^\alpha}(\mathcal{V} + H_s)\rangle \equiv P_\ell + P_s, (1.6)$$

where the pressure P_ℓ results from the interaction of the atoms in the lattice:

$$P_\ell = -\frac{1}{3V}\sum_{f,\alpha} f^\alpha \frac{\partial}{\partial f^\alpha}\langle \mathcal{V}(\dots \vec{R}_f \dots)\rangle, \qquad (1.6a)$$

and P_ℓ is the pressure arising from the dependence of the exchange integral on the coordinates of atoms:

$$P_{\mathrm{s}} = \frac{1}{6V} \sum_{f,g} (f^{\alpha} - g^{\alpha}) \langle (\vec{S}_f \cdot \vec{S}_g) \frac{\partial}{\partial R_f^{\alpha}} J(\vec{R}_f - \vec{R}_g) \rangle . \quad (1.6b)$$

The equation (1.6) determines the equilibrium parameters of the lattice, which depend on the temperature, external pressure and the magnetization (external field). The magnetic term (1.6b) gives a substantial contribution only in the temperature range below the Curie temperature.

Now we expand the potential energy of the crystal and the exchange integral in a Taylor series in powers of the atomic displacements u_f^{α}:

$$V(.. \vec{R}_f ..) = \exp \left\{ \sum_f (\vec{u}_f \cdot \nabla_f) \right\} V_0(.. \vec{f} ..)$$

$$= \sum_{n=0}^{\infty} \frac{1}{n!} \sum_{1...n} \Phi_{1...n} u_1 ... u_n , \quad (1.7)$$

$$J(\vec{R}_f - \vec{R}_g) = \exp \{ \vec{u}_{fg} \cdot \nabla_f \} J_0(\vec{f} - \vec{g})$$

$$= \sum_{n=0}^{\infty} \frac{1}{n!} \sum_{1...n} J_{fg}{}^{1...n} u_1 ... u_n , \quad (1.8)$$

where, for the sake of brevity, we have used the notations: $u_1 = u_{f_1}{}^{\alpha_1}$, etc., $\vec{u}_{fg} = \vec{u}_f - \vec{u}_g$, $\nabla_f^{\alpha} = \partial / \partial f^{\alpha}$. We have also introduced the function of the phonon-phonon interaction:

$$\Phi_{1...n} = \nabla_1 ... \nabla_n V_0(.. \vec{f} ..)$$

$$= \frac{\partial}{\partial f_1^{\alpha_1}} ... \frac{\partial}{\partial f_1^{\alpha_1}} V_0(.. \vec{f} ..), \quad (1.7a)$$

and the function of the spin-phonon interaction:

$$J_{fg}{}^{1...n} = \nabla_1 ... \nabla_n J_0(\vec{f} - \vec{g})$$

$$= \prod_{i=1}^{n} \{ (\delta_{i,f} - \delta_{i,g}) \nabla_f^{\alpha_1} ... \nabla_f^{\alpha_n} \} J_0(\vec{f} - \vec{g}) . \quad (1.8a)$$

The values of the exchange integral for the equilibrium positions of atoms determine the spin-spin interaction in the absence of phonons ($\eta = 0$ in equation (1.8)):

$$J_{fg} = J_0(\vec{f} - \vec{g}) . \quad (1.8b)$$

Therefore, the Hamiltonian of an anharmonic ferromagnetic crystal can be written in the form

$$H = \sum_f \frac{\vec{p}_f^2}{2M} + \sum_{n=0}^{\infty} \frac{1}{n!} \sum_{1..n} \Phi_{1...n} u_1 ... u_n$$

$$- h \sum_f S_f^z - \frac{1}{2} \sum_{n=0}^{\infty} \frac{1}{n!} \sum_{f,g} \sum_{1...n} J_{fg}^{1...n} u_1 ... u_n \vec{S}_f \cdot \vec{S}_g . \quad (1.9)$$

The dynamical variables $\vec{p}_f$, $\vec{u}_f$ and $\vec{S}_f$ obey the usual commutation relations, e.g.:

$$[u_f^\alpha, p_{f'}^\beta] = i\nabla(\vec{f} - \vec{f}')$$

$$(1.10)$$

$$[S_f^x, S_{f'}^y] = iS_f^z (\vec{f} - \vec{f}'), \qquad \hbar = 1.$$

In the following sections we shall discuss the equations of motion for the Green functions of the system described by the Hamiltonian (1.9).

2. SELF-CONSISTENT ELEMENTARY EXCITATIONS IN A FERROMAGNETIC ANHARMONIC CRYSTAL [17]

In this section we shall first of all consider the elementary excitations in a ferromagnetic anharmonic crystal in the mean phonon field approximation. This approximation can be built up most simply on the basis of the Bogoliubov variational principle. We assume that in the system described by the Hamiltonian (1.9) the mixed spin-phonon excitations (bound states) [2,3] do not exist, so that it can be described by the trial Hamiltonian of non-interacting spin and phonon subsystems:

$$H_0 = \tilde{H}_\ell + \tilde{H}_s = \sum_f \frac{\vec{p}_f^2}{2M} + \frac{1}{2} \sum_{\substack{f,g \\ \alpha,\beta}} \tilde{\Phi}_{fg}^{\alpha\beta} u_f^\alpha u_g^\beta$$

$$- h \sum_f S_f^z - \frac{1}{2} \sum_{f,g} \tilde{J}_{fg} \vec{S}_f \cdot \vec{S}_g, \quad (2.1)$$

where the effective dynamical matrix $\tilde{\Phi}_{fg}^{\alpha\beta}$ and the effective exchange integral are the variational parameters. The trial free energy can be written in the form

$$F_1 = F_0 + \langle H - H_0 \rangle_0, \quad (2.2)$$

$$F_0 = - \theta \ln \mathrm{Sp}(e^{-H_0/\theta}) = F_{0\ell} + F_{0s} \qquad (2.3)$$

$$F_{0\ell} = - \theta \ln \mathrm{Sp}(e^{-\tilde{H}_\ell/\theta}) = \theta \sum_{\vec{q}j} \ln\left(2sh\ \frac{\omega_{\vec{q}j}}{2\theta}\right) \qquad (2.3a)$$

$$F_{0s} = - \theta \ln \mathrm{Sp}(e^{-\tilde{H}_0/\theta})$$

$$= - \theta \ln \mathrm{Sp}\left\{\frac{1}{\theta}\left[h \sum_f S_f^z + \frac{1}{2} \sum_{fg} \tilde{J}_{fg}\vec{S}_f\cdot\vec{S}_g\right]\right\} . \qquad (2.3b)$$

We have written the free energy of the phonon subsystem in the usual way, introducing the frequencies $\omega_{\vec{q}j}$ and the polarization vectors $\vec{e}_{qj}$ as the solutions of the eigenvalue equations [25]:

$$e_{\vec{q}j}{}^\alpha \omega_{\vec{q}j}{}^2 = \frac{1}{MN} \sum_{f,f'} \vec{e}_{qj}{}^\alpha \tilde{\Phi}_{ff'}{}^{\alpha\beta} e^{-i\vec{q}(\vec{f}-\vec{f}')}. \qquad (2.4)$$

The free energy of the spin subsystem (2.3b) may be evaluated by means of any approximate method, e.g. with the help of the molecular field approximation. The evaluation of the expectation values $\langle ... \rangle$ in equation (2.2) with the trial Hamiltonian (2.1) is performed independently for the phonon and the spin subsystem:

$$\langle H - H_0\rangle_0 = \tilde{v}(\ .. \ \vec{f} \ .. \ ) - \frac{1}{2} \sum_{ij} \tilde{\Phi}_{ij}\langle u_i u_j\rangle_0$$

$$- \frac{1}{2} \sum_{f,g} \{\langle J(\vec{R}_f - \vec{R}_g)\rangle_0 - \tilde{J}_{fg}\}\langle \vec{S}_f\cdot\vec{S}_g\rangle_0 \qquad (2.5)$$

where the averages over the phonon variables in the harmonic approximation may be evaluated in the usual way (see, e.g., [25]):

$$\tilde{v}(\ .. \ \vec{f} \ .. \ ) = \langle \exp \sum_f (u_f\cdot\nabla_f)\rangle_0 v_0(\ .. \ \vec{f} \ .. \ )$$

$$\qquad (2.6a)$$

$$= \exp \{\frac{1}{2} \sum_{i,j} \langle u_i u_j\rangle_0 \nabla_i \nabla_j\}\cdot v_0(\ .. \ \vec{f} \ .. \ )$$

$$\langle J(\vec{R}_f - \vec{R}_g)\rangle_0 = \exp \left\{\frac{1}{2} \sum_{\alpha,\beta} \langle u_{fg}{}^\alpha u_{fg}{}^\beta\rangle_0 \nabla_f{}^\alpha \nabla_f{}^\beta\right\}\cdot J_0(\vec{f} - \vec{g}). \qquad (2.6b)$$

The displacement-displacement correlation function (the pair correlation function) in the trial harmonic approximation has the standard form [25]:

$$\langle u_f^\alpha u_g^\beta \rangle_0 = \frac{1}{MN} \sum_{\vec{q}j} \frac{e_{\vec{q}j}^\alpha e_{\vec{q}j}^\beta}{2\omega_{\vec{q}j}} \coth\left(\frac{\omega_{\vec{q}j}}{2\theta}\right) e^{i\vec{q}(\vec{f}-\vec{g})}. \tag{2.7}$$

We determine now the variational parameters from the condition that the free energy (2.2) is stationary with respect to variations of $\tilde{\Phi}_{ij}$ and $\tilde{J}_{fg}$, or equivalently, with respect to variations of the correlation functions $\langle u_i u_j \rangle_0$ and $\langle S_f \cdot S_g \rangle_0$:

$$\frac{\delta F_1}{\delta \langle u_f^\alpha u_g^\beta \rangle_0} = 0, \tag{2.8a}$$

$$\tilde{\Phi}_{fg}^{\alpha\beta} = \nabla_f^\alpha \nabla_g^\beta \{\tilde{V}(\,..\, \vec{f} \,..\,) - \tfrac{1}{2} \sum_{f'g'} \tilde{J}_{f'g'} \langle \vec{S}_{f'} \cdot \vec{S}_{g'} \rangle_0 \},$$

$$\frac{\delta F_1}{\delta \langle \vec{S}_f \cdot \vec{S}_g \rangle} = 0, \tag{2.8b}$$

$$\tilde{J}_{fg} = \langle J(\vec{R}_f - \vec{R}_g) \rangle_0 .$$

Therefore, in the mean phonon field approximation, according to equations (2.4,8), we receive the following system of self-consistent equations for the frequencies of trial phonons:

$$e_{\vec{q}j}^\alpha \omega_{\vec{q}j}^2 = \frac{1}{MN} \sum_{f,f'} e_{\vec{q}j}^\beta e^{-i\vec{q}(\vec{f}-\vec{f}')} \{\nabla_f^\alpha \nabla_{f'}^\beta \tilde{V}(\,..\, \vec{f} \,..\,)$$

$$- \sum_g (\delta_{ff'} - \delta_{f'g}) \langle S_f \cdot S_g \rangle \nabla_f^\alpha \nabla_f^\beta \tilde{J}_{fg} \}, \tag{2.9}$$

where the effective dynamical matrix on the effective spin-spin interaction $\tilde{J}_{fg}$ (2.8b), on the displacement correlation function (2.7) and on the spin correlation function:

$$\langle S_f \cdot S_g \rangle_0 = \frac{\mathrm{Sp}\{e^{-\tilde{H}_s/\theta} (\vec{S}_f \cdot \vec{S}_g)\}}{\mathrm{Sp}(e^{-\tilde{H}_s/\theta})} . \tag{2.10}$$

The upper bound for the free energy of the real system with the Hamiltonian (1.9), according to the Bogoliubov variational principle, is given by (2.2),

$$F \leqslant F_1.$$

The internal energy of the system becomes

$$E = \langle H \rangle_0 = \sum_{\vec{q}j} \frac{\omega_{\vec{q}j}}{4} \coth\left(\frac{\omega_{\vec{q}j}}{2\theta}\right) + \tilde{V}(\;..\; \bar{F} \;..\;)$$

$$- hN\langle S_f^z \rangle_0 - \frac{1}{2} \sum_{f,g} J_{fg}\langle \vec{S}_f \cdot \vec{S}_g \rangle_0. \qquad (2.11)$$

The equation of state in the mean phonon field approximation, according to (1.5), can be written as

$$P = -\frac{1}{3V} \sum_{f,\alpha} f^\alpha \nabla_f^\alpha \tilde{V}(\;..\; \vec{F} \;..\;) + \frac{1}{\sigma v} \sum_{f,g} f^\alpha \langle S_f \cdot S_0 \rangle \nabla_f^\alpha J_{f,0},$$

$$\qquad (2.12)$$

$$v = \frac{V}{N}.$$

We notice, that though the trial Hamiltonian (2.1), the corresponding free energy (2.2-5), and the internal energy (2.11)
have additive form with respect to the phonon and the spin variables, nevertheless the effective parameters of interaction $\tilde{\Phi}_{ij}$
(2.8a) and $\tilde{J}_{f,g}$ (2.8b,6b), take into account the mutual influence of the spin and phonon subsytems. For example, the phonon
frequencies (2.9) depend on the state of the spin subsytem, and
near the Curie temperature have a singularity which is the result of the singularity of the spin correlation function:

$$\left(\frac{\partial}{\partial T} \omega_{\vec{q}j}^{\;2}\right)_{T \to T_c} \sim \left(\frac{\partial}{\partial T} \langle \vec{S}_f \cdot \vec{S}_g \rangle\right)_{T \to T_c} \to \infty. \qquad (2.13)$$

On the other hand, the spectrum of magnetic excitations and the
magnetization of the spin subsystem, which depend on the effective exchange integral (2.6b), acquire the additional temperature dependence due to the displacement correlation function,
and, in the case of the constant pressure, due to thermal expansion of the lattice. The coefficient of thermal expansion of
the lattice depends also on the state of the spin subsystem, and
for $T \to T_c$, according to (2.12), has the singularity of the type:

$$\alpha = \frac{1}{V}\left(\frac{\partial V}{\partial T}\right)_P = -\frac{1}{V} \frac{\left(\frac{\partial P}{\partial T}\right)_V}{\left(\frac{\partial P}{\partial V}\right)_T}$$

$$\sim \left[\frac{\partial}{\partial T} \langle \vec{S}_f \cdot \vec{S}_g \rangle\right]_{T \to T_c} \to \infty. \qquad (2.14)$$

Therefore, we see that the mean phonon field approximation
allows us to consider in a self-consistent manner the magnetic,

mechanical and thermal properties of a ferromagnetic crystal. However, this approximation does not take into account the damping of phonons, and therefore it is not satisfactorily justified. More consistent is the approach based on the Green functions method.

3. THE PHONON EXCITATIONS IN FERROMAGNETIC CRYSTAL [18,19]

In order to describe the one-phonon excitations, which are connected with the translational motion of the atoms in the lattice, we introduce the displacement-displacement Green function defined as follows [14,26]:

$$G_{ii'}(t - t') = \langle\langle u_i(t); u_{i'}(t') \rangle\rangle$$

$$= - i\theta(t - t')\langle [u_i(t), u_{i'}(t')] \rangle$$

$$= \int_{-\infty}^{\infty} \frac{d\omega}{2\pi} G_{ii'}(\omega) e^{-i\omega(t-t')} \qquad (3.1)$$

$$G_{ii'}(\omega) \equiv \langle\langle u_i | u_{i'} \rangle\rangle_\omega$$

where the displacement operators $u_i = u_{f_i}{}^{\alpha i}$ are written in the Heisenberg picture with the Hamiltonian (1.9): $u_i(t) = e^{iHt} u \, e^{-iHt}$. Now, we derive the equations of motion for the Green function (3.1). Taking twice the derivative of (3.1) with respect to the time t, and using the equations of motion for the displacement $u_i(t)$ and the momentum $p_i(t)$ operators, we get

$$- M \frac{\partial^2}{\partial t^2} G_{ii'}(t - t') = \delta_{ii'}\delta(t - t')$$

$$+ \sum_{n=1}^{\infty} \frac{1}{n!} \sum_{1...n} \Phi_{i1...n} G_{1...n,i'}(t - t')$$

$$(3.2)$$

$$- \frac{1}{2} \sum_{n=0}^{\infty} \frac{1}{n!} \sum_{f,g,1...n} J_{fg}{}^{1...n}$$

$$\times \langle\langle u_1(t)...u_n(t)\vec{S}_f(t)\cdot\vec{S}_g(t); u_{i'}(t') \rangle\rangle.$$

In order to separate the renormalization of phonons in the mean phonon field, we apply the method of the irreducible Green functions, developed in the theory of the highly anharmonic crystal [26].

The irreducible Green functions with respect to the phonon operators taken at the time t are defined as follows:

$$\langle\!\langle\, \{u_1(t)...u_n(t)\vec{S}_f(t)\cdot\vec{S}_g(t)\}^{ir};A(t')\rangle\!\rangle$$

$$= \langle\!\langle\, \{u_1(t)...u_n(t)\vec{S}_f(t)\cdot\vec{S}_g(t) - \langle u_1...u_n\vec{S}_f\cdot\vec{S}_g\rangle\};A(t')\rangle\!\rangle$$

$$- \sum_{m=0}^{n-1} C_n^m\langle u_{m+1}...u_n\rangle\, \langle\!\langle\, \{u_1...u_m(t)\vec{S}_f(t)\vec{S}_g(t)\}^{ir};A(t')\rangle\!\rangle$$

$$- \sum_{m=1}^{n} C_n^m\langle u_{m+1}...u_n\vec{S}_f\cdot\vec{S}_g\rangle\, \langle\!\langle\, \{u_1(t)...u_m(t)\}^{ir};A(t')\rangle\!\rangle\,, \qquad (3.3)$$

where

$$C_n^m = \frac{n!}{(n-m)!m!}\,.$$

By this definition, the irreducible Green function of order n is connected with the lower order $m \leqslant n - 1$ Green functions. The relation (3.3) can be rewritten in the form of an expansion of the total (exact) Green function in irreducible Green functions:

$$\langle\!\langle\, \{u_1(t)...u_n(t)\vec{S}_f(t)\cdot\vec{S}_g(t) - \langle u_1...u_n\vec{S}_f\cdot\vec{S}_g\rangle\};A(t')\rangle\!\rangle$$

$$= \sum_{m=0}^{n} C_n^m\langle u_{m+1}...u_n\rangle\, \langle\!\langle\, \{u_1...u_m\vec{S}_f\cdot\vec{S}_g\}^{ir};A\rangle\!\rangle \qquad (3.4)$$

$$+ \sum_{m=1}^{n} C_n^m\langle u_{m+1}...u_n\vec{S}_f\cdot\vec{S}_g\rangle\, \langle\!\langle\, \{u_1...u_m\}^{ir};A\rangle\!\rangle\,.$$

For the sake of brevity, we have omitted the time arguments of the operators on the right hand side of expressions (3.3,4) and we have used the symmetry property of the Green functions with respect to the permutations of commuting operators $\{u_1...u_n\}$. The definition of the pure phonon irreducible Green function and the corresponding expansion of the exact phonon Green function in irreducible Green functions are obtained from (3.3,4) by excluding the terms with the spin correlation functions $\langle u_1...u_n\vec{S}_f\cdot\vec{S}_g\rangle$ [26]. As follows from the definition (3.3), the irreducible Green functions cannot be simplified by means of pairing (contraction) of the phonon operators which have the same time arguments. Therefore, these functions contain no diagrams of the mean phonon field. We now insert the expansion given by (3.4) and the similar expansion for the phonon Green function,

$$\mathcal{G}_{1\ldots n,i'}(t - t') = \langle\langle u_1(t)\ldots u_n(t); u_{i'}(t')\rangle\rangle,$$

into equation (3.2). We perform the summation over n for every irreducible Green function of the m-th order, $n \geqslant m$, and after doing this, we sum over all irreducible Green functions. Finally, taking the Fourier transform with respect to the time, according to (3.1), we get

$$\omega^2 G_{ii'}(\omega) = \delta_{ii'} + \sum_j \tilde{\Phi}_{ij} G_{ji'}(\omega)$$

$$+ \sum_{n=2}^{\infty} \frac{1}{n!} \sum_{1\ldots n} \tilde{\Phi}_{i1\ldots n} G_{1\ldots n,i'}{}^{ir}(\omega) \tag{3.5}$$

$$- \frac{1}{2} \sum_{n=0}^{\infty} \frac{1}{n!} \sum_{f,g,1\ldots n} \tilde{J}_{fg}{}^{i1\ldots n} \langle\langle u_1\ldots u_n \vec{S}_f \cdot \vec{S}_g | u_{i'}\rangle\rangle_{\omega}{}^{ir}.$$

In equation (3.5) we have introduced the effective phonon interaction:

$$\tilde{\Phi}_{1\ldots n} = \sum_{n'=0}^{\infty} \frac{1}{n'!} \sum_{1'\ldots n'} \Phi_{1\ldots n,1'\ldots n'} \langle u_{1'}\ldots u_{n'}\rangle \tag{3.6}$$

$$- \frac{1}{2} \sum_{n'=0}^{\infty} \frac{1}{n'!} \sum_{f,g,1'\ldots n'} J_{fg}{}^{1\ldots n,1'\ldots n'} \langle u_{1'}\ldots u_{n'} \vec{S}_f \cdot \vec{S}_g\rangle$$

$$= \nabla_1\ldots\nabla_n\{\langle \mathcal{V}(\;..\;\vec{R}_f\;..\;)\rangle - \frac{1}{2} \sum_{f,g} \langle J(\vec{R}_f - \vec{R}_g)\vec{S}_f \cdot \vec{S}_g\rangle\}$$

and the effective spin-spin interaction:

$$\tilde{J}_{fg}{}^{1\ldots n} = \sum_{n'=0}^{\infty} \frac{1}{n'!} \sum_{1'\ldots n'} J_{fg}{}^{1\ldots n,1'\ldots n'} \langle u_{1'}\ldots u_{n'}\rangle$$

$$= \nabla_1\ldots\nabla_n \langle J(\vec{R}_f - \vec{R}_g)\rangle. \tag{3.7}$$

The correlation functions, which determine the renormalization of the interaction in equations (3.6,7), may be presented in the form of expansions in the irreducible correlation functions. The latter may be determined directly from the irreducible Green functions by using the spectral theorem:

$$\langle AB \rangle = \int_{-\infty}^{\infty} d\omega n(\omega) \left[-\frac{1}{\pi} \, \text{Im} \langle\langle B | A \rangle\rangle_{\omega+i\epsilon} \right] , \qquad (3.8)$$

where

$$n(\omega) = (e^{\omega/\theta} - 1)^{-1}.$$

From (3.4,8) it follows that

$$\langle u_1 ... u_n \rangle = \sum_{m=1}^{n-1} C_{n-1}^m \langle u_{m+2} ... u_n \rangle (\langle u_1 ... u_{m+1} \rangle^{ir}), \qquad (3.9)$$

$$\langle u_1 ... u_n \vec{S}_f \cdot \vec{S}_g \rangle = \langle u_1 ... u_n \rangle \langle \vec{S}_f \cdot \vec{S}_g \rangle$$

$$+ \sum_{m=1}^{n} C_n^m \langle u_{m+1} ... u_n \rangle (\langle u_1 ... u_m \vec{S}_f \cdot \vec{S}_g \rangle^{ir}), \qquad (3.10)$$

where

$$\langle u_1 ... u_n \rangle^{ir} = \int_{-\infty}^{\infty} d\omega n(\omega) \left[-\frac{1}{\pi} \, \text{Im} \langle\langle u_2 ... u_n | u_1 \rangle\rangle_{\omega+i\epsilon}^{ir} \right] , \quad (3.9a)$$

$$\langle u_1 ... u_n \vec{S}_f \cdot \vec{S}_g \rangle^{ir}$$

$$= \int_{-\infty}^{\infty} d\omega n(\omega) \left[-\frac{1}{\pi} \, \text{Im} \langle\langle u_2 ... u_n \vec{S}_f \cdot \vec{S}_g | u_1 \rangle\rangle_{\omega+i\epsilon}^{ir} \right] . \quad (3.10a)$$

Therefore, the system of equations (3.5-10) is self-consistent with respect to the irreducible Green functions. Making use of the expansions (3.9,10) we obtain the following expressions for the average potential energy:

$$\langle \mathcal{V}(\,.. \vec{R}_f \,.. \,) \rangle$$

$$= \exp \left\{ \sum_{n=3}^{\infty} \frac{1}{n!} \sum_{1...n} \langle u_1 ... u_n \rangle^{ir} \nabla_1 ... \nabla_n \right\} \tilde{\mathcal{V}}(\,.. \vec{f} \,.. \,), \quad (3.11)$$

$$\tilde{\mathcal{V}}(\,.. \vec{f} \,.. \,)$$

$$= \exp \left\{ \frac{1}{2} \sum_{1'2'} \langle u_{1'} u_{2'} \rangle \nabla_{1'} \nabla_{2'} \right\} \mathcal{V}_0(\,.. \vec{f} \,.. \,), \qquad (3.11a)$$

and for the average value of the exchange integral:

$$\left\langle J(\vec{R}_f - \vec{P}_g) \right\rangle$$

$$= \exp\left\{ \sum_{n=3}^{\infty} \frac{1}{n!} \sum_{1\ldots n} \langle u_1 \ldots u_n \rangle^{ir} \nabla_1 \ldots \nabla_n \right\} \tilde{J}(\vec{f} - \vec{g}), \qquad (3.12)$$

$$\tilde{J}(\vec{f} - \vec{g}) = J_{fg} = \exp\left\{ \frac{1}{2} \sum_{\alpha,\beta} \langle u_{fg}^{\alpha} u_{fg}^{\beta} \rangle \nabla_f^{\alpha} \nabla_f^{\beta} \right\} J_0(\vec{f} - \vec{g}). \quad (3.12a)$$

In the first order of the self-consistent phonon theory (cf. [26]) one takes only into account the pair correlation functions, i.e., the renormalization has the form given by equations (3.11a, 12a). The effective phonon interaction (3.6), in the first approximation with respect to the spin-phonon interaction, according to (3.10), becomes

$$\Phi_{1\ldots n}^{(1)} = \nabla_1 \ldots \nabla_n \left\{ \tilde{v}(\;.. \; \vec{f} \; .. \;) - \frac{1}{2} \sum_{f,g} \tilde{J}_{fg} \langle \vec{S}_f \cdot \vec{S}_g \rangle \right\}. \qquad (3.6a)$$

A similar expression has been obtained in the preceding section by variational method (cf. (2.8a)). In the first order approximation of the self-consistent phonon field, the lattice vibrations are described by the first term of the right hand side of equation (3.5) and the inelastic processes of phonon scattering are neglected. For this reason it is convenient to choose as the 'zero Green function' G^0, the Green function in the mean phonon field approximation, defined by the equation:

$$M\omega^2 G^0_{ii'}(\omega) = \delta_{ii'} + \sum_j \tilde{\Phi}_{ij} G^0_{ji'}(\omega). \qquad (3.13)$$

Making use of the above equation we can re-write equation (3.5) in the form

$$G_{ii'}(\omega) = G^0_{ii'}(\omega) + \sum_j G^0_{ij}(\omega) \sum_{n\,2} \frac{1}{n!} \sum_{1\ldots n} \tilde{\Phi}_{j1\ldots n} G_{1\ldots n,i'}{}^{ir}(\omega) \tag{3.14}$$

$$- \sum_j G^0_{ij}(\omega)\frac{1}{2} \sum_{f,g,1\ldots n} \tilde{J}_{fg}^{1\ldots n} \langle\langle u_1 \ldots u_n \vec{S}_f \cdot \vec{S}_g | u_{i'} \rangle\rangle_\omega{}^{ir}.$$

The system of equations of the first order (3.13,6a,11a,12a) is a closed one. It is equivalent to variational approach which have been developed in the preceding section.

We consider now the inelastic processes of the phonons scattering, which are described by the last two sums with irreducible Green functions in equation (3.14). We write down the equations of motion for irreducible Green functions of the form

$$G_{A,i'}{}^{ir}(t - t') = \langle\!\langle \{A(t)\}^{ir}; u_{i'}(t') \rangle\!\rangle , \qquad (3.15)$$

where

$$A^{ir} = \{u_1 ... u_n\}^{ir} \qquad or \qquad A^{ir} = \{u_1 ... u_n \vec{S}_f \vec{S}_g\}.$$

Taking twice the derivative of (3.15) with respect to time t', we get the following equation of motion:

$$- M \frac{\partial^2}{\partial t'^2} G_{A,i'}{}^{ir}(t - t')$$

$$= - \delta(t - t')\langle [A, iP_{i'}] \rangle$$

$$+ \sum_{n'=1}^{\infty} \frac{1}{n'!} \sum_{1'...n'} \Phi_{i'1'...n'} \langle\!\langle A(t); u_{1'}(t')...u_{n'}(t') \rangle\!\rangle$$

$$- \tfrac{1}{2} \sum_{n'=0}^{} \frac{1}{n'!} \sum_{f',g',1'...n'} J_{f'g'}{}^{i'1'...n'}$$

$$\times \langle\!\langle A(t); u_{1'}(t')...u_{n'}(t')\vec{S}_{f'}(t')\vec{S}_{g'}(t') \rangle\!\rangle . \qquad (3.16)$$

It can be shown by direct substitution of the definitions of the irreducible Green functions (3.3) that the averaged commutator in the right hand side of equation (3.16) is equal to zero:

$$\langle [\{u_1 ... u_n\}^{ir}, ip_{i'}] \rangle = 0, \qquad n \geqslant 2,$$

$$\langle [\{u_1 ... u_n \vec{S}_f \cdot \vec{S}_g\}^{ir}, ip_{i'}] \rangle = 0, \qquad (3.17)$$

Expanding, in analogy with (3.3,4), the exact Green functions appearing on the right hand side of equation (3.16) in irreducible Green functions with respect to the operators which have the time argument t', we re-write equation (3.16) for the Fourier transformed Green function (3.15) in the form

$$M\omega^2 \langle\!\langle A | u_{i'} \rangle\!\rangle_\omega = \sum_{n'=1}^{\infty} \frac{1}{n'!} \sum_{1'...n'} \tilde{\Phi}_{i'1'...n'} \langle\!\langle A | u_{1'}...u_{n'} \rangle\!\rangle_\omega{}^{ir} - (Contd)$$

(Contd)

$$- \frac{1}{2} \sum_{n'=0} \frac{1}{n'!} \sum_{f',g',1'\ldots n'} \tilde{J}_{f'g'}{}^{i'1'\ldots n'} \langle\!\langle A | u_{1'} \ldots u_{n'} \vec{S}_{f'} \cdot \vec{S}_{g'} \rangle\!\rangle_\omega{}^{ir}.$$

(3.18)

The first term on the RHS with $n' = 1$ describes the propagation of the undamped excitations described by the 'zero Green function' G^0 (3.13). Multiplying the equation (3.18) from the left by $G^0{}_{i'j'}(\omega)$, summing up over j' with the use of equation (3.13), we obtain the following expression for the Green function (3.15):

$$G_{A,i'}{}^{ir}(\omega) = \qquad\qquad\qquad\qquad\qquad\qquad\qquad\qquad (3.19)$$

$$\sum_{j'} G^0{}_{i'j'}(\omega) \Bigg\{ \sum_{n'=2}^{\infty} \frac{1}{n'!} \sum_{1'\ldots n'} \tilde{\Phi}_{j'1'\ldots n'} \langle\!\langle A | u_{1'} \ldots u_{n'} \rangle\!\rangle_\omega{}^{ir}$$

$$- \frac{1}{2} \sum_{n'=0} \frac{\tilde{}}{n'!} \sum_{f',g',1'\ldots n'} \tilde{J}_{f'g'}{}^{j'1'\ldots n'} \langle\!\langle A | u_{1'} \ldots u_{n'} \vec{S}_{f'} \cdot \vec{S}_{g'} \rangle\!\rangle_\omega{}^{ir} \Bigg\}.$$

If we substitute into equation (3.14) the expression given by equation (3.19) (cf. also (3.15)), we get the equation for the one-phonon Green function in the form

$$G_{ii'}(\omega) = G^0{}_{ii'}(\omega) + \sum_{jj'} G^0{}_{ij}(\omega) P_{jj'}(\omega) G^0{}_{j'i'}(\omega), \qquad (3.20)$$

where the scattering operator $P_{jj'}(\omega)$ is given by

$$P_{jj'}(\omega) =$$

$$\sum_{n,n'=2} \frac{1}{n!n'!} \sum_{\substack{1\ldots n \\ 1'\ldots n'}} \tilde{\Phi}_{j1\ldots n} \tilde{\Phi}_{j'1'\ldots n'} \langle\!\langle u_1 \ldots u_n | u_{1'} \ldots u_{n'} \rangle\!\rangle_\omega{}^{ir}$$

$$+ \frac{1}{4} \sum_{n,n'=0} \frac{1}{n!n'!} \sum_{\substack{f,g,1\ldots n \\ f',g',1'\ldots n'}} J_{fg}{}^{j1\ldots n} J_{f'g'}{}^{j'1'\ldots n'}$$

(Contd)

$$\times \langle\!\langle u_1 \ldots u_n \vec{S}_f \cdot \vec{S}_g | u_{1'} \ldots u_{n'} \vec{S}_{f'} \cdot \vec{S}_{g'} \rangle\!\rangle_\omega{}^{ir} -$$

(Contd)

$$-\frac{1}{2}\sum_{n=2}^{\infty}\frac{1}{n!}\sum_{n'=0}^{\infty}\frac{1}{n'!}\sum_{\substack{1\ldots n\\f',g',1'\ldots n'}}\tilde{\Phi}_{j1\ldots n}\tilde{J}_{f'g'}{}^{j'1'\ldots n'}$$

$$\times\left\langle\!\left\langle u_1\ldots u_n|u_1\cdot\ldots u_{n'}\vec{S}_{f'}\cdot\vec{S}_{g'}\right\rangle\!\right\rangle_\omega^{ir}$$

$$-\frac{1}{2}\sum_{n'=2}^{\infty}\frac{1}{n'!}\sum_{n=0}^{\infty}\frac{1}{n!}\sum_{\substack{f,g,1\ldots n\\1'\ldots n'}}\tilde{J}_{fg}{}^{j1\ldots n}\tilde{\Phi}_{j'1'\ldots n'}$$

$$\times\left\langle\!\left\langle u_1\ldots u_n S_f\cdot S_g|u_{1'}\ldots u_{n'}\right\rangle\!\right\rangle_\omega^{ir}.$$

The equation for the one-phonon Green function (3.20) can be re-written in the form of the Dyson equation

$$G_{ii'}(\omega) = \{\omega^2 M\delta_{ii'} - \tilde{\Phi}_{ii'} - \Pi_{ii'}(\omega)\}^{-1}, \qquad (3.22)$$

where the mass operator $\Pi_{ii'}(\omega)$ is defined by the equation:

$$P_{ii'}(\omega) = \Pi_{ii'}(\omega) + \sum_{jj'}\Pi_{ij'}(\omega)G^0{}_{j'j}(\omega)P_{ji'}(\omega). \quad (3.23)$$

From equation (3.23) it follows that the mass operator $\Pi_{ii'}(\omega)$ is equal to the compact (c) part of the scattering operator $P_{ii'}(\omega)$:

$$\Pi_{ii'}(\omega) = P_{ii'}{}^{(c)}(\omega).$$

The compact part $P_{ii'}{}^{(c)}(\omega)$ cannot be split into two simpler parts connected by the single G^0 line. Therefore, the equations (3.20-23) give the exact representation for the one-phonon Green function.

We now consider the second order approximation with respect to the effective interaction in the framework of the self-consistent phonon theory. In this case it is sufficient to evaluate the Green functions on the right hand side of equation (3.21) in the simplest approximation, which takes into account only the first free terms of the equations. They are determined by means of the two-time decoupling of the appropriate correlation functions:

$$\left\langle u_1(t)\ldots u_n(t)|u_{1'}\ldots u_{n'}\right\rangle^{ir} \approx \delta_{n,n'}n!\prod_{i=1}^{n}\left\langle u_i(t)u_{i'}\right\rangle, \quad (3.24a)$$

$$\langle u_1(t)...u_n(t)\vec{S}_f(t)\cdot\vec{S}_g(t)|u_1\text{'}...u_n\text{'}\vec{S}_f\text{'}\cdot\vec{S}_g\text{'}\rangle$$

$$\approx \langle \vec{S}_f(t)\cdot\vec{S}_g(t)|\vec{S}_f\text{'}\cdot\vec{S}_g\text{'}\rangle^{ir}\delta_{n,n\text{'}}n! \prod_{i=1}^{n} \langle u_i(t)u_i\text{'}\rangle, \qquad (3.24b)$$

$$\langle u_1(t)...u_n(t)\vec{S}_f(t)\cdot\vec{S}_g(t)|u_1\text{'}...u_n\text{'}\rangle \approx 0, \qquad (3.24c)$$

where we have used the symmetry properties of the correlation functions. Making use of the spectral representations for the Green functions in equation (3.21) and taking into account the approximations given by equation (3.24), we get the following expression for the mass operator in the second order approximation:

$$\Pi_{ii\text{'}}^{(2)}(\omega) = \int_{-\infty}^{\infty} \frac{d\omega\text{'}}{\omega - \omega\text{'}} (e^{\omega\text{'}/\theta} - 1)\int_{-\infty}^{\infty} \frac{dt}{2\pi} e^{-i\omega\text{'}t}$$

$$\times \left\{ \frac{1}{4} \sum_{\substack{f,g \\ f,\text{'}g\text{'}}} \langle (\vec{S}_f(t)\cdot\vec{S}_g(t))(\vec{S}_f\text{'}\cdot\vec{S}_g\text{'})\rangle^{ir} \right.$$

$$\times \exp(\sum_{jj\text{'}} \langle u_j(t)u_j\text{'}\rangle\nabla_j\nabla_j\text{'})\tilde{J}_{fg}{}^i\tilde{J}_{f\text{'}g\text{'}}{}^{i\text{'}}$$

$$+ \sum_{n=2}^{\infty} \frac{1}{n!} (\sum_{jj\text{'}} \langle u_j(t)u_j\text{'}\rangle\nabla_j\nabla_j\text{'})^n$$

$$\left. \times \tilde{\Phi}_i(\,.. f_j ..\,)\tilde{\Phi}_{i\text{'}}(\,.. f_j\text{'} ..\,) \right\}, \qquad (3.25)$$

where the operators ∇_j and $\nabla_j\text{'}$ act on the functions:

$$\tilde{J}_{fg}{}^i = \nabla_i\tilde{J}(\vec{f} - \vec{g}),$$

$$\tilde{\Phi}_i = \tilde{\Phi}_i(\,.. f_j ..\,) = \nabla_i\tilde{\Phi}(\,.. f_j ..\,),$$

and

$$\tilde{J}_{f\text{'}g\text{'}}{}^{i\text{'}} = \nabla_i\text{'}\tilde{J}(f\text{'} - g\text{'}),$$

$$\tilde{\Phi}_{i\text{'}} = \tilde{\Phi}_{i\text{'}}(\,.. f_j\text{'} ..\,) = \nabla_i\text{'}\tilde{\Phi}(\,.. f_j\text{'} ..\,),$$

respectively. The irreducible spin correlation function, according to the definitions (3.3,10a), has the form

$$\langle(\vec{S}_f(t)\cdot\vec{S}_g(t))|\vec{S}_{f'}\cdot\vec{S}_{g'}\rangle^{ir} =$$

$$\langle\{(\vec{S}_f(t)\cdot\vec{S}_g(t)) - \langle\vec{S}_f\cdot\vec{S}_g\rangle\}\{(\vec{S}_{f'}\cdot\vec{S}_{g'}) - \langle\vec{S}_{f'}\cdot\vec{S}_{g'}\rangle\}\rangle. \qquad (3.26)$$

To evaluate this function it is necessary to consider the equations for the spin Green functions. The phonon correlation functions $\langle u_j(t)u_{j'}\rangle$ are determined by means of the full phonon Green function (3.22) with the mass operator (3.25), according to (3.8). This procedure leads to the closed system of equations with respect to the phonon variables.

The real and imaginary parts of the mass operator,

$$\nabla_{ii'}(\omega) = Re\Pi_{ii'}(\omega), \qquad \Gamma_{ii'}(\omega) = - Im\Pi_{ii'}(\omega + i\epsilon), \qquad (3.27)$$

describe, respectively, the additional renormalization of the frequencies and the damping of the atoms vibrations in the crystal lattice, due to the inelastic processes of scattering of phonons on phonons (the last term in (3.25)), as well as of phonons on the magnetic excitations of the spin subsystem (the first term in (3.25)). These inelastic processes are not included in consideration in the framework of the variational approach, developed in section 2, but their contribution may be important in certain cases. More detailed estimation of the range of value of the damping needs consideration of the appropriate model of the crystal, and this problem is beyond the scope of the present lectures.

4. THE MAGNETIC EXCITATIONS IN AN ANHARMONIC FERROMAGNETIC CRYSTAL

The mean phonon field approximation, which has been discussed by means of the variational method in section 2, leads only to the renormalization of the exchange integral due to the thermal vibration of the crystal lattice, but is unable to take into account the inelastic processes of the spin-phonon interaction. In this section we shall carefully consider the magnetic excitations in the system with the Hamiltonian (1.9). Those excitations are described by the spin Green function depending on the transverse components of the spin operators:

$$G_{ff'}(t - t') = \langle\langle S_f^+(t); S_{f'}^-(t')\rangle\rangle$$

$$= - i\theta(t - t')\langle[S_f^+(t), S_{f'}^-(t')]\rangle = \text{(Contd)}$$

(Contd) $$= \int_{-\infty}^{\infty} \frac{d\omega}{2\pi} e^{-i\omega(t-t')} G_{ff'}(\omega),$$ (4.1)

where we have used the usual notation $S_f^{\pm} = S_f^x \pm iS_f^y$, etc.. Making use of the equations of motion for the spin operators in the Heisenberg picture with the Hamiltonian (1.9), we get for (4.1) the following equation of motion:

$$\left[i \frac{\partial}{\partial t} - h\right] G_{ff'}(t - t') = 2\langle S_f^z\rangle \delta_{ff'} \delta(t - t')$$

$$+ \sum_{n=0}^{\infty} \frac{1}{n!} \sum_{g,1...n} J_{fg}^{1...n}$$ (4.2)

$$\times \langle\langle u_1(t)...u_n(t)(S_f^+(t)S_g^z(t) - S_f^z(t)S_g^+(t)); S_{f'}^-(t')\rangle\rangle .$$

Further, we take into account the renormalization of the exchange interaction in the mean phonon field by introduction, in analogy with (3.3), the irreducible Green functions with respect to the phonon operators:

$$\langle\langle \{u_1(t)...u_n(t)\}^{ir} S_f^+(t)S_g^z(t); S_{f'}^-(t')\rangle\rangle =$$

$$\langle\langle u_1(t)...u_n(t)S_f^+(t)S_g^z(t); S_{f'}^-(t')\rangle\rangle$$ (4.3)

$$- \sum_{m=0}^{n-1} C_n^m \langle u_{m+1}...u_n\rangle \langle\langle \{u_1(t)...u_m(t)\}^{ir} S_f^+(t)S_g^z(t); S_{f'}^-(t')\rangle\rangle .$$

As distinct from (3.3), here the spin-phonon correlation function $\langle u_1... u_n S_f^+ S_g^z\rangle$ is equal to zero. Substituting the expansion for the spin-phonon Green functions from equation (4.2) and summing up over n for every irreducible Green function of the m-th order, we write equation (4.2) for the Fourier transform of the Green function in the form

$$(\omega - h)G_{ff'}(\omega) = 2\delta_{ff'}\langle S_f^z\rangle$$

$$+ \sum_{g} \tilde{J}_{fg} \langle\langle (S_f^+ S_g^z - S_f^z S_g^+)|S_{f'}^-\rangle\rangle_{\omega}$$ (4.4)

$$+ \sum_{n=1}^{\infty} \frac{1}{n!} \sum_{g,1...n} \tilde{J}_{fg}^{1...n} \langle\langle (u_1...u_n)^{ir}(S_f^+ S_g^z - S_f^z S_g^+)|S_{f'}^-\rangle\rangle .$$

The interaction $J_{fg}^{1...n}$, renormalized in the mean phonon field, has been defined in (3.7).

The first sum on the right hand side of equation (4.4) describes the direct spin-spin interaction in the mean phonon field and the second one determines the inelastic processes of the spin-phonon scattering.

It is convenient to introduce the energy of magnetic excitations in the mean spin field approximation, as has been proposed in [27], by means of irreducible, with respect to the spin operators, Green functions, which are defined as follows:

$$\langle\!\langle \{S_f^+ S_g^z - S_f^z S_g^+\}^{ir}; S_{f'}^- \rangle\!\rangle$$

$$= \langle\!\langle (S_f^+ S_g^z - S_f^z S_g^+); S_{f'}^- \rangle\!\rangle$$

$$- A_{fg}\langle\!\langle S_f^+; S_{f'}^- \rangle\!\rangle - A_{gf}\langle\!\langle S_g^+; S_{f'}^- \rangle\!\rangle, \qquad (4.5)$$

where the coefficients A_{fg} are chosen from the condition that the inhomogeneous term in the equation for the irreducible Green function should be equal zero:

$$\langle [\{S_f^+ S_g^z - S_f^z S_g^+\}^{ir}, S_{f'}^-] \rangle = 0. \qquad (4.5a)$$

From this condition, taking into account (4.5), for coefficients A_{fg} with $f \neq g$, we get

$$A_{fg} = A_{gf} = \frac{1}{2\langle S^z \rangle} \{2\langle S_f^z S_g^z \rangle + \langle S_f^- S_g^+ \rangle\} \qquad (4.6)$$

These coefficients determine the first order Green function in the mean field approximation. It is convenient to choose this function as 'the zero Green function' G^0:

$$(\omega - h)G^0_{ff'}(\omega)$$

$$= 2\delta_{ff'}\langle S^z \rangle + \sum_g \tilde{J}_{fg}A_{fg}\{G^0_{ff'}(\omega) - G^0_{gf'}(\omega)\}. \qquad (4.7)$$

The solution of the above equation can be written, with the aid of the Fourier transformation, in the form

$$G^0_{ff'}(\omega) = \frac{1}{N} \sum_{\vec{v}} e^{i\vec{v}(\vec{f}-\vec{f}')} \frac{2\langle S^z \rangle}{\omega - E(\vec{v})}, \qquad (4.8)$$

where the energy of the spin excitations in the mean field is equal to

$$E(\vec{v}) = h + \frac{1}{N} \sum_{f,g} \tilde{J}_{fg} A_{fg} (1 - e^{i\vec{v}(\vec{f}-\vec{g})})$$

$$= h + \langle S^z \rangle \{ \tilde{J}(0) - \tilde{J}(\vec{v}) \} + \frac{1}{N} \sum_{\vec{v}'} \{ \tilde{J}(\vec{v}') - \tilde{J}(\vec{v} - \vec{v}') \} N_{v'}$$

$$+ \frac{1}{\langle S^z \rangle} \frac{1}{N} \sum_{\vec{v}'} \{ \tilde{J}(\vec{v}') - \tilde{J}(\vec{v} - \vec{v}') \} K_{\vec{v}'}{}^{zz}. \tag{4.9}$$

Here, we have used the spatial Fourier transform for the renormalized exchange integral (3.7) and for the correlation functions in equation (4.6), such that:

$$\tilde{J}_{fg} = \langle J(\vec{R}_f - \vec{R}_g) \rangle = \frac{1}{N} \sum_{\vec{v}} \tilde{J}(\vec{v}) e^{i\vec{v}(\vec{f}-\vec{g})}, \tag{4.10}$$

$$K_{fg}{}^{-+} = \langle S_f^- S_g^+ \rangle = \frac{2\langle S^z \rangle}{N} \sum_{\vec{v}} N_{\vec{v}} e^{i\vec{v}(\vec{f}-\vec{g})}, \tag{4.11a}$$

$$K_{fg}{}^{zz} = \langle (S_f^z - \langle S_f^z \rangle)(S_g^z - \langle S_g^z \rangle) \rangle,$$

$$= \frac{1}{N} \sum_{\vec{v}} K_{\vec{v}}{}^{zz} e^{i\vec{v}(\vec{f}-\vec{g})}. \tag{4.11b}$$

The expression (4.9) (but without renormalization in the mean phonon field) had been obtained in the framework of the first order Green functions decoupling in the paper [28]. It can be also derived by means of the diagrammatic technique for the spin operators [29,30] in the first order with respect to the interaction [29]. The first term in equation (4.9) corresponds to the Random Phase Approximation (Tyablikov decoupling), the second and the third ones take into account, respectively, the renormalization of the excitation energy due to the elastic scattering on spin waves ($\sim N_v$) and on the fluctuations of the z-component of the spin ($\sim K_v{}^{zz}$) (cf. the discussion in [31]).

Taking into account now the definition of the 'zero Green function' (4.7), we re-write the equation (4.4) in the form

$$G_{ff'}(\omega) = G^0{}_{ff'}(\omega)$$

$$+ \sum_{\ell} G^0{}_{f\ell}(\omega) \frac{1}{2\langle S^z \rangle} \left\{ \sum_g \tilde{J}_{\ell g} \langle \langle \{ S_\ell^+ S_g^z - S_\ell^z S_g^+ \} \dot{\iota} r \, S_{f'}^- \rangle \rangle_\omega \right\} +$$

$$\tag{4.12}$$

(Contd)

(Contd)

$$(4.12)$$

$$+ \sum_{n=1}^{\infty} \frac{1}{n!} \sum_{g,1...n} \mathcal{J}_{fg}^{1...n} \langle\langle \{u_1...u_n\}^{ir}(S_\ell^+ S_g^z - S_\ell^z S_g^+) | S_{f'}^- \rangle\rangle_\omega .$$

To determine the irreducible Green functions of the type $\langle\langle A(t); S_{f'}^-(t') \rangle\rangle$ appearing on the right hand side of the equation, which describe both the effects of the spin-spin ($A = \{S_\ell^+ S_g^z - S_\ell^z S_g^+\}^{ir}$) and of the inelastic spin-phonon ($A = [\{u_1...u_n\}^{ir} \times (S_\ell^+ S_g^z - S_\ell^z S_g^+)]$) interactions, we differentiate this Green function with respect to the second time argument t' and we obtain

$$\left[-i \frac{\partial}{\partial t'} - h \right] \langle\langle A(t); S_{f'}^-(t') \rangle\rangle$$

$$= \langle [A, S_{f'}^-] \rangle \delta(t - t') + \sum_{n'=0}^{\infty} \frac{1}{n'!} \sum_{g',1'...n'} J_{f'g'}^{1'...n'} \qquad (4.13)$$

$$\times \langle\langle A(t); u_{1'}(t')...u_{n'}(t')(S_{g'}^z(t')S_{f'}^-(t') - S_{f'}^z(t')S_{g'}^-(t')) \rangle\rangle .$$

Now we introduce the irreducible Green functions with respect to the phonon operators in analogy with equation (4.3) and, with respect to the spin ones (for $t' = 0$) in analogy with equation (4.5), both kinds of operators having the time argument t'. Taking next the Fourier transform of the equation (4.13) we get

$$\langle\langle A | S_{f'}^- \rangle\rangle_\omega = \frac{1}{2\langle S^z \rangle} \sum_{\ell'} G^0_{f'\ell'}(\omega) \qquad (4.13a)$$

$$\times \left\{ \sum_{g'} \mathcal{J}_{\ell'g'} \langle\langle A | (S_{g'}^z S_{\ell'}^- - S_{\ell'}^z S_{g'}^-)^{ir} \rangle\rangle_\omega \right.$$

$$\left. + \sum_{n'=1}^{\infty} \frac{1}{n'!} \sum_{g',1'...n'} J_{fg}^{1'...n'} \langle\langle A | (u_{1'}...u_{n'})^{ir} S_{g'}^z S_{f'}^- - S_{f'}^z S_{g'}^-) \rangle\rangle_\omega \right\} .$$

We have taken into account that, according to the definition of the irreducible functions (4.3,5), the inhomogeneous term in equation (4.13) is equal to zero: $\langle [A, S_{f'}^-] \rangle = 0$. Substituting the equation (4.13a) into equation (4.12), we obtain the equation for the spin Green function (4.1) in the form

$$G_{ff'}(\omega) = G^0_{ff'}(\omega) + \sum_{\ell\ell'} G^0_{f\ell}(\omega) P_{\ell\ell'}(\omega) G^0_{\ell'f'}(\omega), \qquad (4.14)$$

where we have introduced the operator

$$(2\langle S^z \rangle)^2 P_{ff'}(\omega) =$$

$$\sum_{n,n'=1}^{\infty} \frac{1}{n!n'!} \sum_{g,1...n} \sum_{g',1'...n'} J_{fg}^{1...n} J_{f'g'}^{1'...n'}$$

$$\times \langle\langle (u_1...u_n)^{ir}(S_f^+ S_g^z - S_f^z S_g^+) |$$

$$|(u_{1'}...u_{n'})^{ir}(S_{g'}^z S_{f'}^- - S_{f'}^z S_{g'}^-) \rangle\rangle_\omega$$

$$+ \sum_{gg'} \tilde{J}_{fg}\tilde{J}_{f'g'} \langle\langle (S_f^+ S_g^z - S_f^z S_g^+)^{ir} | (S_{f'}^- S_{g'}^z - S_{f'}^z S_{g'}^-)^{ir} \rangle\rangle_\omega$$

$$+ \sum_{gg'} \sum_{n=1}^{\infty} \frac{1}{n!} \sum_{1...n} \tilde{J}_{fg}^{1...n} \tilde{J}_{f'g'} \langle\langle (u_1...u_n)^{ir}(S_f^+ S_g^z - S_f^z S_g^+) |$$

$$|(S_{f'}^- S_{g'}^z - S_{f'}^z S_{g'}^-)^{ir} \rangle\rangle_\omega$$

$$+ \sum_{gg'} \sum_{n'=1}^{\infty} \frac{1}{n'!} \sum_{1'...n'} \tilde{J}_{fg}\tilde{J}_{f'g'}^{1'...n'} \qquad (4.15)$$

$$\times \langle\langle (S_f^+ S_g^z - S_f^z S_g^+)^{ir} | (u_1...u_{n'})^{ir}(S_{f'}^- S_{g'}^z - S_{f'}^z S_{g'}^-) \rangle\rangle_\omega.$$

The mass operator of the spin Green function $M_{ff'}(\omega)$, defined by the Dyson equation:

$$G_{ff'}(\omega) = G^0_{ff'}(\omega) + \sum_{\ell\ell'} G^0_{f\ell}(\omega) M_{\ell\ell'}(\omega) G_{\ell'f'}(\omega), \qquad (4.16)$$

is connected with the operator $P_{ff'}(\omega)$ by means of the equation:

$$P_{ff'}(\omega) = M_{ff'}(\omega) + \sum_{\ell\ell'} M_{f\ell}(\omega) G^0_{\ell\ell'}(\omega) P_{\ell'f'}(\omega), \qquad (4.17)$$

i.e., it represents the compact part (c) of the operator $P_{ff'}(\omega)$, which contains no parts connected by one G^0 line. Taking the lattice Fourier transform, by analogy with (4.8), we obtain the solution of the Dyson equation in the form

$$G_{ff'}(\omega) = \frac{1}{N} \sum_{\vec{v}} e^{i\vec{v}(\vec{f}-\vec{f}')} \frac{2\langle S^z \rangle}{\omega - E(\vec{v}) - 2\langle S^z \rangle M(\vec{v},\omega)}, \qquad (4.18)$$

where the mass operator, according to equations (4.15,17), is

$$M(\vec{v},\omega) = \frac{1}{N} \sum_{ff'} e^{-i\vec{v}(\vec{f}-\vec{f}')} P_{ff'}(c)(\omega). \qquad (4.19)$$

The spectral density of the magnetic excitations with the wave vector $\vec{v}$ is determined by the imaginary part of the Green function (4.18) and has the form

$$g(\vec{v},\omega) = -\frac{1}{\pi} \text{Im}G(\vec{v},\omega + i\varepsilon)$$

$$= \frac{2\langle S^z\rangle \Gamma_{\vec{v}}(\omega)}{[\omega - E(\vec{v}) - \Delta_{\vec{v}}(\omega)]^2 + [\Gamma_{\vec{v}}(\omega)]^2}, \qquad (4.20)$$

where the frequency dependent shift $\nabla_{\vec{v}}(\omega)$ and the half-width $\Gamma_{\vec{v}}(\omega)$,

$$\Delta_{\vec{v}}(\omega) = 2\langle S^z\rangle \text{Re}M(\vec{v},\omega),$$

$$\Gamma_{\vec{v}}(\omega) = -2\langle S^z\rangle \text{Im}M(\vec{v},\omega + i\varepsilon), \qquad (4.21)$$

have been introduced, The function $g(\vec{v},\omega)$ determines directly the inelastic magnetic scattering of slow neutrons.

Let us derive now the approximate expression for the mass operator (4.19,15), in the second order with respect to the spin-phonon interaction. In this approximation the last two terms in equation (4.15) give no contribution, and for the first term we can use the two-time decoupling of the spin-phonon correlation functions, analogously with (3.24):

$$\langle \{u_{1'}(t)...u_{n'}(t)\}^{ir} S_{g'}{}^z(t) S_{f'}{}^-(t) | \{u_1...u_n\}^{ir} S_g{}^z S_f{}^+\rangle \qquad (4.22)$$

$$\approx \langle \{u_{1'}(t)...u_{n'}(t)\}^{ir} \{u_1...u_n\}\rangle^{ir} \langle S_{g'}{}^z(t) S_{f'}{}^-(t) S_g{}^z S_f{}^+\rangle$$

$$\approx \delta_{n,n'} n! \prod_{i=1}^{n} \langle u_{i'}(t)u_i\rangle \langle S_{g'}{}^z(t) S_{f'}{}^-(t) S_g{}^z S_f{}^+\rangle.$$

(Remember that the equal-time pairing of the phonon operators has been in fact included by introducing the irreducible Green functions (4.3)). Therefore, in the second order approximation with respect to the spin-phonon interaction, for the mass operator we get the following expression:

$$2\langle S^z\rangle M(\vec{v},\omega) = \qquad \text{(Contd)} \qquad (4.23)$$

(Contd)

$$= \frac{1}{2\langle S^z \rangle N} \sum_{ff'} e^{-i\vec{v}(\vec{f}-\vec{f}')} \int_{-\infty}^{\infty} \frac{d\omega'}{\omega - \omega'} (e^{\omega'/\theta} - 1) \int_{-\infty}^{\infty} \frac{dt}{2\pi} e^{-i\omega't}$$

$$\times \left\{ \sum_{gg'} \tilde{J}_{fg}\tilde{J}_{f'g'} \langle \{S_{f'}^{-}(t)S_g^{z}(t) - S_{f'}^{z}(t)S_g^{-}(t)\}^{ir} \right.$$

$$\times \{S_f^+ S_g^z - S_g^+ S_f^z\}^{ir} \rangle$$

$$+ \sum_{gg'} \langle \{S_{f'}^{-}(t)S_g^{z}(t) - S_{f'}^{z}(t)S_g^{-}(t)\}\{S_f^+ S_g^z - S_f^z S_g^+\} \rangle$$

$$\times \sum_{n=1}^{\infty} \frac{1}{n!} \left[\sum_{\alpha\beta} \langle u_{f'g'}^{\alpha} t u_{fg}^{\beta} \rangle \nabla_{f'}^{\alpha} \nabla_f^{\beta} \right]^n \tilde{J}_{f'g'}\tilde{J}_{fg} \right\} , \qquad (4.23)$$

where, in the last term, the operators $\nabla_{f'}^{\alpha}$ and ∇_f^{β} act on the functions $\tilde{J}_{f'g'} = \tilde{J}(\vec{f}' - \vec{g}')$ and $\tilde{J}_{fg} = \tilde{J}(\vec{f} - \vec{g})$, respectively. We have also used the abbreviation $u_{fg}^{\alpha} = u_f^{\alpha} - u_g^{\alpha}$.

Further simplification of the expression (4.23) for the mass operator can be obtained if we evaluate approximately the four-spin time correlation functions. The first term contains the irreducible correlation function, defined by equation (4.5). Taking into account equation (4.6), we have in the simplest approximation:

$$\langle \{S_{g'}^{z}(t)S_{f'}^{-}(t)\}^{ir}\{S_f^+ S_g^z\}^{ir} \rangle$$

$$\approx \langle \{(S_{g'}^{z}(t) - \langle S_{g'}^{z}\rangle)S_{f'}^{-}(t)\}\{S_f^+(S_g^z - \langle S_g^z\rangle)\} \rangle$$

$$\approx K_{g'g}^{zz}(t)K_{f'f}^{-+}(t). \qquad (4.24)$$

This approximation takes into account the inelastic scattering of the spin excitations on the fluctuations of the spin z-component (longitudinal fluctuations). If we employ next in (4.24) the static approximation for the correlation function $K_{g'g}^{zz}(t)$:

$$K_{g'g}^{zz}(t) \approx K_{g'g}^{zz}(0) \equiv K_{g'g}^{zz}, \qquad (4.25)$$

and, if we evaluate the correlation function $K_{f'f}^{-+}(t)$ with the help of the Green function (4.18), taken in the pole approximation:

$$-\frac{1}{\pi} \text{Im}G(\nu, \omega + i\varepsilon) \approx 2\langle S^z \rangle \delta(\omega - h - \varepsilon(\vec{v})), \qquad (4.26)$$

we get the contribution to the mass operator of the form

$$2\langle S^z \rangle M_{s-s}^{(1)}(\vec{v}, \omega) = \frac{1}{N} \sum_{\nu'} \frac{[\tilde{J}(\vec{v}') - \tilde{J}(\vec{v}' - \vec{v})]^2}{\omega - h - \varepsilon(\vec{v} - \vec{v}')} , \qquad (4.27)$$

which coincides with the expression obtained in the first order
by Vaks *et al.* [29]. Making use, in the evaluation of the irreducible correlation functions, appearing in (4.23), of the expansion

$$S_f{}^z = S - \frac{1}{2S} S_f{}^- S_f{}^+ + \dots ,$$

we can obtain, in the second order, the contribution to the mass
operator coming from the inelastic scattering on the transverse
spin components (spin waves). In the case of the low temperatures we can easily obtain the contribution to the mass operator
of the wellknown form [29]:

$$2\langle S^z \rangle M_{\text{S-S}}{}^{(2)}(\vec{\vartheta},\omega)$$

$$= - \frac{1}{2N^2} \sum_{\vec{p},\vec{q}} \frac{[\tilde{J}(\vec{p}) + \tilde{J}(\vec{q}) - \tilde{J}(\vec{p} - \vec{\vartheta}) - \tilde{J}(\vec{q} - \vec{\vartheta})]^2}{h + \varepsilon(\vec{p}) + \varepsilon(\vec{q}) - \varepsilon(\vec{p} + \vec{q} - \vec{\vartheta}) - \omega}$$

$$[(1 + N_{\vec{p}} + N_{\vec{q}})N_{\vec{p}+\vec{q}-\vec{\vartheta}} - N_{\vec{p}}N_{\vec{q}}], \qquad (4.28)$$

where in evaluation of the correlation function $K_{ff'}{}^{-+}(t)$ we
have used the pole approximation (4.26).

The second term in (4.23) describes the inelastic spin-phonon
scattering ($n \geqslant 1$). The spin correlation function in this term
may be approximately represented as follows:

$$\langle S_{g'}{}^z(t)S_f{}^-(t)S_g{}^z S_f{}^+ \rangle \approx \{\langle S^z \rangle^2 + K_{g'g}{}^{zz}(t)\}K_{f'f}{}^{-+}(t). \qquad (4.29)$$

In this approximation the spin-phonon part of the mass operator
(4.23) takes the form

$$2\langle S^S \rangle M_{\text{S-ph}}(\nu,\omega) = \int_{-\infty}^{\infty} \frac{d\omega'}{\omega - \omega'} (e^{\omega'/\theta} - 1) \int_{-\infty}^{\infty} \frac{dt}{2\pi} e^{\prime i\omega' t}$$

$$\times \sum_{ff'gg'} e^{-i\vec{\nu}(\vec{f}-\vec{f'})} \sum_{n=1}^{\infty} \frac{1}{n!} \left(\sum_{\alpha\beta} \langle u_{f'g'}{}^{\alpha}(t)u_{fg}{}^{\beta} \rangle \nabla_{f'}{}^{\alpha}\nabla_f{}^{\beta} \right)^n \tilde{J}_{f'g'} \tilde{J}_{fg}$$

$$\times \frac{1}{2\langle S^z \rangle} \{\langle S^z \rangle^2 [K_{f'f}{}^{-+}(t) + K_{g'g}{}^{-+}(t) - K_{f'g}{}^{-+}(t) - K_{g'f}{}^{-+}(t)]$$

$$+ [K_{g'g}{}^{zz}(t)K_{f'f}{}^{-+}(t) + K_{f'f}{}^{zz}(t)K_{g'g}{}^{-+}(t)$$

$$- K_{g'f}{}^{zz}(t)K_{f'g}{}^{-+}(t) - K_{f'g}{}^{zz}(t)K_{g'f}{}^{-+}(t)] . \qquad (4.30)$$

The first term in the brace brackets determines the scattering
of the magnetic excitations with the emission and absorption of
the phonons in the mean field of the z-component of the spins
($\sim\langle S^z \rangle^2$). The second term takes into account the fluctuations
of the longitudinal spin components ($\sim K_{ff'}^{zz}(t)$) and the additional momentum transfer to the spin subsystem. Taking into account in (4.30) the one-phonon ($n = 1$) inelastic scattering
only, we can easily obtain the results of the work [14] (see
also [21]), based on the rather complicated diagrammatic technique for the spin-phonon system. It is not difficult to make
some estimation in (4.30) of the multiphonon scattering processes, which may be important in the high temperature region.

5. DISCUSSION

The main feature of the theory of the spin-phonon interaction
we have presented in these lectures is the self-consistent account of the mutual influence of the spin subsystem on the lattice vibrations, and the reciprocal action of the lattice vibrations on the spectrum of the magnetic excitations and the magnetization. This account, in its simplest form, is done in the mean
phonon field approximation with the aid of the Bogoliubov variational principle (section 2). The more consistent approach is
given in the framework of the Green function method. The self-
consistency of the calculations within the Green function method
is achieved by evaluating the correlation functions, which determine the mass operator, with the aid of the full Green function (cf. (3.25),(4.23,30)). This procedure leads to the self-
consistent system of non-linear equations, which does not contain the zero order Green functions, describing the propagation
of the free excitations, which do not describe correctly the
system with strong interactions (cf. [26]). In this respect,
the self-consistent method is equivalent to construction of the
perturbation theory in the framework of the diagrammatic technique with respect to the skeleton diagrams, which contain only
the full Green functions [32]. At the same time it is not difficult to set up a correspondence between a given decoupling in
the mass operator (3.24), (4.22,24) and the definite set of the
diagrams in the diagrammatic technique, and thus to estimate a
chosen approximation [33].
   The second characteristic feature of the theory developed
consists in taking into consideration all terms in the expansion
of the crystal potential energy as well as of the exchange interaction in thermal displacement of the atoms in the Hamiltonian (1.9), This procedure allows us to consider in a consistent
manner the spin-phonon interaction in the highly anharmonic
crystals.
   The example of such a system are: the solid He^3, where the
exchange of nuclei in the crystal lattice sites is described by
the Heisenberg Hamiltonian [34], and some hydrogen bonded ferro-
electrics, which are described by the so called quasi-spin model

[35]. The spin-phonon interaction in this model is of principal importance [36]. The amplitudes of the atomic vibrations in such crystals are large, so that the simple harmonic approximation is inapplicable.

Let us notice finally that the account of the compressibility of a ferromagnetic crystal, which can be, in the simplest way, described by the equation (1.16), plays an important rôle in the determination of the type of the magnetic phase transitions, and in the evaluation of the temperature dependence of the spectrum of magnetic excitations and of the magnetization [37].

REFERENCES

1. Akhiezer, A.J., Baryakhtar, V.G. and Kaganov, M.J. (1960). *Usp. Fiz. Nauk*, **71**, 533; **72**, 3; [(1961). *Sov. Phys. Usp.*, 3, 567, 661].

2. Akhiezer, A.J., Baryakhtar, V.G. and Peletminski, S.W. (1968). *Spin Waves*, (North Holland, Amsterdam).

3. Izyumov, Yu.A. (1961). *Fiz. Met. Metalloved.*, **12**, 20.

4. Kashcheev, V.N. and Krivoglaz, M.A. (1961). *Fiz. Tverd. Tela*, 3, 1541; [(1961).*Sov. Phys.*, *Solid State*, 3].

5. Elliot, R.J. and Stern, H. (1961). *Proceedings of The Conference on Inelastic Scattering of Neutron in Solids and Liquids*, (IAEA, Vienna).

6. Szaniecki, J. (1962). *Acta. Phys. Polon.*, **22**, 3, 9, 21.

7. Yakovlev, E.N. (1962). *Fiz. Tverd. Tela*, **4**, 594; [(1962). *Sov. Phys.*, *Solid State*, **4**, 433].

8. Pytte, E. (1965). *Ann. Phys. (N.Y)*, **32**, 377.

9. Bennet, H.S. and Pytte, E. (1967). *Phys. Rev.*, **155**, 553.

10. Tani, K. and Tanaka, H. (1970). *J. Phys.*, **C2**, L90; Tani, K. and Mori, H. (1968). *Prog. Theor. Phys.*, **39**, 876.

11. Silberglitt, R. (1969). *Phys. Rev.*, **188**, 786.

12. Kashcheev, V.N. (1971). *Phys. Stat. Sol.*, **43**, 51.

13. Kashchenko, M.P., Balakhonov, N.F. and Kurbatov, L.V. (1972). *Fiz. Met. Metalloved.*, **33**, 18.

14. Tyablikov, S.V. and Konwent, H. (1968). (Preprint, JINR P4-3794), (Dubna); (1968). *Phys. Lett.*, **A27**, 130.

15. Konwent, H. (1968). (Preprint, JINR P4-4028), (Dubna); (1968). *Phys. Lett.*, **A28**, 237.

16. Konwent, H. and Plakida, N.M. (1970). *Teor. Mat. Fiz.*, **3**, 135.

17. Plakida, N.M. (1970). *Phys. Lett*, **A32**, 134.

18. Konwent, H. and Plakida, N.M. (1971). *Teor. Mat. Fiz.*, **8**, 119.

19. Konwent, H. and Plakida, N.M. (1971). *Phys. Lett.*, **A37**, 173.

20. Konwent, H. (1972). *The Effects of the Spin-Phonon Interaction in Anharmonic Crystals*, (University of Wrocław, Institute of Theoretical Physics), (Preprint, No.247), (in Polish).

21. Konwent, H., Plakida, N.M. and Vukajlovič, F.R. (1973). *Teor. Mat. Fiz.*, (Preprint, JINR P4-7145), (Dubna), (to be

published).

22. Meissner, G. (1970). *Z. Phys.*, **237**, 272.

23. Boccara, N. (1971). *Phys. Stat. Sol.*, **43**, K11.

24. Zagrebnov, V.A. and Fedyanin, V.K. (1972). *Teor. Mat. Fiz.*, **10**, 127.

25. Leibfried, G. (1955). Gittertheorie der Mechanischen und Thermischen Eigenschaften der Kristalle; in *Handbuch der Physik, Vol VII/I*, (ed. Flugge, S.), (Springer Verlag), p. 104.

26. Plakida, N.M. (1970). *Teor. Mat. Fiz.*, **5**, 147; (1972). *Teor. Mat. Fiz.*, **12**, 135; (1973) Green Function Method in the Theory of Anharmonic Crystals, in *Statistical Physics and Quantum Field Theory*, (ed. Bogoliubov, N.N.), (Nauka, Moskva), (in Russian).

27. Plakida, N.M. (1973). *Phys. Lett.*, **A43**, 481.

28. Mubayi, V. and Lange, R.V. (1969). *Phys. Rev.*, **178**, 882; Kenan, R.P. (1970). *Phys. Rev.*, **B1**, 3205.

29. Vaks, V.G., Larkin, A.J. and Pikin, S.A. (1967). *Zh. Eksp. Teor. Fiz.*, **53**, 281, 1053; [(1967). *Sov. Phys. JETP*, **26**, 188, 647].

30. lzyumov, Yu.A. and Kassan-Ogly, F.A. (1968). *Fiz. Met. Metalloved.*, **26**, 385; (1970). *Fiz. Met. Matalloved.*, **30**, 225.

31. Rudoy, Yu.G. and Tserkovnikov, Yu.A. (1973). *Teor. Mat. Fiz.*, **14**, 102.

32. Abrikosov, A.A., Gorkov, L.P. and Dzyaloshinski, J.E. (1965). *Methods of Quantum Field Theory in Statistical Physics*, (2nd edn.), (Prentice Hall, London).

33. Vukajlovič, F.R. (1973). (Preprint, JINR P4-7291), (Dubna).

34. Trickey, S.B., Kirk, W.P. and Adams, L.D. (1972). *Rev. Mod. Phys.*, **44**, 668.

35. De Gennes, P.G. (1963). *Solid State Commun.*, **1**, 132; Blinc, R. (1960). *J. Phys. Chem. Sol.*, **13**, 204; Blinc, R. and Zeks, B. (1972). *Adv. Phys.*, **21**, 693.

36. Kabayashi, K.K. (1968). *J. Phys. Soc. Jap.*, **24**, 497; Stasyuk, J.V. and Levitsky, R.R. (1970). *Ukrain. Fiz. Zh.*, **15**, 460; Stasyuk, J.V. (1971). *Teor. Mat. Fiz.*, **9**, 431; Moore, M.A. and Williams, H.C.W.L. (1972). *J. Phys.*, **C5**, 3168.

37. Vukajlovič, F.R. (1973). *Phys. Lett.*, **A44**, 231.

THE SYMMETRY PROPERTIES OF MAGNETOELASTIC WAVES AND THE DETERMINATION OF SELECTION RULES FOR MAGNON-PHONON INTERACTIONS

ARTHUR P. CRACKNELL

*Carnegie Laboratory of Physics,
University of Dundee, Dundee DD1 4HN, Scotland*

1. INTRODUCTION

Ever since the classic paper by Bouckaert, Smoluchowski and Wigner [1], group theory has been used extensively in connection with the electronic energy bands of crystalline solids. As a result of a well known theorem owing to Wigner, the wave function of an electron in a given energy band in some crystalline material must belong to, that is it must transform according to, one of the irreducible representations of the space group of the material. Consequently, the irreducible representations of the space groups have been used extensively in labelling electronic energy bands, in simplifying the forms of electronic wave functions, and in determining selection rules for processes involving electronic transitions (for references, see, e.g., the bibliography given by Bradley and Cracknell [2]). It is only much more recently that these ideas have begun to be applied to other particles or quasi-particle excitations in crystalline solids, such as phonons and magnons.

The early experimental measurements of phonon dispersion relations by means of inelastic neutron scattering were mostly performed for important directions in crystals of high symmetry. Each normal mode could then be described in a meaningful way as either longitudinal acoustic (LA), transverse acoustic (TA), longitudinal optic (LO), or transverse optic (TO). However, as measurements were extended to directions of lower symmetry and to crystals with more complicated structures, it became apparent that this simple scheme was inadequate for classifying the normal modes of vibration in many crystals. Consequently the use

of space group representations, which had previously been used
for several decades in labelling electronic energy bands, has
now also been adopted for labelling the phonon dispersion relations of crystals and in studying the degeneracies of these phonon dispersion relations (see, for example, [3-8]). It is also
possible to classify the various branches of the spin wave dispersion relations of a magnetic crystal into acoustic magnon
branches (MA) and optic magnon branches (MO). The physical significance of such a classification is not very great (see section 3) and so group theoretical labels have now also been introduced for labelling magnon dispersion relations. In this case
the dispersion relations are labelled in terms of the irreducible
corepresentations of the Heesch-Shubnikov space group, or magnetic space group, that describes the symmetry of the magnetically
ordered structure. Particular examples have been considered by
a number of authors [9-13] while the general principles were described in my lectures here at the VI-th Winter School [14]. So
far the assignment of group theoretical labels to the various
branches of the phonon or magnon dispersion relations has been
performed for only a rather limited number of crystalline materials, although very recently a computer program has been developed to perform this analysis for phonons [15].

Originally, magnon-phonon interactions, for small k, were investigated by ultrasonic techniques. More recently the effect
of magnon-phonon interactions on the dispersion relations for
these quasi-particle excitations in magnetic crystals has come
to be of considerable interest, as a result of refinements in
the techniques involved in inelastic neutron scattering experiments [16]. From such experiments it became apparent that, at
least for the special lines of symmetry in the Brillouin zone, a
given magnon may interact appreciably with certain phonons but
not with others; this has been observed, for example, in Tb [17].
In these lectures we shall examine the group theoretical aspects
of the magnon-phonon interactions for various wave vectors, k.
Some brief mention of the use of group theory in determining selection rules for magnon-phonon interactions has been made by
one or two authors [18], (see also p. 233 of the article by
Mackintosh and Møller [19]); more detailed discussions have been
given for ferromagnetic hcp metals [20] and for the antiferromag-
tic difluorides MnF_2, FeF_2 and CoF_2 [21,22].

Following a brief discussion of the symmetries of magnetoelastic waves in section 2, we shall use group theoretical arguments
in section 3 for the determination of general selection rules for
magnon-phonon interactions, without reference to any particular
form of the interaction. It is our view that these arguments are
valid in their own right, as a consequence of the direct application of Wigner's theorem (see chapter 11 of [23]) to the magnetoelastic waves. After considering the application to some important common magnetic crystals in sections 4,5 we shall return in
section 6 to a discussion of the determination of selection rules
for various particular interaction mechanisms and of the relation
of such selection rules to those determined only from group theor-

etical considerations, since this also is a matter of current
interest (see, for example, the discussion at the end of [24]).

2. MAGNETOELASTIC WAVES

It is common to assume that, to a very good first approximation, the lattice vibrations of a crystal are unaffected by the
behaviour of any magnetic moments which may be associated with
the atoms or ions in the crystal. Similarly, in attempting to
describe the motions of the spins in a magnetically ordered system, that is in discussing spin waves, it is also traditional to
assume that the atoms or ions with which the magnetic moments
are associated are frozen in their equilibrium positions and not
involved in any translational motions. This approximation is in
the same spirit as the adiabatic or Born-Oppenheimer approximation which is commonly used when calculating the electronic band
structure of a crystalline solid; in this approximation it is assumed that the electronic band structure can be calculated neglecting the motions of the nuclei, or ions, and assuming them to
be frozen in their equilibrium positions.

If the non-interacting approximation is abandoned, the situation in figure 1 may arise. The Hamiltonian of the system will
consist of a lattice part and a spin part and at least one term
which involves both the displacements of the nuclei and the orientations of the spins. The collective excitations of the system described by this Hamiltonian are therefore neither pure
lattice vibrations nor pure spin waves, but rather they are
coupled *magnetoelastic waves*. A magnetoelastic wave involves
displacements both in the positions of the atoms in a crystal
and in the orientations of the spin vectors associated with the
atoms. The existence of such magnetoelastic waves has been
known, at least from a theoretical point of view, for a long
time (see, for example, chapter 4 of [25]). However, it is only
recently that it has become possible to study the dispersion
relations for such magnetoelastic waves experimentally in any
great detail. Instead of the two intersecting curves AD and BC
describing lattice vibrations and spin waves respectively in
figure 1, the two curves no longer cross at k_0 and two non-in-
tersecting curves AC and BD are obtained, each of which describes
a magnetoelastic wave. For very small k, the quasi-particle excitation described by AC is almost entirely a pure lattice vibration, whilst for very large k near the Brillouin zone boundary
it is almost a pure spin wave; it is only in the vicinity of k_0
that this quasi-particle excitation involves substantial displacements both in the atomic positions and in the spin orientations.
That is why the curve AC is described in figure 1 as being 'phonon-like' near A and 'magnon-like' near C. Similarly, curve BD
is magnon-like for small k and phonon-like for large k. It will
need a little ingenuity to invent a term to describe the quantum
of energy of a magnetoelastic wave near k_0 involving both substantial phonon-like and magnon-like behaviour.

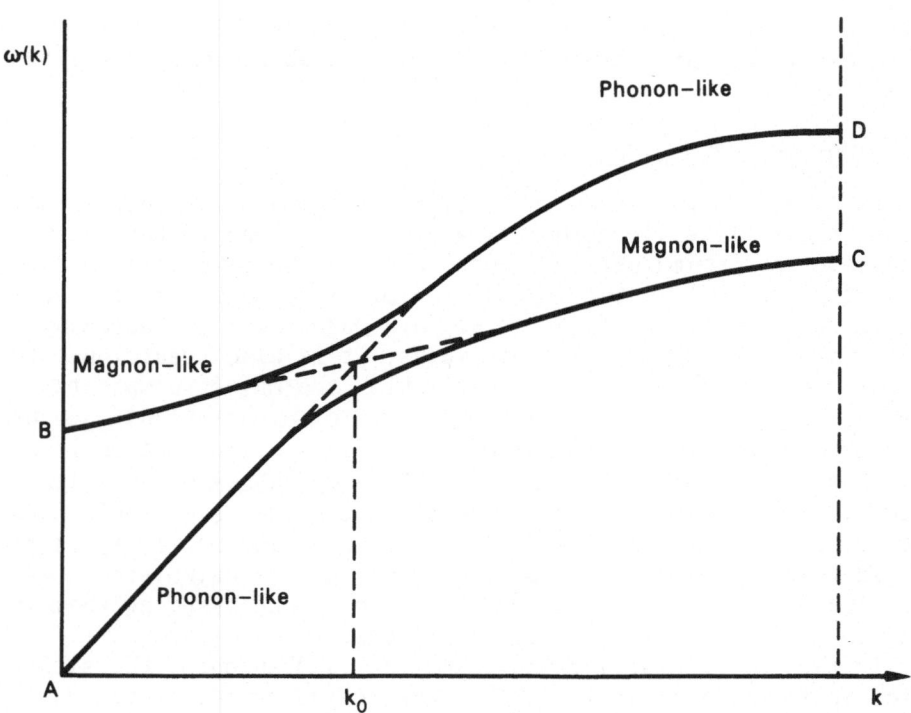

Figure 1 - Sketch of the dispersion relations for coupled
magnetoelastic waves in a hypothetical magnetic crystal
involving one acoustic phonon branch and one magnon branch.
The broken lines indicate the dispersion relations in the
absence of any coupling between spin waves and lattice
vibrations.

3. THE SYMMETRIES OF MAGNETOELASTIC WAVES AND THE DETERMINATION OF SELECTION RULES FOR MAGNON-PHONON INTERACTIONS

The various branches of the lattice vibration spectrum of a
given magnetic crystal are commonly labelled by using the irreducible representations of the classical space group, G, which
describes the symmetry of the configuration of the atomic positions and neglects the existence of the magnetic moments associated with the atoms. In a similar manner the various branches
of the spin wave dispersion spectrum of a magnetic crystal may
be labelled by using the irreducible corepresentations of the
magnetic space group which describes the structure, including
the magnetic ordering pattern. It is, perhaps, important to
relate these labelling schemes to the older labelling schemes
in terms of LA, TA, LO, and TO for phonons and MA and MO for
magnons.

The question of the definition of longitudinal and transverse
waves in complex crystals is not trivial, and simple ideas based on
structures like rocksalt and zincblende are inadequate for the

general case. Suppose we have an acoustic wave with a wave vector **k** corresponding to a long but finite wavelength, that is **k**
is small but non-zero. The centre of mass of each primitive cell
is moving periodically in time, and if the displacement of the
centre of mass is in a straight line parallel to **k** the mode is
LA; if it is in a straight line or an ellipse lying in a plane
normal to **k** the mode is TA.   If neither of these situations applies, the classification into LA and TA is not valid. A similar argument applies to infrared active optic modes, if for the
displacement of the centre of mass one substitutes the electric
dipole moment of the primitive cell. It follows that non-primi-
tive translations for space group elements are irrelevant, as
the phase shift of the wave from one cell to the next can be
made arbitrarily small by allowing **k** to tend towards zero in a
well defined direction. (In the same way macroscopic tensors
for a crystal are governed by its point group, not by its space
group). Suppose the group of the wave vector **k** is derived from
elements of the form $\{R|\mathbf{v}\}$ in the Seitz notation. When one ap-
plies the operation $\{R|\mathbf{v}\}$ to the acoustic mode the displacement
of the centre of mass suffers a rotation R; therefore the acous-
tic mode must transform under $\{R|\mathbf{v}\}$ in the same way that a Eu-
clidian vector transforms under the point group operation R,
provided that we can neglect the small phase shift produced by
v. The three acoustic modes of wave vector **k** are mutually orthogonal, and it follows that they transform like the three orthogonal components of a polar vector in Euclidian space. It
should be noted that these arguments apply strictly only in the
limit of long waves, where the acoustic frequencies are determined by the elastic constants. In the neighbourhood of the
Brillouin zone boundary the distinction between TA and LA can
be quite meaningless, and cases exist where a branch is LA near
the zone centre, passes smoothly through the zone boundary and
becomes TA near the centre of the next zone (see, for example,
figure 6 of [26]). We prefer to avoid making any attempt to
classify magnons as being longitudinal or transverse.

We should also consider the validity, in general, of the distinction between acoustic modes (A) and optic modes (O) for
phonons and magnons. For any crystal there must be three lattice vibrational modes which can be described approximately in
terms of the macroscopic elastic constants of the crystal for
long wavelength, that is for small **k**. The group theoretical
labels for the acoustic modes can easily be identified from the
fact that the acoustic modes transform like the components of a
polar vector. The acoustic modes will, for small values of **k**,
have a velocity which is independent of wavelength. Therefore,
for small **k** the dispersion relation $\omega(\mathbf{k})$ against $|\mathbf{k}|$ for each of
these modes will take the form of a straight line passing through
the origin. If we adopt the rule that all the remaining normal
modes for any given crystal are described as optic modes it can
be seen that the classification into acoustic and optic modes for
phonons is of general validity. Whereas the existence of the

acoustic modes is a general feature of the phonon dispersion relations for any crystal, a casual inspection of a selection of
published magnon dispersion relations reveals that the existence
of modes with the characteristic 'acoustic mode' properties,
namely with the simple linear dispersion relation $\omega_j(\mathbf{k}) = C_1|\mathbf{k}|$
in the vicinity of $\mathbf{k} = 0$ where $\omega_j(\mathbf{k})/2\pi$ is the magnon frequency
and C_1 is a constant, is by no means universal. In particular
the dependence of $\omega_j(\mathbf{k})$ on $|\mathbf{k}|$ is very often not linear and there
may also be a gap, that is $\omega_j(\mathbf{k}) \neq 0$ at $\mathbf{k} = 0$. The gap is often sufficiently large that the energies of the magnons in the
vicinity of $\mathbf{k} = 0$ correspond to infrared frequencies, or very
high ultrasonic frequencies, rather than to ordinary acoustic
frequencies. Nevertheless, in spite of the reduced physical significance of the distinction, it is quite common to adopt for
magnons a rule that is similar to that used for phonons: that is,
we describe the lowest spin wave mode at each $\mathbf{k}$ as an acoustic
magnon mode (MA) and all the rest as optic magnon modes (MO).
Since we are dealing with only one degree of freedom, the number
of acoustic magnon branches to the dispersion relations for any
given crystal is one and not three.

The same general theorem, namely Wigner's theorem, which lies
behind the use of group theory in labelling the electronic band
structure, the phonon dispersion relations, and the magnon dispersion relations, also applies to the magnetoelastic waves.
That is to say, if $\mathcal{H}$ is the Hamiltonian used to describe the
coupled magnetoelastic vibrations of a magnetically ordered crystal, any given magnetoelastic wave must belong to one or other
of the irreducible representations of the group of $\mathcal{H}$, which in
turn will be determined by the symmetry of the crystal itself.
The question of the existence of accidental degeneracies for the
dispersion relations of a particle, or of quasi-particle excitations, in a crystalline solid was studied by Herring [27], who
showed that the occurrence of an accidental degeneracy between
two branches at the same $\mathbf{k}$ and belonging to the same representation
is vanishingly improbable (except for certain very special simple
forms of Hamiltonian, such as for example that used in the 'empty
lattice' approximation in band theory). The importance of this
result has long been appreciated in connection with electronic
band structures. Along a line of symmetry we therefore expect to
find intersections of electronic energy bands, that is accidental degeneracies between energy bands at certain $\mathbf{k}$, for energy
bands belonging to different irreducible representations, but we
do not expect to find intersections between two bands belonging
to the same representation. If we consider the electronic energy bands along an arbitrary (i.e. non-symmetry) direction of
$\mathbf{k}$ from the centre to the boundary of the Brillouin zone, there
is only one irreducible space group representation, $\Gamma_1{}^{\mathbf{k}}$, and all
the bands therefore belong to $\Gamma_1{}^{\mathbf{k}}$. This means that for an arbitrary direction of $\mathbf{k}$ we should expect to find no intersections
among the energy bands. These arguments were originally given
for electronic energy bands and, in the past, they have princip-

ally been used in this connection. However, the arguments are general and can be applied equally well to the dispersion relations for any particle or quasi-particle excitation in a crystalline solid. In particular we are concerned at present with their relevance to magnetoelastic waves.

The above discussion implies that we start with a Hamiltonian $\mathcal{H}$ with terms describing the vibrations of the atoms, the oscillations of the spins, and the interactions between the motions of the atoms and the orientations of the spins. However, we have seen in the previous section that it is profitable to regard magnetoelastic waves as arising from a relatively weak coupling between the lattice vibrations and the spin waves. This suggests that it would be profitable to consider the symmetries of coupled magnetoelastic waves in terms of the symmetries of those phonons and magnons which become coupled together. We first note that since the dispersion curve for a magnetoelastic wave, or for any other quasi-particle excitation, is a quasi-continuous function of k this dispersion curve must belong to the same irreducible representation for all values of k between Γ, the centre of the Brillouin zone, and a point on the boundary of the Brillouin zone. The situation illustrated in figure 1 applies for an arbitrary direction of k in the Brillouin zone. All the phonon branches in this particular direction belong to Γ_1^k and all the magnon branches also belong to Γ_1^k. Similarly, all the magnetoelastic waves will also belong to Γ_1^k. Therefore, in this direction any given magnon branch which intersects some phonon branch can interact with it to form a coupled magnetoelastic wave. This means that, for any arbitrary direction in the Brillouin zone, there is no selection rule preventing the interaction of phonons and magnons. When we speak about 'selection rules' in this context what we mean is the possibility of the lifting of an accidental degeneracy between a phonon and a magnon with the same value of k, the cause of the lifting of this degeneracy being the magnon-phonon interactions.

We now consider the case of a special line of symmetry in the Brillouin zone. Each branch of the dispersion relations for coupled magnetoelastic waves is a quasi-continuous function of k. Thus, everywhere along this line of symmetry, any given branch of these dispersion relations will be labelled by the same representation, or corepresentation, of the space group of the crystal. This leads to the selection rule:

> If k is on a line of symmetry the degeneracy between a magnon and a phonon can only be lifted if the magnon and the phonon have the same symmetry, that is they must belong to the same irreducible representation, or corepresentation, of the space group of the crystal.

For the purpose of this selection rule we do not regard the two members of a pair of complex conjugate representations as 'different'. In practice this statement of the selection rule may need modification by replacing the word 'different' by the word

'incompatible'; this is because for any given magnetic crystal
it is quite common to use different space groups in describing
the phonon symmetries and the magnon symmetries. For the phonon symmetries it is common to use the (classical) space group,
say G, which describes the symmetry of the atomic positions in
the crystal and neglects the magnetic moments, whereas for the
magnon symmetries it is common to use the magnetic space group,
M, of the magnetically ordered crystal or, possibly, the spin
space group G_s [28,29]. In testing the compatibilities it will
then be necessary to use the grey group derived from G for the
phonons.

Let us assume that group theory has been used to determine
which irreducible representations describe the phonons and the
magnons in a given crystal and also to determine the forms of
their eigenvectors. By constructing and diagonalizing suitable
Hamiltonians, possibly with some parameters that have to be determined experimentally, it may be possible to distribute these
irreducible representations uniquely among the observed branches
of the dispersion relations both for the magnons and for the
phonons. If this has been done then it will be possible to see
by inspection whether any given magnon branch which intersects
a certain phonon branch in the non-interacting approximation
will be expected to couple, or mix, with that phonon branch when
magnon-phonon interactions are included. Alternatively, if the
phonon or magnon symmetries have not been distributed unambiguously among the observed branches of the dispersion relations,
the study of the interactions among the various phonon and magnon branches may be useful to assist in determining these assignments unambiguously. We shall illustrate the use of these selection rules for some examples in the next two sections, based on
the previous papers by Cracknell [20,21] and Montgomery and
Cracknell [22].

4. SELECTION RULES FOR FERROMAGNETIC hcp METALS

The space group of a paramagnetic hexagonal close packed
metal is $P6_3/mmc(D_{6h}^4)$ or, if time reversal symmetry is included,
$P6_3/mmc1'$. As basis vectors of the hexagonal Bravais lattice we
take

$$t_1 = (0,-a,0),$$

$$t_2 = (\tfrac{1}{2}\sqrt{3}a,\tfrac{1}{2}a,0), \qquad (1)$$

$$t_3 = (0,0,c),$$

when the basis vectors g_1, g_2 and g_3 of the reciprocal lattice
will be

$$g_1 = \frac{2\pi}{a} \left(\frac{1}{\sqrt{3}} , -1, 0 \right),$$

$$g_2 = \frac{2\pi}{a} \left(\frac{2}{\sqrt{3}} , 0, 0 \right), \qquad\qquad (2)$$

$$g_3 = \frac{2\pi}{c} (0, 0, 1).$$

This is the notation used by Bradley and Cracknell [2]. In addition to the pure translations of the Bravais lattice the space group $P6_3/mmc1'$ contains the operations

$\{E\|0\}$	1	$\{I\|0\}$	13	$\{\sigma_h\|\tau\}$	16,1	$\{C_2\|\tau\}$	4,1
$\{C_3^+\|0\}$	3	$\{S_6^-\|0\}$	15	$\{S_3^+\|\tau\}$	18,1	$\{C_6^-\|\tau\}$	6,1
$\{C_3^-\|0\}$	5	$\{S_6^+\|0\}$	17	$\{S_3^-\|\tau\}$	14,1	$\{C_6^+\|\tau\}$	2,1
$\{C_{21}'\|\tau\}$	10,1	$\{\sigma_{d1}\|\tau\}$	22,1	$\{\sigma_{v1}\|0\}$	19	$\{C_{21}''\|0\}$	7
$\{C_{22}'\|\tau\}$	8,1	$\{\sigma_{d2}\|\tau\}$	20,1	$\{\sigma_{v2}\|0\}$	23	$\{C_{22}''\|0\}$	11
$\{C_{23}'\|\tau\}$	12,1	$\{\sigma_{d3}\|\tau\}$	24,1	$\{\sigma_{v3}\|0\}$	21	$\{C_{23}''\|0\}$	9

together with their products with θ, the operation of time inversion. Each operation is denoted first by its Seitz symbol and secondly by the label used by Miller and Love [30]. In the Seitz symbols the notation is that of Altmann and Bradley [31], except that we have moved the origin of the coordinate axes to the inversion centre so that $\tau = (0,0,\frac{1}{2}c)$.

The symmetries of the phonons for an hcp metal can be obtained from table XXIV of [7]. At the point Γ at the centre of the Brillouin zone the phonons belong to

$$\Gamma: \quad A_{2-}'' \oplus A_{2+}'' \oplus E_+' \oplus E_-', \qquad\qquad (3)$$

where the irreducible representations are labelled in the notation of [31]. This can be translated into the notation of [30] using table 7 of [32], when we find that the normal modes at Γ belong to

$$\Gamma: \quad \underset{(1)}{\Gamma}_2^- \oplus \underset{(2)}{\Gamma}_6^- \oplus \underset{(1)}{\Gamma}_3^+ \oplus \underset{(2)}{\Gamma}_5^+, \qquad\qquad (4)$$

where the degeneracies are indicated in brackets. Of these normal modes Γ_2^- and Γ_6^- are acoustic modes, while Γ_3^+ and Γ_5^+ are optic modes. The phonon symmetries along the lines of symmetry Δ, Σ, and Λ (or T), which pass through Γ can easily be identified using the compatibility tables given by Miller and Love [30] between these lines and Γ. By studying the transformation

TABLE I

Phonon Symmetries Along Δ, Λ *(or T) and* Σ *for*
hcp Metals

Γ	Δ		Λ (or *T*)		Σ	
Γ_2^-	Δ_1	LA	Λ_3	TA	Σ_3	TA
Γ_6^-	Δ_6	TA	$\begin{cases}\Lambda_1 \\ \Lambda_4\end{cases}$	$\begin{array}{l}\text{LA} \\ \text{TA}\end{array}$	$\begin{cases}\Sigma_1 \\ \Sigma_2\end{cases}$	$\begin{array}{l}\text{LA} \\ \text{TA}\end{array}$
Γ_3^+	Δ_4	O†	Λ_2	O†	Σ_3	TO
Γ_5^+	Δ_5	O†	$\begin{cases}\Lambda_1 \\ \Lambda_4\end{cases}$	$\begin{array}{l}\text{LO} \\ \text{TO}\end{array}$	$\begin{cases}\Sigma_1 \\ \Sigma_2\end{cases}$	$\begin{array}{l}\text{LO} \\ \text{TO}\end{array}$

† L/T classification does not apply.

properties of a vector (ξ,η,ζ) where ζ is along the direction
of **k** and ξ and η are normal to **k**, it is also possible to assign
the appropriate label L or T to each of the normal modes, except
where the L/T classification is not meaningful. The results are
given in table I. Since we shall be concerned with some of the
black and white space groups in connection with the discussion
of magnon symmetries in terms of the corepresentations of $P6_3/$
$mmc1'$ rather than in terms of the representations of $P6_3/mmc$.
However, for Γ, Δ, Σ, and Λ (or T) all the irreducible representations of $P6_3/mmc$ lead to case (a) corepresentations of $P6_3/$
$mmc1'$ (see for example, p.373 of Bradley and Cracknell [2]) so
that the inclusion of time reversal symmetry causes no change in
the degeneracies of the phonons for these wave vectors.

To study selection rules for magnon-phonon interactions in a
ferromagnetic hcp metal we have to consider the magnon symmetries,
too. For this we use the appropriate Heesch-Shubnikov space
group. For the ferromagnetic hcp metals, there are two magnetic
atoms per unit cell and therefore there are two branches to the
spin wave dispersion relations, namely one acoustic branch (MA)
and one optic branch (MO). The group theoretical labels for the
acoustic and optic branches can be identified by considering a
Heisenberg Hamiltonian for two sublattices *A* and *B* corresponding
to the different sites in the unit cell (see, for example, pp.
195-6 of the article [*19*]). The symmetry of the acoustic magnon

branch and the optic magnon branch can be identified by studying the transformation properties of the operators

$$\alpha_{1\mathbf{k}}{}^\dagger = \Lambda(\mathbf{k})\{a_{\mathbf{k}}{}^\dagger + b_{\mathbf{k}}{}^\dagger\} \tag{5}$$

and

$$\alpha_{2\mathbf{k}}{}^\dagger = \Lambda(\mathbf{k})\{a_{\mathbf{k}}{}^\dagger - b_{\mathbf{k}}{}^\dagger\}, \tag{6}$$

respectively, where $\Lambda(\mathbf{k})$ is some function of $\mathbf{k}$ that does not affect the symmetry arguments,

$$\left. \begin{aligned} a_{\mathbf{k}}{}^\dagger &= N^{-\frac{1}{2}} \sum_{\ell} \exp(-i\mathbf{k}\cdot\mathbf{R}_\ell)S_\ell{}^- \\[2mm] b_{\mathbf{k}}{}^\dagger &= N^{-\frac{1}{2}} \sum_{m} \exp(-i\mathbf{k}\cdot\mathbf{R}_m)S_m{}^- \end{aligned} \right\} \tag{7}$$

where S_ℓ and S_m are the spin operators for the sites on the A
and B sublattices respectively, and N is the number of unit cells
in the crystal. The actual magnetic space group that one has to
use will depend on the orientation of the magnetization, M, relative to the crystallographic axes [13,14]. The translational
symmetry of a ferromagnetic hcp metal will be the same as in the
paramagnetic phase, because all the spins of the metal ions are
parallel to each other and there are no displacements of the
atoms, since the ferromagnetic phase transition is assumed to be
a second order phase transition. The shape of the Brillouin
zone for the magnetically ordered crystal, neglecting any distortions which may occur below the Curie temperature, is therefore
identical with the shape of the Brillouin zone for the paramagnetic phase. However, the reduction in symmetry due to the magnetic ordering means that this Brillouin zone may be the Brillouin zone of a Bravais lattice with less symmetry, for example
an orthorhombic Bravais lattice, but with special values of the
axial ratios determined by the fact that the symmetry of the array of atomic positions is that of the hcp structure, see figure
2.

    If the magnetization M is in a completely arbitrary orienta-
tion relative to the crystallographic axes the symmetry will be
that of the rather trivial triclinic space group $P\bar{1}$ which, apart
from the lattice translations, contains just the identity $\{E|0\}$
and the space inversion $\{I|0\}$. If M is in an arbitrary direction
$[uv\dagger0]$ in the basal plane the magnetic space group is $P2_1{}'/m'$
containing the elements

$$\{E|0\}, \qquad \{I|0\}, \qquad \theta\{C_2|\tau\}, \qquad \theta\{\sigma_h|\tau\}. \tag{8}$$

For other special orientations of the magnetization some less

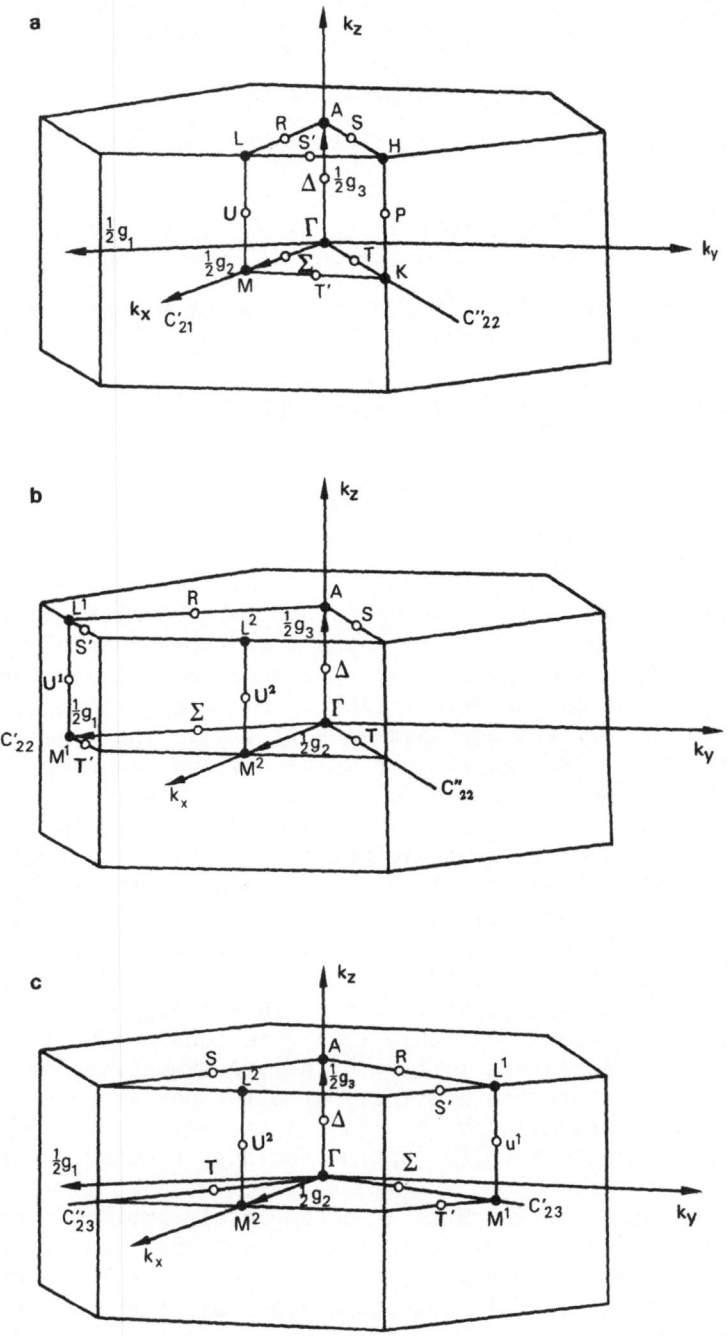

Figure 2 - (a) The Brillouin zone of a ferromagnetic hcp
metal magnetized parallel to [0001] ($P6_3/mm'c'$); (b) the
Brillouin zone of a ferromagnetic hcp metal magnetized
parallel to [10$\bar{1}$0] ($Cm'cm'$); (c) the Brillouin zone of a
ferromagnetic hcp metal magnetized parallel to [11$\bar{2}$0]
($Cmc'm'$).

TABLE II

The Magnetic Space Group of a Ferromagnetic hcp Metal

Direction of M	Magnetic space group	Unitary subgroup				Anti-unitary elements ($\times\ \theta$)			
[0001]	$P6_3/mm'c'$	$\{E\|0\}$	$\{C_3{}^+\|0\}$	$\{C_3{}^-\|0\}$		$\{C_{21}'\|\tau\}$	$\{C_{22}'\|\tau\}$	$\{C_{23}'\|\tau\}$	
		$\{I\|0\}$	$\{S_6{}^-\|0\}$	$\{S_6{}^+\|0\}$		$\{\sigma_{d1}\|\tau\}$	$\{\sigma_{d2}\|\tau\}$	$\{\sigma_{d3}\|\tau\}$	
	⫿^{270}_{194}	$\{C_2\|\tau\}$	$\{C_6{}^+\|\tau\}$	$\{C_6{}^-\|\tau\}$		$\{C_{21}''\|0\}$	$\{C_{22}''\|0\}$	$\{C_{23}''\|0\}$	
		$\{\sigma_h\|\tau\}$	$\{S_3{}^-\|\tau\}$	$\{S_3{}^+\|\tau\}$		$\{\sigma_{v1}\|0\}$	$\{\sigma_{v2}\|0\}$	$\{\sigma_{v3}\|0\}$	
[10$\bar{1}$0]	$Cm'cm'$ ⫿^{464}_{63}	$\{E\|0\}$	$\{C_{22}'\|\tau\}$	$\{I\|0\}$	$\{\sigma_{d2}\|\tau\}$	$\{C_{22}''\|0\}$	$\{C_2\|\tau\}$	$\{\sigma_{v2}\|0\}$	$\{\sigma_h\|\tau\}$
[11$\bar{2}$0]	$Cm2'm'$ ⫿^{463}_{63}	$\{E\|0\}$	$\{C_{23}''\|0\}$	$\{I\|0\}$	$\{\sigma_{v3}\|0\}$	$\{C_{23}'\|\tau\}$	$\{C_2\|\tau\}$	$\{\sigma_{d3}\|\tau\}$	$\{\sigma_h\|\tau\}$
[$uvt0$]	$P2_1'/m'$ ⫿^{54}_{11}	$\{E\|0\}$	$\{I\|0\}$			$\{C_2\|\tau\}$	$\{\sigma_h\|\tau\}$		
[$uvtw$]	$P\bar{1}$ ⫿^{4}_{2}	$\{E\|0\}$	$\{I\|0\}$			——			

trivial magnetic space group will apply, see table II. Examples of ferromagnetic hcp metals in which the easy axes of magnetization are the first three given in table II are, respectively,

[0001]   Co;   Gd   (232 K $\leqslant T \leqslant$ 293 K)

[10$\bar{1}$0]   Tb;

[11$\bar{2}$0]   Dy.

To help to clarify the situation the relative positions of several important directions in the basal plane are identified in figure 3. We shall now consider each of the orientations of the magnetization M in table II in turn. In each case we shall identify the phonon symmetries for the appropriate magnetic space group and make use of the known magnon symmetries [13] to determine the selection rules for magnon-phonon interactions along those lines of symmetry that pass through Γ.

First we consider the case when the magnetization M is parallel to [0001] and the symmetry of the ferromagnetic crystal is described by the space group $P6_3/mm'c'$ (⫿^{270}_{194}). Since $P6_3/mm'c'$ is a subgroup of the grey group $P6_3/mmc1'$, it is possible to determine the phonon symmetries in $P6_3/mm'c'$ by the process of subduction, using the phonon symmetries given in table I. That is, we establish the compatibility relations between the irreducible

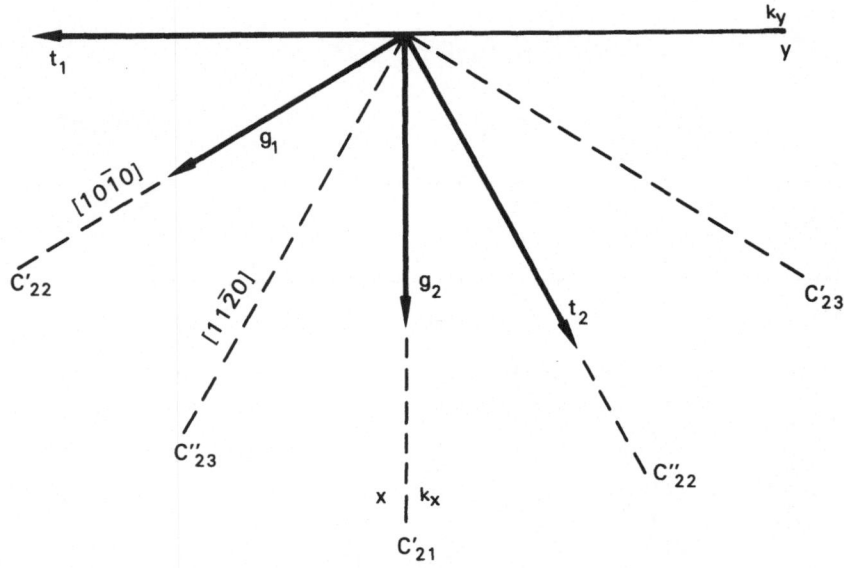

Figure 3 - Identification of some of the principal directions for the hcp structure

corepresentations of $P6_3/mmc1'$ and those of $P6_3/mm'c'$ for a given
k. These compatibilities are not given explicitly in the tables
of Miller and Love [30] but they can be derived from those tables
quite easily using the characters and the matrix representatives
given for the two groups $P6_3/mmc$ and $P6_3/mm'c'$. For example, if
we consider the line of symmetry Δ we find the following compatibilities for those representations of $P6_3/mmc$ to which the phonons belong

$P6_3/mmc$	$P6_3/mm'c'$
Δ_1 (LA)	Δ_1
Δ_6 (TA)	$\Delta_4 \oplus \Delta_6$
Δ_4 —	Δ_2
Δ_5 —	$\Delta_3 \oplus \Delta_5$.

By studying the transformation properties of the operators $\alpha_{1\mathbf{k}}^\dagger$
and $\alpha_{2\mathbf{k}}^\dagger$ in equations (5,6) under the operations of $P6_3/mm'c'$
it is possible to show that the acoustic magnon branch and the
optic magnon branch belong to Δ_6 and Δ_5, respectively, of $P6_3/$
$mm'c'$. Therefore the selection rules for interactions between
magnons and phonons along Δ are easy to see:

  (1) The phonon branches labelled by Δ_1 and Δ_4 in $P6_3/mmc$

cannot interact with either magnon branch of the dispersion relations;

(2) The phonon branch labelled by Δ_5 or Δ_6 of $P6_3/mmc$ can interact with one of the magnon branches, namely Δ_5 or Δ_6 respectively, of $P6_3/mm'c'$, but not with the other magnon branch, namely Δ_6 or Δ_5, respectively, of $P6_3/mm'c'$.

It will, of course, be realized that even though an interaction may be allowed group theoretically this does not give any guarantee that such an interaction will occur. That is, these selection rules give necessary, but not sufficient, conditions for an interaction to occur between a given magnon branch and a given phonon branch. We have gone into some detail in the consideration of the line Δ just to illustrate what is involved. We shall not repeat the arguments for the other lines of symmetry passing through Γ but simply quote the results in table III. From table III it can be seen that the selection rules are less restrictive along Λ or Σ than along the line Δ.

The arguments which we have given above for $P6_3/mm'c'$, that is for M parallel to [0001], can be repeated for the space groups $Cm'cm'$ and $Cmc'm'$ which correspond to the other special orientations of M. The details have already been given elsewhere [20] and so they will not be included here; we simply note the results which are given in table III. Finally, if M is in some arbitrary direction $[uv\bar{t}0]$ in the basal plane or in some completely arbitrary direction $[uv\bar{t}w]$, it is possible to show that no selection rule is expected for magnon-phonon interactions for any line of symmetry passing through Γ in the hcp Brillouin zone (for details see [20]); any magnon is able to interact with any phonon in these directions if both quasi-particles have common values of both energy and wave vector.

It is perhaps interesting to note from table III that, for each of the three most important magnetization directions in ferromagnetic hcp metals and for each of the lines of symmetry Δ, Λ (or T) and Σ, the LA-MA interaction is forbidden, although in several cases the LA-MO interaction is not forbidden. However, it should be stressed that this result is only demonstrated for these particular materials; it should not be taken to apply to other materials without further justification.

The two ferromagnetic hcp metals which have so far proved most profitable from the point of view of studying magnon-phonon interactions experimentally are Tb and Dy, for both of which the magnetization, M, is in the basal plane. Little evidence of any significant magnon-phonon interactions was observed in Gd in which the magnetization M is parallel to [0001] [33] while Co appears not to have been studied extensively in the search for magnon-phonon interactions. We have already noted in table III that when M is parallel to [0001] these interactions are certainly not all ruled out group theoretically. The non-existence of these interactions in Gd is probably not for any group-theoretical reason but because the appropriate magnon and phonon branches

TABLE III

Selection Rules for Magnon-Phonon Interactions in Ferromagnetic hcp Metals

Phonon $P6_3/mmc$	M∥[0001] $P6_3/mm'c'$ Phonon	MA	MO	M∥[10Ī0] $Cm'cm'$ Phonon	MA	MO	M∥[11Ž0] $Cmc'm'$ Phonon	MA	MO
		Δ_6	Δ_5		Λ_2	Λ_1		Λ_2	Λ_2
Δ_1 LA	Δ_1	×	×	Λ_1	×	✓	Λ_1	×	×
Δ_6 TA	Δ_4	×	×	Λ_1	×	✓	Λ_1	×	×
	Δ_6	✓	×	Λ_2	✓	×	Λ_2	✓	✓
Δ_4 O	Δ_2	×	×	Λ_2	✓	×	Λ_1	×	×
Δ_5 O	Δ_4	×	×	Λ_1	×	✓	Λ_1	×	×
	Δ_5	×	✓	Λ_2	✓	×	Λ_2	✓	✓
		Λ_2	Λ_2		Σ_2	Σ_1		Σ_2	Σ_1
Λ_3 TA	Λ_2	✓	✓	Σ_1	×	✓	Σ_2	✓	×
Λ_1 LA	Λ_1	×	×	Σ_1	×	✓	Σ_1	×	✓
Λ_4 TA	Λ_1	×	×	Σ_2	✓	×	Σ_2	✓	×
Λ_2 O	Λ_2	✓	✓	Σ_2	✓	×	Σ_1	×	✓
Λ_1 LO	Λ_1	×	×	Σ_1	×	✓	Σ_1	×	✓
Λ_4 TO	Λ_1	×	×	Σ_2	✓	×	Σ_2	✓	×
		Σ_2	Σ_2		Δ_2	Δ_2		Δ_2	Δ_2
Σ_3 TA	Σ_2	✓	✓	Δ_2	✓	✓	Δ_1	×	×
Σ_1 LA	Σ_1	×	×	Δ_1	×	×	Δ_1	×	×
Σ_2 TA	Σ_1	×	×	Δ_2	✓	✓	Δ_2	✓	✓
Σ_3 TO	Σ_2	✓	✓	Δ_2	✓	✓	Δ_1	×	×
Σ_1 LO	Σ_1	×	×	Δ_1	×	×	Δ_1	×	×
Σ_2 TO	Σ_1	×	×	Δ_2	✓	✓	Δ_2	✓	✓
(1)	(2)	(3)	(4)	(5)	(6)	(7)	(8)	(9)	(10)

Notes: (i) The phonons are labelled in the first column according to the representations of $P6_3/mmc$, the space group of the positions of the atoms in the crystal, neglecting spins. The L/T classification is not relevant to Δ_4, Δ_5 or Λ_2. The phonons are also re-labelled in columns 2, 5, and 8 according to the corepresentations of the appropriate magnetic space group. The magnons are also labelled according to the corepresentations of the appropriate magnetic space group. For both phonons and magnons the notation is that of Miller and Love [30]. (ii) A tick indicates that an interaction is allowed (i.e. a gap may appear) and a cross indicates that an interaction is forbidden (i.e. there is no gap). (iii) For M parallel to $[uvt0]$ or $[uvtw]$ there are no restrictions (see text).

which would be allowed to interact were already widely separated from one another in energy at all **k** before the interactions were considered.

The magnon-phonon interactions in Tb have been studied principally along the line Δ in the Brillouin zone [*17,18,34*]. Interactions were detected between the acoustic magnon branch (MA) and the transverse phonon branch (TA,Δ_6) and also between the acoustic magnon branch (MA) and the two-fold degenerate optic phonon branch (Δ_5), where the phonon symmetries are labelled in the notation for the space group $P6_3/mmc$. The optic magnon branch (MO) is forbidden by symmetry to interact with the non-degenerate Δ_4 optic phonon (see table III); also in this particular material the MO branch lies above the two-fold degenerate optic phonon branch Δ_5 for all **k** along Δ. Therefore the question of interaction between the MO branch and the optic phonon branches does not arise. The situation is illustrated in figure 4. The experimental results obtained for Dy by Nicklow and

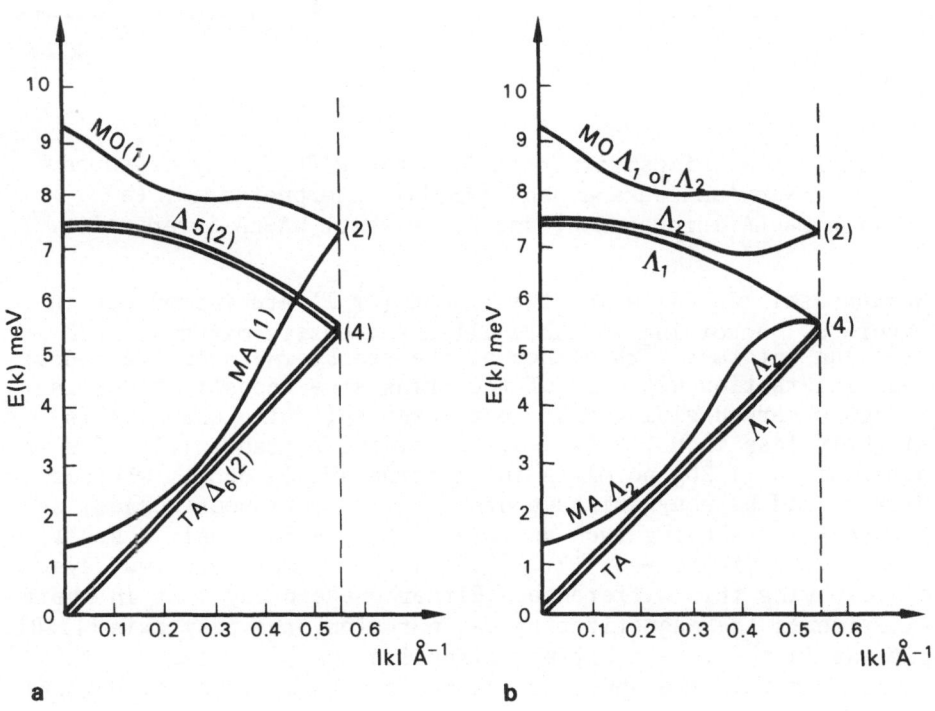

Figure 4 - Dispersion relations for magnons and for Δ_5 and Δ_6 phonons in hcp ferromagnetic metals with **M** in the basal plane: (a) in the absence of magnon-phonon interactions; (b) with interactions included. (b) applies for **M** ‖ [10$\bar{1}$0] and **M** ‖ [11$\bar{2}$0], the only difference between the two cases being in the group-theoretical label for MO: (b) also applies for **M** ‖ [uvt0] if all the branches are labelled as Λ_1.

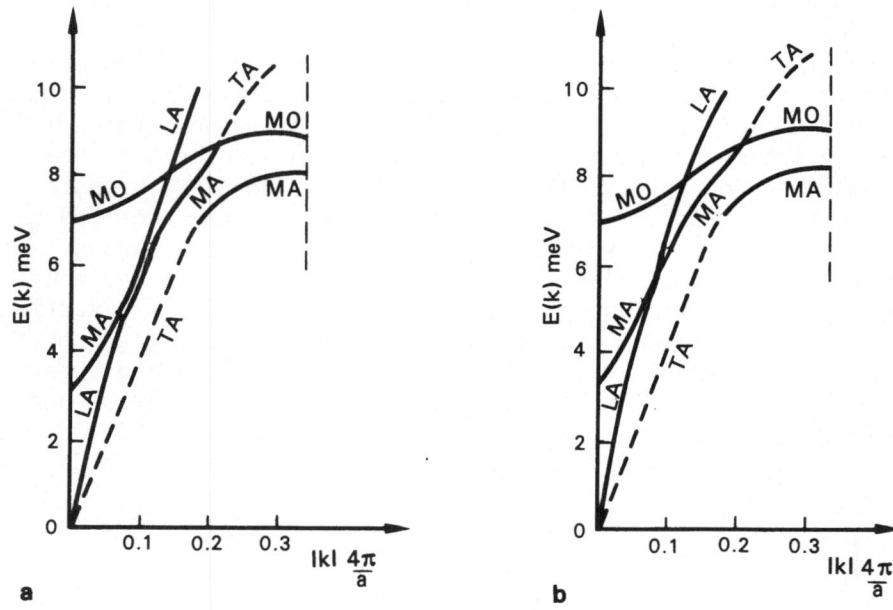

Figure 5 - Dispersion relations along ΓK for Dy measured
by Nicklow and Wakabayashi [35,36] interpreted as: (a)
with MA-LA interaction, and (b) without MA-LA interaction.

Wakabayashi [35,36] along the line Λ (or T) are reproduced in
figure 5. According to table III the acoustic magnon branch
(Σ_2) can interact with either of the transverse acoustic modes;
this interaction with one of the transverse acoustic modes is
indeed observed giving rise to a large splitting when $|\mathbf{k}|$ is
slightly less than 0.2 ($4\pi/a$). According to table III, if **M** is
parallel to [11$\bar{2}$0] no MA-LA interaction should occur, whereas
Nicklow and Wakabayashi have drawn their continuous curves, see
figure 5(a), showing such an interaction, albeit only a small
one, in the vicinity of $|\mathbf{k}| = 0.1$ ($4\pi/a$). There are two ways
of resolving this difference. Either one can say that in their
experiments the magnetization was not along the easy axis [11$\bar{2}$0],
but was in the more arbitrary direction [$uv\dagger0$] or [$uv\dagger w$]; this
would mean that the specimen was subject to an external magnetic
field during their experiments. Alternatively one can say that
since there is no MA-LA interaction when **M** is parallel to [11$\bar{2}$0]
the curves shown in figure 5(a) for these two branches should
cross as shown in figure 5(b).

5. SELECTION RULES FOR SIMPLE ANTIFERROMAGNETS WITH THE RUTILE STRUCTURE.

There are, at present, only a few antiferromagnetic structures for which the symmetries of both the phonons and the mag-

nons have been studied in detail. As an example we consider
antiferromagnetic FeF_2 [21,22] because magnon-phonon interactions in this material have recently been studied experimentally [24].

Antiferromagnetic FeF_2 has the rutile structure with the direction of the sublattice magnetization parallel or anti-parallel to the c axis; its symmetry is described by the magnetic
space group $P4_2'/mnm'$. This is the same as the structure of
antiferromagnetic CoF_2 or MnF_2 [37]. The classical space group
which describes the symmetry of the positions of the atoms in
the crystal, ignoring their spins, is $P4_2/mnm$ (D_{4h}^{14}). The
Brillouin zone is the same for these two space groups and is illustrated in figure 6.

The symmetries of the spin waves in FeF_2 will be the same as
those of MnF_2 which were identified in terms of the corepresentations of $P4_2'/mnm'$ by Loudon [9], see table IV, where the notation is that of Dimmock and Wheeler [37]. We prefer to avoid using the acoustic/optic classification (MA/MO) for the magnons in
antiferromagnetic FeF_2 for two reasons. First, there is a gap
at Γ, $\mathbf{k} = 0$, so that the lowest spin wave frequency at Γ is in
the infrared range rather than the acoustic range of frequen-
cies. Secondly, the two spin-wave branches are degenerate at Γ
and they are also degenerate at a number of special points on the
Brillouin zone boundary too; therefore a classification based on
the idea of a high frequency branch and a low frequency branch
would not be very meaningful.

The phonon symmetries have been identified for the rutile
structure; in terms of the classical space group $P4_2/mnm$, by
Katiyar [38]. It would then be possible to determine the phonon
symmetries in terms of the corepresentations of $P4_2'/mnm'$ by using Katiyar's results together with the compatibility tables
given by Dimmock and Wheeler [37]. However, we are primarily
concerned with the acoustic phonons. Their symmetries, together
with their assignments in the LA/TA classification scheme, can
easily be determined by studying the transformation properties
of a polar vector. For example, consider the line Δ in the Brillouin zone, where $\mathbf{k}$ has components $(0,\beta,0)$. Dimmock and Wheeler
[37] give the character table for Δ and its limiting form at very
small β is

	E	C_{2y}	σ_h	σ_{vx}	
Δ_1	1	1	1	1	$\mathbf{j}$
Δ_2	1	-1	1	-1	$\mathbf{i}$
Δ_3	1	-1	-1	1	$\mathbf{k}$
Δ_4	1	1	-1	-1	

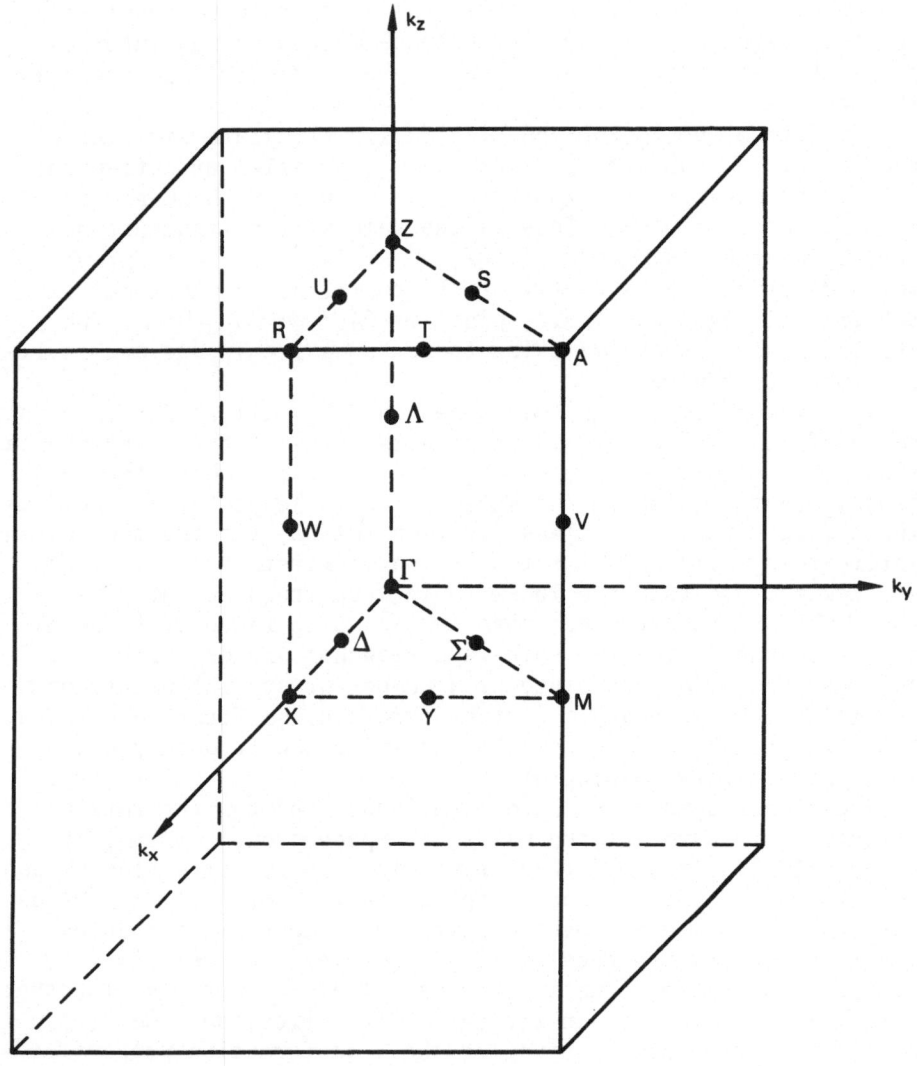

Figure 6 - Brillouin zone for antiferromagnetic FeF$_2$.

The unit vectors **i,j,k** transform as Δ_2, Δ_1 and Δ_3 respectively, and these must be the representations of the acoustic branches. Δ_1 is LA, while Δ_2 and Δ_3 are TA. Similar arguments can be applied to the lines Σ ([1$\bar{1}$0]) and Δ ([001]).

Having determined the labels both for the magnons and for the phonons in FeF$_2$ in terms of the space group $P4_2'/mnm'$, it is quite straightforward to write down the selection rules for magnon-phonon interactions in this material. The results are given in table V. From table V we see that for the Δ direction one of the magnon branches, Δ_4, should not be expected to interact with any of the acoustic phonon branches in FeF$_2$ and the other magnon branch, Δ_3, should be expected to interact with one of the TA

TABLE IV

Magnon Symmetries in Antiferromagnetic FeF$_2$

Point	Magnons	
Γ	$\Gamma_3^+ + \Gamma_4^+$	$D\Gamma_{3,4}^+$
M	M_3^+, M_4^+	DM_3^+, DM_4^+
Z	Z_2	DZ_2
A	A_2	DA_2
R	R_1^+	DR_1^+
X	X_2	DX_2
Δ	Δ_3, Δ_4	$D\Delta_3, D\Delta_4$
U	U_1	DU_1
Λ	$\Lambda_3 + \Lambda_4$	$D\Lambda_{3,4}$
V	V_3, V_4	DV_3, DV_4
Σ	$2\Sigma_2$	$2D\Sigma_2$
S	S_1, S_2	DS_1, DS_2
Y	Y_3, Y_4	DY_3, DY_4
T	T_1	DT_1
W	W_1	DW_1

Note: In column two the magnon symmetries are identified in terms of
the irreducible representations of
Pnnm [*37*], and in column three they
are identified in terms of the irreducible corepresentations of $P4_2'/mnm'$
(see section 7.7 of [*2*]).

phonon branches, but not with the other TA phonon branch and not
with the LA phonon branch. This is precisely what is observed in
practice in the experimental results [*24*], see figure 7. Similar

TABLE V

*Selection Rules for Magnon–Phonon Interactions
in Antiferromagnetic FeF$_2$*

Phonon $P4_2/mnm$		Phonon $P4_2'/mnm'$	Magnon $P4_2'/mnm'$	
			Δ_3	Δ_4
Δ_1	LA	Δ_1	×	×
Δ_2	TA	Δ_2	×	×
Δ_3	TA	Δ_3	✓	×
			Σ_2	Σ_2
Σ_1	LA	Σ_1	×	×
Σ_2	TA	Σ_1	×	×
Σ_4	TA	Σ_2	✓	✓
			Λ_3	Λ_4
Λ_1	LA	Λ_1	×	
Λ_5	TA	Λ_3 Λ_4	✓	

Notes: (i) The magnons are labelled according to the irreducible representations of *Pnnm*, the unitary subgroup of $P4_2'/mnm'$, where the notation is that of [37]. (ii) The bracket } or ⌣ is used to indicate that Λ_3 and Λ_4 become degenerate when the anti-unitary operations of $P4_2'/mnm'$ are included.

agreement with experiment is also obtained for the Σ ([110]) and Λ ([001]) directions.

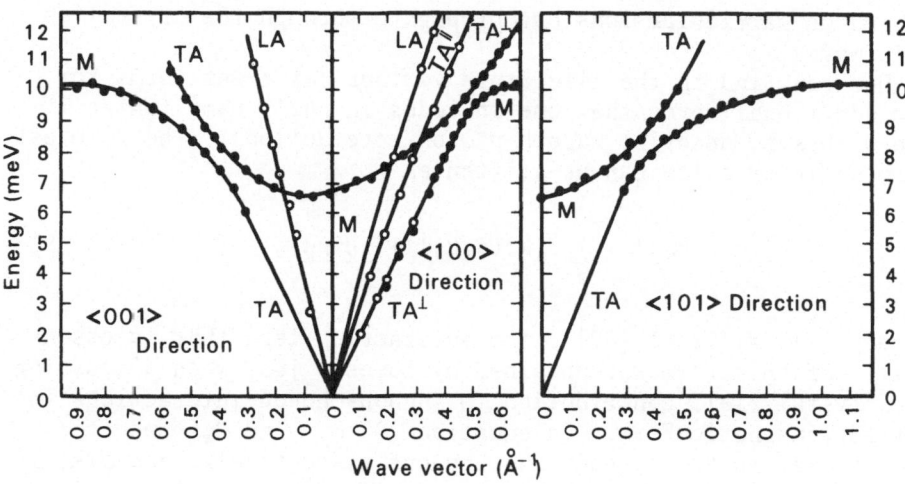

Figure 7 - Dispersion relations showing a magnon-phonon
interaction in antiferromagnetic FeF_2 [24]. M labels
the 'magnon-like' parts of the modes.

6. MECHANISMS

The selection rules that we have given in table III for ferromagnetic hcp metals and in table V for antiferromagnetic MnF_2,
FeF_2 or CoF_2 are general requirements that are independent of
the mechanism that is postulated as being responsible for the
interaction. However, even if the interaction is allowed group
theoretically there is no guarantee that any particular given
mechanism will lead to a detectable interaction. Interactions
between lattice vibrations and spin waves may arise from several
different physical origins; these include (i) magnetostriction,
(ii) changes in the crystalline electric field, (iii) a more complicated interaction proposed by Liu [18] involving the annihilation of a phonon and the creation of a magnon via the (virtual)
excitation of an electron-hole pair. We shall confine our attention to the first two of these mechanisms. The existence of magnetostriction, a change in the shape of a specimen resulting from
a change in its magnetization, is well known as a static phenomenon and the mechanism (i) simply involves applying this to a
dynamic situation. This is, the lattice vibrations lead to local
changes in the strain of the crystal which therefore lead to local changes in the magnetization of the crystal. Resolving the
displacements of the atoms and the variations in the magnetization in terms of normal modes in each case we have the corresponding picture in terms of magnons and phonons. In the second
mechanism the lattice vibrations are regarded as causing changes
in the crystalline electric field experienced by the magnetic
ions. These may lead to transitions involving changes in the
spin states of the ions; then, because of the strong coupling between the magnetic ions, such spin deviations do not remain loc-

alized on particular ions but propagate through the crystal as spin waves.

One can find in the literature various different forms for the additional terms that one includes in one's Hamiltonian if one wishes to describe magnon-phonon interactions. The simplest form of interaction can be written as

$$\mathcal{H}_{mp}' = \sum_{\mathbf{k}} U(\mathbf{k})(\alpha_{\mathbf{k}}\beta_{\mathbf{k}}{}^\dagger + \alpha_{\mathbf{k}}{}^\dagger\beta_{\mathbf{k}}), \qquad (9)$$

(see p. 74 of Kittel [39]). An interaction term that is essentially of this form was obtained by Lovesey [40] when discussing the crystal field contribution to magnon-phonon interactions in antiferromagnetic FeF_2. In equation (9) $\alpha_{\mathbf{k}}{}^\dagger$ and $\beta_{\mathbf{k}}{}^\dagger$ are creation operators for magnons and phonons respectively, and $U(\mathbf{k})$ is a function representing the magnitude of the interaction. A typical term in equation (9) such as

$$V_1 = U(\mathbf{k})\alpha_{\mathbf{k}}{}^\dagger\beta_{\mathbf{k}} \qquad (10)$$

corresponds to the annihilation of a phonon with wave vector $\mathbf{k}$ and the creation of a magnon with wave vector $\mathbf{k}$. When considering magnon-phonon interactions in Tb Jensen [17] obtained a more complicated interaction term of the form

$$\mathcal{H}_{mp}' =$$

$$\sum_{\mathbf{k}} W(\mathbf{k})(\alpha_{\mathbf{k}}{}^\dagger + \alpha_{-\mathbf{k}})(\beta_{\mathbf{k}} + \beta_{-\mathbf{k}}{}^\dagger) + \sum_{\mathbf{k}_1,\mathbf{k}_2} \{U(\mathbf{k}_2,\mathbf{k}_1)\alpha_{\mathbf{k}_1+\mathbf{k}_2}{}^\dagger\alpha_{\mathbf{k}_1}$$

$$+ \tfrac{1}{2}V(\mathbf{k}_2,\mathbf{k}_1)(\alpha_{\mathbf{k}_1+\mathbf{k}_2}{}^\dagger\alpha_{-\mathbf{k}_1}{}^\dagger + \alpha_{\mathbf{k}_1}\alpha_{-\mathbf{k}_1-\mathbf{k}_2})\}(\beta_{\mathbf{k}_2} + \beta_{-\mathbf{k}_2}{}^\dagger), \qquad (11)$$

where $W(\mathbf{k})$, $U(\mathbf{k}_2,\mathbf{k}_1)$ and $V(\mathbf{k}_2,\mathbf{k}_1)$ are functions representing the magnitudes of the interactions and depend on the exchange, crystal field, and magnetoelastic parameters of the crystal. The $W(\mathbf{k})$ terms are of the same form as the terms in Lovesey's Hamiltonian. The $U(\mathbf{k}_2,\mathbf{k}_1)$ and $V(\mathbf{k}_2,\mathbf{k}_1)$ terms are more complicated. A typical term would be

$$V_2 = U(\mathbf{k}_2,\mathbf{k}_1)\alpha_{\mathbf{k}_1+\mathbf{k}_2}{}^\dagger\alpha_{\mathbf{k}_1}\beta_{\mathbf{k}_2}, \qquad (12)$$

which corresponds to the creation of one magnon with wave vector $\mathbf{k}_1 + \mathbf{k}_2$ by annihilating one magnon with wave vector $\mathbf{k}_1$ and one phonon with wave vector $\mathbf{k}_2$, or, in other words, the scattering of a magnon from $\mathbf{k}_2$ to $(\mathbf{k}_1 + \mathbf{k}_2)$ with the absorption of a phonon with wave vector $\mathbf{k}_2$. Huberman and Burstein [41] considered a third kind of interaction, namely a two-phonon-one-magnon process,

involving the scattering of a phonon from k_1 to $k_1 + k_2$ with the
absorption (or emission) of a magnon with wave vector k_2 (or
$-k_2$). A typical term in this case would take the form

$$V_3 = U'(k_2,k_1)\beta_{k_2+k_1}{}^{\dagger}\beta_{k_1}\alpha_{k_2}, \tag{13}$$

where $U'(k_2,k_1)$ represents the magnitude of the interaction.

To determine selection rules for magnon-phonon interactions
for a particular mechanism, rather than on general considerations that were described in section 3 and exploited in sections
4 and 5, one would proceed as follows. The interaction must
first be expressed as some term added to the Hamiltonian that
was used to describe the system in the absence of magnon-phonon
interactions. This additional term must then be expressed in
terms of magnon and phonon creation and annihilation operators
as in equations (9-13). From this stage onward the arguments
are group-theoretical ones again. For example, suppose we consider the term V_1 in equation (10) which is added to the Hamiltonian $\mathcal{H}$ of a crystal with the symmetry of the space group G.
If $\alpha_k{}^{\dagger}$ belongs to $\Gamma_i{}^k$ of G and β_k belongs to $\Gamma_j{}^k$ of G, the term
$U(k)\alpha_k{}^{\dagger}\beta_k$ belongs to the Kronecker product representation

$$\Gamma_i{}^k \boxtimes \Gamma_j{}^k \tag{14}$$

of G since $U(k)$ is just a number and therefore belongs to $\Gamma_1{}^+$,
the identity representation of G. The Hamiltonian $\mathcal{H}$ must possess the full symmetry of the space group G, of the crystal,
that is it belongs to $\Gamma_1{}^+$ of G. Therefore, unless the Kronecker
product $\Gamma_i{}^k \boxtimes \Gamma_j{}^k$ contains $\Gamma_1{}^+$ the term $U(k)\alpha_k{}^{\dagger}\beta_k$ will not possess the symmetry required of the Hamiltonian $\mathcal{H}$ and therefore
$U(k)$ will have to be zero. This means that unless $\alpha_k{}^{\dagger}$ and β_k belong to complex conjugate representations, or to the same repre-
sentation if all the characters are real, interactions between
this phonon branch and this magnon branch will be forbidden. In
other words an interaction of the form in equation (10) leads to
exactly the same selection rule as we obtained from general considerations in section 3.

If we consider a different form for the interaction, such as
V_2 in equation (12) or V_3 in equation (13), we may obtain some
different selection rules. Let us consider V_2 for example. V_2
must belong to the product representation

$$\Gamma_p{}^{k_1+k_2*} \boxtimes \Gamma_q{}^{k_1} \boxtimes \Gamma_r{}^{k_2} \tag{15}$$

where $\alpha_{k_1+k_2}{}^{\dagger}$, α_{k_1} and β_{k_2} belong to $\Gamma_p{}^{k_1+k_2*}$, $\Gamma_q{}^{k_1}$ and $\Gamma_r{}^{k_2}$ respectively. Therefore, unless this triple Kronecker product
contains the identity representation Γ_1 of G at $k = 0$, the inter-

action described by V_2 will be forbidden. In general, this can be expected to lead to a different set of selection rules from those obtained from the interaction V_1. The procedure for the reduction of the triple Kronecker products of the form of (15) has been discussed by a number of authors: for further details see section 4.7 of the book by Bradley and Cracknell [2] in which many more specialized references are cited.

7. CONCLUSION

Our discussion of magnon and phonon symmetries and of selection rules for magnon-phonon interactions should assist in interpreting the results of inelastic neutron scattering experiments on these materials. Let us suppose that the orientation of the magnetization relative to the crystallographic axes in a given magnetic crystal is already known. Then the correct magnetic space group can be assigned to the structure. Let us assume that group theory has been used to determine which irreducible representations (or corepresentations) describe the phonons and the magnons in a given crystal and also to determine the forms of their eigenvectors. By constructing and diagonalizing suitable Hamiltonians, possibly with some parameters that have to be determined experimentally, it may be possible to distribute these irreducible representations uniquely among the observed branches of the dispersion relations both for the magnons and for the phonons. It should then be possible to ensure that the correct degeneracies and connectivities are assigned to each branch of the dispersion relations. Especially one can establish which phonon branches are allowed to couple with any given magnon branch and which ones are not allowed to couple with that magnon branch. Alternatively, if the phonon or magnon symmetries have not been distributed unambiguously among the observed branches of the dispersion relations, the experimental investigation of the interactions among the various phonon and magnon branches may be useful to assist in determining these assignments unambiguously.

Another interesting possibility is that of changing the selection rules for magnon-phonon interactions by the application of an external magnetic field. Suppose that the magnetization, or sublattice magnetization, **M** is spontaneously in some given special direction (the easy axis or direction) relative to the crystal axes. Then a certain set of selection rules apply and these can be enumerated in detail by the procedure that we have described and illustrated. If now an external magnetic field is applied, and in response to this field the magnetization becomes oriented in some arbitrary direction, these selection rules will no longer apply. Thus we have the interesting possibility of introducing selection rule breaking by using an external magnetic field to swing the magnetization away from one of the high-symmetry directions (in real space) into some arbitrary direction. In practice the values of the anisotropy make this much more likely to be observable in ferromagnets than in antiferromagnets.

REFERENCES

1. Bouckaert, L.P., Smoluchowski, R. and Wigner, E. (1936). *Phys. Rev.*, **50**, 58.
2. Bradley, C.J. and Cracknell, A.P. (1972). *The Mathematical Theory of Symmetry in Solids: Representation Theory for Point Groups and Space Groups*, (Clarendon Press, Oxford).
3. Cowley, R.A. (1964). *Phys. Rev.*, **134**, A981.
4. Johnson, F.A. and Loudon, R. (1964). *Proc. Roy. Soc.*, **A281**, 274.
5. Poulet, H. (1965). *J. Phys. (Paris)*, **26**, 684.
6. Maradudin, A.A. and Vosko, S.H. (1968). *Rev. Mod. Phys.*, **40**, 1.
7. Warren, J.L. (1968). *Rev. Mod. Phys.*, **40**, 38.
8. Montgomery, H. (1969). *Proc. Roy. Soc.*, **A309**, 521.
9. Loudon, R. (1968). *Adv. Phys.*, **17**, 243.
10. Joshua, S.J. and Cracknell, A.P. (1969). *J. Phys.*, **C2**, 24.
11. Daniel, M.R. and Cracknell, A.P. (1969). *Phys. Rev.*, **177**, 932.
12. Cracknell, A.P. and Joshua, S.J. (1969). *Proc. Camb. Phil. Soc.*, **66**, 493.
13. Cracknell, A.P. (1970). *J. Phys., Met. Phys. Suppl.*, **C3**, S175.
14. Cracknell, A.P. (1969). *Lectures, Sixth Annual Winter School of Theoretical Physics, Karpacz, February 16 - March 1, 1969*; (1969). *Rep. Prog. Phys.*, **32**, 633.
15. Worlton, T.G. and Warren, J.L. (1972). *Comput. Phys. Commun.*, **3**, 88.
16. Dolling, G. and Cowley, R.A. (1966). *Phys. Rev. Lett.*, **16**, 683.
17. Jensen, J. (1971). *Int. J. Mag.*, **1**, 271.
18. Liu, S.H. (1972). *Phys. Rev. Lett.*, **29**, 793.
19. Mackintosh, A.R. and Møller, H.B. (1972). Spin Waves, in *Magnetic Properties of Rare Earth Metals*, (ed. Elliot, R.J. Plenum Press, London), p. 187.
20. Cracknell, A.P. (1974). *J. Phys.*, **F4**, 466.
21. Cracknell, A.P. (1973). *J. Phys.*, **C6**, 1054.
22. Montgomery, H. and Cracknell, A.P. (1973). *J. Phys.*, **C6**, 3156.
23. Wigner, E.P. (1959). *Group Theory and Its Application to The Quantum Mechanics of Atomic Spectra*, (Academic Press, New York).
24. Rainford, B.D., Houmann, J.G. and Guggenheim, H.J. (1972). *Neutron Inelastic Scattering 1972. Proceedings of a Symposium, Grenoble 6-10 March 1972*, (IAEA, Vienna), p. 655.
25. Akhiezer, A.I., Bar'yakhtar, V.G. and Peletminskii, S.V. (1968). *Spin Waves*, (North-Holland, Amsterdam).

26. Montgomery, H. and Dolling, G. (1972). *J. Phys. Chem. Solids*, **33**, 1201.
27. Herring, C. (1937). *Phys. Rev.*, **52**, 365.
28. Brinkman, W.F. and Elliott, R.J. (1966). *J. Appl. Phys.*, **37**, 1457.
29. Brinkman, W.F. and Elliott, R.J. (1966). *Proc. Roy Soc.*, **A294**, 343.
30. Miller, S.C. and Love, W.F. (1967). *Tables of Irreducible Representations of Space Groups and Co-Representations of Magnetic Space Groups*, (Pruett Press, Boulder, Colorado).
31. Altmann, S.L. and Bradley, C.J. (1965). *Rev. Mod. Phys.*, **37**, 33.
32. Sedaghat, A.K. and Cracknell, A.P. (1974). *Physica*, **71**, 615.
33. Koehler, W.C., Child, H.R., Nicklow, R.M., Smith, H.G., Moon, R.M. and Cable, J.W. (1970). *Phys. Rev. Lett.*, **24**, 16.
34. Møller, H.B., Houmann, J.G., Jensen, J., and Mackintosh, A.R. (1972). *Neutron Inelastic Scattering 1972. Proceedings of a Symposium, Grenoble 6-10 March 1972*, (IAEA, Vienna), p. 603.
35. Nicklow, R.M. and Wakabayashi, N. (1971). *AIP Conference Proceedings, No.5; Magnetism and Magnetic Materials*, p. 1446.
36. Nicklow, R.M. and Wakabayashi, N. (1972). *Neutron Inelastic Scattering. Proceedings of a Symposium, Grenoble 6-10 March 1972*, (IAEA, Vienna), p. 611.
37. Dimmock, J.O. and Wheeler, R.G. (1962). *Phys. Rev.*, **127**, 391.
38. Katiyar, R.S. (1970). *J. Phys.*, **C3**, 1087.
39. Kittel, C. (1963). *Quantum Theory of Solids*, (Wiley, New York).
40. Lovesey, S.W. (1972). *J. Phys.*, **C5**, 2769.
41. Huberman, B.A. and Burstein, E. (1971). *AIP Conference Proceedings, No.5: Magnetism and Magnetic Materials*, p. 1350.

SOUND ABSORPTION IN MAGNETICS

V.G. KAMENSKY

*Landau Institute for Theoretical Physics,
Academy of Sciences of the USSR, Moscow*

1. INTRODUCTION

One of the important aspects of the dynamics of magnets is
the sound propagation near the critical point. A number of
theories have been developed for the case of ultrasonic attenuation at magnetic phase transitions (see for instance the paper
by B. Luthi, T.J. Moran and R.J. Pollina [1] and references
quoted there). The spin-phonon interaction, responsible for the
critical effects, arises in most cases via the strain modulation
of the exchange interaction (volume magnetostrictive coupling).
As a result, the attenuation is proportional to the space-time
Fourier Laplace transform of a four spin correlation function.
This result is the starting point of all further calculations.
Most of the present theories, however, consider temperatures a-
bove T_c, and therefore come down to the calculation of the characteristic decay time, t_e, of the spin fluctuations. Using the
various expressions for t_e, one gets the temperature dependence
for the attenuation. As a result one finds for the attenuation
coefficient $\gamma \sim \omega^2 \tau^{-\eta}$ ($\tau = (T - T_c)/T_c$). The critical changes
in sound wave attenuation follow a quadratic frequency depen-
dence, whilst the temperature dependence varies widely for different substances. The variation of the critical exponent can
be characterized by the following parameters: spin structure
(ferro- or antiferromagnets), degree of magnetic anisotropy of
the spin interaction, and range of the exchange interaction.

Definite interest is, however, attached to the case $T < T_c$,
when magnetic damping of sound is determined by the interaction
of phonons with the spin waves and spin fluctuations. We can

601

obtain the expression for the damping of sound in isotropic and anisotropic magnets near the transition point by using the diagram technique for a system with spin-spin and spin-phonon interaction, developed in [2-4]. As an example of the technique we discuss the case of the anisotropic ferromagnet [3].

2. MODEL AND DIAGRAM TECHNIQUE

We consider an anisotropic Heisenberg ferromagnet with an arbitrary exchange interaction between the spins, with a Hamiltonian in the form

$$\hat{H} = -\mu H \sum_{\vec{r}} S_{\vec{r}}^{z}$$

$$-\frac{1}{2} \sum_{\vec{r} \neq \vec{r}'} V(\vec{r} - \vec{r}')[aS_{\vec{r}}^{z}S_{\vec{r}'}^{z} + b(S_{\vec{r}}^{x}S_{\vec{r}'}^{x} + S_{\vec{r}}^{y}S_{\vec{r}'}^{y})]. \quad (1)$$

Here $\vec{S}_{\vec{r}}$ is the spin operator of the atom, which is assumed to be fixed in the crystal lattice site; $\vec{r}$ is the coordinate of site; $V(\vec{r} - \vec{r}')$ is the effective potential of the interaction between the spins; H is the external magnetic field and is directed along the z-axis; μ is the Bohr magneton; a and b the arbitrary constants. ($a = b$ yields the isotropic Heisenberg model; $a = 0$, $b \neq 0$ the xy-model and $b = 0$, $a \neq 0$ the Ising model). The summation is over all lattice sites, and the interaction radius may be large.

We shall consider the intermediary case $a \neq b \neq 0$ on condition that the average spin $\langle S \rangle$ is directed along the z-axis. That means that for $b > a$ there must be [5]

$$\mu H \geqslant (b - a)\langle S \rangle V_0, \quad (2)$$

where

$$V_0 = \sum_{\vec{r}} V(\vec{r}).$$

In order to obtain successive approximations, it is convenient to separate in (1) the interaction with the average spin $\langle \vec{S} \rangle \equiv \langle S^z \rangle$. The Hamiltonian then takes on the form

$$\hat{H} = \frac{1}{2}aNV_0\langle S \rangle^2 - \sum_{\vec{r}} S_r^z(\mu H + aV_0\langle S \rangle)$$

$$-\frac{1}{2} \sum_{\vec{r} \neq \vec{r}'} V(\vec{r} - \vec{r}') \, a(S_{\vec{r}}^z - \langle S_{\vec{r}}^z \rangle)(S_{\vec{r}'}^z - \langle S_{\vec{r}'}^z \rangle)$$

$$+ b(S_{\vec{r}}^+ S_{\vec{r}'}^- + S_{\vec{r}}^- S_{\vec{r}'}^+). \quad (1')$$

Here N is the number of sites, $S^{\pm} = (S^x \pm iS^y)/\sqrt{2}$.

The zeroth self-consistent field approximation, referred to as the molecular-field approximation, is obtained when the last term in (1) is neglected. In this case the average spin $\langle S^z \rangle$ is obtained from the condition that the free energy be a minimum

$$\frac{\partial F}{\partial y} \equiv \frac{\partial}{\partial y}\left[-\frac{1}{\beta N} \ln \mathrm{Sp}(-\beta\hat{H}) \right] = 0,$$

where $y = \beta(V_0\langle S^z \rangle + \mu H)$, $\beta = 1/T$.

One gets

$$\langle S^z \rangle = \frac{y - \beta\mu H}{\alpha\beta V_0} = B(y). \qquad (3)$$

Here $B(y)$ is a function connected with the well known Brillouin function $B_S(y)$:

$$B(y) = \frac{[\mathrm{Sp}S^z e^{yS^z}]}{\mathrm{Sp}e^{yS^z}} = SB_S(y) = (S + \tfrac{1}{2})\coth[y(S + \tfrac{1}{2})] - \tfrac{1}{2}\coth\tfrac{1}{2}y.$$

It follows from (3) that, in the absence of a magnetic field, a second order phase transition into a ferromagnetic state takes place at a temperature T_c

$$T_c = \frac{1}{3} \alpha V_0 S(S + 1). \qquad (4)$$

In order to find the next approximation, we shall use the temperature diagram technique developed by Vaks, Larkin and Pikin [2]. This technique constitutes a series of successive approximations of the self-consistent field method, and is suitable for ferromagnetic with a large interaction radius. The corrections to the thermodynamic and correlation functions are represented by a set of different connected diagrams, each of which can be represented in the form of single-cell blocks connected by the interaction lines $V(\vec{r} - \vec{r}')$. Each interaction line joins the vertices of different blocks, either S^z with S^z, or S^+ with S^-. A definite frequency and momentum correspond to each of these lines. The conservation laws are satisfied in each block. The Fourier component of a single-cell block with n outgoing interaction lines is given by the expression

$$\Gamma_n^{\alpha_1,\alpha_2,\ldots,\alpha_n}(\omega_1,\omega_2,\ldots,\omega_n)$$

$$= T^n\int_0^\beta \prod_{j=1}^n d\tau_j\exp(i\omega_j\tau_j)\times \qquad \text{(Contd)}$$

(Contd)

$$\times [\langle \hat{T} \prod_{j=1}^{n} S^{\alpha_j}(\tau_j) \rangle - \sum_{m_1+m_2+..+m_k=n} \Gamma_{m_1}{}^{\alpha_1\cdots}...\Gamma_{m_k}{}^{\cdots\alpha_n}]. \qquad (5)$$

Here $S^\alpha(\tau) = \exp(\tilde{H}_0\tau)S^\alpha\exp(-\tilde{H}_0\tau)$, $\tilde{H}_0 = -yTS^z$, T is the T-ordering symbol, $i\omega_m = 2\pi i mT$ are the imaginary frequencies of the temperature diagram technique, the mean value $\langle ... \rangle$ denotes $\mathrm{Sp}\rho_0(\cdots)$ with $\rho_0 = \exp(-\beta\tilde{H}_0)/\mathrm{Sp}\exp(-\beta\tilde{H}_0)$. The second term in the right hand side of (5) represents the sum of products of all the possible blocks of smaller order.

We present the analytic expressions for the simplest vertices, which were obtained in [2]:

$$\Gamma_2{}^{zz}(\omega_1,\omega_2) = B'(y)\delta(\omega_1)\delta(\omega_2),$$

$$\Gamma_2{}^{+-}(\omega_1,\omega_2) = B(y)G(\omega_1)\delta(\omega_1 - \omega_2), \qquad (6)$$

$$\Gamma_3{}^{+-z}(\omega_1,\omega_2,\omega_3) = -B(y)G(\omega_1)G(\omega_2)\delta(\omega_1 - \omega_2 + \omega_3),$$

where $\delta(\omega_k)$ is the Kronecker symbol of the corresponding frequency difference, and $G(\omega_k) = (y - i\beta\omega_k)^{-1}$.

In calculating the contributions corresponding to different diagrams, one sums over the internal frequencies and momenta.

The result of the summation over the momentum is an expression proportional to $r_0^{-3} = (d/R_0)^3$, where R_0 is the average interaction radius and d is the cell dimension, so that the correction of order ℓ in the expansion in r_0^{-3} will be represented by the aggregate of all possible connected diagrams containing ℓ closed loops. It is more convenient for further purposes to introduce the effective interactions which take into account the particle correlation and the presence of spin waves. In the first approximation in the self-consistent field, they have the form:

$$V^{zz}(\vec{k},i\omega_n) = \frac{aV_{\vec{k}}}{1 - \beta aB'(y)V_{\vec{k}}\delta(\omega_n)}, \qquad (7)$$

$$V^{+-}(\vec{k},i\omega_n) = \frac{bV_{\vec{k}}}{1 - \beta bB(y)V_{\vec{k}}^2 G(i\omega_n)},$$

where

$$V_{\vec{k}} = \sum_{\vec{r}} e^{ik\vec{r}}V(\vec{r}).$$

The interaction V^{zz} connects the S^z vertices of a single block
or of different blocks and will be shown by a wavy line. The
interaction V^{+-} connects an S^+ vertex with an S^- vertex and will
be shown by a full drawn line. In the case of the spin-phonon
interaction the technique contains also blocks Γ_{nm}, which contain
n spin operators and m phonon operators. It can be shown [3]
that the expressions for the Fourier components of such blocks
are obtained from the expressions for the blocks Γ_n of the same
order in n by making a suitable interchange of frequencies. If
in the block Γ_{nm} there are operators $S^{\alpha j}(\tau_j)$ and $\tilde{a}(\tau_j)$ with co-
inciding time dependent arguments, then, in order to obtain its
Fourier component, it is necessary to replace ω_j in (5) by $\omega_j +$
Ω_j, where Ω_j is the frequency of the corresponding phonon line.

3. SOUND WAVE DAMPING NEAR THE TRANSITION POINT

   Now we consider the sound propagation below T_c. The lattice
vibrations induce an oscillation of the distance $\vec{R}_{ij}$ between the
sites of spin i and j, and lead to a distortion of the exchange
integral $V(\vec{r}_i - \vec{r}_j)$. This distortion produces an interaction of
the sound with the spins, giving rise to an attenuation of the
sound. The connection between the spins and lattice vibrations
in the Hamiltonian (1) is given by the terms that arise when the
interaction potential is expended in terms of the displacements
of the sites:

$$V(\vec{r} - \vec{r}') = V^0(\vec{r} - \vec{r}') + [\nabla_{\vec{r}} V(\vec{r} - \vec{r}')]^0 \vec{u}_{\vec{r}}$$

$$+ [\nabla_{\vec{r}'} V(\vec{r} - \vec{r}')]^0 \vec{u}_{\vec{r}'}, \qquad (8)$$

where the zero index corresponds to quantities taken at the equi-
librium values of $\vec{r}$ and $\vec{r}'$, and $\vec{u}_{\vec{r}}$ is the displacement of the
corresponding lattice site. We shall henceforth omit the zero
index where there is no danger of misunderstanding. Substituting
(8) in (1') and recognizing that

$$V(\vec{r} - \vec{r}') = V(\vec{r}' - \vec{r}),$$

$$\sum_{\vec{r} \neq \vec{r}'} \vec{S}_{\vec{r}} (u_{\vec{r}} \nabla_{\vec{r}} V(\vec{r} - \vec{r}')) = \sum_{\vec{r}} \vec{S}_{\vec{r}} (u_{\vec{r}} \nabla_{\vec{r}} \sum_{\vec{r}'} V(\vec{r} - \vec{r}')) \equiv 0,$$

we obtain for the complete Hamiltonian

$$\hat{H} = \hat{H}_{ph} + \hat{H}_m + \hat{H}_{int}, \qquad (9)$$

where $\hat{H}_{ph}$ is the ordinary Hamiltonian of the non-interacting
phonons,

$$\hat{H} = \tfrac{1}{2} a N V_0 \langle S \rangle^2 - \sum_{\vec{r}} S_{\vec{r}}^z (\mu H + a V_0 \langle S \rangle), \qquad \text{(Contd)}$$

(Contd) $\hat{H}_{int} = -\frac{1}{2} \sum_{\vec{r} \neq \vec{r}'} V(\vec{r} - \vec{r}')[a(S_{\vec{r}}^{z} - \langle S_{\vec{r}} \rangle)(S_{\vec{r}'}^{z} - \langle S_{\vec{r}'} \rangle)$

$$+ b(S_{\vec{r}}^{+}S_{\vec{r}'}^{-} + S_{\vec{r}}^{-}S_{\vec{r}'}^{+})]$$

$$- \sum_{\vec{r} \neq \vec{r}'} a\langle S \rangle u_{\vec{r}} \nabla_{\vec{r}} V(\vec{r} - \vec{r}')(S_{\vec{r}'}^{z} - \langle S_{\vec{r}'} \rangle)$$

$$- \sum_{\vec{r} \neq \vec{r}'} u_{\vec{r}} \nabla_{\vec{r}} V(\vec{r} - \vec{r}')[a(S_{\vec{r}}^{z} - \langle S_{\vec{r}} \rangle)(S_{\vec{r}'}^{z} - \langle S_{\vec{r}'} \rangle)$$

$$+ b(S_{\vec{r}}^{+}S_{\vec{r}'}^{-} + S_{\vec{r}}^{-}S_{\vec{r}'}^{+})].$$

The corrections to the Green's function of the phonons

$$\mathcal{D}(\vec{r},\tau;\vec{r}',\tau') = - \langle \hat{T}\vec{u}(\vec{r},\tau)\vec{u}(r,\tau') \rangle,$$

where $\vec{u}(\vec{r},\tau)$ is the displacement operator, will be represented in the form of single-cell blocks connected by the interaction lines $V(\vec{r} - \vec{r}')$ and $\nabla_{\vec{r}}V(\vec{r} - \vec{r}')$ and by the phonon lines. The Dyson equation for the phonon Green's function (we shall consider now longitudinal phonons) has, in the momentum representation, the form

$$\mathcal{D}^{-1}(\vec{k},i\Omega_m) = \mathcal{D}^{(0)-1}(\vec{k},i\Omega_m) - \Pi(\vec{k},i\Omega_m), \tag{10}$$

where $\Pi(\vec{k},i\Omega_m)$ is the irreducible self-energy part and

$$\mathcal{D}^{(0)}(\vec{k},i\Omega_m) = - \frac{1}{M[\Omega_m^2 + \Omega_0^2(k)]}$$

is the Green's function of the free phonons (M is the mass of the cell, $\Omega_0(k) = ck$, c is the velocity of the sound).

Assuming the average interaction radius to be large compared with the inter-atomic distance, let us consider the first term in the expansion of the self-energy part $\Pi(\vec{k},i\Omega_m)$ in powers of r_0^{-3}. In this case the irreducible self-energy part $\Pi_1(\vec{k},i\Omega_m)$ will be equal to the sum of the expressions corresponding to diagrams with only one loop.

Figure 1 shows these diagrams. The lines marked in figure 1 by crosses correspond to the interaction $\nabla_{\vec{r}}V(\vec{r} - \vec{r}')$ the dashed lines correspond to phonons. Generally speaking, the complete set of the diagrams for $\Pi_1(\vec{k},i\Omega_m)$ must include also the similar diagrams with the single-cell blocks Γ_2^{zz}, Γ_3^{zzz}, Γ_{21}^{zz}, Γ_{22}^{zz} and Γ_{31}^{zzz}.

However, in order to study the dependence Π of Ω, one needs to investigate the difference $\Pi(\vec{k},i\Omega) - \Pi(\vec{k},0)$, where $\Pi(\vec{k},0) \equiv$

Figure 1

$\pi(\vec{k},i\Omega_m)|_{m=0}$. The blocks Γ_2^{zz}, Γ_3^{zzz}, Γ_{21}^{zz} Γ_{22}^{zz} and Γ_{31}^{zzz} contain $\delta(\omega_j)$ as a factor, where ω_j is the frequency corresponding to $V^{zz}(\vec{p},i\omega_j)$. It is obvious that inasmuch as this frequency coincides with the phonon frequency Ω_m, such diagrams need not to be taken into account. As a result, in order to obtain $\Pi_1(\vec{k},i\Omega_m) - \Pi_1(\vec{k},0)$, it is necessary to take into account only the diagrams which are shown in figure 1.

Summing all expressions which correspond to them, carrying out summation over the inner frequency and performing the analytic continuation $i\Omega_m \to \Omega + i\delta$, we obtain

$$\Pi_1(\vec{k},\Omega) = \frac{3\sqrt{3}}{16\pi^3} \frac{v_c T}{R_0^5 T_c} \int d^3 p \left[a\langle S\rangle k V_{\vec{k}} + p\cos\alpha\, bB(y) V_{\vec{p}} \right.$$

$$\left. - (p\cos\alpha + k)bB(y)V_{\vec{p}+\vec{k}} \right]^2$$

$$\times \frac{\coth\left(\dfrac{\varepsilon_{\vec{p}}}{2T}\right) - \coth\left(\dfrac{\varepsilon_{\vec{p}+\vec{k}}}{2T}\right)}{\varepsilon_{\vec{p}+\vec{k}} - \varepsilon_{\vec{p}} - \Omega - i\delta} \, . \qquad (11)$$

where v_c is the volume of the elementary cell (the summation over $\vec{p}$ was transformed to an integral), α is the angle between the vector $\vec{p}$ and the phonon wave vector $\vec{k}$,

$$k = |\vec{k}|R_0, \qquad p = |\vec{p}|R_0,$$

and

$$\varepsilon_{\vec{p}} = [\mu H + (a - b)B(y)V_0] + 3bB(y)V_0k^2, \qquad [3],$$

The absorption of the sound will be determined by the imaginary part of $\Pi_1(k,\Omega)$. Obviously, $\mathrm{Im}\Pi_1(\vec{k},\Omega)$ exists only for $\Omega < \varepsilon_{\vec{p}+\vec{k}} - \varepsilon_{\vec{p}} \sim bB(y)kR_0V_0$. In our case $\Omega = c|\vec{k}|$ and $bB(y)V_0 \sim (bT_c/a)\sqrt{|\tau|}$. Therefore, if $\sqrt{|\tau|} > ac/bR_0T_c$ and k is very small, the principal rôle in the attenuation of the sound will be played by the long wave magnons, which must meet the requirement of the existence of the spin waves.

Let us calculate the damping of the sound, $\gamma = \mathrm{Im}\Pi_1(k,\Omega)/2M\Omega$ below T_c in the case $\alpha + \mu Hm \ll |\tau|$. Calculating the imaginary part of equation (11) and keeping only the singular term, we obtain

$$\gamma \sim \frac{\Omega}{r_0^4\sqrt{|\tau|}} \frac{\omega_D}{Mc^2}, \qquad (12)$$

where ω_D is the Debye frequency.

Analogous calculations give in the case $\alpha + \mu Hm \gtrsim |\tau|$,

$$\gamma \sim \frac{\Omega\sqrt{|\tau|}}{r_0^4} \frac{\omega_D}{Mc^2} \frac{a\alpha^2}{b(\alpha + \mu Hm)^2}. \qquad (13)$$

The calculation of the next approximations does not change the qualitative form of the obtained results.

In the case of the two-sublattice antiferromagnetic one can obtain in similar way [4]

$$\gamma \sim \Omega^2\tau^{-3/2}. \qquad (14)$$

4. CONCLUSION

It should be noted that the model considered is very rough and the results obtained have a rather qualitative character. First of all, the interaction radius of the actual magnets, which we assumed to be large, is usually small, therefore the parameter of the expansion $r^{-3}|\tau|^{-\frac{1}{2}}$ is not small. The forms of the Hamiltonian are also highly conditional. One can only hope that the results obtained give a correct qualitative description. However, it is interesting to compare our results with the present experiments.

The heavy rare earth elements are known to have localized magnetic moments composed of the $4f$ electrons and to have a long range exchange interaction. It makes these rare earth metals good candidates for our description. It was shown [6] that for Dy, Tb, Ho, η lies between 1.0 and 1.6 and is comparable with (14). The results of our work do not pretend to describe the small area, near the transition point, where the spin fluctuations are essential. We have another mechanism for sound attenuation than that of the theories of reference [1], therefore our results differ from theirs and describe another temperature area. From the experimental results for RbMnF$_3$ [1] one can clearly see two power law regions with $\eta = 0.28$ for $\tau < 4.10^{-2}$ and $\eta = 0.7$ for $\tau > 4.10^{-2}$. Such a two power law region has been observed in other substances, e.g. MnF$_2$ and CoO. The theories quoted above do not explain this picture. One can suppose that it implies the existence of two different mechanisms of the sound attenuation.

It is necessary to note that all the investigations show the tendency of the reduction of the singularity by the strong anisotropy.

In conclusion, we would like to say that such simple models can be useful for the qualitative investigations of the various phenomena, and in special cases can give satisfactory quantitative results.

REFERENCES

1. Luthi, B., Moran, T.J. and Pollina, R.J. (1970). *J. Phys. Chem. Solids*, **31**, 1741.

2. Vaks, V.G., Larkin, A.I. and Pikin, S.A. (1967). *Zh. Eksp. Teor. Fiz.*, **53**, 281, [(1968). *Sov. Phys. JETP*, **26**, 188, 647].

3. Kamensky, V.G. (1970). *Zh. Eksp. Teor. Fiz.*, **59**, 2244; [(1971). *Sov. Phys. JETP*, **32**, 1214]; (1973). *Physica*, **67**.

4. Kamensky, V.G. and Tigane, A.A. (1971). *Izv. Akad. Nauk ESSR*, **20**, 20.

5. Turov, E.A. (1965). *Physical Properties of Magnetically Ordered Crystals*, (Academic Press, New York).

6. Pollina, R.J. and Luthi, B. (1968). *Phys. Rev.*, **177**, 841.

CONTRIBUTORS

Abrikosov, A. A.   Landau Institute for Theoretical Physics,
                   Academy of Sciences of the USSR, Moscow,
                   USSR

Brandt, U.   Institut für Physik, Universität Dortmund,
                   46 Dortmund, West Germany

Cooper, B. R.   General Electric Co. Research and Development,
                   P.O. Box 8, Schenectady, New York 12301, USA

Coqblin, B.   Laboratoire de Physique des Solides, Université
                   Paris-Sud, Centre d'Orsay, 91405 Orsay,
                   France

Cracknell, A. P.   Carnegie Laboratory of Physics, University of
                   Dundee, Dundee, DD1 4HN, Scotland, UK

Davis, H. L.   Solid State Division, Oak Ridge National
                   Laboratory, Oak Ridge, Tennessee 37830, USA

Enz, C. P.   Département de Physique Théorique, Université
                   de Genève, 1211 Geneva 4, Switzerland

Glasser, M. L.   Battelle Memorial Institute, 505 King Avenue,
                   Columbus, Ohio 43201, USA

Goodenough, J. B. Lincoln Laboratory, Massachusetts Institute of
                   Technology, Lexington, Massachusetts 02173,
                   USA

Götze, W.   Physik-Department der Technischen Universität
                   München, and Max Planck-Institut für Physik,
                   München, West Germany

Irkhin, Yu. P.   Institute of Metal Physics, Sverdlovsk, USSR

Jullien, R.   Laboratoire de Physique des Solides, Université
                         Paris-Sud, Centre d'Orsay, 91405 Orsay, Franc

Kamensky, V. G.   Landau Institute for Theoretical Physics,
                         Academy of Sciences of the USSR, Moscow, USS

Konwent, H.   Institute of Chemistry, University of Wrocław,
                         Wrocław, Poland

Kopayev, Yu. V.   Landau Institute for Theoretical Physics,
                         Academy of Sciences of the USSR, Moscow, USS

Kühnel, A.   Sektion Physik, Karl-Marx-Universität, Leipzig,
                         East Germany

Lam, D. J.   Argonne National Laboratory, Argonne, Illinois
                         60439, USA

Leschke, H.   Institut für Physik, Universität Dortmund,
                         46 Dortmund, West Germany

Levy, P. M.   New York University, New York, New York 10003,
                         USA

Lindgård, A. P.   Danish Atomic Energy Commission, 4000 Roskilde,
                         Denmark

Morkowski, J.   Laboratorium Ferromagnetyczne, Instytut Fizyki,
                         PAN, 60-179 Poznan, Poland

Natoli, C. R.   C.N.R.S., Grenoble, France, and C.N.E.N.,
                         Frascati, Rome, Italy

Plakida, N. M.   Laboratory of Theoretical Physics, Joint
                         Institute of Nuclear Research, Dubna, USSR

Ranninger, J.   Institut Laue-Langevin, Grenoble, France

Schweitzer, J. W.   Department of Physics and Astronomy, University
                         of Iowa, Iowa City, Iowa 52242, USA

Sivardière, J.   Département de Recherche Fondamentale, Centre
                         d'Etudes Nucléaires, BP 85, 38041 Grenoble
                         CEDEX, France

Stachowiak, H.   Institute of Low Temperatures and Structure
                         Research, Pl. Katedralny 1, Wrocław, Poland

Stevens, K. W. H.   Department of Physics, University of Nottingham,
                     Nottingham, England

Thellung, A.   Institut für Theoretische Physik, Universität
                     Zürich, Switzerland

INDEX

Acoustic magnon modes 577
Actinides
  band structure calculations
    details 303
    electron occupation 299
    results 305,315
  bonding mechanisms 299
  coupling schemes 291
  crystal field theory 290,
    293
  electronic structure 302
  exploratory band structure
    295-318
  hybridization 334, 336
  magnetic properties 287-
    294, 319-350
    band approach 325
    structure 288
    theory 334
    variety of 319
  physical properties 295
Actinide impurities
  hyperfine fields on 330
  transition based alloys
    with 330
Actinide metals
  hybridization 328
  magnetic properties 319,
    325
Actinide monocarbides 296
Actinide monochalcogenides
    296
Actinide monopnictides 296
Actinide systems
  nearly magnetic
    resistivity at high
      temperatures 338
Alkali metals
  ground state energy in 28

Alloys
  coherent potential approx-
    imation applied to 111
  enhanced magnetic see
    Enhanced magnetic al-
      loys
  narrow band magnetism of
    91-130
  Slater-Pauling curves 55
  transition metal see
    Transition metal al-
      loys
Americium
  magnetic properties 287,
    288
Anderson model 511
Andrews critical point 394
Anisotropic exchange in
    rare earths
  paramagnetic susceptibil-
    ity anisotropy 199
Anisotropy constants of
    rare earths 198
Anisotropy in RCo_5 compounds
    200
Antiferromagnets
  axial 480
    hydrodynamics 490,498
  Heisenberg 460
    hydrodynamics 498
    magnon resonance state
      in 453
  hydrodynamics of 488
  insulating
    magnon light scattering
      435-462
  K_2NiF_4
    window condition 476
  MnF_2

window condition 476
magnon light scattering
    in
    anisotropy fields 450
    calculation of cross
        section 445
    exchange constants 450
    mechanism 442
    resonance state 453
magnon-photon interaction
    selection rules 590
planar
    hydrodynamics 489, 498
window condition 476
Antiferromagnetic coupling
    43, 46, 55, 70
Antiferromagnetic FeF_2
    magnon symmetries in 592,
    593
Antiferromagnetic ordering
    temperature 42
Antiferromagnetic spin
    density wave 48
Antiferromagnetic state
atomic moments 49
Antiferromagnetism
    impurity coupled 156
    spontaneous 41
    super-exchange interac-
        tions 82
Antimony $^cNMn_3^f$ compounds
    78
Arsenic $^cNMn_3^f$ compounds. 78
Atomic moments 48,53
    general expressions for
        56
    in metal alloys 58
Augmented plane wave method
    204

Band structure 299
    band width 299
    calculations 306
        details 303
        NpSb 311
        PuN 310
        results 305,315
        ThP 306
        US 313
    electron occupation 299

model 302
    photoemission data 314
    Schrödinger equations 302
    wavefunction character
        308
Band theories 1
Bloch states 95
Bloch-Stoner model 514
Blume-Emery-Griffiths model
    396, 407
Bogoliubov approximation
    229
Bogoliubov variational prin-
    ciple 547
Born-Oppenheimer approxima-
    tion 575
Brillouin zones 298, 589
    antiferromagnetic FeF_2 592
    shape of 583, 584
    symmetry in 579

Carbides
    energy bands 65
Cerium bismuthide and anti-
        monide 258
Cerium monopnictides 226
    anisotropy 249, 256
    anomalous magnetic proper-
        ties 246
    magnetic structures 247,
        249, 259
    self-consistent molecular
        field 254
    susceptibility behaviour
        246, 251, 256
Cesium monopnictides
    susceptibility behaviour
        261
Chromium
    paramagnetism in 55
    Slater-Pauling curves for
        59
Cobalt
    Slater-Pauling curves for
        59
    spontaneous ferromagnetism
        in 55
Coherent potential approxim-
        ation 91, 92
    applied to magnetic alloys
        111

density of states and 98,
    99
single-site 94
Conceptual phase alloys
  iron-aluminium 63
Conceptual phase diagrams
    35-90
  $n_\ell = 1$ 40
  $n_\ell = \frac{1}{2}$ 45
  atomic moments 48
  iron-silicon alloys 63
  magnetic order 50
  manganese-copper alloys 61
  manganese-gold alloys 61
  nitrides and carbides
      $M^C X M_3'^f$ 65
  ordered alloys 61
  preliminary considerations
      36
  Slater-Pauling curve 52
  substitutional binary
      alloys 55
Copper
  Fermi surfaces of 191
  outer electron configura-
      tion 61
  Slater-Pauling curves for
      59
Copper $^C N M n_3{}^f$ compounds 73
Correlation energy 20, 27,
    37
Coulomb correlation effects
  in disordered alloys 101,
      102, 103, 111, 123
Crystal
  magnetic phases of the
      Hubbard model for 101
Crystals
  ferromagnetic *see Ferro-*
      magnetic crystals
  magnetic *see Magnetic*
      crystals
Crystal field theory
  application to actinides
      290, 293
  of rare earths 198
Cubic transition metal-ion
    compounds 427

Debye-Waller exponent 139

De Gennes formula 194, 197
Density of states 98, 99
  Hubbard model 106, 118
  positron annihilation
      curves and 183
Dipolar ordering in magnetic
    crystals 411
  geometrical description
      411
  origin of interaction 412
  parameters 413
  phase transitions 415
    in $S = 1, 3/2$ models
      419
    phenomenological de-
        scription 427
  spin-lattice coupling 425
Dipole-dipole interaction
    485, 488, 494
Dysprosium
  exchange interaction 218
Dysprosium antimonide 266

Electron
  conduction
    interaction with $4f$
        electrons 206
  correlation
    short range inter-atomic
        460
    exchange interaction 504
    $4f$
      effect of magnetic order-
          ing on 208
      exchange interaction
          with conduction elec-
          trons 206
  gas
    magnetic properties of
        15
  localisation centres 504
Electron-phonon interaction
    36, 41
Electron structure of rare
    earths 193
Energy bands
  exchange interaction in
      heavy rare earth met-
      als calculated by
      203-223

group theory and 573
nitrides and carbides 65
Ruderman-Kittel-Kasuya-
    Yosida exchange inter-
    action with 206
Enhanced magnetic alloys
narrow band magnetism 113
Equation of motion technique
double time Green's func-
    tion and 375-383
Erbium
exchange interaction 218
Euler's constant 31
Europium chalcogenides 297
Exchange energy 20, 26
Exchange interaction
between conduction and $4f$
    electrons 206
calculation in rare earth
    metals
    augmented plane wave
        method 204
    numerical methods 212
effect of magnetic order-
    ing 208, 217
in ordered phase 209
s-f
    magnitude of 211
Exclusion principle 4, 46

Ferromagnet
axial
    dipole-dipole interac-
        tion 485, 494
    hydrodynamics 480
Heisenberg
    arbitrary spin and 386
    dipole-dipole interac-
        tion 488
    Green's function theory
        of 385-391
    hydrodynamics 487
    magnon fluid of 464
    quadratic magnon spec-
        trum 463
    sound absorption 602
    sound wave damping in
        605
isotropic see Ferromagnet,
    Heisenberg

planar
    hydrodynamics of 465
    hydrodynamic modes of
        472
Ferromagnetic coupling, 43,
    46, 156
Ferromagnetic anharmonic
    crystal
    magnetic excitations in
        560
    self-consistent element-
        ary excitations 547
    spin-phonon interaction
        543-571
Ferromagnetic crystal
    phonon excitations in 551
    single-domain isotropic
        544
Ferromagnetic hcp metals
    selection rules for 580
Ferromagnetic metals
    spin-disorder resistivity
        342
Ferromagnetic nickel
    anisotropy 179, 183
    electronic structure of
        179
    positron annihilation
        177-192
    interpretation 183
Ferromagnetic solutions,
    magnetization and electron
        densities 122, 123,
        124, 125
Ferromagnetism
    band 503, 511
    itinerant electron 47
    magnon interaction in
        159-176
    random phase approxima-
        tion 159
    origin 503, 512
    resonance 504, 524
Ferro nickel
    studies by positron an-
        nihilation 177-192

Gadolinium
    exchange interaction 218
Gallium $^CNMn_3^f$ compounds 76

Germanium $^{C}MNn_3^f$ compounds 78

Gold
  outer electron configuration 61
Goldstone theorem 272, 275, 282, 285
Green's function 16, 17, 19, 21, 92, 96, 97, 242, 327, 354, 552, 555
  double time
    approach to 375
    definition 381
  one-electron 111
  of Heisenberg ferromagnet 385-391
  spin operators and 383, 385
Ground state
  Hubbard model 101, 110
Ground state energy 15
  exchange energy 20
  inverse dielectric function 21
  kinetic energy 26
  leading terms 31
  random phase approximation 22
Ground state function
  correlation energy 20, 27, 37
  kinetic energy 26, 28, 30
Group theory 3, 573

de Haas-van Alphen terms 32
Hamiltonian
  actinides 321
  as transition operator in Green's function 440
  crystal field 291
  effective magnon 163, 169, 172
  exchange contact 131
  ferromagnetism 513
  free electron 16
  Heisenberg 450, 544
  Hubbard 92, 160, 351
  impurity 132, 137
  Ising 419
  magnetoelastic waves 575
  magnon-phonon interaction 173

many-electron unperturbed 2
$(N - 2n)$-site Heisenberg chain 357
non-degenerate narrow band model 91
one-electron 111, 127
  perturbation 207
planar ferromagnet 465
rare earth 279, 285
rare earth compounds 266, 267
singlet ground state magnetism 226, 232
spin boson 150
spin one Ising 395, 397, 407
structural and magnetic phase transition 266
typical solid 1
unperturbed 1, 2
  definition of 6
variations 2
Hankel's formula 24
Hartree-Fock
  approximation 20, 29, 30, 101, 111, 126, 321, 325, 327, 363, 369, 445
  equations 28
  ground state 28
  hypothesis 37
Hartree-Fock-Slater wave functions 305
Heesch-Shubnikov space 574, groups 574, 582
Heisenberg
  antiferromagnets see Antiferromagnets, Heisenberg
  chain
    Hamiltonian 357
    thermodynamics of 360
  equation 151
  ferromagnet see Ferromagnet, Heisenberg
  Hamiltonian 450, 544
  magnet
    spin ½ and 386
Heusler alloys 62
  ferromagnetism in 75

Hilbert space 537
Hubbard
  alloy analogy approxima-
      tion 113
  Hamiltonian 92
  model 160, 511
    correlation effects in
        104
    density of states 106,
        118
    ground state 110
    magnetic properties
        101
    itinerant electron
        ferromagnetism 160
    magnetic phases of 101
    magnetic properties 105
    one-dimensional *see One-
        Dimensional Hubbard
        model*
    perturbation theory and
        352

Impurity ferromagnetism in
    non-magnetic semicon-
    ductors 501-525
Impurity spin susceptibility
  of Kondon alloys 131-157
    soft boson singularities
        138, 157
    spin boson model 132
Indium $^CNMn_3^f$ compounds 76
Indium semiconductors 501
  ferromagnetism in 504
Insulators
  localised moments in 2
Inverse dielectric function
    21
Iron
  magnetic properties of 54
Iron-aluminium alloys 63
Iron-silicon alloys 63
Iron-titanium alloy
  spontaneous magnetism 59,
    60
$Iron_{1-x}^CNi_x^CNFe_3^f$ system 68
Ising Hamiltonian 419
  spin one 395
Ising model
  spin half 393, 394

spin one 394
Ising transverse field prob-
    lem 237
Itinerant electron magnetism
  conceptual phase diagrams
    in 35
Itinerant electron model of
    ferromagnetism
  Hubbard model 160
  magnon interaction in 159-
    176
  random phase approximation
    159
    improvements of theory
        162

Jahn-Teller transitions 266,
    415, 429

Kondo
  alloys
    impurity spin suscepti-
        bility of 131-157
  susceptibility 145
Korringa-Kohn-Rostoker
    method 304
Korringa law 152
Kramers-Kronig relation 342

Landau
  method 416, 424
  theory 430
Lattice effect
  magnetic behaviour of rare
    earths and 225
Light scattering
  magnon *see Magnon, light
    scattering*
Lindhard function 216
Linear response theory 145
Localised moments
  many electron theory 1-14

Magnesium-copper-manganese
    interaction 75
Magnetic crystals
  coupling in 543

dipolar and quadrupolar
  ordering in 411-433
  classification of
    structure 415
  geometrical description
    411
  origin of interaction
    412
  parameters 413
  phase transitions 415,
    427
  spin-lattice coupling
    425
  fluids in 464
  ordered state of spins in
    463
Magnetic exciton behaviour
    229, 238
  temperature dependence
    241, 244
Magnetic fields
  correlation energy 27
  exchange energy 26
  high field quantum limit
    25, 29, 30
    correlation energy 28
    exchange energy 27
    kinetic energy 26
  intermediate 25, 29
    correlation energy 27
    exchange energy 27
    kinetic energy 26
  kinetic energy 26, 28, 30
  quantum mechanics of free
    electrons in 15
  regions of strength 25
  strength of 15
  weak 25, 29
    correlation energy 27
    exchange energy 27
    kinetic energy 26
Magnetic order 50
Magnetic ordering
  effect on 4 electrons 208
  effect on exchange inter-
    action 208, 217
Magnetics
  sound absorption in 601-
    609
Magnetism
  spontaneous 42, 50, 523

disappearance of 44
  in metal alloys 59, 60
  iron, cobalt and niob-
    ium 54
  suppression of 45
Magnetization
  low temperature 386
Magnetoelastic waves 575,
    see also Magnons
  symmetry properties of
    573-600
Manganese
  spontaneous ferromagnet-
    ism in 55
Manganese-copper alloys 61
Manganese-gold alloys 61
Manganese $^{C}N_{1-x}C_xMn_3^f$ sys-
    tem 82
Magnons 463
  acoustic modes 577, 578
  antiferromagnetic FeF_2
    symmetries 593
  bound states of 165, 172
  drift 464, 470
  effective Hamiltonian 163,
    165, 173
  elementary theory 160
  energy 162, 385
    density 468
    dipolar corrections 169
  heat diffusion 476
  light scattering in insul-
    ating antiferromagnets
    435-462
    anisotropy fields 450
    calculation of cross
      section 445
    dynamical behaviour 449
    exchange constants 450
    mechanism 442
    quantitative descrip-
      tion 440
  longitudinal adiabatic
    velocity 478
  momentum balance 468
  optical modes 577 578
  transverse isothermal
    velocity 494
Magnon creation operator
    161, 162
  definition 163

Magnon fluid
  hydrodynamics of 464
  thermal equilibrium 475
Magnon interaction
  elementary theory of 160
  in itinerant electron
    model of ferromagnet-
    ism 159-176
Magnon-phonon interaction
  effective Hamiltonian 173
  selection rules for 573-
    600
    antiferromagnetic FeF_2
    594
    antiferromagnets 590
    ferromagnetic hcp metals
    580
    mechanism 595
Magnon relaxation 165, 170
Magnon wave vectors 169
Many-electron states 5
Many-electron theory 1-14
  exclusion principle 4
Many-electron unperturbed
  Hamiltonian 2
Metals
  positron annihilation in
    177
Metal alloys
  atomic moments in 58
  non-transitional
    Slater-Pauling curves in
    55
  transitional
    Slater-Pauling curves in
    57
Metal-ion compounds
  cubic transition 427
Metamagnetic
  behaviour 48
  transitions 51
Molecular field approxima-
    tion 363
Molecular field theory 228,
    236
Multi-electron function 193,
    195

Néel temperatures 250, 256,
    258, 260

Neptunium
  dilution in transition
    hosts 330
  magnetic properties 287,
    288, 289
  resistivities 345, 346
    at high temperature 340
Neutron scattering 203
Nickel
  electronic structure at
    higher temperatures
    187
  ferromagnetic
    anisotropy 179, 183
    electronic structure of
    179
    studies by positron an-
    nihilation 177-192
  paramagnetic
    electronic structure of
    187
Niobium
  Slater-Pauling curves for
    59
Niobium $^cNMn_3^f$ compounds 71
Nitrides
  energy bands of 65
Non-transitional metal al-
    loys
  Slater-Pauling curves in
    55
Nucleons
  interaction between 284

One-dimensional Hubbard
    model
  thermodynamic properties
    of 351-362
Opechowski and Guccione
    method 416
Ordered systems
  two-fluid hydrodynamic de-
    scription of 463-499

Palladium
  Fermi surface of 189, 191
Paramagnetic susceptibility
    anisotropy of rare
    earths 198
Paramagnetism in chromium 55

Peierls' equation 530
  solution by perturbation
    theory 535
Perovskites 435, 460
Perturbation theory 7, 284,
    352
  approach to double time
    Green's function 375
  decoupling 8
    within P 11
  double exchange 51
  exclusion principle 46
  mean free time of phonon
    and 535
  solution of Peierls' equa-
    tion by 535
  stages 8
  super-exchange 46
Phase diagrams
  Hubbard model for a per-
    fect crystal 102, 103
Phonons *see also Magnon-*
    phonon interactions
  acoustic 591
  dispersion relations 573
  excitations in ferromag-
    netic crystal 551
  hydrodynamics
    conditions 527
    in solids 527-542
  in equilibrium 528
  interactions 528
  mean free time approxima-
    tion 531, 535
  optical 589
  Peierls' equation 530
    solution 535
  perturbation theory and
    535
  Poiseuille flow of gas
    528, 533
  second sound 528, 533
    damping of 535
  spin interaction in an-
    harmonic ferromagnet
    crystals 543-571
  symmetries 581, 582, 585
    rutile structure and
    590, 591
  thermal equilibrium 529
Plutonium

dilution in transition
    hosts 330
  magnetic properties 287,
    288
  resistivities 345, 347
    at high temperature 338,
    340
Polarons 37
Positron annihilation
  in metals 177
Praesodymium metal system
    229
  magnetic exciton spectrum
    239, 243
Pytte and Stevens model 430

Quadrupolar ordering in mag-
    netic crystals 411
  classification of struct-
    ure 415
  origin of interaction 412
  parameters 413
  phase transitions 415
    in $S = 1, 3/2$ models
    419
    phenomenological de-
      scription 427
Quantum oscillations 503,
    509

Random Phase Approximation
    22, 159
  improvements of theory
    162
Rare earth arsenates 266
Rare earth compounds
  anisotropy 268, 271
  band structure calcula-
    tions 297, 300, 301
  bonding mechanism 299,
    300
  collective excitation be-
    haviour 234
  crystalline field proper-
    ties 428
  decoupling 280
  elementary excitation
    spectra 272
  spin one 272

two-site excitations
276, 285
Hamiltonian 266, 267, 279,
285
higher degree pair inter-
actions 265-286
lattice effects 225
magnetic behaviour
discontinuity in 234
lattice effects and 225-
263
magnetic exciton behaviour
229, 238
magnetic phase transitions
in 415
perturbation theory 284
phase transitions 428
quadrupolar coupling 279,
285
Rare earth metals
electron behaviour 1
exchange interactions in
203-223
calculation by augmented
plane wave method 204
calculation by numerical
methods 212
low lying states of 3
spin-orbit coupling 203
Rare earth monopnictides
225, 228
Rare earth phosphates 266
Rare earth pnictides 266
single ground state mag-
netism 226
Rare earth vanadates 266
Rare earths
anisotropy constants of
198
electron structure of 193
magnetic properties of
193-201
orbital contributions
193
paramagnetic susceptibil-
ity anisotropy 198
anisotropic exchange
199
crystal field theory 198
Resistivity 343, 344
Rock salt structures 435

Roth
two-pole approximation of
105
Ruderman-Kittel-Kasuya-
Yosida (RKKY) exchange
interaction 9, 203,
289
effect of magnetic order-
ing 208, 217
in ordered phase 209
with realistic energy
bands 206
Russell-Saunders coupling
scheme 291
Rutiles 435, 460
Rutile structure
phonon symmetries and 590,
591

Schrödinger equation
band structure 302
Self-consistency
approximation scheme 364
definitions and problems
364
derivable approximations
366
stationarity principle
366
weak 363-373
Self-consistency relation
369
Semiconductors
non-magnetic
impurity ferromagnetism
in 501-525
Shubnikov-de Haas oscillations 501
Silver $^cNMn_3^f$ compounds 73
Singlet ground state magnetism 225, 226
pseudo-spin representation
232
Slater-Pauling curve 52
elements 54
model 52
substitutional binary
alloys 55
transitional metal alloys
57

Soft boson singularities 138, 157
  physical relevance of 140
Solids
  magnetic properties 35
  phonon hydrodynamics in 527-542
Sound absorption in magnetics 601-609
Sound wave damping 605
Spin boson model
  impurity spin susceptibility 132
Spin fluctuation systems 348
Spin half
  Ising model 393, 394
Spin one
  Ising Hamiltonian 395, 397, 407
  Ising model 394
  lattice-gas model 393-410
    application to binary fluids 398
      tricritical points 407
    application to simple fluids 398
    application to ternary fluids 406
      tricritical points 407
    equations of state 397
Spin operators 560
  Green's function and 383, 385
Spin pairing 44
Spin-phonon interaction
  in anharmonic ferromagnet crystals 543-571
  theory 569
Spin relaxation in metals
  Korringa law 152
Spin waves see Magnons
Stationarity principle 366, 370
Stoner theory 112
Substitutional binary alloys
  Slater-Pauling curves in 55

Tellurium in non-magnetic semiconductors 501, 503

Temperature
  electronic structure of nickel and 187
  nearly magnetic actinide systems and 338
Terbium
  exchange interaction 218
  magnon-photon interaction in 589
Thorium
  resistivity 346
Tin $^cNMn_3^f$ compounds 78
Tomonagons 138, 152
Transition based alloys, actinide impurities in 330
Transition metal
  alloys
    band occupancies 87
    narrow band magnetism 113
    Slater-Pauling curves in 57
  carbides 39
  compounds 297
    bonding mechanism 299, 300
  -ion compounds
    phase transitions in 415
  nitrides 39
Two-pole approximation 105

Ultrasonic attenuation
  at magnetic phase transitions 601
Uranium
  alloys 287
  magnetic properties 287, 289
  monopnictides 289

Wannier
  representation 92, 97
  states 95
Wigner's theorem 578
Window condition 475
Wolff model 115

Zeeman energy 430
Zinc $^{c}NMn_3^{f}$ compounds 76
Zubarev Green's function
        145, 151